D0722136

FOURTH EDITION

PHYSICS OF THE ATOM

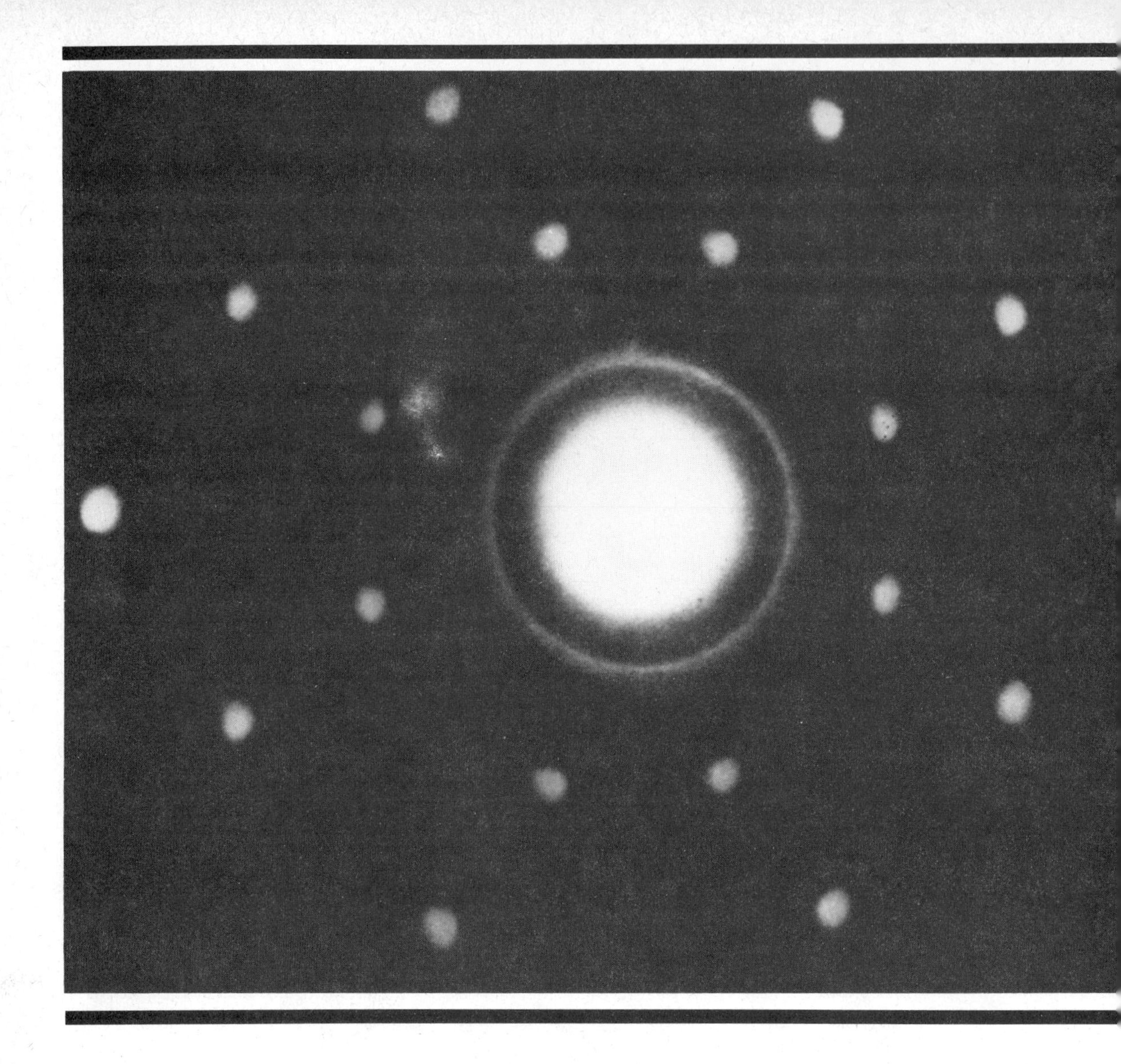

ADDISON-WESLEY PUBLISHING COMPANY
Reading, Massachusetts / Menlo Park, California
London / Amsterdam / Don Mills, Ontario / Sydney

M. RUSSELL WEHR

Drexel University

JAMES A. RICHARDS, JR.

SUNY Agricultural and Technical College

THOMAS W. ADAIR, III

Texas A & M University

FOURTH EDITION

PHYSICS OF THE ATOM

Cover photograph: Courtesy of C. G. Shull and E. O. Wollan

Library of Congress Cataloging in Publication Data

Wehr, M. Russell (Mentzer Russell), 1902–
Physics of the atom.

Includes index.
1. Matter—Constitution. 2. Nuclear physics.
I. Richards, James Austin, 1916– II. Adair, Thomas W. III. Title.
QC173.W42 1984 539.7 83-7072
ISBN 0-201-08878-9

Reprinted with corrections, January 1985

DEFGHIJ–HA–89876

PREFACE

The fourth edition of *Physics of the Atom* is designed, as were the first three editions, to meet the modern need for a better understanding of the atomic age. It is an introduction suitable for any student with a background in college physics and mathematical competence at the level of calculus. Some parts of the book will be better appreciated by those with a knowledge of chemistry. This book is designed to be an extension of the introductory college physics course into the realm of atomic physics: It should give the student a proficiency in this field comparable to that developed in mechanics, heat, sound, light, and electricity.

The approach has been to deal with a logical series of topics leading to the practical conversion of mass-energy into kinetic energy. The development of modern physics is such a logical sequence of discoveries that we have had little difficulty in deciding what topics to include. The order of presentation is usually chronological. Mechanics can be presented in a from-simple-to-complex manner because mechanics is an organization of observations with which most of us are familiar. Atomic physics, on the other hand, involves concepts that are quite foreign to general experience, and it is both helpful and stimulating for students to feel that they are growing in understanding as others grew before them. This book capitalizes on the fact that the story has a plot.

This plot is developed as simply and directly as possible. It is difficult to know how a new idea comes into being. Even the person who creates a new idea cannot fully describe that moment when the light dawned. But we, as "Monday-morning quarterbacks," can state the problem, present the relevant information, and, we hope, make the idea that originally required genius appear obvious. With informal discussion and many analogies the student may be led to understand new and difficult concepts. Although any serious student of atomic physics must go on to a more elaborate mathematical treatment, we feel that even at this level the most convincing argument is a careful mathematical development and the presentation of original data. We have not hesitated to use calculus where it is appropriate, and we have clearly indicated the weakness of our argument on those few occasions when the argument is indeed weak.

Many who use this book will not have access to a laboratory. Thus we have described a few projects that most students can do for themselves.

The book is designed for a one-term course in either a liberal arts or an engineering curriculum. Basic physics is basic physics whether the student is motivated by curiosity or possible application. Our preference for the SI system of units is due to the fact that this system is rapidly becoming *the* system in engineering and physics, and because this system includes the units all scientists use to measure electrical quantities. Except in equations that include units peculiar to our subject, like the electron volt, our equations are valid in any consistent rationalized system of units.

The references at the end of each chapter include sources that have been helpful to us in preparing the text and to which students may *profitably* refer. Many of our references to original articles are to reprints of source materials. Our thought is that these compilations and translations are more convenient for students to use. In some cases we have made remarks designed to facilitate the student's choice of reference reading.

A few of the problems at the end of each chapter may involve mere substitution of numbers into formulas. In these cases the purpose is to give the student a feeling for the magnitudes of atomic quantities. Most of the problems, however, require thoughtful analysis. Some are stated in general terms instead of numbers. The order of the problems is the same as the order in which the theory is developed. Some of the problems in the earlier editions have been retained without change, others have been rewritten, and over 40 new problems have been added, bringing the total to 500. Review questions have also been added at the end of each chapter. These questions should help students determine if they have understood the main topics.

Appendix 1 is a chronology of discoveries important to the atomic view of nature. It was impossible to formulate any sharp criterion for selecting the items to be included, and so, in general, we have taken a broad view. The purpose of this chronology is to convey a sense of the development of a natural science. We think it will also serve to stress the international character of scientific achievement and to emphasize that many of the great scientists did their first important work early in life.

The principal changes in the second edition of this book were the rewriting of the chapters on waves and particles and the addition of a new chapter on matter waves, both prepared by I. A. Miller of Drexel University. In the third edition the principal changes were the addition of numerous new sections—for example, those on laser applications, stellar aberration, general relativity, Miller indices, integrated circuits, superconductivity, the Josephson effect, and conservation of parity. Significant revisions were made in the sections on blackbody radiation, radiation detectors, reactors, fusion, and elementary particles. The significant changes in the fourth edition include the following: a new section on Boltzmann's law; a new section on cosmic blackbody radiation; an expanded section on lasers and laser light, including a discussion of the Einstein coefficients; the section on relativity revised to include gravitational lenses and black holes; a new section on extraterrestrial x-rays; a revised discussion of wave groups; the chapter on quantum mechanics expanded to include a discussion of the barrier potential and an expanded discussion of the quantum mechanical model of the atom; a new discussion of solar cells and light emitting diodes; a new discussion of low temperature physics; a new section on nuclear models; a revised section on fusion; and an updated discussion of particle physics to include quarks and the basic interactions.

This textbook is dedicated to the students who will use it and to the former students whose questions, comments, and reactions have largely determined our presentation. For the first edition, we gratefully acknowledge the assistance of Elliot H. Weinberg of North Dakota State College. We also wish to thank our former colleagues at Drexel University, especially Henry S. C. Chen, Dennis H. Le Croissette, and Irvin A. Miller, for their constructive suggestions. For assistance with the second edition we wish to thank A. Capecelatro, D. E. Charlton, E. M. Corson, T. T. Crow, E. I. Howell, A. P. Joblin, G. E. Jones, P. Kaczmarczik, A. Meador, M. W. Minkler, T. J. Parmley, L. E. Peterson, D. J. Prowse, F. W. Sears, F. O. Wooten, and M. W. Zemansky. And for very useful comments about the third edition we are pleased to thank Gordon E. Jones, Richard Madey, Philip N. Parks, Kenneth W. Rothe, and R. A. Atneosen. We are extremely grateful for the very helpful comments on the fourth edition by Richard D. Haracz, Samuel Olanoff, and Philip Parks. In addition, we wish to thank Minnette Bilbo for typing the many preliminary drafts of the third and fourth editions and the preparation of both final drafts. The writing of these editions was made much easier because of her expertise and dedicated hard work.

This book is neither a treatise nor a survey. It is a textbook that bridges the gap between classical physics and the present frontiers of physical investigation.

Havertown, Pennsylvania
Delhi, New York
College Station, Texas
November 1983

M. R. W.
J. A. R., Jr.
T. W. A. III

CONTENTS

1 THE ATOMIC VIEW OF MATTER 1

2 THE ATOMIC VIEW OF ELECTRICITY 37

9 THE ATOMIC VIEW OF SOLIDS 311

10 NATURAL RADIOACTIVITY 365

11 NUCLEAR REACTIONS AND ARTIFICIAL RADIOACTIVITY 409

1 THE ATOMIC VIEW OF MATTER

1.1 INTRODUCTION

The ancient Greeks speculated about almost everything. Democritus, for example, theorized that not only matter but also the human soul consists of particles. His statements, made about twenty centuries before the advent of experimental science, can be regarded primarily as demonstrating the fertility of his imagination. Still, it would be a mistake to discredit Democritus completely, for although he was no student of atomic physics, he had characteristics that every student needs.

Because atomic physics is built from big ideas about very small things, its development often leads along paths that run counter to common sense. As we consider things and events that are orders of magnitude removed from everyday experience, the difficulty of understanding their nature increases. Our common sense enables us to understand the relationship between a brick and a house. Conceiving of the earth as round may involve a little uncommon sense, but for most people it presents no great difficulty. However, the relationship between water and a water molecule is more difficult. While we can see the earth, whether flat or round, we cannot see a water molecule even with the best of instruments. All of our information about single water molecules is of an indirect kind, yet it would be a very unsophisticated chemist for whom the concept of a single water molecule would not be a part of his or her common sense. As a person's knowledge expands, more and more facts assume the aspect of "common sense." Certain velocity relationships are common sense. To an observer in a moving car, the velocity of another moving car appears different than it does to an observer standing beside the highway. In fact, a very young child once observed when the car in which he was traveling was passed by another, "We are backing up from the car ahead." However, the statement made by Albert Einstein that the velocity of light is the same for all observers regardless of their own velocities is very uncommon sense. In a later chapter we will attempt to show that his statement is reasonable and can appropriately be incorporated into our common sense. The conflict between the earth's actual roundness and its apparent flatness is resolved conceptually, i.e., by imaginative understanding, with the realization that the earth is a very big sphere. Somewhat similarly, the apparent conflict between our statements about relative velocities is resolved conceptually with the realization that the velocity of light is a very large velocity. Democritus, who could propose an atomic theory in the fifth century B.C., would today have the courage and imagination to face the ideas that lie before us.

It is the business of philosophers to discuss the nature of reality. It is the business of physicists (once called natural philosophers) to discuss the nature of physical reality. Philosophy, therefore, includes all of physics and a lot more besides. It is natural, then, that physics should have a continuing influence on philosophy. As physical discovery is quickly put into engineering practice and made to bear on our physical environment, so it also affects the formulation of philosophical theory and bears directly on our outlook and interpretation of life.

The old or classical physics of Newton was extraordinarily successful in dealing with events observed in his day. Using Newton's methods, it is simple to equate the earth's gravitational force on the moon to the centripetal force and obtain verifiable relationships about the behavior of the moon. The same methods can be extended to orbits that cannot be regarded as circular. In fact, three observations of a new comet enable astronomers to foretell with great accuracy the entire future behavior of the comet. Given a certain amount of specific data known as initial or boundary conditions, classical Newtonian mechanics enables us to determine future events in a large number of situations. It is easy to move a step further and argue that what Newton has demonstrated to be true often, is true always, and that given sufficient initial data and boundary conditions, laws may be found that show every future event to be determined. The motion of a falling leaf or the fluctuations in the price of peaches may be very complex phenomena. It may require tremendous amounts of data and the application of very complicated laws that we do not yet understand to be able to make predictions in these cases. The important philosophical consequence of classical mechanics was not that every problem had been solved but that a point of view had been established. It was felt that each new discovery would fall into the Newtonian mechanistic framework. Philosophical questions became more pressing. Do we humans make decisions that alter the course of our lives or are we, like the bodies of the solar system, acting according to a set of inflexible laws and in accordance with a set of boundary conditions? Are we free or is our apparent ability to make decisions an illusion? Is everything we do beyond our responsibility, having been determined at the time of creation? Although mechanistic philosophy is rather repulsive when applied to ourselves, we nevertheless lean heavily upon it in interpreting things that go on about us. Indeed, the whole argument over whether human behavior is influenced more by heredity or environment is based on the assumption that human behavior is determined by some combination of the two.

To the extent that this mechanistic philosophy is based on classical physics, it is due for revision. Upon examination of events that are either very large or very small, we find that classical physics begins to fail. When a new theory or a modified theory has had to be applied in order to describe experimental observations, it has often resulted that the new theory is very different from classical physics. The method of attack, the mathematical techniques, and the form of the solution are often quite different. At one point we shall show that the observations of natural phenomena are inherently *uncertain*. It becomes evident, then, that if some circumstance had led to the development of atomic physics before classical physics, the influence of atomic physics on philosophy would have been against mechanism rather than for it.

Atomic physics has given us electronics and all that that word implies, including radio, radar, television, computers, etc. Atomic physics has given us nuclear energy. The new physics is as successful with submicroscopic events as classical physics was with large-scale events. But it may be that the most important benefits that can result from the study of atomic physics are philosophical rather than technical.

1.2 CHEMICAL EVIDENCE FOR THE ATOMIC VIEW OF MATTER

The speculations of Democritus and of the Epicurean school, whose philosophy was based on atomism, were not the generally accepted views of matter during the Middle Ages and the Renaissance. The prevailing concepts were those of Aristotle and the Stoic philosophers, who held that space, matter, and so on were continuous, and that all matter was one primordial stuff that was the habitat of four elementary principles — hotness, coldness, dryness, and wetness. Different materials differed in the degree of content of these principles. The hope of changing the amount of these principles in the various kinds of matter was the basis of alchemy. Not until the development of quantitative chemistry in the last half of the eighteenth century did the experimental evidence needed for evaluating the conflicting speculations about the constitution of matter begin to appear.

Antoine Lavoisier of France was outstanding among the early chemists. He evolved the present concept of a chemical element as "the last point which analysis is capable of reaching." He also concluded from his observations on combustion that matter was conserved in chemical reactions. In 1799 the French chemist J. L. Proust stated the law of definite or constant proportions, which summed up the results of his studies of the substances formed when pairs of elements are combined. The law states that *in every sample of any compound substance, formed or decomposed, the proportions by weight of the constituent elements are always the same*. This statement actually defines chemical compounds, because it differentiates them from solutions, alloys, and other materials that do not have definite composition.

The principal credit for founding the modern atomic theory of matter goes to John Dalton, a teacher in Manchester, England. His concern with atoms seems to have originated with his speculations about the solubilities of gases in water and with his interest in meteorology, which led him to try to explain the fact that the atmosphere is a homogeneous mixture of gases. Eventually, he believed that an element is composed of atoms that are both *physically* and *chemically* identical, and that the atoms of different elements differ from one another. In a paper he read in 1803 at a meeting of the Manchester Literary and Philosophical Society, Dalton gave the first indication of the quantitative aspect of his atomic theory. He said, "An enquiry into the relative weights of the ultimate particles of bodies is a subject, as far as I know, entirely new: I have lately been prosecuting this enquiry with remarkable success." This was followed by his work on the composition of such gases as methane (CH_4), ethylene (C_2H_4), carbon monoxide (CO), carbon dioxide (CO_2), and others, which led him to propose the law of multiple proportions in 1804. This law states that *if substance A combines with substance B in two or more ways, forming substances C and D, then if mass A is held constant, the masses of B in the various products will be related in proportions which are the ratios of small integers.* The only plausible interpretation of this law is that when elementary substances combine, they do so as discrete entities or atoms. Dalton emphasized the importance of relative weights of atoms to serve as a guide in obtaining the composition of other substances, and stressed that a chemical symbol means not only the element but also a fixed mass of that element. The introduction of the concept of atomic weights (strictly, atomic masses) was Dalton's greatest contribution to the theory of chemistry, because it gave a precise quantitative basis to the older vague idea of atoms. This concept directed the attention of quantitative chemistry to the determination of the relative masses of atoms.

An important law pertaining to volumes of gases was announced by Gay-Lussac in 1808. He said that *if gas A combines with gas B to form gas C, all at the same temperature and pressure, then the ratios of the volumes of A, B, and C will all be ratios of simple integers.* Two examples of this law are (a) the combining of two volumes of hydrogen and one volume of oxygen to form two volumes of water vapor, and (b) the union of one volume of nitrogen and three volumes of hydrogen to produce two volumes of ammonia. The following are symbolic forms of these reactions:

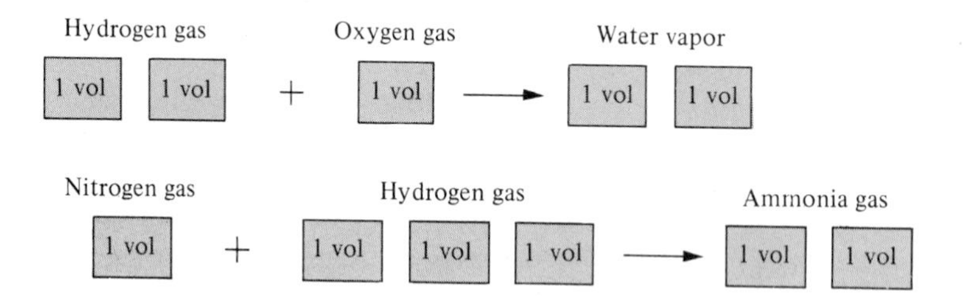

It is obvious that Gay-Lussac's law, like the law of multiple proportions, implies that the substances that participate in these reactions participate in discrete or corpuscular amounts. The ratio between the number of shoes worn to the number of people wearing them is almost an exact integer, namely two, showing that both people and shoes are discrete entities. The ratio of the number of tomatoes used per serving of tomato soup is quite a different kind of situation, and if the ratio is integral it is only by coincidence.

Gay-Lussac's law supported the work of Dalton, but it also raised difficult questions about the composition of an element in the gaseous state. In the case of the first reaction given, does each atom in the given oxygen gas divide to spread through the two volumes of water vapor? If so, the indivisibility of atoms must be abandoned. Or does each entity in the oxygen gas consist of a multiplicity of atoms? If so, how many atoms are grouped together? Similar questions can be raised about each of the gases in the two reactions given. It is evident that the numerical values of the relative weights of the atoms determined from these reactions will depend on the answers to these questions.

In 1811, Avogadro, an Italian physicist, proposed the existence of different orders of small particles for the purpose of correlating the works of Dalton and Gay-Lussac. He postulated the existence of "elementary molecules" (atoms) as the smallest particles that can combine to form compounds, and the existence of "constituent molecules" (molecules of an element) and "integral molecules" (molecules of a compound) as the smallest particles of a body that can exist in the free state. He went on to state (without proof) a very important generalization, known as Avogadro's law, that *at the same temperature and pressure equal volumes of all gases contain the same number of molecules*. From this law and his concepts of atoms and molecules, Avogadro showed that the ammonia-producing reaction required that nitrogen gas consist of diatomic molecules and that oxygen must also be diatomic to account for the water-vapor reaction. He further concluded that water must consist of a union of two atoms of hydrogen and one atom of oxygen.

Unfortunately, the ideas advocated by Avogadro received little notice even when revived by Ampere in 1814. The notion that hydrogen and other gases were composed of diatomic molecules was ridiculed by Dalton and others, who would not conceive of a combination of atoms of the same kind. They asked, "If two hydrogen atoms in a container filled with this gas can cling together, why do not all cling together and condense to a liquid?" This is indeed a very good question. Science was not able to give a satisfactory answer until over a century later, after the introduction of the Bohr theory of the atom.

In the next two sections in this chapter, we will describe some of the methods that were and still are used to determine the relative weights of atoms. The results obtained by the analytical chemists using these several methods during the first half of the nineteenth century were often contradictory. They frequently obtained different values for the atomic mass of the same element. By the 1850s, inconsistencies were so numerous that many felt that the atomic theory of matter would have to be discarded. However, the contradictions were resolved in 1858 by the Italian chemist Cannizzaro, who had an intimate knowledge of the then known methods for determining atomic masses and a broad grasp of the whole field of chemistry. He showed that Avogadro really had provided a rational basis for finding atomic masses, and that the inconsistent results obtained by various experimenters resulted from a lack of clear distinction between atomic masses, equivalent masses, and molecular masses. The views of Cannizzaro received the approval of the scientific world when they were adopted by the international conference on atomic masses, which met in Karlsruhe, Germany, in 1860. This, then, is the year in which the fundamental ideas of modern chemistry were widely accepted.

1.3 MOLECULAR MASSES

After Cannizzaro had clarified and established some of the basic definitions in chemistry, Avogadro's law opened the door to one of the methods for determining molecular weights. No one had any idea what a single molecule weighed, but once there was a way of isolating equal numbers of different kinds of molecules, the relative masses could be determined. The hydrogen molecule was found to be the lightest molecule, and the hydrogen atom proved to be the lightest atom. In 1815, Prout had proposed that the relative atomic mass of hydrogen be arbitrarily taken as one. On this basis, most other light atoms and molecules had relative masses that were nearly integers. But, for reasons to be discussed later, it turned out that the atomic masses of many of the heavier atoms were not very nearly integers. Hydrogen appeared to be a poor basis for the system, and more nearly integral atomic masses for all atoms could be obtained by making a heavier atom the basis of the system. Originally, oxygen was chosen and its atomic mass was arbitrarily set at exactly 16. This was the basis for many years until, in 1961, it was replaced by setting the atomic mass of carbon at exactly 12. (This will be discussed further in Chapter 2 when we consider isotopes.) Under either plan, the atomic mass of hydrogen is not exactly unity, although it is nearly so.

These relative molecular and atomic masses are all dimensionless ratios. If about 4 parts by weight of hydrogen were combined with 32 parts by weight of oxygen, about 36 parts by weight of water vapor can be formed, according to the familiar equation, $2H_2 + O_2 = 2H_2O$. In the carbon-12 isotope, there are exactly 12 nucleons (i.e., 12 neutrons plus protons) per atom. We define a mole of carbon as the amount of carbon in 12 grams of the substance. Note that for carbon the number of grams in a mole equals the atomic mass. More generally, a mole is defined implicitly in accordance with two rules: (1) the number of grams in a mole of any element equals that element's atomic mass; (2) the number of grams in a mole of any compound equals that compound's molecular mass. It is a simple consequence that a mole contains a definite number of atoms or molecules, whatever the element or compound. The number of atoms or molecules in the mole is called Avogadro's number or the Avogadro constant, N_A, and it is of basic importance in physics and physical chemistry. These definitions emphasize that a mole is a fixed number of particles of a material and it is evident that the number of moles in an amount of a compound is the mass in *grams* of the compound divided by its formula (molecular) mass. (Note that the mole and the Avogadro constant have been defined in terms of 12 grams of carbon. This is the modern practice. However, these definitions do not always conform to the SI system of units. Therefore, we will on occasion use the kilomole, kmole, which is 1000 moles. Obviously a kilomole of carbon 12 has a mass of 12 kilograms.)

The value of the Avogadro constant was of relatively minor importance to chemistry in the early nineteenth century and its magnitude was not even estimated until Loschmidt did so in 1865. We will discuss Perrin's method of determining it later in this chapter. Here is an interesting case where knowing the existence of a number was more important than knowing its magnitude as, for example, in determining the relative masses of the atoms involved in the ammonia-producing hydrogen-nitrogen reaction previously described. The value of the Avogadro constant is almost inconceivably large; by modern measurements it is

$$N_A = (6.0220943 \pm 0.0000063) \times 10^{23}$$

particles per mole. Only after the magnitude of Avogadro's number was known could the absolute mass of an atomic particle be computed.

It follows from Avogadro's law that the volume of a mole of a gas under given conditions of temperature and pressure is the same for all gases. The normal volume of a perfect gas or the standard molar volume of an ideal gas, V_0, is the volume occupied by a mole of the gas at a pressure of 1 standard atmosphere and a temperature of 0°C. The value of V_0 is

$$(2.241383 \pm 0.000070)^{\dagger} \times 10^{-2}\ \text{m}^3 \text{ per mole.}$$

† If it seems strange that this and some other constants are given with the uncertainty expressed to more than one significant figure, refer to the article, "Probable Values of the General Physical Constants," by R. T. Birge [*Phys. Rev. Suppl.* **1** (1929), p. 6]. We note here only that the concepts of probable errors and significant figures do not correspond completely. If the probable error can be determined to less than 10 percent of itself, then more than one significant figure is required to express it.

Principal articles containing the values of various constants are: "Values for the Physical Constants Recommended by NAS-NRC," *Nat. Bur. Std. (U.S.) Tech. News Bull.*, **47** (1963), p. 175; "World Sets Atomic Definition of Time," *Nat. Bur. Std. (U.S.) Tech. News Bull.*, **48** (1964), p. 209; E. A. Mechtly, *The International System of Units*, NASA SP-7012. Washington, D.C.: U.S. Government Printing Office (1964); E. R. Cohen and J. W. M. Du Mond, "Our Knowledge of the Fundamental Constants of Physics and Chemistry in 1965," *Rev. Mod. Phys.*, **37** (1965), p. 537; B. N. Taylor, W. H. Parker, and D. N. Langenberg, "Determination of e/h Using Macroscopic Quantum Phase Coherence in Superconductors: Implications for Quantum Electrodynamics and the Fundamental Physical Constants," *Rev. Mod. Phys.*, **41** (1969), p. 375; T. G. Trippe et al., "Review of Particle Properties," *Rev. Mod. Phys.*, **48**, No. 2, Part II (1976), p. 61; B. N. Taylor and E. R. Cohen, Proceedings of the Fifth International Conference on Atom Masses and Fundamental Constants (AMOC-5), Paris (1975); B. N. Taylor and E. R. Cohen, "Present Status of the Fundamental Constants," edited by J. H. Sanders and A. H. Wapstra, *Atomic Masses and Fundamental Constants*, **5** (1976), p. 663.

1.4

ATOMIC MASSES

Avogadro's law provided a systematic method for determining molecular weights, but a large amount of quantitative data on the formation of various compounds was required before the atomic masses of the known elements could be determined. The situation is somewhat like the following: Suppose that man A pays man B \$1.00 in coin, using no coin smaller than quarters, and that we wish to determine how this is done. Man A may do this in any one of four ways:

1. one \$1.00 coin,
2. two 50¢ coins,
3. one 50¢ and two 25¢ coins, or
4. four 25¢ coins.

Assuming man B had no money earlier, if he now pays man C 25¢, possibilities (1) and (2) are eliminated, but there is still a doubt as to how the original transaction was made. By careful observation of further transactions of those who spend the original \$1.00, it could be determined just what coins man A must have had originally.

An aid to the solution of this puzzle was the empirical discovery by Dulong and Petit in 1819 that for most elements in the solid state the atomic mass times the specific heat at constant volume is about 6 cal/mole · K. The law of Dulong and Petit permits a rough independent determination of atomic mass by dividing this constant by the specific heat. We shall discuss the theoretical basis of this law in Chapter 9.

1.5

PERIODIC TABLE

Probably the most significant discovery in all chemistry, aside from the atomic nature of matter, was the periodic properties of the elements, now depicted in the familiar periodic table of the elements (Appendix 3). The chemical properties of this table are probably familiar to most readers of this book; the physical properties will be discussed later. The table was proposed independently by Meyer and by Mendeléev in 1869. Its usefulness lay both in its regularities and in its irregularities. One interesting irregularity in the original table was that in order to have the elements fall in positions consistent with their chemical properties, it was necessary to leave numerous spaces unoccupied. Mendeléev suggested that these spaces would be filled with as yet undiscovered elements. Using his table, he was able to describe in considerable detail the properties these elements could be expected to have when they were discovered. It was nearly 100 years before all the predictions that Mendeléev made were fulfilled.

Reflect, for a moment, on the vast simplification that the chemical discoveries here outlined provide. Looking about us, we see innumerable kinds of materials. The atomic view indicates that these materials are of discrete kinds whose number, however large, is not infinite. The discovery of elements is a further simplification in that the many materials we encounter are shown to be composed of only about 100 chemically distinguishable materials, many of which are rare. It turns out that even these elements are not a heterogeneous group but are subject to further classification into a periodic table. The problems of chemistry are many; however, it is easy to see that things are much simpler than might at first appear.

1.6 PHYSICAL EVIDENCE FOR THE ATOMIC VIEW OF MATTER

In our discussion thus far, all atomic properties have been inferred from studies of gross matter. In 1827, the English botanist Robert Brown observed that microscopic pollen grains suspended in water appear to dance about in random fashion. At first the phenomenon was ascribed to the motions of living matter. In time, however, it was found that any kind of fine particles suspended in a liquid performed such a perpetual dance. Eventually it was realized that the molecules of a liquid are in constant motion and that the suspended particles recoiled (Brownian movement) when hit by the molecules of the liquid. However, long before the equations for Brownian movement were derived early in the twentieth century, the particles of matter were thought of as moving about in a random manner and undergoing frequent collisions. Such processes are decidedly in the domain of physics. How can the principles of mechanics be applied to molecular collisions?

The simplest state of matter to consider was a gas. The ideal gas law, for n moles of a gas, is $pV = nRT$, where R is the universal gas constant per mole and p, V, and T are the pressure, volume, and temperature, respectively. This law was a well-established *empirical* relationship, and its derivation was one of the objectives of physics. The application of classical physics to the mechanics of gases is called the *kinetic theory of gases*. Although Daniel Bernoulli had some success in developing this theory as early as 1738, the principal contributions that led to its establishment were made between 1850 and 1900 by Clausius, Maxwell, Boltzmann, and Gibbs.

1.7 KINETIC THEORY OF GASES. MOLAR HEAT CAPACITY

Early in our study of physics, we investigated the mechanics of bodies that can be regarded as particles. The study of extended bodies was treated by introducing certain averages, and the translational problem of extended bodies was solved by

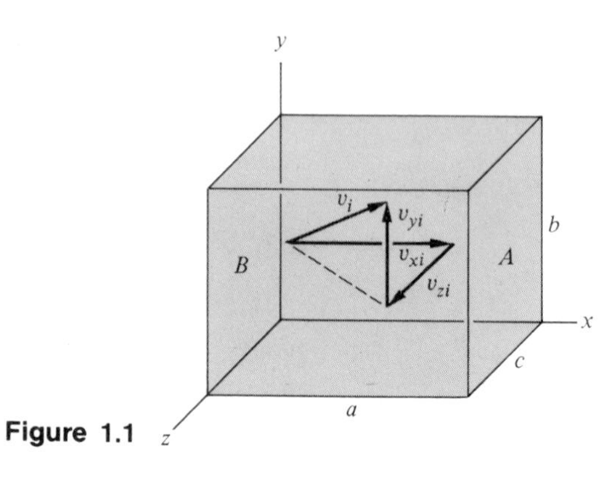

Figure 1.1

introducing the concept of a *center of mass* that moves as though it were a particle. The study of rotational properties of extended bodies was similarly facilitated by the introduction of another average property of the body, its *moment of inertia*. In the kinetic theory of gases, we assume that pressure, temperature, etc., are *averages* of properties of all the molecules of a gas. Kinetic theory is a large and elegant subject. We can convey its spirit by deriving the ideal gas law and a few other relationships.

Consider a rectangular container, the edges of which are parallel to the x-, y-, and z-axes and have dimensions a, b, and c, as shown in Fig. 1.1. There are N identical particles, each of mass m, in the box moving in random directions with a wide range of speeds. These identical particles may be, but are not necessarily, atoms or molecules.

We assume that the particles are very small so collisions between them are rare compared to collisions with the plane walls of the container. Neglecting minor forces such as gravity and intermolecular forces, we shall consider that the only forces acting on the particle are those resulting from collisions with the walls. We number the particles $1, 2, \ldots, i, \ldots, N$. Figure 1.1 shows the ith particle, whose velocity is v_i. This velocity may be broken into the rectangular components v_{xi}, v_{yi}, and v_{zi}, as shown. We assume collisions to be perfectly elastic such that when a particle strikes a wall, the velocity component related to the axis that is perpendicular to that wall is reversed in direction but unchanged in magnitude. The other two velocity components remain unchanged. Thus if the particle strikes side A, the x-component of its momentum is changed from $+mv_{xi}$ to $-mv_{xi}$. The net change in the x-component of its momentum is

$$(-mv_{xi}) - (+mv_{xi}) = -2mv_{xi}.$$

Since collision with the wall causes the particle to change its momentum, the wall experiences an impulsive force. This impulsive force is unknown because we cannot estimate the time of contact in a meaningful way. Fortunately, it is not the impulsive force but the average force due to repeated hits that we seek. Since collisions with the top, bottom, far, and near sides have no effect on v_{xi}, and since collisions with ends A and B merely reverse the direction of v_{xi}, we see that the time interval between successive hits on side A is the total x-distance, $2a$, divided by the x-component of the velocity or $2a/v_{xi}$. By applying Newton's second law, we find the average force F_i of the wall on the ith particle to be

$$F_i = \frac{\Delta(mv)}{\Delta t} = \frac{-2mv_{xi}}{2a/v_{xi}} = -\frac{mv_{xi}^2}{a}. \quad \textbf{(1.1)}$$

This force is equal in magnitude and opposite in direction to the force of the particle on the wall and thus the particle produces an average pressure on side A given by

$$p_i = \frac{-F_i}{\text{area}} = \frac{-F_i}{bc} = \frac{mv_{xi}^2}{abc} = \frac{mv_{xi}^2}{V}, \quad \textbf{(1.2)}$$

where V is the volume of the container.

The pressure we have computed is due to but one, the ith, particle. The pressure due to each particle is computed in the same way. Adding the pressures due to the N identical particles, we have

$$p = \sum_{i=1}^{N} p_i = \frac{m}{V}\sum_{i=1}^{N} v_{xi}^2. \quad \textbf{(1.3)}$$

To evaluate the sum on the right-hand side of the equation we note (see Fig. 1.1) that

$$v_i^2 = v_{xi}^2 + v_{yi}^2 + v_{zi}^2. \quad \textbf{(1.4)}$$

Since this equation holds for each of the particles, we can add the corresponding equations and obtain

$$\sum_{i=1}^{N} v_i^2 = \sum_{i=1}^{N} v_{xi}^2 + \sum_{i=1}^{N} v_{yi}^2 + \sum_{i=1}^{N} v_{zi}^2. \quad \textbf{(1.5)}$$

We now define the *mean square velocity*, $\overline{v^2}$, to be the average of the sum of the squares of the velocities; therefore

$$\overline{v^2} = \left(\sum_{i=1}^{N} v_i^2\right)\Big/N. \quad \textbf{(1.6)}$$

Applying this definition to all terms in Eq. (1.5), we find that it becomes

$$\overline{v^2} = \overline{v_x^2} + \overline{v_y^2} + \overline{v_z^2}, \quad \textbf{(1.7)}$$

and, substituting terms from Eq. (1.6) into Eq. (1.3), we get

$$p = \frac{m}{V}N\overline{v_x^2}. \quad \textbf{(1.8)}$$

Since we assume these velocities to be completely random in direction and magnitude, the three mean square velocity components must be equal or $\overline{v_x^2} = \overline{v_y^2} = \overline{v_z^2}$. This assumption enables us to deduce from Eq. (1.7) that

$$\overline{v^2} = 3\overline{v_x^2} = 3\overline{v_y^2} = 3\overline{v_z^2}. \quad \textbf{(1.9)}$$

The square root of the quantity $\overline{v^2}$ is called the *root-mean-square speed, or velocity*, v_{rms}. Substituting $\overline{v_x^2}$ from Eq. (1.9) into Eq. (1.8), we get

$$p = \frac{Nm}{V}\frac{\overline{v^2}}{3}$$

or

$$pV = \tfrac{1}{3}Nm\overline{v^2}. \tag{1.10}$$

Assume that particles in the container are all identical gas molecules. Then the number of moles n in the container equals the number of molecules in the container N divided by the number of molecules per mole, N_A. Thus, n equals N/N_A, or N equals nN_A. Since the product of a mass of a molecule m and the Avogadro constant N_A is the molar mass M, we can express the total mass of a gas in the box as $Nm = nN_Am = nM$. When this resúlt is substituted in Eq. (1.10) it becomes

$$pV = \tfrac{1}{3}nM\overline{v^2}. \tag{1.11}$$

This is not the result we sought, $pV = nRT$, so we have as yet no justification for the many assumptions we have made. The result is interesting, however, because it contains the pV term, and the term $\frac{1}{3}nM\overline{v^2}$ has a familiar look. If we write

$$\tfrac{1}{3}nM\overline{v^2} = \tfrac{2}{3}\left(\tfrac{1}{2}nM\overline{v^2}\right), \tag{1.12}$$

the quantity in parentheses is clearly the total *translational* kinetic energy of the molecules. This energy must be the internal energy of the gas U because it was assumed that no forces of attraction or repulsion exist that could give rise to molecular potential energy. By combining Eqs. (1.11) and (1.12), we obtain

$$pV = \tfrac{2}{3}U. \tag{1.13}$$

If we could show that $U = \frac{3}{2}nRT$, then Eq. (1.13) would become $pV = nRT$, the ideal gas law.

By 1850 Joule had demonstrated that heat is a form of energy. When a gas is heated at constant volume, the heat energy supplied causes a temperature change that must increase the energy of the gas. If we assume that the relation $U = \frac{3}{2}nRT$ is correct, then the change in internal energy with respect to temperature can be given by

$$dU/dT = \tfrac{3}{2}nR. \tag{1.14}$$

The change in internal energy with respect to temperature of one mole of an ideal gas at constant volume is called the *molar heat capacity* C_v. Therefore we obtain

$$C_v = \tfrac{3}{2}R, \tag{1.15}$$

and its value is

$$C_v = \frac{3}{2}R = \frac{3}{2} \times \frac{8.31\ \text{J}}{\text{mole} \cdot \text{K}} \times \frac{1\ \text{cal}}{4.18\ \text{J}} = 2.97\ \text{cal/mole} \cdot \text{K}.$$

Table 1.1 C_v of gases

Gas	C_v, cal / mole · K
Helium, He	3.00
Argon, A	3.00
Mercury, Hg	3.00
Hydrogen, H_2	4.82
Oxygen, O_2	4.97
Chlorine, Cl_2	6.01
Ether, $(C_2H_5)_2O$	30.8

The experimental values of C_v for several gases are given in Table 1.1. Note that three values agree very closely with the computed value but that the others are quite different. Both the agreements and the disagreements are interesting. The values that agree are those of monatomic gases and those that disagree are not. We shall have more to say about the apparent disagreements. The point here is to recall that in our discussion of kinetic theory we assumed that our molecules were isolated elastic spheres. We should not expect our result to apply to diatomic dumbbells or to complicated molecules. Thus the evidence is strong that our assumption that $U = \frac{3}{2}nRT$ is correct and we have from Eq. (1.13) the gas law equation we sought:

$$pV = nRT. \tag{1.16}$$

1.8 EQUIPARTITION OF ENERGY

The agreement we have observed for monatomic molecules would not have been possible had there not been the number 3 in the expression $C_v = 3R/2$. Looking back over our derivation, we find that the 3 entered into the calculation from the statement $\overline{v^2} = 3\overline{v_x^2}$, that is, because the molecule was free to move in three-dimensional space. The expression for the average kinetic energy of translation of the molecules of a mole of gas is composed of three equal parts, $R/2$ per degree of absolute temperature associated with each coordinate. The principle of *equipartition of energy* states that if a molecule can have energy associated with several coordinates, the average energy associated with each coordinate is the same. Because of this principle, the number of coordinates necessary to specify the position and configuration of a body is called the number of its *degrees of freedom*.

A monatomic molecule requires three coordinates to specify its position. A rigid diatomic molecule requires three position coordinates and two more are necessary to specify its configuration. If we assume the second atom is at a fixed distance from the first, its location is specified as being on a sphere with the first atom at its center.

It requires but two additional coordinates to specify where on this sphere the second atom lies. Thus the addition of a second atom adds two degrees of freedom to the molecule. If our derivation for molar heat capacity had been based on diatomic rather than monatomic molecules, we would have obtained $5R/2$ instead of $3R/2$. We find that $5R/2 = 4.95$ cal/mole · K, which agrees closely with the measured molar heat capacities of such diatomic molecules as hydrogen and oxygen, as shown in Table 1.1. Six coordinates are enough to specify the position of any *rigid* molecule, however complex, but if the molecules are composed of vibrating atoms, then the number of degrees of freedom may become very large. This accounts for the large molar heat capacity of ether. A fuller discussion of heat capacities requires the introduction of quantum theory but classical kinetic theory reveals much, both qualitatively and quantitatively.

With the help of independent data from molar heat capacities, we have found that the kinetic theory of matter provides a quantitative mechanical model for both the ideal gas law and molar heat capacities. A very useful relation can be obtained from one of the derived equations and the ideal gas law. Having shown that $pV = \frac{1}{3}nM\overline{v^2}$ and knowing that $pV = nRT$, we see that $\frac{1}{3}nM\overline{v^2} = nRT$. This may be written as

$$\frac{1}{2}M\overline{v^2} = \frac{3}{2}RT. \tag{1.17}$$

This equation shows that the kinetic energy of translation of the molecules of a gas depends only on the absolute temperature, and that at a given temperature, the lighter molecules have the greater speeds. The root-mean-square speed of hydrogen molecules at room temperature is about 1800 m/s, or more than 1 mi/s.

1.9 MAXWELL'S SPEED DISTRIBUTION LAW

We have found that $\overline{v^2}$ can be computed from the temperature of a gas. The speed thus determined is one of the important average properties of a gas. But an average can be misleading. Consider first that a man who earns one million dollars a year is riding in a taxi, alone except for the driver. The average income of the occupants of the car is about \$500,000. This true statement tends to imply that there are two wealthy men in the car. Similarly for a large group of people, a statement of the average income fails to disclose that some incomes are far from average. A fuller economic picture is conveyed if we know what fractions of the populace have incomes within certain ranges. We might find, for example, 1% below \$1000, 5% between \$1000 and \$2000, etc. Such data determine what is called a distribution function.

Thus it is reasonable to ask how individual molecular speeds are distributed about the average. Maxwell solved the following problem: What fraction of the

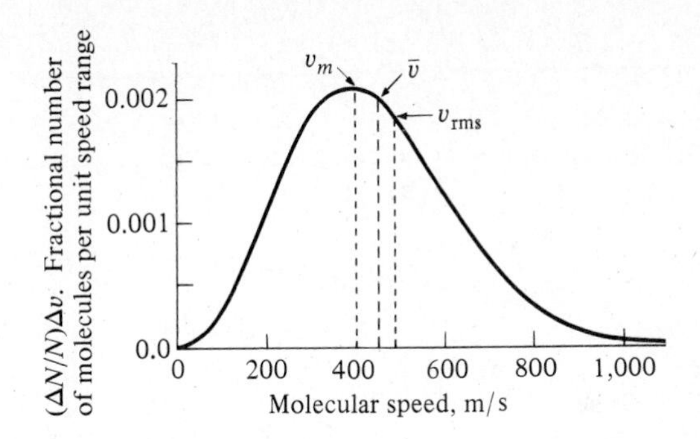

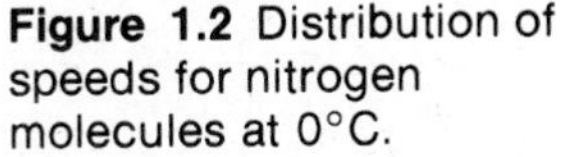
Figure 1.2 Distribution of speeds for nitrogen molecules at 0°C.

molecules of a gas have speeds between v and $v + dv$? Maxwell's solution is too difficult to present here, but the result is

$$\frac{dN}{N} = \frac{4}{\sqrt{\pi}}\left(\frac{M}{2RT}\right)^{3/2} v^2 e^{-Mv^2/2RT}\, dv. \tag{1.18}$$

In this equation, dN/N is the fractional number of molecules that have speeds in the interval dv, M is the molecular mass of the gas, T is its absolute temperature, and R is the gas constant.

This Maxwellian distribution of speeds is shown graphically for a particular case in Fig. 1.2. The curve shows that the molecular speeds range from zero to infinity. The speed for which the curve is maximum is the *most probable speed*, v_m. If we were to pick molecules at random, this speed would be found most often. The average speed obtained from the pressure calculation was $\sqrt{\overline{v^2}}$, or the root-mean-square speed v_{rms}. It is larger than the most probable speed because it is based on the square of the speeds, which gives high speeds more relative importance than low ones. Intermediate between these is the *average speed* $\bar{v}$.

Equation (1.18) can be used to obtain v_m, $\bar{v}$, and v_{rms} (note that $\overline{v^2}$ is the mean square velocity discussed in Section 1.7). The coefficient of dv in Eq. (1.18) is the probability of finding a molecule with speed v. Therefore the most probable speed can be obtained by differentiating this quantity and setting it equal to zero:

$$\frac{d}{dv}\left[\frac{4}{\sqrt{\pi}}\left(\frac{M}{2RT}\right)^{3/2} v^2 e^{-Mv^2/2RT}\right] = 0. \tag{1.19}$$

Solving for v gives

$$v_m = \sqrt{\frac{2RT}{M}}. \tag{1.20}$$

The average speed $\bar{v}$ is found by adding the speeds of all molecules and dividing by the total number of molecules. If there are ΔN_1 molecules with speed v_1, ΔN_2

Table 1.2 Values of $\int_0^\infty x^n e^{-ax^2}\,dx$†

n	$\int_0^\infty x^n e^{-ax^2}\,dx$
0	$\frac{1}{2}\sqrt{\frac{\pi}{a}}$
1	$\frac{1}{2a}$
2	$\frac{1}{4}\sqrt{\frac{\pi}{a^3}}$
3	$\frac{1}{2a^2}$
4	$\frac{3}{8}\sqrt{\frac{\pi}{a^5}}$
5	$\frac{1}{a^3}$

† For evaluation of the integral, see Appendix A in D. C. Kelly, *Thermodynamics and Statistical Physics*. New York: Academic Press (1973).

molecules with speed v_2, and so on,

$$\bar{v} = \frac{v_1\,\Delta N_1 + v_2\,\Delta N_2 + \cdots}{N} = \sum_{i=1}^{q} \frac{v_i\,\Delta N_i}{N},$$

where q is the number of different speeds. In integral form this becomes

$$\bar{v} = \int v\frac{dN}{N} = \frac{4}{\sqrt{\pi}}\left(\frac{M}{2RT}\right)^{3/2}\int_0^\infty v^3 e^{-Mv^2/2RT}\,dv. \tag{1.21}$$

The integral is of the form $\int_0^\infty x^n e^{-ax^2}\,dx$ where $n = 3$ and $a = M/2RT$. From Table 1.2 we find that the value for the integral is $\frac{1}{2}(2RT/M)^2$ and

$$\bar{v} = \sqrt{\frac{8RT}{\pi M}}. \tag{1.22}$$

In a similar way we can obtain $v_{\text{rms}} = \sqrt{\overline{v^2}}$. Considering the square of velocities this time we have†

$$\overline{v^2} = \frac{v_1^2\,\Delta N_1 + v_2^2\,\Delta N_2 + \cdots}{N} = \sum_{i=1}^{q} \frac{v_i^2\,\Delta N_i}{N} \tag{1.23}$$

or

$$\overline{v^2} = \int v^2\frac{dN}{N} = \frac{4}{\sqrt{\pi}}\left(\frac{M}{2RT}\right)^{3/2}\int_0^\infty v^4 e^{-Mv^2/2RT}\,dv. \tag{1.24}$$

† The form of this equation is different from that of Eq. (1.6) because here we are summing over the number of different speeds, while for Eq. (1.6) we summed over the number of particles.

The value of the integral is taken from Table 1.2 and

$$\overline{v^2} = 3\frac{RT}{M}; \tag{1.25}$$

therefore,

$$v_{\text{rms}} = \sqrt{\overline{v^2}} = \sqrt{\frac{3RT}{M}}. \tag{1.26}$$

The relations between the characteristic speeds of gas molecules can be obtained by comparing Eqs. (1.20), (1.22), and (1.26) and are as follows:

$$\begin{aligned} v_{\text{rms}} &= \sqrt{\frac{3RT}{M}}, \\ v_m &= \sqrt{\frac{2}{3}}\, v_{\text{rms}} = 0.817 v_{\text{rms}}, \\ \bar{v} &= \sqrt{\frac{8}{3\pi}}\, v_{\text{rms}} = 0.921 v_{\text{rms}}. \end{aligned} \tag{1.27}$$

EXAMPLE Calculate (a) the root-mean-square speed of the molecules of nitrogen under standard conditions, and (b) the kinetic energy of translation of one of these molecules when it is moving with the most probable speed in a Maxwellian distribution. (The required physical data can be found in Appendixes 3, 7, and 8.)

Solution (a) Using Eq. (1.17) and considering a kmole of the gas, we have

$$\tfrac{1}{3}M\overline{v^2} = RT,$$

or

$$\overline{v^2} = 3RT/M$$

$$\overline{v^2} = 3 \times 8.31 \times 10^3 \frac{\text{J}}{\text{K} \cdot \text{kmole}} \times 273\ \text{K} \times \frac{1\ \text{kmole}}{28\ \text{kg}}$$

$$= 2.43 \times 10^5 \frac{\text{m}^2}{\text{s}^2};$$

$$v_{\text{rms}} = \sqrt{\overline{v^2}} = \sqrt{2.43 \times 10^5\ \text{m}^2/\text{s}^2} = 492\ \text{m/s}.$$

This result can also be obtained from Eq. (1.11):

$$pV = \tfrac{1}{3}nM\overline{v^2};$$

$$\overline{v^2} = \frac{3pV}{nM}$$

$$= 3 \times 1.013 \times 10^5 \frac{N}{\text{m}^2} \times 22.42 \frac{\text{m}^3}{\text{kmole}} \times \frac{1\ \text{kmole}}{1 \times 28\ \text{kg}}$$

$$= 2.43 \times 10^5 \frac{\text{m}^2}{\text{s}^2};$$

$$v_{\text{rms}} = \sqrt{\overline{v^2}} = \sqrt{2.43 \times 10^5\ \text{m}^2/\text{s}^2} = 492\ \text{m/s}.$$

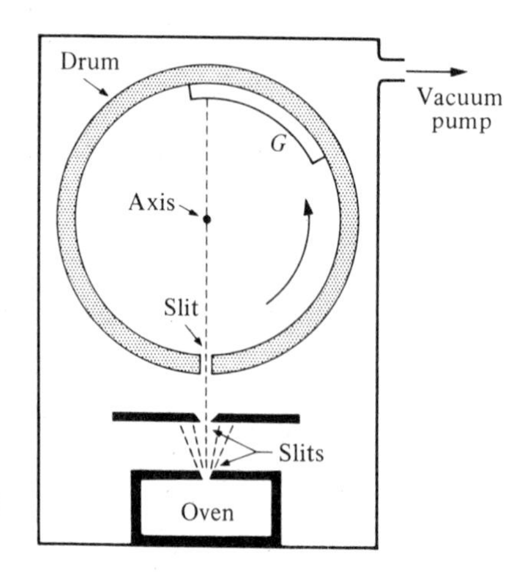

Figure 1.3 Diagram of apparatus used by Zartman and Ko.

(b) $E_k = \frac{1}{2}mv_m^2$.

The mass of a nitrogen molecule can be calculated from its molecular mass and the Avogadro constant:

$$\begin{aligned} m &= M/N_A \\ &= 28\frac{\text{kg}}{\text{kmole}} \times \frac{1\ \text{kmole}}{6.02 \times 10^{26}\ \text{molecules}} \\ &= 4.65 \times 10^{-26}\ \text{kg/molecule}. \end{aligned}$$

From Eq. (1.27) we have

$$v_m = \sqrt{\tfrac{2}{3}}\, v_{\text{rms}},$$

$$\begin{aligned} \therefore E_k &= \tfrac{1}{2} \times 4.65 \times 10^{-26}\ \text{kg} \times \tfrac{2}{3} \times 2.43 \times 10^5\ \text{m}^2/\text{s}^2 \\ &= 3.76 \times 10^{-21}\ \text{J}. \end{aligned}$$

Maxwell's distribution of speeds was employed to calculate other gas properties and was indirectly verified in terms of these secondary properties. A direct experimental verification was obtained by Zartman and Ko in 1930. They used an oven, shown in Fig. 1.3, containing bismuth vapor at a known high temperature (827°C). Bismuth molecules streamed from a slit in the oven into an evacuated region above.† The beam was made unidirectional by another slit, which admitted only properly directed molecules. Above the slit was a cylindrical drum that could be rotated in the vacuum about a horizontal axis perpendicular to the paper. A slit along one side of the drum had to be in a particular position to enable the beam of molecules to enter it. When the drum was stationed so that the beam could enter, the beam moved

† Fast-moving molecules escape from the oven more often than slow ones. Computation shows that if the oven is at a temperature T, the root-mean-square speed of escaping molecules is the same as the root-mean-square speed within an oven at a higher temperature, $4T/3$.

along a diameter of the drum and was deposited on a glass plate G mounted on the inside surface of the drum opposite the slit. During the experiment, the drum was rotated at a constant angular velocity so that short bursts of molecules were admitted on each rotation. Because the speeds of the molecules varied, some crossed the diameter quickly and others took much more time and, since the drum was turning while the molecules were moving across it, they struck the glass plate at different places. Thus the distribution of speeds was translated by the apparatus into a distribution in space around the inside of the drum, as indicated by the variation in the darkening of the glass where the bismuth was deposited. The thickness of the deposit was measured optically, and comparison of the experimental distribution of speeds with Maxwell's theoretical distribution expression showed excellent agreement.

1.10 COLLISION PROBABILITY. MEAN FREE PATH

If molecules were truly geometrical points, no collisions would take place between them. Actual molecules, however, are of finite size, and for the purposes of this discussion we are assuming that a molecule is a rigid, perfectly elastic sphere. A collision between two molecules is considered to take place whenever one molecule makes contact with another. Let us refer to one of the colliding molecules as the target molecule, of radius r_t, and to the other as the bullet molecule, of radius r_b. Then a collision occurs whenever the distance between the centers of the molecules becomes equal to the sum of their radii, $r_t + r_b$, as in Fig. 1.4(a).

When we are considering collisions of molecules of a given gas with other molecules of the same gas, the radii r_t and r_b are equal and there is no difference

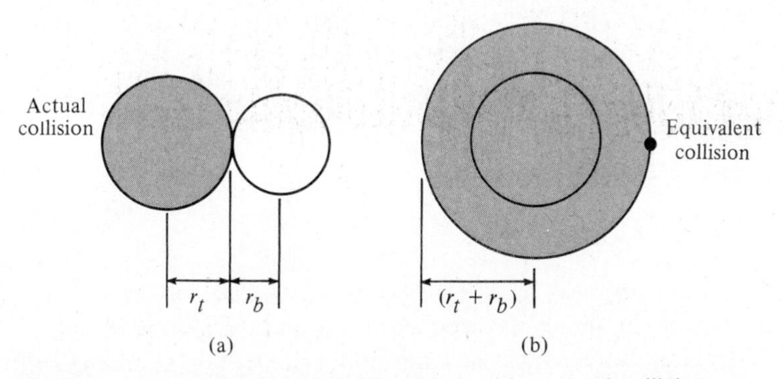

Figure 1.4 For mathematical convenience, the actual collision depicted in part (a) may be represented by the equivalent collision shown in (b).

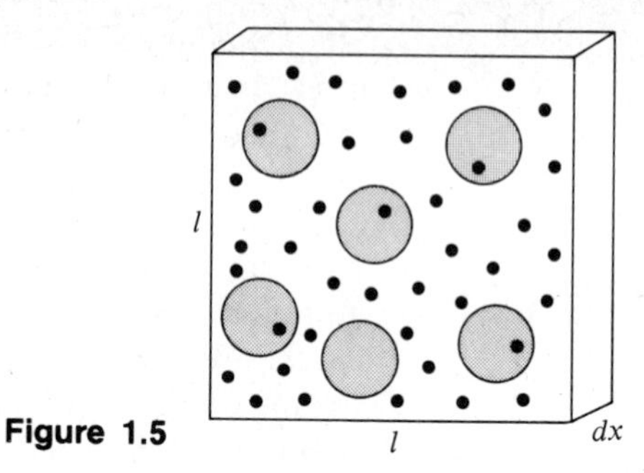

Figure 1.5

between target molecules and bullet molecules. In many instances, however, we wish to consider collisions between different kinds of particles, and so we shall speak of the target molecules as though they differed from the bullet molecules.

Since it is only the center-to-center distance that determines a collision, it does not matter whether the target is large and the bullet is small, or vice versa. We may therefore replace an actual collision with the equivalent collision shown in Fig. 1.4(b), in which the bullet molecule has been considered to shrink to a geometrical point and the target molecule to expand to a sphere of radius $r_t + r_b$.

Now consider a thin layer of material of dimensions l, l, and dx. The layer contains (equivalent) target molecules only, and to begin with we assume that these are at *rest*. We then imagine that a very large number N of bullet molecules are incident normally on the face of the layer like a blast of pellets from a shotgun, in such a way that they are distributed over the face. If the thickness of the layer is so small that no target molecule can *hide* behind another, the layer presents to the bullet molecules the appearance shown in Fig. 1.5, where the shaded circles represent the target molecules and the black dots the bullet molecules.

Most of the bullet molecules will pass through the layer, but some will collide with target molecules. The ratio of the number of collisions, dN, to the total number of bullet molecules, N, is equal to the ratio of the area presented by the target molecules to the total area presented by the layer:

$$\frac{dN}{N} = \frac{\text{target area}}{\text{total area}}.$$

The target area σ of a single (equivalent) molecule is

$$\sigma = \pi(r_t + r_b)^2. \tag{1.28}$$

This area is called the *microscopic collision cross section* of one (equivalent) molecule. The total target area is the product of this and the number of molecules in the layer. If there are n target molecules per unit volume, this number is $nl^2\,dx$, so the total target area is

$$n\sigma l^2\,dx.$$

The total area of the layer is l^2, so

$$\frac{dN}{N} = \frac{n\sigma l^2\, dx}{l^2} = n\sigma\, dx. \tag{1.29}$$

The quantity $n\sigma$ is called the *macroscopic cross section* of the group of (equivalent) molecules. Note that the unit of macroscopic cross section is reciprocal length, not area.

In the preceding equation, the quantity dN/N is the fractional number of molecules that undergo collisions and therefore this ratio is simply the probability of a collision. (Strictly, this should have a negative sign because dN molecules are removed from the stream of bullets.) In the beginning of this discussion the cross section was thought of as an actual area presented by a target molecule, but this was soon replaced by an equivalent area. The microscopic cross section can now be defined in a more realistic way through Eq. (1.29), i.e., that the probability of an interaction between two molecules is directly proportional to σ.

If N_0 bullet molecules per unit area are incident normally on the face of a layer of material containing *stationary* molecules having the macroscopic cross section $n\sigma$, then N, the number transmitted per unit area through a finite thickness x, can be found by integrating Eq. (1.29). We then have

$$\int_{N_0}^{N} -\frac{dN}{N} = \int_0^x n\sigma\, dx, \tag{1.30}$$

and obtain

$$\ln\frac{N}{N_0} = -n\sigma x$$

or

$$N = N_0 e^{-n\sigma x}. \tag{1.31}$$

This exponential equation is plotted as a solid line in Fig. 1.6.

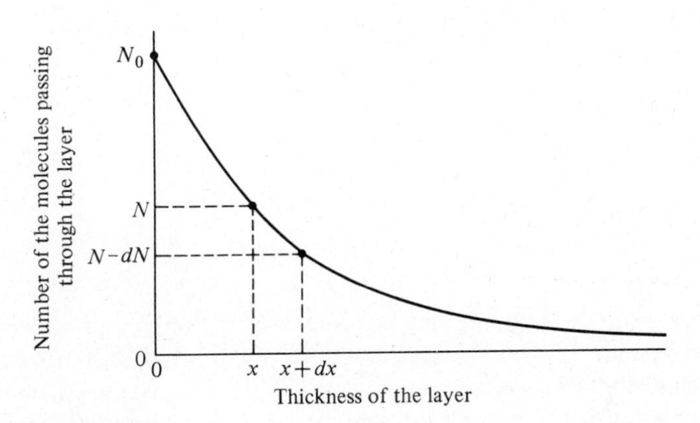

Figure 1.6

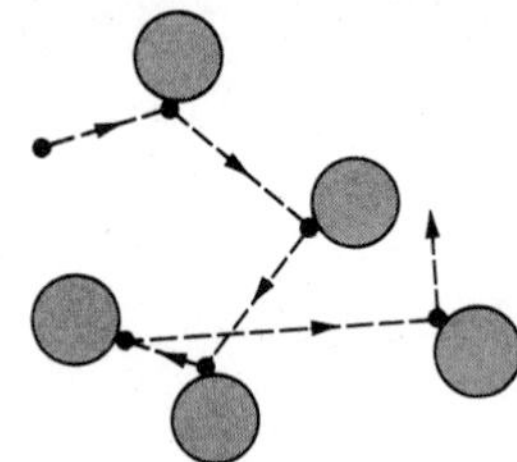

Figure 1.7 Molecular free paths.

Let us next follow in imagination a single bullet molecule as it makes its way through a material along the zigzag path shown in Fig. 1.7. We wish to obtain an expression for the average distance traveled between collisions, known as the *mean free path*, L. This can be deduced from the results above by a type of reasoning that is common and useful in problems of this sort.

When molecules are passing through the thin layer of material in Fig. 1.5, the number removed from the beam by collisions is small compared with the original number, and we can say that N molecules have each traversed a thickness dx of material and that in the process a number dN of collisions have taken place. The total distance traveled by all of the N molecules is then $N\,dx$. We now make the hypothesis that the number of collisions made by a single molecule in traversing the same total distance $N\,dx$ is equal to the number of collisions made by N molecules, each traversing a distance dx. Then from Eq. (1.29), the total number of collisions made by the single molecule in a total path length $N\,dx$ is

$$dN = N\,n\sigma\,dx. \tag{1.32}$$

The mean free path of the molecule is equal to the total path length divided by the number of collisions, or

$$L = \frac{\text{total path length}}{\text{total number of collisions}}.$$

From the expressions above for the total path length and the total number of collisions, we have

$$L = \frac{N\,dx}{N\,n\sigma\,dx}$$

or

$$L = \frac{1}{n\sigma}. \tag{1.33}$$

The concept of mean free path may be visualized by thinking of someone shooting a rifle aimlessly into a forest. Most of the bullets will hit trees, but some bullets will travel much farther than others. It is easy to see that the average distance the bullets go will depend inversely on both the denseness of the woods and the size of the trees.

The expression for the mean free path of a bullet molecule can also be found from Eq. (1.31). Obviously, L is the mean or average distance $\bar{x}$ traveled by the bullet molecules before collisions with the targets in a layer of material so thick that no molecule goes all the way through it. Therefore, what we seek is the numerical average of the distances traveled by the bullets as their number decreases from N_0 to 0. The required relation can be found by considering the curve in Fig. 1.6. Let N be the number of bullets that have traveled a distance x, and $N - dN$ those moving onward beyond $x + dx$; then dN is the number of bullets that collided with targets in the distance dx. Therefore, the combined ranges of the bullets in this group at the time of collision is $x\,dN$. The average distance traveled by a bullet will be the sum of the combined ranges of all the groups from N_0 to 0 divided by the total number of bullets. Mathematically, this is

$$\bar{x} = L = \frac{\int_{N_0}^{0} x\,dN}{\int_{N_0}^{0} dN}. \tag{1.34}$$

This is most easily integrated if dN in the numerator is replaced by its equivalent in terms of dx. From Eq. (1.31) we get

$$N = N_0 e^{-n\sigma x},$$

and the derivative of this is

$$dN = -N_0 n\sigma e^{-n\sigma x}\,dx. \tag{1.35}$$

The new limits are obtained from the conditions that when $N = N_0$, $x = 0$; and when $N = 0$, $x = \infty$. Substituting the expression for dN from Eq. (1.35) into the numerator of Eq. (1.34) and integrating the denominator, we have

$$\bar{x} = L = \frac{\int_0^\infty - xN_0 n\sigma e^{-n\sigma x}\,dx}{-N_0} = n\sigma \int_0^\infty x e^{-n\sigma x}\,dx. \tag{1.36}$$

We integrate this by parts, obtaining

$$L = -\left[\frac{x + 1/n\sigma}{e^{n\sigma x}}\right]_0^\infty.$$

This is indeterminate at the upper limit but, by using l'Hospital's rule, we find the result is

$$L = \frac{1}{n\sigma}. \tag{1.37}$$

This is identical to the previous expression for the mean free path. (We will obtain equations of the same form as Eq. (1.31) in later chapters, especially in the discussions of x-rays and of radioactivity. Note that the derivation of Eq. (1.37) shows that the mean value, over the range 0 to ∞, of the variable in the exponent in Eq. (1.31) is equal to the negative of the reciprocal of the coefficient of that variable.)

In the above analysis, we assumed that the target molecules were at rest. This assumption is valid for a bullet molecule going through a solid, and we will meet this situation in a later chapter when we consider neutrons passing through a solid moderator in a nuclear reactor. If, however, we consider a gas in which both the target and the bullet molecules are moving randomly, the mean free path will decrease because now there are not only head-on collisions as before, but also "sideswipes" with targets moving across the line of travel of the bullet. It is found that the mean free path of a molecule of an ideal gas having a Maxwellian distribution of speeds is

$$L = \frac{0.707}{n\sigma}. \tag{1.38}$$

1.11 FARADAY'S LAW OF ELECTROLYSIS — SKEPTICISM

Another line of argument supporting the atomic view of matter came from the work of Faraday. In 1833, he observed that if the same electric charge is made to traverse different electrolytes, the masses of the materials deposited on the electrodes are proportional to the chemical equivalent weights of the materials. The quantity of electricity required to deposit a mole of univalent ions in electrolysis is called the Faraday constant, F, and is equal to 9.64870×10^4 coulombs. Like the law of multiple proportions proposed by Dalton, this also implied atomicity of matter. Faraday's law, however, brings electricity into the picture and implies that both electricity and matter are atomic.

We have traced a few highlights of the development of the atomic view of matter through most of the nineteenth century, but since no one had ever seen a molecule, the entire theory was still regarded with skepticism. Maxwell, who proposed the distribution of speeds already discussed, did his greatest work in electrical theory. It was he who found the relationship between electricity and light, and it is because of his work that we often call light "electromagnetic radiation." In his comprehensive book on electricity and magnetism (1873), after explaining Faraday's laws of electrolysis on the basis of the atomic theory of matter and electricity, Maxwell says, "It is extremely improbable that when we come to understand the true nature of electrolysis we shall retain in any form the theory of molecular charges, for then we shall have obtained a secure basis on which to form a true theory of electric currents and so become independent of these provisional theories."

As late as 1908 the physical chemist Wilhelm Ostwald and the physicist Ernst Mach opposed the atomic theory of matter. Their skepticism is an interesting question in epistemology. These scientists were unwilling to accept purely indirect evidence. Mach makes their position clear in the following analogy: A long elastic rod held in a vise may be made to execute slow, perceivable vibrations. If the rod is shortened, the vibrations become a blur in which individual motions of the rod cannot be followed. If the rod is shortened further, the blur may be visually unobservable but a tone is heard. If the rod is made so short that we no longer experience a physical sensation from its behavior, we may still think of it as vibrating when struck. This, according to Mach, is a safe extrapolation of our ideas because it proceeds from the *directly* observable to the *indirectly* observable. Those who were skeptical about the atomic theory objected to the fact that the evidence was *entirely indirect*. The experiments described in the next section provided the observable events that made the indirect evidence we have given acceptable to everyone.

1.12 PERRIN'S VERIFICATION OF THE ATOMIC VIEW OF MATTER

Credit for removing the remaining skepticism of atomic theory goes to the French physical chemist Perrin. He tested the hypothesis that the suspended particles that dance about in a stationary liquid in Brownian movement behave like large gas molecules. For his experiments, Perrin prepared a water suspension of particles which met a stringent set of requirements. The particles had to be large enough to be seen individually but small enough to have an appreciable thermal motion that could be measured; they also had to be of known uniform size and mass. Further, the concentration of the particles in the suspension had to be so low that the force effects between them could be neglected. In short, the particles had to be directly observable and conform to the assumptions of the kinetic theory of gases. Perrin was able to obtain a suspension of particles that met these requirements by centrifuging a water mixture of powdered gamboge, a gum resin more dense than water. The centrifuge separated the particles according to size. After drawing off a portion of the mixture where the magnitude of the particle size was suitable for his experiments, he could centrifuge again and again until the size of the remaining particles was nearly uniform. Although gamboge is more dense than water, Perrin observed that the particles did not settle out of still water. They assumed a distribution in height, with more particles per unit volume near the bottom of the container than at the top. He measured this distribution as a function of height.

To derive the distribution equation, consider a vertical column of gas (Fig. 1.8) that has a cross-sectional area a and is at a uniform temperature. Through this column, let us take a horizontal slice of thickness dy. If the weight of the gas above this slice is $\mathbf{w}$, then the weight of the gas above the bottom of the slice will be

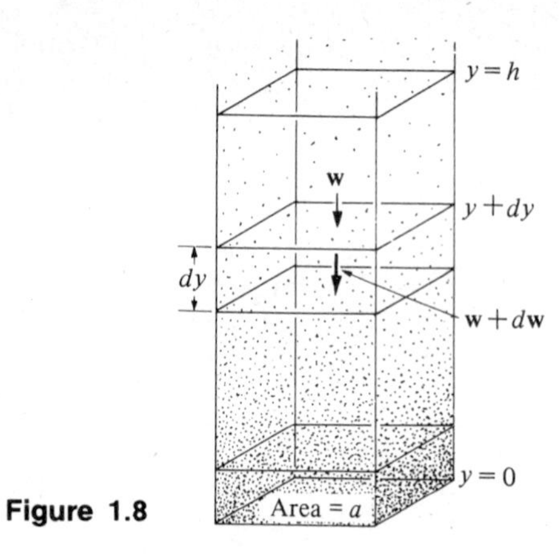

Figure 1.8

$\mathbf{w} + d\mathbf{w} = \mathbf{w} + mgna\,dy$, where mg is the weight of a molecule, n is the average number of molecules per unit volume, and $a\,dy$ is the volume of the slice. (Note that we are employing an atomic view of the gas.) The difference of these weights per unit area, $mgn\,dy$, is the pressure difference due to the gas in the slice, that is, $dp = -mgn\,dy$. The minus sign denotes that the pressure decreases as the height increases. Since the molecules in the column of gas have weight, the number of molecules per unit volume at a low level is greater than at a higher one. Because of this difference in concentration, there is a corresponding difference in the number of molecular collisions per unit time at the two levels, and the column of gas comes to dynamic equilibrium. In this state, the weight of the molecules in any layer is just balanced by the net upward force caused by the difference between the number of molecular impacts per unit time on the lower and upper horizontal surfaces of the layer. We shall next obtain an expression for the difference in pressure between two levels produced by a difference in molecular concentrations.

The equation $pV = RT$ holds for one mole of any gas that can be regarded as ideal. According to atomic theory, the number of molecules in a mole is the Avogadro constant N_A. We can obtain an expression that contains the concentration of the molecules by dividing the ideal gas law equation by the Avogadro constant. Thus we have

$$\frac{pV}{N_A} = \frac{RT}{N_A}$$

or

$$p = \left(\frac{N_A}{V}\right)\left(\frac{R}{N_A}\right)T.$$

The quantity N_A/V is the molecular concentration n. (Note that here n is not the number of moles as earlier in this chapter.) This concentration is a function of the pressure and therefore, since R, N_A, and T are constant, $dp = (R/N_A)T\,dn$. When this expression for the difference of pressure due to the difference in molecular concentrations in the layer of height dy is equated to the difference due to the weight of the layer, we obtain

$$\frac{R}{N_A} T\,dn = -mgn\,dy,$$

or

$$\frac{dn}{n} = -\frac{N_A mg}{RT}\,dy. \tag{1.39}$$

The relation between the molecular concentration n_0 at $y = 0$ and n at $y = h$ can be found by integrating Eq. (1.39). We have

$$\int_{n_0}^{n} \frac{dn}{n} = \int_0^h -\frac{N_A mg}{RT}\,dy,$$

which gives

$$n = n_0 e^{-(N_A mgh/RT)}. \tag{1.40}$$

Equation (1.40) may also be written in the form

$$n = n_0 \exp\left[-\frac{N_A mgh}{RT}\right]. \tag{1.41}$$

Since the pressure of a gas is directly proportional to the number of molecules per unit volume, Eq. (1.41) can be rewritten in terms of pressures. The resulting equation is then called the *law of atmospheres*, since it gives the distribution in height of the pressure in a column of gas at constant temperature and subject to the force of gravity.

Equation (1.41) must be modified slightly to make it applicable to the study of Brownian movement. The effective weight of a particle suspended in a fluid is the resultant of its weight and Archimedes' buoyant force. The volume of a particle of mass m and density ρ is m/ρ and the mass of an equal volume of liquid having a density ρ' is $\rho' m/\rho$. Therefore, the buoyant force on this particle when submerged in the liquid is $mg\rho'/\rho$, and its effective weight becomes

$$mg - mg\frac{\rho'}{\rho} = mg\left(\frac{\rho - \rho'}{\rho}\right).$$

When we replace the actual weight mg in Eq. (1.41) by the effective weight, we obtain

$$n = n_0 \exp\left[-\frac{N_A mg(\rho - \rho')h}{\rho RT}\right]. \tag{1.42}$$

This is the equation for *sedimentation* equilibrium of a suspension as a result of Brownian movement.

Perrin measured a series of n's at a series of h's in a very dilute suspension. The results verified the sedimentation equation. Equally important, his measurements yielded a value for N_A, since all other quantities in the equation were known. The very existence of the Avogadro constant implies the correctness of the atomic theory.

It is interesting that Einstein, the greatest contributor to the development of modern physics, also had a part in the final establishment of the atomic theory. In 1905 he derived an equation that describes how a suspended particle should migrate in a random manner through a liquid. His expression involved the Avogadro constant, and Perrin's verification of the Einstein formula confirmed the value of this constant. The results showed that small particles suspended in a stationary liquid do move about in the manner predicted by the molecular-kinetic theory of gases.

Altogether, Perrin used four completely independent types of measurements, each of which was an observable verification of atomic theory and each of which gave a quantitative estimate of the Avogadro constant. Since the publication of his results in 1908, no one has seriously doubted the atomic theory of matter.

1.13 BOLTZMANN CONSTANT

Once N_A was determined, it was possible to evaluate the frequently recurring quantity R/N_A. This is the universal gas constant *per molecule*, and is known as the *Boltzmann constant*, k. Its value is 1.380×10^{-23} J/K. Earlier in this chapter we computed the translational kinetic energy of the molecules of a gas from $\frac{1}{2}M\overline{v^2} = \frac{3}{2}RT$ (Eq. 1.17), where M and R are the mass and gas constant per mole, respectively. Replacing M by N_Am and R by N_Ak in this equation, we obtain

$$\frac{1}{2}m\overline{v^2} = \frac{3}{2}kT, \tag{1.43}$$

where m and k are the mass and gas constant *per molecule*, respectively. Thus the average translational kinetic energy of any molecule of a gas depends only on its temperature (if indeed the word temperature has meaning for a single molecule), and the *energy per degree of freedom* is $kT/2$.

Let us now return to Eq. (1.40), the law of atmospheres. In this equation N_A/R equals $1/k$ and mgh is the gravitational potential energy E_p of a particle. Substituting in Eq. (1.40), we get

$$n = n_0 e^{-E_p/kT}. \tag{1.44}$$

This expression yields the relative number of particles per unit volume in two different energy states in terms of their difference of gravitational potential energy, E_p. It turns out that this same equation may be used for any case where the potential energy is a function of position. For two energy states of electrified particles, E_p would be the difference in their electrostatic potential energies. When all types or forms of energy are considered, we obtain the *Boltzmann distribution law*

$$n = n_0 e^{-E/kT}, \tag{1.45}$$

where E is the difference in energy between two states. We will need this equation in Chapter 3, when we discuss the distribution of energy in a group of harmonic oscillators.

1.14 MAXWELL REVISITED

We have not *derived* Boltzmann's law; we have gotten it as an extension of the law of atmospheres. But we can use Boltzmann's law to get Maxwell's law, (Eq. 1.18).

In an ideal monatomic gas, a particle has kinetic energy of translation only, $E_k = \frac{1}{2}mv^2$. The mass m is the molecular mass divided by the Avogadro constant so that $E_k = \frac{1}{2}Mv^2/N_0$. Thus Boltzmann's law for molecules of an ideal gas is

$$N = N_0 e^{-1/2\,Mv^2/N_0kT} = N_0 e^{-Mv^2/2RT}, \tag{1.46}$$

where we have used the fact that $N_0 k = R$.† This exponential equation indicates that there are more molecules with small velocity than with large.

Maxwell's law gives a distribution of *speeds* without regard for direction. To go from velocities to speeds, consider a three-dimensional-velocity space in which all molecular velocities are translated so that all the vector "tails" are at the origin (Fig. 1.9). The velocity vectors look like sparks from a firecracker, randomly directed and distributed in length according to Boltzmann's law.

All vectors having a narrow range of speeds between v and $v + dv$ lie in a spherical shell between two spheres of "radius" v and $v + dv$. Since the area of a sphere is $4\pi r^2$, the "volume" of this shell in *velocity space* is $4\pi v^2\,dv$. All velocity vectors that end in this shell have *speeds* between v and $v + dv$. The number of vectors ending in the shell is the "volume" times the number "at that distance" or

$$dN = 4\pi N_0 v^2 e^{-Mv^2/2RT}\,dv. \tag{1.47}$$

The total number of molecules must be $N = \int dn$, or

$$N = 4\pi N_0 \int_0^\infty v^2 e^{-Mv^2/2RT}\,dv. \tag{1.48}$$

This integral may be evaluated from Table 1.2 with $x = v$, $n = 2$, and $a = M/2RT$,

$$N = N_0\left(\frac{2\pi RT}{M}\right)^{3/2}. \tag{1.49}$$

† The N_0 in the exponent must be Avogadro's number. The N_0 after the equal sign need not be Avogadro's number. It will cancel later.

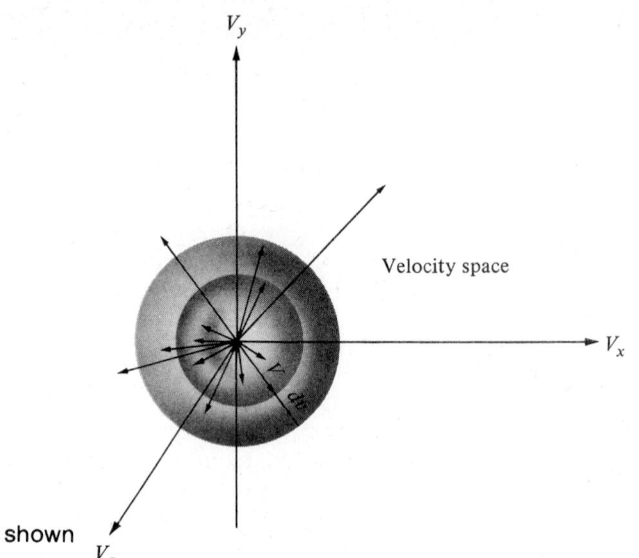

Figure 1.9 Velocity space shown in three dimensions.

Combining Eq. (1.47) with Eq. (1.49), we have finally

$$\frac{dN}{N\,dv} = \frac{4}{\sqrt{\pi}}\left(\frac{M}{2RT}\right)^{3/2} v^2 e^{-Mv^2/2RT}, \tag{1.50}$$

which is the same as Eq. (1.18).

This shows that given Boltzmann's law, Maxwell's law can be derived; and Maxwell's law has been tested, as we have seen. This exercise increases our confidence in Boltzmann's law.

FOR ADDITIONAL READING

American Institute of Physics, D. E. Gray, editor, *American Institute of Physics Handbook*. 3rd ed. New York: McGraw-Hill, 1972. A very comprehensive book.

Bridgman, P. W., *The Logic of Modern Physics*. New York: Macmillan, 1938.

Brown, T. B., editor, *The Lloyd William Taylor Manual of Advanced Undergraduate Experiments in Physics*. Reading, Mass.: Addison-Wesley, 1959. Discusses theory and apparatus required for numerous experiments.

Conant, James B., *On Understanding Science; An Historical Approach*. New Haven, Conn.: Yale University, 1947.

Einstein, Albert, *Investigation on the Theory of the Brownian Movement*. R. Fürth, translator. New York: Dover Publications, 1948.

Frank, Philipp, *Philosophy of Science*. Englewood Cliffs, N.J.: Prentice-Hall, 1957.

Hodgeman, C. D., editor, *Handbook of Chemistry and Physics*. 63rd ed. Cleveland, Ohio: Chemical Rubber Publishing Co., 1982. A good reference book for chemical and physical data.

Holton, Gerald, *Introduction to Concepts and Theories in Physical Science*. 2nd ed. Revised and with new material by Stephen G. Brush. Reading, Mass.: Addison-Wesley, 1973. This is an excellent historical account of how the scientific method has been employed in investigating the physical world. This book should be read by every student in the sciences.

Jeans, James, *Physics and Philosophy*. New York: Macmillan, 1943.

Kelly, Don C., *Thermodynamics and Statistical Physics*. New York: Academic Press, 1973. See Appendix A: Gaussian Integrals.

Laue, Max Von, *History of Physics*. R. E. Oesper, translator. New York: Academic Press, 1950. A short but comprehensive book.

Lee, J. F., F. W. Sears, and D. L. Turcotte, *Statistical Thermodynamics*. Reading, Mass.: Addison-Wesley, 1963.

Leicester, H. M., and H. S. Klickstein, *A Source Book in Chemistry*. New York: McGraw-Hill, 1952. A collection of the principal papers of the great men in chemistry. Every student of atomic theory should read the original works of Dalton, Avogadro, Gay-Lussac, and Cannizzaro.

Mehra, Jagdish, editor, *The Physicist's Conception of Nature*. Boston: D. Reidel Publishing Co., 1973.

Perrin, Jean, *Atoms*. D. L. Hammick, translator. London: Constable, 1923.

Planck, Max, *The Philosophy of Physics*. New York: Norton, 1936.

Sambursky, S., *The Physical World of the Greeks*. Morton Dagut, translator. New York: Macmillan, 1956.

Sarton, George, *A History of Science: Ancient Science through the Golden Age of Greece*. Cambridge, Mass.: Harvard University Press, 1952.

Sears, F. W., and G. L. Salinger, *Thermodynamics, Kinetic Theory and Statistical Thermodynamics*. 3rd ed. Reading, Mass.: Addison-Wesley, 1975.

Segré, Emilio, *From X-Rays to Quarks*. San Francisco: W. H. Freeman, 1980. An excellent book about the main discoveries of modern physics and how they were accomplished. Contains an exceptional bibliography.

Taylor, F. Sherwood, *A Short History of Science and Scientific Thought*. New York: Norton, 1949.

Zemansky, M. W., *Temperatures Very Low and Very High*, Momentum Book No. 6. Princeton, N.J.: Van Nostrand, 1954.

REVIEW QUESTIONS

1. What is the evidence for the atomic view of matter?
2. What is Avogadro's number?
3. What characteristics of the elements allow them to be arranged in the periodic table?
4. What are the three speeds, v_m, $\bar{v}$, v_{rms}?

5. What is meant by mean free path?
6. What is Boltzmann's constant?

PROBLEMS

1.1 Describe one or more experiments that show that the earth is essentially spherical.

1.2 Some early chemists wish to determine the atomic mass of nitrogen. They assume that the atomic mass of oxygen is exactly 16 and they prepare four oxides of nitrogen that are distinctly different compounds (see data below).

Nitrogen, parts by wt.	Oxygen, parts by wt.	Product oxide
86.3	197	*A*
500	285	*B*
300	343	*C*
108.2	186	*D*

(a) Show that these data demonstrate the law of multiple proportions. [*Hint:* First find the masses of one of these elements that unite with a unit mass of the other.] (b) From the above data the chemical formulas of the products cannot be determined completely but, by assuming that nature is simple, one may propose several possible sets of product formulas. Write out several possible sets of product formulas. (c) Calculate the atomic mass of nitrogen for each set.

1.3 The specific heats of the elements are given in Appendix 4. Choose eight of these that are solids at room temperature and distributed over the range of atomic masses. Calculate the molar heat capacity of each in cal/mole · K Plot these results against atomic mass as the abscissa.

1.4 The specific heat of helium is about 0.75 calorie/gram K; that of nitrogen is about 0.18 calorie/gram K. How closely do these agree with expectations?

1.5 The specific heat of lead is approximately 0.03 calorie/gram K at room temperature. That of diamond at room temperature is 0.10 calorie/gram K. How closely do they satisfy the Dulong-Petit law?

1.6 (a) What would happen if the collisions between the molecules of a gas were not perfectly elastic? (b) What would happen to the shape of an air-filled spherical toy balloon if, on the average, the velocities of the gas molecules in one direction were to become greater than those in another?

1.7 A dart board with area A hangs on a vertical wall. Darts are thrown horizontally with a speed v and stick in the dart board. If the mass of each dart is m and r darts strike the board in each second, derive an equation for the average pressure p exerted on the board.

1.8 Compute the molar heat capacity at constant volume of a gas composed of molecules that are rigid three-dimensional structures.

1.9 Show that if ρ represents the density of a gas, then

$$v_{rms} = \sqrt{3p/\rho}\,.$$

1.10 The velocity of sound in a monatomic gas is equal to $[5P/3\rho]^{1/2}$, where P is the pressure and ρ is the density. Find the ratio of the root-mean-square velocity of the molecule to the velocity of sound.

1.11 A horizontal beam of mercury atoms emerges from an oven whose temperature is 1000 K. The atoms travel a horizontal distance of 1 m. How far do they fall during this travel? Assume that the atoms travel at the rms speed corresponding to the oven temperature (see footnote page 18).

1.12 Repeat the derivation of the ideal gas law assuming a mixture of particles having two different masses, m_1 and m_2. From this derivation generalize to the law of partial pressures, which states that the pressure of a mixture of gases is the sum of the pressures due to each gas separately.

1.13 (a) Compute the arithmetic mean speed and the root-mean-square speed for each of the following distributions of the speeds of eight particles: (1) all eight have speeds of 10 m/s; (2) two have speeds of 3 m/s, four have speeds of 6 m/s, and two have speeds of 10 m/s; (3) one has a speed of 3 m/s, three have speeds of 6 m/s, and four have speeds of 10 m/s; (4) four are at rest and four have speeds of 10 m/s. (b) In each case decide whether the shape of the graph of the speed distribution would be the same as that of the translational kinetic energy distribution, assuming that each particle has the same mass.

1.14 The speed distribution function of a group of N particles is given by $dN_v = k\,dv$, where dN_v is the number of particles that have speeds between v and $v + dv$ and k is a constant. No particle has a speed greater than the value V and the range of speeds is from 0 to V. (a) Draw a graph of the distribution function, that is, plot (dN_v/dv) versus v. (b) Find the constant k in terms of N and V. (c) Compute the average and the root-mean-square speeds in terms of V. (d) What percent of the particles have speeds between the average speed and V? between the root-mean-square speed and V?

1.15 The speed distribution function of a group of N particles is given by $dN_v = kv\,dv$ where dN_v is the number of particles that have speeds between v and $v + dv$ and k is a constant. No particle has a speed greater than V and the speeds range from 0 to V. (a) Draw a graph of the distribution function, that is, plot (dN_v/dv) versus v. (b) Find the constant k in terms of N and V. (c) Compute the average speed, the root-mean-square speed, and the most probable speed in terms of V. (d) What percent of the particles have speeds between the average speed and V? between the root-mean-square speed and V?

1.16 For which of the following gases do the molecules have (a) the highest most probable speed and (b) the lowest most probable speed at a given temperature: CO, H_2, O_2, Ar, NO_2, Cl_2, He?

1.17 An object can escape from the surface of the earth if its speed is greater than $\sqrt{2gR}$, where g is the acceleration due to gravity and R is the radius of the earth. (a) Using a radius of 6.4×10^6 m, calculate this escape speed. (b) Explain why oxygen and nitrogen remain in the earth's atmosphere while hydrogen does not.

1.18 Consider a container of helium gas that is a mixture of the two isotopes, ^{3}He and ^{4}He, at $T = 300$ K and atmospheric pressure. What is the ratio of the average speeds of the two different isotopes?

1.19 Uranium hexafluoride, UF_6, is a gaseous compound at 100°C. Given a mixture of two such hexafluorides, one formed of isotope ^{235}U and the other of isotope ^{238}U, find the ratio of the average speeds of the two molecules. The atomic mass of F is 19. The fact that this ratio is not equal to one formed the basis for separation of ^{235}U by gaseous diffusion for use as fuel in nuclear reactors.

1.20 The speed of propagation of a sound wave in air at 27°C is about 348 m/s. Find the ratio of this speed to the rms speed of nitrogen molecules at this temperature.

1.21 (a) At what temperature will the rms speed of oxygen molecules be twice their rms speed at 27°C? (b) At what temperature will the rms speed of nitrogen molecules equal the rms speed of oxygen molecules at 27°C?

1.22 Find the rms speed, the average speed, and the most probable speed of the molecules of gaseous hydrogen at a temperature of (a) 20°C and (b) 120°C.

1.23 Find the rms speed, the average speed, and the most probable speed of the molecules of gaseous hydrogen at a constant temperature of 20°C and a pressure of (a) 1 atm and (b) 100 atm.

1.24 (a) To what temperature would an ideal gas in which the "particles" are baseballs have to be heated so their rms speed in a Maxwellian distribution would equal that of a fast ball having a speed of 30.5 m/s? (The mass of a baseball is 144 g.) (b) What is the microscopic cross section of these "particles"? (The circumference of a baseball is 23 cm.)

1.25 The drum of a Zartman-Ko apparatus, Fig. 1.3, has a radius of 8 cm and rotates at 6000 rpm. The oven contains mercury atoms at a temperature of 600 K. Two atoms of mercury, one with the most probable speed at oven temperature and the other with the rms speed at the same temperature, leave the oven and enter the rotating drum. These two atoms are then deposited on the glass plate at the far side of the drum. What is the separation of these two atoms on the glass plate?

1.26 A certain lecture hall measures 10 m × 20 m × 30 m and the temperature in the hall is 0°C. (a) Calculate the total kinetic energy of the air assuming that the pressure is 1 atmosphere. (Note that the diatomic molecules, O_2 and N_2, have five degrees of freedom at this temperature.) (b) Calculate the force on the wall of the room that is 10 m × 30 m. (c) How many degrees of freedom would be involved in calculating the pressure used in part (b) from the kinetic theory of gases?

1.27 When the atoms in a deuterium gas have an average translational kinetic energy of 12×10^{-14} J, they can approach one another so closely that nuclear fusion will occur. (a) What is the speed of a deuterium atom having this kinetic energy? (b) To what temperature would the deuterium gas have to be heated so that the rms speed of the atoms would equal the speed in the preceding part? Deuterium is hydrogen having an atomic mass of 2.014.

1.28 (a) What is the total kinetic energy of translation of the atoms in 4 moles of helium at a temperature of 27°C? (b) What would be the answer for the same amount of any other ideal gas?

1.29 What is the total kinetic energy of the molecules in 2 moles of argon gas at a temperature of (a) 300 K and (b) 301 K? (c) Calculate the internal energy change of 2 moles of argon when the temperature is increased 1 K, using the specific heat at constant volume, and compare your answers with the change in kinetic energy.

1.30 The microscopic cross section for a certain bullet and particle is σ when they are electrically neutral. Would the value of σ increase or decrease if the bullet and particle carried electric charges (a) of like sign and (b) of unlike sign?

1.31 (a) Show that n, the number of molecules per unit volume of an ideal gas, is given by $n = pN_A/RT$ where N_A is the Avogadro constant. (b) Find the number of molecules in 1 m^3 of an ideal gas under standard conditions. (c) What is the number of molecules in 1 m^3 of an ideal gas at a pressure of two atmospheres and a temperature of 47°C?

1.32 The diameters of molecules can be computed from an equation derived by using the kinetic theory to explain the viscosity of a gas. The results show that the molecular diameters are about 2×10^{-10} m for all gases. Find (a) the microscopic cross section and (b) the macroscopic cross section of the molecules of hydrogen at a temperature of 20°C and a pressure of 1 atm.

1.33 Derive Eq. (1.37) from Eq. (1.36).

1.34 (a) If the pressure is kept constant, at what temperature will the mean free path of the molecules of a given mass of an ideal gas be twice that at 27°C? (b) If the temperature is kept constant, at what pressure in millimeters of mercury will the mean free path of the molecules of a given mass of an ideal gas be 1000 times greater than that at a pressure of 1 atm?

1.35 Find the mean free path for a Maxwellian distribution of speeds (Eq. 1.38) of the molecules of hydrogen gas when at a pressure of 1 atm and a temperature of (a) 20°C and (b) 120°C; and (c) when at a pressure of 100 atm and a temperature of 20°C. (Data for calculating the macroscopic cross section are given in Problem 1.32.) (d) How many collisions per second would a molecule that is always moving with the average speed in a Maxwellian distribution make in each of the preceding cases? (The time of contact during collisions is negligible.) (e) What is the ratio of the mean free path in part (a) to the wavelength of green light, $\lambda = 5500 \times 10^{-10}$ m?

1.36 A beam of bullet particles is incident normally on a layer of material containing stationary target particles. Find (a) the fraction of the incident beam transmitted and the fraction that experienced collisions in a layer whose thickness equals the mean free path, and (b) the thickness of the layer in terms of the mean free path required to reduce the transmitted beam to one-half the intensity of the incident beam.

1.37 Show that the kinetic energy of translation of a molecule having the most probable speed in a Maxwellian distribution is equal to kT.

1.38 At room temperature (300 K), what is the value of kT? The surface of the sun is about 6000 K. What is the value of kT for this temperature?

1.39 A neutron is a fundamental particle. Like ordinary gas molecules, neutrons have a distribution of speeds, and this distribution is of prime importance in the theory of nuclear reactors. Quantitatively, a thermal neutron is usually defined as one having the most probable speed of a Maxwellian distribution at 20°C. Find (a) the kinetic energy of translation and (b) the speed of a thermal neutron. (c) A thermal neutron is sometimes called a "kT neutron." Why?

1.40 Assume that the mean free path of nitrogen molecules in a 5-liter spherical container is 3 mm at room temperature. If enough of the molecules are pumped out of the container to reduce the pressure to a pressure of one fifth the original pressure, what is the mean free path?

1.41 If the thermal conductivity of a gas composed of helium mass isotope four is $\lambda = 0.030$ watt/K cm at 800 K, estimate the thermal conductivity of the gas at 200 K. Estimate the thermal conductivity of a gas of helium mass isotope three at 800 K (assume cross sections for He^3 and He^4 are the same).

1.42 Assume that hydrogen atoms in the atmosphere of the sun obey the Maxwellian speed distribution. Calculate (a) the kinetic energy of one of these atoms moving with the most probable speed in the distribution given that the temperature is 6000 K, and (b) the speed of this atom.

1.43 Consider nitrogen molecules at a temperature of 0°C. (a) What is the average translational kinetic energy of a molecule? (b) What is the average rotational kinetic energy of a molecule? (c) What is the average total kinetic energy of a molecule?

1.44 (a) Show that when the Maxwellian speed distribution is expressed in terms of the mass m of a molecule the result is

$$\frac{dN}{N} = \frac{4}{\sqrt{\pi}}\left(\frac{m}{2kT}\right)^{3/2} v^2 e^{-mv^2/2kT}\, dv.$$

(b) Assuming that the energy E of a molecule is only translational kinetic energy, show that the fractional number of molecules that have energies in the range dE is

$$\frac{dN}{N} = \frac{2}{\sqrt{\pi}} \left(\frac{1}{kT} \right)^{3/2} E^{1/2} e^{-E/kT} \, dE.$$

(c) From the energy distribution of part (b) show that the most probable energy is $\frac{1}{2}kT$. (d) What is the ratio of the average translational kinetic energy to the most probable translational kinetic energy?

1.45 In one of his experiments, using a water suspension of gamboge at 20°C, Perrin observed an average of 49 particles per unit area in a very shallow layer at one level and 14 particles per unit area at a level 60 microns higher. The density of the gamboge was 1.194 g/cm^3 and each particle was a spherical grain having a radius of 0.212 micron. (1 micron = 10^{-6} m.) Find (a) the mass of each particle, (b) the Avogadro constant, and (c) the molecular weight of a particle if each grain is regarded as a single giant molecule. Use the results from parts (a) and (b) to calculate (c).

1.46 Starting with Eq. (1.47), derive Eq. (1.50).

2 THE ATOMIC VIEW OF ELECTRICITY

2.1 ELECTRICAL DISCHARGES

We have already considered how Faraday's law of electrolysis implies that both matter and electricity are atomic. In spite of Maxwell's skepticism, it is very difficult to explain the fact that the passage of one faraday of electricity through an electrolyte liberates or deposits an equivalent weight of a substance, except by assuming that both matter and electricity exist in units that preserve their identity throughout the process.

In order to learn more about "particles of electricity," we turn to another line of investigation and consider the passage of electricity through gases. Although Benjamin Franklin's very dangerous experiment with kite and key was not a particularly convincing one, it nevertheless led to the correct conclusion that lightning is the discharge of electricity through a gas (air). Every electric spark is an example of this process. Since sparks are one of the most dramatic electric effects, it is natural that they should have been a subject of early study.

The passage of electricity through gases is a very complicated process and a great deal has been learned from it. There are many ways in which the character of an electrical discharge can be altered, but here we shall direct our attention to the effect of gas pressure. A typical discharge tube is shown in Fig. 2.1. This system has a gauge that measures the gas pressure and a pumping system that varies the pressure. Electrodes are sealed into the ends of the tube so that an electric field can be established between them.

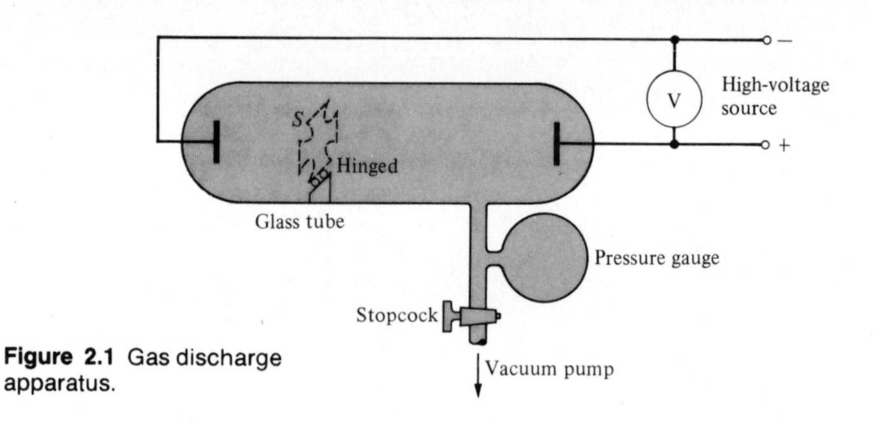

Figure 2.1 Gas discharge apparatus.

When the pressure in the tube is atmospheric, a very large electric field is required to produce a discharge (about 3×10^6 volts/m for air). The discharge is a violent spark as the gas suddenly changes from being an excellent insulator to being a good conductor. As the pressure is reduced, the discharges are more easily established (Fig. 2.2), until, at very low pressures, they again become difficult to start. Discharges start most easily at a pressure of about 2 mm of mercury (although this will depend on the kind of gas and the geometry of the electrodes). As the pressure is reduced, the discharge changes in character. With air in the tube, the bright spark changes to a purple glow filling the whole tube, and with neon, one obtains the red glow seen in many advertising signs. On further lowering of the pressure, the glow assumes a remarkable and complicated structure, with striations and dark spaces. At very low pressures the glow of the gas becomes dim and a new effect appears — the glass itself begins to glow. If the bulb has within it a device that is hinged so that it can be made to move into or out of the region between the electrodes by tipping the entire bulb (S in Fig. 2.1), then another effect may be seen.

Figure 2.2 Typical starting potential curve of gaseous discharge.

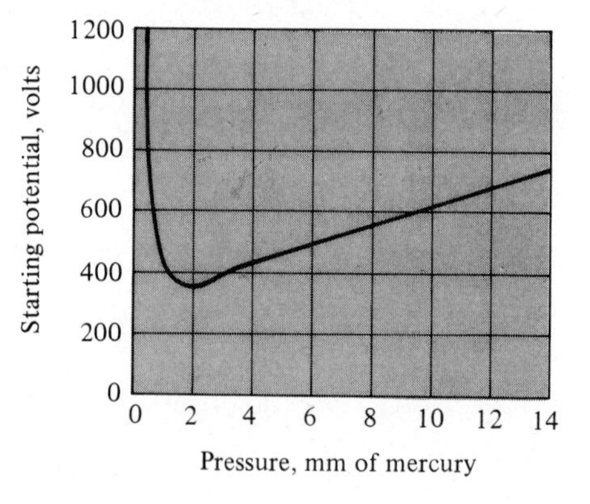

The greenish glow of the glass, which appears everywhere between the electrodes when the object S is out of the way, is partly obliterated when S is swung between the electrodes. If the object S has some distinctive shape, it may be seen clearly that it is casting a shadow. The shadow is on the side of S that is away from the negative electrode or cathode. If the cathode is small, the shadow is rather sharp. It is a simple deduction that the greenish glow is caused by some kind of rays from the cathode that cannot penetrate the obstruction S. These rays are called *cathode rays*. Many years ago it was observed that these rays could be deflected by both electric and magnetic fields, and the direction of these deflections showed that the rays were negatively charged.

Sir J. J. Thomson undertook a quantitative study of cathode rays in 1897. He was able to show that all cathode rays or corpuscles possess a common property. He showed that the ratio of their charge to their mass, q/m, was a constant. His measurements did not establish that all the rays have identical charges or identical masses, although this is the simplest interpretation of his results. He did, however, discover a unique characteristic of these rays and he is regarded as the discoverer of a fundamental particle of electricity, the electron.

2.2 CHARGED-PARTICLE BALLISTICS

Before discussing one of the methods by which q/m can be measured, let us review some basic facts of electricity and magnetism. When a particle having charge $+q$ is in a uniform electric field of intensity E, the particle experiences a force in the direction of the field of magnitude

$$F = qE \qquad [\text{in vector notation } \mathbf{F} = q\mathbf{E}]. \tag{2.1}$$ [†]

If all other forces on the particle are negligible compared with this one, the particle will undergo uniformly accelerated motion, and we have

$$ma = qE. \tag{2.2}$$

Note that if the particle has a component of its velocity perpendicular to E, that component remains *unchanged*, whereas the component of velocity in the direction of E *changes* with an acceleration a.

If a particle having charge q moves in a uniform magnetic field where the magnetic induction is B with a velocity v that is perpendicular to B, the particle experiences a force F that is *perpendicular* to both v and B and has a magnitude

[†] This equation and those following are valid in any consistent system of units. No conversion factors for units need be introduced provided *all* are electrostatic units, *all* are electromagnetic units, or *all* are meter-kilogram-second-ampere units. This will be true of all equations in this book except in cases where units peculiar to atomic physics, such as angstroms or electron volts, are specified.

given by

$$F = qvB \qquad [\text{in vector notation } \mathbf{F} = q(\mathbf{v} \times \mathbf{B})]. \tag{2.3}$$

If all other forces on the particle are negligible compared with this one, the particle will change its direction of motion but not its speed. After an infinitesimal change of direction, the particle has the same speed and is still moving perpendicular to B. Thus the particle experiences a force of constant magnitude and changing direction that causes it to move in a circular path in a plane perpendicular to the magnetic induction or flux density. Since the force required to maintain a mass m, with tangential velocity v, in a circular path of radius r is

$$F = mv^2/r,$$

we obtain

$$mv^2/r = qvB, \tag{2.4}$$

or

$$mv = qBr. \tag{2.5}$$

2.3 THOMSON'S MEASUREMENT OF q/m

We are now ready to consider how Thomson measured the ratio of charge to mass, q/m, for what he called "cathode corpuscles." His apparatus (Fig. 2.3) consisted of a highly evacuated glass tube into which several metal electrodes were sealed. Electrode C is the cathode from which the rays emerged. Electrode A is the anode, which was maintained at a high positive potential so that a discharge of cathode rays passed to it. Most of the rays hit A, but there was a small hole in A through which some of the rays passed. These rays were further restricted by an electrode A' in

Figure 2.3 Thomson's apparatus for measuring the ratio q/m for cathode rays.

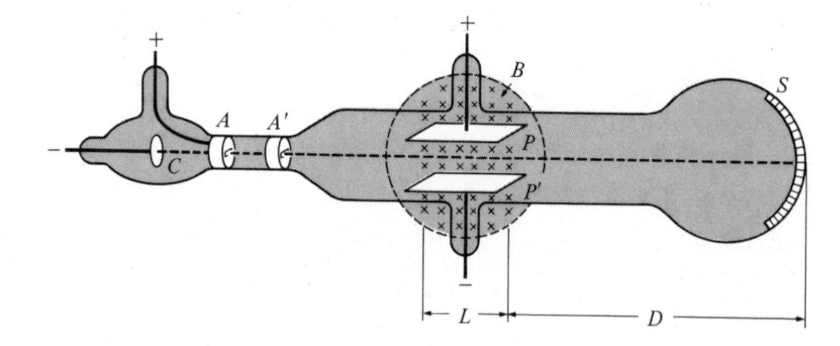

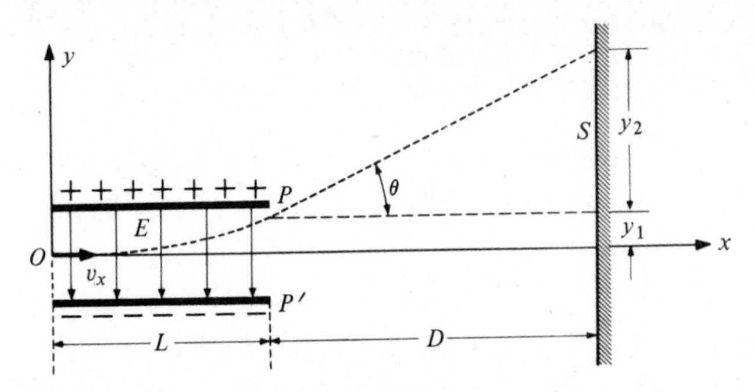

Figure 2.4 Electrostatic deflection of cathode rays.

which there was another hole. Thus a narrow beam of the rays passed into the region between the two plates P and P'. After passing between the plates, the rays struck the end of the tube, where they caused fluorescent material at S to glow. (We will discuss how this happens when we consider fluorescence in Chapter 4.)

The deflection plates P and P' were separated a known amount, so that when they were at a known difference of potential the electric field between them could be computed. We shall assume that the field was uniform for a distance L between the plates and zero outside them. When the upper plate P was made positive, the electric field deflected the negative cathode rays upward.

The trajectory of the cathode corpuscle is obtained in the same way that trajectories were found for projectiles in the gravitational field of force in mechanics. The only difference is that here the constant electric force is upward and is limited to the region between the plates. The gravitational force mg and the electrostatic force between two corpuscles are so small that they may be neglected.

In Fig. 2.4, if the cathode rays enter the region between the plates at the origin O with a velocity v_x, this velocity will continue to be the horizontal component of the velocity of the rays. The general equation for displacement in uniformly accelerated motion is

$$s = s_0 + v_0 t + \tfrac{1}{2}at^2. \tag{2.6}$$

Applying Eq. (2.6) to the horizontal direction, we obtain

$$x = v_x t. \tag{2.7}$$

Between the plates, however, the rays experience an upward acceleration

$$a_y = \frac{qE}{m}, \tag{2.8}$$

obtained from Eq. (2.2). In this case E is constant, since fringing of the electric field is neglected, and it is equal to the potential difference between the deflection plates divided by their separation. Hence the general displacement equation becomes

$$y = \frac{qEt^2}{2m}. \tag{2.9}$$

Elimination of t between Eqs. (2.7) and (2.9) yields the equation for the parabolic trajectory:

$$y = \frac{qEx^2}{2mv_x^2}. \tag{2.10}$$

The quantity y_1, defined in Fig. 2.4, is the value of y when $x = L$.

Beyond the plates, the trajectory is a straight line because the charge is then moving in a field-free space. The value of y_2 is $D \tan \theta$, where D and θ are defined as in Fig. 2.4. The slope of this straight line is

$$\tan \theta = \left(\frac{dy}{dx}\right)_{x=L} = \left(\frac{qEx}{mv_x^2}\right)_{x=L} = \frac{qEL}{mv_x^2}. \tag{2.11}$$

The total deflection of the beam, y_E, is $y_1 + y_2$, so that

$$y_E = y_1 + y_2 = \frac{qEL^2}{2mv_x^2} + \frac{qELD}{mv_x^2} = \frac{qEL}{mv_x^2}\left(\frac{L}{2} + D\right). \tag{2.12}$$

If q/m is regarded as a single unknown, then there are two unknowns in this equation. The initial velocity of the rays, v_x, must be determined before q/m can be found. We need another equation involving the initial velocity v_x, so that this unknown velocity can be eliminated between the new equation and Eq. (2.12).

Thomson obtained another equation by applying a magnetic field perpendicular to both the cathode-corpuscle beam and the electric field. It is represented in Fig. 2.3 as being into the page and uniform everywhere within the x-marked area. Thus the electric and magnetic forces acted on the cathode rays in the same geometric space.

Figure 2.5 shows the situation when the magnetic field alone is present. The negatively charged rays experience a force that is initially downward. This force is not constant in direction but is always normal to both the field and the direction of motion of the rays. Therefore the cathode corpuscles move in a circular path according to Eq. (2.5). The center of curvature of the trajectory is at C, and the radius of curvature of the path is

$$R = mv_x/qB, \tag{2.13}$$

where v_x is the initial velocity of the rays in the x-direction. Referred to the origin O, the equation of this circular path is

$$x^2 + (R + y)^2 = R^2. \tag{2.14}$$

Solving for R, we get

$$R = -\frac{x^2 + y^2}{2y} \approx -\frac{x^2}{2y}. \tag{2.15}$$

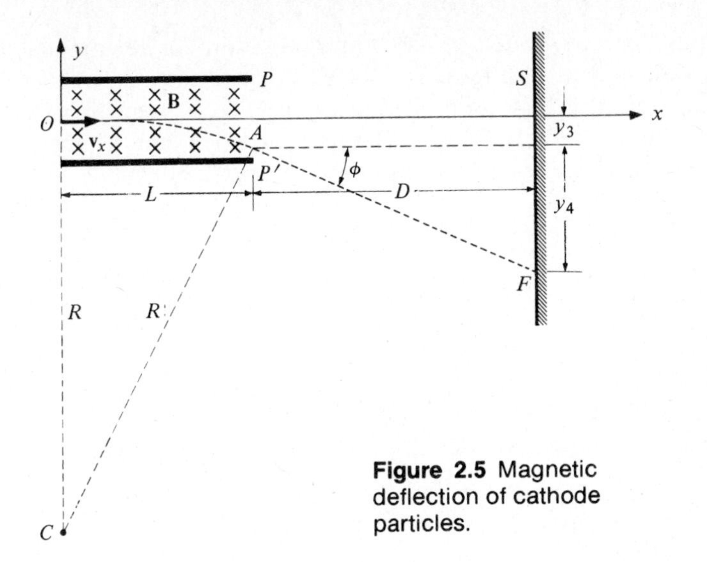

Figure 2.5 Magnetic deflection of cathode particles.

The approximation is good if the deflection is small compared with the distance the rays have moved into the magnetic field, that is, when $y^2 \ll x^2$.

Since the radius of curvature is difficult to measure, we eliminate R between Eqs. (2.13) and (2.15), and obtain

$$y = -\frac{qBx^2}{2mv_x}. \tag{2.16}$$

Therefore, for small deflections, the circular path may be approximated by the parabolic path of Eq. (2.16). The minus sign indicates that the curve is concave downward.

Just as in the electric case, we find that y_3 is the value of y for $x = L$. The rays again move in a straight line through the field-free region, so that

$$y_4 = D \tan \phi = D\left(\frac{dy}{dx}\right)_{x=L} = -\frac{DqBL}{mv_x}. \tag{2.17}$$

For the total magnetic deflection y_B, we have

$$y_B = y_3 + y_4,$$

or

$$y_B = -\frac{q}{m}\left(\frac{BL^2}{2v_x} + \frac{BLD}{v_x}\right) = -\frac{qBL}{mv_x}\left(\frac{L}{2} + D\right). \tag{2.18}$$

Equation (2.18) is very similar to Eq. (2.12). It contains q/m and v_x together with measurable quantities, so that v_x can be eliminated and q/m found. It is interesting, however, to follow Thomson's procedure for determining v_x by considering the simultaneous application of the electric and the magnetic fields. If these are adjusted so that there is no deflection on the screen, then the force of the electric field on the charged particle is balanced by that of the magnetic field. For this condition of balance, we find from Eqs. (2.1) and (2.3) that

$$F = qE - qv_x B = 0, \tag{2.19}$$

or, in terms of v_x,

$$v_x = \frac{E}{B}. \tag{2.20}$$

This result can also be derived from Eqs. (2.12) and (2.18). When y_E and y_B are equal and opposite, the resultant deflection of the cathode-ray beam is zero, and we then have

$$y_E + y_B = \frac{q}{m}\frac{EL}{v_x^2}\left(\frac{L}{2} + D\right) - \frac{q}{m}\frac{BL}{v_x}\left(\frac{L}{2} + D\right) = 0. \tag{2.21}$$

This expression is easily reduced to $E/v_x^2 = B/v_x$ or $v_x = E/B$.

For this particular ratio of the fields, the particle goes straight through both fields. It is undeflected, and therefore the measurement of v_x does not depend on the geometry of the tube. Since $y = 0$ at all times, the approximation in Eq. (2.15) is avoided. The velocity thus determined may be substituted into Eq. (2.12), which was derived without approximation.

Thomson measured q/m for cathode rays and found a unique value for this quantity that was independent of the cathode material and the residual gas in the tube. This independence indicated that cathode corpuscles are a common constituent of all matter. The modern accepted value of q/m is $(1.758796 \pm 0.000019) \times 10^{11}$ coulombs per kilogram. Thus Thomson is credited with the discovery of the first subatomic particle, the electron. Because it was shown later that electrons have a unique charge e, the quantity he measured is now denoted by e/m_e. He also found that the velocity of the electrons in the beam was about one-tenth the velocity of light, much larger than any previously measured material particle velocity.

It was fortunate that the electrons Thomson studied had nearly equal velocities. If this had not been the case, the spot on the end of his experimental tube would have been seriously smeared. Both Eqs. (2.12) and (2.18) include the electron velocity and if all electrons had not had the same velocity, each would have undergone a different deflection. Thomson could tell that the electrons had a uniform velocity when he observed an undeflected spot upon proper adjustment of **E** and **B**.

It is interesting to explore why the electrons in Thomson's apparatus had nearly uniform velocities. The electrons he studied came from the cathode and were accelerated toward the anode by a potential difference that we can call V. Since the energy required to separate the electrons from the cathode is negligibly small, the work done by the electric field on the charges went into kinetic energy.

In general, from the law of conservation of energy, the change of kinetic energy plus the change of electrical potential energy of a charge as it goes from point 1 to point 2 must equal zero because no work is done by external forces.† Therefore we have

$$\left(\tfrac{1}{2}mv_2^2 - \tfrac{1}{2}mv_1^2\right) + \left(qV_2 - qV_1\right) = 0, \tag{2.22}$$

or

$$-q(V_2 - V_1) = \frac{m}{2}(v_2^2 - v_1^2). \tag{2.23}$$

The quantity $V_2 - V_1$ is the potential of the second electrode relative to the first.

If we apply Eq. (2.23) to Thomson's experiment, noting that the charges are negative, the accelerating voltage $V_2 - V_1$ is V, the initial velocity v_1 is zero, and the final velocity v_2 is v_x, we obtain

$$qV = \frac{m}{2}v_x^2,$$

or

$$q/m = e/m_e = v_x^2/2V. \tag{2.24}$$

This is another equation relating e/m_e and v_x. It could have been used with Eq. (2.18) to give e/m_e. Thomson could have measured the potential difference between the cathode and anode and been spared either the electric or magnetic deflection of the beam in the vicinity of P and P'. Indeed, other methods of measuring e/m_e utilize this principle.

Cathode-ray tubes such as those Thomson used have been developed into important modern electronic components. Electrostatic deflection of an electron beam is used in the cathode-ray tube of modern oscilloscopes. Such tubes usually have two sets of deflecting plates (Fig. 2.6), so that the electron beam can be deflected right and left as well as up and down. These tubes utilize the fact that the deflection is proportional to the electric field between the plates, as shown by Eq. (2.12). Television tubes, on the other hand, commonly utilize magnetic deflection to cause the beam to sweep over the face of the picture area.

Anyone can demonstrate that electric and magnetic fields deflect electron beams. Holding a strong permanent magnet near the face of a television picture produces weird distortions. Rubbing the face of a picture tube or even the plastic protective window with wool, silk, or nylon will produce strong electric fields when the humidity is low. Neither the magnetic nor electric fields thus produced are uniform or perpendicular to the beam, and the deflections they produce are striking in their unpredictability.

†Strictly speaking, this is true only if we ignore radiation energy. We are not always justified in doing so, but in many experimental situations we may do so to a good approximation.

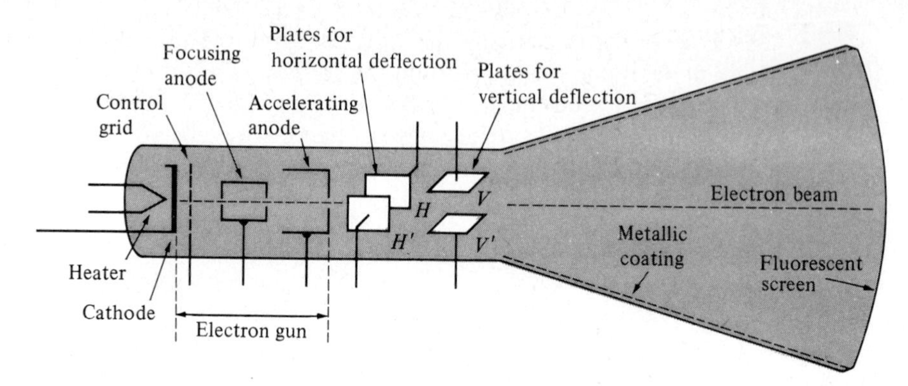

Figure 2.6 Basic elements of a cathode-ray tube.

EXAMPLE (a) Find the speed an electron acquires if it is accelerated from rest through a potential difference of 565 volts.

This electron now moves in a vertical plane with its acquired speed and enters a region where there is a uniform electric field of 35 V/cm directed downward. Find (b) the coordinates of the electron 5×10^{-8} s after it passes through the point of entry along a course directed at an angle of 30° below the horizontal (Fig. 2.7), and (c) the direction of its velocity at this time.

Solution (a) Recall Eq. (2.23):

$$-q(V_2 - V_1) = \frac{m}{2}\left(v_2^2 - v_1^2\right).$$

Since $v_1 = 0$, we obtain

$$v_2 = \sqrt{\frac{-2e(V_2 - V_1)}{m_e}} = \sqrt{-2 \times (-1.76 \times 10^{11}\ \mathrm{C/kg}) \times 565\ \mathrm{J/C}}$$

$$= \sqrt{1.99 \times 10^{14} \frac{\mathrm{kg \cdot m^2/s^2}}{\mathrm{kg}}} = 1.41 \times 10^7\ \mathrm{m/s}.$$

(b) The field is uniform and directed downward, so there is a constant force on the electron directed upward along the positive y-axis. Therefore the acceleration of the

Figure 2.7

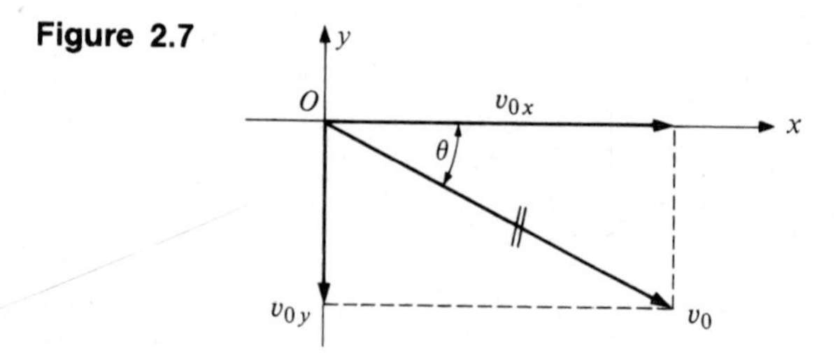

electron is constant and its value is

$$a = \frac{F}{m} = \frac{Eq}{m} = \frac{Ee}{m_e}$$

$$= 35 \frac{\text{J/C}}{\text{cm}} \times 10^2 \frac{\text{cm}}{\text{m}} \times 1.76 \times 10^{11} \frac{\text{C}}{\text{kg}}$$

$$= 6.16 \times 10^{14} \frac{\text{m}}{\text{s}^2}.$$

Referred to the point of entry O as the origin of coordinates, the equations of motion for the electron are

$$x = v_0 \cos \theta t = 1.41 \times 10^7 \frac{\text{m}}{\text{s}} \times 0.866 \times 5 \times 10^{-8}\ \text{s}$$

$$= 0.61\ \text{m}$$

and

$$y = -v_0 \sin \theta t + \tfrac{1}{2} a t^2$$

$$= -1.41 \times 10^7 \frac{\text{m}}{\text{s}} \times 0.500 \times 5 \times 10^{-8}\ \text{s}$$

$$+ \tfrac{1}{2} \times 6.16 \times 10^{14} \frac{\text{m}}{\text{s}^2} \times (5 \times 10^{-8}\ \text{s})^2$$

$$= -0.35\ \text{m} + 0.77\ \text{m} = 0.42\ \text{m}.$$

(c) $$v_x = v_0 \cos \theta$$

$$= 1.41 \times 10^7 \frac{\text{m}}{\text{s}} \times 0.866 = 1.22 \times 10^7 \frac{\text{m}}{\text{s}}.$$

$$v_y = -v_0 \sin \theta + at$$

$$= -1.41 \times 10^7 \frac{\text{m}}{\text{s}} \times 0.500 + 6.16 \times 10^{14} \frac{\text{m}}{\text{s}^2} \times 5 \times 10^{-8}\ \text{s}$$

$$= -0.71 \times 10^7 \frac{\text{m}}{\text{s}} + 3.08 \times 10^7 \frac{\text{m}}{\text{s}}$$

$$= 2.37 \times 10^7 \frac{\text{m}}{\text{s}}.$$

Let ϕ_x be the direction of motion with respect to the x-axis; then

$$\tan \phi_x = \frac{v_y}{v_x} = \frac{2.37 \times 10^7\ \text{m/s}}{1.22 \times 10^7\ \text{m/s}} = 1.94,$$

and

$$\phi_x = 62°\ 45'.$$

The direction of motion can also be obtained from the equation of the trajectory. Combining the x- and y-coordinate equations in part (b) by eliminating t gives

$$y = -x \tan \theta + \frac{1}{2} \frac{Ee}{m_e} \frac{x^2}{v_0^2 \cos^2 \theta}.$$

It is left to the reader to show that the slope of this curve at $x = 0.61$ m is 1.94. ▬

2.4 ELECTRONIC CHARGE

Although the measurement of e/m_e indicated the identity of electrons, another measurement was required before e and m_e could be known separately. This was first made with precision in 1909 by R. A. Millikan, who perfected a technique suggested by J. J. Thomson and H. A. Wilson.

Both the charge e and the mass m_e of an electron are incredibly small quantities. The mass of any body can be determined from the measurement of the force acting on it when it is accelerated. Even if a single electron could be isolated for study, no instrument could measure its mass directly. Similarly, the charge on a body can be determined by measuring the force it experiences in an electric field. This method does not require the isolation of a single electron and, since very intense electric fields can be created, a measurable force can be produced.

An experiment to measure e must be carried out with a body having so few charges that the change of one charge makes a noticeable difference. Since the experiment must be done with very little charge, the force the body experiences will be small even though a large electric field is utilized. If the force on the charged body is very small, then the body itself must be very light. The force of gravity is always with us, and if the small electric force is not to be masked by a large gravitational force, then the mass of the body must be both small and known. If the body is small enough that the electric force on its charges is of the same order of magnitude as the gravitational force it experiences, then it may be that the gravitational force will be a useful standard of comparison rather than an annoying handicap.

Millikan used a drop of oil as his test body. It was selected from a mist produced by an ordinary atomizer. The drop was so small that it could not be measured optically, but with a microscope it could be seen as a bright spot because it scattered light from an intense beam, like a minute dust particle in bright sunlight.

When such a drop falls under the influence of gravity, it is hindered by the air it passes through. The way in which the fall of a small spherical body is hindered by air had been described by Stokes, who found that such a body experienced a resisting force **R** proportional to its velocity, or

$$R = kv. \tag{2.25}$$

The proportionality constant k was found by Stokes to be

$$k = 6\pi\eta r, \tag{2.26}$$

where η is the coefficient of viscosity of the resisting medium and r is the radius of the body. (This law assumes that the resisting medium is homogeneous. A more complicated law must be used if the size of the body is of the same order of magnitude as the mean free path of the molecules of the medium.) This is a friction equation very different from that introduced in mechanics to describe the force between two sliding bodies. In that case we assumed that the friction force depended only on the nature of the sliding surfaces and the normal force pressing the surfaces together. Hence in mechanics we discussed a force that did not depend on the speed of the motion. In the problem of a box sliding against friction down an inclined plane, the friction produced a constant force opposing the motion, but the acceleration was constant and the velocity increased continuously. A body subject to a frictional force like that given by Stoke's law will behave very differently.

A falling droplet of oil is acted on by its weight w, the buoyant force B of the air, and the resisting force $R = kv$ (Fig. 2.8). The resultant downward force F is

$$F = w - B - kv. \tag{2.27}$$

Initially, the velocity v is zero, the resisting force is zero, and the resultant downward force equals $w - B$. The drop therefore has an initial downward acceleration. As its downward velocity increases, the resisting force increases and eventually reaches a value such that the resultant force is zero. The drop then falls with a constant velocity called its *terminal velocity*, v_g. Since $F = 0$ when $v = v_g$, we have from Eq. (2.27),

$$w - B = kv_g. \tag{2.28}$$

Let ρ be the density of the oil and ρ_a the density of the air. Then

$$w = \tfrac{4}{3}\pi r^3 \rho g, \qquad B = \tfrac{4}{3}\pi r^3 \rho_a g, \tag{2.29}$$

and inserting the value of k from Eq. (2.26), we get

$$\tfrac{4}{3}\pi r^3(\rho - \rho_a)g = 6\pi\eta r v_g. \tag{2.30}$$

All of the quantities in this equation except r are known or measurable. We can therefore solve for the drop radius r and hence can express the proportionality constant k in terms of known or measurable quantities. The result is

$$k = 18\pi\left[\frac{\eta^3 v_g}{2g(\rho - \rho_a)}\right]^{1/2}. \tag{2.31}$$

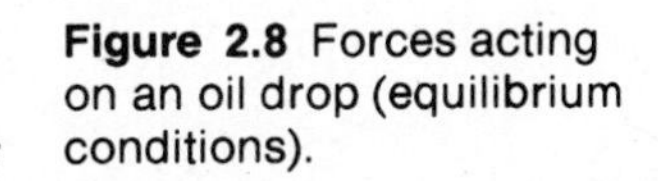

Figure 2.8 Forces acting on an oil drop (equilibrium conditions).

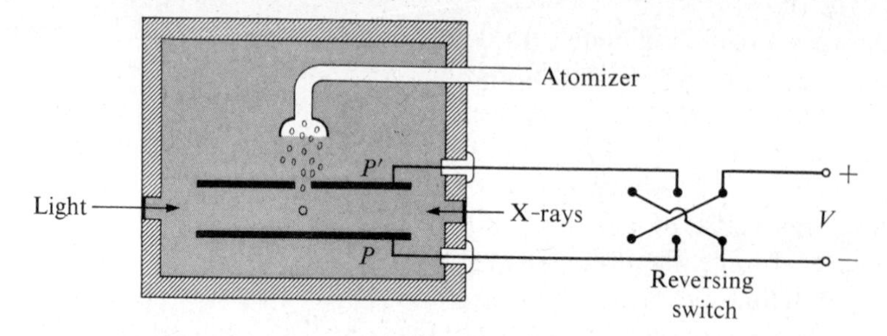

Figure 2.9 Millikan's oil-drop experiment.

In the experiment, the oil drop is situated between two horizontal plates where a known strong electric field may be directed upward or downward or may be turned off (Fig. 2.9). The droplet has a small electric charge q that may be minus or plus, depending on whether it has an excess or deficiency of electrons. The droplet gets this charge from rubbing against the nozzle of the atomizer and from encounters with stray charges left in the air by cosmic rays or deliberately produced by x-rays or by bringing a radioactive material nearby. In the electric field, the drop will experience a force $q\mathbf{E}$, which can always be directed upward by the proper choice of the direction of $\mathbf{E}$. The experimenter must manipulate $\mathbf{E}$ so that the drop rises and falls in the region between the plates but never touches either.

The microscope with which the drop's movements are followed is equipped with two horizontal hairlines whose separation represents a known distance along the vertical line in which the drop travels. By timing the trips of the drop over this known distance, the terminal velocities of the drop are found. The velocities of fall, $\mathbf{v}_g$, are all the same, since oil does not evaporate and therefore the weight of the drop is constant. The velocity of rise, however, depends on the charge q. The resultant force on the drop while it is rising is

$$F = qE + B - w - kv. \tag{2.32}$$

When the terminal velocity $\mathbf{v}_E$ is reached, the resultant force is zero, so

$$qE = w - B + kv_E. \tag{2.33}$$

But from Eq. (2.28), $w - B = kv_g$, so finally

$$q = \frac{k}{E}(v_g + v_E). \tag{2.34}$$

Since these terminal velocities are constant, they are relatively easy to measure.

Equation (2.34) permits the evaluation of q, the charge on the drop. In the oil-drop experiment, the value of v_g is determined for a particular drop with the electric field off, and a whole series of v_E's for the same drop is observed with the field on. If we knew that the electronic charge was unique and that there was only one charge on the drop, then Eq. (2.34) would give the value of this charge at once. Since the nature of the electronic charge was not known, Millikan repeated the experiment with many different charges on the drop. This provided a set of q's that he found to be integral multiples of one charge, which he took to be the ultimate unit of charge, e. Thus he established the *law of multiple proportions* for electric charges and concluded from it that electricity must be atomic in character.

Millikan made observations on oil drops of different sizes and also on drops of mercury. In one instance a drop was watched continuously for eighteen hours. The sets of observations always gave the same magnitude of the electronic charge or "atom" of electricity. The best modern determination of e is $(1.6021892 \pm 0.0000046) \times 10^{-19}$ coulomb.

2.5 MASS OF THE ELECTRON. AVOGADRO CONSTANT

Since e/m_e and e are now known, it is only simple arithmetic to find the mass of the electron to be

$$m_e = (9.109534 \pm 0.000047) \times 10^{-31} \text{ kg}.$$

Still another basic atomic constant may now be calculated with precision by using the value of the electronic charge. The Faraday constant is the amount of charge required to transport one atomic (molecular) mass of a univalent ion of a material through an electrolyte. Dividing the Faraday constant by e gives the number of electrons that have participated in this transport, or the Avogadro constant. The result agrees with Perrin's value, which had finally established the atomic view of matter.

2.6 POSITIVE RAYS

After the particle of negative electricity, the electron, had been identified, it was reasonable to ask about positive electricity. The search was made in a discharge tube very similar to that which disclosed cathode rays. In 1886, Goldstein observed that if the cathode of a discharge tube had slots in it, there appeared streaks of light in the gas on the side away from the anode. These channels of light, first called "canal rays," were easily shown to be due to charged particles. They moved in the direction of the electric field that was producing the discharge, and they were deflected by electric and magnetic fields in directions that proved that their charge was positive. Attempts were made to measure q/m, the ratio of the charge to the mass, of these *positive rays*. It was soon discovered that q/m for positive rays was much less than for electrons and that it depended on the kind of residual gas in the tube. The

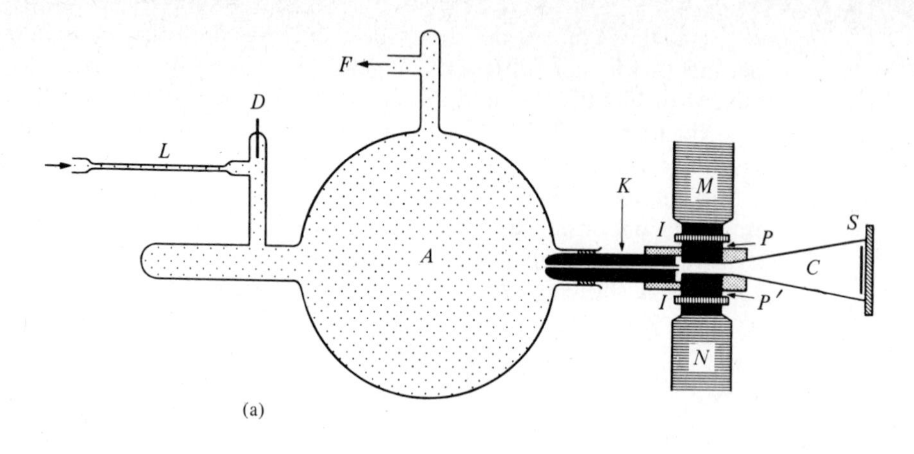

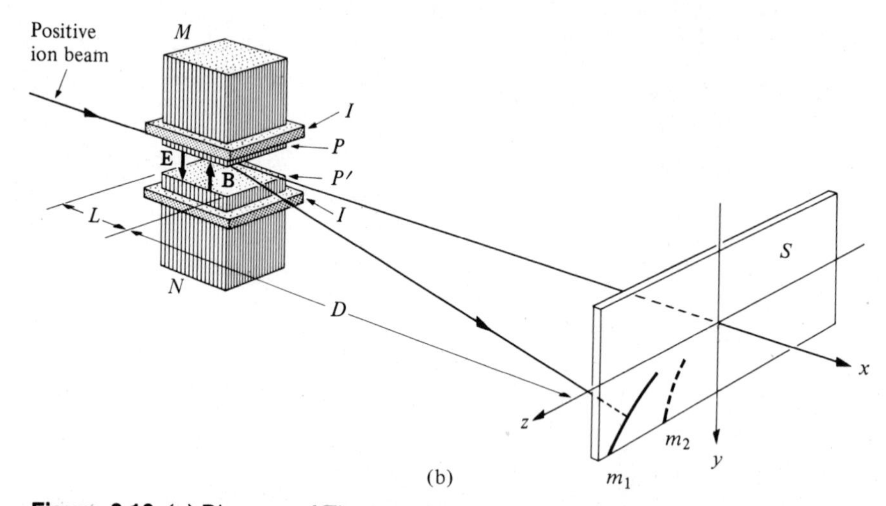

Figure 2.10 (a) Diagram of Thomson's apparatus for positive-ray analysis. (b) Formation of positive-ray parabolas.

velocities of these positive rays were found to be nonuniform and much less than electron velocities.

Thomson devised a different method for measuring q/m of these positive rays having nonuniform velocities. Figure 2.10(a) shows the apparatus he used. The main discharge took place in the large bulb A at the left, where K is the cathode and D is the anode. The gas under study was slowly admitted through the tube at L and was simultaneously pumped out at F. Thus a very low gas pressure was maintained. Most of the positive rays produced in the bulb hit the cathode and heated it. The cathode had a "canal" through it, so that some of the positive rays passed into the right half of the apparatus. Just to the right of the cathode are M and N, the poles of an electromagnet. The pole pieces of this magnet were electrically insulated by sheets, I, so that the magnetic pole pieces could also be used as the plates of a capacitor for the establishment of an electric field. With neither electric nor magnetic fields, the positive rays passed straight through the chamber C to the sensitive layer at S. This layer was either the emulsion on a photographic plate or a fluorescent screen. The beam was well defined because of the narrow tunnel in the cathode through which it had to pass. Instead of crossed fields as in the electron apparatus, this apparatus has its fields perpendicular to the rays but parallel to each other. The electric field is directed downward and the magnetic induction is upward, so that in Fig. 2.10(b) the electric force is toward the bottom of the page along the y-axis and the magnetic force is out of the page toward the reader along the z-axis.

Let a positively charged particle of unknown q/m enter the region between P and P' in Fig. 2.10(b) with an unknown velocity v_x along the x-axis. Then, according to Eq. (2.12), the deflection of the particle on the screen S due to the electric field is

$$y = \frac{qEL}{mv_x^2}\left(\frac{L}{2} + D\right), \tag{2.35}$$

and, according to Eq. (2.18), the deflection at S due to the magnetic field is

$$z = \frac{qBL}{mv_x}\left(\frac{L}{2} + D\right). \tag{2.36}$$

These two equations together are the parametric equations of a parabola, where v_x is the parameter. Since v_x is different for different particles of the same type, the pattern on the screen is not a point but a locus of points. Elimination of v_x between these equations leads to

$$z^2 = \frac{q}{m}\frac{B^2L}{E}\left(\frac{L}{2} + D\right)y, \tag{2.37}$$

which is the equation of a parabola.

Some actual parabolas obtained by Thomson's method are shown sketched in Fig. 2.11. Examination of the figure reveals several things: Positive rays have distinct values of q/m, as is shown by the fact that the traces are clearly parabolas. That a single experiment discloses several values of q/m is evident from the fact that there

Figure 2.11 The parabolas of neon.

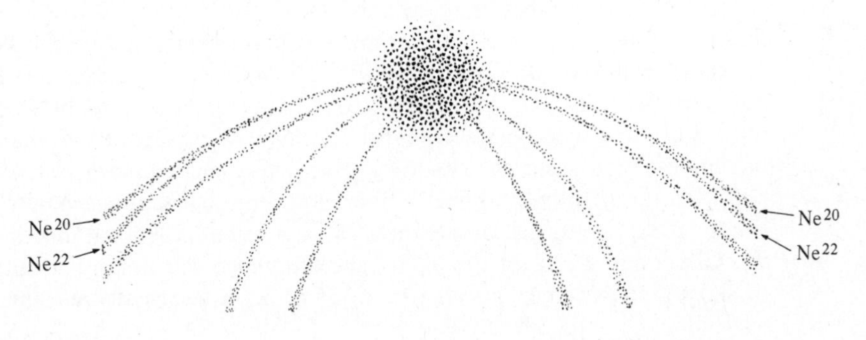

are several parabolas. It is apparent that the method is not capable of great precision because the parabolas are not sharp.

Thomson assumed that each particle of the positive rays carried a charge equal and opposite to the electronic charge, and he attributed the divergent parabolas to differences in mass. He assumed that the positive rays were positive because each had lost one electron. Thomson could identify particular parabolas with particular ions (charged atoms or molecules are called *ions*). Thus for atomic hydrogen, he could verify that the q/m he measured was equal to the value one would expect from dividing the electronic charge by the mass per atom (the atomic mass of hydrogen divided by Avogadro's number). The reason that positive rays move more slowly than electrons and have lower values of q/m than electrons is now clear: The positive rays are much more massive. The largest q/m for positive rays is that for the lightest element, hydrogen. From the value of q/m it was found that the mass of the *hydrogen ion* or *proton* is 1836.15 times the mass of an electron. Electrons contribute only a small amount to the mass of material objects.

2.7 ISOTOPES

The most striking thing that was shown by the Thomson parabolas was that certain chemically pure gases had more than one value of q/m. Most notable was the case of neon, of atomic mass 20.2. Neon exhibited a parabola situated to correspond to a particle of atomic mass 20, but it also had a parabola that indicated an atomic mass of 22. Since the next heavier element, sodium, has an atomic mass of 23.0, efforts to explain away the unexpected value of q/m failed at first. Finally, it was concluded that there must be two kinds of neon, with different masses but chemically identical. The proof of this interpretation was given by Aston, one of Thomson's students.

Aston used a principle that we discussed in Chapter 1. We pointed out there that the average kinetic energy of a molecule in a gas is $3kT/2$. Different gas molecules mixed together in a container must be at the same temperature, and hence the average kinetic energy of each kind of molecule must be the same. If the two gases have different molecular masses, the lighter molecules must have the higher average velocity, and these will make more collisions per unit time with the walls of the container than the heavier molecules. Therefore if these molecules are allowed to diffuse through a porous plug from a container into another vessel, the lighter molecules will have a higher probability of passing through than the heavier, slower ones. Aston took chemically pure neon gas and passed part of it through such a plug. Since one such pass accomplishes only a slight separation, the process had to be repeated many times. He ended with two very small amounts of gas. One fraction had been through the plug many times and the other had been "left behind" many times. He measured the atomic mass of each fraction and found values of 20.15 for the former and 20.28 for the latter. The difference was not great, but it was enough to show that there are indeed two kinds of neon. Many other elements have since been shown to exist in forms that are chemically identical but different in mass. Such forms of an element are called *isotopes*. Thus Dalton's belief that all of the atoms of an element were physically identical in every way was not correct.

The discovery of iostopes solved several problems. It explained the two parabolas observed by Thomson. It also gave a logical explanation of the fact that the atomic mass of neon, 20.2, departs so far from an integral value. If chemical neon is a mixture of neon of atomic mass 20 and of neon of atomic mass 22, then there is some proportion of the two that will mix and have an average atomic mass of 20.2.

2.8 MASS SPECTROSCOPY

A detailed search for the isotopes of all the elements required a more precise technique. Aston built the first of many instruments called mass spectrographs in 1919. His instrument had a precision of one part in 10,000, and he found that many elements have isotopes. Rather than discuss his instrument, however, we shall describe an elegant one built by Bainbridge. The Bainbridge mass spectrograph (Fig. 2.12) has a source of ions (not shown) situated above S_1. The ions under study pass through slits S_1 and S_2 and move down into the electric field between the two plates P and P'. In the region of the electric field there is also a magnetic induction B, perpendicular to the paper. Thus the ions enter a region of crossed electric and magnetic fields like those used by Thomson to measure the velocity of electrons in his determination of e/m_e. Those ions whose velocity is E/B pass undeviated through this region, but ions with other velocities are stopped by the slit S_3. All ions that emerge from S_3 have the same velocity. The region of crossed fields is called a *velocity selector*. Below S_3 the ions enter a region where there is another magnetic

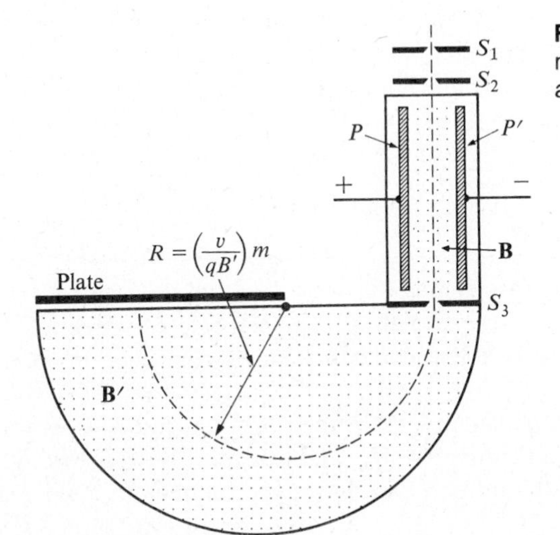

Figure 2.12 Bainbridge's mass spectrometer, utilizing a velocity selector.

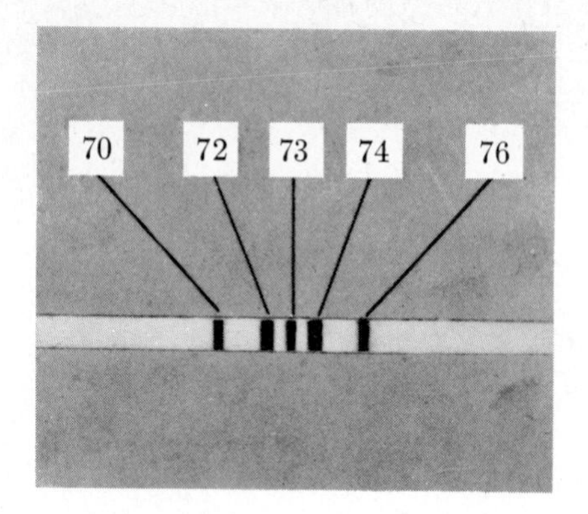

Figure 2.13 The mass spectrum of germanium, showing the isotopes of mass numbers 70, 72, 73, 74, 76.

field B', perpendicular to the page, but no electric field. Here the ions move in circular paths of radius R. From Eq. (2.5), we find that

$$m = \frac{qB'R}{v}. \tag{2.38}$$

Assuming equal charges on each ion, we find, since B' and v are the same for all ions, that the masses of the ions are proportional to the radii of their paths. Ions of different isotopes are converged at different points on the photographic plate. The relative abundance of the isotopes is measured from the densities of the photographic images they produce. Figure 2.13 shows the mass spectrum of germanium. The numbers shown beside the isotope images are not exactly the relative masses or atomic masses of each of the atoms but the integers nearest the relative masses. The integer closest to the isotopic mass is called the *mass number A*, and isotopes are written with the mass number as a superscript to the chemical symbol. Thus the isotopes shown are written Ge^{70}, Ge^{72}, etc., or, in another notation, ^{70}Ge, ^{72}Ge, etc.

As in the case of neon, the discovery of the isotopes of the various elements largely accounted for the fact that many chemical atomic masses are not integers. If germanium has mass numbers 70, 72, 73, 74, and 76, it is no wonder that a mixture of isotopes of germanium has a chemical atomic mass of 72.6.

2.9 HELIUM LEAK DETECTOR

A special use for a mass spectrometer is in the locating of small leaks in high vacuum systems. For this purpose a special mass spectrometer, designed to detect only helium ions, is attached to the vacuum system that is to be investigated for leaks (Fig. 2.14).

Assume that there is a small leak in the system at point A. Molecules from the air (mostly N_2 and O_2 but some H_2, He, and others) will enter through this hole and some will travel down the tube to the ionization chamber B of the mass spectrometer. They will be ionized and some will be accelerated through the slit at C and into the region of the magnetic field at D. The heavier ions like N_2 and O_2 will be bent less than the helium ions, while the lighter ions like H_2 will be bent more than those of He. All except the helium ions will be stopped by the baffles at E. The helium ions continue to the target at F, which is attached to the electrometer tube detector causing a signal that is amplified and measured on an ammeter. The ammeter reading is proportional to the number of helium ions arriving at the target.

In order to find the leak at A it is only necessary to spray a fine stream of helium gas in the suspected area. If the leak exists many helium molecules will enter

Figure 2.14

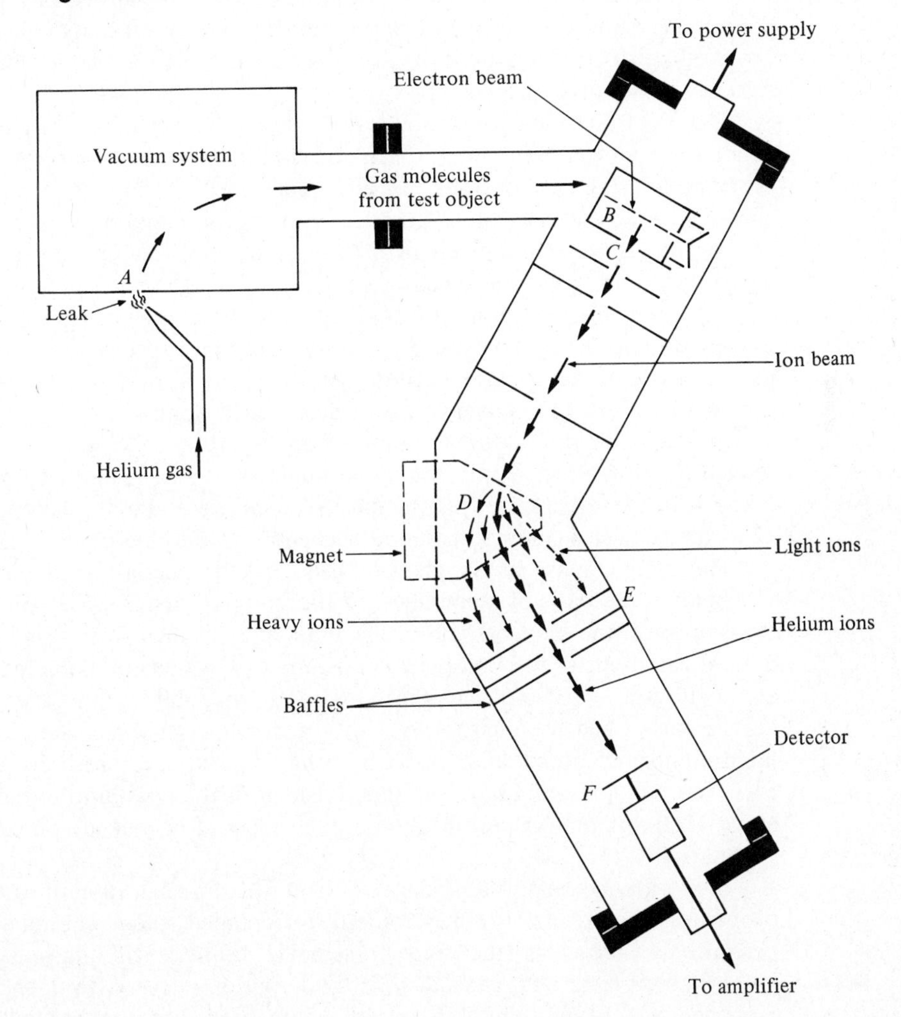

the vacuum chamber and make their way to the mass spectrometer. There will be many more helium ions in the beam and, therefore, many more arriving at the target with a corresponding increase in the signal.

2.10 ISOTOPIC MASS. UNIFIED ATOMIC MASS UNIT

Before approximately 1930, both chemists and physicists had assigned an atomic mass of 16 to natural oxygen and thus used this as the basis for the scale of atomic masses. However, the discovery that natural oxygen is a mixture of three isotopes, ^{16}O, ^{17}O, and ^{18}O, and a somewhat variable mixture at that, led physicists to assign the number 16 to the isotope of oxygen having the lowest mass, whereas chemists, not needing so precise a definition, continued to use 16 as the atomic mass of the mixture of the three isotopes of oxygen.

This dual basis led to two tables of relative masses of atoms that differed by about 275 ppm (parts per million). In addition, the Faraday constant, the Avogadro constant, and the gas constant had different values, depending on which basis was chosen. A great deal of confusion resulted from this situation. For example, values of the various constants given in tables could be misleading unless the scale on which they were based was given.

Further confusion followed from the fact that although physicists were using the oxygen isotope 16 as the base of their system, mass spectroscopists had found it more convenient to use the carbon isotope 12 as a standard because this atom provides a series of reference points in mass spectrograms.

In 1957, the International Commission on Atomic Weights initiated steps to resolve the difficulties arising from the dual system. In 1961, the final action was taken, which resulted in setting the mass of the most abundant isotope of carbon at 12 as the base of a common system for chemistry and physics.

The atomic mass of an element used by the chemist is the average value, weighted on the basis of abundance, of the isotopic masses, based on carbon 12, of the isotopes of the element. The atomic mass of an element varies slightly because of natural variations in the isotopic composition of the element. Some observed ranges are hydrogen, ± 0.00001; oxygen, ± 0.0001; and sulfur, ± 0.003. Chemists use atomic masses because their samples of a material, even when small, contain a very large number of atoms and for their techniques isotopes are indistinguishable. If they were to use isotopic masses and consider all the possibilities, they would find about 1200 different values of the molecular mass of chemically pure hydrated zinc sulfate, $ZnSO_4 \cdot 6H_2O$.

In physics the relative abundance of isotopes, the separation of isotopes, and the properties of particular isotopes come in for detailed study. The relative masses of *neutral atoms* are called the *isotopic masses*. The unit of atomic masses, called the *unified atomic mass unit*, *u*, or *amu*, is by definition one-twelfth of the mass of C^{12}. Since the mass of an atom is its isotopic mass divided by the Avogadro constant, it follows from the definition that the u is equal to the reciprocal of the Avogadro constant, or $(1.6605518 \pm 0.000001) \times 10^{-27}$ kg. Thus the discovery of isotopes not only accounted for those chemical atomic masses that were far from integral, but also provided a new basis in terms of which the isotopic masses are very near to being integers.

FOR ADDITIONAL READING

Anderson, D. L., *The Discovery of the Electron*, Momentum Book No. 3. Princeton, N.J.: Van Nostrand, 1964.

Aston, F. W., *Isotopes*. 2nd ed. London: Arnold, 1924. An account of the first precision determinations of isotopic masses.

Bailey, P. T., "Discovery of the Electron," *Phys. Today* **19**, No. 7 (July 1966), 12. Brief discussion of and complete references to the determinations of q/m for cathode rays prior to the work of J. J. Thomson.

Glasstone, Samuel, *Sourcebook on Atomic Energy*. 3rd ed. New York: Van Nostrand/ Reinhold, 1967. A comprehensive source of basic information in atomic and nuclear physics.

Harnwell, G. P., and J. J. Livingood, *Experimental Atomic Physics*. New York: McGraw-Hill, 1933. An excellent book on experimental methods and relevant theory.

Loeb, Leonard B., *Fundamental Processes of Electric Discharge in Gases*. New York: Wiley, 1939.

Magie, William F., *A Source Book in Physics*. Cambridge, Mass.: Harvard University Press, 1963. A collection of the principal papers of the great men in physics. A valuable source for learning how some of the important discoveries were made. Among the papers included are the following: "The Cathode Discharge," by William Crookes; "Laws of Electrolysis," by Michael Faraday; "The Canal Rays," by Eugen Goldstein; "The Cathode Discharge," by Johann W. Hittorf; "The Negative Charges in the Cathode Discharge," by Jean Perrin; "The Electron," by Joseph J. Thomson; "The Canal Rays," by Wilhelm Wien.

Millikan, Robert A., *The Electron*. 2nd ed. Chicago, Ill.: University of Chicago, 1924. A detailed account of the determination of the electronic charge.

Nier, A. O. C., "The Mass Spectrometer," *Scientific American* (March 1968).

Shamos, M. H., editor, *Great Experiments in Physics*. New York: Holt, 1959. Includes the original accounts of 24 experiments that laid the foundation for modern physics, with annotations by the editor. An excellent book.

REVIEW QUESTIONS

1. How did J. J. Thomson measure the ratio of charge to mass for cathode-ray corpuscles? What did he conclude from the results obtained?
2. What did the results of the Millikan oil-drop experiment show?

3. What was the unexpected result of positive ray analysis?
4. What are isotopes?
5. What is mass number? What is an atomic mass unit?
6. What are the essential features of a mass spectrometer?

PROBLEMS

2.1 What is the ratio of the electric force on a charged particle in an electric field of 20 V/cm to the force of gravity on the particle if it is (a) an electron and (b) a proton? (c) Is the weight negligible compared with the electric force?

2.2 An electron moving in a vertical plane with a speed of 5.0×10^7 m/s enters a region where there is a uniform electric field of 20 V/cm directed upward. Find the electron's coordinates referred to the point of entry and the direction of its motion at a time 4×10^{-8} s later if it enters the field (a) horizontally, (b) at 37° above the horizontal, and (c) at 37° below the horizontal.

2.3 If the charged particle in Problem 2.2 were a proton instead of an electron, what must be the magnitude and direction of the electric field so that the answers for the proton would be the same as they were for the electron?

2.4 The dimensions of some parts of a typical commercial cathode-ray tube are given in Fig. 2.15. If electrons start from rest at the cathode, what is their velocity v_x at the origin O for an accelerating voltage of 1136 V between the anode and cathode?

2.5 Given that the potential difference between the deflecting plates P and P' is 50 V in the cathode-ray tube in Problem 2.4, (a) find the y-coordinate and the direction of motion of the electrons when $x = L$. (b) What is the total deflection on the screen S? (c) What would be the magnitude and direction of a magnetic field that would cause the particles to go between plates P and P' undeflected?

2.6 A large plane metal plate is mounted horizontally at a distance of 0.80 cm above another similar horizontal plate. The plates are charged to a potential difference of 40 V, the upper plate being positive. An electron is projected horizontally with a velocity of 10^6 m/s from a point O, which is midway between the plates. (a) Find the x-coordinate of the point at which the electron strikes a plate. (b) Compute the tangent of the angle that gives the direction of the electron's motion as it strikes the plate. (c) What is the change in kinetic energy of the electron in going from O to the plate? (d) What would be the answer to part (c) if the electron had no initial velocity at O?

Figure 2.15

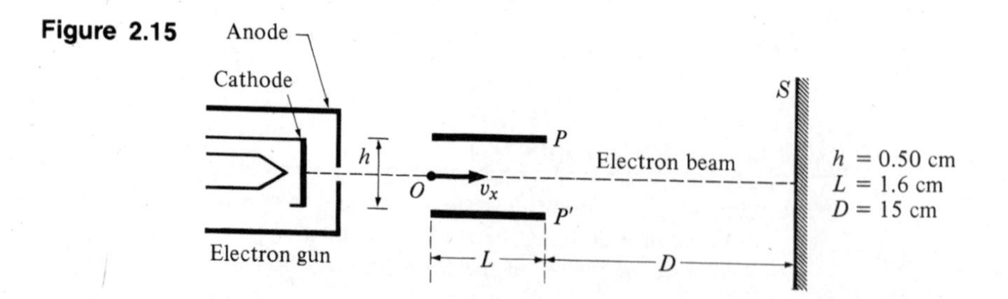

2.7 An electron beam consists of electrons that have been accelerated from rest through a potential difference of 1000 V. The beam current is 50 mA. The beam is incident on a metal surface that stops the electrons. Calculate the force exerted on the metal plate by the electron beam.

2.8 A particle having a mass of 1 g carries a charge of -3×10^{-8} C. The particle is given an initial horizontal velocity in the plane of the paper of 6×10^4 m/s in the gravitational field of the earth. What is the magnitude and direction of the minimum magnetic induction that will keep the particle moving along the same straight line?

2.9 An electron is accelerated from rest through a potential difference of 50 V. The electron then enters a uniform electric field of intensity 3000 N/C, the electron moving in the same direction as the electric field lines. (a) How many seconds elapse between the time the electron enters the field and the instant it comes to rest? (b) How far into the field does the electron travel before it comes to rest? (c) Is it possible to use a magnetic field in place of an electric field to stop the electron?

2.10 (a) Through what potential difference would a deuterium ion have to be accelerated from rest in a vacuum so that it would have a speed of 8.47×10^6 m/s? (Refer to Problem 1.27.) (b) What would have to be the magnitude and direction of the smallest magnetic induction that would constrain the moving deuterium ion to a circular path in an evacuated toroidal tube 1 m in diameter?

2.11 A uniform electric field of intensity 40×10^4 V/m is perpendicular to a uniform magnetic field of flux density 2×10^{-2} T. (A tesla T is a weber per square meter, Wb/m^2.) An electron moving perpendicularly to both fields experiences no net force. (a) Show in a diagram the relative orientation of the electric field vector, the magnetic induction vector, and the velocity of the electron. (b) Calculate the speed of the electron. (c) What is the radius of the electron orbit when the electric field is removed?

2.12 An electron is accelerated from rest to a speed of 10^7 m/s by an electron gun. The electrons leave the gun and strike a screen 4 cm from the front of the gun. When a uniform magnetic field, perpendicular to the electron's velocity, is established between the gun and the screen a deflection of 2 cm is noted on the screen. (a) What is the accelerating potential in the gun? (b) What is the magnetic induction?

2.13 A charged particle enters the region between two very large parallel metal plates, the particle's velocity being parallel to the plates when the particle enters. The plates are separated a distance of 2 cm and the potential difference between the plates is 2000 V. When the particle has penetrated 5 cm into the space between the plates, it is found that the particle has been deflected 0.6 cm. If a perpendicular magnetic field of flux density 0.1 T is impressed simultaneously with the electric field, the particle is found to undergo no deflection at all. Calculate the charge to mass ratio for this particle.

2.14 A uniform magnetic field whose induction is 4×10^{-4} T acts on the electron beam over the distance L in the cathode-ray tube in Problem 2.4. The field is normal to the plane of the trajectory and directed outward. (a) Find the y-coordinate and the direction of motion of the electron when $x = L$. (b) What is the total deflection on the screen S? (c) What would be the deflection on the screen computed both from the approximate and the exact expressions for the radius of curvature if the magnetic induction extended the whole distance, $L + D$, from O to the screen?

2.15 The cathode-ray tube of Problem 2.4 is mounted horizontally and oriented so that the screen faces north. Given that the earth's magnetic induction is 5×10^{-5} T at a dip of 53°, find the magnitude and direction of the horizontal component and of the vertical component of the displacement of the electron beam with respect to the original position on the screen when the tube is turned 90° so that it faces west. Assume that the earth's field extends only over the distance $L + D$ and that the deflections are small.

2.16 Given a cathode-ray tube of the form shown in Problem 2.4, (a) show in general terms that the total deflection on the screen due to an electric field between P and P' is equal to that which the electrons would undergo if they traveled along the axis from O to the point $x = L/2$ and then were deflected at the angle θ given in Eq. (2.11). (b) Would a similar relation hold if a magnetic field instead of an electric field extended over the length L?

2.17 A cathode-ray tube is placed in a uniform magnetic induction B with the axis of the tube parallel to the lines of force. If electrons emerging from the gun with a velocity v make an angle θ with the axis as they pass through the origin O, show (a) that their trajectory is a helix, (b) that they will touch the axis again at the time $t = 2\pi m_e/Be$, (c) that the coordinate of the point of touching is $x = 2\pi m_e v \cos\theta/Be$, and (d) that for small values of θ, the coordinate of the point of crossing or touching the axis is independent of θ. (e) The arrangement in this problem is called a magnetic lens. Why? (f) How do the trajectories of the electrons passing through the origin at an angle θ above the axis differ from those directed at an angle θ below the axis?

2.18 Electrons are accelerated through a potential difference of 1000 V in an electron gun and leave the narrow hole in the anode as a narrow diverging beam. What magnitude of axial magnetic induction is required to focus the beam on a screen 50 cm from the hole? [*Hint:* See Problem 2.17.]

2.19 What is the final velocity of an electron accelerated through a potential difference of 1136 V if it has an initial velocity of 10^7 m/s?

2.20 Two large, plane metal plates are mounted vertically 4 cm apart and charged to a potential difference of 200 V. (a) With what speed must an electron be projected horizontally from the positive plate so that it will arrive at the negative plate with a velocity of 10^7 m/s? (b) With what speed must it be projected from the positive plate at an angle of 37° above the horizontal so that the horizontal component of its velocity when arriving at the negative plate is 10^7 m/s? (c) What is the magnitude of the y-component of the velocity when arriving at the negative plate? (d) What is the electron's time of transit from one plate to the other in each case? (e) With what speed will the electron arrive at the negative plate if it is projected horizontally from the positive plate with a speed of 10^6 m/s?

2.21 Two positive ions having the same charge q but different masses, m_1 and m_2, are accelerated horizontally from rest through a potential difference V. They then enter a region where there is a uniform electric field E directed upward. (a) Show that if the ion beam entered the field along the x-axis, then the value of the y-coordinate for each ion at any time t is $y = Ex^2/4V$. (b) Can this arrangement be used for isotope separation?

2.22 Two positive ions having the same charge q but different masses, m_1 and m_2, are accelerated horizontally from rest through a potential difference V. They then enter a region where there is uniform magnetic induction B normal to the plane of the motion. (a) Show that if the beam entered the magnetic field along the x-axis, the value of the y-coordinate for a small deflection of each at time t is

$$y = Bx^2(q/8mV)^{1/2}.$$

(b) Can this arrangement be used for isotope separation?

2.23 In a cathode-ray tube of the form shown in Problem 2.4, electrons are accelerated from rest through a potential difference V. These then enter a vertical electric field E that extends over the distance L, and also a magnetic field having flux density B normal to the paper over the distance $L + D$. If both fields are adjusted so that there is no deflection on the screen, show that

$$\frac{e}{m_e} = \frac{E^2L^2}{2VB^2}\frac{(L+2D)^2}{(L+D)^4}.$$

2.24 Particles with charge q and mass m are injected into a homogeneous magnetic field having induction B. When their velocities are initially perpendicular to the field, the particles travel in circular orbits. Derive an expression for the frequency of revolution of the particles and show that the frequency is independent of the velocity.

2.25 A beam of protons is passed through a velocity selector consisting of crossed electric and magnetic fields perpendicular to the beam. In this selector $E = 2.4 \times 10^5$ volts/meter and $B = 6 \times 10^{-2}$ T. (a) What is the velocity of the protons? (b) If the protons then move in a region where only the field B acts on them, what is the radius of their orbits?

2.26 The following data were obtained in a Millikan oil-drop experiment:

Plate separation	0.016 m
Voltage across the plates	5,085 V
Distance of fall	1.021×10^{-2} m
Viscosity of air	1.824×10^{-5} N s/m^2
Density of oil	0.92×10^3 kg/m^3
Density of air	1.2 kg/m^3
Average time of fall (no field)	11.88 s
Successive times of rise (with field)	1. 22.37 s
	2. 34.80 s
	3. 29.25 s
	4. 19.70 s
	5. 42.30 s

(a) Compute the radius of the oil drop. (b) Find the charge on the drop for all five cases. (c) Obtain an average value of the electronic charge e from these results.

2.27 A charged oil drop falls 4.0 mm in 16.0 s at constant speed in air in the absence of an electric field. The relative density of the oil is 0.80, that of the air is 1.30×10^{-3}, and the viscosity of the air is 1.81×10^{-5} N · s/m^2. Find (a) the radius of the drop and (b) the mass of the drop. (c) If the drop carries one electronic unit of charge and is in an electric field of 2000 V/cm, what is the ratio of the force of the electric field on the drop to its weight?

2.28 Derive the expression for k in Eq. (2.31) from Eqs. (2.25), (2.29), and (2.30). What are the units of k in the SI system?

2.29 When the oil drop in Problem 2.27 was in a constant electric field of 2000 V/cm, several different times of rise over the distance of 4.0 mm were observed. The measured times were 36.1, 11.5, 17.4, 7.55, and 23.9 s. Calculate (a) the velocity of fall under gravity, (b) the velocity of rise in each case, and (c) the sum of the velocity in part (a) and each velocity in part (b). (d) Show that the sums in part (c) are integral multiples (two significant figures) of some number and interpret this result. (e) Calculate the value of the electronic charge from these data.

2.30 Show that the electric field E necessary to raise an oil drop of mass m and charge q with a speed that is twice the speed of fall of the drop when there is no field is $E = 3mg/q$, given that the buoyant force of the air is negligible.

2.31 In an experiment to count and "weigh" atoms, it is found that a current of 0.800 A flowing through a copper sulfate solution for 1800 s deposits 0.473 g of copper. The atomic weight of copper is 63.54, its valence is 2, and the electronic charge is 1.60×10^{-19} C. Using only the data *given in this problem*, find (a) the number of electronic charges carried by the ions that deposited as copper atoms, (b) the number of copper atoms deposited, (c) the mass of a copper atom, (d) the number of atoms in a gram-atomic weight of copper, (e) the number of electronic charges carried by a gram-equivalent weight of copper ions, (f) the number of coulombs required to deposit a gram-equivalent weight of copper, and (g) the mass of a hydrogen atom given that its atomic weight is 1.008.

2.32 What must be the direction of the electric field E and the magnetic induction B in Fig. 2.10(b) so that the segment of the positive-ion parabola will be in (a) the lower right quadrant, (b) the upper right quadrant, and (c) the upper left quadrant, as viewed from the right of the diagram?

2.33 (a) If the ion beam in Fig. 2.10(b) contains two types of ions having equal charges but different masses, which of the two parabolic segments will have those of greater mass? (b) If the masses are equal but the charges different, which segment will contain those having the larger charge?

2.34 For a particular parabola in Thomson's mass spectrograms, what physical quantity is different for the ions that land close to the origin than for those landing farther away? Why does this difference exist, since the accelerating voltage is the same for all the ions?

2.35 In the Thomson positive-ray experiment of Fig. 2.10, what must be the value of v_x for an ion so that $y = 0$ and $z = 0$?

2.36 In a mass spectrometer, ions having the same charge q but different masses m_1 and m_2 are accelerated from rest through a potential difference V. A narrow beam of these ions then enters a magnetic field having a magnetic induction B that is perpendicular to the motion of the particles. (a) Derive a simple expression in terms of the *given* quantities for the *ratio* of the radii of the trajectories of the two types of ions in the magnetic field assuming the following:

Case 1. All m_1 ions have the same velocity, and all m_2 ions have the same velocity, which is not necessarily equal to that of the m_1 ions.
Case 2. There is a distribution of velocities in the ion beam but, before entering the magnetic field B, the beam passes through an effective velocity selector having an electric field E and magnetic induction B_1.

(b) In which of the preceding cases is the resolving power of the mass spectrometer greater?

2.37 If the electric field between the plates PP' in Fig. 2.12 is 100 V/cm and the magnetic induction in both magnetic fields is 0.2 T, (a) what is the speed of an ion that will go undeviated through the slit system? (b) Given that the source produces singly charged ions of the carbon isotopes C^{12} and C^{13}, find the distance between the center of the lines formed by them on the photographic plate. (Assume that the atomic masses of these isotopes are equal to their mass numbers.) (c) If the slit S_3 is 1 mm wide, will the images on the plate overlap?

2.38 A ^{3}He–^{4}He dilution refrigerator can be used to cool samples to 0.010 K by mixing ^{3}He with ^{4}He in a special chamber. A mass spectrometer is used to measure the ratio of these two isotopes in the refrigerator system. (a) If the instrument described in Problem 2.37 were used to do this, what would be the separation of the lines on the photographic plate? (b) Do you think this instrument would do the job? Explain. (c) If not, how would you modify the mass spectrometer so that it would?

2.39 A deuteron is an ionized hydrogen isotope with mass number 2. Protons and deuterons are accelerated through a potential difference of 150 V, pass through a small slit, and then enter a uniform magnetic field where the magnetic induction is 0.010 T. The field causes the particles to move in a circular path. What is the separation of the beams after completing a semicircle?

2.40 The text does not tell why in Bainbridge's mass spectrometer the ion beam is caused to execute a semicircle before striking the plate. In Fig. 2.12 the beam passing down through S_3 is slightly divergent. Use a compass (or a 25-cent coin) to show that divergent beams through S_3 that have a common radius of curvature are in best focus at the plane of the plate.

2.41 Copper has two isotopes whose masses are 62.9 and 64.9, respectively. What is the percent abundance of each in ordinary copper having an atomic mass of 63.5?

2.42 The isotopic mass of ^{12}C on the former ^{16}O scale was 12.003816. (a) By what percent must this value of the mass of ^{12}C be reduced to make it exactly 12 (the ^{12}C base)? (b) Calculate the isotopic mass of the former oxygen 16 base on the ^{12}C scale.

2.43 Uranium isotopes of mass number 235 and 238 are to be separated from a piece of uranium by using a mass spectrometer that will deflect them through 180° into two collectors 4 cm apart. If the singly charged ions have energies of 2 KeV when entering the magnetic field, calculate (a) the magnetic flux density necessary to achieve this separation and (b) the radii of the paths of the ions.

2.44 A certain piece of equipment is designed to contain neon. Design a leak detector that can be used to find leaks in this apparatus.

3 THE ATOMIC VIEW OF RADIATION

3.1 INTRODUCTION

All sources of light consist of matter that is excited in one way or another. The firefly excites its body matter by some obscure chemical process; the matter of the sun is excited by heat. But ever since Heinrich Hertz demonstrated the validity of Maxwell's theory of electromagnetic radiation, we have known that the ultimate source of radiation is an accelerated electric charge. We cannot begin the story of radiation with Maxwell, however, if we are to appreciate one of the most dramatic demonstrations of the scientific method.

3.2 PARTICLES OR WAVES

In ancient times, some Greek theorists argued that since the blind reach out to feel with their hands, the sighted must reach out with their eyes. They thought of light as a kind of tentacle emitted by the eye, yet retaining contact with the eye so that information about objects touched was conveyed to the mind. Such a view obviously failed to explain why a person cannot see at night unless there is an outside source of illumination.

It was realized long ago that light consists of something that goes out from certain "sources," bounces off objects, and may finally enter the eye. In the seventeenth century, there were two views on the nature of the "something" that was bouncing about. Newton defended the premise that light consists of a stream of fast-moving elastic particles of very small mass. His view accounted for the law of reflection, which states that the angle of incidence is equal to the angle of reflection. (This is the way perfectly elastic balls bounce from the sidewalk.) He accounted for the law of refraction by arguing that when particles of light are very near any optically dense medium like glass, they are attracted to it, and this attraction increases the component of the velocity of light in a direction perpendicular to the surface. Thus, according to Newton, the light travels through the medium *faster* than it does in free space and has its direction altered *toward* the normal. Christian Huygens, on the other hand, supported the view that light consists of waves. The most impressive argument in his favor at that time was that two light beams can cross through each other without "colliding." He too explained reflection and refraction. His explanation of refraction was that when a wavefront penetrates an optically dense medium at an angle, the wave moves more *slowly*. This slowing of the wavefront causes the wave's direction of advance to be altered *toward* the normal.

3.3 ELECTRICITY AND LIGHT

After a century of neglect, the undulatory theory of light was revived by the versatile English scientist Thomas Young. In 1801, he showed that only the principle of interference of waves could explain the colors of thin films and of striated surfaces. During the next half century, further experimental work, especially by the French physicists Fresnel, Arago, Malus, Cornu, Fizeau, and Foucault, showed that the particle theory of light was not tenable. This work reached its culmination in 1864 when James Clerk Maxwell announced the results of his efforts to put the laws of electricity into good mathematical form. He had succeeded in this formulation and found in addition an important by-product: The laws could be combined into the mathematical form of the wave equation for electromagnetic waves. He showed, furthermore, that the velocity of these waves is the *velocity of light*! Thus in one dramatic move he put the theory of electricity in order and incorporated all optics into that theory.

Huygens's view completely displaced Newton's when Foucault found the velocity of light in an optically dense medium *less* than its velocity in free space, and Maxwell's theory was verified in 1888 when Hertz demonstrated that oscillating currents in an electric circuit can radiate energy through space to another similar circuit. Hertz used a circuit containing inductance and capacitance, hence capable of oscillating. Whenever a spark jumped across a gap in the active (transmitting) circuit, electromagnetic waves were radiated from the region in which the electric discharge occurred. (Modifications of this first transmitter were used for radio communication until the advent of vacuum tubes.) The passive or receiving circuit was a loop of wire containing a gap. When energy was transferred from one circuit to the other, sparks jumped across the receiver gap. Hertz's experiments showed that the radiation generated by electric circuits obeyed the known laws of optics. It thus appeared that the theory of light was in a satisfactory and elegant state.

Yet the last word on this subject had not been said. Hertz noted that the induced spark was more easily produced when the terminals of the receiving gap were illuminated by light from the sparks in the transmitter gap. This effect was studied more fully by one of Hertz's students, Hallwachs, who showed that a negatively charged clean plate of zinc loses its charge when illuminated by ultraviolet light. Thus Hertz's verification of Maxwell's wave theory of light led almost simultaneously to the discovery of the photoelectric effect which, as we shall see, led in turn to a profound reinterpretation of the wave theory of radiation.

3.4 ELECTRODYNAMICS

It is unfortunate that we do not have time to develop here the methods of Maxwell's electrodynamics. We can, however, develop qualitatively the idea that the ultimate source of radiation is an accelerated electric charge.

Every electric charge produces an electric field whose lines of force extend radially from the charge through all space. When this charge is in motion, a magnetic field is produced in accordance with Ampère's law and the magnetic field lines are circles concentric with the current. According to the viewpoint of Maxwell's field theory, it is the motion of the electric lines of force that sets up the magnetic field transverse to them. A steady electric current is accompanied by steady electric and magnetic fields; however, a varying current (i.e., one composed of accelerated charges) will produce changes in both fields associated with it. These changes are propagated outward from the accelerated charges through space with the speed of light. The acceleration of a charge produces a pulse of electromagnetic radiation that consists of an electric field component and a magnetic field component that are perpendicular to each other. If the charge oscillates with a simple periodic motion, an electromagnetic wave like that in Fig. 3.1 will be produced.

The energy transmitted by the electromagnetic waves in a radiation field may be specified either in terms of the intensity or of the energy density of the wave motion. The intensity of the radiation is defined as the energy transmitted in unit time through a unit area normal to the direction of propagation of the waves. The SI unit of intensity is watts per square meter. The energy density or volume density of the radiation is defined as the amount of radiant energy in a unit volume of space. The SI unit of energy density is joules per cubic meter. It is evident that the energy density is equal to the intensity divided by the velocity of propagation of the wave.

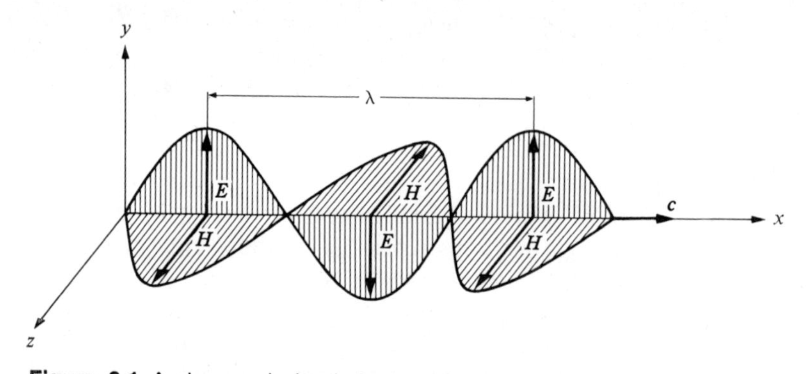

Figure 3.1 A plane-polarized electromagnetic wave of wavelength λ showing the relation of the vectors **E**, **H**, and **c**.

The term *energy density* is particularly useful in discussing the radiation within a heated enclosure.

Our discussion has implied a linear acceleration but, according to Maxwell's theory, radiation occurs whenever an electric charge is accelerated in any manner. For example, a charge moving with constant speed in a circular path will be a source of radiation.† This case is equivalent to two mutually perpendicular simple periodic motions with equal amplitudes and frequencies, but with a phase difference of 90°.

Every radio transmission is a refined repetition of Hertz's original experiments verifying Maxwell's proposition. Instead of physical motion of charged bodies, a radio transmitter causes electrons to move back and forth in an antenna that is physically at rest. These accelerated electrons produce radiation. The electric field in this radiation will exert force on any charges it encounters. Thus the electrons in a receiving antenna respond to the radiation by being accelerated, and their motion constitutes an electric current. A modern radio receiver amplifies these currents and makes them easy to observe. Hertz had no amplifiers, so the emf induced in the receiver had to be large enough to create visible sparks.

3.5 THE UNITY OF RADIATION

From previous studies, the readers of this book are aware that the many forms of radiation — heat, light, radar, radio, etc. — differ from one another in frequency but not in kind. The so-called "kinds" of radiation are characterized by the techniques used to produce and detect them; actually, they all travel through free space with the same velocity and are generally understood in terms of the same theory. The tremendous range of the electromagnetic spectrum is shown in Fig. 3.2 (photon

† This is the source of synchrotron radiation. See Section 11.5.

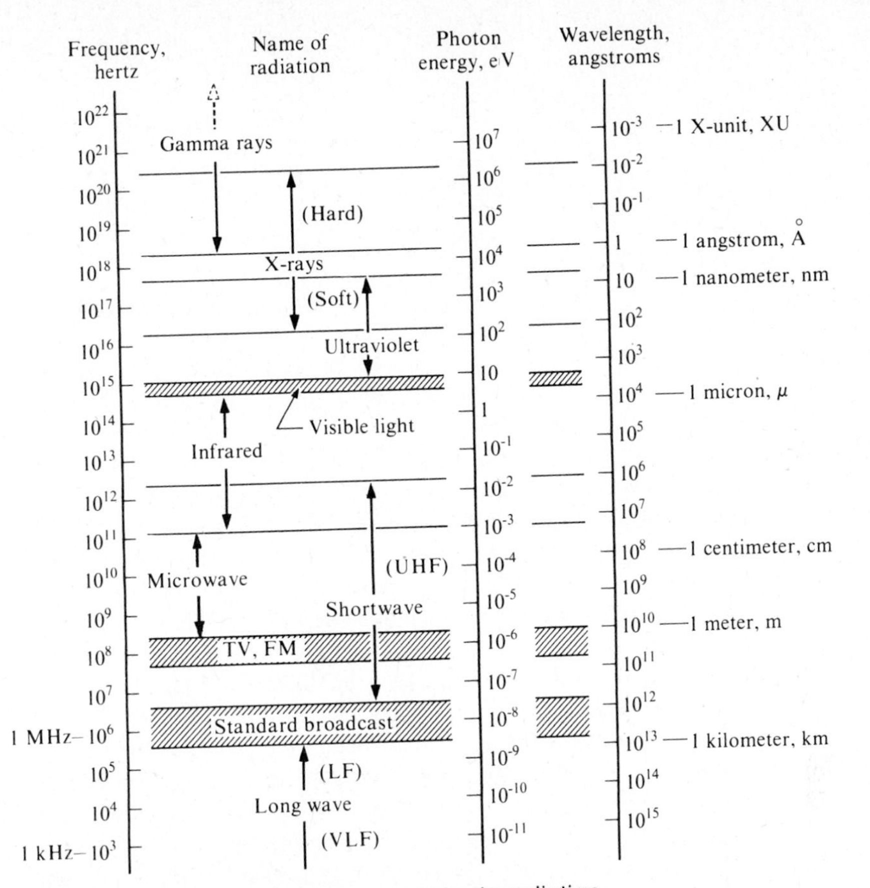

Figure 3.2 The spectrum of electromagnetic radiation.

energy, shown in the figure, will be discussed later in this chapter). The classical theory of Maxwell applies to all these radiations and all are due ultimately to the acceleration of electrical charges. Except for differences due to frequency, an observation made on one "kind" of radiation is usually assumed to be true of all other kinds.

3.6 THERMAL RADIATION

Information about the nature of all radiation may be obtained from a study of any of the "kinds" of radiation. We shall now consider the radiation from heated bodies, since that investigation has proved particularly fruitful. We all know that a body will

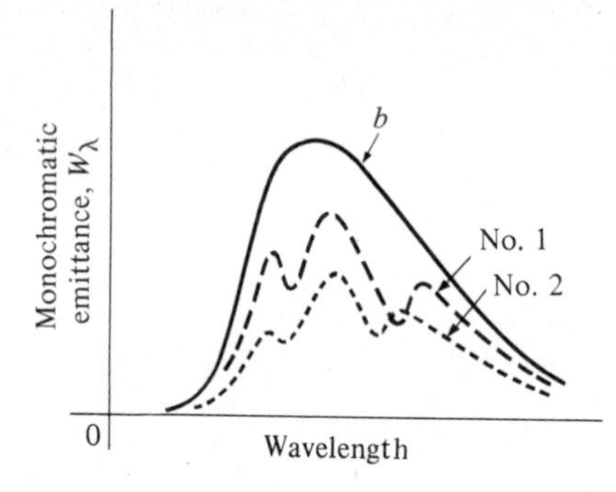

Figure 3.3 The radiation spectrum of several hot bodies.

emit visible radiation if it is hot enough. A close relation between temperature and radiation is further implied by the fact that a white-hot body is hotter than a red-hot one. We might explore this matter further by passing the radiation from a hot body through some dispersive instrument such as a prism or grating spectrometer. If we measure the radiant energy emitted by a hot body for a whole series of radiant frequencies, we might obtain a graph similar to the dashed curve (No. 1) in Fig. 3.3. The ordinate of this curve is called the *monochromatic emittance*, W_λ, which is the amount of energy radiated per unit time per unit area of emitter in a wavelength range $d\lambda$; the abscissa is the wavelength rather than the frequency. Repeating the same experiment for another body of a different material but at the same temperature, we might now obtain the dashed curve (No. 2) of Fig. 3.3. It is clear from the figure that at most wavelengths the first body is a more efficient emitter at the given temperature than the second. Although the two curves differ, they have the same general character. They come to their highest points at about the same wavelength. Upon studying a great variety of substances all at the same temperature, we would obtain a great variety of emission curves, but none of these would ever have a greater monochromatic emittance, $W_{\lambda b}$, than the envelope curve, shown as a solid line in Fig. 3.3. It appears that this curve may have a significance that does not depend on the nature of the emitting material. Let us attempt to find or make an emitter that has an emission curve identical to the solid curve of Fig. 3.3.

3.7 EMISSION AND ABSORPTION OF RADIATION

It may be wondered why it is, if the surfaces of all bodies are continually emitting radiant energy, that all bodies do not eventually radiate away all their internal energy and cool down to a temperature of absolute zero. The answer is that they would do so if energy were not supplied to them in some way. In the case of the filament of an electric lamp, energy is supplied electrically to make up for the energy radiated. As soon as this energy supply is cut off, these bodies do, in fact, cool down very quickly to room temperature. The reason that they do not cool further is that their surroundings (the walls and other objects in the room) are also radiating, and some of this radiant energy is intercepted, absorbed, and converted into internal energy. The same thing is true of all other objects in the room — each is both emitting and absorbing radiant energy simultaneously. If any object is hotter than its surroundings, its rate of emission will exceed its rate of absorption. There will thus be a net loss of energy and the body will cool down unless heated by some other method. If a body is at a lower temperature than its surroundings, its rate of absorption will be larger than its rate of emission and its temperature will rise. When the body is at the same temperature as its surroundings, the two rates become equal, and there is no net gain or loss of energy and no change in temperature.

Figure 3.3 shows that the emittance of a surface W is different at different wavelengths, and the paragraph above implies that it is greater for higher temperatures. To simplify our next discussion, consider an infinitesimal band of wavelengths and several opaque bodies in thermal equilibrium with each other and their surroundings. Because the bodies are opaque they will not transmit radiation and therefore, in general, part of the incident radiation will be reflected and the remainder will be absorbed. The fraction absorbed, called the *absorptance* a, plus the fraction reflected, called the *reflectance* r, must be unity or, since the surfaces may be different,

$$a_1 + r_1 = 1, \qquad a_2 + r_2 = 1, \quad \text{etc.} \tag{3.1}$$

Since the bodies are in thermal equilibrium with their surroundings, they will be "bathed" in radiation of uniform intensity, I. If this is not obvious, it may be helpful to think of the bodies as being tiny specks near each other but too small to cast shadows on each other. We can now write the last sentence quantitatively. The total radiation in a time Δt from body No. 1, which has an area ΔA_1 and a radiant emittance W_1, is $W_1 \Delta A_1 \Delta t$. The absorption by the same body in the same time is $a_1 I \Delta A_1 \Delta t$. For the condition of thermal equilibrium to exist these must be equal. Therefore we have

$$W_1 \Delta A_1 \Delta t = a_1 I \Delta A_1 \Delta t, \tag{3.2}$$

and similarly for another body,

$$W_2 \Delta A_2 \Delta t = a_2 I \Delta A_2 \Delta t, \quad \text{etc.} \tag{3.3}$$

Dividing Eq. (3.2) by Eq. (3.3), we obtain

$$\frac{W_1}{W_2} = \frac{a_1}{a_2} \qquad \text{or} \qquad \frac{W_1}{a_1} = \frac{W_2}{a_2}, \quad \text{etc.} \tag{3.4}$$

Since the number of specks or kinds of surface has not been restricted, it becomes evident that W/a for any substance must be a constant (which may, of course, still depend on wavelength and temperature).

We have just proved that a body or surface that is a good emitter (high value of W) must be a good absorber (high value of a) and conversely. If we could find a perfect absorber, we would necessarily find the best possible emitter, the graph of which is shown as b in Fig. 3.3.

3.8 BLACKBODY RADIATION

In acoustics, an open window is taken to be a perfect absorber of sound, since an open window reflects virtually no sound back into the room. In optics, there are few things darker than the keyhole of a windowless closet, since what little light gets into the closet bounces around against absorbing surfaces before it is redirected out the keyhole. Painting the inside of the closet black may increase the darkness of the keyhole, but the essential darkness of the hole is due to the geometry of the cavity rather than to the absorptivity of its surfaces. A small hole in a cavity of opaque material is the most perfect absorber of radiant energy that can be found. Conversely, a small hole in a cavity is the most perfect emitter that has been devised. We can conclude this from the proof above or we can understand it more thoroughly from the following. If we look into a hole in a heated cavity, we can see the radiation from the inside wall just opposite to the hole. In addition, we see some radiation from other parts of the inside of the cavity that was directed toward the spot of wall we are looking at and is reflected to us by that spot. The absorption and emission of radiation by a hole in a hollow tungsten cylinder is shown in Fig. 3.4(a). The light streak across the center of the figure is an incandescent filament maintained at a constant color temperature for comparison purposes. When the cylinder is cold, the hole is darker than any other part and actually appears black, but when the cylinder is heated sufficiently, the hole is brighter than the body of the tube and matches the reference filament. Such a hollow absorber-emitter is called a blackbody. Figure 3.4(b) shows the appearance of the dark design on a piece of white porcelain when it is cold, and the reversal of the pattern when the piece is heated to a high temperature. It is obvious that the same kind of radiation is emitted from a surface as is absorbed by that surface. Since a is equal to unity for a blackbody, from Eq. (3.4) we obtain

$$\frac{W_1}{a_1} = \frac{W_2}{a_2} = \frac{W_b}{1} = W_b. \tag{3.5}$$

This relation is called Kirchhoff's law of radiation: *The ratio of the radiant emittance of a surface to its absorptance is the same for all surfaces at a given temperature and is equal to the radiant emittance of a blackbody at the same temperature.*

We shall now discuss the multiple reflection situation of a radiant cavity quantitatively. Consider the radiation traveling back and forth in an isothermal cavity formed between two plane parallel sheets of different materials, as shown in

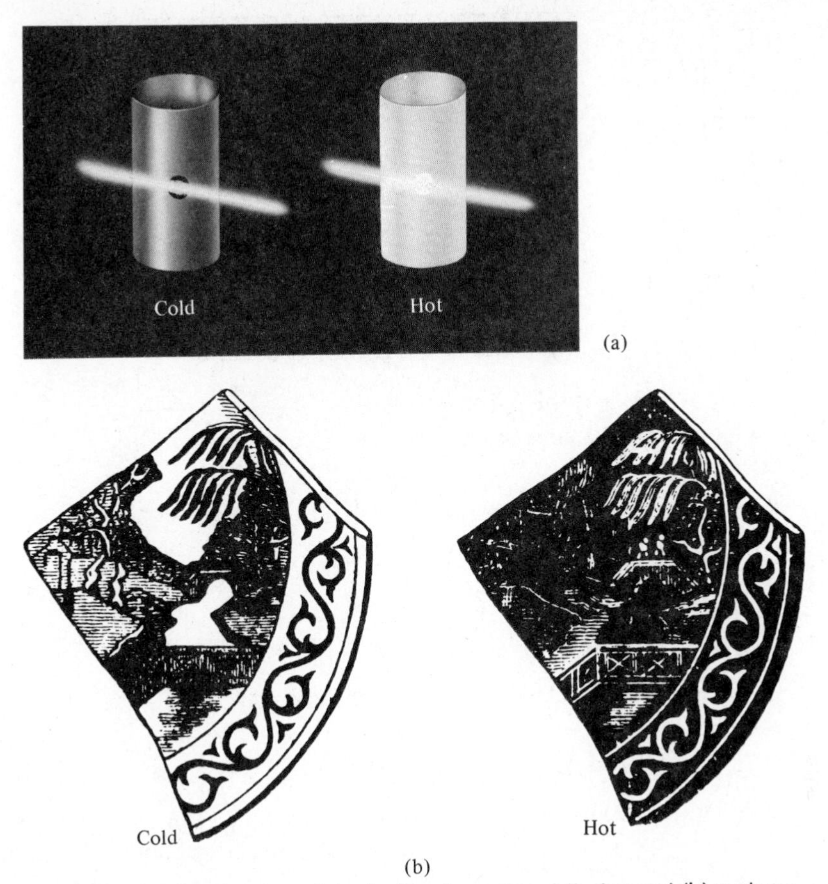

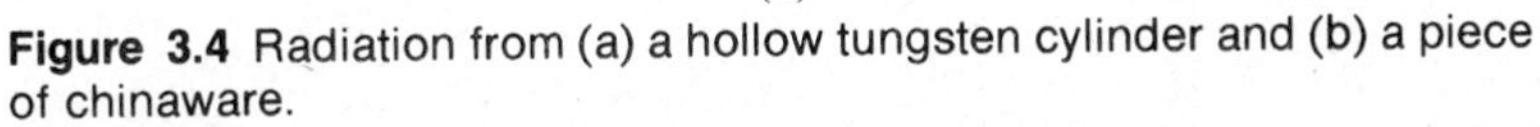

Figure 3.4 Radiation from (a) a hollow tungsten cylinder and (b) a piece of chinaware.
From Edser, *Heat for Advanced Students*. London: Macmillan, 1913.

Fig. 3.5. Let us now follow the history of the radiation emitted from a unit area of each face in a time interval Δt, which is just long enough to permit the radiation to travel across the space to the other face. It is reflected there with some loss of energy, returns to the first face, is reflected there with further loss, and so on. Table 3.1 gives the values and the direction of travel of the components in the radiation streams per unit area between the walls after several intervals of Δt have elapsed.

While these successive traverses of the initial radiation from each face are occurring, both faces continue to emit. Therefore when the steady state has been reached, there are simultaneous columns of thermal radiation going back and forth in the space between the faces. The total radiation streaming in one direction, say to the right, is simply the sum of all the components in the direction $\rightarrow$. Therefore, the

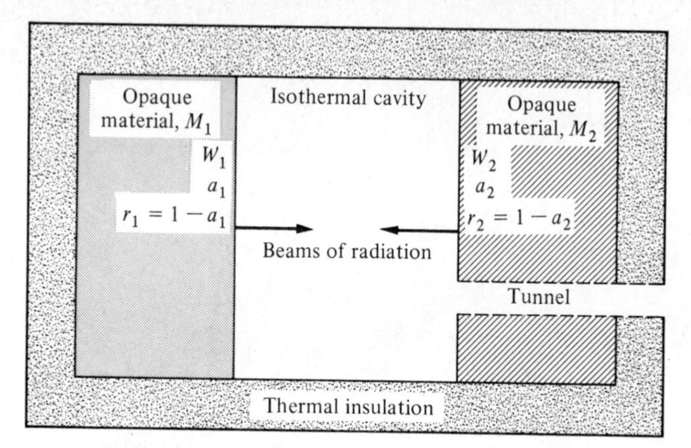

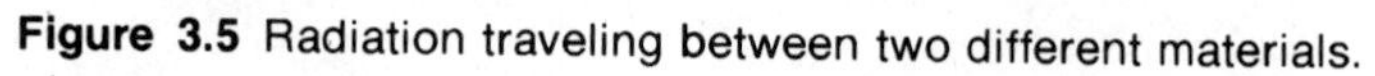
Figure 3.5 Radiation traveling between two different materials.

total effective emittance toward the right is given by

$$\begin{aligned} W_r\,\Delta t = {} & W_1\,\Delta t + (1 - a_1)(1 - a_2)W_1\,\Delta t + (1 - a_1)^2(1 - a_2)^2 W_1\,\Delta t \\ & + \cdots + (1 - a_1)W_2\,\Delta t + (1 - a_1)^2(1 - a_2)W_2\,\Delta t \\ & + (1 - a_1)^3(1 - a_2)^2\,\Delta t + \cdots . \end{aligned} \tag{3.6}$$

Let $x = (1 - a_1)(1 - a_2)$. Then substituting this in the preceding equation we obtain

$$W_r = W_1[1 + x + x^2 + \cdots] + W_2(1 - a_1)[1 + x + x^2 + \cdots]. \tag{3.7}$$

The series within the pairs of brackets is a simple geometric progression whose limit, since $0 < x < 1$, is $1/(1 - x)$. In terms of the absorptances, this limit is

$$\begin{aligned} \frac{1}{1 - x} &= \frac{1}{1 - (1 - a_1)(1 - a_2)} = \frac{1}{1 - (1 - a_1 - a_2 + a_1 a_2)} \\ &= \frac{1}{a_1 + a_2 - a_1 a_2}. \end{aligned} \tag{3.8}$$

Table 3.1

Initially emitted by M_1	Initially emitted by M_2
$\rightarrow W_1\,\Delta t$	$\leftarrow W_2\,\Delta t$
$\leftarrow (1 - a_2)W_1\,\Delta t$	$\rightarrow (1 - a_1)W_2\,\Delta t$
$\rightarrow (1 - a_1)(1 - a_2)W_1\,\Delta t$	$\leftarrow (1 - a_1)(1 - a_2)W_2\,\Delta t$
$\leftarrow (1 - a_1)(1 - a_2)^2 W_1\,\Delta t$	$\rightarrow (1 - a_1)^2(1 - a_2)W_2\,\Delta t$
$\rightarrow (1 - a_1)^2(1 - a_2)^2 W_1\,\Delta t$	$\leftarrow (1 - a_1)^2(1 - a_2)^2 W_2\,\Delta t$
$\leftarrow (1 - a_1)^2(1 - a_2)^3 W_1\,\Delta t$	$\rightarrow (1 - a_1)^3(1 - a_2)^2 W_2\,\Delta t$
$+ \cdots$	$+ \cdots$

Because the system is isothermal, we can obtain the following relations from Kirchhoff's radiation law (Eq. 3.5):

$$W_1 = a_1 W_b \quad \text{and} \quad W_2 = a_2 W_b. \tag{3.9}$$

Substituting the values from Eqs. (3.8) and (3.9) in Eq. (3.7), we have

$$W_r = \frac{a_1 W_b + a_2 W_b (1 - a_1)}{a_1 + a_2 - a_1 a_2} = \frac{W_b (a_1 + a_2 - a_1 a_2)}{a_1 + a_2 - a_1 a_2} = W_b. \tag{3.10}$$

This equation shows that the radiation to the right (it could just as well have been in any other direction) is effectively radiated from the left surface as though from a blackbody. If a tunnel, which is so small it does not subtract a significant portion of the radiation in the cavity, is bored through the right-hand face, then *the leakage radiation will be blackbody radiation*. It is to be noted that the derivation contained no assumptions about either the nature of thermal radiation or of the kinds of surfaces inside the enclosure.

We now know how to make a blackbody and have achieved the goal we set at the end of Section 3.6.

Readers can demonstrate for themselves that blackbodies can even be made from bright objects. A pile of razor blades at least one sixteenth of an inch thick looks black when viewed from the sharp side. The incident radiation is completely absorbed as a result of all the partial absorptions experienced at the many successive partial reflections it undergoes in traveling down into the relatively deep, narrow spaces between the blades.

We now return to the question of the spectrum of the radiation emitted by a hot body. If we take a blackbody as our sample, we can measure the emission from the hole as we did the material samples in getting the data for Fig. 3.3. This experiment shows that the emission of the blackbody gives at once the smooth solid curve of Fig. 3.3, which, unlike the other curves in the figure, is independent of the material used to make the emitter. This confirms what we might have suspected before, that the solid curve portrays a general characteristic of thermal radiation at a given temperature. A study of this curve should give information about radiation itself. With consideration of the material composing the cavity eliminated, the remaining important variable is the temperature of the radiation source. Mathematically, the total energy radiated per unit time per unit area of emitter is proportional to the area under the curve, and Stefan found empirically that this area is directly proportional to the fourth power of the absolute temperature,

$$W = \sigma T^4 \tag{3.11}$$

where

$$\sigma = (5.66961 \pm 0.00096) \times 10^{-8}\ W/m^2K^4.$$

(Note that σ is not a cross section as in Chapter 1.) Equation (3.11) is called the *Stefan-Boltzmann law* or the "fourth-power law." It was derived theoretically by Boltzmann, who used a thermodynamic argument. Wien found that as the temperature of any blackbody is changed the curve retains its general shape, but that the maximum of the curve shifts with temperature so that the wavelength of the most intense radiation is inversely proportional to the absolute temperature, or

$$\lambda_{max} = \text{const}/T$$

or

$$\lambda_{max} T = \text{const.} \qquad (3.12)$$

This is a special case of *Wien's displacement law*, which states that at corresponding wavelengths the monochromatic energy density of the radiation in the cavity of a blackbody varies directly as the fifth power of the absolute temperature. The relation defining corresponding wavelengths at temperatures T_1 and T_2 is

$$\lambda_1 T_1 = \lambda_2 T_2. \qquad (3.13)$$

The displacement law enables us to predict the entire curve at *any* temperature, given the entire curve at *one* particular temperature. Neither of these radiation laws, however, treats the basic problem of why the energy radiated from a blackbody has this particular wavelength distribution.

3.9 WIEN AND RAYLEIGH-JEANS LAWS

A comparison of the blackbody radiation curves of Fig. 3.6 and the Maxwell distribution of speeds in a gas shown in Fig. 1.2 shows a remarkable similarity. Wien

Figure 3.6 The distribution of energy in the spectrum of the radiation from a blackbody at different temperatures.

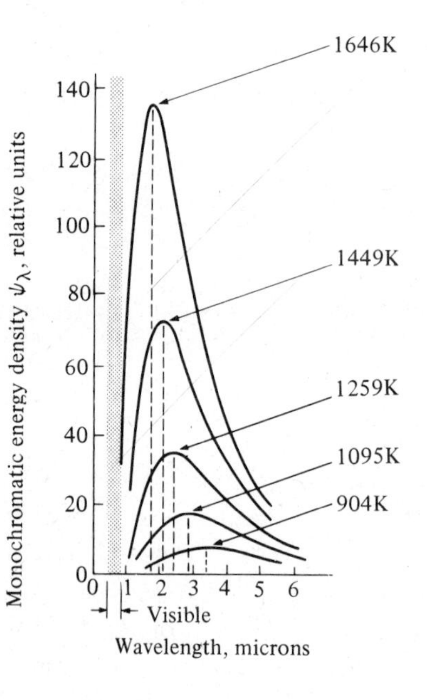

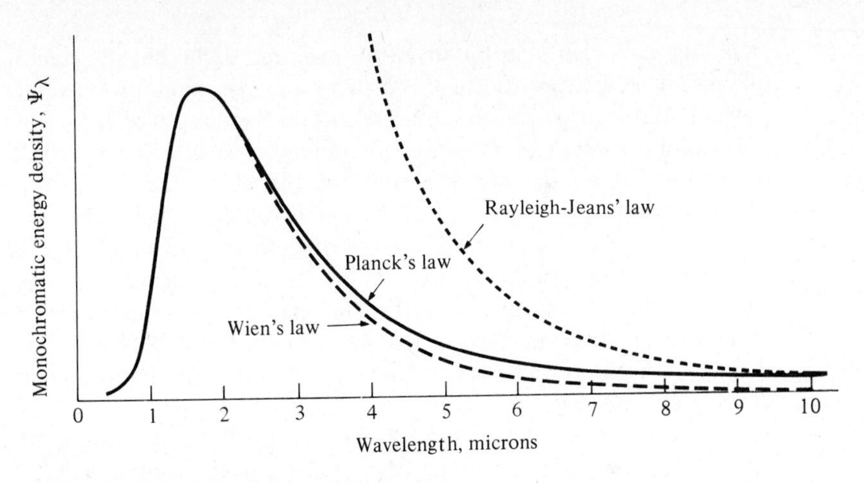

Figure 3.7 Graphs of the radiation laws.

noted this similarity and tried to fit a function such as Maxwell had derived for the speed distribution to the blackbody wavelength distribution. There is more than the similarity of the curves to justify this approach. If the molecules of the blackbody are thermally agitated, then their distribution of speeds may be somewhat like that derived by Maxwell. The accelerations of these molecules should be related to their velocities. These molecules contain charges that are therefore thermally accelerated, and we have shown that classical electrodynamics indicates that radiation results from accelerated charges. This argument is hardly rigorous, but it is a plausible explanation of the relationship between the similar curves. The expression that Wien obtained for the monochromatic energy density Ψ_λ within an isothermal blackbody enclosure in the wavelength range λ to $\lambda + d\lambda$ is

$$\Psi_\lambda \, d\lambda = \frac{c_1 \lambda^{-5}}{e^{c_2/\lambda T}} d\lambda, \tag{3.14}$$

where λ is the wavelength, T is the absolute temperature, and e is the base of natural logarithms. This formula is essentially empirical and contains two adjustment constants c_1 and c_2, called the first and second radiation constants, respectively. Wien chose these constants so that the fit he obtained was rather good except at long wavelengths. The graph of Wien's law is shown dashed in Fig. 3.7. But a "rather good" fit is not good enough, and a formula that is essentially empirical tells us nothing about the nature of radiation.

Lord Rayleigh set out to derive the radiation distribution law in a rigorous way. We shall not repeat his argument here except to mention that he concentrated attention on the radiation itself. He said that the electromagnetic radiation inside an isothermal blackbody cavity is reflected back and forth by the walls to form a system of standing waves for each frequency present. Thus these waves should resemble the standing waves on a violin string or, even more closely, the standing sound waves within an acoustic cavity. Just as a string can vibrate to produce a fundamental and a whole series of overtones, so there should be many modes of vibration present in the standing waves of radiation in the cavity space. It is a small sample of this radiation that streams out of a hole in a blackbody for spectrum analysis. Because the system is in thermal equilibrium, the radiation from the cavity absorbed by the interior walls must equal that emitted to the cavity by the atomic oscillators in the walls. Each mode of vibration introduces two degrees of freedom, one for the potential energy of the oscillator and one for its kinetic energy. This will be discussed in detail in Section 9.2. In Section 1.13 we found that the energy per degree of freedom was $\frac{1}{2}kT$, and thus each mode of vibration has a total energy kT associated with it.

We cannot reproduce here Rayleigh's involved derivation of the number of modes of vibration within the cavity. The result he obtained for the number of modes of vibration, dn_λ, per unit volume of space in the wavelength range λ to $\lambda + d\lambda$ is

$$dn_\lambda = 8\pi \frac{d\lambda}{\lambda^4}. \tag{3.15}$$

If we accept this result, we need only multiply the energy per mode of vibration, kT, by the number of modes per unit volume of space within the wavelength interval $d\lambda$ to obtain the Rayleigh-Jeans law. The result is

$$\Psi_\lambda \, d\lambda = 8\pi kT \frac{d\lambda}{\lambda^4}. \tag{3.16}$$

This equation is also plotted in Fig. 3.7. At first glance, this law appears vastly inferior to Wien's. Although it fits well for long wavelengths, at short wavelengths or high frequencies it heads toward infinity in what has been dramatically called the "ultraviolet catastrophe." Theoretically, however, the Rayleigh law must be taken far more seriously than Wien's. It was derived rigorously on the basis of classical physics. It involves no arbitrary constants, and where it does fit the experimental curve, it fits exactly. Whereas the failure of the Wien law was "too bad," the failure of the Rayleigh law presented a crisis. It indicated that classical theory was unable to account for an important experimental observation. This was the situation to which Max Planck directed himself.

3.10 PLANCK'S LAW. EMISSION QUANTIZED

Planck's first step was essentially empirical. He found that by putting a mere -1 into the denominator of the Wien formula to obtain

$$\Psi_\lambda \, d\lambda = \frac{c_1 \lambda^{-5}}{e^{c_2/\lambda T} - 1} d\lambda \tag{3.17}$$

and by adjusting Wien's constants, he could get a formula that reduced to the Rayleigh formula at long wavelengths and that fitted the experimental curve everywhere. He knew that he had found a correct formula and that it should be derivable. Planck's position was a little like that of a student who has peeked at the answer in the back of the book and is now faced with the task of showing how that answer can be logically computed. Planck tried by every method he could conceive to derive this correct formula from classical physics. He was finally forced to conclude that there was no flaw in Rayleigh's derivation and that the flaw must lie in classical theory itself.

Planck had to eliminate the "ultraviolet catastrophe" that came into Rayleigh's derivation because of the assumption that the radiation standing waves had a fundamental and also an infinite number of harmonic modes of vibration. Each of these was assumed to have an average energy kT. Instead of taking the average energy per mode to be kT directly, Planck examined this matter in more detail.

Let us, as Planck did, assume that each mode can receive energy only in discrete steps, u. The energy content of one mode will be 0, u, $2u$, $3u$, etc., or in general mu, where m is an integer. (The method at this point involves following a procedure used by Boltzmann in 1877 to determine the distribution of kinetic energy among the molecules in a gas.) The number of oscillators having the energy mu, as given by the Boltzmann distribution law (Eq. 1.45), is

$$n_m = n_0 e^{-mu/kT}. \tag{3.18}$$

The energy contributed by the n_m oscillators is obviously

$$mun_m = mun_0 e^{-mu/kT}. \tag{3.19}$$

Therefore the average energy $\bar{w}$ of an oscillator is

$$\bar{w} = \frac{\sum_{m=0}^{\infty} mun_0 e^{-mu/kT}}{\sum_{m=0}^{\infty} n_0 e^{-mu/kT}}. \tag{3.20}$$

Since m is an integer, Eq. (3.20) becomes

$$\bar{w} = \frac{0 + ue^{-u/kT} + 2ue^{-2u/kT} + 3ue^{-3u/kT} + \cdots}{1 + e^{-u/kT} + e^{-2u/kT} + e^{-3u/kT} + \cdots}. \tag{3.21}$$

Let $x = e^{-u/kT}$. Then Eq. (3.21) can be written

$$\bar{w} = ux\frac{1 + 2x + 3x^2 + \cdots}{1 + x + x^2 + \cdots}. \tag{3.22}$$

The limits of these convergent series can be found by the usual methods (note that $x < 1$). The convergence limit of the series in the numerator is $1/(1 - x)^2$. This can be checked by expanding $(1 - x)^{-2}$ according to the binomial theorem. The denominator is a simple geometric progression converging to $1/(1 - x)$. Substituting these limits in Eq. (3.22), we have

$$\overline{w} = ux\frac{1/(1-x)^2}{1/(1-x)} = \frac{ux}{1-x} = \frac{u}{(1/x)-1}. \tag{3.23}$$

When x is replaced by its equivalent, the result is

$$\overline{w} = \frac{u}{e^{u/kT}-1}. \tag{3.24}$$

If we now multiply Eq. (3.24) by the number of modes of vibration in a unit volume of cavity space, from Eq. (3.15), we obtain the energy density in a wavelength range $d\lambda$,

$$\Psi_\lambda \, d\lambda = \frac{8\pi}{\lambda^4}\frac{u}{e^{u/kT}-1}\,d\lambda. \tag{3.25}$$

Recall that in this derivation the energy of an oscillator has been assumed to be an integer, m, times some small energy, u. Classical physics says the energy may have any value. This is equivalent to saying that u may be exceedingly small and, in the limit, approach zero. If we set $u = 0$, Eq. (3.24) is indeterminate, $0/0$. If we apply l'Hospital's rule, differentiating both numerator and denominator with respect to u before letting $u = 0$, we find that

$$\overline{w} = kT, \tag{3.26}$$

which is in complete agreement with Rayleigh's classical assumption. As we have seen, however, this assumption does not lead to the correct radiation law.

The relation given in Eq. (3.25) begins to look like Wien's law, Eq. (3.14), if u is not zero. Indeed, the denominators of these two equations become identical (except for the minus one) if a value of u is chosen so that the powers of the exponential terms are the same. To obtain this value of u, we let

$$\frac{c^2}{\lambda T} = \frac{u}{kT}, \tag{3.27}$$

or

$$u = \frac{c_2 k}{\lambda} = \frac{c_2 k}{c}\nu. \tag{3.28}$$

In this last equation, c is the free-space velocity of light and ν is the frequency of the oscillator and therefore also the frequency of the radiation it emits. If we replace the constants $(c_2 k/c)$ by another constant h, we have

$$u = \frac{hc}{\lambda} = h\nu. \tag{3.29}$$

When the value of u from Eq. (3.29) is substituted in Eq. (3.25), we obtain Planck's law for the energy density of blackbody or cavity radiation. This law is

$$\Psi_\lambda \, d\lambda = \frac{8\pi ch\lambda^{-5}}{e^{ch/\lambda kT}-1}\,d\lambda. \tag{3.30}$$

This equation does agree with the experimental results. It is plotted in Fig. 3.7. This expression for Planck's law also contains the Stefan-Boltzmann law and the Wien displacement law (see Section 3.11).

The new constant h is called the *Planck constant*. We have seen that it could be determined from Wien's constant c_2 but it can also be evaluated from the photoelectric effect discussed later in this chapter. Its value is $h = (6.6256 \pm 0.0005) \times 10^{-34}$ J · s. (Note that the units are those of angular momentum.)

Thus Planck was led to his startling, nonclassical assumption that the energy states of an oscillator must be an *integral* multiple of the product of the constant h and the frequency ν of the electromagnetic radiation it emits. If E represents the smallest permissible energy change, Planck's famous quantum† equation is

$$E = h\nu. \tag{3.31}$$

Planck introduced the quantum concept in 1900, and it eventually led to the conclusion that radiation is not emitted in continuous amounts but in discrete bundles of energy each equal to $h\nu$. These bundles or packets of radiant energy are now called *quanta* or *photons*. This was the beginning of the atomic theory of radiation, which has grown to become the quantum theory. It is obvious, however, that quanta of radiation of different frequencies have different "sizes" (energies), and that they are atomic only in the sense that they are discrete. Planck thought at first that his *ad hoc*‡ hypothesis applied only to the oscillators and, possibly, to the emitted radiation in their immediate neighborhood and that, at most, it was a slight modification of Maxwell's theory of radiation. However, we shall see that he initiated a series of events that have changed our whole concept of the interaction of electromagnetic radiation with matter.

EXAMPLE What is the energy in a quantum of radiation having a wavelength of 5000 Å?

$$E = h\nu = \frac{hc}{\lambda} = 6.63 \times 10^{-34}\ \text{J}\cdot\text{s} \times 3 \times 10^{8}\ \frac{\text{m}}{\text{s}} \times \frac{1}{5000\ \text{Å}} \times \frac{1\ \text{Å}}{10^{-10}\ \text{m}}$$

$$= 3.98 \times 10^{-19}\ \text{J}$$

3.11 STEFAN-BOLTZMANN LAW AND WIEN DISPLACEMENT LAW

As was pointed out in the previous section, the Stefan-Boltzmann law and the Wien displacement law are contained in the Planck law and thus may be derived from Eq. (3.30). The Planck equation gives the energy density of the blackbody. The intensity

†*Quantum* is the Latin word for *how much* or *how great.*

‡*Ad hoc* means literally *to this* and is used to describe a hypothesis that is applicable to only one (the present) situation.

of the emitted radiation is equal to $c/4$ times this energy density (see Problem 3.16). Thus, multiplying Eq. (3.30) by $c/4$ and integrating from $\lambda = 0$ to $\lambda = \infty$, we have

$$W = \frac{c}{4}\int_0^\infty \frac{8\pi ch\lambda^{-5}}{e^{ch/\lambda kT} - 1}\,d\lambda. \tag{3.32}$$

Let $x = ch/\lambda kT$ to obtain

$$W = \frac{2\pi k^4 T^4}{c^2 h^3}\int_0^\infty \frac{x^3\,dx}{e^x - 1}. \tag{3.33}$$

The definite integral is equal to $\pi^4/15$; therefore,

$$W = \frac{2\pi^5 k^4}{15c^2 h^3}T^4 \tag{3.34}$$

or

$$W = \sigma T^4, \tag{3.35}$$

with

$$\sigma = \frac{2\pi^5 k^4}{15c^2 h^3}. \tag{3.36}$$

When the values of k, c, and h are inserted in Eq. (3.36), the calculated and experimental values of σ agree to within the experiment error.

The Wien displacement law is obtained by finding the wavelength for maximum total energy radiated per unit time per unit area at a certain temperature (in other words, finding the maximum of the solid curve in Fig. 3.7). To do this, it is necessary to differentiate

$$\frac{d}{d\lambda}\left[\frac{8\pi ch\lambda^{-5}}{e^{ch/\lambda kT} - 1}\right] = 0. \tag{3.37}$$

This yields

$$5 - \frac{ch}{\lambda_m kT} = 5e^{-ch/\lambda_m kT}.$$

Then if we let

$$y = \frac{ch}{\lambda_m kT}$$

we have

$$5 - y = 5e^{-y}.$$

A solution to this is very nearly $y = 5$. Taking $y = 5$ in the exponential but not in $(5 - y)$, gives

$$y = 4.97$$

or

$$\frac{ch}{\lambda_m kT} = 4.97$$

and finally

$$\lambda_m T = \frac{ch}{4.97k}. \tag{3.38}$$

Inserting the appropriate values of the speed of light, Planck's constant and Boltzmann's constant will give $2.90 \times 10^{-3}\ m \cdot K$ for the Wien displacement constant.

3.12 COSMIC BLACKBODY RADIATION

Until recent years, there were two major competing theories of cosmology — the big bang and the steady state. According to the big bang theory, all of the matter and energy of the universe was at one point in time contained in a single point. A gigantic explosion then sent all matter rushing away from all other matter in the universe. This model was first proposed by A. G. Lemaitre in 1927 and details were worked out by George Gamow. In contrast is the steady state model, proposed by Bondi, Gold, and Hoyle. This theory states that the universe has always been and will always remain as it exists today.

Over the years evidence has leaned in favor of the big bang theory; however, it has not been overpowering. By studying the light coming from distant galaxies, astronomers have been able to determine their distances as well as their velocities. The evidence is strong that the galaxies farthest from us are moving faster than those near us. These data are consistent with the big bang model because matter initially traveling faster should be farther away than that traveling more slowly.

The big bang theory of Gamow described a primeval nucleus of radiation that was originally extremely hot and dense. As the universe expanded and cooled, the leftover radiation permeated the entire universe. According to this theory, such an expansion and cooling for billions of years would have left electromagnetic radiation corresponding to a blackbody at a temperature of about 3 K.

Significant experimental evidence in favor of the big bang theory came in 1965 with the observation of the background radiation by Penzias and Wilson. They were using a conventional radio telescope at the surface of the earth and continually saw background microwave radiation that appeared to be coming from all directions. The radiation they observed was consistent with the radiation that would be coming from a blackbody at 3 K. However, their data were only from a very small part of the electromagnetic spectrum. Conclusive evidence for the shape of the curve was given by Woody, Mather, Inshioka, and Richards when they measured the radiation on the high frequency side of the blackbody radiation peak. Their experiment covered the spectrum from a wavelength of 5 mm to a wavelength of 0.6 mm. They showed that the peak was at 1.6 mm, which is consistent with a blackbody radiation curve for 2.9 ± 0.1 K.

3.13 PHOTOELECTRIC EFFECT

We now turn from thermal radiation to another portion of the electromagnetic spectrum and consider an effect that is due to radiation of higher frequency. We mentioned earlier that even before the discovery of the electron, Hallwachs observed that zinc irradiated with ultraviolet light lost negative charge. He proposed that somehow the radiation caused the zinc to eject negative charge. In 1899, Lenard showed that the radiation caused the metal to emit electrons.

This phenomenon, called the photoelectric effect, can be studied in detail with the apparatus shown in Fig. 3.8. In this figure, S is a source of radiation of variable and known frequency ν and intensity I, E is an emitting electrode of the material being studied, and C is a collecting electrode of the same material. Both electrodes are contained in an evacuated glass envelope with a quartz window that permits the passage of ultraviolet and visible light. The electric circuit allows the electrodes to be maintained at different known potentials and permits the measurement of any current between the electrodes. We first make the collecting electrode positive with respect to the emitting electrode, so that any electrons ejected will be quickly swept away from the emitter. About 10 volts is enough to do this but not enough to free electrons from the negative electrode by positive ion bombardment as was the case in the early cathode-ray tubes. If the tube is dark, no electrons are emitted and the microammeter indicates no current. If ultraviolet light is allowed to fall on the emitting electrode, electrons are liberated and the current is measured by the microammeter. It is found that the rate of electron emission is proportional to the light intensity. By holding the frequency ν of the light and the accelerating potential V constant, we can obtain data like that represented in Fig. 3.9.

It is hardly surprising that if a little light liberates a few electrons, then more light liberates many. If we vary either the frequency of the light or the material irradiated, only the slope of the line changes.

We can now experiment by keeping the light intensity constant and varying the frequency. The graphs of these data are shown in Fig. 3.10, where A and B represent

Figure 3.8 Apparatus for investigating the photoelectric effect.

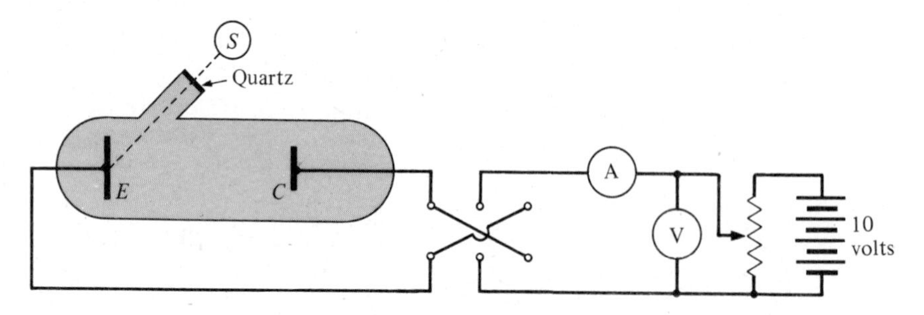

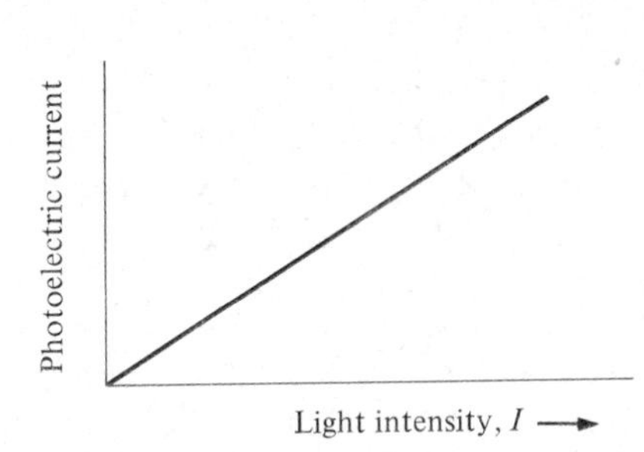

Figure 3.9 Photoelectric current as a function of the intensity of the light. The frequency of the light and the accelerating potential are kept constant.

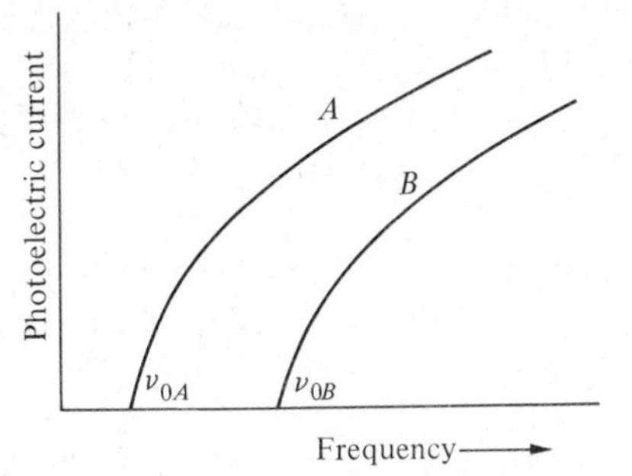

Figure 3.10 Photoelectric current for two materials as a function of the frequency of the light. The intensity of the light and the accelerating potential are kept constant.

two different irradiated materials. The significant thing about these curves is that for every substance irradiated there is a limiting frequency below which no photoelectrons are produced. This frequency, called the *threshold frequency*, ν_0, is a characteristic of the material irradiated. The wavelength of light corresponding to the threshold frequency is the *threshold wavelength*, λ_0. No photoelectrons are emitted for wavelengths greater than this.

The existence of a threshold frequency is difficult to explain on the basis of the wave theory of light. If we think of light as consisting of a pulsating electromagnetic field, we can imagine that that field is sometimes directed so as to tend to eject electrons from a metallic surface. We might even feel it reasonable that certain frequencies of light would resonate with the electrons of the metal so that, for a particular metal, there might be preferred light frequencies that would cause emission more efficiently. The striking thing about these data is that for each material there is a frequency below which no photoelectrons are emitted and above which they are emitted. This effect is independent of the intensity of the light.

In 1905, Einstein proposed a daring but simple explanation. He centered attention on the energy aspect of the situation. Whereas Planck had proposed that radiation is composed of energy bundles only in the neighborhood of the emitter, Einstein proposed that these energy bundles *preserve their identity throughout their life*. Instead of spreading out like water waves, Einstein conceived that the emitted energy bundle stays together and carries an amount of energy equal to $h\nu$. For Einstein, the significance of the light frequency was not so much an indication of the frequency of a pulsating electric field as it was a measure of the energy of a bundle of light called a *photon*. His interpretation of the data of Fig. 3.10 would be that a quantum of light below the threshold frequency just does not have enough energy to remove an electron from the metal, but light above that frequency does.

The threshold frequency is dependent on the nature of the material irradiated because there is for each material a certain minimum energy necessary to liberate an electron. The *photoelectric work function* or *threshold energy*, W_0, of a material is the *minimum* energy required to free a photoelectron from that material.

In a third photoelectric experiment, let us hold both the frequency and intensity of the light constant. The variable is the potential difference across the photoelectric cell. Starting with the collector at about 10 volts positive, we reduce this potential to zero and then run it negative until the photocurrent stops entirely. Curve I_1 of Fig. 3.11 shows the type of curve we might expect for this particular substance. This curve requires careful interpretation.

When the potential difference across the tube is about 10 volts or more, *all* the emitted electrons travel across the tube. This stream of charges is called the saturation current, and it is obvious that an increase in the potential of the collector cannot cause an increase in current. As the accelerating potential is reduced from positive values through zero to negative values, the tube current reduces because of the applied retarding potential. Eventually this potential is large enough to stop the current completely.

The *stopping potential* V_s is the value of the retarding potential difference that is just sufficient to halt the *most energetic* photoelectron emitted. Therefore the product of the stopping potential and the electronic charge, $V_s e$, is equal to the maximum kinetic energy that an emitted electron can have. Since this stopping potential has a definite value, it indicates that the emitted electrons have a definite upper limit to their kinetic energy. Doubling the intensity of the light doubles the current at each potential, as in I_2 of Fig. 3.11, but the stopping potential is *independent* of the intensity.

If, however, the experiment is repeated with a series of different light frequencies, it is found that the stopping potential increases linearly with the frequency. This is best shown by plotting the stopping potential against the frequency, as shown in Fig. 3.12. Below the threshold frequency no electrons are emitted and the stopping potential is of course zero; however, as the frequency is increased above the threshold, the stopping potential increases linearly with the frequency.

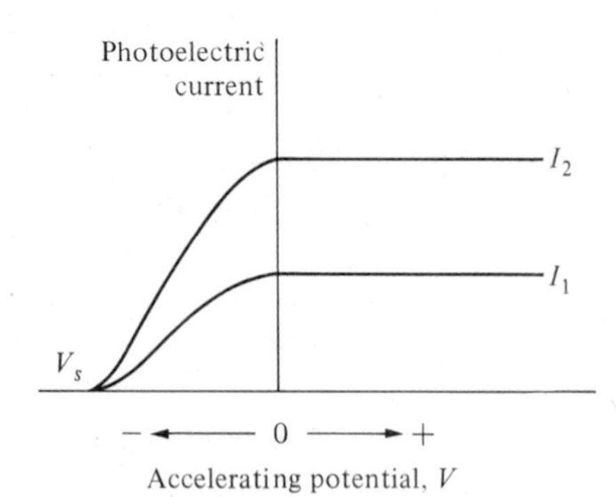

Figure 3.11 Photoelectric current as a function of the accelerating potential for light of different intensities having a 2-to-1 ratio. The frequency of the light is constant.

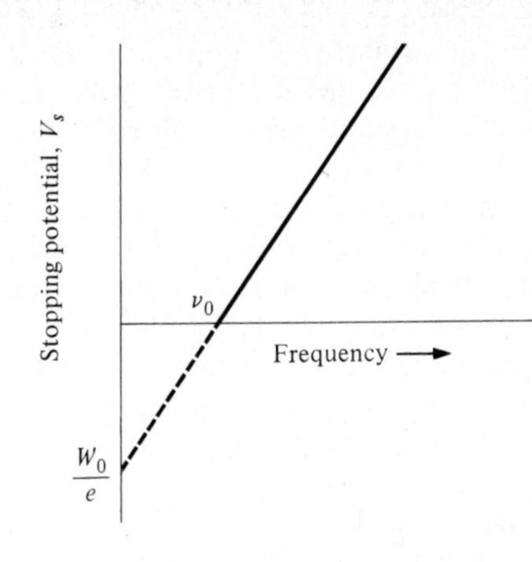

Figure 3.12 Stopping potential as a function of the frequency of the light. Results are independent of intensity.

To see how fully the data of Fig. 3.12 confirm the Einstein photon interpretation of the photoelectric effect, we now interpret the graph of Fig. 3.12 as he would have. For light frequency between zero and the threshold frequency there are no photoelectrons produced, since the incident photons have less energy than the work function of the material. For light above the threshold frequency, photoelectrons are emitted. The energy of these emitted electrons may vary greatly. According to the Einstein view, however, there must be an upper limit to the energy of the emitted photoelectrons. No photoelectrons can have energy in excess of the energy of the incoming photon less the minimum energy to free an electron, the work function. Since the photon energy is proportional to frequency and since the stopping potential is a measure of the maximum kinetic energy of the emitted photoelectrons, the graph of stopping potential against frequency should be a straight line. This is precisely what Fig. 3.12 shows. The quantitative check of the Einstein interpretation is that the slope of the straight line provides a method of determining the way in which photon energy depends on photon frequency, which is the Planck constant.

A final and decisive experiment consists of making the irradiating light extremely weak. In this case, the number of photoelectrons is very small and special techniques are required to detect them. The significant result of this experiment is that dim light causes emission of photoelectrons which, however few, are emitted almost instantaneously and with the same maximum kinetic energy as for bright light of the same frequency.

According to the wave theory of light, pulsating electromagnetic fields spread out from their source. Dim light corresponds to waves of small amplitude and small energy. If dim light spreads over a surface, conservation of energy requires that either no photoelectrons should be emitted or the electrons must store energy over long periods of time before gathering enough energy to become free of the metal. The fact that high-energy photoelectrons appear almost immediately can be explained only by assuming that the light energy falls on the surface in concentrated bundles. According to Einstein, the dim light consists of a few photons each having energy depending only on the light frequency. This energy is not spread over the surface uniformly as required by the wave theory. A photon that is absorbed gives all its energy to one electron, and that electron will be emitted violently even though the number of such events is small.

If someone dives into a swimming pool, the swimmer's energy is partly converted into waves that agitate other swimmers in the pool. If it were observed that when someone dives into a pool another swimmer is suddenly ejected from the pool onto a diving board, we would be forced to conclude that the energy provided by the diver did not spread out in an expanding wavefront but was somehow transferred in concentrated form to the ejected swimmer. A swimming party held in such a superquantum pool would be a very odd affair compared with one in a classical pool.

To summarize Einstein's interpretation of the photoelectric effect, we equate the energy of an incident photon of frequency ν to the sum of the work function of the emitter/collector material ($W_0 = h\nu_0$) and the maximum kinetic energy that an ejected photoelectron can acquire.† We then have

$$h\nu = h\nu_0 + \tfrac{1}{2}m_e v_{max}^2, \qquad \text{or} \qquad \tfrac{1}{2}m_e v_{max}^2 = h\nu - h\nu_0,$$

or

$$V_s e = h\nu - W_0. \tag{3.39}$$

This is a linear equation. It is the equation of the graph in Fig. 3.12. It is evident that the slope of the curve V_s plotted as a function of ν is equal to h/e, that ν_0 is the ν-axis intercept, and that W_0/e is the intercept of the extrapolated curve on the V_s-axis. Note that if $\nu < \nu_0$, v is imaginary. The physical meaning is that photoelectric emission is then impossible. Einstein proposed his photoelectric equation in 1905. It was first verified with precision eleven years later by Millikan, who made careful measurements of the photoemission from many different substances.

3.14 SUMMARY OF THE ATOMIC VIEW OF RADIATION

We introduced this chapter by outlining the disagreement between Newton and Huygens over the nature of light. We described how Maxwell strengthened Huygens's wave theory when he showed that electromagnetic waves were a consequence of the laws of electricity and magnetism. We reported that Hertz demonstrated that electric circuits could be made to produce the electromagnetic waves Maxwell predicted. But

† By choosing the collector material to be the same as the emitter material, we have avoided the problem of the contact potential difference between the two materials. If the emitter and collector are made of different materials, the work function that we would need to use would be the work function of the collector material. See the paper on the photoelectric effect by Rudnick and Tannhauser referenced at the end of this chapter.

we also mentioned that Hertz observed that he could produce sparks more easily when his spark gap was illuminated. Thus Hertz's work, which supported Maxwell's theory, also contained the first observation of the photoelectric effect. Although the wave theory of light accounts beautifully for many optical phenomena, it fails to account for either the blackbody radiation or the photoelectric effect, where light appears to possess marked particle aspects. It is hardly satisfactory to regard light as a wave motion part of the time and a particle phenomenon at other times. We shall return to the question of the resolution of this conflict of viewpoint in Chapter 7. At this point it is clear, however, that the resolution of this paradox can never eliminate the idea that light is emitted and absorbed in bundles of energy called photons, and that now radiation must join matter and electricity in having a basically atomic character. The fact that radiant energy is quantized is a radical departure from classical physics and will require us to re-examine the whole energy concept from the quantum point of view. Establishing this fact has been the main business of this chapter. Before closing the chapter, however, we shall consider two related topics.

3.15 THE ELECTRON VOLT

In the Einstein equation, we measured the maximum kinetic energy of the emitted electrons by noting the potential energy difference (eV_s), which was equivalent to the electron kinetic energy. This method of determining and expressing electron energies is a particularly convenient one, and it suggests a new unit of energy. This new unit of energy is called the *electron volt*, eV, which is defined as the amount of energy equal to the change in energy of one electronic charge when it moves through a potential difference of one volt. Since the electron volt is an energy unit, it is in the same category as the foot-pound, the British thermal unit, and the kilowatt-hour.

To convert energy in joules to energy in electron volts we multiply by 1 electron volt/1.6×10^{-19} joules. Less formally, we divide by $e_c = 1.60 \times 10^{-19}$. In this case e_c is *not a charge* but a *conversion factor* having the units of joules per electron volt. The Einstein equation, Eq. (3.39), is valid in any consistent system of units. If we choose SI units and divide Eq. (3.39) by the factor e_c, we obtain the same relation in electron volts:

$$E_{k(\max)}(\textit{numerically}\text{ equal to } V_s) = \frac{h\nu}{e_c} - \frac{W_0}{e_c}. \quad (3.40)$$

In words, this equation states that the maximum kinetic energy of a photoelectron in electron volts equals the energy of the photon in electron volts minus the work function in electron volts.

When the electron volt is too small a unit, it is convenient to use 10^3 eV = 1 keV (kilo-electron volt), and 10^6 eV = 1 MeV (million-electron volt). Other prefixes used to form multiples of units are listed in Appendix 7.

It is also useful to express photon energies in electron volts. In terms of photon frequency,

$$E = \frac{h\nu}{e_c} = 6.63 \times 10^{-34}\,\text{J}\cdot\text{s} \times \frac{1\ \text{eV}}{1.60 \times 10^{-19}\,\text{J}} \times \nu$$
$$= 4.14 \times 10^{-15}\,\nu\ (\text{s}\cdot\text{eV}). \tag{3.41}$$

Or, since $\nu = c/\lambda$, we get

$$E = 4.14 \times 10^{-15}\,\text{s}\cdot\text{eV} \times 3 \times 10^{8}\,\frac{\text{m}}{\text{s}} \times \frac{1}{\lambda}$$
$$= \frac{1.24 \times 10^{-6}}{\lambda}\ (\text{m}\cdot\text{eV}). \tag{3.42}$$

If the wavelength is expressed in angstroms instead of meters, we have

$$E = \frac{1.24 \times 10^{-6}}{\lambda}\ (\text{m}\cdot\text{eV}) \times \frac{1\ \text{Å}}{10^{-10}\ \text{m}},$$

or

$$E = \frac{1.24 \times 10^{4}}{\lambda}(\text{eV}\cdot\text{Å}). \tag{3.43}$$

We shall use Eq. (3.43) *frequently.*

EXAMPLE Light having a wavelength of 5000 Å falls on a material having a photoelectric work function of 1.90 eV. Find (a) the energy of the photon in eV, (b) the kinetic energy of the most energetic photoelectron in eV and in joules, and (c) the stopping potential.

Solution (a) From Eq. (3.43),

$$E = \frac{1.24 \times 10^{4}\ \text{eV}\cdot\text{Å}}{\lambda} = \frac{12{,}400\ \text{eV}\cdot\text{Å}}{5000\ \text{Å}} = 2.47\ \text{eV}.$$

(b) The law of conservation of energy gives

Maximum kinetic energy = photon energy − work function

or

$$E_k = 2.47\ \text{eV} - 1.90\ \text{eV} = 0.57\ \text{eV}.$$

Also

$$E_k = 0.57\ \text{eV} \times \frac{1.60 \times 10^{-19}\ \text{J}}{1\ \text{eV}} = 9.11 \times 10^{-20}\ \text{J}.$$

(c) $$V_s = \frac{0.57\ \text{eV}}{1\ \text{electronic charge}} = 0.57\ \text{V}.$$

3.16 THERMIONIC EMISSION

We have already considered two ways in which electrons can be released from a metal. In Chapter 2 we discussed the discharge of electricity through gases. The electrons that participate in cold-cathode discharges are obtained from the cathode while it is bombarded by the positive ions produced in the residual gas in the tube. In this chapter we have considered another emission process, called photoelectric emission. There is still another kind of electron emission, which we shall now discuss briefly.

If a metal is heated, the thermal agitation of the matter may give electrons enough energy to exceed the work function of the material. Thus the space around a heated metal is found to contain many electrons. A study of this effect shows that the *thermionic work function* is very nearly the same as the photoelectric work function — a most satisfying result.

Since a fine wire can be heated easily by passing an electric current through it, thermionic emission is one of the most convenient electron sources. Most television picture tubes use a heated cathode as their electron source. Thermal emission of negative charges from a hot wire in a vacuum was first observed by Edison in 1883 when he was making incandescent lamps, and such thermal emission is called the *Edison effect*. In 1899, J. J. Thomson showed that the thermions in this effect are electrons.

FOR ADDITIONAL READING

Arons, A. B., and M. B. Peppard, "Einstein's Proposal of the Photon Concept — A Translation of the *Annalen der Physik* Paper of 1905," *Am. J. Phy.* **33**, No. 367 (1965). This is the only English translation of the famous paper proposing the concept of photons and the photoelectric law.

Beth, R. A., "Mechanical Detection and Measurement of the Angular Momentum of Light," *Phys. Rev.* **30**, No. 115 (1936).

Carruthers, P., "Resource Letter QSL-1 on Quantum and Statistical Aspects of Light," *Am. J. Phys.* **31**, No. 5 (May 1963); and the booklet, "Quantum and Statistical Aspects of Light, Selected Reprints," New York: American Institute of Physics.

Gamow, George, *Thirty Years that Shook Physics*. Garden City, N.Y.: Doubleday, 1966. A lively account of the twentieth-century revolution in physics.

Goldhaber, A. S., and M. M. Nieto, "The Mass of the Photon," *Scientific American* **234**, No. 86 (May 1976).

Hertz, Heinrich, *Electric Waves*. D. E. Jones, translator. London: Macmillan, 1893. Read pages 68–69 for the account of how Hertz's careful observations revealed an unexpected phenomenon and led to the "accidental" discovery of the photoelectric effect.

Klein, M. J., *The Natural Philosopher*, Vol. 2. New York: Blaisdell, 1963. This contains a critical analysis and the historical background of Einstein's first paper on photons.

Magie, William F., *A Source Book in Physics*. Cambridge, Mass.: Harvard University Press, 1963. This contains reprints of the following papers: "Electric Discharge by Light," by Wilhelm Hallwachs; "Temperature Radiation," by Josef Stefan.

Muller, R. A., "The Cosmic Background Radiation and the New Aether Drift," *Scientific American* **238**, p. 64 (May 1978).

Richtmyer, F. K., E. H. Kennard, and T. Lauristen, *Introduction to Modern Physics*. 5th ed. New York: McGraw-Hill, 1955. Chapter 1, brief history of the evolution of physics; Chapter 3, photoelectric and thermionic effects; and Chapter 4, origin of the quantum theory.

Rudnick, J., and D. S. Tannhauser, "Concerning a Widespread Error in the Description of the Photoelectric Effect," *Am. J. Phy.* **44**, No. 3-796 (1976).

Scully, M. O., and M. Sargent, III, "The Concept of the Photon," *Physics Today* **25**, No. 3 (1972), p. 38.

Simon, Ivan, *Infrared Radiation*, Momentum Book No. 12. Princeton, N.J.: Van Nostrand, 1966.

REVIEW QUESTIONS

1. What are the similarities and differences between the radiations of the electromagnetic spectrum?
2. What is a blackbody radiator?
3. What was the so-called ultraviolet catastrophe? How did Planck eliminate the problem?
4. What is the difference between Planck's quantization to explain the blackbody radiator and Einstein's quantization to explain the photoelectric effect?
5. What is the significance of the observation shown in Fig. 3.4(b)?
6. What are the blackbody characteristics of clean snow?
7. What is an electron volt?
8. In the photoelectric effect, what can be said about the time interval between the arrival of the light at the surface of the metal and the ejection of the electron?

PROBLEMS

3.1 The cavity of a blackbody radiator is in the shape of a cube measuring 2 cm on a side and the blackbody is at a temperature of 1500 K. (a) Calculate the number of modes of vibration per unit volume in the cavity in the wavelength band between 4995 Å and 5005 Å. (b) What is the total radiant energy in the cavity in this wavelength band?

3.2 Show that the area under the Planck radiation curve is proportional to the fourth power of the absolute temperature.

3.3 At the surface of the earth, a 1-cm^2 area oriented at right angles to the sun's rays receives about 0.13 J of radiant energy each second. Assume that the sun is a blackbody radiator.

What is the surface temperature of the sun? (The radius of the sun is about 7×10^8 m and the earth is about 1.49×10^8 km from the sun.)

3.4 A tungsten sphere 0.5 cm in radius is suspended within a large evacuated enclosure whose walls are at 300 K. Tungsten is not a blackbody but has an average emissive power that is 0.35 that of a blackbody. What power input is required to maintain the sphere at a temperature of 3000 K, if heat conduction along the supports is neglected?

3.5 Find the percent change in the total energy radiated per unit time by a blackbody if the temperature of the blackbody is increased by (a) 100%, (b) 10%, (c) 1%, and (d) 0.1%.

3.6 Show that the Planck radiation law is consistent with the special case of Wien's displacement law, as discussed in the text material.

3.7 At what wavelength does the maximum intensity of the radiation from a blackbody occur at a temperature of (a) 300 K, (b) 1000 K, and (c) 10,000 K?

3.8 Using your answer to Problem 3.3, calculate the wavelength for the maximum radiated intensity from the sun. In what range of the electromagnetic spectrum would this wavelength be found?

3.9 A star is determined to have a temperature of 3000 K and radiates 400 times as much energy per unit time as the sun. What is the radius of this star in sun radii? (Use 6000 K for the temperature of the sun.)

3.10 One way of estimating the temperature of a stellar surface is to determine the wavelength corresponding to the maximum of the radiation intensity in a spectrum whose composition is sufficiently close to that of a blackbody. This wavelength is 0.55 micron for the sun, 0.43 micron for the North Star, and 0.29 micron for Sirius. Calculate the corresponding star temperatures.

3.11 A spherical blackbody 0.10 m in diameter is held at a constant temperature. What is this temperature if the power radiated by the body is 15 kcal/min?

3.12 Predict from the curves in Fig. 3.6 the sequence of colors that would be seen if a piece of iron were heated from 20°C to 2500°C in a dark room.

3.13 Radiant energy with wavelengths longer than visible radiation (infrared region) is often called "heat radiation." (a) Can shorter wavelengths heat a body on which the radiation falls? (b) To what extent is it proper to call infrared rays heat rays? (See Fig. 3.6.)

3.14 If 5% of the energy supplied to an incandescent light bulb is radiated as visible light, how many visible quanta are emitted per second by a 100-watt bulb? Assume the wavelength of all the visible light to be 5600 Å.

3.15 A 100-watt light bulb operates at 3000 K. Assuming it loses energy only by radiation, calculate the total surface area of the filament. Is your answer reasonable?

3.16 Show that for blackbody radiation the intensity of the emitted radiation is equal to $c/4$ times the energy density: $W_\lambda = (c/4)\Psi_\lambda$.

3.17 In order that an object be visible to the naked eye, the intensity of light entering the eye from the object must be at least 1.5×10^{-11} J/m$^2 \cdot$ s. What is the minimum rate at which photons must enter the eye so that an object is visible, given that the diameter of the pupil is 0.7 cm? Assume a wavelength of 5600 Å.

3.18 The directions of emission of photons from a source of radiation are random. According to the wave theory, the intensity of radiation from a point source varies inversely as the square of the distance from the source. Show that the number of photons from a point source passing out through a unit area is also given by an inverse square law.

3.19 Show that Planck's radiation law, Eq. (3.30), reduces to Wien's law for short wavelengths, and to the Rayleigh-Jeans law for long ones. [*Hint:* Express the exponential term as a series to obtain the second of these laws.]

3.20 Show that the energy density of blackbody radiation expressed in terms of frequency is

$$\Psi_\nu \, d\nu = \frac{8\pi h \nu^3}{c^3 (e^{h\nu/kT} - 1)} \, d\nu .$$

3.21 Since it has been shown that matter, electricity, and radiant energy are discrete or "atomic" in character, is it strictly correct to write their differentials dm, dq, and dE, respectively? Explain.

3.22 What is the energy in eV of a photon having a wavelength of 912 Å?

3.23 If for a certain diatomic material the maximum wavelength of a photon needed to separate it into two atoms is 3000 Å, what is the energy holding the molecule together (binding energy)?

3.24 What is the energy of each photon leaving the antenna of your favorite AM radio station? FM radio station?

3.25 A radio station operating at a frequency of 105 MHz has a power of 100 kW. Determine the rate of emission of quanta from the station.

3.26 Using Eq. (3.36), obtain the numerical value of the Stefan-Boltzmann constant.

3.27 Show that Planck's constant h has the same physical units as angular momentum.

3.28 A mass of 10 g hangs from a spring with a force constant of 25 N/m. Assume this oscillator is quantized just as the radiation oscillators are. (a) What is the minimum energy that can be supplied to the mass? (b) If the mass at rest absorbs the minimum energy of part (a), what is the resulting amplitude? (c) How many quanta must the mass absorb in order to have an amplitude of 10 cm?

3.29 Particles of a certain system can have energies of E, $2E$, or $3E$, where $E = 0.025$ eV. (a) What are the ratios of the number of particles in each of the upper states to the number of particles in the lowest state when the system is in equilibrium at 290 K? (b) What is the average energy of a particle in the equilibrium distribution of these states?

3.30 (a) In Problem 1.39, it was found that a thermal neutron has a velocity of 2200 m/s. What is a thermal neutron's kinetic energy in eV? (b) What is the increase in kinetic energy in eV of each water molecule in a stream that goes down a 450-ft (137.1-m) waterfall?

3.31 The fissioning of a ^{235}U atom yields 200 MeV of energy. How many such atoms must fission to provide an amount of energy equal to that required to lift a mosquito one inch? The mass of an average mosquito is 0.90 mg.

3.32 (a) Calculate the energy released in eV per atom or molecule in the complete combustion of each of the compounds listed in the accompanying table. (b) The fissioning of a ^{235}U atom releases 200 MeV of energy. What is the ratio of this energy to that released per molecule of the compounds in the first part of this problem? (c) What is the ratio of the energy released per unit mass of uranium to the energy released per unit mass of each of the compounds?

Compound	Atomic or molecular weight	Heat of combustion
1. Coal	12 (carbon)	14,000 Btu/lb (7,780 kcal/kg)
2. Ethyl alcohol	46	327.6 kcal/mole
3. Starch	162 (assumed)	4,179 kcal/kg
4. Trinitrotoluene (TNT)	227	821 kcal/mole
5. Gasoline	100 (n-heptane)	20,750 Btu/lb (11,530 kcal/kg)

3.33 At 200 MeV per fission per atom of ^{235}U find, on the basis of the energy released in complete combustion (data given in Problem 3.32), (a) the number of grams of uranium needed to furnish a megawatt-day (24 h) of energy, (b) the number of tons of coal equivalent to 1 lb of uranium, and (c) the number of pounds of uranium that equal the blast effect of a megaton of TNT. [*Note:* Only about 30% of the energy of combustion of TNT goes into its blast or explosive effect. Assume that the blast efficiency of the uranium is 100%.] It is very useful in nuclear engineering to remember the relations or conversion factors calculated in parts (a) and (b).

3.34 In the photoelectric effect, the electrons are emitted instantaneously even if the irradiating light is made extremely weak (which is explained by the Einstein postulates). Assume that 10^{-5} W/m^2 falls on sodium and assume the radius of the sodium atom to be about 2 Å. Classically, how long would it take for the atom to absorb 2 eV of energy, which is approximately what would be needed to free an electron?

3.35 The light-sensitive compound on most photographic films is silver bromide, AgBr. We will assume that a film is exposed when the light energy absorbed dissociates this molecule into its atoms. (The actual process is more complex, but the quantitative result does not differ greatly.)

The energy of or heat of dissociation of AgBr is 23.9 kcal/mole. Find (a) the energy in eV, (b) the wavelength, and (c) the frequency of the photon that is just able to dissociate a molecule of silver bromide. (d) What is the energy in eV of a quantum of radiation having a frequency of 100 MHz? (e) Explain the fact that light from a firefly can expose a photographic film, whereas the radiation from a TV station transmitting 50,000 watts at 100 MHz cannot. (f) Will photographic films stored in a light-tight container be ruined (exposed) by the radio waves constantly passing through them? Explain. (g) After a class in which this problem was discussed one student came to the instructor to remark "But I have photographed TV towers." Comment!

3.36 When a certain photoelectric surface is illuminated with light of different wavelengths, the following stopping potentials are observed:

λ, A	3660	4050	4360	4920	5460	5790
V_s, V	1.48	1.15	0.93	0.62	0.36	0.24

Plot the stopping potential as ordinate against the frequency of the light as abscissa. Determine (a) the threshold frequency, (b) the threshold wavelength, (c) the photoelectric work function of the material, and (d) the value of the Planck constant h (the value of e being known).

3.37 The photoelectric work function of potassium is 2.0 eV. If light having a wavelength of 3600 Å falls on potassium, find (a) the stopping potential, (b) the kinetic energy in eV of the most energetic electrons ejected, and (c) the velocities of these electrons.

3.38 What will be the change in the stopping potential for photoelectrons emitted from a surface if the wavelength of the incident light is reduced from 4000 Å to 3980 Å? (Assume that the decrease in wavelength is so small that it may be considered a differential.)

3.39 The threshold wavelength for photoelectric emission from a certain material is 6525 Å. Find the stopping potential when the material is irradiated with (a) light having a wavelength of 4000 Å and (b) light having twice the frequency and three times the intensity of that in the previous part. (c) If a material having double the work function were used, what would then be the answers to parts (a) and (b)?

3.40 Light of wavelength 4000 Å liberates photoelectrons from a certain metal. The photoelectrons now enter a uniform magnetic field having an induction of 10^{-4} T. The electrons move normal to the field lines so that they travel circular paths. The largest circular path has a radius of 5.14 cm. Find the work function for the metal.

3.41 A surface is irradiated with monochromatic light of variable wavelength. Above a wavelength of 5000 Å, no photoelectrons are emitted from the surface. With an unknown wavelength, a stopping potential of 3 V is necessary to eliminate the photoelectric current. What is the unknown wavelength?

3.42 In a certain radio tube, the electron current from the tungsten cathode to the plate is 200 mA. The thermionic work function of tungsten is 4.5 eV. (a) At what rate must energy be supplied, in watts, to get the electrons outside of the cathode? (b) Assuming a Maxwellian distribution of the velocities of the electron "gas" in tungsten, what must the temperature of the cathode be if the energy of the electron with the most probable velocity is to equal 4.5 eV? Does the answer seem reasonable? (c) What assumptions were made about the molecules of a gas when the kinetic theory equations were derived in Chapter 1? To what extent are these fulfilled by the electron "gas" in this problem?

4 THE ATOMIC MODELS OF RUTHERFORD AND BOHR

4.1 INTRODUCTION

We have traced how matter, electricity, and radiation came to be regarded as atomic in character. We have established the existence of some elementary particles that are more fundamental than the chemical elements. Electrons, for example, are common to all elements and are a common building block of all matter. Our discussion of positive rays and mass spectroscopy showed that matter also has positive constituents that are much more massive than electrons. Thomson, who made the first quantitative measurements on electrons and positive rays, assumed that a normal chemical atom consists of a mixture of constituents. This mixture came to be called the "plum-pudding" atomic model: The atom was regarded as a heavy positive sphere of charge seasoned with enough electron plums to make it electrically neutral.

4.2 THE RUTHERFORD NUCLEAR ATOM

A very different atomic model was indicated by an experiment performed by Rutherford in 1911.

We shall discuss radioactivity at some length in Chapter 10, but in order to be able to discuss the Rutherford experiments a few observations need to be made now. Certain atoms are unstable and fly apart of their own accord. The nature of these

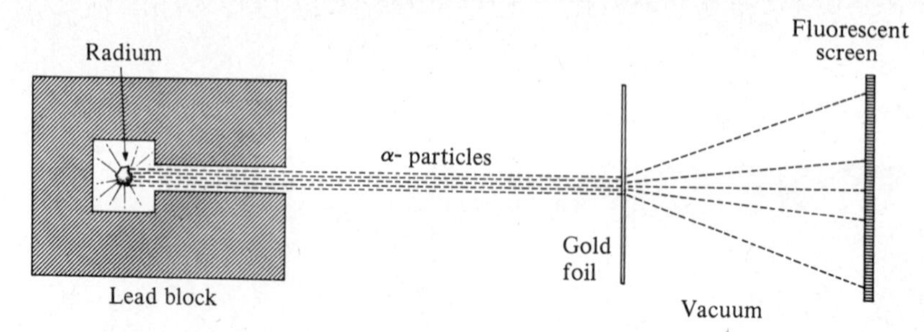

Figure 4.1 Schematic diagram of Rutherford's alpha-particle scattering apparatus.

disintegrations depends on the element that is disintegrating, but in every case the fragments ejected consist either of electrons, here called beta rays, or of doubly ionized helium atoms, called alpha particles. These disintegration fragments usually are ejected with high energies from a radioactive substance and are often accompanied by very short-wavelength photon radiation called gamma rays. Radium, for example, is an excellent source of high-energy alpha particles. These alpha particles can travel through a few centimeters of air before they are stopped, and in a vacuum they travel long distances without losing energy. When they strike certain materials, they cause visible fluorescent light flashes.†

Rutherford studied how these alpha particles from radium were absorbed by matter. He found that they were absorbed by sheets of metal a few hundredths of a millimeter thick, but that they could readily pass through gold foil several ten thousandths of a millimeter thick. Rutherford's apparatus is shown schematically in Fig. 4.1. Radium was placed in a cavity at the end of a narrow tunnel in a lead block. Alpha particles were emitted in random directions by this source and the lead absorbed all except those emitted along the axis of the tunnel. In this way Rutherford obtained a collimated beam of alpha particles that streamed toward the gold foil. Particles that went through the foil produced flashes on the fluorescent screen.

Many of the alpha particles did go straight through the foil or were deflected only by very small amounts. Amazingly, however, some alpha particles were deflected through very large angles. A few even returned to the side of the gold foil from which they came. Rutherford's astonishment at this is evident in his comment, "It was quite the most incredible event that has ever happened to me in my life. It was

†Radiolight watch dials are painted with a paste containing fluorescent material and a trace of a radioactive material. Under a microscope the glow of the dial can be seen to be a multitude of flashes, called scintillations, that resemble the twinkling of the stars on a summer night. The effect may be seen through a 4-power magnifier, but it is better to use two or three times this magnification. The light is less likely to appear continuous if there is very little radioactive material on the dial. Observations must be made in a completely darkened room and it may be necessary to wait about five minutes for the eyes to become dark-adapted. This time delay will also permit any phosphorescence to fade and eventually die out. Both fluorescence and phosphorescence will be discussed in more detail in Section 4.11.

almost as incredible as if you fired a 15-inch shell at a piece of tissue paper and it came back and hit you." This observed scattering through large angles was contrary to predictions based on the Thomson model of the atom. Only small deflections were expected for the following reasons. The effect of electrons can be neglected. Although there are electrostatic forces between the electrons and the alpha particles, it is the electrons, rather than the much more massive alpha particles, that are appreciably deflected. The deflection of the alpha particle is small, even when it passes near or through a positive sphere of charge. The particle experiences only a small resultant force because it is repelled in many directions by other nearby charges. The haphazard deflections of an alpha particle as it passes through the foil nearly cancel, and therefore the net deviation is nearly zero. Thus the plum-pudding model provides no mechanism to account for large deflections. Rutherford replaced the plum-pudding model by another that correctly explained his experimental results. He assumed that the positive, massive part of the atom was concentrated in a very small volume at the center of the atom. This core, now called the *nucleus*, is surrounded by a cloud of electrons, which makes the entire atom electrically neutral. This model accounts for alpha-particle scattering in the following way. The forces between electrons and the alpha particles may be neglected as before, but now an alpha particle passing near the center of an atom is subject to an increasingly large coulomb repulsion as the separation of the two particles decreases. Because the atom is mostly empty space, many of the alpha particles go through the foil with practically no deviation. But an alpha particle that passes close to a nucleus experiences a very large force exerted by the massive, positive core, and is deflected through a large angle in a single encounter.

The significant interaction, then, was between the doubly charged alpha particle ($+2e$) and the positive core having an integral number Z of positive charges of electronic magnitude ($+Ze$). Rutherford assumed that these two charges acted on each other with a coulomb force that in this case was repulsive. The equation for this force in rationalized units is

$$F = \frac{1}{4\pi\varepsilon_0}\frac{2e \cdot Ze}{r^2} = \frac{2Ze^2}{4\pi\varepsilon_0 r^2}, \tag{4.1}$$

where ε_0 is the permittivity of free space ($1/4\pi\varepsilon_0 = 9 \times 10^9$ SI units).

This force is inversely proportional to the square of the distance between the bodies. In advanced mechanics it is shown that such a force always results in a path that is a conic section. When the force is attractive, like the gravitational force between the sun and planets of the solar system, the paths may be parabolas, ellipses, hyperbolas, or, in special cases, circles or straight lines. When the inverse square force is repulsive, however, the conic section must be the far branch of a hyperbola or its degenerate form, a straight line.

Of course the alpha particle and the atom core of the scattering metallic foil both experience the force given by Eq. (4.1). But since the gold atom is much more massive than the alpha particle, we shall assume that the gold atom remains fixed.

Consider, now, an alpha particle aimed directly at the core of an atom of a metal. The repelling force tends to slow the alpha particle, and since this force becomes greater as the particle nears the core, the alpha particle must finally stop. After this momentary pause, the alpha particle will be accelerated away from the atom along the same line by which it approached. In this case the path of the particle is a straight line. Obviously, this case is rare, since it requires that the alpha particle move directly toward the atom core.

In general, the alpha particle will not be perfectly aimed. If an alpha particle is aimed so as to pass near the core (Fig. 4.2), then it will begin its approach at A, moving parallel to a line that is one asymptote of the hyperbolic path. Because the repulsive force is radial from the nucleus whereas the particle's motion is not, the alpha particle is forced to move away from this asymptote of its trajectory. At its point of nearest approach, B, the velocity of the alpha particle is normal to the radius from the nucleus, and beyond this point the path straightens out, becoming parallel to another asymptote through C. The alpha particle is scattered through the angle θ between these two asymptotes. In the rare case of perfect aim, θ is 180°, whereas for the most common case of a wide miss, it is 0°. Between these extremes the angle of scattering depends on the initial speed of the alpha particle, the charge on the core, and the aiming of the alpha particle.

In a given experimental situation the first two of these variables are constant, while the third is impossible to measure or control. But the very randomness of the aim makes it possible to treat this variable statistically, and Rutherford could compute the angular distribution of the alpha particles emerging from the foil.

In 1913, Geiger and Marsden carried out an exhaustive test of the Rutherford theory, using gold and silver foils. The agreement between the theory and their experimental results was excellent. Their technique was not very sensitive to the core charge Z, but they estimated that it was about half the atomic weight. Later, Chadwick made more refined measurements with copper, silver, and platinum foils. He showed that the integer Z is very near the atomic number. In Chapter 6 we shall discuss the work of Moseley, who showed that the number of core charges is *exactly* the atomic number.

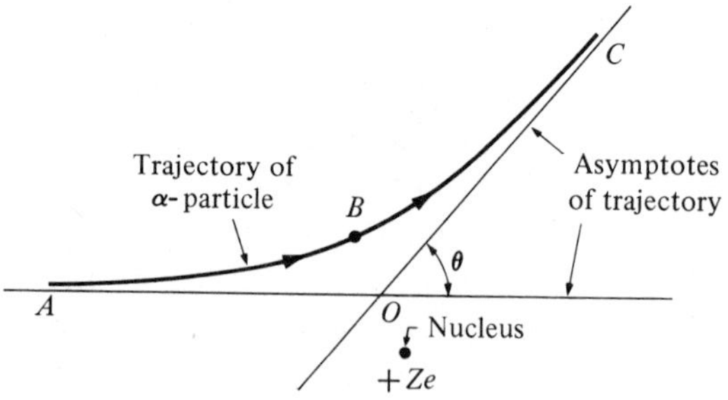

Figure 4.2 Scattering of an alpha particle by the Rutherford nuclear atom.

The Rutherford theory assumes that alpha particles are scattered by stationary point charges. This theory also gives a way of estimating the size of the "point." The required equation for the special case of an alpha particle aimed directly at the nucleus is readily derived from the law of conservation of energy. Let M be the mass of the alpha particle, $+2e$ its charge, and v its initial velocity. At its distance of closest approach d to the core charge $+Ze$, the particle is momentarily at rest, so that all of its kinetic energy must then have been converted into electrical potential energy. Thus we have

$$\frac{1}{2}Mv^2 = \frac{1}{4\pi\varepsilon_0}\frac{2e \cdot Ze}{d}, \tag{4.2}$$

or

$$d = \frac{Ze^2}{\pi\varepsilon_0 Mv^2}. \tag{4.3}$$

For the metals studied, the "radius" of the nucleus, d, is about 10^{-14} m. This is far less than the radius of the space the atom occupies as computed from the density of the metal and the Avogadro constant or from the kinetic theory of matter. This atomic "radius" is about 10^{-10} m. Furthermore, the distance of nearest approach is only an upper limit on the size of the core.

Since aluminum is a light metal with a rather low atomic number, aluminum should repel alpha particles less readily than the heavier metals do. Equation (4.3) shows that for smaller Z the distance of nearest approach is less. For aluminum, d can be as small as 0.8×10^{-14} m. But it has been found that the Rutherford formulas do not fit perfectly for aluminum. The deviations can be explained by assuming that at very small distances from the force center, the force of repulsion is actually less than that computed from Coulomb's law. This suggests that a very short-range attractive force is competing with the repulsive electric force. This force cannot be gravitational. Not only is the gravitational force many magnitudes too small, but also it is an inverse square force that varies with distance, exactly as the coulomb force does. This new force must depend on distance more strongly than does an inverse square force. The size of the region where this new short-range force is significant, compared with the coulomb force, can be investigated with very high-energy particles. When such particles are sent through thin foils, the analysis of the distribution of the scattering angles leads to the discovery that the radius R of the nucleus (assumed spherical) is proportional to the cube root of the mass number of the scattering atoms. The relation is

$$R = R_0 A^{1/3}, \tag{4.4}$$

where R_0 is equal to about 1.4×10^{-15} m. The value of R_0 is somewhat different for different incident particles — alpha particles, protons, or electrons — but it is always of the order of 10^{-15} m. A unit of length that is sometimes used in giving nuclear dimensions is the *fermi* f which is 10^{-15} meters. Equation (4.4) shows that the cube of the nuclear radius is proportional to A; thus the nuclear volume is proportional to the mass number. Therefore, since the density of a substance is its mass divided by its volume, the density of the nucleus is about the same for all atoms.

The success of Rutherford's theory of alpha-particle scattering gives him the distinction of having discovered the *atomic nucleus*. He probed matter with alpha particles and found it mostly empty space. He found the most massive part of the atom concentrated in a region of density about 10^{17} kg/m^3 (relative density about 10^{14}). This massive part contains positive particles that must repel one another with large electric forces. But nuclear stability and the fact that the Coulomb law is not accurate for short distances indicate that there is some entirely *new short-range force* that binds nuclear particles together. The Rutherford model raised many questions about how atoms are held apart and how solid bodies can retain their rigid structure. The answers to these questions come from the structure of the electronic "mist" outside the nucleus. To study this electron arrangement, we next direct our attention to atomic spectra.

4.3 SPECTRA

Most readers of this book have studied light and know that spectrographs are instruments that analyze light according to its distribution of frequency or color. These instruments always have an entrance slit and lens, a dispersive component that may be a prism or grating, and an optical system that forms an image of the slit on a detector that is usually a photographic plate. The instrument forms an image of the slit for each frequency of light present, so that light that is continuous in its frequency distribution forms a wide image that is a continuous succession of slit images. Light that is discontinuous in frequency distribution forms a discrete set of slit images called spectral lines.

In the earlier study of light, attention was particularly directed toward the theory of operation of spectrographs, but it was pointed out at that time that the light from any element in gaseous form produces a discontinuous line spectrum. Each element has its own characteristic frequency distribution or spectrum, so that each element can be identified by the light that it emits. The most dramatic instance of such identification occurred when the element helium was "discovered" in the spectrum of the sun before it was chemically isolated here on earth.

Our present interest in spectroscopy goes far deeper than an interest in instruments or a technique for analysis. Our concern lies in what the light emitted by an element can tell about the structure of that element. Our situation is like that of a visitor from Mars who attempts to learn the structure of a piano by analyzing the sounds it can make.

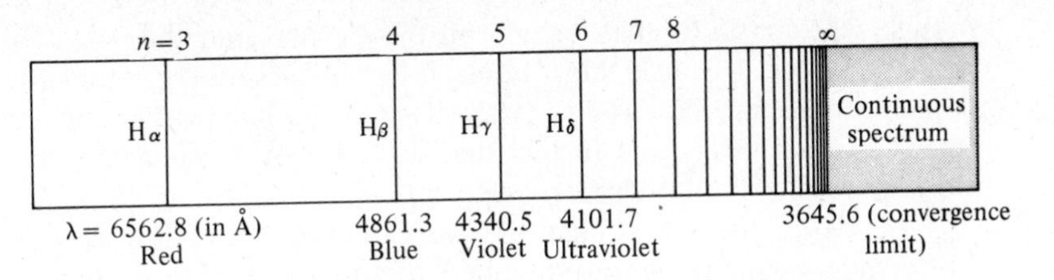

Figure 4.3 Diagram of the Balmer series of atomic hydrogen. (The wavelengths are the values in air.)

4.4 THE HYDROGEN SPECTRUM

The obvious place to start the study of spectra is with the spectrum of hydrogen. It is not surprising that this lightest element has the simplest spectrum and probably the simplest structure. The hydrogen spectrum is shown in Fig. 4.3. The regularity of the spectral lines is immediately evident, and it appears obvious that there is some interrelationship among them. In 1884, a Swiss high-school mathematics teacher by the name of Balmer took the wavelengths of these lines as a problem in numbers. He set out to find a formula that would show their interrelation. He hit upon a formula that could be made to give these wavelengths very precisely. The Balmer formula is

$$\lambda \text{ (angstroms)} = \frac{3645.6n^2}{n^2 - 4}. \tag{4.5}$$

Each different wavelength is obtained by putting into the formula different values of the running integers n, which are $n = 3$, $n = 4$, $n = 5$, etc.

The success of the Balmer formula led Rydberg to attempt a formulation that would apply to heavier elements. He proposed an equation of the form

$$\bar{\nu} = \frac{1}{\lambda} = A - \frac{R}{(n + \alpha)^2}, \tag{4.6}$$

where $\bar{\nu}$ is the *wave number*,† R is the *Rydberg constant*, which is equal to $(1.097373143 \pm 0.00000010) \times 10^7\ \text{m}^{-1}$, and n is a running integer. A and α are

† The wave number is the number of waves per unit length, that is, reciprocal wavelength. One might suppose that the logical quantity to use for the reciprocal form of wavelength would be the frequency, c/λ. In spectroscopy, the wave number is used because in order to compute the frequency without losing the remarkable precision of wavelength measurements, it would be necessary to know the velocity of light to an equal precision. Wave numbers can be computed without knowing the velocity of light and so they retain all the accuracy of spectroscopic wavelength measurements. However, the wave number is not an absolute constant for a given spectral line because its wavelength depends on the index of refraction of the medium in which the measurements are made. The wavelength in air is corrected to vacuum by means of the relation $\lambda_{vac} = \mu\lambda_{air}$, where μ is the index of refraction of air for the particular wavelength. In the visible region, λ_{vac} is approximately 2.5×10^{-2} percent greater than λ_{air}.

adjustment constants that depend on the element and the part of the spectrum or spectral series to which the formula is applied. Rydberg found that this formula, which can be regarded as a generalization of the Balmer formula, could be fitted to many spectral series, and further that the value of R was nearly the same when the formula was applied to different elements.

In 1908, Ritz noted that the wave numbers of many spectral lines are the differences between the wave numbers of other spectral lines, and that the A term of the Rydberg formula was really a particular value of a term, like the second term of the Rydberg formula. Using this "combination principle," Ritz rewrote the Rydberg formula as

$$\bar{\nu} = \frac{R}{(m+\beta)^2} - \frac{R}{(n+\alpha)^2}, \tag{4.7}$$

where α and β are adjustment constants that depend on the element. For different spectral series of a given element, m takes on different integral values. The different lines within a series are computed by changing the running integer n. It is easily shown that when $\alpha = \beta = 0$ and $m = 2$, Eq. (4.7) reduces to the Balmer formula for hydrogen.

Also in 1908, Paschen found another hydrogen series of lines in the infrared region to which Eq. (4.7) could be fitted by making $\alpha = \beta = 0$, $m = 3$, and $n = 4, 5, 6$, etc. Thus, both the then-known hydrogen series could be represented by

$$\bar{\nu} = R\left(\frac{1}{m^2} - \frac{1}{n^2}\right). \tag{4.8}$$

This gives the Balmer series when $m = 2$ and $n = 3, 4, 5$, etc., and correctly predicts the Paschen series for $m = 3$ and $n = 4, 5, 6$, etc.

4.5 THE BOHR MODEL AND THEORY OF THE ATOM

Equation (4.8) represented the entire known hydrogen spectrum with great precision, but it was an empirical formula. At this point we have a correct but underived formula for the hydrogen spectrum. In 1913, Niels Bohr succeeded in deriving this important relation.

Bohr extended Rutherford's model of the atom. He retained the small core or nucleus of the atom and proposed, in addition, that there were electrons moving in orbits around the nucleus. In the case of hydrogen, Bohr proposed that the nucleus consisted of one proton with one electron revolving about it. This is a planetary model of the atom where the heavy positive nucleus is like the sun and the light, negative electron is like the planet earth. In this model, hydrogen is a tiny, one-planet solar system with the gravitational force of the solar system replaced by the electrostatic force of attraction between the oppositely charged particles. The general equations for the gravitational force and the electrostatic force are, respectively,

$$F = G\frac{MM'}{r^2}$$

and

$$F = \frac{1}{4\pi\varepsilon_0}\frac{qq'}{r^2}. \tag{4.9}$$

Both forces are inversely proportional to the square of the distance between the particles. The planets of the solar system have elliptical orbits that are nearly circular. Bohr assumed that the planetary electron of hydrogen moves in a circular orbit, which makes the analysis of the classical aspects of the problem straightforward. Let v be the tangential speed of a mass M' that is revolving around a very large mass M in a circular orbit of radius r. Revolution occurs around the center of mass of the system which, in effect, is at the center of the large, massive body. The centripetal force acting on M' is the gravitational force of attraction due to M. Thus we have

$$F = G\frac{MM'}{r^2} = M'a = \frac{M'v^2}{r}, \tag{4.10}$$

from which we obtain

$$v^2 = \frac{GM}{r}. \tag{4.11}$$

In Bohr's model of the atom, an electron of charge e, mass m_e, and tangential speed v revolves in a circular orbit of radius r around a massive nucleus having a positive charge Ze. In this case, too, the center of the orbit is essentially at the center of the heavy nucleus. The centripetal force acting on the orbiting electron is the electrostatic force of attraction of the nuclear charge, and therefore the force equation is

$$F = \frac{1}{4\pi\varepsilon_0}\frac{Ze \cdot e}{r^2} = m_e a = \frac{m_e v^2}{r}. \tag{4.12}$$

From this equation we find that

$$v^2 = \frac{Ze^2}{4\pi\varepsilon_0 m_e r}. \tag{4.13}$$

(For hydrogen, the atomic number Z equals one. We include Z for generality.) Equations (4.11) and (4.13) both provide a relationship between the variables v and r. If one is known, the other can be found. In the gravitational case, any pair of values of v and of r that satisfy Eq. (4.11) may actually occur. In the electrical case, classical physics imposes no limitation on the number of solutions there can be for Eq. (4.13). For the case of the hydrogen atom, Bohr introduced a restrictive condition that is known as the first Bohr postulate. He assumed that not all the possible orbits that can be computed from Eq. (4.13) are found in hydrogen. Bohr's *first postulate* is that *only those orbits occur for which the angular momenta of the planetary electron are integral multiples of* $h/2\pi$, that is, $nh/2\pi$.† Here n is any integer and h is Planck's constant. Bohr's first postulate introduces the integer idea that appears in the Ritz formula and also introduces Planck's constant, which we have seen plays an important role in the atomic view of radiation. Stated mathematically, this first postulate is

$$I\omega = \frac{nh}{2\pi}, \tag{4.14}$$

where I is the moment of inertia of the electron about the center, ω is its angular velocity, and $n = 1, 2, 3, \ldots$. For the revolving electron, the quantity $I\omega = mr^2\omega = mvr$. From this we obtain an equation that expresses Bohr's first postulate in a very useful form:

$$m_e vr = \frac{nh}{2\pi}. \tag{4.15}$$

The product $m_e vr$ is also called the moment of momentum of the electron.

The orbiting electron in hydrogen must simultaneously satisfy the conditions expressed by Eqs. (4.13) and (4.15). After eliminating v between these two equations, we find that the orbits that exist or are "permitted" in the hydrogen atom are only those that have radii

$$r = \frac{\varepsilon_0 h^2 n^2}{\pi m_e Z e^2}. \tag{4.16}$$

Since we must next consider the energy of the planetary electron and since it may seem odd that it proves convenient to consider the electron energy as being negative, we now review some basic energy concepts. The energy concept is useful in calculations only when there is an exchange of energy. Fundamentally, every energy calculation is the result of an integration that always involves either an initial and a final state (evaluation of a definite integral) or an arbitrary constant (evaluation of an indefinite integral). As a matter of convenience we arbitrarily assign a certain energy to a particular state. Thus, in considering kinetic energy, we usually say that a body at rest has no kinetic energy. A man on a moving train has no kinetic energy relative to himself, but an observer on the ground regards the man on the train as moving and having kinetic energy. Each is correct in terms of *his* arbitrary definition of zero energy. This arbitrary choice of reference level is more familiar in the case of

† The 2π in this expression has no particular physical significance. This factor could have appeared in quantum theory, although this was not pointed out in our discussion of it. Planck said the energy of a photon is $E = h\nu$. If instead of using the frequency in cycles per second, he had used the angular frequency, $\omega = 2\pi\nu$, in radians per second, then his constant would have been $h/2\pi$, a quantity often written $\hbar$ and read "h-bar." In terms of $\hbar$, the energy of a photon is $E = \hbar\omega$, and the angular momentum of Bohr's first postulate is $n\hbar$. (Similarly, $\bar{\lambda}$ means $\lambda/2\pi$.)

potential energy. When we say the gravitational potential energy of a mass m is mgy, it is necessary to state what is meant by $y = 0$.

In discussing the energy of a planetary electron, we shall use the usual convention of field theory, namely, that the electron has no energy when it is at rest infinitely far from its nucleus. Because an electron can do work as it moves nearer the positive nucleus, it loses electric potential energy. Since the electron starts from rest at infinity with zero energy, its potential energy must become more negative as it approaches the nucleus.

To obtain an expression for the electric potential energy of an electron, we note that the potential at a point which is at a distance r from a nucleus having a charge Ze is

$$V = \frac{1}{4\pi\varepsilon_0}\frac{Ze}{r}.$$

The potential energy of a negative electronic charge at this point is $E_p = V(-e)$ or

$$E_p = -\frac{Ze^2}{4\pi\varepsilon_0 r}. \tag{4.17}$$

Note that the potential energy of the electron in this case is zero at infinity and negative elsewhere. We can use Eq. (4.13) to find its kinetic energy:

$$E_k = \tfrac{1}{2}m_e v^2 = \frac{Ze^2}{8\pi\varepsilon_0 r}. \tag{4.18}$$

The total energy of the planetary electron is the sum of the potential and kinetic energies:

$$E = E_k + E_p = \frac{Ze^2}{8\pi\varepsilon_0 r} - \frac{Ze^2}{4\pi\varepsilon_0 r} = -\frac{Ze^2}{8\pi\varepsilon_0 r}. \tag{4.19}$$

We have computed the total energy as a function of r. But we have seen that r can have only those values given by Eq. (4.16). Using this equation to eliminate r, we find that

$$E_n = -\frac{m_e e^4 Z^2}{8\varepsilon_0^2 h^2 n^2}, \tag{4.20}$$

where $n = 1, 2, 3, \ldots,$ for the energy states that it is possible for the electron to have.† The integer n is called the *total* or *principal* quantum number and it can have

† In calling E the energy of the electron, we are not precise. We have assumed that the electron does all the moving while the nucleus remains at rest. Since the mass M of the proton is 1836.15 times the mass m_e of the electron, the latter has most of the kinetic energy of the atomic system. A detailed treatment would require us to consider the movement of all particles about their common center of mass. There is a theorem in mechanics that states that in a two-body problem such as this, the motion of one body may be neglected if the mass of the other body is taken to be the "*reduced mass*," which is the product over the sum of the two masses, $m_e M/(m_e + M) = m_e/(1 + m_e/M)$. If, in Eq. (4.20) and elsewhere, we *replace* the electron mass m_e by the reduced mass, then our equations correctly describe the atomic system *as a whole*.

any of the series of values, 1, 2, 3, ... The values of n determine the energies of the states. When n is large, the energy is large, that is, less negative than for small integers. The energy required to remove an electron from a particular state to infinity is called the *binding energy* of that state. It is numerically equal to E_n.

We now consider how Bohr used this set of energies to account for the hydrogen spectrum. In Chapter 3, we described how classical electrodynamics predicts that energy will be radiated whenever a charged particle is accelerated. We were careful to point out that the acceleration could be due to a change of direction of motion as well as due to a change of speed. According to classical theory, an orbital electron should radiate energy because of its centripetal acceleration. In order to preserve his atomic model of planetary electron orbits, Bohr had to devise a theory that would violate this classical prediction since, according to it, any electron that separated from the nucleus would soon radiate away its energy and fall back into the nucleus. Bohr's second break with classical physics is contained in his *second postulate*, which states that *no electron radiates energy so long as it remains in one of the orbital energy states; and that radiation occurs only when an electron goes from a higher energy state to a lower one, the energy of the quantum of radiation, $h\nu$, being equal to the energy difference of the states*. Let the quantum number $n = n_2$ represent a higher energy state and $n = n_1$ represent a lower energy state ($n_1 < n_2$); then the second Bohr postulate can be written as

$$h\nu = E_{n_2} - E_{n_1}. \tag{4.21}$$

Substituting for the energies from Eq. (4.20), we have for the frequency of the emitted radiation

$$\nu = \frac{m_e e^4 Z^2}{8\varepsilon_0^2 h^3}\left(\frac{1}{n_1^2} - \frac{1}{n_2^2}\right) \tag{4.22}$$

or, in terms of the wave number,

$$\bar{\nu} = \frac{1}{\lambda} = \frac{\nu}{c} = \frac{m_e e^4 Z^2}{8\varepsilon_0^2 h^3 c}\left(\frac{1}{n_1^2} - \frac{1}{n_2^2}\right) \tag{4.23}$$

where c is the speed of light in a vacuum. Comparing Eq. (4.23) with Eq. (4.8) shows that both have the same form.

Equally impressive is the fact that the constant factor of the Bohr formula is the Rydberg constant, R. Again comparing Eqs. (4.8) and (4.23), we find that, since $Z = 1$ for hydrogen,

$$R = \frac{m_e e^4}{8\varepsilon_0^2 h^3 c} = 1.0973731 \times 10^7 \text{ m}^{-1}. \tag{4.24}$$

The R given here is R_∞, which would be correct if the mass of the nucleus were infinite compared with the mass of an electron. If the motion of the nucleus is taken

Table 4.1 The spectral series of hydrogen

Values of n_1	Name of Series	Values of n_2
1	Lyman	2, 3, 4, etc.
2	Balmer	3, 4, 5, etc.
3	Paschen	4, 5, 6, etc.
4	Brackett	5, 6, 7, etc.
5	Pfund	6, 7, 8, etc.

into account, m_e must be replaced by the reduced mass. Therefore, in general, $R = R_\infty/(1 + m_e/M)$. This accounts for the slight variation of R from element to element noted by Rydberg. It is a triumph of the Bohr model and theory that the slight differences between the spectra of ordinary hydrogen and its isotope, heavy hydrogen (deuterium), can be attributed to the influence of the nuclear mass. In fact, heavy hydrogen was discovered spectroscopically by Urey in 1932.

The Bohr formula gives the Balmer series for $n_1 = 2$ and the Paschen series for $n_1 = 3$, as we have seen before. But the Bohr theory places no restrictions on n_1 and his result suggested that there might be additional hydrogen series not yet found experimentally. In 1916, Lyman found a series in the far ultraviolet, in 1922 Brackett found a new series in the infrared, and in 1924 Pfund located another in the same region. Table 4.1 summarizes the five hydrogen series.

4.6 EVALUATION OF THE BOHR THEORY OF THE ATOM

Bohr's successful derivation of the Rydberg–Ritz formula opened the door to an understanding of the extranuclear structure of atoms that is now virtually complete. The student may feel that we can all get the right answer to a problem if we are permitted to write our own postulates and break the rules of the game — especially when we know the answer before starting. Upon reflection, however, it is remarkable that Bohr could get the right answer without being more arbitrary than he was. There is no harm in introducing a new constant into physics as Planck did, but it is a great achievement to find that an empirical constant like the Rydberg constant is expressible in terms of basic constants already known. It is true that Bohr did work backwards in order to know that in his first postulate he should let the angular momentum be $nh/2\pi$. The impressive thing is that he found something so simple. There was no need to set the angular momentum to n times an arbitrary number. In view of later developments, it is surprising that Bohr could retain as much classical theory as he did.

4.7 ENERGY LEVELS

Bohr's second postulate — that there are only certain discrete energy levels in the hydrogen atom — is especially important because it has been found to have wide application throughout atomic physics. Energy levels are most conveniently expressed in electron volts. Equation (4.20) gives the energy states of a one-electron atom in basic units. If E_n is in joules, it may be converted into electron volts by dividing by the conversion factor e_c, 1.60×10^{-19} J/eV. For this case we have

$$E(\text{eV}) = \frac{E_n}{e_c} = -\frac{1}{e_c} \cdot \frac{m_e e^4 Z^2}{8\varepsilon_0^2 h^2} \cdot \frac{1}{n^2}. \tag{4.25}$$

Figure 4.4 Energy-level diagram of the hydrogen atom.

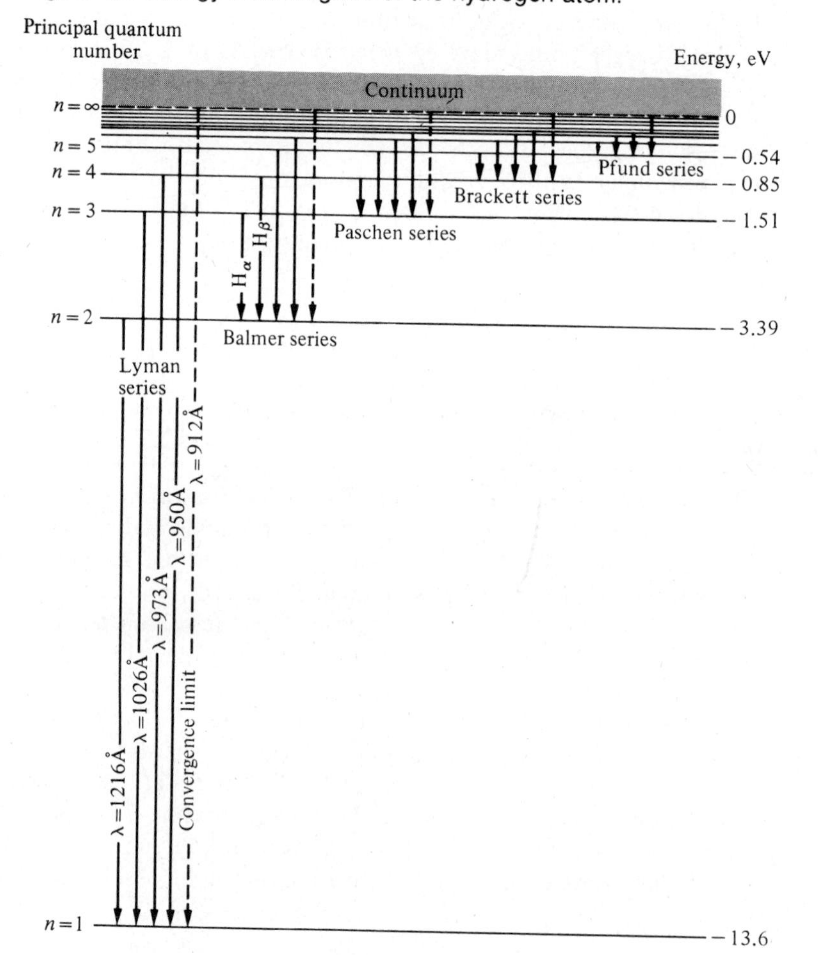

Substituting the values of the constants in this equation and letting $Z = 1$, we find that the energy levels of the hydrogen atom are given by

$$E(\text{eV}) = -\frac{13.6}{n^2}. \tag{4.26}$$

These energy levels can now be represented graphically as shown in Fig. 4.4. The quantum numbers are shown at the left and the corresponding energies of hydrogen in electron volts are given at the right. In this array of energies, the higher (less negative) energies are at the top while the lower (more negative) are toward the bottom. In a normal unexcited hydrogen atom, the electron is in its lowest energy state at the bottom, with $n = 1$. An electron in this *ground state* is stable and moves in this orbit continuously without emitting or absorbing energy. The "excitement" begins when the electron absorbs energy in some way. There are a variety of ways in which this may be brought about. If the hydrogen is in an electric discharge, a free electron that has been accelerated by the electric field may hit it.† If the hydrogen is heated, the electron may be excited by a thermal-motion collision. If the hydrogen is illuminated, it may absorb energy from a photon. Suppose the electron in hydrogen absorbs about 20 eV of energy in one of these ways. This is enough energy to lift the electron to $n = \infty$ (13.6 eV) with 6.4 eV left over. In this case, the electron is made entirely free of its home nucleus and is given 6.4 eV of kinetic energy besides. If the electron absorbs just 13.6 eV, it is merely freed from its home nucleus and drifts about with only its thermal kinetic energy. In either of these cases the remaining nucleus is an ion. If the energy of a bombarding electron is less than that required for ionization but equal to or greater than that needed to raise an electron in an atom to one of its permitted energy levels, then the atomic electron will absorb just enough energy to put it into some higher energy state. After the bombarding electron has transferred enough energy to the atom to excite it, the electron will leave the encounter, carrying away any excess as kinetic energy. The *excitation energy* of any level in electron volts is *numerically* equal to the *excitation potential* of that state in volts.

After excitation, the atomic electron returns to its normal state. If it was excited to $n = 4$, it may jump from 4 to 1 in one step. It may also go 4, 2, 1 or 4, 3, 1 or 4, 3, 2, 1. In each step of the return trip, the electron must lose an amount of energy equal to the difference of the energy levels. The only mechanism available for this energy loss is through the emission of electromagnetic radiation. Thus, in Fig. 4.4 we have represented graphically the second Bohr postulate, given in Eq. (4.21). When we see the light from a hydrogen discharge, we are "seeing" the electrons go from excited states to lower states.

The electron transitions that end on $n = 1$ constitute the Lyman series, on $n = 2$, the Balmer series, on $n = 3$, the Paschen series, etc. From the energy-level

† We use the word "hit" loosely. We saw in the discussion of alpha-particle scattering that a collision between charged bodies does not involve physical contact in the usual sense.

diagram we can see that the Lyman transitions involve the largest changes of energy, produce the highest frequencies, and provide the "bluest" (ultraviolet) light.

The shaded region at the top of Fig. 4.4 represents the fact that electrons at infinity may have kinetic energy, so that their energy there is not zero but positive. If the electron of hydrogen is completely removed from its nucleus, then one of these electrons at infinity having *any* energy may fall into any one of the energy levels. Such an electron undergoes a change of energy equal to its energy at infinity minus the negative energy of the level to which it falls. The "double negative" in the last sentence enables us to conclude that the energy radiated by such a transition is the sum of the electron's kinetic energy at infinity and that involved in the transition from $n = \infty$ to the final level. The energy radiated will have a value greater than that involved in the transition from $n = \infty$ to the final level. Since there is a wide distribution of the energies among the electrons at infinity, there is a continuous spectrum below the short-wavelength convergence limit of any series.

4.8 IONIZATION POTENTIALS

Confirmation of the energy level concept is convincingly given by a consideration of ionization potential. Consider first a tube; Fig. 4.5, which contains only a filament-heated cathode and a plate as anode. When the plate is positive with respect to the cathode, electrons will move across the tube to the anode. This current is limited by two factors. First, the number of electrons emitted per unit area from a cathode depends on cathode composition and temperature. In the remainder of this discussion we shall assume that the cathode is operated hot enough so that the tube current is not significantly limited by cathode emission. The second factor is the effect of the electrons in the region between the electrodes on those emerging from the cathode. The concentration of negative charge in the interelectrode region is called *space charge*, and it lowers the potential in the vicinity of the emitting surface. Indeed, the potential in this region usually falls below the potential of the emitter. Thus, although the plate or accelerating potential is still positive, this decrease in potential due to space charge will reduce the electron current because the potential barrier will turn back electrons emitted with low kinetic energy. (The electrons in

Figure 4.5 Apparatus for determining ionization potential.

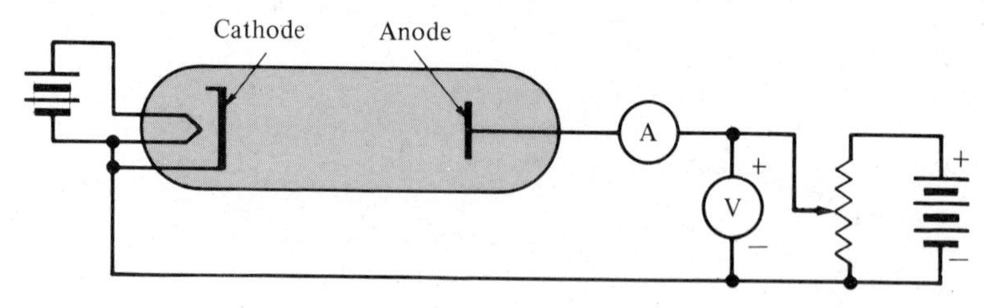

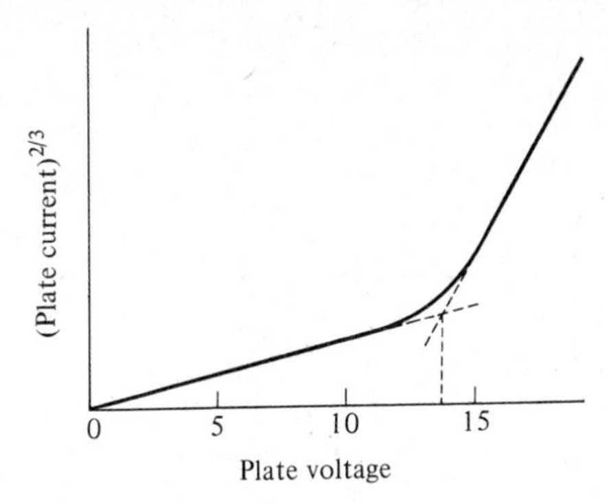

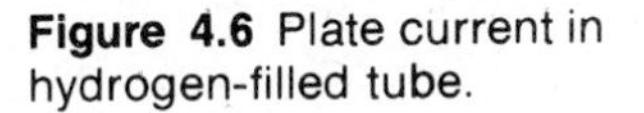
Figure 4.6 Plate current in hydrogen-filled tube.

thermionic emission have a distribution of speeds similar to that of the molecules of a gas.) However, those high-energy electrons that get beyond this barrier caused by space charge arrive at the plate with the same energy they would have had if the space charge had been absent, since the total potential difference V between the electrodes is independent of the space charge. The space charge limited current is not a linear function of the potential difference as in the case of an ohmic resistor; it is found to be proportional to the three-halves power of the potential difference. This is known as the Child-Langmuir law. In mathematical form, it is $I = kV^{3/2}$. The value of k depends on the geometry of the tube and the volume density of the charges between the electrodes.

The ionization potential of a gas is determined by introducing some of the gas into a tube such as that shown in Fig. 4.5, and then measuring the plate current as a function of plate voltage. As the potential difference is increased, it is found that above some particular value of the potential the current increases much more rapidly than it does below that value, as shown in Fig. 4.6. When the plate potential reaches this critical value, the electrons arriving at the anode have acquired enough energy to knock electrons off the atoms of the gas close to this electrode. The positive ions produced when the voltage equals or exceeds the critical value neutralize some of the negative space charge. Thus ionization causes a marked increase in the tube current. If the gas under study is hydrogen, the ionization potential is found to be 13.6 volts. It is a remarkable confirmation of the energy-level idea that there should be excellent agreement for the ionization potential as measured by the very different techniques of spectroscopy and electronics. Since the electrons are emitted from the cathode with an initial velocity distribution, some of them acquire enough energy to produce ionization at lower accelerating potentials than others. This accounts for the curved section joining the two straight-line portions of the graph in Fig. 4.6. Except for this curved part, the break in the line is quite abrupt. This sudden change of slope would not occur if several low-energy electrons could often combine in their "efforts" to ionize the atom of hydrogen.

EXAMPLE Consider a *hypothetical* one-electron atom that we assume does not have the hydrogen energy levels but that obeys Bohr's second postulate. The wavelengths of the first four lines of the spectral series terminating on $n = 1$ are 1200 Å, 1000 Å, 900 Å, and 840 Å. The short wavelength limit of this series is 800 Å. (a) Find the values of the first five energy levels of this atom in eV and construct the energy-level diagram. (b) What is the ionization potential? (c) What is the wavelength of the line emitted for the transition from the energy level $n = 3$ to $n = 2$? (d) What is the minimum energy that must be supplied to the electron in the ground state so that it can make the transition in part (c)?

Solution (a) The energy-level transitions that produce the lines of the series are shown in Fig. 4.7. The energy in a quantum of radiation of given wavelength is found from Eq. (3.43). Thus the energy for the first line is

$$E = \frac{1.24 \times 10^4}{\lambda}(\text{eV} \cdot \text{Å}) = \frac{1.24 \times 10^4}{1200\ \text{Å}}(\text{eV} \cdot \text{Å}) = 10.3\ \text{eV}.$$

Similarly, the energies for the other wavelengths are found to be 12.4 eV, 13.8 eV, 14.8 eV, and 15.5 eV, respectively.

Because the energy in the quantum radiated when the electron goes from the zero energy level ($n = \infty$) down to the lowest level ($n = 1$) is 15.5 eV, the energy of the ground state is $E_1 = -15.5$ eV. The difference in energy between the levels $n = 2$ and $n = 1$ is 10.3 eV. Therefore the energy of the state $n = 2$ is found from $E_2 - E_1 = 10.3$ eV, or

$$E_2 = E_1 + 10.3\ \text{eV} = -15.5\ \text{eV} + 10.3\ \text{eV} = -5.2\ \text{eV}.$$

Figure 4.7

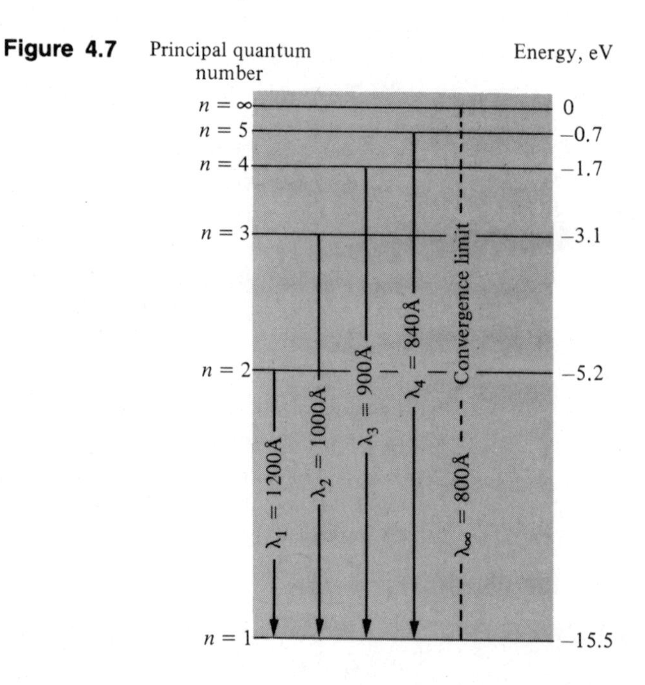

In the same way, we find that

$$E_3 = -3.1\text{ eV}, \qquad E_4 = -1.7\text{ eV}, \qquad \text{and} \quad E_5 = -0.7\text{ eV}.$$

(b) 15.5 volts.
(c) From Eq. (3.43) we have

$$\lambda = \frac{1.24 \times 10^4}{E_3 - E_2}(\text{eV} \cdot \text{Å}) = \frac{1.24 \times 10^4}{(-3.1\text{ eV}) - (-5.2\text{ eV})}(\text{eV} \cdot \text{Å})$$

$$= \frac{1.24 \times 10^4}{2.1}(\text{Å}) = 5900\text{ Å}.$$

(d) $E_3 - E_1 = (-3.1\text{ eV}) - (-15.5\text{ eV}) = 12.4\text{ eV}.$ ▬

4.9 RESONANCE POTENTIALS

The experiment just described was set up to measure the potential through which electrons must be accelerated before they can lift orbital electrons from their lowest energy state (ground state) to infinity. This ionization was detected by an increase in the current through the tube. But before the bombarding electrons have enough energy to take the atom apart by removing an electron, they have enough energy to lift an electron to an excited state. Orbital electrons in the gas can be transferred from their lowest energy state to any of the higher states. As implied earlier, the quantum conditions that require an orbital electron to emit only certain frequencies as radiation apply also to the absorption process. These electrons can absorb only energies represented by transitions between energy levels. If an orbital electron is hit by a bombarding electron with insufficient energy to produce an energy transition, the orbital electron absorbs no energy from the bombarding electron and *the collision is perfectly elastic*. If the orbital electron is hit by a high-energy bombarding electron, then the orbital electron can absorb energy by making a transition. This leaves the bombarding electron with that much less energy. *Such a collision is inelastic*, since the bombarding electron is left with less energy than it had before the collision. Such an inelastic collision puts the orbital electron in one of the excited states, and hence it can radiate energy in returning to a lower state.

Consider again the ionization experiment. As the potential difference across the tube is slowly increased, the electrons from the heated cathode are accelerated to higher and higher velocities. At low speeds, these electrons make completely elastic collisions with the electrons of the gas, so that they are deviated but not slowed by the collision process. As the potential difference across the tube is increased, however, a potential is reached where energy can be transferred to an orbital electron. For the purpose of discussion consider hydrogen, which has an ionization potential of 13.6 volts. A look at Fig. 4.4 discloses that the least amount of energy the orbital electron in the ground state can absorb is 13.6 eV − 3.4 eV or 10.2 eV. A bombarding electron with 10.2 eV of energy can "resonate" with hydrogen and transfer its energy to the hydrogen. This produces no ionization, so the current through the tube is not changed, but after making such collisions, the bombarding electrons proceed more slowly and the hydrogen shows its "excitement" by radiating. The hydrogen will not glow visibly because this resonance radiation is one line of the Lyman series, which is in the ultraviolet region; however, ultraviolet spectroscopy confirms that the radiation is there.

In order to demonstrate the resonance phenomenon electronically, we need a more elaborate tube, such as that used by Franck and Hertz, who first performed the experiment in 1913 using mercury vapor as the gas. The principal parts of such a tube are shown schematically in Fig. 4.8. From the standpoint of electronics, the effect of resonance is that the bombarding electrons are slowed down, so we need a device that will measure the energy of the bombarding electrons after they have made collisions. Suppose that the anode of the ionization tube is perforated or made of wire mesh. In this case, some of the bombarding electrons will pass through the electrode rather than hit it. We now need to know the energy with which the bombarding electrons arrive at the anode.

In our consideration of the photoelectric effect, we measured the energy of photoelectrons by making them move against the force action of an electric field, and the energy of the photoelectrons was given by the stopping potential. Here we use much the same technique and insert into the tube another electrode beyond the anode. This collector electrode is maintained less positive than the anode, say 0.5 V, so that any electrons that pass through the anode will be slowed by the field between the electrodes. Electrons that reach this last electrode must have passed the anode with an energy of at least 0.5 eV.

The experimental procedure consists of measuring the collector current as a function of the anode potential with respect to the cathode, and typical results are shown in Fig. 4.9. From $V = 0$ to $V = 0.5$ V, there is no collector current, since no electrons can reach the anode if they have less than 0.5 eV of energy. Above $V = 0.5$ V, the collector current rises because the number of electrons having at least this minimum energy increases. When V reaches the resonance potential of the gas, the collector current begins to decrease because some of the bombarding electrons are

Figure 4.8 Apparatus for determining resonance potential of gas.

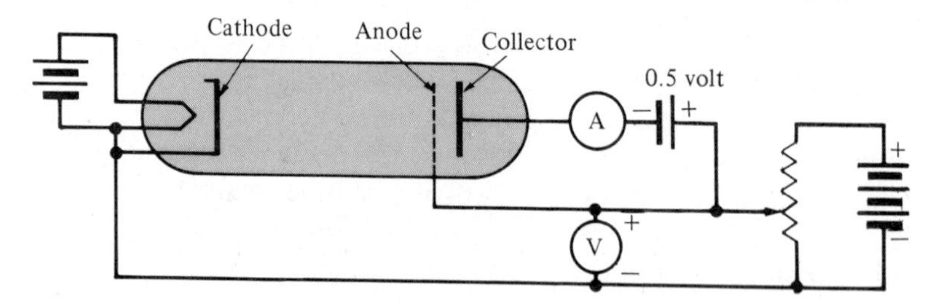

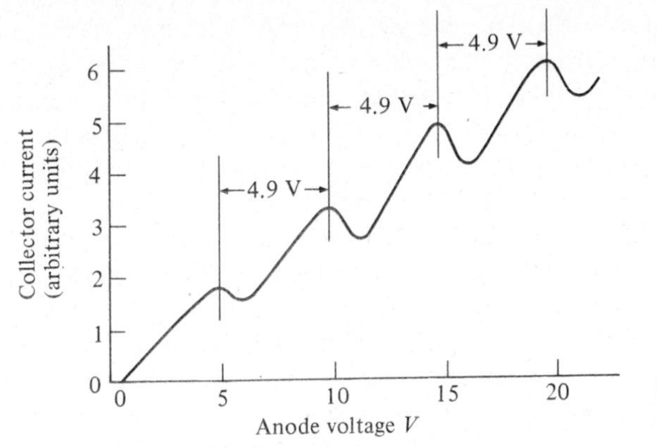

Figure 4.9 Resonance potential curve for mercury.

slowed by inelastic collisions with orbital electrons in the gas. The current rises again as V is further increased since, in the stronger field, bombarding electrons can make inelastic collisions early and still undergo enough acceleration to surmount the 0.5 V barrier.

The second dip is not due to a new energy transition, because very few electrons ever get enough energy to excite the next transition. The second dip occurs at twice the resonance potential and is caused by the bombarding electrons suffering two inelastic collisions of the same kind. Thus each peak of the curve signifies more collisions and each peak is an integral multiple of the resonance potential. The separation of successive resonance peaks is 4.9 V for mercury. (Since mercury ionizes at 10.4 V, the third resonance peak will be masked by other effects if the tube shown in Fig. 4.8 is used. Actually, a more complicated tube which differentiates the resonance and ionization effects is used. This important experimental detail does not in any way alter the principle discussed in this section. Mercury vapor is chosen because it is monatomic whereas hydrogen is ordinarily a diatomic molecule whose dissociation energy is 4.5 eV. If the gas were hydrogen, most of the bombarding electrons would give up their energy to excite molecular energy states and also to dissociate the molecules after the accelerating potential reached 4.5 V. This complex situation would conceal the effects of atomic hydrogen in this type of experiment.)

The data from the Franck–Hertz experiment corroborate the spectrographic observations and further support the concept of discrete energy levels.

4.10
PHOTON ABSORPTION

In the previous two sections we have discussed how an atom can absorb discrete amounts of energy from bombarding electrons. An atom may also absorb energy from photons, but there is an important difference. Absorbed photons disappear entirely. A photon with more energy than the ionization energy of an atom can always be absorbed because the excess energy will appear as kinetic energy of the photoelectron. A photon with less than the ionization energy cannot be absorbed unless its energy is equal to one of the excitation energies of the absorbing atom. (Some of these statements about photon absorption will be slightly modified when we discuss the scattering of high-energy x-ray quanta in Chapter 6.) Consider hydrogen again. Ordinarily, the probability of finding a hydrogen atom in an excited state is very small; therefore we *assume* that it is always in its ground state. Thus hydrogen can *only* absorb photons whose wavelengths correspond to those emitted in the far ultraviolet, the Lyman series. Hydrogen atoms are therefore transparent to visible and infrared light. If we pass radiation of all wavelengths through hydrogen and analyze the transmitted light by means of a spectrograph, we find the transmitted intensity *reduced* for the Lyman wavelengths. Such a spectrum, having a bright background and *dark lines*, is called an *absorption spectrum*. Because the atoms that have been excited by the absorption of radiation re-emit photons in *random* directions upon returning to the ground state, there is a decrease of intensity along the direction of the transmitted radiation. The absorption lines observed are actually very faint bright lines that appear dark by contrast.

There are certain advantages to studying absorption spectra. For many atoms, as for hydrogen, the absorption spectrum is simpler than the emission spectrum. For hydrogen, one can usually observe absorption for only one of the five emission series. This one series is sufficient to establish the entire energy-level diagram, and these levels permit an explanation of all the emission series.

The determination of elements on the sun is a dramatic example of absorption spectroscopy. The sun is a hot body that emits a continuous spectrum of photons. As these photons pass through the outer atmosphere of the sun, wavelengths that are characteristic of the gases present are absorbed. Thus the continuous spectrum of the light from the sun is crossed with (relatively) dark lines that were first observed by Fraunhofer in 1815. Most of the Fraunhofer lines correspond to the wavelengths of elements found on the earth. The absorption lines of the Balmer series of hydrogen are especially prominent in the spectrum of the sun. The Balmer lines of hydrogen and the visible-region lines of other elements have rather long wavelengths and their photons can be absorbed *only* by excited atoms of the respective elements. These higher energy states are produced in the following way. The solar atmosphere's temperature is several million degrees. Therefore many solar atmospheric atoms are excited by collision. Thus these atoms are raised above the lowest energy level to states where they can absorb wavelengths longer than the ultraviolet. For many years one set of Fraunhofer lines in the visible region could not be associated with any known element. It was presumed to be due to a new "sun element" that was appropriately named helium. This hypothesis was confirmed when helium was isolated on the earth and its emission spectrum was found to correspond with the previously unidentified Fraunhofer lines.

Many of the spectral series that are characteristic of molecular structure are in the infrared region. Absorption spectroscopy is the only feasible way of investigating the structure of those molecules that would be dissociated by being excited in an electric arc in an attempt to produce emission lines.

Molecules have the three degrees of freedom of translation at very low temperatures, and additional degrees of freedom due to rotation and vibration at higher temperatures. These additional degrees of freedom produce many quantized energy states. Since the energy differences between these states are small, the emitted wavelengths associated with them are long and so, for the most part, they lie in the infrared or beyond. Molecular spectra have many lines and are very complex. The rotational and vibrational states are characteristic of the molecule. These states differ even in the case of isomers, which are molecules having the same atomic composition but different arrangement or structure.

By means of molecular spectra it is possible to measure the frequencies of rotations and vibrations of the atoms that join together to form compounds. In this way we can learn about moments of inertia, spring constants, and the angles such as those between the lines joining the three atoms in H_2O. But the mechanical motions of atoms in molecules are much slower than the activity of atomic electrons, and few molecular studies can be carried out with radiation in the visible region. Progress in molecular research has led from the visible region into the infrared, the microwave (radar), and the radio region. An interesting application of one of the results of this work (to be discussed in Section 4.12) is the use of one of the molecular frequencies of ammonia as the "pendulum" for controlling a so-called atomic clock. Such clocks have very high precision. The problems of molecular structure are still under active study, especially in the field of microwave spectroscopy.

4.11 FLUORESCENCE AND PHOSPHORESCENCE

Another application of the energy-level concept is in the explanation of fluorescence. The fluorescent lamps used for modern lighting work in the following way. The electric discharge within the lamp is through mercury vapor. The spectrum of mercury has some lines in the visible region, but most of the emitted radiation is concentrated in a line in the ultraviolet. If the tube were made of quartz, which can transmit ultraviolet light, this radiation would be able to get out of the tube, where it could be used for air sterilization or to produce "sunburn." But a clear tube is a poor source of visible light. Therefore the inside of a fluorescent lamp is coated with a material that absorbs the invisible ultraviolet light. Thus the atoms of this fluorescent material become excited. If the excited electrons fell back to their normal state in one step, they would re-emit the ultraviolet light, which would still be invisible. But the excited electrons return to their normal state in more than one step. Each step produces radiation of less energy than the original excitation, so that the energy of the ultraviolet light is converted into visible light. The various tints that different lamps have are controlled by the nature of the fluorescent material used.

Some materials have what are called *metastable* states. When an electron is excited into one of these states, it does not return to its normal state at once, but may remain excited for an appreciable time. Such materials have a persistent light, called *phosphorescence*, which may last several hours after all external excitation is removed. These materials are sometimes used on the screens of cathode-ray tubes, and they are sometimes used to make light switches glow, so that they may be found in the dark. Most fluorescent tubes have some phosphorescence, which may be observed in a dark room a few minutes after the light is turned off.

4.12 MASERS AND LASERS

We now have the background to understand a whole group of new devices called *masers* [*m*olecular (formerly microwave) *a*mplification by *s*timulated *e*mission of *r*adiation]. We shall discuss the underlying principle of these devices and then consider their functional and practical differences.

In our discussion of the Bohr model of the hydrogen atom, we made the too-simple assumption that *all* atoms are normally in the lowest or ground state. This is not *quite* true. If we assume that we have monatomic hydrogen gas at room temperature (chemically impossible), then it is easy to calculate the relative numbers of atoms in higher energy states. Thermal collisions will raise some atoms to higher energy states just as thermal collisions cause "air" molecules to rise physically high in the atmosphere. In Chapter 1 we derived the law of atmospheres, Eq. (1.41), which we generalized to the Boltzmann distribution law, Eq. (1.45). To compare the numbers of atoms at two different energies in thermal equilibrium we write this distribution law as

$$n_2/n_1 = \exp[-(E_2 - E_1)/kT]. \tag{4.27}$$

For hydrogen at room temperature this becomes

$$\frac{n_2}{n_1} = \exp\left(-\left\{\frac{-[-3.39-(-13.6)]e_c}{k \cdot 293}\right\}\right),$$

where the energy levels have been introduced in electron volts and room temperature has been taken to be 20°C or 293 K. Since $e_c = 1.60 \times 10^{-19}$ J/eV and k is Boltzmann's constant ($= 1.38 \times 10^{-23}$ J/K), we find that

$$\frac{n_2}{n_1} = \exp\left[-\left(\frac{10.2 \times 1.60 \times 10^{-19}}{1.38 \times 10^{-23} \times 293}\right)\right] = e^{-404} = 10^{-176},$$

which is incredibly small. This calculation shows that we were *almost* correct in stating that all atoms are in the ground state. Note that if the temperature had been much higher (as in the atmosphere of the sun, 6000 K) the proportion of atoms in the excited state would then have been significant, $n_2/n_1 = 10^{-8.6}$. Note too that the fraction of excited atoms would have been great, even at room temperature, if the separation of the energy levels had been much less. Note especially that e^0 equals 1 so that no matter how small the energy level difference or how high the temperature, the number of atoms in the higher state cannot exceed the number in the lower state (this whole discussion assumes the hydrogen to be in thermal equilibrium).

Having explained that some atoms (however few) may "normally" be in an excited state, we now state an additional fact. Whereas we have explained previously that a 10.2 eV photon can raise a hydrogen atom from the $n = 1$ state to the $n = 2$ state with the absorption of the photon, we now state that a 10.2 eV photon interacting with an excited hydrogen atom can cause a transition from the $n = 2$ state to the $n = 1$ state *with the emission of an additional photon* that must have the same energy. When 10.2 eV photons pass through hydrogen, absorption of photons is obviously the observed effect. But in a *few* instances the reverse effect of *stimulated emission* takes place. For future reference it is important to mention that stimulated photons and the stimulating photon are coherent (that is, in phase with one another). The three processes that can occur in the hydrogen then are: (1) a 10.2 eV photon can be absorbed by a hydrogen atom in the ground state giving an excited hydrogen

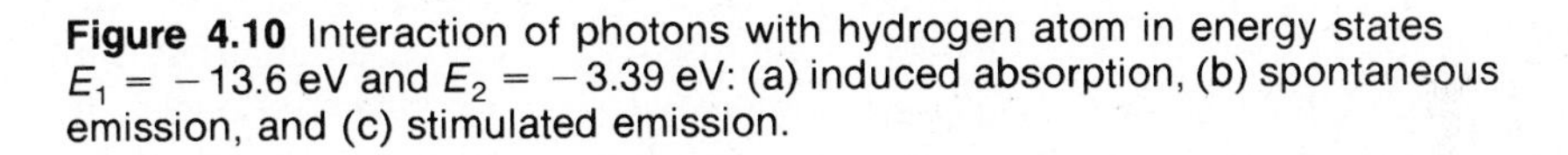

Figure 4.10 Interaction of photons with hydrogen atom in energy states $E_1 = -13.6$ eV and $E_2 = -3.39$ eV: (a) induced absorption, (b) spontaneous emission, and (c) stimulated emission.

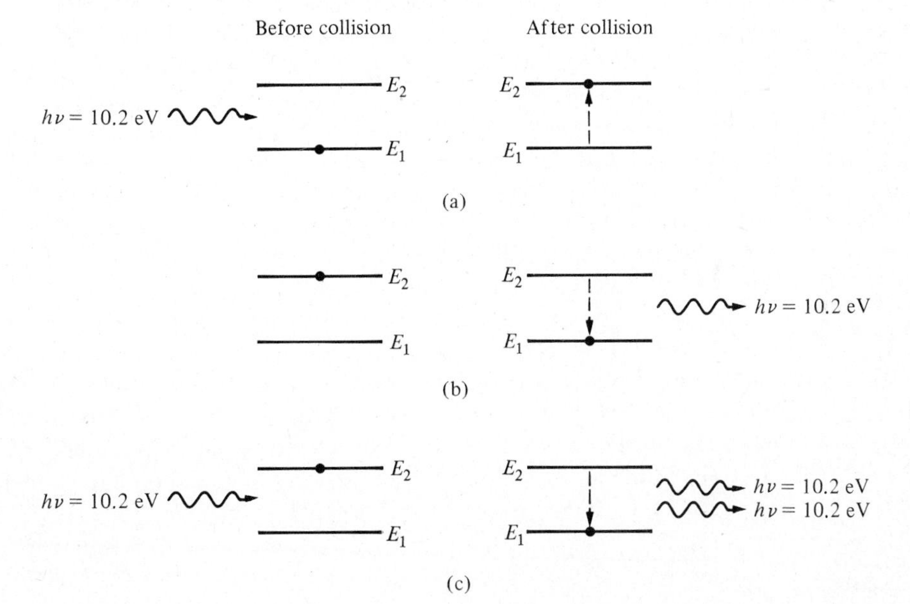

atom in the $n = 2$ state; (2) an excited hydrogen atom in the $n = 2$ state decays to the $n = 1$ state with the emission of a 10.2 eV photon; and (3) a hydrogen atom in the excited state interacts with a 10.2 eV photon and returns to the ground state with the emission of a second 10.2 eV photon in phase with the first photon. These possibilities are summarized in Fig. 4.10.

Lasers and masers are devices in which these stimulated photons are cleverly enhanced so they outnumber absorbed photons.

Neither lasers nor masers can be readily made with hydrogen, but we will develop the idea one step more before we leave our consideration of this familiar substance.

Suppose that a chamber of atomic hydrogen were subjected to *very* intense radiation of 10.2 eV photons. Absorptions would tend to increase the number of atoms in the excited state. However, the more atoms there are in the excited state, the more likelihood there is of stimulated emission to the ground state. The probability of one process is intrinsically equal to that of the other, so that if the number of excited atoms became equal to the number in the lower state, the number of stimulated emissions would become equal to the number of absorptions. Thus radiation disturbs thermal equilibrium. We have calculated that hydrogen "in the dark" has only a very small fraction of its atoms excited. Hydrogen in intense radiation of appropriate energy will tend to have equal numbers of atoms in the higher and lower states, although the tendency toward thermal equilibrium always prevents the number in the higher state from quite equaling the number in the lower.

To illustrate this, let us consider a box containing grains of rice not more than one layer deep. Grains on the bottom of the box correspond to hydrogen atoms in the ground state. If we shake the box, some grains hit the top, a situation that corresponds to the atoms in the excited state. Shaking the box is intrinsically as likely to bang a grain down as up, but gravity (as in the law of atmospheres) corresponds to thermal equilibrium and tends to keep the grains at the bottom. If the box is shaken violently, the number of grains at the top and bottom tend to become equal, but there are always more at the bottom than at the top. Thus mere intense radiation can never cause more stimulated emission than absorption. If we could invent a way to establish a condition in which more atoms were in the higher state, then stimulated emission would exceed absorption. We would get out more than we "put in," and amplification would result.

Before we discuss the devices, let us look at the relationships between the transition rates for absorption and emission. Three processes are of interest — induced absorption, spontaneous emission, and stimulated emission (see Fig. 4.10). Einstein defined coefficients for these processes in 1917. He postulated that the induced absorption transition rate was proportional to the number of atoms with electrons in the lower state and to the density of radiation energy incident on these atoms or

$$\left.\frac{dN_{12}}{dt}\right|_I \propto N_1\rho_{21}.$$

N_1 is the number of atoms with electrons in the $n = 1$ state and ρ_{21} is the density of electromagnetic radiation with energy equal to the energy difference between the two states. Thus,

$$\left.\frac{dN_{12}}{dt}\right|_I = B_{12}\rho_{21}N_1,$$

where B_{12} is the Einstein coefficient for induced absorption.

In spontaneous emission (Fig. 4.10b), an electron in an energy level $n = 2$ spontaneously drops into a lower energy state $n = 1$. Einstein postulated that the spontaneous emission transition rates were proportional to the number of atoms with electrons in the upper state

$$\left.\frac{dN_{21}}{dt}\right|_S \propto N_2$$

or

$$\left.\frac{dN_{21}}{dt}\right|_S = A_{21}N_2,$$

where A_{21} is the Einstein coefficient for spontaneous emission. The reciprocal of A_{21} is the time for this transition to take place and is called the transition life time for the upper state.

Finally, the transition rate for stimulated emission was postulated to be proportional to N_2 and to the density of radiation incident on the atoms with energy equal to the energy difference between the two states

$$\left.\frac{dN_{21}}{dt}\right|_S \propto N_2\rho_{21}.$$

This becomes

$$\left.\frac{dN_{21}}{dt}\right|_S = B_{21}\rho_{21}N_2,$$

where B_{21} is the Einstein coefficient for stimulated emission.

After making the above assumptions, Einstein showed that for thermal equilibrium the coefficients of induced absorption and stimulated emission are equal:

$$B_{12} = B_{21}.$$

He also showed that the relationship between the coefficient of spontaneous emission and the coefficient of stimulated emission is

$$A_{21} = \frac{8(E_2 - E_1)}{h^2c^3}B_{21},$$

where $(E_2 - E_1)$ is the energy difference between the two states, h is the Planck constant, and c is the speed of light. These two equations are known as the Einstein relations.

In 1955 Gordon, Zeiger, and Townes operated the first maser. They used ammonia molecules instead of the hydrogen we have been discussing. The ammonia molecule (NH_3) has energy states like hydrogen, but instead of the first excited state being 10.2 eV above the ground state, it is excited by a mere 10^{-4} eV and thus at 20°C we have

$$\frac{n_2}{n_1} = \exp - \frac{10^{-4} \times 1.60 \times 10^{-19}}{1.38 \times 10^{-23} \times 293} = \exp\left[-\frac{1}{252}\right] \approx 1.$$

In thermal equilibrium the relative fraction of excited atoms is nearly unity. There are "naturally" almost as many molecules excited as unexcited. The remarkable achievement of Townes and his associates was their invention of a molecule sorter. They passed ammonia through a jet into a vacuum. The stream of molecules passed through an electrode structure where, because of their differing electrical properties, excited molecules were not deflected and unexcited molecules were deflected. The excited molecules were allowed to pass on into a chamber. Now, the energy of the excited state is 10^{-4} eV, which corresponds to a wavelength of 1.24 cm and a frequency of 23,870 MHz. This is in the microwave region. By making the chamber of proper size it became a resonant cavity for any radiation produced. If one molecule went from its excited state to its ground state, it was likely to stimulate another, and since initially all atoms were excited, this process could quickly build up to significant proportions. As excited molecules were converted to absorbing molecules, new excited molecules were introduced to sustain the production of photons. Thus the cavity "sang" with microwave radiation like a sounding organ pipe. It was an oscillator.

The maser just described has many remarkable properties. One of these is its stability. The frequency of the oscillations is determined by the nature of the ammonia molecule itself. One excited ammonia molecule is indistinguishable from another. They all cause radiation of the same frequency. By electronic techniques, this oscillator can be made to govern the frequency of slower oscillators. A succession of such frequency reductions provides a low frequency oscillator that ticks like a clock but has all the stability built into ammonia molecules. Such a clock is called an *atomic clock* and its precision is such that it will neither gain nor lose more than one second in 1000 years! Whereas this gas maser is superb as a stable oscillator, another type, the *solid-state maser*, makes a better amplifier.

The most common solid-state maser is made from ruby. A ruby is basically clear alumina made red by a small concentration of chromium. It is the chromium "impurity" that is the active atom of the maser. When the ruby is in a steady magnetic field, the chromium acquires energy states, three of which are represented schematically in Fig. 4.11. In thermal equilibrium the number of atoms in the three states obeys Boltzmann's law, as shown graphically in Fig. 4.12(a). The ruby material is irradiated with photons from an external source whose frequency (energy) corresponds to the energy difference, $E_3 - E_1$. This causes absorption transitions from E_1 to the *metastable* state E_3 and stimulated transitions from E_3 to

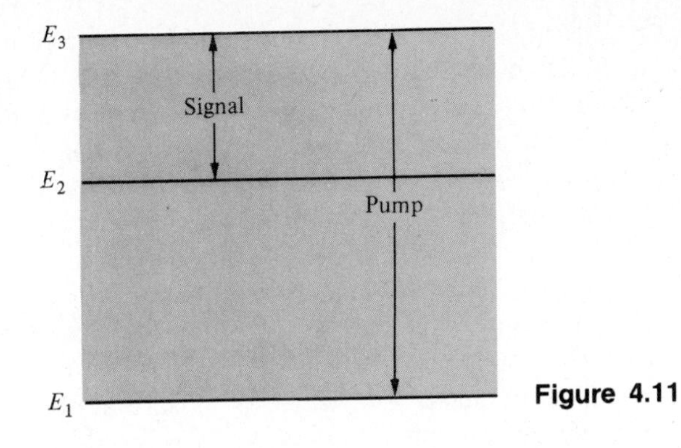

Figure 4.11

E_1. Although these latter transitions are stimulated, they themselves do not account for the maser action. The effect of these transitions is, as we have seen, to make the concentration of chromium atoms in states E_1 and E_3 tend to become equal. The concentration of atoms in E_2 remains substantially unchanged, as shown in Fig. 4.12(b). This *optical pumping* of atoms from E_1 to E_3 causes a population inversion between states E_2 and E_3, which is enhanced because E_3 is a metastable state of chromium. If the ruby material is now exposed to photons whose energy is $E_3 - E_2$, transitions will occur both up and down between these two levels. The significant fact is that with this inverted population of these states, there are more stimulated emissions than there are absorptions. Amplification results.

It is true that signal frequency photons (with power in the microwatt range) tend to make the populations of states 2 and 3 become equal, but the overwhelming pumping between states 1 and 3 (milliwatts) keeps state 3 more populated than state 2.

Figure 4.12

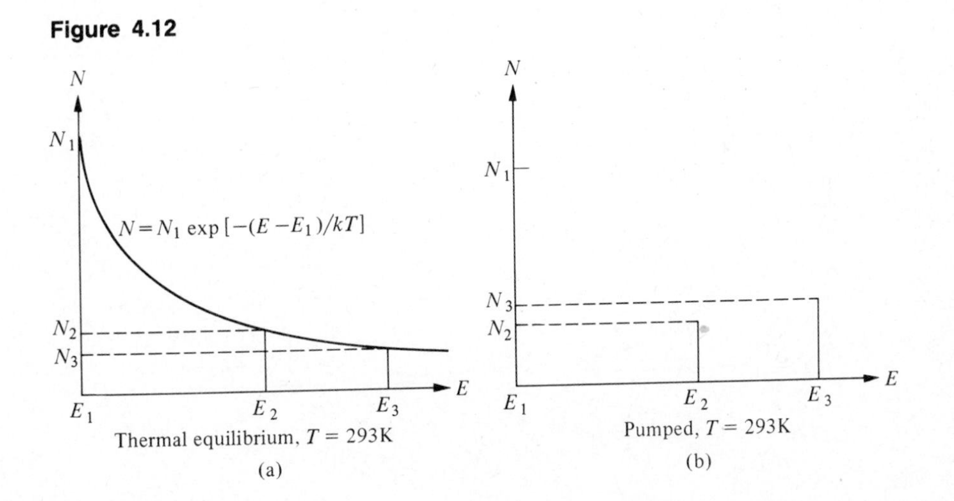

Spontaneous transitions from state 3 to state 2 do occur, as you would expect from our discussion of the Bohr theory of hydrogen. Photons thus produced are random in origin and constitute the ultimate defect of all amplifiers, noise. But these random transitions are so remarkably rare that this amplifier has less noise and can amplify weaker signals than amplifiers of any other type. One application of maser amplifiers is in detecting and measuring extremely weak microwave signals from outer space.

The frequency of microwaves that this maser can amplify is determined by the energy difference $E_3 - E_2$. This difference can be varied by changing the magnetic field in which the ruby material is placed. Thus, unlike the frequency stability of the ammonia maser, the solid state maser has versatility as an amplifier.

Ruby masers are always operated at low temperatures — frequently at the temperature of liquid helium, 4.2 K. Comparing Figs. 4.13(a) and (b) to the earlier figures shows how cooling enhances the inversion between states 2 and 3.

Lasers (*L*ight *a*mplification by *s*timulated *e*mission of *r*adiation) are, in principle, no different from the maser just described. A material is used that has an energy diagram like that shown in Fig. 4.11, except that the energy differences are much greater, and thus the wavelengths are in the visible region instead of in the microwave region. The pumping is done with light and the laser produces light. Of course the pumping light must be "bluer" than the laser light.

A typical laser can be a cylindrical crystal or a gas-filled cylinder with optically parallel ends, one of which is fully silvered and the other partly silvered. The pumping light is admitted through the sides, and the laser light emerges from the partly transmitting end. Light from a laser differs from light from most other sources in four basic ways. Laser light is much more directional, monochromatic, and coherent; it can also be produced as polarized light.

Figure 4.13

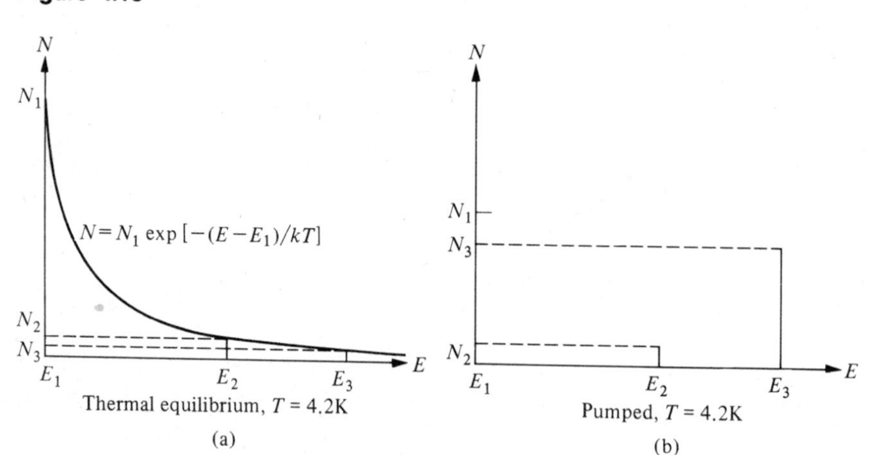

First, let's discuss directionality. Light from a laser diverges very little. The reason for this is that the active material is in a cylindrical resonant cavity. Light is reflected back and forth through this cylindrical cavity such that light traveling parallel to the axis finally exits and becomes the laser beam. Any light that is traveling in a direction other than parallel to the axis of the cylinder is scattered out of the beam. Hence, after the beam leaves the laser the only divergence results from diffraction.[†]

A quantity that measures the directionality of the laser beam (or any other light beam) is the full angle beam divergence. This quantity is the angle between the two outer edges of the beam where the edge is defined as the location in the beam where the intensity decreases to $1/e$ of that at the center. This angle tells us how much the beam will spread as it travels through space. A typical divergence for a small laser is 10^{-3} rad, which means that the laser beam increases by about 1 mm for every meter it travels.

In Section 3.3, light was described as an electromagnetic wave. The frequency of this wave is the number of oscillations per second the electric field (or magnetic field) makes at a point in space. If the electromagnetic wave or light has only one frequency of oscillation for this electric field, the light is said to be monochromatic.[‡] It is not possible to produce light with only one frequency. However, it is possible to have light with a very narrow band of frequencies. This narrow band would be called the light's linewidth, $\Delta\nu$. The linewidth for white light, which would contain all of the visible frequencies, would be about 3×10^{14} Hz. If this white light is sent through a filter made of colored glass (for example, red) the remaining band would have a linewidth of perhaps 3×10^{13} Hz. If we take the light coming from a hydrogen discharge lamp as was discussed in Section 4.4 and send it through a prism retaining only the red spectral line, the linewidth would be approximately 10^{10} Hz. In a good laser, it is possible to have monochromaticities of 10^2 Hz.

Light from a laser is coherent. By "coherent" we mean that there is a connection between the amplitude and phase of the light at one point and time and the amplitude and phase of the light at another point and time. Coherence can be broken into two concepts: spatial and temporal. For our discussion of spatial coherence, consider two waves at some time, say t_1, where the electric field at P_1 is E_1 and at P_2 is E_2 (see Fig. 4.14a). We can see that the phase difference between the two electric vectors is zero. Now consider the same two waves at some later time t_2 (see Fig. 4.14b). Now E_1 at P_1 and E_2 at P_2 are different than they were at time t_1. However, the phase between them is still zero. If the phase difference between the two electric vectors remains zero for all times, then the two waves are spatially coherent.

To discuss temporal coherence, consider the electric field at one point in space at two different times, t_1 and t_2 where $t_2 - t_1 = \tau$ (see Fig. 4.15). The phase

[†]See any basic optics book.

[‡]In Greek, *monos* means *single* and *chroma* means *color.*

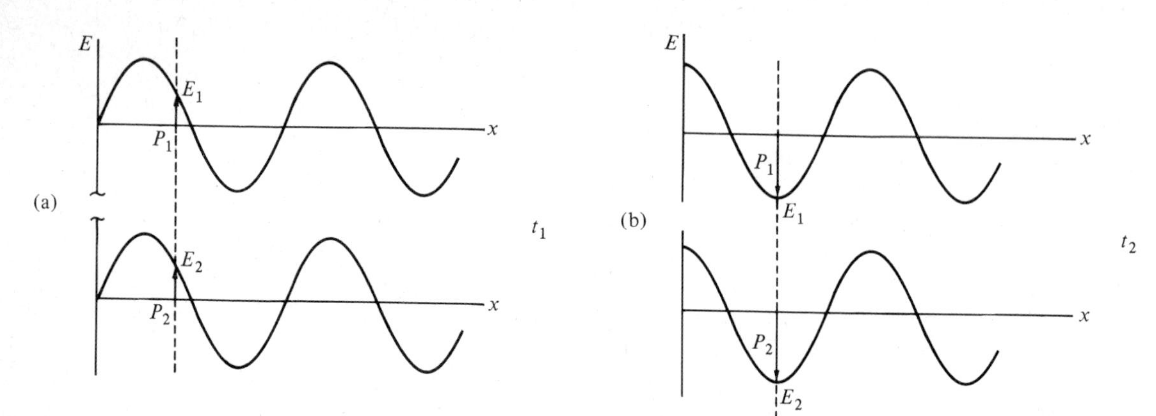

Figure 4.14

difference between the field at t_1 and the field at t_2 is ϕ_1, as can be seen by noting the position of the dashed curve. Now consider the electric fields at later times t_3 and t_4, where $t_4 - t_3 = \tau$. The phase between the fields at these two different times is ϕ_2, as is shown in the figure. If ϕ_2 is equal to ϕ_1 and this is true for any two times that differ by τ, the wave satisfies the requirement of being temporally coherent.

Light that eventually emerges from a laser is coherent because at each interaction of a photon with an excited atom the photon stimulates the transition of an electron from an upper state to a lower state and the atom emits a photon in phase with the initial photon. These two photons eventually interact with two other excited atoms, each causing the emission of a photon in phase with each stimulating photon. In a pulsed laser, this interaction takes place between photons and many excited atoms over a short period of time. The pulse of light that comes out is then coherent. In a continuous laser, the interaction takes place continuously with each stimulated

Figure 4.15

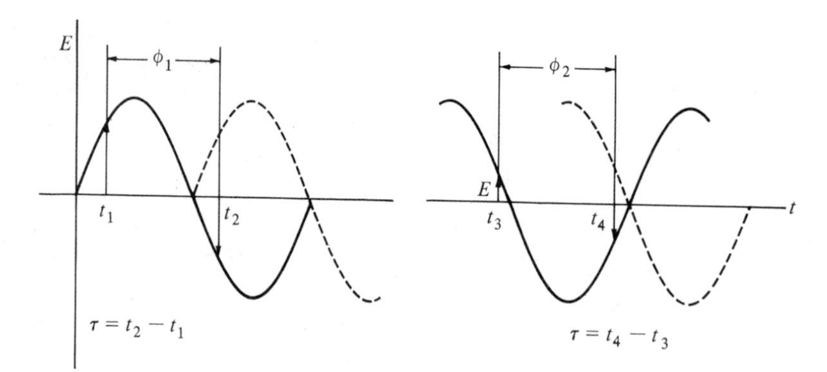

photon being in phase with the stimulating photon. Thus the continuous beam of light is coherent.

Light from most lasers is plane polarized to some extent. Figure 3.1 in Chapter 3 is a diagram of a plane polarized electromagnetic wave. By plane polarized we mean the orientation of the E-vector remains in one plane (in that figure the *xy*-plane) while its magnitude changes with time. A typical laser has a cylindrically shaped laser media between two mirrors. One of the mirrors is totally reflecting and the other is partially reflecting. Often the ends of the lasing material (or the container) are cut at a special angle called the Brewster angle. Light incident on a surface at the Brewster angle will be reflected or transmitted depending on its polarization. Light polarized parallel to the plane of incidence will be transmitted and not reflected. Light polarized perpendicular to the plane of incidence will be transmitted and reflected. Therefore in the laser cavity with ends cut at the Brewster angle, light polarized perpendicular to the plane of incidence will be reflected out of the cavity away from the mirrors. The remaining light will reflect back and forth between the mirrors and produce the laser beam predominantly polarized parallel to the plane of incidence.

Various types of materials can be made to lase. The first laser was built by T. H. Maiman in 1960 using a ruby crystal. Ruby, of course, is a naturally occurring precious stone but can also be formed in the laboratory. A typical ruby laser is constructed from a rod a few millimeters in diameter and several centimeters long. Pumping is accomplished with a xenon flash lamp helically wrapped around the rod. The ruby laser is an example of the general category doped-insulator lasers.

Gas is another media that can be used to construct a laser. The gas is contained in a glass tube and placed between appropriate mirrors. The most common of this type is the helium-neon laser that uses ten parts helium and one part neon. The neon atoms provide the energy states for the transitions while the helium provides a mechanism for efficiently exciting these neon atoms to upper metastable states. Figure 4.16 shows the energy level diagram for the important states in the helium and neon. When an electric field is applied across the laser, electrons collide with helium atoms and drive them into the upper states labeled He_2 and He_3. These are metastable states. Thus these atoms remain in these excited states long enough to interact with neon atoms in the ground state. This interaction excites the neon atoms to their metastable states, labeled Ne_4 and Ne_6, where they stay until laser transitions take place from these states to the states labeled Ne_3 and Ne_5, respectively. The lasing transitions that take place and are reinforced to produce the laser beam depend on the construction of the glass envelope and mirrors. This is controlled by choosing a mirror with maximum reflectivity at the wavelength desired. These lasers are continuous because the collision process maintains the energy states Ne_6 and Ne_4 at larger population densities than the lower states. This continued population inversion gives a continuous lasing action.

A dye laser is another type of laser that we will only mention. The lasing media is an organic dye dissolved in a solvent. The primary advantage of a dye laser is its

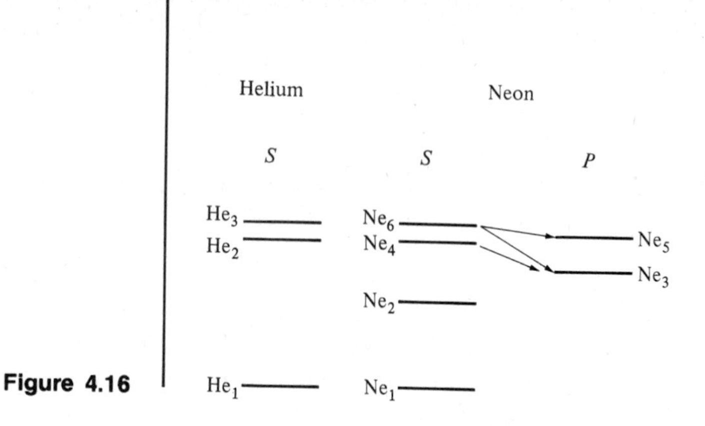

Figure 4.16

tunability. By varying the dye, the pumping light, and the angles of the mirrors, the frequency of the laser radiation can be varied over a broad range.

We shall delay our discussion of another type of laser, the semiconductor laser, until we have introduced semiconductors in Chapter 9.

4.13 LASER APPLICATIONS

The development of the laser has opened many avenues of investigation and improved many others. The important characteristics that have contributed to its usefulness in these many areas are the coherence, monochromaticity, and great intensity of its radiation.

The laser light can be focused to an extremely high energy density. With suitable lenses, all energy from a typical laser can be concentrated into a diffraction limited image that is only one micron (10^{-6} m) in diameter. Because of the small size of this focused image and because of the precise control over the energy, the lasers are used extensively in microelectronics for the cutting, welding, and drilling of circuits for calculators, computers, and related equipment. Lasers also have been used for making special tools; for example, drilling holes in industrial diamonds used for drawing fine wire. Lasers are used to repair flaws in diamonds of jewelry quality. A small flaw in a diamond can be removed with the local heating of the well-defined focused laser beam.

To obtain energy from a controlled fusion process (see Chapter 12), a temperature on the order of 10^7 K is needed. Because of the ability to obtain extremely high energy densities with lasers they are being used in the attempt to obtain these very high temperatures. Fuel pellets of solid deuterium and tritium are irradiated with a high-energy laser beam that vaporizes the pellet and then heats the gas to very high temperatures. Presently, the necessary temperatures are not being obtained because some of the energy goes to expand the plasma (hot ionized gas). Magnetic fields that could be used to confine the plasma and reduce the energy that goes into this expansion are not yet available. When they are available, the laser may help lead the way to the solution of our energy problem.

Because of its high intensity and exceptional directionality, the laser is an excellent tool for surveying. In order to be certain that a bridge or tunnel is being constructed along the proper line it is only necessary to set up a laser and have it pointed in the proper direction. Then, at each point along the bridge or tunnel, the engineers can measure very accurately using the laser beam as a reference. Also, when excavating for a pipe line a ditch must be dug only to a certain depth. If it is made too deep, the pipe will eventually settle and rupture. The depth of the ditch may be controlled very accurately if the laser is pointed along the top of the ditch and the bottom determined with reference to the laser beam immediately above.

The most interesting surveying measurement for which the laser has been used is the measurement of the distance from the earth to the moon. The old techniques used from the time of Newton until the 1950s for obtaining this distance have been optical parallax and stellar occultations. An optical parallax measurement is a measurement of the angle between the horizontal and the line to the moon at two or more places on earth at the same time or the same measurement from one place at two different times. Stellar occultations (measurement of the time at which the edge of the moon passes over a star) from two different positions on earth also will yield the distance to the moon. The accuracy of each of these measurements is about two miles.

Radar was used for the first time in 1967 to measure the earth-moon distance and the accuracy was 0.7 mile. Then, in 1965, laser light was reflected from the moon and the distance was measured to within an accuracy of 600 feet. The major breakthrough in measuring the distance to the moon came in 1969 when the astronauts placed a lunar laser reflector on the surface of the moon. This reflector enabled the scientist to measure the distance to the moon to within six inches. Very precise measurements will allow extremely accurate determination of the moon orbit and this information will yield valuable data about the distribution of the moon mass. Also, the surface of the earth is believed to be made up of "plates" that are moving with respect to each other. For example, the Pacific plate is believed to be moving toward Japan. By carefully measuring the change in the distance to the moon with a laser set up in Japan and one set up in the United States, the rate of this movement can be obtained. The location of the North Pole is known only to within five feet and, also, is known to trace an almost elliptical path of approximately 200 feet during a year. The new accuracy of the measurements of the earth-moon distance will allow the determination of the location of the North Pole to within six inches and will allow the determination of the exact movement of the

Pole during the period of a year. Finally, it is thought that the gravitational constant may not really be constant but may be decreasing with time due to the expansion of the universe. With the use of the lunar laser reflector, this possibility can be checked.

The information capacity of a communication channel — for example, the microwave system used extensively by the telephone company — is proportional to the width of its band of frequencies. Therefore, the theoretical capacity of a typical visible light wave is approximately 10^5 times greater than that of the microwave. This fact makes communication by laser very promising.

A fast growing use of the laser is in holography. Holography is a technique of producing an interference pattern made by having a direct laser beam and a laser beam reflected from an object fall on a photographic plate. When the interference pattern on the developed photographic plate (the hologram) is illuminated with laser light in the proper way, a three-dimensional, very realistic image of the object is produced. This image can be viewed with the eye or other optical instruments or recorded with a camera. In fact, for optical purposes a hologram of a subject is as useful as the actual subject.

Holography is becoming a very useful technique for microscopic investigations — especially for microscopic investigations of biological subjects. Also, holography is becoming the scene of considerable activity in the area of data processing and storage. (For details of holography and its applications, see Keith S. Pennington's *Scientific American* article referenced at the end of this chapter. Also, for details of general applications of lasers see the book by O'Shea, Callen, and Rhodes referenced at the end of this chapter.)

4.14 MANY-ELECTRON ATOMS

We have stated that the Bohr theory is successful only for one-electron atoms, which pretty well limits quantitative application of the theory to hydrogen, heavy hydrogen, and ionized helium.

In our solar system, the various planets interact with one another only very slightly. The motion of the planets is understood by first considering that each planet experiences only the gravitational force of the sun. When these problems are solved for each planet, the weak planet-to-planet forces are treated as minor perturbations. The solar analogy is fruitful in understanding hydrogen, since hydrogen has but one electron. When an atom has two or more electrons, the interelectron interactions are not small and the simple planetary model breaks down. But one can sense what is in fact found: that a multiplicity of energy levels exists in these cases and so there are many lines in the spectra. Furthermore, the whole treatment so far has been of an *isolated* atom, that is, atoms in the gaseous state. In the liquid and solid states, the interactions between the electronic systems of the closely packed atoms introduce so many additional energy levels that the spectrum is now essentially continuous. Thus if the electrodes for an electric arc are copper rods, the spectrum of the light from the incandescent rods is continuous, whereas that from the gaseous copper ions in the arc between the electrodes is a line spectrum.

For a logical treatment of atomic structure, we should next turn to wave mechanics, which is a form of quantum mechanics. These theoretical techniques permit a more sophisticated understanding of both hydrogen and the heavier elements. But, historically, electronic structure was developed empirically before it was developed theoretically. Because our interest is directed inward toward the nucleus rather than outward to the more complex electronic structure of the heavier elements and molecules, our treatment of these subjects will be somewhat empirical and somewhat superficial.

4.15 QUANTUM NUMBERS

We have computed the spectrum of hydrogen in terms of but one quantum number n, called the principal quantum number. Measurements of extreme precision on hydrogen and of less precision on other elements show that the energy levels have fine structure (spectroscopes used in elementary courses in light are often able to separate the sodium doublet). If the substance under study is placed in a magnetic or an electric field, the energy levels are further subdivided. These are called the *Zeeman* and *Stark* effects, respectively. It turns out, therefore, that additional quantum numbers are needed in order to specify in detail the energy levels of atomic electrons. It is found that four quantum numbers are required.

The Bohr atomic model assumes that the electron orbits are circular. In planetary motion the circular orbit is possible but it is a very special case of an ellipse. An obvious extension of the Bohr model was to allow for the possibility of elliptical orbits. This extension was made by Sommerfeld in 1915.

Bohr's first postulate was that the angular momentum of an orbital electron is an integral multiple of $(h/2\pi)$. Although this integer n is associated only with angular momentum in the case of circular orbits, this association must be modified when elliptical orbits are considered because both the distance of the electron from the nucleus and the angular position of it change during a revolution. Just as the circular path is one extreme of an elliptical orbit, so a straight line is another. Mathematically, at least, we can consider a linear orbit where the electron passes directly through the nucleus. Such an orbit will have an energy associated with it, although it has no angular momentum. We can imagine a whole set of orbits as having the same energy but ranging from the straight-line orbit with no angular momentum to a circular orbit for which the angular momentum is maximum.

Sommerfeld's argument led *ultimately* to the retention of the principal or total quantum number, n, associated with the mean separation of the electron and proton as in our Eq. (4.16), and to the introduction of an orbital quantum number to characterize the angular momentum. The *orbital* quantum number, l, specifies the *number of units of angular momentum* $(h/2\pi)$ associated with an electron in a given orbit. This quantity can be represented vectorially as shown in Fig. 4.17 by a straight line that is parallel to the axis of rotation and whose positive sense is given by the direction of advance of a right-hand screw rotated with the motion. It was found that the integer l could take on only positive values from zero to $(n - 1)$. Thus the electron in the smallest orbit $(n = 1)$ will have no angular momentum, since for it l must equal zero. For a larger orbit with $n = 3$, there are three possible orbital shapes or eccentricities: a straight line through the nucleus without angular momentum for $l = 0$, an ellipse with an angular momentum for $l = 1$, and a "rounder" ellipse with more angular momentum for $l = 2$. It may seem mechanically absurd to think of an electron passing through the nucleus as required for $l = 0$, but this was reasonable in the wave mechanics that had been introduced when this scheme was completed.

The orbital motion of an electron is equivalent to a current in a loop of wire, so each orbit has a magnetic moment. The magnetic moment vector is normal to the plane of the orbit, but *antiparallel* to l because the electron's charge is negative. When the atom is placed in an external magnetic field, each electronic orbit will be subject to a torque that tends to make the l-vector parallel to the field. These vectors are shown in Fig. 4.18, where θ is the angle between l and the external magnetic induction, B. Because of the righting torque of the field on the revolving system, the l-vector will precess about the field for the same general reasons that account for the motion of a spinning top. This motion of the orbit, called the *Larmor precession*,

Figure 4.17 Vector representation of orbital angular momentum.

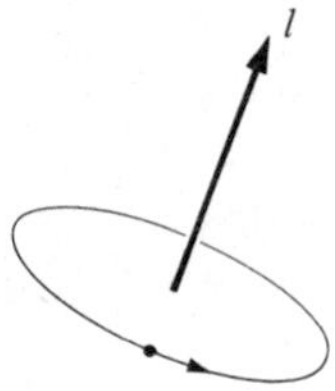

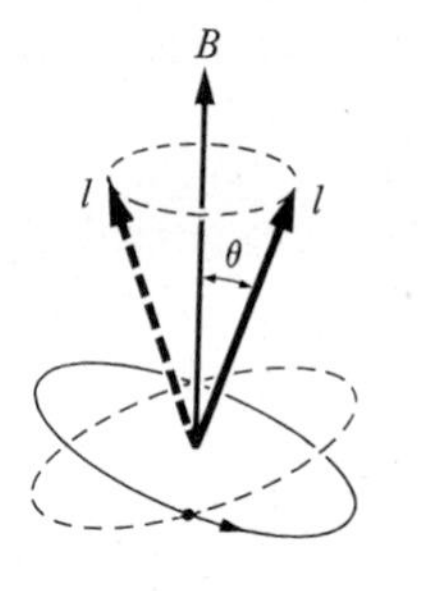

Figure 4.18 Larmor precession of an electron orbit in a magnetic field.

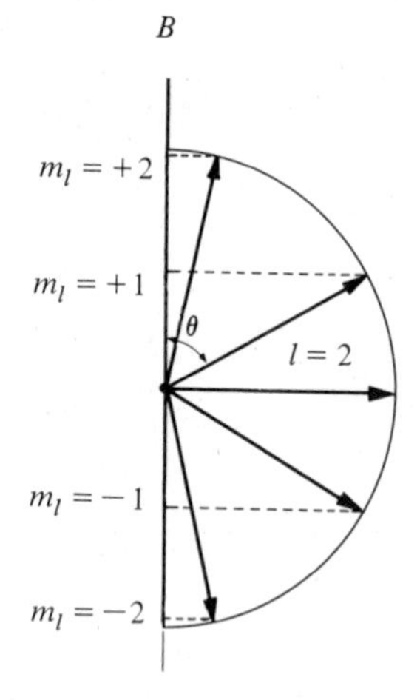

Figure 4.19 Possible orientations of two units of orbital angular momentum in a magnetic field.

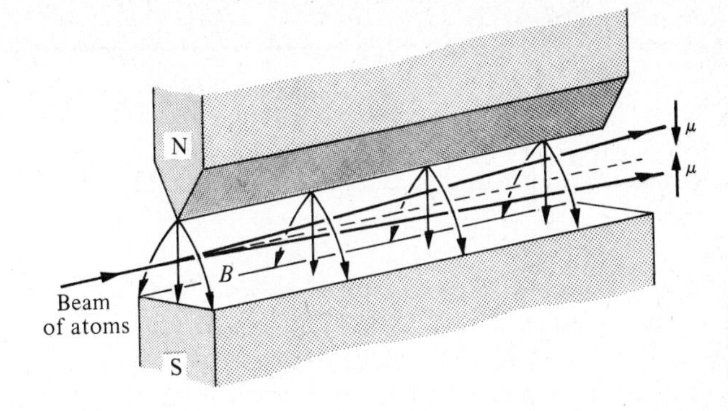

Figure 4.20 Schematic diagram of the Stern-Gerlach experiment.

introduces additional energy states into the atomic system. The amount of energy depends on the precessional velocity and this depends on the righting torque which, in turn, depends on θ. If all values of θ could occur, then there would be an infinite number of new energy states. The Zeeman effect shows, however, only a few additional lines, not a continuous spectrum. Therefore only a few values of θ are "allowed" and these are those for which $l \cos\theta$, the projection of the l-vector on the direction of the magnetic induction, is an integer.

This introduces another quantum number called the *orbital magnetic* quantum number, m or m_l. It can have any integral value from $-l$ to $+l$ inclusive. Since m_l is limited to whole numbers, it means that the component of the orbital angular momentum along the magnetic field is restricted to *integral multiples* of $(h/2\pi)$. This is shown in Fig. 4.19 for $l = 2$. The "allowed" values of θ are given by $m_l = l \cos\theta$. Because of the restrictions on the orientation of the electron orbits, they are said to be space quantized. Space quantization of atoms was first verified in 1921 by Stern and Gerlach by observing the deflection of a narrow beam of atoms in silver vapor in an inhomogeneous magnetic field.

To discuss the Stern–Gerlach experiment, we return to the discussion earlier in this section and note that an orbiting electron is equivalent to a current loop whose magnetic moment vector μ is antiparallel to its angular momentum vector l. When the loop is in a uniform magnetic field, it experiences only a torque, which produces the Larmor precession already mentioned. In a nonuniform field, however, in addition to the torque action there is a resultant force that gives the loop motion of translation. The magnitude of the force depends on the nonuniformity of the magnetic induction B and on the inclination of the magnet moment vector to the field, as specified by the angle θ in Fig. 4.18. The direction of the translational force is either toward or away from the region of increased magnetic field intensity, depending on the direction of the component of μ along a field line. The force is maximum in one direction when $\theta = 0°$, maximum in the opposite direction when $\theta = 180°$, and zero when $\theta = 90°$. In the Stern–Gerlach experiment, a beam of silver atoms was projected through a nonuniform magnetic field, as shown in Fig. 4.20. Classically, one would expect the beam to be "smeared" out in the vertical

Table 4.2 Values of quantum numbers

Principal	$n = 1, 2, 3, 4, \ldots$
Orbital	$l = 0, 1, 2, 3, \ldots, (n-1)$
Orbital magnetic	$m_l = -l, (-l+1), \ldots, (l-1), l$
Spin	$m_s = -\frac{1}{2}, +\frac{1}{2}$

direction, because the atomic current loops would be randomly oriented. However, it was found that the beam split into two distinct parts with nothing between them. This means that the magnetic moment in the case of the silver atom can only be aligned with or against the field. Similar experiments with beams of atoms of other elements also show space quantization of the atoms in a magnetic field.†

In order to account for the fine structure of some spectral lines, Uhlenbeck and Goudsmit introduced a fourth quantum number in 1925. This is the electron spin quantum number, s or m_s. This spin of the electron about its own axis as it revolves about the nucleus is analogous to the rotation of the earth as its moves along its orbit around the sun. The spin momentum has the *numerical value* $\frac{1}{2}(h/2\pi)$. The rotating electron also has a magnetic moment, and spectroscopic observations show that the spin vector is capable of orientation only in either of two ways, parallel or antiparallel to the surrounding magnetic field. Therefore m_s can have only two values, $+\frac{1}{2}$ or $-\frac{1}{2}$. In the absence of a magnetic field, there is no unique assignment to the direction of m_l or m_s.

The order in which the four quantum numbers have been discussed gives, in general, the order of their importance in determining the energy of any particular level. The restrictions on the values of these quantum numbers are summarized in Table 4.2.

The elements whose spectra most closely resemble hydrogen are the alkali metals, lithium, sodium, and potassium. Like hydrogen, these elements display series of lines. These series were given the descriptive names *sharp*, *principal*, *diffuse*, and *fundamental*. When it was learned that the sharp series resulted from electron transitions to the $l = 0$ state, and the principal series resulted from transitions to the $l = 1$ state, etc., electrons in the $l = 0, 1, 2, 3, 4, 5, \ldots$ states came to be described as being in $s, p, d, f, g, \ldots$ (continuing alphabetically) states. This notation is extended by writing the value of n for a state before the letter, and if an atom has more than one electron with particular values of n and l, this number is written as a superscript of the letter. Thus if an atom has six electrons for which $n = 3$ and $l = 2$ the atom is said to have $3d^6$ electrons.

Taking four quantum numbers into consideration greatly increases the complexity of an energy-level diagram and the spectrum from which it is deduced. But the

† The magnetic moment of the silver atom causing the splitting of the beam is associated with the electron spin (see next paragraph).

increase in the complexity of the spectrum is less than might at first be expected because many of the energy transitions that appear possible *do not occur* or occur rarely. There are certain rules, called *selection rules*, that specify between which energy levels an electron does move and, in consequence, which energy differences do appear as radiation. For example, one selection rule is that an atom can undergo no energy transition unless $\Delta l = \pm 1$. Thus $4s$ to $2p$, $2p$ to $1s$, and $5d$ to $2p$ are observed transitions. On the other hand, $4s$ to $1s$, $3d$ to $1s$, and $4f$ to $2p$ transitions are not observed.

4.16 PAULI EXCLUSION PRINCIPLE

Another generalization of utmost importance that was empirically arrived at is known as the *Pauli exclusion principle*. It states that in any one atom no two electrons have identical sets of quantum numbers. The electron in hydrogen does not demonstrate this rule since there is but one electron in this element, but the Pauli principle provides a vital key to the electron structure of the heavier elements. This principle, together with the *usual* energy importance of the quantum numbers, enables us to specify the states of the electrons in unexcited atoms — at least near the beginning of the periodic table.

The electron in hydrogen may have any energy state listed in Table 4.2. When this electron is in its lowest energy state, however, two of the quantum numbers are $n = 1$ and $l = 0$, and it is then in the $1s$-state. It is obvious from Table 4.2 that $m_l = 0$ whenever the electron is in any s-state. Helium has two electrons, one of which has those quantum numbers assigned to hydrogen. The second electron has the same quantum numbers except that its spin has a sign opposite to that of the first. Therefore these two $1s$ electrons have different quantum numbers, as required by the Pauli exclusion principle.

Lithium has three electrons, two of which have the quantum numbers already assigned. These two $1s$ electrons constitute a helium core. The third electron must have considerably more energy than either of the first two. We have exploited both permitted variations of spin. The next lowest energy quantum number, m_l, cannot be increased without increasing l, but since l must remain less than n, which thus far has been 1, we see that we must now increase the most important quantum number n. Thus the lowest set of quantum numbers that can be assigned to the third lithium electron require it to be in the $2s$-state. The $2s$-state has $l = 0$, which requires that $m_l = 0$. But since the spin may be either $+\frac{1}{2}$ or $-\frac{1}{2}$, we find that the fourth electron, which forms beryllium, may also be a $2s$ electron. The electron configuration of beryllium is written $1s^2 2s^2$. Boron has $Z = 5$ and has five electrons. The first four states "used up" all four possible s-states with $n = 2$. But since $n = 2$, then l may equal 1, and the fifth electron is a $2p$ electron. When $l = 1$, then m_l may be -1,

0, or 1, and each value of m_l may have spins $\pm\frac{1}{2}$ so there may be six electrons in the $2p$-state. The states described account for the configurations through neon, whose 10 electrons have the configuration $1s^2 2s^2 2p^6$. With neon we have exploited every possible state with electrons for which $n = 1$ or 2. The eleventh electron in sodium must therefore have $n = 3$, and its configuration is $1s^2 2s^2 2p^6 3s$. The $3s$-state will also accommodate magnesium, but the last electrons added for aluminum through argon must be $3p$ electrons. Table 4.3 summarizes how the Pauli principle enables us to establish atomic electron configurations through potassium where irregularities set in because the l quantum number becomes more energy-important than n, so that it is "cheaper" to increase n rather than l. Although our scheme becomes more complex with potassium, it is possible to extend the table through all the irregulari-

Table 4.3

			Atom in the ground state			
Element	Z	First ionization potential, V	Quantum number of last added electron		Electron configuration	Outermost shell occupied
			n	l		
H	1	13.6	1	0	$1s$	K
He	2	24.6	1	0	$1s^2$	K
					Helium core, 2 electrons	
Li	3	5.39	2	0	$2s$	L
Be	4	9.32	2	0	$2s^2$	L
B	5	8.30	2	1	$2s^2\,2p$	L
C	6	11.3	2	1	$2s^2\,2p^2$	L
N	7	14.5	2	1	$2s^2\,2p^3$	L
O	8	13.6	2	1	$2s^2\,2p^4$	L
F	9	17.4	2	1	$2s^2\,2p^5$	L
Ne	10	21.6	2	1	$2s^2\,2p^6$	L
					Neon core, 10 electrons	
Na	11	5.14	3	0	$3s$	M
Mg	12	7.64	3	0	$3s^2$	M
Al	13	5.98	3	1	$3s^2\,3p$	M
Si	14	8.15	3	1	$3s^2\,3p^2$	M
P	15	10.6	3	1	$3s^2\,3p^3$	M
S	16	10.4	3	1	$3s^2\,3p^4$	M
Cl	17	13.0	3	1	$3s^2\,3p^5$	M
Ar	18	15.8	3	1	$3s^2\,3p^6$	M
					Argon core	
K	19	4.34	4	0	$4s$	N

ties and assign quantum numbers to all the electrons of the unexcited elements. Short as Table 4.3 is, it permits a qualitative understanding of many properties of the atoms.

Table 4.3 shows a distinct periodicity for those elements with $n = 1, 2, 3$. The $n = 1$ group consists of chemically active hydrogen and chemically inactive helium. The $n = 2$ and $n = 3$ groups each begin with active alkalies and end with inert gases. Silicon and carbon have very similar chemistry and we note that, except for n, their most energetic electrons have identical quantum numbers. These groups constitute the rows of the periodic table, and Table 4.3 accounts for the first three rows with two, eight, and eight elements, respectively.

4.17 ELECTRON SHELLS AND CHEMICAL ACTIVITY

From the preceding discussion and from a study of Table 4.3, it is evident that whenever an element is reached for which the vector sum of the magnetic quantum numbers and the sum of the spin numbers are each equal to zero, we have a very stable electron configuration. This is the case for each of the noble gases — He, $Z = 2$; Ne, $Z = 10$; Ar, $Z = 18$, etc. Thus the most stable elements are those with atomic numbers 2, 10, 18, 36, 54, and 86. Whenever these sums are zero, we say that we have completed an *electron grouping* or *shell*. The innermost grouping is called the *K-shell* and is complete for helium, which has two *K*-shell electrons. As we go on adding electrons to the structure (and, of course, an equal number of protons to the nucleus), the grouping takes place farther out than before, in the region called the *L-shell*. This is filled when we arrive at neon. This element has a helium core surrounded by eight electrons in the completed *L*-shell. Continuing the addition of electrons brings us to argon, which has a neon core surrounded by eight electrons in the completed *M-shell*. In a similar way, the *N-*, *O-*, *and P-shells* become populated as we go on in the periodic table. In general, none of these outer shells are filled completely before some electrons enter one or more other shells that are not yet fully populated. The last column of Table 4.3 shows the outermost of each unexcited atom. We will need the electron-shell concept in our discussion of x-rays in Chapter 6.

Atoms having completed shells are very stable. This conclusion is supported by their chemical inactivity, but there is substantiating physical evidence too. The stability of the atomic system can be judged from the first ionization potential, which is a measure of the energy required to remove an electron from it. (The second ionization potential refers to the removal of the second electron and so on.) Table 4.3 shows that this first ionization potential rises to its greatest value, 24.6 V, for helium and then drops sharply to 5.39 V for lithium. This means that it is difficult to disrupt the completed *K*-shell, but easy to remove a lone electron in the *L*-shell. The required potential increases as we go from element to element until it again reaches a maximum in neon. After that, it drops once more to a low value but gradually increases until another maximum is reached in argon. This pattern is repeated periodically as one goes on through the periodic table of elements. Since the *L*-shell electrons of neon are somewhat screened from the attraction of the nucleus by the helium core, the ionization potential of neon is less than that of helium. Similarly, the *M*-shell electrons of argon are partially screened by the neon core. You probably have noticed that the potentials do not increase uniformly in going from beryllium to neon or from sodium to argon. Indeed, some are out of order on the basis of increasing numerical values. It is beyond the scope of this book to discuss these apparent anomalies, but the explanation can be found in more exhaustive books on atomic structure.

The stability of chemical compounds can be inferred from the ionization potentials in Table 4.3. From this it is evident that fluorine does not give up an electron readily; rather, it seeks an additional electron to complete its *L*-shell quota and thus attain the stable neon configuration. It is electron hungry and, therefore, readily becomes a negative ion in solution. On the other hand, sodium has a lone electron in the *M*-shell that is easily removed. If it lost this one, it would have the stable neon structure. Because of this excess electron, sodium readily becomes a positive ion in solution. Thus one expects, and in fact finds, sodium fluoride to be a very stable compound because each atom provides what the other one needs. Now consider carbon, which is halfway between helium and neon. It could revert to the helium arrangement by "giving up" electrons. This is what it does when it forms carbon dioxide, CO_2. On the other hand, it could approach the neon state by "taking on" electrons. This is what happens when it unites with beryllium to form beryllium carbide, Be_2C. In general, elements located just before a noble gas in the periodic table form strongly electronegative ions; those that follow are strongly electropositive; and those in the middle range between two of the noble gases are electronegative in their reactions with some elements but electropositive in others.

4.18 MOLECULES

The tendency for electrons to pair off within an atom so that the vector sum of their quantum numbers is zero has a strong bearing on the tendency of atoms to combine into compounds. A single, unexcited hydrogen atom has one electron with a spin of either $+\frac{1}{2}$ or $-\frac{1}{2}$. If this atom is near another hydrogen atom, the situation becomes more complex. If both atoms have their spin vectors parallel and if they should combine, the total electron spin of the molecule would be -1. Since this is in opposition to the observed tendency, we know these atoms will repel each other and not combine. If, however, the spins are antiparallel, the two atoms attract each other, forming a molecule whose total electron spin is zero. Therefore, we normally find hydrogen as the molecule H_2. Thus we have a stable molecule of an element when the resultant of the electron spins of its constituent atoms is zero. This is the answer to the objection, stated in Chapter 1, of Dalton and others to Avogadro's view that the molecules of hydrogen, nitrogen, oxygen, and many other gases are diatomic.

Helium, on the other hand, has two electrons with oppositely directed spins. Since its two electrons are paired, it has no tendency to react with atoms of its own or other kinds. To make helium chemically active, we would have to reverse the spin of one of its electrons. But we have seen that the Pauli principle requires an increase in the principal quantum number n before the spin can be changed. This requires a large energy that might be supplied by heating. Thermal agitation might excite the helium until it was chemically active, but the hot helium atom is not near other excited helium atoms long enough to react. Any that do react have so much energy that the molecule is unstable and quickly dissociates. Thus helium rarely has compounds and is almost always found in the atomic state.

In carbon, the six electrons are all paired off and carbon is inert at room temperature. The eight oxygen electrons are similarly paired off (though oxygen easily forms O_2, since its last electron can be reversed if m_l becomes unity). If we hold a match to paper, the carbon in the paper is easily excited. Its electrons become unpaired so that it reacts with oxygen. If there is not much oxygen available, the product is carbon monoxide, CO. In this case, one electron of each atom reverses and the two unpaired electrons of each atom attract each other. In CO, each atom has a valence of two. If there is an abundance of oxygen, the reaction product is CO_2. In this case, two carbon electrons reverse, causing all four outer electrons to be unpaired. Each two of these four attracts an oxygen atom to the one carbon atom. In this case, carbon has a valence of four while oxygen still has a valence of two.

Although examples of this kind can be cited almost without limit, it must be admitted that we have only scratched the surface of the problem of molecular structure. But in showing that chemical forces between atoms are determined by their electronic structure, we have suggested an answer to the question raised by Rutherford's nuclear model of the atom. A rigid body is not a cluster of plum-pudding atoms stuck together with glue. A rigid body is mostly empty space. Most of the mass is concentrated in tiny positive nuclei, each of which is surrounded by electrons. These electrons are not uniquely associated with their home nuclei but are shared with neighboring nuclei of the same or different kinds. Thus electronic interactions provide the "glue" that holds matter together, while electrostatic repulsion keeps the nuclei apart.

4.19 THE STATUS OF BOHR'S MODEL AND THEORY OF THE ATOM

In this chapter, we have seen how Rutherford's discovery of the nucleus led Bohr to propose a model for the hydrogen atom and devise a theory that accounted for the spectral lines that hydrogen emits. Unfortunately, the Bohr model of the atom suffers from serious defects. In the first place it applies only to one-electron atoms, principally hydrogen. But even for this simple element, Sommerfeld found it necessary to postulate elliptical orbits and to consider the relativistic mass change (see Chapter 5) of the revolving electron in order to account for the details in the fine structure of the spectral lines. Attempts to apply Bohr theory to heavier elements were moderately successful in a few cases. The next heavier element, helium, has two electrons, but when one of these is removed by ionization, the remaining ion has but one electron and is hydrogen-like. The alkaline metals share with hydrogen the same column in the periodic table and their spectra are also hydrogen-like when ionized until only one orbital electron remains. But the Bohr model is not a general solution to the problem of atomic structure.

Furthermore, the Bohr model does not answer all the questions we can raise about hydrogen. A complete analysis should not only reveal the frequency of the emitted light, but also the *relative intensity* of the different spectral lines. In Chapter 8 we shall discuss quantum mechanics, which has replaced the Bohr analysis. However, we need not be dismayed with Bohr's work simply because it has been replaced by something better. Road maps are far less accurate than topographical survey maps. But on a cross-country trip, road maps are a better guide because the survey maps are cumbersome and contain details that are often irrelevant. The Bohr theory of hydrogen is like a road map that we use even when we know a more precise map exists.

Despite its shortcomings, the Bohr model was a conceptual "breakthrough" that facilitated many empirical observations about the electronic structure of the heavier atoms to which Bohr's work did not apply directly. We shall find Bohr's conceptual scheme very useful in describing the x-ray spectra of heavy elements. Although wave mechanics has replaced the Bohr model, wave mechanics confirms and builds on the energy-level concept that Bohr introduced.

FOR ADDITIONAL READING

Balmer, Johann J., "The Hydrogen Spectral Series," in W. F. Magie, *A Source Book in Physics*, pp. 360–365. Cambridge, Mass.: Harvard University Press, 1935.

Beyer, Robert T., *Foundations of Nuclear Physics*, New York: Dover, 1949, A collection of facsimiles of thirteen fundamental studies as they were originally reported in the scientific journals. Included is "The Scattering of α and β Particles by Matter and the Structure of the Atom" by Rutherford.

Bohr, Niels, *The Theory of Spectra and Atomic Constitution*. 2nd ed. Cambridge, England: University Press, 1924. Bohr's own account of the stages in the development of his theory of atomic structure.

Candler, Chris, *Atomic Spectra and the Vector Model*. 2nd ed. Princeton, N.J.: Van Nostrand, 1964.

Christensen, C. Paul, "Some Emerging Applications of Lasers," *Science* **218**, No. 4568 (October 8, 1982).

Christy, R. W., and Agnar Pytte, *The Structure of Matter*. New York: Benjamin, 1965.

Faller, J. E., and E. J. Wampler, "The Lunar Laser Reflector," *Scientific American* **38** (March 1970).

Friedman, F. L., and L. Sartori, *The Classical Atom*. Reading, Mass.: Addison-Wesley, 1965. An excellent account of the experiments and subsequent theory of the atom prior to quantum mechanics.

Goldenberg, H. M., D. Kleppner, and N. F. Ramsey, "Atomic Hydrogen Maser," *Phys. Rev. Letters* **5**, No. 361 (1960).

Gordon, J. P., "The Maser," *Scientific American* **42** (December 1958).

Herzberg, Gerhard, *Atomic Spectra and Atomic Structure*. 2nd ed. J. W. T. Spinks, translator. New York: Dover, 1944. A good basic book.

Hund, Friedrick, "Paths to Quantum Theory Historically Viewed," *Phys. Today* **19**, No. 8, 23 (August 1966).

King, G. W., *Spectroscopy and Molecular Structure*. New York: Holt, Rinehart, and Winston, 1964.

LaRocca, A. V., "Laser Applications in Manufacturing," *Scientific American* **246**, p. 94 (March 1982).

O'Shea, D. C., W. Russell Callen, and William T. Rhodes, *Introduction to Lasers and Their Applications*. Reading, Mass.: Addison-Wesley, 1977.

Pauling, Linus, *The Nature of the Chemical Bond*. 3rd ed. Ithaca, N.Y.: Cornell University Press, 1960.

Pennington, K. S., "Advances in Holography," *Scientific American* **40** (February 1968).

Richtmyer, F. K., E. H. Kennard, and T. Lauritsen, *Introduction to Modern Physics*. 5th ed. New York: McGraw-Hill, 1955. Chapter 6.

Schawlow, A. L., "Advances in Optical Masers," *Scientific American* **34** (July 1963).

Schawlow, A. L., "Laser Light," *Scientific American* **120** (September 1968).

Svelto, O., and D. C. Hanna, *Principles of Lasers*. New York and London: Plenum Press, 1976.

Thyagarajan, K., and A. K. Khatak, *Lasers, Theory and Applications*. New York and London: Plenum Press, 1981.

Zorn, J. C., *Experiments with Molecular Beams*, Molecular Beams, Vol. 1 and *Atomic and Molecular Beam Spectroscopy*, Molecular Beams, Vol. 2, Resource Letter MB-1 and Selected Reprints. New York: American Institute of Physics, 1965.

REVIEW QUESTIONS

1. What major conclusion resulted from Rutherford's alpha particle scattering experiment?
2. What are the Bohr postulates?
3. How did the Bohr model explain atomic spectra?
4. How can you use the idea of energy states to explain the operation of a laser?
5. How does light from a laser differ from light from a flashlight?
6. What are the conditions on the basic quantum numbers?
7. What is the major consequence of the Pauli exclusion principle?
8. How can you justify the following statement: "Most of the chair on which you are sitting is empty space"?

PROBLEMS

4.1 Derive the equation for the radii of the Bohr orbits, Eq. (4.16), from Eqs. (4.12) and (4.14).

4.2 Show that the tangential speed of an electron in its orbit is

$$v = \frac{Ze^2}{2\varepsilon_0 nh},$$

and its angular speed is

$$\omega = \frac{\pi m Z^2 e^4}{2\varepsilon^2 n^3 h^3}.$$

4.3 (a) Calculate the radii of the first, second, and third "permitted" electron orbits in hydrogen in angstroms. (b) What is the diameter of the hydrogen atom in the ground state?

4.4 The Bohr model for hydrogen shows that the orbital electron can be found only at certain fixed distances from the proton, the larger radii corresponding to higher quantum numbers. Assume that the electron in a hydrogen atom moves outward to larger radii. Which of the following quantities increase and which decrease: angular momentum, total energy, potential energy, kinetic energy, frequency of rotation?

4.5 (a) Calculate the force on the electron in the hydrogen atom due to the electrostatic field of the proton at the distance of the radius of the first Bohr orbit. (b) Calculate the force on an electron due to the gravitational attraction of a proton at this same distance. (c) What is the ratio of these forces?

4.6 The energy loss per unit time due to the acceleration of an electron is given by

$$P = \frac{e^2 a^2}{6\pi\varepsilon_0 c^3},$$

where a is the acceleration of the particle. (a) What would be the energy loss per unit time for an electron in the first Bohr orbit in a hydrogen atom if it were to lose energy by this process? (b) Estimate the time that it would take for the electron in the first Bohr orbit to spiral into the nucleus if it were to lose energy in this way.

4.7 A particle of mass m moves in a circular orbit of radius r under the influence of a force kr directed toward the center (k is a constant). Assuming that Bohr's postulates apply to this system, derive the equation for (a) the radii of the permissible orbits and (b) the energies of these orbits in terms of the quantum number n. (c) Show that the frequency radiated when the particle makes a transition from one orbit to the adjacent orbit is the same as the frequency of the circular motion.

4.8 Suppose one has a tank of atomic hydrogen at a temperature of 20°C and a pressure of 1 atm. If each of the atoms and also the mean free path were expanded until the diameter of each nucleus is 1 inch, what would then be (a) the radius of the smallest electron orbit in feet, and (b) the average distance traveled between collisions in miles? (Assume that the nuclear diameter is 3 fermis, and that the mean free path is 9×10^{-5} cm. This is four times the mean free path calculated for molecular hydrogen in Problem 1.35.) (c) What would be the mean free path on this expanded scale if the gas pressure were reduced to 1 mm of mercury? (d) If the gas were liquefied so that the atoms "touched," would there still be open space in the world of hydrogen atoms?

4.9 Calculate the binding energy of the electron in hydrogen in joules and in eV when $n = 1, 2, 3$, and infinity.

4.10 (a) Show that the average translational kinetic energy of a molecule of hydrogen gas at 27°C is much less than the energy required to raise a hydrogen atom from its ground state to its first excited state. (b) At what temperature will the average molecular kinetic energy of a molecule in hydrogen gas equal the ground-state binding energy of a hydrogen atom?

4.11 An electron and a proton are separated by a large distance and the electron approaches the proton with a kinetic energy of 2 eV. If the electron is captured by the proton to form a hydrogen atom in the ground state, what wavelength photon would be given off?

4.12 An alpha particle having a kinetic energy of 7.68 MeV is projected directly toward the nucleus of a copper atom. What is their distance of closest approach? The mass of an alpha particle is four times that of the proton and the atomic number of copper is 29.

4.13 Evaluate the Rydberg constant for hydrogen from atomic constants, assuming a nucleus of infinite mass.

4.14 Rearrange and alter the Balmer formula, Eq. (4.5), so that the left-hand side is wave number in reciprocal meters. Show that the result agrees with Eq. (4.23) when the Rydberg constant is substituted in the latter equation and when $n_1 = 2$ and $n_2 = n$.

4.15 Calculate (a) the frequency, (b) the wavelength, and (c) the wave number of the H_β line of the Balmer series of hydrogen. This line is emitted in the transition from $n_2 = 4$ to $n_1 = 2$. Assume that the nucleus has infinite mass.

4.16 Using the reduced mass equations, calculate the wavelength of the H_β line (see Problem 4.15) in the Balmer series for the three isotopes of hydrogen: ^1H; deuterium, ^2H; tritium, ^3H.

4.17 An atom of tungsten has all of its electrons removed except one. (a) Calculate the ground-state energy for this one remaining electron. (b) Calculate the energy and wavelength of the radiation emitted when this electron makes a downward transition from $n = 2$ to $n = 1$. (c) In what portion of the electromagnetic spectrum is this photon?

4.18 (a) Calculate the first three energy levels for the electron in Li^{++}. (b) What is the ionization potential of Li^{++}? (c) What is the first resonance potential for Li^{++}?

4.19 The Rydberg constants (reduced mass form) for hydrogen and singly ionized helium are 10967757.6 m^{-1} and 10972226.3 m^{-1}, respectively. For the nuclei of these atoms,

$$M_{He} = 3.9726 M_H.$$

From the given data, calculate the ratio of the mass of the proton to that of the electron to four significant figures.

4.20 Calculate the short wavelength limit of each of the series listed in Table 4.1, and find the energy of the quantum in eV for each.

4.21 Some of the energy levels of a *hypothetical* one-electron atom (*not* hydrogen) are listed in the table below.

n	1	2	3	4	5	∞
E_n, eV	−15.60	−5.30	−3.08	−1.45	−0.80	0

Draw the energy level diagram and find (a) the ionization potential, (b) the short wavelength limit of the series terminating on $n = 2$, (c) the excitation potential for the state $n = 3$, and (d) the wave number of the photon emitted when the atomic system goes from the energy state $n = 3$ to the ground state. (e) What is the minimum energy that an electron will have after interacting with this atom in the ground state if the initial kinetic energy of the electron was (1) 6 eV, (2) 11 eV?

4.22 For a certain *hypothetical* one-electron atom (*not* hydrogen), the wavelengths in angstroms for the spectral lines for transitions originating on $n = p$, and terminating on $n = 1$ are given by

$$\lambda = \frac{1500p^2}{p^2 - 1}, \qquad \text{where} \quad p = 2, 3, 4, \ldots .$$

(a) What are the least energetic and most energetic photons in this series? (b) Construct an energy level diagram for this element showing the energies of the lowest three levels. (c) What is the ionization potential of this element?

4.23 (a) What is the least amount of energy in eV that must be given to a hydrogen atom so that it can emit the H_β line (see Problem 4.15 and Fig. 4.4) in the Balmer series? (b) How many different possibilities of spectral line emission are there for this atom when the electron goes from $n = 4$ to the ground state?

4.24 A system of hydrogen atoms, all in the ground state, is bombarded by a beam of electrons. Through what minimum potential difference must the beam be accelerated in order that (a) only one spectral line is observed, and (b) only three spectral lines are observed?

4.25 In a certain gas discharge tube containing hydrogen atoms, electrons acquire a maximum kinetic energy of 13 eV. What are the wavelengths of all the lines that can be radiated?

4.26 If white light passes through a quantity of atomic hydrogen in the ground state, what wavelengths can be absorbed?

4.27 The energy levels in eV of a *hypothetical* one-electron atom (*not* hydrogen) are given by

$$E_n = -18.0/n^2, \qquad \text{where} \quad n = 1, 2, 3, \ldots .$$

(a) Compute the four lowest energy levels and construct the energy level diagram. (b) What is the excitation potential of the state $n = 2$? (c) What wavelengths in angstroms can be emitted when these atoms in the ground state are bombarded by electrons that have been accelerated through a potential difference of 16.2 V? (d) If these atoms are in the ground state, can they absorb radiation having a wavelength of 2000 Å? (e) What is the photoelectric threshold wavelength of this atom?

4.28 The frequencies ν of the spectral lines emitted by a certain *hypothetical* one-electron atom (*not* hydrogen) are given by the relation

$$\nu = 864 \times 10^{12}\left(\frac{1}{n_1^2} - \frac{1}{n_2^2}\right) Hz,$$

where the n's are the principal quantum numbers. (a) Find the wavelengths in angstroms of the first three lines of the series terminating on the ground state. (b) What is the photoelectric threshold wavelength of this atom? (c) Construct the energy level diagram. Give the values of the energies in eV of the first four levels and show, with labeled arrows, the transitions that cause the emission of the wavelengths in parts (a) and (b). (d) What is the binding energy of the electron when it is in the state $n = 3$? (e) State and clearly explain the possible interactions when a large number of these hypothetical atoms in the ground state are bombarded by (1) a beam of electrons having 2.90 eV of kinetic energy; (2) a beam of 2.90-eV photons.

4.29 Compute the magnitude of (a) the longest wavelength in the Lyman series and (b) the shortest wavelength in the Balmer series. (c) From these wavelengths determine the ionization potential of hydrogen.

4.30 The first ionization potential of helium is 24.6 V. (a) How much energy in eV and in joules must be supplied to ionize it? (b) To what temperature would an atmosphere of helium have to be heated so that an atom of it moving with the most probable speed in a Maxwellian distribution would have just enough kinetic energy of translation to ionize another helium atom by collision?

4.31 Neglecting reduced mass corrections, show that (a) the short-wavelength limit of the Lyman series ($n = 1$) of hydrogen is the same as that of the Balmer series ($n = 2$) of singly ionized helium, and (b) that the wavelength of the first line of this helium series is 1.35 times the wavelength of the first line of the Lyman series of hydrogen. (c) How much energy must be given to singly ionized helium in the ground state so that the first line in the Paschen series will be emitted?

4.32 (a) Show that the frequency of revolution of an electron in its circular orbit in the Bohr model of the atom is $\nu = mZ^2e^4/4\varepsilon_0^2n^3h^3$. (b) Show that when n is very large, the frequency of revolution equals the radiated frequency calculated from Eq. (4.22) for a transition from $n_2 = n + 1$ to $n_1 = n$. (This problem illustrates Bohr's *correspondence principle*, which is often used as a check on quantum calculations. When n is small, quantum physics gives results that are very different from those of classical physics. When n is large, the differences are not significant and the two methods then "correspond.")

4.33 A 10-kg satellite circles the earth once every 2 h in an orbit having a radius of 8000 km. (a) Assuming that Bohr's angular momentum postulate applies to satellites just as it does to an electron in the hydrogen atom, find the quantum number of the orbit of the satellite. (b) Show from Bohr's first postulate and Newton's law of gravitation that the radius of an earth-satellite orbit is directly proportional to the square of the quantum number, $r = kn^2$, where k is the constant of proportionality. (c) Using the result from part (b), find the distance between the orbit of the satellite in this problem and its next "allowed" orbit. (d) Comment on the possibility of observing the separation of the two adjacent orbits. (e) Do quantized and classical orbits correspond for this satellite? Which is the "correct" method for calculating the orbits?

4.34 On the average, an atom will exist in an excited state for a shake, 10^{-8} s, before it makes a "downward" transition and emits a photon. Assuming that the electron in hydrogen is in the state $n = 2$, how many revolutions about the nucleus are made before the electron "jumps" to the ground state?

4.35 What volume, in cubic miles, of *atomic* hydrogen at a pressure of 1 atmosphere and a temperature of 20°C is required so that, according to the Boltzmann distribution, it will contain one atom in the energy state $n = 2$?

4.36 (a) What would the temperature of the atmosphere of the sun have to be so that thermal agitation alone would put one millionth of the hydrogen atoms there into the necessary energy state for the absorption of the Balmer series wavelengths, that is, into the state $n = 2$? (b) What other means would be available in the atmosphere of the sun for pumping up hydrogen to this higher energy state? (c) Compare the answers to parts (a) and (b) with the results obtained in Problem 1.41 in which the temperature given is about the actual value for the atmosphere of the sun.

4.37 Consider an electron in the hydrogen atom moving in the first Bohr orbit. Calculate the magnetic field produced by this electron at the center of its orbit.

4.38 (a) Show that the magnetic moment of a circular Bohr orbit in hydrogen is given by $n(h/4\pi)(e/m_e)$. (The magnetic moment of a current-carrying loop of wire is equal to the product of the current and the area bounded by the loop.) (b) Calculate the magnetic moment of the orbit in hydrogen for which $n = 1$. (This particular value of the magnetic moment is called the Bohr magneton, μ_β.)

4.39 How can you account for the fact that lithium and sodium have similar chemical properties?

4.40 (a) Show that the maximum number of electrons that can be accommodated in a shell specified by the quantum number n is $2n^2$. (b) How many elements would there be if the electronic shells through $n = 7$ were completely occupied?

4.41 Show that the angular momentum has the same dimensions as Planck's constant.

4.42 On 20 July 1969 the astronauts of Apollo 11 placed a reflector on the surface of the moon that with the telescope and electronic equipment on earth allows measurements of the time of flight of laser pulses from earth to the moon and back. Scientists are able to measure the time of flight with an accuracy of one nanosecond (10^{-9} s). What is the accuracy of the distance-to-moon measurements?

4.43 A typical helium-neon laser emits light with $\lambda = 6328$ Å. How many photons per second would be emitted by a one milliwatt helium-neon laser?

4.44 A typical ruby laser emits radiation of $\lambda = 6940$ Å because of transitions between the energy levels of the impurity chromium ions. If a ruby 5 cm long and 1 cm in diameter contains 10^{19} of these ions per cubic cm, what is the maximum energy a pulse of radiation emitted by the laser could have? If the pulse lasts for 5×10^{-9} sec, what is the average power of the laser during the pulse?

4.45 As was discussed in the text, the angular spread of a laser beam is ideally diffraction limited. If so the spread would be given by $\theta_o = 1.22\lambda/D$ where θ_o can be considered the angle between a line parallel to the center of the beam and a line parallel to the "edge" of the beam, where λ is the wavelength and where D is the diameter of rod. For a ruby laser of diameter 1 cm and wavelength of 7000 Å, (a) what would be the radius of a spot projected on a screen at a distance of 1 km? (b) What would be the radius of a spot projected on the moon (distance to moon $\approx$ 240,000 miles)?

5 RELATIVITY

5.1 INTRODUCTION

Before we can continue the discussion of atomic theory, we need to obtain some of the results of Einstein's remarkable theories of relativity. This discussion will enable us to proceed to the topics of Compton scattering, pair production, radioactivity, and especially the behavior of charged particles in high energy accelerators. We shall begin our discussion with Einstein's special theory of relativity that he developed in 1905 to relate events in reference frames moving with uniform motion. We shall conclude the chapter with the discussion of the general theory of relativity, a theory Einstein published in 1917 for nonuniform motion and, as it turned out, a theory of gravitation.†

5.2 IMPORTANCE OF VIEWPOINT

What one observes often depends on one's viewpoint. An election will be interpreted differently by Democrats and Republicans. A change in college rules will be viewed differently by students and administrators. What scientists observe also depends on

†Some people wrongly believe Einstein's work replaced Newton's. Einstein's views were clearly expressed in 1948. "No one must think that Newton's great creation can be overthrown by this (relativity) or any other theory. His clear and wide ideas will forever retain their significance as the foundation on which our modern conceptions of physics have been built."

viewpoint. Consider the interpretations of physical events that might be made by a woman of high IQ born and brought up on a merry-go-round. She would experience a force somewhat like the force of gravity. It would be down at the center of the merry-go-round, and, because of the centrifugal component, it would be directed down and out at points away from the center. Our observer could learn to live happily with this force, and might even seek to write a quantitative description of it and express its nature in terms of mathematical equations. Though we express the force of gravity we experience as $F = mg$ (always down), our observer would have a harder time of it, since her force is not constant in either magnitude or direction. If our merry-go-round observers were to have the genius to devise a whole system of mechanics, the mechanics she would devise would not be the mechanics of Newton. Newton's mechanics, which incorporates the concept of inertia introduced by Galileo, includes the law that a moving body subject to no forces will continue to move with a constant speed in a constant direction. But an airplane that flew "straight" over the merry-go-round would appear to move in a curved path to our special observer.

The situation described appears farfetched, but in fact we were all born and brought up on a merry-go-round — the earth! The acceleration due to gravity is complicated by the rotation of the earth, and there are many other effects, like cyclonic storms, that are a direct result of the rotation. Newton knew of this complication. The earth turns rather slowly compared with a regular merry-go-round, and the effects of the earth's rotation are small enough that most of the time we can ignore them. But, strictly speaking, Newton's laws just do not hold precisely for an observer who is earthbound. For Newton's laws to be valid precisely, we must observe events from what is called an *inertial frame* or a Galilean–Newtonian coordinate system. Such a system is one that has no acceleration relative to a system of coordinates known to be fixed absolutely. This raises the question: What, *if anything*, constitutes a fixed system?

The whole structure of classical physics, then, is based on the assumption that we interpret all events as they would be interpreted by an observer whose viewpoint is in an inertial reference frame. Riders on a merry-go-round would have a physics far from Newtonian. By stepping onto the earth they would find this physics more nearly Newtonian. The genius of Newton is, in part, that although he never could step off the earth physically, he did step off the earth mentally. He interpreted events as though he had no acceleration. Because of this shift in his viewpoint, he was able to write his laws of mechanics in the particularly simple form that he did.

But Newton never really knew to where he projected himself. He excluded the earth as a vantage point because the earth not only rotates on its axis but revolves about the sun. The sun offered possibilities, but even the sun moves and is probably accelerated through space. The stellar constellations were named by the ancients and the stability of their arrangement led to their being called the "fixed" stars. Yet it would be the strangest of coincidences if the "fixed" stars really were fixed. Any motion looks small when seen from far enough away, and from modern astronomical measurement we know that the constellations are slowly changing.

It would seem, however, that if we locate a frame of reference so that it is fixed relative to the stars, this will be sufficiently steady for Newton's laws to serve well enough for every practical purpose. Such a frame is good enough for the practical people who want to fly aircraft, build rocket engines, or communicate via television. But for those whose primary concern is the understanding of the nature of things and whose goal is the discovery of truth, uncertainty about the frame of reference represented a serious flaw in the logical structure of classical physics. *It is an amazing fact that concern over this seemingly minor point led ultimately to the discovery of how to release the energy of the nucleus of the atom.*

5.3 THE SEARCH FOR A FRAME OF REFERENCE — THE ETHER

The search for something more fixed than the stars went something like this. As we pointed out in Chapter 3, James Clerk Maxwell demonstrated that electricity and light are related phenomena. Starting with known properties of electricity and magnetism, Maxwell derived equations that are identical in form to the equations that describe many wave phenomena. He could demonstrate, furthermore, that the velocity of the waves he discovered was the same as the velocity of light. He could derive many other properties of light, and it was soon accepted that he had put the wave theory of light on a firm foundation. In this theory, light is an electromagnetic wave motion.

Every wave motion has something that "waves." Sound waves have air and water waves have water. Surely, it was argued, light waves must involve the waving of something even in free space. No one knew what it was, but it was given the name "luminiferous ether," a light-transmitting medium.

Light passes through many kinds of materials. It passes through relatively heavy materials like glass and it passes through the nearly perfect vacuum that must lie between the stars and the earth. Thus ether must permeate all of space. Light is a transverse wave motion. This comes out of Maxwell's theory and from many experimental observations, particularly those on polarized light. This implies that the ether is a solid. Transverse waves involve shear forces and can occur only in solids that can support shear. Sound waves in air must be longitudinal because of this fact. Furthermore, the ether must be a rigid solid. The propagation velocity of mechanical waves in various materials depends on the elastic constants of the material. These are much greater for steel than for air. The very great velocity of light thus implies that the ether must have a very large shear modulus. It is rather hard on the imagination to suppose that all space is filled with this rigid solid and that all material objects move through this solid without resistance, yet it was supposed to exist. However fanciful it may seem to us, physicists felt that this ether might be just the thing to which to attach a Newtonian coordinate system. It was conceived that

Newton's laws would hold exactly for an observer moving without acceleration *relative to the ether.*[†]

If the ether is assumed to be at rest, then the interesting question is, How fast are we moving through the ether? Since all speculations about the ether stem from its properties as a medium for carrying light, an optical experiment is indicated. It is not hard to compute how sensitive the apparatus must be in order to measure the ether drift. Assuming, for the sake of argument, that the sun has no ether drift, the velocity of the earth through the ether must be its orbital velocity. If the sun has an ether drift, then the drift of the earth will be even greater than its orbital velocity at some seasons. Knowing that the radius of the earth's orbit is about 93 million miles, we can find the orbital velocity to be about 18.5 mi/s. By performing the experiment at the best season of the year, we know that we should be able to find an ether drift of at least 18.5 mi/s. The velocity of light is 186,000 mi/s. Great as our orbital velocity is, it is only about 10^{-4} times the velocity of light; so it is evident that a very sensitive instrument is required.

5.4 THE MICHELSON INTERFEROMETER

A device of sufficient sensitivity was made and used in the United States by Michelson and Morley in 1887. The principle of their apparatus is brought out by the following analogy. Suppose two equally fast swimmers undertake a race in a river between floats anchored to the river bed. Two equal courses, each having a total length $2L$, are laid out from the starting point, float A, as shown in Fig. 5.1. One course is AD, parallel to the flow of the river relative to the earth, and the other is AC, perpendicular to it. How will the times compare if each of the swimmers goes out and back on his or her course? Let the speed of each swimmer relative to the water be c, and let the water drift or velocity with respect to the earth be v. When the swimmer on the parallel course goes downstream, his velocity will add to that of the water, giving him a resultant velocity of $(c + v)$ with respect to the earth. The time required for him to swim the distance L from A to D is $L/(c + v)$. On his return, he must overcome the water drift. His net velocity then is $(c - v)$, and his return time is $L/(c - v)$. His total time is the sum of these two times. This is seen to depend on the velocity of the water, and is given by

$$t_{\parallel} = \frac{L}{c+v} + \frac{L}{c-v} = \frac{2Lc}{c^2 - v^2}. \tag{5.1}$$

[†]Some physicists must have had doubt about the existence of the ether. Dr. F. G. Swann (1884–1962), a lively speaker, while discussing the contradictions of the existence of the ether stated that "the luminiferous ether is a hypothetical and tenuous medium by means of which physicists propagate their misconceptions from point to point in space."

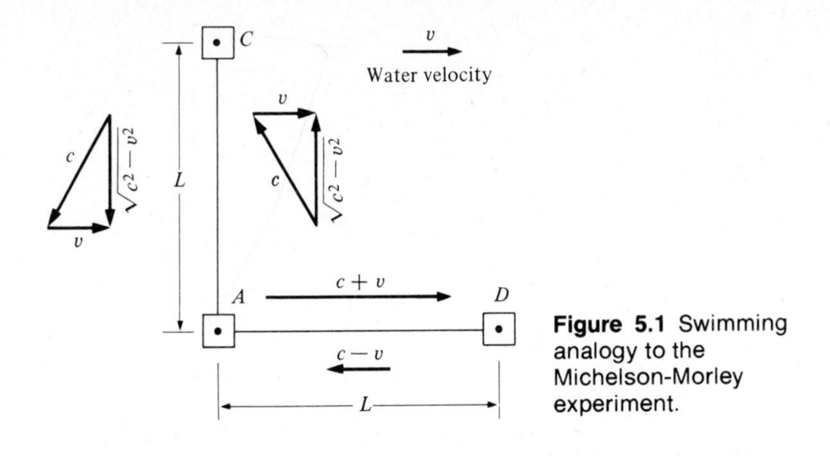

Figure 5.1 Swimming analogy to the Michelson-Morley experiment.

The other swimmer, going perpendicular to the water drift, spends the same time on each half of her trip, but she must head upstream if she is not to be carried away by the current. The component of her velocity that carries her toward her goal is $\sqrt{c^2 - v^2}$ with respect to the earth. The total time for her trip also depends on the water drift, and is

$$t_{\perp} = \frac{2L}{\sqrt{c^2 - v^2}}. \tag{5.2}$$

To see how these two times compare, we divide the parallel course time, Eq. (5.1), by the perpendicular course time, Eq. (5.2), and obtain

$$\frac{t_{\parallel}}{t_{\perp}} = \frac{2Lc}{c^2 - v^2} \cdot \frac{\sqrt{c^2 - v^2}}{2L} = \frac{1}{\sqrt{1 - (v^2/c^2)}}. \tag{5.3}$$

In still water $v = 0$, the ratio of the times is unity, and the race is a tie, as we would expect. In slowly moving water, the ratio is greater than unity and the swimmer on the perpendicular course wins; or, put differently, if the swimmers are stroking in phase when they leave float A, they will be out of phase when they return to it. If the velocity of the river increases to nearly that of the swimmers, then the ratio tends toward infinity. If the river velocity exceeds the swimmer velocity, the entire analysis breaks down. The ratio becomes imaginary and both swimmers are swept off the course by the current. The point is that, by observing the race, the water velocity relative to the anchored floats can be measured.

The optical equivalent of the above situation is to have a race between two light rays over identical courses, one parallel and one perpendicular to the ether drift. The instrument used is shown schematically in Fig. 5.2.

Let us follow a ray of light that enters the apparatus from an extended source at the left. At A it is incident on a glass plate that is half-silvered on its right side. This

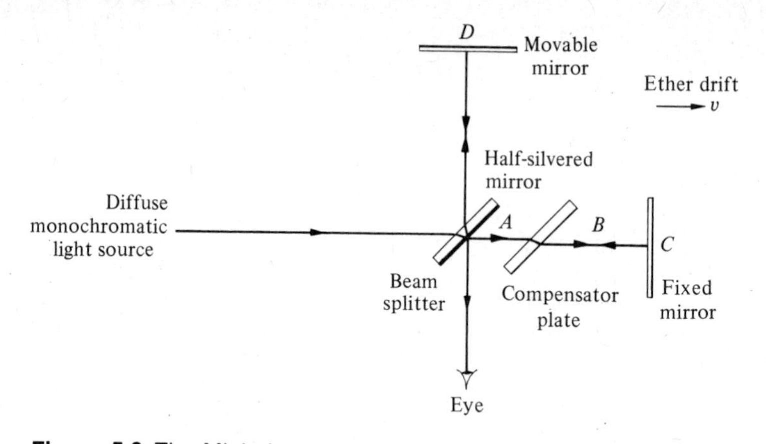

Figure 5.2 The Michelson apparatus.

surface reflects half of the light up toward D, while the other half goes on to C. Both C and D are full-silvered, front-surface mirrors that reflect their beams back toward A. The beam from D is partly reflected at A, and the remainder goes on through to the eye of the observer. A portion of the beam from C is reflected at A to the observer, and the rest goes through the glass plate and is lost. The plate of glass at B has the same thickness and inclination as that at A, so that the two light paths from source to observer pass through the same number of thicknesses of glass. If the light from the source did not diverge and remained very narrow in going through the apparatus, the observer would see a line of light. The brightness of this line would depend on the difference in the optical length† of the two light paths. If these differed by any whole number of wavelengths of the light (including zero), the line would be bright. If the paths differed by an odd number of half-wavelengths, then the line would be dark. Between these extremes every brightness gradation would be observed. In practice, the light does diverge in the apparatus, and there are a great many slightly different paths being traversed simultaneously. Consequently, the observer does not see one line but a multiplicity of lines. The loci of points where the paths differ by whole wavelengths are bright, and where the paths differ by an odd number of half-wavelengths there is darkness. Thus, as one path length is varied, the observer sees fringes, like the teeth of a comb, move across the field, rather than a single line becoming lighter and darker. It is fortunate that the optical system works as it does, since it is easier for the eye to detect differences in position than differences in intensity.

The precision of this device is remarkable. If yellow light from sodium is used, the wavelength is 5.893×10^{-7} m. Moving the mirror C away from A one-half this

† Two paths have the same optical length if light traverses both in the same time. The optical lengths of the interferometer paths can be changed by changing their physical length, by changing the index of refraction of the region through which the light passes, or, if the swimming analogy applies, by moving the apparatus relative to the light-carrying medium.

distance will increase one path length by a whole wavelength and cause the pattern to move an amount equal to the separation of two adjacent dark lines. If we can estimate to hundredths of fringes, then the smallest detectable motion is only 2.9×10^{-9} m. Upon moving a mirror 0.1 mm, 340 fringes would go by. (One way of defining a meter is to say that it is the length of 1,650,763.73 wavelengths in vacuum of the orange-red line of krypton-86.)

The similarity between the apparatus for the Michelson experiment and the swimming race should be evident. Light corresponds to the swimmers and has the free-space velocity, c, with respect to its ether medium. The ether drift corresponds to the water current drift and has the velocity v with respect to the earth. Just as we could learn about the river flow by seeing the outcome of the swimmers' race, so we wish to measure the ether drift by conducting a "light race" over equal paths parallel and perpendicular to the ether drift.

Suppose that instead of taking the ratio of the times for the two paths of the river race we now take their difference; then

$$\Delta t = \frac{2Lc}{c^2 - v^2} - \frac{2L}{\sqrt{c^2 - v^2}} = \frac{2L}{c}\left[\left(1 - \frac{v^2}{c^2}\right)^{-1} - \left(1 - \frac{v^2}{c^2}\right)^{-1/2}\right]. \quad \textbf{(5.4)}$$

Using the first two terms of the binomial expansion, we have, to a good approximation if $v \ll c$, that

$$\Delta t = \frac{2L}{c}\left[\left(1 + \frac{v^2}{c^2}\right) - \left(1 + \frac{v^2}{2c^2}\right)\right] = \frac{Lv^2}{c^3}. \quad \textbf{(5.5)}$$

In the interferometer, the time difference should appear as a fringe shift from the position the fringes would have if there were *no* ether drift. The distance light moves in a time Δt is $d = c\,\Delta t$ and if this distance represents n waves of wavelength λ, then $d = n\lambda$. Therefore the fringe shift would be

$$n = \frac{Lv^2}{\lambda c^2}. \quad \textbf{(5.6)}$$

Thus if the light race is carried out with light of speed c and wavelength λ in an interferometer whose arms are of length L, one of which is parallel to the ether drift of velocity v, then Eq. (5.6) gives the number of fringes that should be displaced because of the motion of the earth through the ether compared with their positions if the earth were *at rest* in the ether.

5.5 THE MICHELSON–MORLEY EXPERIMENT

The apparatus used was large and had its effective arm length increased to about 10 m by using additional mirrors to fold up the path (see Fig. 5.3). The entire apparatus was floated on mercury so that it could be rotated at constant speed

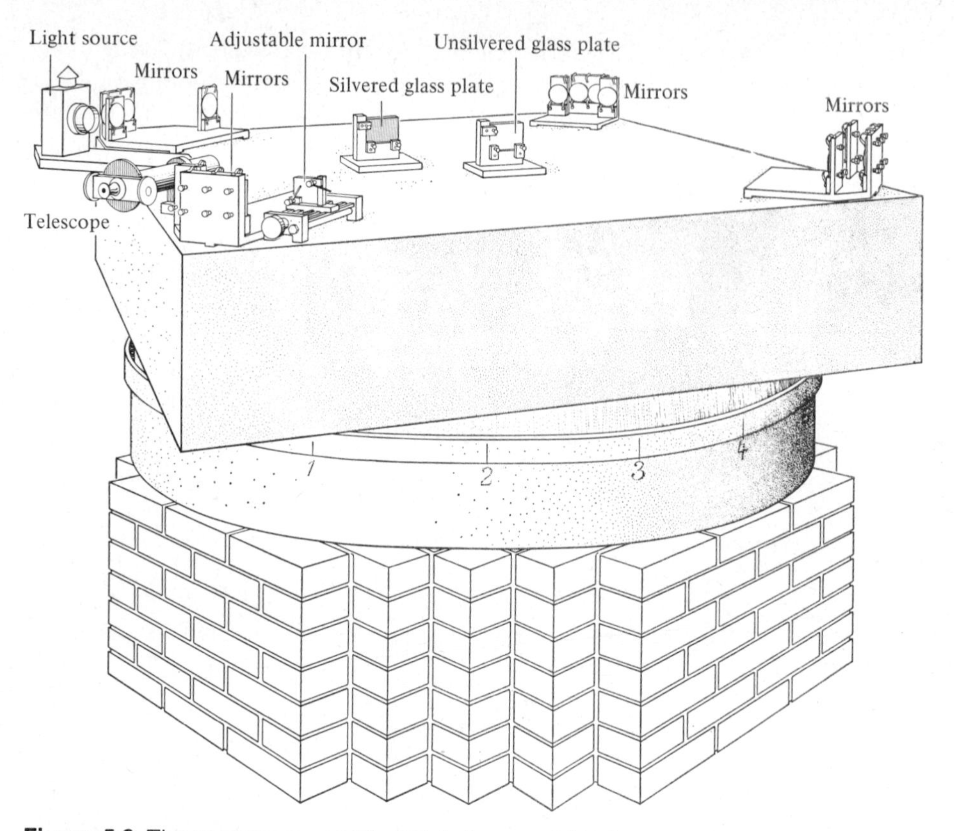

Figure 5.3 The apparatus used by Michelson and Morley.

From "Michelson-Morley Experiment" by R. S. Shankland.

without introducing strains that would deform the apparatus. *Rotation was necessary* in order to make the fringes shift, and by rotating through 90°, first one arm and then the other could be made parallel to the drift, thereby *doubling* the fringe displacement of Eq. (5.6). We can now estimate whether this instrument should be sensitive enough to detect the ether drift. Recall that at some time of the year, the ether drift v was expected to be at least the orbital velocity of the earth, which is about 10^{-4} c. Thus we expect v/c to be at least 10^{-4}. Using light of wavelength 5×10^{-7} m, the computed shift is $\Delta n = 0.4$ fringe. Michelson and Morley estimated that they could detect a shift of one-hundredth of a fringe. Sensitivity to spare!

Measurements were made over an extended period of time at all seasons of the year, but no significant fringe shift was observed. Thinking that the earth might drag a little ether along with it just as a boat carries a thin layer of water when it glides, Michelson and Morley took the entire apparatus to a mountain laboratory in search of a site that would project into the drifting ether. Again a diligent search failed to measure an ether drift. The experiment "failed."

5.6 STELLAR ABERRATION

One interpretation of the negative results of the Michelson–Morley experiment could be that there is an ether but it is centered on the earth; in other words, it moves with the earth in its motion through space. An argument against this interpretation is the phenomenon of stellar aberration. Stellar aberration was first observed in 1727 by James Bradley, an English astronomer, who noted that the stars appear to have a circular motion in the sky with a period of one year. This phenomenon is the apparent displacement of a star in the direction of the earth's motion in its orbit around the sun. The angular diameter of this apparent stellar motion is 41 seconds of arc.

This observation can be understood with the assistance of Fig. 5.4. Figure 5.4(a) shows the alignment of a telescope with a star near the zenith if the earth and, therefore, the telescope are not moving relative to the star. In this case, obviously, the telescope is vertical. If the earth and telescope are moving relative to the star and the telescope is held in the vertical position, the path of light down the tube will be displaced causing the image to miss the eyepiece of the telescope. The reason for this displacement is that the telescope has moved during the time it takes the light to travel from the top of the telescope to the eyepiece. To adjust for this displacement of the image and thus cause the light to fall on the eyepiece, the telescope must be tilted in the direction of motion as shown in Fig. 5.4(b). This tilting causes the image of the star to be displaced away from the actual star in the direction of the earth's motion. The angle θ that the telescope must be tilted can be obtained from

$$\text{tangent } \theta = \frac{vt}{ct} = \frac{v}{c}.$$

The velocity of the earth in its orbit around the sun is approximately 30 km per

Figure 5.4 Aberration of starlight.

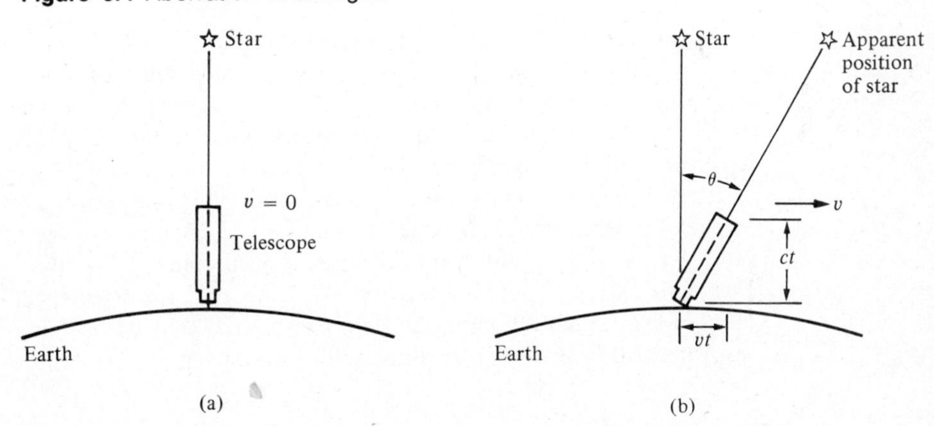

second; therefore,

$$\text{tangent } \theta = \frac{30 \times 10^3 \text{ m/s}}{3 \times 10^8 \text{ m/s}} = 10^{-4}.$$

Since for small angles the tangent θ is approximately equal to the angle, $\theta \simeq 10^{-4}$ radian, which is about 20.5 seconds of arc. Twice this value would be the circular diameter expected for aberration and is in agreement with the measured value of 41 seconds.

An important conclusion from the observation of the stellar aberration is that the ether, if it were to exist, is not dragged around with the earth. If the ether were at rest with respect to the telescope, the light would be dragged along with the ether. In this case, the ether in Fig. 5.4(b) would be moving to the right with velocity v and the light would be pulled along with it so there would be no need to correct for the earth's motion by tilting the telescope. The Michelson–Morley experiment thus can be interpreted as showing that there is no ether.

5.7 FITZGERALD–LORENTZ CONTRACTION

Few experimental "failures" have been more stimulating than this. The negative result of the Michelson–Morley experiment presented a challenge to explain its failure. Fitzgerald proposed an ad hoc explanation in an attempt to save the idea of an ether. He pointed out that there might be an interaction between the ether and objects moving relative to it such that the object became shorter in all its dimensions parallel to the relative velocity. Recall that in the flowing-river analogy, the ratio of the times of the swimmers in Eq. (5.3) was

$$\frac{1}{\sqrt{1 - v^2/c^2}}.$$

If the route parallel to the flow had been shorter by this factor, then the ratio of the times would have been one and the race would have been a tie. A similar shortening of the parallel arm would account for the tie race Michelson and Morley always observed. The shortening could never be measured because any rule used to measure it would also be moving relative to the ether and would shorten also. Lorentz developed the mathematical expressions for the Fitzgerald explanation of the Michelson–Morley experiment. His formulas became a part of the Einstein theory of relativity, although with very different interpretation. Whether you accept the Fitzgerald–Lorentz contraction hypothesis or not, the Michelson–Morley experiment indicates that all observers who measure the velocity of light will get the same result regardless of their own velocity through space.

5.8 THE CONSTANT SPEED OF LIGHT

Speed trials of cars, boats, and airplanes are never official unless they are made in the following way. The record contender must drive the craft in opposite directions over a measured course and the speed attained is calculated to be the double distance divided by the total time spent between markers. This technique is used to make sure that any wind, water, or other conditions that may be helpful in one direction are canceled out by their hindrance on the reverse trip. Measurements of the speed of light are similarly made by timing a flash of light as it goes to a distant mountain and returns. This technique is rather good and, *to a first approximation*, the influence of a moving medium does cancel out. But the cancellation is not perfect and *to a second approximation*, the effect of a moving medium does *not* cancel. This becomes obvious if the medium moves faster than the speed under test. In this case, the test cannot be made, since the thing tested is carried away. The fact that Fizeau, in France, and others obtained consistent results for the velocity of light at a variety of times, places, and in different directions was in itself evidence that the speed of the earth through the ether (if any) was small compared with the velocity of light. It is highly significant that the Michelson–Morley interferometer was sensitive enough to detect the second-order term. Referring to Eq. (5.5), you will note that the first terms (the ones) cancel out and the significant result remains only because the second terms do not cancel. The Michelson–Morley experiment was sensitive to the second-order terms because, instead of trying to measure the times of transit of light through their apparatus, they measured the *difference* of times. Michelson and Morley found that the speed of the earth through space made *no difference* in the speed of light relative to them. The inference is clear: either the earth moves in some way through the ether space more slowly than it moves about the sun, or *all observers must find that their motion through space makes no difference in the speed of light relative to them.*

The consideration of electromagnetic phenomena led Einstein to conclude that the speed of light does not depend on the motion of the observer and that there is no preferred reference frame for the laws of physics. We quote the beginning of his first paper on relativity, which is titled "On the Electrodynamics of Moving Bodies."†

> It is known that Maxwell's electrodynamics — as usually understood at the present time — when applied to moving bodies leads to asymmetries which do not appear to be inherent in the phenomena. Take, for example, the reciprocal electrodynamic action of a magnet and a conductor. The observable phenomenon here depends only on the relative motion of the conductor and the magnet,

† Excerpt from A. Einstein, "Zur Electrodynamik bewegter Körper," *Annalen der Physik* **17**, 891 (1905), translated by W. Perrett and G. B. Jeffery, in *The Principle of Relativity* by Einstein, Lorentz, Minkowski, and Weyl. New York: Dover.

> whereas the customary view draws a sharp distinction between the two cases in which either the one or the other of these bodies is in motion. For if the magnet is in motion and the conductor at rest, there arises in the neighborhood of the magnet an electric field with a certain definite energy, producing a current at the places where parts of the conductor are situated. But if the magnet is stationary and the conductor in motion, no electric field arises in the neighborhood of the magnet. In the conductor, however, we find an electromotive force, to which in itself there is no corresponding energy, but which gives rise — assuming equality of relative motion in the two cases discussed — to electric currents of the same path and intensity as those produced by the electric forces in the former case.
>
> Examples of this sort, together with the unsuccessful attempts to discover any motion of the earth relatively to the "light medium," suggest that the phenomena of electrodynamics as well as of mechanics possess no properties corresponding to the idea of absolute rest. They suggest rather that, as has already been shown to the first order of small quantities, the same laws of electrodynamics and optics will be valid for all frames of reference for which the equations of mechanics hold good. We will raise this conjecture (the purport of which will hereafter be called the "Principle of Relativity" to the status of a postulate, and also introduce another postulate, which is only apparently irreconcilable with the former, namely, that light is always propagated in empty space with a definite velocity c which is independent of the state of motion of the emitting body. These two postulates suffice for the attainment of a simple and consistent theory of the electrodynamics of moving bodies based on Maxwell's theory for stationary bodies. The introduction of a "luminiferous ether" will prove to be superfluous inasmuch as the view here to be developed will not require an "absolutely stationary space" provided with special properties, nor assign a velocity-vector to a point of the empty space in which electromagnetic processes take place.
>
> The theory to be developed is based — like all electrodynamics — on the kinematics of the rigid body, since the assertations of any such theory have to do with the relationships between rigid bodies (systems of coordinates), clocks, and electromagnetic processes. Insufficient consideration of this circumstance lies at the root of the difficulties which the electrodynamics of moving bodies at present encounters.

It should be noted at this point that the theory of relativity (both the special and general theories that are to be discussed) are just theories that are consistent with all experiments performed to date. For example, the special theory of relativity is only one explanation for the results of the Michelson–Morley experiment. It does not follow logically or necessarily from the results of the experiment and other interpretations are possible. With that comment we shall proceed to discuss the special and then the general theory of relativity, each being the most accepted theory to explain the appropriate experiments performed.

5.9 CLASSICAL RELATIVITY

Let us first consider the pre-Einstein relativity of physical quantities in classical or Galilean–Newtonian physics, and ask how events in one system, S, moving with constant linear velocity, appear from another system S', also moving with constant linear velocity.

No generality will be lost if one of the systems, say S, is regarded as being at rest, and the other, S', as moving with a uniform velocity v. Our problem, then, is like that of comparing the observations of someone on the ground with those of someone on a uniformly moving train.

In Fig. 5.5, the earth observer considers himself to be at the origin O of his system S and he chooses an x-axis parallel to the track. The train observer in S' likewise measures distances from himself at O' and chooses his x'-axis parallel to the track. Let us measure time from the moment when the two observers are exactly opposite each other. Suppose that at some later time t each observer decides to measure the separation of two birds, B_1 and B_2, which happen to be hovering over the track. The observer at O' observes that the positions of B_1 and B_2 are the small distances x_1' and x_2'. The observer at O finds the positions of B_1 and B_2 are the larger distances x_1 and x_2. The observer at O can compute the O' observations by noting that the observer at O' has moved a distance vt, with the result

$$x_1' = x_1 - vt \quad \text{and} \quad x_2' = x_2 - vt. \tag{5.7}$$

The observer on the train (who may think the train is at rest with the earth moving under it) can account for the difference in their observations by observing that O has

Figure 5.5 Coordinate systems S and S'.

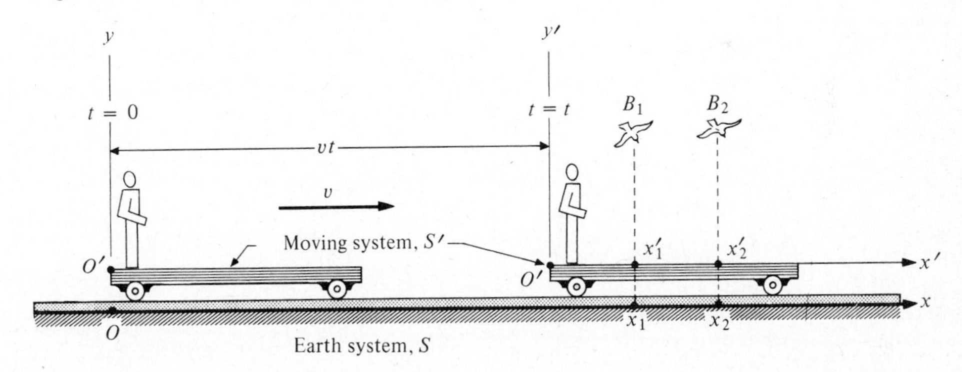

drifted away from him a distance vt. He obtains

$$x_1 = x_1' + vt \quad \text{and} \quad x_2 = x_2' + vt. \tag{5.8}$$

These are transformation equations in that they transform observations from one system to the other. By solving either set of equations for the separation of the birds, we get

$$x_2 - x_1 = x_2' - x_1' \quad \text{or} \quad \Delta x = \Delta x'. \tag{5.9}$$

This shows that the two observers agree on how far apart the two birds are. Note that the relative velocity v of the observers need not be known. Similarly, the time t since the two observers were opposite each other need not be known, but all observations must be *simultaneous* — even if the birds cooperate by staying the same distance apart.

What we have just shown is that *distance* or length is an *invariant* quantity when transformed from one Newtonian coordinate system to another.

We now compare results by the same two observers for a velocity measurement. Suppose that one bird flies from position B_1 at time t_1 to position B_2 at time t_2. The transformation equations of the positions of the bird are of the same form except for the difference of time, thus

$$x_1' = x_1 - vt_1, \qquad x_2' = x_2 - vt_2. \tag{5.10}$$

To solve for the average observed velocity u', we take the difference between the two transformation equations and divide by the time interval, obtaining

$$\begin{aligned} x_2' - x_1' &= (x_2 - vt_2) - (x_1 - vt_1) \\ &= (x_2 - x_1) - v(t_2 - t_1), \end{aligned} \tag{5.11}$$

or

$$\Delta x' = \Delta x - v\,\Delta t. \tag{5.12}$$

Therefore the velocity relation is

$$u' = \frac{\Delta x'}{\Delta t} = \frac{\Delta x}{\Delta t} - v = u - v. \tag{5.13}$$

Thus the *velocities* measured by the two observers are *not* the same. They are *not invariant* under a transformation between Galilean–Newtonian coordinate systems.

Suppose the bird again cooperates and swoops over the train with different velocities at positions 1 and 2. Each observer could measure the velocity at each position. Using the result just derived, we find that the velocity transformation equations are

$$u_1' = u_1 - v \quad \text{and} \quad u_2' = u_2 - v. \tag{5.14}$$

If the observers also measured the time Δt it took the bird to get from one position to the other, they could solve for the average acceleration. Taking the difference between the velocity transformations gives

$$u_2' - u_1' = u_2 - u_1. \tag{5.15}$$

Dividing by Δt, we have $a' = a$, which shows that velocity difference and *acceleration* are *invariant* under transformation between Newtonian inertial frames of reference.

The transformation equations we have just derived formally are fairly obvious. The most complicated result, the *variant* velocity transformation, is used without formal proof in the study of impact and Doppler effect, and was used in our recent discussion of the Michelson–Morley experiment. In fact, the problem presented by the negative result of that experiment was that, contrary to the Newtonian transformation, observers *must* find the velocity of light the *same* whether the observers are moving or not.

The classical transformation equations just derived apply only for reference frames moving with *constant linear velocity* with respect to one another. It is their failure in cases involving uniform motion that leads us into Einstein's special theory of relativity. We have not treated the classical transformation equations that deal with accelerated or rotating frames of reference, since these are much more complicated.

In the derivations above, we have tacitly made the *classical assumption* that time intervals are the same for all observers. Actually, there is no a priori reason for assuming that time or any other physical quantity is invariant under a transformation of coordinates.[†] Whether or not an assumption is correct is determined solely by the experimental verification of the results predicted with its aid. We will find later in this chapter that the invariance of the time interval will have to be abandoned, when the invariance of the free-space velocity of light is assumed.

5.10 EINSTEINIAN RELATIVITY

In the special theory of relativity, proposed in 1905,[‡] Einstein treats problems that arise when one frame of reference moves with a *constant linear velocity* relative to another. This theory is sufficient for discussing most atomic and nuclear phenomena. It contains much "uncommon sense" and many ideas that tease the imagination. But the mathematics is simple enough that we can derive several interesting

[†] An a priori reason is one deduced from previous assumptions or known causes.

[‡] This work was published in a 1905 volume of *Annalen der Physik* and was one of five papers by Einstein appearing in that volume, which led Max Born to say, "One of the most remarkable volumes in all of scientific literature is Volume 17, Series 4 of *Annalen der Physik*, 1905." Another physicist, Witkowski, after reading the special relativity article, said "A new Copernicus has been born."

relationships that are basic to an understanding of the material that follows.

The Michelson–Morley experiment was carried out to measure the velocity of the earth through the ether in the hope that the ether would provide a fixed frame of reference relative to which Newton's laws would hold exactly. Einstein assumed that all experiments designed to locate a fixed frame of reference would fail. Since he assumed that a fixed frame of reference could never be found, he went to the other extreme and postulated that the laws of physics should be so stated that they apply relative to any frame of reference.

If the laws of physics took different forms for different moving observers, then the different forms of the laws would provide a method for determining the nature of an observer's motion. This would take us back to the Newtonian view that there is a unique frame of reference for which the laws of physics are Newton's laws. Going back to our observer on the merry-go-round, it is hard to see how the laws of motion could ever seem the same to her as they would seem to a person standing on the earth. However, this is a problem in general relativity. For special or restricted relativity, we need only face the problem of making the laws of physics take the same form for all observers *translating* relative to each other with *constant* velocity. Limiting ourselves to special relativity, we may state its two postulates as follows:

1. *The laws of physics apply equally well for any two observers moving with constant linear velocity relative to each other or, in other words, the observations on one reference frame are not preferred above those on any other.*
2. *All observers must find the same value of the free-space velocity of light regardless of any motion they may have.*

5.11 RELATIVISTIC SPACE-TIME TRANSFORMATION EQUATIONS

The equations we are about to derive were first obtained by Lorentz as he successively refined electromagnetic theory to conform with the results of experiments. As stated earlier, both Fitzgerald and Lorentz obtained an ad hoc explanation of the negative result of the Michelson–Morley experiment by supposing that the interferometer shortened along its velocity vector through the ether. But we shall derive the relativistic transformation equations as consequences of Einstein's two postulates.

We return to our two observers at O and O' in systems S and S'. We ask the observer at O' to go back and pass the observer at O again. This time — just as O is opposite O' — we fire a photographic flashbulb. Thus at time $t = 0$, each observer is at the source of a spherical light wave. If we endow each observer with a supernatural power so that each can "see" the light spread out into space, then *each* observer must feel that *he* is at the center of the growing sphere of light. This must be the case, since we are now imposing the condition that the velocity of light is the same for all observers even if the velocity of an object is not.

The equation of the expanding sphere seen by the observer at O is

$$x^2 + y^2 + z^2 = c^2t^2, \tag{5.16}$$

where c is the velocity of light. Similarly, the observer at O' writes for his equation of the sphere

$$x'^2 + y'^2 + z'^2 = c^2t'^2. \tag{5.17}$$

Clearly, the relativistic transformation equations must be different from the classical ones if both observers are *each* to seem to be at the center of the same sphere.

As we consider what the new transformation equations are to be, we find that we are somewhat limited. These new equations must be linear. Any quadratic equation has two solutions and higher-order equations have more solutions. Surely any observations from the system S must have a unique interpretation in the system S'. There must be a "one-to-one" correspondence between what each observer "sees." The transformation equations must be linear in the space coordinates and in the times. Since the classical transformation equations were found to be $x' = x - vt$ and $x = x' + vt'$, let us here assume the next simplest linear equations,

$$x' = \gamma(x - vt) \qquad \text{and} \qquad x = \gamma'(x' + vt'), \tag{5.18}$$

where the γ's are transformation quantities to be determined. Note that our notation t and t' admits the possibility that time intervals may not be identical for the two observers. Note further that γ and γ' must be equal if, as required by Einstein's first postulate, there is to be no preferred reference system. Hereafter we shall let γ represent both γ and γ'.

We now seek to transform Eq. (5.17) so that the observer at O' "sees" what the observer at O would see. Mathematically, this requires that we eliminate x', y', z', and t' by using the assumed transformation equations. The coordinates y, y', z, and z' are perpendicular to the relative velocity of the observers, so that $y' = y$ and $z' = z$.† We have assumed a transformation equation for x', but we need some manipulation before we can eliminate t'. To get t' in terms of unprimed quantities, we work with our two assumed transformation equations in Eq. (5.18). Solving the second for t' and using the first to eliminate x', we obtain

$$t' = \gamma\left[\frac{x}{v}\left(\frac{1}{\gamma^2} - 1\right) + t\right]. \tag{5.19}$$

We can now transform the observation from O' into the observation from O by

† If this conclusion seems hasty, one can write $y' = \gamma_y(y - v_y t)$. Since the only relative velocity is along the x-axis, we have $v_y = 0$. This leaves $y' = \gamma_y(y)$, which we could carry into our subsequent derivation. If we did so, it would be obvious immediately that $\gamma_y = 1$. A similar argument applies to the z-transformation. In order to make these simplifications in our derivations, we assumed the relative velocity to be parallel to the x-axis.

eliminating the primed quantities from Eq. (5.17). We obtain

$$\gamma^2(x - vt)^2 + y^2 + z^2 = c^2\gamma^2\left[\frac{x}{v}\left(\frac{1}{\gamma^2} - 1\right) + t\right]^2. \tag{5.20}$$

Upon expanding and collecting terms, we have

$$\left[\gamma^2 - \frac{c^2\gamma^2}{v^2}\left(\frac{1}{\gamma^2} - 1\right)^2\right]x^2 - \left[2v\gamma^2 + \frac{2c^2\gamma^2}{v}\left(\frac{1}{\gamma^2} - 1\right)\right]xt + y^2 + z^2$$
$$= \left[c^2\gamma^2 - v^2\gamma^2\right]t^2. \tag{5.21}$$

This equation represents the expanding sphere of light as seen by the observer at O' whose observations have been interpreted (transformed) by the observer at O. This equation must be identical to the direct observation of the expanding sphere as seen from O, namely Eq. (5.16). For this to be the case, it is necessary that the quantities within brackets in Eq. (5.21) equal 1, 0, and c^2, respectively.†

Since the quantity within the third pair of brackets appears the simplest, we set it equal to c^2, obtaining

$$c^2 = c^2\gamma^2 - v^2\gamma^2 = \gamma^2(c^2 - v^2). \tag{5.22}$$

From this we get

$$\gamma^2 = \frac{c^2}{c^2 - v^2}$$

or

$$\gamma = \frac{1}{\sqrt{1 - v^2/c^2}}. \tag{5.23}$$

We choose the positive value of the square root, so that relativistic and classical physics correspond at low velocities.

Setting the quantity in the second pair of brackets in Eq. (5.21) equal to zero and solving for γ, we again get the same result. This is encouraging. It appears possible that the transformation we have assumed can indeed allow each observer to consider himself to be at the center of the expanding sphere of light. The final check is to determine whether the value of γ we have obtained makes the coefficient of x^2 become unity. Upon substituting for γ, we find that the quantity within the first pair

† In the $x = 0$ plane, the sphere becomes a circle, so that the sphere of Eq. (5.16) degenerates to $y^2 + z^2 = c^2t^2$ and the same sphere described by Eq. (5.21) degenerates to $y^2 + z^2 = [c^2\gamma^2 - v^2\gamma^2]t^2$. Since these equations describe the same circle, they must be identical, and to be identical, the quantity within the brackets must equal c^2.

of brackets becomes

$$\gamma^2 - \frac{c^2\gamma^2}{v^2}\left(\frac{1}{\gamma^2} - 1\right)^2 = \frac{c^2}{c^2 - v^2} - \frac{c^2c^2}{v^2(c^2 - v^2)}\left(\frac{c^2 - v^2}{c^2} - 1\right)^2$$

$$= \frac{c^2}{c^2 - v^2} - \frac{c^4}{v^2(c^2 - v^2)}\left(\frac{c^2 - v^2 - c^2}{c^2}\right)^2$$

$$= \frac{c^2}{c^2 - v^2} - \frac{v^2}{c^2 - v^2} = 1.$$

The consistency of these results means that mathematically it is possible to devise linear transformation equations that permit the transformation of one velocity to be invariant. We have chosen that one velocity to be the free-space velocity of light by assuming that our two observers are "watching" an expanding light wave. We need modify only the classical transformation equations by the inclusion of $\gamma = 1/\sqrt{1 - v^2/c^2}$ and note that time, which was the same for all Newtonian observers, must now be transformed along with the space coordinates. Knowing γ, we can simplify Eq. (5.19) to

$$t' = \gamma\left[t - \frac{vx}{c^2}\right]. \tag{5.24}$$

We can now summarize both classical and relativistic transformation equations for two observers having a relative velocity v parallel to the x-axis as follows:

Galileo–Newton (classical)	Lorentz–Einstein (relativistic)
$x' = x - vt$	$x' = \gamma(x - vt)$
$y' = y$	$y' = y$
$z' = z$	$z' = z$
$t' = t$	$t' = \gamma\left(t - \dfrac{vx}{c^2}\right),$

where

$$\gamma = \frac{1}{\sqrt{1 - v^2/c^2}}. \tag{5.25}$$

Note that if $v \ll c$, then γ is nearly 1 and the relativistic transformation equations reduce to the Newtonian forms. Thus Newtonian physics can be regarded as a special case of relativistic physics. Recall that the tremendous velocity of the earth about the sun is 18 mi/s, but that this is still only one ten-thousandth of the velocity of light. Thus the relativity correction for the observation of positions from the earth compared with observations from the sun is only about 5×10^{-7} percent. Relativity makes no significant difference for "ordinary" engineering applications.

The relativistic transformation equations, however, point the way to important philosophical advances. The portion of special relativity we have developed applies only to observers not accelerated relative to each other. Note, however, that neither of these observers has a *preferred* viewpoint. If the transformation equations for primed quantities in terms of unprimed quantities are solved for the unprimed in terms of the primed ones, we obtain the *inverse transformations*. The resulting forms are the same except that the sign of their relative velocity, v, changes. [Thus $x = \gamma(x' + vt')$.] This one difference is expected, since if O' moves north relative to O, then O must move south relative to O'. A step has been taken toward the goal of general relativity, namely, that the laws of physics shall take on the same form for *all* observers.

5.12 LENGTH CONTRACTION

With the aid of the Lorentz–Einstein transformation equations we can now "explain" the Michelson–Morley experiment. The ad hoc factor of Fitzgerald and Lorentz is the factor γ, which has been derived from the basic relativistic assumption that all observers must get the same result if they measure the speed of light. Under relativity, space is not a rigid fixed thing but changes in size depending on the motion of the observer. This is very different from Newton's view. He said, "Absolute space, in its own nature, without regard to anything external, remains always similar and immovable."

Let us now consider a rod placed in the O' or primed reference frame (Fig. 5.5), with its length parallel to the x-axis. Let the coordinate of the left end be x_1' and that of the right end be x_2'. The length of the rod as measured by an observer at rest with respect to the rod is called its *proper length*. Its value is $(x_2' - x_1')$ in this case. What is the length of the rod measured by an observer in the O, or unprimed, frame when the primed frame is moving in the positive x-direction with speed v? From Eq. (5.25) we find that

$$x_1' = \gamma(x_1 - vt_1) \quad \text{and} \quad x_2' = \gamma(x_2 - vt_2).$$

Subtracting the first of these expressions from the second, we obtain

$$(x_2' - x_1') = \gamma[(x_2 - x_1) - v(t_2 - t_1)]. \tag{5.26}$$

This equation shows that the measured distance between the ends of the rod in the O-frame can have many different values that depend on the choice of t_1 and t_2, the times when the ends of the rod are observed. Because of this, we define the length of a moving rod as the measured distance between its ends obtained when the two ends are observed simultaneously. Then $t_1 = t_2$ and Eq. (5.26) reduces to

$$(x_2' - x_1') = \gamma(x_2 - x_1) \tag{5.27}$$

or

$$L' = \gamma L = (1 - v^2/c^2)^{-1/2} L.$$

Since γ is always greater than unity, $(x_2 - x_1)$ will always be less than the proper length, $(x_2' - x_1')$, and therefore it is said that the rod has contracted. We now see that the Fitzgerald–Lorentz contraction discussed in Section 5.7 is mathematically the same as the relativistic length contraction given by Eq. (5.27). However, the two equations are based on significantly different concepts. In the Fitzgerald–Lorentz contraction, v is the speed of the rod relative to the ether, whereas in the relativistic equation, it is the speed of the rod relative to an observer.

Assuming sufficient visual acuity, could one see that a moving body is contracted in the direction of its motion? It turns out that one could not if the moving body subtends a small angle at the observer. The situation here is not the same as that when we obtained Eq. (5.27), the length contraction equation. When we see an object, we have a retinal image produced by light quanta that arrive simultaneously at the retina from different points of the object. Therefore these light quanta could not have been emitted by every point on the object at the same time. The points farther away from the observer must have emitted their part of the image earlier than the closer points. In this case we cannot let $t_1 = t_2$ in Eq. (5.26) and obtain Eq. (5.27). When the differences in the times of emission are taken into account, it is found that a moving body that subtends a small angle at the observer will appear to have undergone a rotation but not a contraction. It is not at all difficult to derive this result, but it is too long for inclusion here. The method of arriving at this interesting conclusion is given in a paper by Terrell and summarized and discussed in one by Weisskopf. For the case when the moving object subtends a large angle at the observer, then, under suitable conditions that are stated in a paper by Scott and Viner, the length contraction will be visible. The reader is urged to study these three articles. The references to them are given at the end of this chapter.

5.13 TIME DILATION AND CAUSAL SEQUENCE

The classical concept of time is contained in Newton's idea that absolute, true, and mathematical time, of itself and from its own nature, flows equably without relation to anything external. There is implicit in this that information can be transmitted at velocities so enormous compared with the relative velocity of observers on different reference frames that no question about defining simultaneity arises. When, however, we assume that light signals are the fastest possible way of communicating, we can expect to encounter difficulties in describing the motion of objects having velocities approaching that of light.

How is the time interval between two events in one reference frame related to the interval between events as observed from another frame moving with respect to the first? Again, as in Fig. 5.5, let the x-axes of the frames be parallel and the relative velocity of the origins O and O' be v. Consider an event that occurs at time

t_1 at place x_1 in the unprimed frame and a later event in the same frame at t_2, x_2. The transformations to t', obtained from Eq. (5.25), are

$$t_1' = \gamma\left(t_1 - \frac{vx_1}{c^2}\right) \quad \text{and} \quad t_2' = \gamma\left(t_2 - \frac{vx_2}{c^2}\right).$$

By taking the difference between these expressions, we find the time interval in the primed frame to be

$$(t_2' - t_1') = \gamma\left[(t_2 - t_1) - \frac{v}{c^2}(x_2 - x_1)\right]. \tag{5.28}$$

To discuss this equation, let us start by considering the case when both events in the O-frame occur at the same place, that is, $x_1 = x_2$. The time interval between two such events measured in the coordinate frame in which the two events occur at the same place is called the *proper time* or local time. In the case under discussion, $(t_2 - t_1)$ is the proper time and Eq. (5.28) becomes

$$(t_2' - t_1') = \gamma(t_2 - t_1). \tag{5.29}$$

Since γ is greater than unity, the time interval between two events measured by the observer in the O'-frame will be longer than the proper time, which is the value of the time interval between the same two events obtained by the observer in the O-frame. This is *time dilation* and we say that the "clock" in the O-frame, the moving clock, runs more slowly than the one in the O'-frame. It is to be noted that although $x_1 = x_2$, the coordinates x_1' and x_2' are not the same because of the relative displacement of the reference frames during the time $t_2 - t_1$. If two events had occurred at the same place in the primed system at times t_1' and t_2', then the inverse transformation shows that the unprimed system would also have reported time dilation given by

$$t_2 - t_1 = \gamma(t_2' - t_1').$$

We can arrive at this equation by using another method that may give more insight into time dilation. Consider our passenger in the moving system of Fig. 5.5, except that now a mirror has been added, as shown in Fig. 5.6(a). A light that can be controlled by our observer is placed below the mirror. The observer pushes a button and a flash of light travels up to the mirror and back to the source. If we call the time that the light is flashed t_1' and the time that the light returns to the source t_2', then we have the time of flight (the time interval between the two events) being

$$t_2' - t_1' = \frac{2D}{c}. \tag{5.30}$$

What would another observer (not shown) in the stationary frame S observe? As is shown in Fig. 5.6(b), the individual would note that at the time the light flashed the car would be in Position 1, and when the light reached the mirror the car would be in Position 2. When the light finally reached the source, the car would be in Position 3. If we call the time in this rest frame when the light is flashed t_1 and the time when

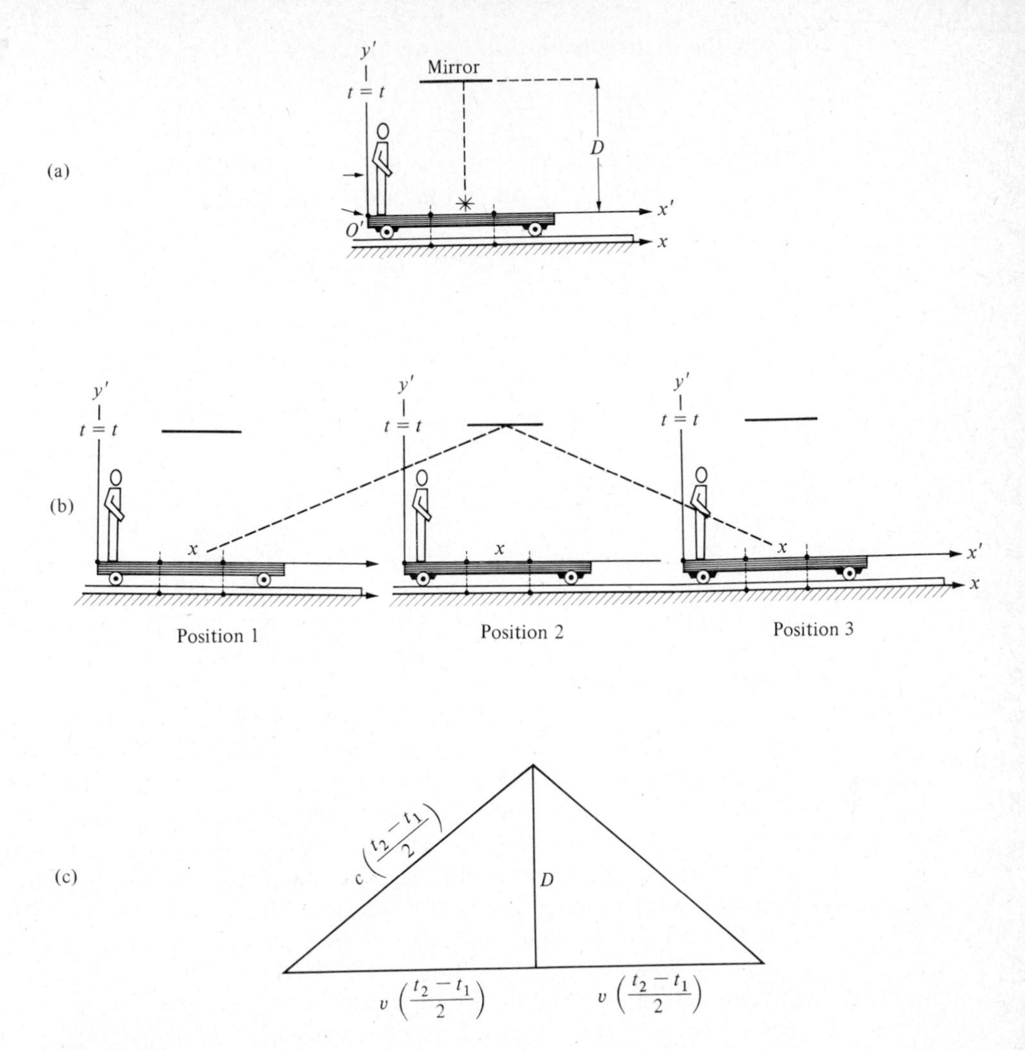

Figure 5.6

the light again reaches the source t_2, the interval will be $t_2 - t_1$. Half of this time interval will be the time necessary for the light to go from the source to the mirror and the car to go from Position 1 to Position 2. We know that the distance the light travels will be the time interval it is traveling times the speed of light. At the same time, we know that the distance the car travels will be the speed of the car times this time interval. Therefore we are able to construct a triangle using the three distances — the distance the light travels from the source to the mirror, the distance the car travels while the light is traveling from the source to the mirror, and the distance D, the distance between the mirror and the light. This triangle is shown in Fig. 5.6(c).

Using the Pythagorean theorem, we then have

$$\left[\frac{c}{2}(t_2 - t_1)\right]^2 = D^2 + \left[\frac{v}{2}(t_2 - t_1)\right]^2. \tag{5.31}$$

The distance between the source and the mirror will be the same for each observer. We can combine Eq. (5.30) and Eq. (5.31) and obtain

$$D^2 = \left[\frac{c}{2}(t_2' - t_1')\right]^2 = \left[\frac{c}{2}(t_2 - t_1)\right]^2 - \left[\frac{v}{2}(t_2 - t_1)\right]^2$$

or solving for $t_2 - t_1$

$$(t_2 - t_1) = \frac{t_2' - t_1'}{\sqrt{1 - \dfrac{v^2}{c^2}}} = \gamma(t_2' - t_1').$$

Returning to Eq. (5.29), we see that if the events in the unprimed frame are simultaneous, that is, $t_2 - t_1 = 0$, then we also have $t_2' - t_1' = 0$. Therefore two observers will agree on the simultaneity of two events if they occur simultaneously at the *same* place in either system of coordinates. Otherwise, complications arise that will be discussed in the next section. Finally, note that if the speed of light were infinite, then $\gamma = 1$ and the time intervals in each frame would be equal or absolute, as Newton assumed.

Direct confirmation of time dilation is found in an experiment with muons. These are subatomic particles that can be created in the laboratory by high-energy particle accelerators. These particles are unstable and it has been found that they disintegrate at an exponential rate such that one-half of them remain unchanged after 3.1×10^{-6} s, measured in the reference frame in which the muons are at rest. These particles are also formed high in the earth's atmosphere by cosmic ray bombardment and are projected toward the earth's surface with a very high velocity. Consider a beam of these particles traveling down toward the earth at a speed of $0.9c$. (For this speed, $\gamma = [1 - (0.9c/c)^2]^{-1/2} = [0.19]^{-1/2} = 2.3$.) This beam would traverse a distance of 840 m in 3.1×10^{-6} s. Therefore one would expect that if a meson counter were placed 840 m above another one in the atmosphere, then the count observed at the lower level would be half of that at the upper one. But it is found to be significantly greater than one-half the count at the upper level. The error in the prediction arose because the earthbound observer used the proper half-life instead of the dilated one in the computations. The dilated half-life of the muons, calculated from Eq. (5.29), is $2.3 \times 3.1 \times 10^{-6}$ s $= 7.2 \times 10^{-6}$ s. Therefore the earth observer will obtain a higher count at a lower level than was first expected because this dilated half-life is much longer than 3.1×10^{-6} s, the time of flight between the counters. Let the earth observer increase the distance between counters to 1920 m, the product of the dilated half-life and the relative velocity of the reference frames. This observer will now find the lower-level count to be half of that at the higher level. What does the observer riding along with the muons report as the ratio of the counts? From the observer's point of view, the separation of the counters is not 1920 m but, because of length contraction, 840 m. This contracted length divided by $0.9c$ gives the time of flight between counters to be 3.1×10^{-6} s, which is equal to the proper half-life. Therefore, the observer in the muons frame will also find the lower-level count to be half of the upper-level one. Thus, both observers do agree on the relative number of muons that survived the trip between the counters, although the basic data they used in their calculations were quite different.

Let us now discuss the transformed time interval when the two events in the unprimed system do *not* occur at the same place. In this case the value of $t_2' - t_1'$ is given by Eq. (5.28). If the quantity within the brackets in this equation is equal to zero, then the two events are simultaneous in the primed frame. If the quantity is positive, then events are observed in the same order in the primed frame as in the unprimed one. If it is negative, then the two events in O should appear in the reverse order from O'. To find the mathematical condition for a reversal, we will rewrite Eq. (5.28) as follows:

$$t_2' - t_1' = \gamma(t_2 - t_1)\left[1 - \frac{v}{c}\frac{x_2 - x_1}{c(t_2 - t_1)}\right].$$

This shows that the quantity within brackets will be negative only if

$$\frac{x_2 - x_1}{c(t_2 - t_1)} > \frac{c}{v}.$$

This inequality can be realized only if the distance between x_2 and x_1 is greater than the distance traveled by light in the time $t_2 - t_1$. But if the separation of the two places were that large, then a light signal leaving Event 1 could not reach Event 2 soon enough to cause Event 2. Thus if two events appear in different time order to two observers, one event cannot be the cause of the other. Therefore we must conclude that cause and effect will never appear in the reverse order to different observers.

Is time *really* dilated and is the length of a moving rod *really* contracted in the direction of its motion? The answers depend on what is meant by *really*. In the physical sciences what is real is what is measured. Only through measurement can one obtain the information needed for assigning properties to a clock, to a rod, to an atom, and so forth. Time dilation and length contraction are real in this sense. Proper time and proper length have nothing of an absolute nature about them. Time, length, area, volume, and other quantities are all relations between an observed body and the observer.

5.14 SIMULTANEOUS EVENTS

What is meant by simultaneous? When two individuals are side by side it is very easy for them to agree on a definition of simultaneous. Let us use as an example lightning striking the ground. If two bolts of lightning strike the ground in front of two individuals, then they will agree that the lightning struck simultaneously or at the same time. No problem!

Now let's say that the two individuals are separated by hundreds of miles, a woman in Houston and a man in Chicago (we shall use the same time zone for convenience). When lightning struck in front of the woman in Houston, she noted the time on her clock, and when lightning struck in front of the man in Chicago, he noted the time on his clock — and both individuals could come to some agreement about whether the two events were simultaneous. Or could they? They could if they were careful about two points: (1) The clocks were synchronized at some time prior to the events, and (2) the clocks ran at the same rate after synchronization.

The clocks must first be synchronized. If we could send signals instantaneously, this would present no difficulty. When the clock in Houston would read a certain time, the man in Chicago would set his clock to coincide with the Houston clock. However, signals cannot be sent instantaneously, and thus we must devise a technique that will synchronize the clocks. For example, since we know the speed of light, the woman in Houston could send a radio signal to the man in Chicago when the Houston clock read 12:00. The man in Chicago would then set his clock to 12:00 plus the time for the signal to travel from Houston to Chicago (distance divided by speed of light). A simpler technique is the technique that Einstein suggested in his discussion of these problems. An individual exactly halfway between Houston and Chicago could send a signal and the individuals in Chicago and Houston could set their clocks to a predescribed time when they received the signal. Then the clocks in Houston and Chicago would be synchronized. Now if they were to run at the same rate from that point on, the individuals could determine if the lightning had struck at the same time.

Our definition of simultaneous for two observers separated by a great distance has been satisfactory. But our example has concerned two individuals at rest with respect to each other. What if they are moving relative to each other? Now we shall see that we still have a problem. In fact, we shall show that two events that are simultaneous for one observer are not necessarily simultaneous for another.

Consider two individuals, a man standing on the 50-yard line of a typical football field and a woman riding in the middle of a 100-yard-long rocket ship (measured by the man on the ground) flying above the field. As the rocket ship passes immediately overhead, lightning strikes the front of the rocket ship and the goal line at that end and lightning strikes the rear of the rocket ship and the goal line at that end. The man standing in the middle of the football field says that the two events were simultaneous. His reasoning goes something like this: Since I am at the 50-yard line and thus an equal distance from each goal and since the flashes from the two events have reached me at the same moment, the lightning occurred simultaneously.

The woman in the middle of the rocket ship will observe something quite different. Lightning strikes the front and back of the rocket ship and during the time that the light travels from each end of the rocket ship to the observer, the rocket ship

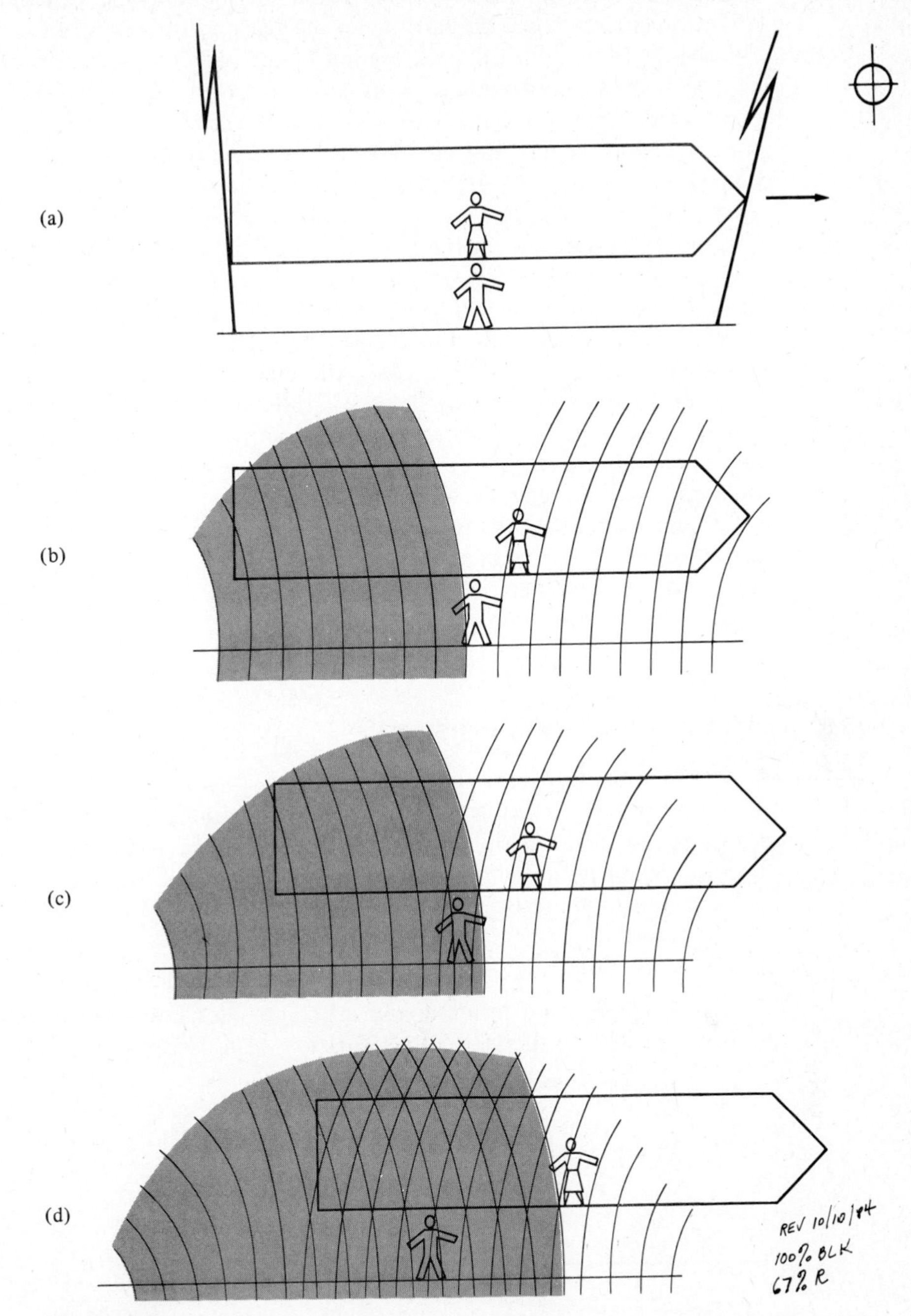

Figure 5.7

has moved. First the observer will be aware that lightning has struck the front of the rocket. Sometime later, the light from the back will catch up to her and she will observe the light from the back section of the rocket. Since she knows that she is halfway between the front and the back of the rocket ship and since she observes the event at the front of the rocket ship before the event at the back, she will be convinced that the lightning struck the front of the rocket ship and after a period of time struck the back of the rocket ship. In other words, she believes that the two events were not simultaneous.

Figure 5.7 summarizes this situation. In Fig. 5.7(a) the rocket ship is precisely over the football field, and the lightning strikes its front and rear and the two goal lines. In Fig. 5.7(b), the information from the lightning striking the front of the rocket ship has reached the rocket passenger. The observer in the football field has observed nothing yet. In Fig. 5.7(c), the man in the middle of the football field observes the lightning striking the rocket in the front and rear and the two goal lines simultaneously. In Fig. 5.7(d), the signal from the lightning striking the rear of the rocket ship finally catches up to the passenger and she is then aware that both the front and back of the rocket ship have been struck.

Which individual is correct? Were the events simultaneous or were they not? The correct answer is that both observers are right and that the events were simultaneous for one observer and not simultaneous for the other.

5.15 THE RELATIVISTIC VELOCITY TRANSFORMATION

To illustrate further how the transformation equations of space and time are used, let us repeat a calculation we made classically. We found the classical velocity transformation for observers with relative velocity v to be $u' = u - v$. To get the corresponding relativistic expression, we use the defining equations

$$u'_x = \frac{dx'}{dt'} \quad \text{and} \quad u_x = \frac{dx}{dt}.$$

We now express u' in terms of the differentials of unprimed quantities obtained from the relativistic transformation equations. The result is

$$u'_x = \frac{dx'}{dt'} = \frac{\gamma(dx - v\,dt)}{\gamma(dt - v\,dx/c^2)} = \frac{dx/dt - v}{1 - (v/c^2)(dx/dt)},$$

or

$$u'_x = \frac{u_x - v}{1 - u_x v/c^2}. \tag{5.32}$$

In a similar way we find the velocity components parallel to the other two axes to be

$$u'_y = \frac{u_y}{\gamma(1 - u_x v/c^2)} \quad \text{and} \quad u'_z = \frac{u_z}{\gamma(1 - u_x v/c^2)}. \tag{5.33}$$

Note that the velocity components that are transverse to the relative motion of the reference frames depend on the relative motion but that transverse distances do not.

Equation (5.32) is the relativistic rule for transforming velocities when the observed velocity is *parallel* to the relative velocity of the observers. We see that as in the classical case, the velocity transformation is not invariant. Consider, however, what happens when the velocity under observation is the velocity of light, c. If one observer measures the velocity of light and gets $u_x = c$, what will another observer moving relative to the first obtain? The equation yields

$$u'_x = \frac{u_x - v}{1 - (u_x v/c^2)} = \frac{c - v}{1 - (cv/c^2)} = \frac{c - v}{(c - v)/c} = c. \tag{5.34}$$

Thus, regardless of their own relative velocity v, any two observers will agree that the velocity of light is c. In relativity, the velocity of light is *invariant*. This result should not surprise us, since it was used as a basic assumption for the derivation of the transformation equations. It is one example of the "uncommon" sense we promised to discuss at the very beginning of this book.

Although the result just derived follows logically from the assumptions of relativity and therefore should not surprise us, a further comment may be helpful. Suppose observer O measures the distance light travels during an interval of time and computes the velocity of light, obtaining c. Let another observer O' moving relative to O perform the same experiment. Intuitively we feel O' cannot get the same result, c. But if O and O' both seem to have "rubber" rulers and "defective" clocks, we see that their results may have any values — including c. The transformation equations of the special theory of relativity are a description of how "rubber" rulers and "defective" clocks vary so that the velocity of light can be unique, so that velocities small compared with c behave classically, and so that intermediate cases lead to no contradictions.

If this were the end of the matter, there is no doubt relativity would have been forgotten or remembered only as a fantastic speculation of a fertile mind. No one would make such a tremendous break with traditional thought merely to "explain" the Michelson–Morley experiment. Before it can be taken seriously, the inductive structure of relativity or any other theory must be subjected to deductive verification. If relativity is to be "true" or useful, it must correctly foretell some unforeseen events that can be subjected to experimental tests.

There are several such verifications of relativity like the decay of mesons discussed in Section 5.13. Some come from general relativity, which will be discussed later. But the most striking and important deductive consequence of relativity comes from the special theory. This is the mass transformation equation. To introduce mass into the picture, we next consider an impact situation.

5.16
RELATIVISTIC MASS TRANSFORMATION

Consider two basketballs that are spherical, perfectly elastic, and have identical masses when compared by an observer at rest relative to them. We give one ball to an observer O' on a railroad train, moving relative to the ground with a constant translational velocity v. We give the other ball to an observer O on the ground at a distance d from the railroad track. The observers have not yet passed each other and we tell both to throw his ball with a certain velocity in a direction perpendicular to the track in such a way that the basketballs bounce perfectly off each other just at the moment when the two observers pass each other. (Deflections due to gravity are irrelevant to this discussion.)

Assuming sufficient skill, this experiment could be carried out and fully understood on the basis of classical mechanics. Each observer sees his ball hit and return to him just as though it had bounced from a perfectly elastic wall with kinetic energy conserved. Each observer must anticipate the moment of passing, and the movements of the balls are as shown in Fig. 5.8.

If we now treat the situation just described from the standpoint of relativity, a new idea emerges. Whereas each thrower gives *his* ball a velocity of magnitude U_y (his U_x being zero), each observes the y-component of the velocity of the other's ball as given by Eq. (5.33), namely,

$$U_y' = \frac{U_y}{\gamma\left(1 - \dfrac{0 \cdot v}{c^2}\right)} = \frac{U_y}{\gamma}. \tag{5.35}$$

Since γ is always greater than unity, it appears to each observer that the other person has thrown his ball slower than he was told to.

Figure 5.8 Perfectly elastic collision of two identical basketballs as viewed from systems in relative motion.

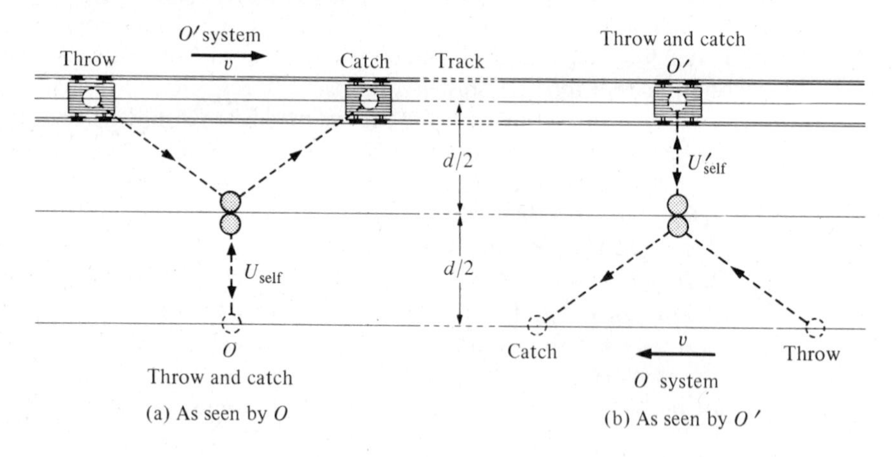

We are now faced with a dilemma. If we assume that the masses remain the same as when they were compared without relative velocity, then conservation of momentum is violated. If we keep conservation of momentum, then the relative motion that made a difference in the observed velocities of the basketballs also resulted in an observed difference in their masses. In relativity it is assumed that *momentum is conserved*. Therefore observer O, who sees his ball change in momentum an amount $2m_{\text{self}}U_{\text{self}}$, equates this change to the change he observes for the other ball. He writes

$$2m_{\text{self}}U_{\text{self}} = 2m_{\text{other}}U'_{\text{other}}. \tag{5.36}$$

When the value of U'_{other} from Eq. (5.35) is substituted in Eq. (5.36), it becomes

$$2m_{\text{self}}U_{\text{self}} = 2m_{\text{other}}\frac{U'_{\text{self}}}{\gamma}, \tag{5.37}$$

which, since the throws were equal such that $U_{\text{self}} = U'_{\text{self}}$, reduces to

$$m_{\text{other}} = \gamma m_{\text{self}}. \tag{5.38}$$

Thus observer O concludes that the masses of the balls are not equal and that there must be a transformation equation for mass in addition to those for the other quantities we have discussed.

In this derivation we have considered an especially simple situation in order to get our result as quickly as possible. We have computed the observed change in mass by considering the effect of a relative x-velocity on observed y-velocities. We could neglect the changes in mass each observer produced in his own ball when he threw it because both observers threw in the same way ($U'_{\text{self}} = U_{\text{self}}$). A general derivation would have led to a result of the same form, namely,

$$m = \gamma m_0 = \frac{m_0}{\sqrt{1 - (v^2/c^2)}}, \tag{5.39}$$

where m_0 (called the rest mass) is the mass of a body at rest relative to the measurer, and m (called the mass) is the observed mass of the body when it has a velocity v relative to the measurer. The system in which the observer is located is usually called the laboratory system.

This result, which has been deduced from the space transformation equations, can be tested experimentally. What is required is the measurement of the mass of a body moving at great speed relative to an observer. J. J. Thomson was startled when he computed the speed of electrons in his cathode-ray tubes. These speeds were greater than any previously measured by humans and, by modern methods, can be made significant compared with the free-space velocity of light, c. Furthermore, an e/m_e experiment is a measure of m_e if we assume that the charge e of the particle is constant. In principle, then, J. J. Thomson provided a technique for the testing of the mass-transformation equation. The first adequate test and verification was made in

1908 by Bucherer, using electrons from a radioactive source, and since then the mass-transformation equation has been confirmed innumerable times. It is now a cornerstone of atomic physics.

5.17 MASS–ENERGY EQUIVALENCE

The derived and experimentally verified mass-transformation equation has important consequences. Whereas in classical physics we considered that the kinetic energy of a body could be increased only by increasing its velocity, we now must take mass variation into account. Suppose we increase the kinetic energy of a body an amount dE_k by exerting a force through a distance. The change of kinetic energy equals the work done, or

$$dE_k = F\,ds.$$

Since we cannot regard mass as constant in relativity, we write Newton's second law in its more general form — force equals the time rate change of momentum:

$$F = \frac{d(mv)}{dt}.$$

The change of kinetic energy then is

$$dE_k = \frac{d(mv)}{dt}\,ds. \tag{5.40}$$

But since $ds/dt = v$, we can write the preceding equation as

$$dE_k = v\,d(mv) = v^2\,dm + mv\,dv. \tag{5.41}$$

This expression can be simplified by using the mass-transformation equation, Eq. (5.39), which after squaring and rearranging becomes

$$m^2c^2 = m^2v^2 + m_0^2c^2. \tag{5.42}$$

We now differentiate, noting that both m_0 and c are constant, and obtain

$$2mc^2\,dm = 2mv^2\,dm + 2m^2v\,dv. \tag{5.43}$$

When $2m$ is divided out of this equation, the right side becomes identical with our expression for dE_k in Eq. (5.41), so we have, finally,

$$dE_k = c^2\,dm. \tag{5.44}$$

This famous equation shows that in relativity a change in kinetic energy can be expressed in terms of mass as the variable. This equation is valid for a change of energy in any form whatever, although it was derived here for a change of kinetic energy.

Since the kinetic energy is zero when $v = 0$, then it is also zero when $m = m_0$. Therefore we integrate, and obtain

$$E_k = \int_0^{E_k} dE_k = c^2 \int_{m_0}^{m} dm = c^2(m - m_0)$$

or

$$E_k = mc^2 - m_0c^2. \tag{5.45}$$

This is the relativistic expression for kinetic energy. The classical relation $E_k = \frac{1}{2}mv^2$ does *not* give the correct value of the kinetic energy even when the relativistic values of the mass and the velocity are used.

Like the other transformation equations, this expression for E_k should reduce to the classical expression for $v \ll c$. If we transform m in Eq. (5.45), we obtain

$$E_k = \frac{m_0c^2}{\sqrt{1 - (v^2/c^2)}} - m_0c^2. \tag{5.46}$$

Now, since $v \ll c$, we can say

$$\frac{1}{\sqrt{1 - (v^2/c^2)}} = \left(1 - \frac{v^2}{c^2}\right)^{-1/2} = \left(1 + \frac{v^2}{2c^2} + \frac{3}{8}\frac{v^4}{c^4} + \cdots\right), \tag{5.47}$$

where we have carried three terms of the binomial expansion. We then have, from Eq. (5.46),

$$\begin{aligned} E_k &= m_0c^2\left(1 + \frac{v^2}{2c^2} + \frac{3}{8}\frac{v^4}{c^4} + \cdots - 1\right) \\ &= \frac{1}{2}m_0v^2 + \frac{3}{8}m_0\frac{v^4}{c^2} + \cdots . \end{aligned} \tag{5.48}$$

The first term of this is the classical expression for the kinetic energy. Obviously, it is the only significant term for low velocities.

Equation (5.44) suggests a broader interpretation of Eq. (5.45). Equation (5.44) implies that, in addition to the familiar forms of energy, such as kinetic, potential, and internal, there is yet another kind, *mass-energy*. Just as Joule's constant relates heat to energy by telling us the number of joules per calorie, so Eq. (5.44) relates mass to energy by telling us the number of joules per kilogram. Just as heat is energy and energy can be observed as heat, so mass is energy and energy can be observed as mass. According to this view, Eq. (5.45) can be written as

$$E = mc^2 = m_0c^2 + E_k, \tag{5.49}$$

where E and mc^2 are the *total* energy, m_0c^2 is the *rest mass-energy*, and E_k is the kinetic energy of a body.

We can now obtain an important relation between the total mass-energy E of a body and its momentum p from Eq. (5.42). If we multiply through this equation by

c^2 and make the appropriate substitutions from $E = mc^2$ and $p = mv$, we get

$$E^2 = p^2c^2 + m_0^2c^4. \tag{5.50}$$

This equation will be used several times in later chapters.

The equivalence of mass and energy brings a satisfying consequence. The broad principle of conservation of energy now takes unto itself another broad conservation principle, the conservation of mass. The identity of mass and energy resulting in a unified concept of mass-energy is certainly the most "practical" consequence of relativity. We shall see in Chapter 11 how nuclear reactions illustrate the fact that neither mass nor energy as conceived classically is conserved separately. In relativity we see that it is mass-energy that is conserved. In any interaction the *total mass*, m (which includes any kinetic energy in mass units), is *the same before and after* the interaction. Similarly, in any interaction, the *total energy* E (which includes any rest mass in energy units), is *the same before and after* the interaction. We are now witnessing the growth of the whole new field of nuclear engineering, which is based on E equals mc^2.

Most of the mass-energy we use comes from the sun, where rest mass-energy is converted into thermal mass-energy. Today we are converting rest mass-energy into thermal mass-energy here on the earth. We shall see that nuclear fission and fusion are the techniques for this conversion.

5.18 THE UPPER LIMIT OF VELOCITY

An examination of Eq. (5.39), the mass-transformation equation, shows that for $v = c$, $m = \infty$. Thus, as the velocity of a body increases toward the free-space velocity of light, the mass of the body increases toward infinity. An infinite mass moving at a snail's pace would have infinite energy. Since it is absurd for any body with finite rest mass m_0 to have infinite energy, we must conclude that it is impossible for such a body to move with the free-space velocity of light. Thus in relativistic mechanics, the free-space velocity of light is the greatest velocity that can be given to any material particle. If a tiny particle is given equal increments of energy, the first increments increase the velocity significantly and the mass insignificantly. But as the particle gains more and more energy, the velocity changes become less as the velocity approaches the velocity of light. As the velocity changes become insignificant, the mass changes become significant.

Expressing the same argument another way, we have two equations for the kinetic energy of a body, both of which are correct relativistic expressions. Equation (5.44) expresses E_k in terms of m_0 and v. For v small compared with c, this is the convenient formula, especially since it reduces to the first term of Eq. (5.48), the classical expression. The other equation is $E_k = c^2(m - m_0)$, Eq. (5.45). This equation expresses E_k in terms of m_0 and the variable mass, m. When the velocity is near c, the mass is the more convenient variable. We can express E_k in terms of either m or v, since we have the mass-variation equation, Eq. (5.39), relating m and v.

Since we argued that the free-space velocity of light is an upper limit for material particles, it is fair to ask whether any velocities equal or exceed this velocity. We have been careful in this discussion to emphasize that the limit is the *free-space* velocity of light. Light itself often travels more slowly than its free-space velocity. The index of refraction of a medium is the ratio of the free-space velocity of light to the actual velocity in the medium. In water, for example, a particle may travel faster than light *in that medium*. An example of this will be discussed when we consider Cherenkov radiation in Chapter 12.

Photons in free space obviously travel with the free-space velocity of light. We have found that photons have energy $h\nu$, and since we have shown that mass and energy are equivalent, we can conclude that each photon has a mass $h\nu/c^2$. Here then is a mass particle actually traveling with the free-space velocity of light. But there is no contradiction here. A photon has no rest mass m_0. If we stop a photon to measure its mass, all its mass-energy is transferred to something else and the photon no longer exists. Perhaps the main difference between a "material" particle and a photon particle is that one has rest mass while the other does not.

There are times when one may encounter velocities greater than the free-space velocity of light, but on these occasions the thing moving is a mathematical function rather than a physical reality. The transmission of a "dot" in radio telegraphy may be regarded as the resultant of the transmission of several frequencies having different velocities. These velocities, called phase velocities, may exceed the free-space velocity of light. The motion of the resultant "dot" is called a group velocity, and it can never exceed the free-space velocity of light. (These two types of velocities associated with waves will be discussed in Chapter 7.) We may state with complete generality that no observer can find the *velocity of mass-energy in any form to exceed the free-space velocity of light.*

5.19 EXAMPLES OF RELATIVISTIC CALCULATIONS

The basic mass unit of nuclear physics is the unified atomic mass unit, u, which is 1/12 of the rest mass of carbon-12. One u = 1.66×10^{-27} kg, and this is nearly the proton's mass. In relativity, this mass-energy may be expressed in energy units as

$$E = m_0c^2 = 1.66 \times 10^{-27}\ \text{kg} \times (3.00 \times 10^8)^2 \frac{\text{m}^2}{\text{s}^2}$$

$$= 1.49 \times 10^{-10}\ \text{J}.$$

Or 1 u may be expressed in eV$/c^2$ and MeV$/c^2$:

$$1\text{ u} = \frac{E}{e_c c^2} = 1.49 \times 10^{-10}\text{ J} \times \frac{1\text{ eV}/c^2}{1.60 \times 10^{-19}\text{ J}}$$
$$= 9.31 \times 10^8\text{ eV}/c^2 = 931\text{ MeV}/c^2.$$

We have now expressed the relativistic rest mass-energy of one u in five ways:

$$1\text{ u} = \tfrac{1}{12}\,{}^{12}\text{C} = 1.66 \times 10^{-27}\text{ kg} = 1.49 \times 10^{-10}\text{ J}/c^2$$
$$= 9.31 \times 10^8\text{ eV}/c^2 = 931\text{ MeV}/c^2.$$

This same quantity could also be expressed in any other units of mass or energy — slugs, calories, grams, kWh, ft · lb, etc. We could have made similar calculations with any quantity of mass or energy. We chose the u as an important example. It is very useful to *remember* that 1 u = 931 MeV$/c^2$. (To six significant figures, 1 u equals 931.441 MeV$/c^2$ on the ^{12}C isotopic mass scale.)

Consider next the velocity that one u would have if it has a kinetic energy equal to twice its rest mass-energy. This kinetic energy can be expressed as 2 u, 3.32×10^{-27} kg, 2.98×10^{-10} J, 1.86×10^9 eV, or 1.86 GeV.

The total mass of the particle is the sum of its rest mass and kinetic energy. Thus its mass, m, is 1 + 2 = 3 u, 4.47×10^{-10} J$/c^2$, etc.

We now have everything necessary to calculate the relativistic velocity. The mass-velocity equation, Eq. (5.39), solved for the velocity is

$$v = c\sqrt{1 - (m_0/m)^2}\,.$$

We may substitute for masses in any consistent units. The velocity will have the same units as those used for c. Using the masses in u, we have

$$v = c\sqrt{1 - (\tfrac{1}{3})^2} = 0.942c = 2.83 \times 10^8\,\frac{\text{m}}{\text{s}}.$$

This is a very large velocity, and we may expect that a classical calculation will be seriously in error. To show this, we next calculate v on the basis of classical physics. We know that the kinetic energy is $E_k = 2.98 \times 10^{-10}$ J, and that the mass is $m = 1.66 \times 10^{-27}$ kg. Solving $E_k = \frac{1}{2}mv^2$ for v, we have

$$v = \sqrt{\frac{2E_k}{m}} = \sqrt{2 \times 2.98 \times 10^{-10}\text{ J} \times \frac{1}{1.66 \times 10^{-27}\text{ kg}}}$$
$$= 6.0 \times 10^8\,\frac{\text{m}}{\text{s}} = 2.0c.$$

Not only is this result very different from the relativistic one, but it is also greater than the velocity of light. Whenever the two methods differ significantly, the relativistic method must be used.

We next seek the kinetic energy of a 15 kg earth satellite traveling with a velocity of 18,600 mi/h. Since the velocity of light in English units is 186,000 mi/s, the ratio of the satellite's velocity to the velocity of light, v/c, is

$$\frac{v}{c} = 18{,}600\,\frac{\text{mi}}{\text{h}} \times \frac{1\text{ h}}{3600\text{ s}} \times \frac{1\text{ s}}{186{,}000\text{ mi}} = \frac{1}{36{,}000}.$$

Strictly speaking, we should compute the satellite's mass from Eq. (5.39), $m = m_0/\sqrt{1-(v/c)^2}$, and then the kinetic energy from Eq. (5.49), $E_k = (m - m_0)c^2$. But the satellite's velocity is so small compared with c that calculations using Eq. (5.48), $E_k = \frac{1}{2}m_0v^2 + \frac{3}{8}m_0(v^4/c^2)\ldots$, would be much more convenient. In this example, the second term on the right is negligible. This shows that a classical calculation is sufficient for a body moving so slowly.

5.20 PAIR PRODUCTION

Certainly one of the most dramatic instances of the relativistic change of energy from one form to another is in the phenomenon known as pair production. *Pair production* is the process in which a photon becomes an electron and a new positive particle. This new particle was anticipated by Dirac in 1928 from his relativistic wave-mechanics theory of the energy of an electron and was experimentally observed by Anderson in 1932.

The discovery was made in the course of cosmic-ray research with a cloud chamber. We shall discuss cloud chambers in Chapter 10. All we need say now is that a cloud chamber is a device that makes the paths of charged particles visible and subject to photographic recording. Figure 5.9 is Anderson's most famous cloud-chamber photograph. It shows a charged particle moving with high speed and with its path curved by a magnetic field directed into the paper. The particle is seen to traverse a sheet of lead 6 mm thick in the middle of the chamber. From the beady nature of the track it was established that the particle was electronlike. Since the lead sheet could only have slowed the particle, its direction of motion must have been from the region of low curvature to the region of high curvature. From the observed direction of curvature and the known direction of the motion and of the magnetic field, it was concluded that the charge of the particle was positive. On the basis of this picture and others similar to it, Anderson announced the discovery of a new particle that he called the *positron* or positive electron.† The positron is just like the electron except for the sign of its electric charge.

The source of the positron is not visible in Fig. 5.9 but several sources are apparent in Fig. 5.10. Note the three sets of diverging tracks leaving the lower side of the lead plate. These are tracks originating at the lead but with no corresponding tracks coming into the lead on the upper side. Evidently these pairs of particles are not produced by other charged particles but by something else that cannot produce tracks, namely, photons. The curvature of the paths is due to a magnetic field normal to the plane of the paper. Since the tracks of the particles of the pair are oppositely curved, the individuals must have opposite charges. In each case, one

†At the same time it was proposed that the electron be called the *negatron* or negative electron. These names have never come into general use.

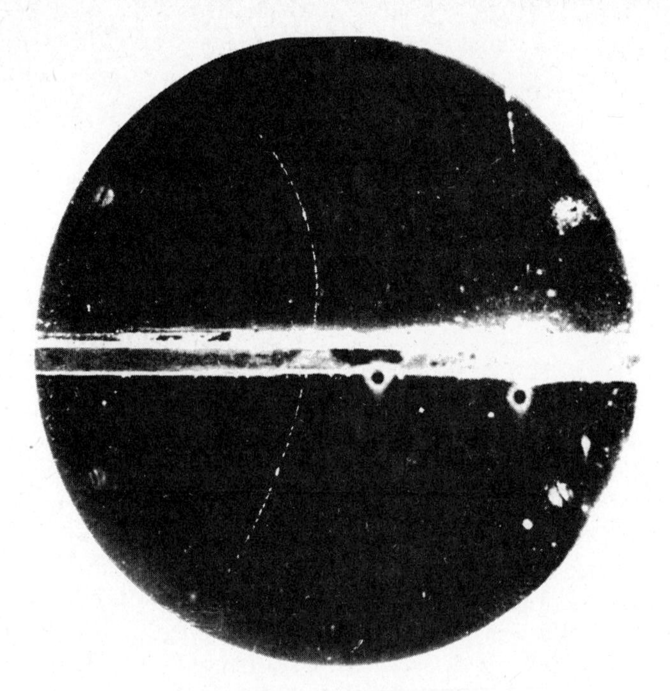

Figure 5.9 Cloud-chamber track of a positive electron (positron) in a magnetic field.
Courtesy of C. D. Anderson, California Institute of Technology.

particle is a positron and the other an ordinary electron. A quantum of radiant energy coming from above the lead sheet has changed into matter forming a *pair* consisting of a positron and an electron. In the process of *pair production* we have the *materialization* of energy.

It is easy to show that the rest mass of an electron is equivalent to 0.51 MeV/c^2. Thus, to create a pair, a photon must have at least enough energy to convert into two electron masses or 1.02 MeV/c^2. A photon could never convert into just one electron or positron, since this would violate the law of conservation of electric charge. The photons that produced the pairs shown in Fig. 5.10 must each have had more than 1.02 MeV of energy, since these pairs were not only created but also were given considerable kinetic energy.

A photon cannot produce a pair just anywhere. The process must take place in an intense field such as that close to the nucleus of an atom. Since a nucleus must be involved in pair production, conservations of energy and of momentum become a bit complex. If the nucleus involved is massive, it can carry away its share of momentum without taking away an appreciable amount of kinetic energy. In this case pair production can occur if the energy of the photon is just a little above the mass-required minimum of 1.02 MeV. If the pair is produced near a light nucleus, on the

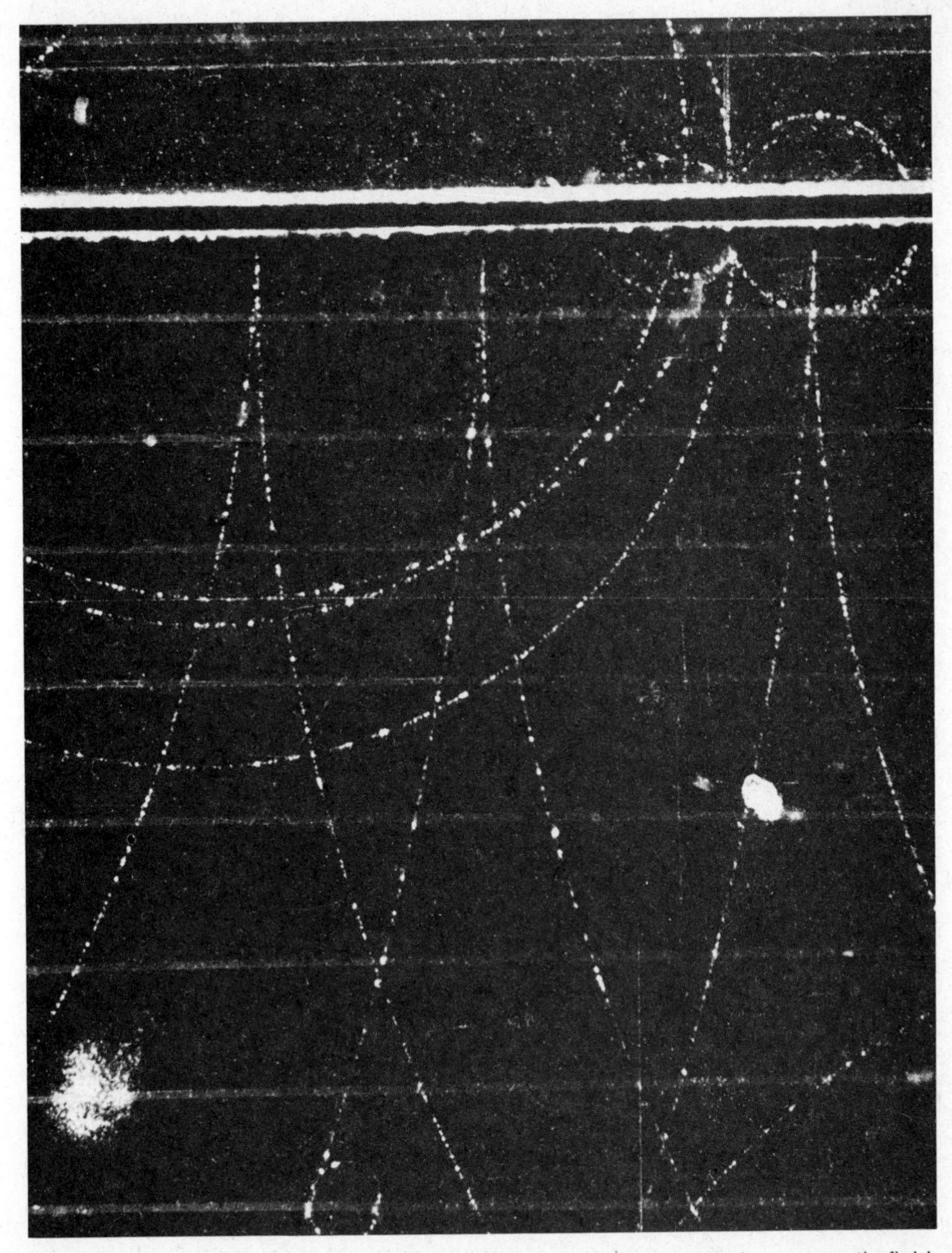

Figure 5.10 Cloud-chamber tracks of three electron-positron pairs in a magnetic field. Three gamma-ray photons entering at the top materialize into pairs within a lead sheet. The coiled tracks are due to low-energy photoelectrons ejected from the lead.

Courtesy of the Radiation Laboratory, University of California.

other hand, the light nucleus carries away appreciable kinetic energy at the expense of the originating photon and thus this photon must have considerably more than 1.02 MeV of energy. For 20-MeV photons, the nuclei of the air are about as effective as those of lead in "catalyzing" pair production. Since pair production occurs near the nucleus of an atom, the repelled positron emerges with a somewhat higher velocity than the attracted electron.

Because electrons are common in all matter, the lives of positrons are very short. Once a positron has lost most of its kinetic energy in passing near charged particles, it will move quickly toward an attracting electron. Before union occurs, however, these particles sometimes revolve momentarily about their center of mass in a semistable configuration called *positronium*.

Positronium is a non-nuclear "element" with an average existence of less than 10^{-7} s. Despite its short life, its spectrum has been measured. These measurements show that positronium is hydrogenlike, as expected. The frequencies of its spectral lines are about half those of the corresponding hydrogen lines because the reduced mass of the electron in positronium is about half its value in hydrogen.

When the electron and positron finally join, they annihilate each other and become electromagnetic radiation. The production of *annihilation radiation* is the converse of the materialization of energy. Just as pair production requires a photon of at least 1.02 MeV of energy, so annihilation produces photons whose energies add up to at least 1.02 MeV. Annihilation resulting in two photons is by far the most common, but the three-photon case has been observed. This rare event is discussed in advanced books. The laws of conservation of mass-energy, of momentum, and of charge hold in all these processes.

Pair production is not limited to the positron and electron. In 1955, the *antiproton* was produced in experiments with high-energy particles from accelerators. The antiproton and the proton have equal charges of opposite sign and equal masses. It requires a minimum of 1.88 GeV of energy to produce a proton-antiproton pair. The antiproton is usually quickly captured by a proton. The subsequent proton-antiproton annihilation usually produces a number of subatomic particles. The positronium analog of this pair has not been observed.

The positron and the antiproton are often called *antiparticles*. They may be thought of as plane mirror images of the electron and proton, respectively. Since the discovery of these pairs, there has been speculation about the possibility of the existence of *antihydrogen*. This would have an antiproton nucleus and an orbital positron. One can indeed imagine the existence of the *antimatter* form of each of the elements. No such mirror-image forms have been found. Note that the term "antimatter" does not mean negative mass.

The processes of pair production and annihilation are both inconceivable without relativity. These are cases where an appreciable amount of radiant energy without rest mass and particles with appreciable rest mass are alternately materialized and destroyed.

5.21 GENERAL THEORY OF RELATIVITY

The special theory of relativity is special in the sense that it applies only to situations involving inertial frames of reference, that is, frames of reference in uniform linear motion. It does not matter which reference frame is considered fixed; the conclusions will be the same. In considering the case of a train moving uniformly past a fixed observer, we will apply to any question the same physics that would be applied if we considered the train fixed and the observer moving. In summary, uniform motion is relative.

This leads us to the obvious next question: Is nonuniform motion also relative? Can we devise an experiment that will indicate that one accelerating reference frame is preferred over another? In the case of a spaceship blasting off, does it make a difference whether we consider the earth fixed and the rocket accelerating or whether we consider the rocket fixed and the earth accelerating away from the rocket? Astronauts in a rocket ship exert an enormous force on their seats. This force, an inertial force, is caused by the rocket's acceleration. If indeed accelerated motion is relative, it should be possible to choose the rocket ship as a fixed frame of reference with the earth and the entire cosmos regarded as moving backward away from the rocket. If so, how can the inertial forces that act on the astronauts be explained? This force pressing them against the seat seems to indicate that the rocket moves, not the earth. Another example that Newton accepted as proof that motion was not relative is the rotation of a bucket of water about its vertical axis. The water will be forced up the sides of the bucket due to the centrifugal force. If the bucket is considered fixed and the cosmos rotating about it, then what causes the water to rise up the sides of the bucket? Newton argued that the bucket's rotation was absolute.

Most physicists, after the publication of the special theory of relativity, were willing to believe that uniform motion was relative but that nonuniform motion was not relative. Einstein was not able to accept this thinking and he worried about the problem for many years. In 1916, he published the general theory, so called because it is a generalization of the special theory to include reference frames in nonuniform motion.

At the very heart of Einstein's general theory is the Mach principle of equivalence: The inertial force and the gravitational force are indistinguishable, that is, gravity and inertia are two different words for exactly the same thing. There was a clue that had been around for many hundreds of years, that gravity and inertia were connected: the proportionality of mass and weight. Other things being equal, a heavy football player is more valuable than a light one. When a heavy player blocks, tackles, or runs with the ball, he is harder to accelerate or decelerate than a light player. The heavy player has more inertia or inertial mass, m_I. The force to produce a given acceleration is, from Newton's second law, $F = m_I a$. The heavy football player also experiences a greater gravitational force. He is harder to carry off the field when he is hurt; his cleats make deeper dents in the mud; a stronger bench is required for him when he is not playing. This second property of the player (almost incidental to the game of football) depends on the gravitational mass of the player. The property we are discussing is, of course, the weight of the players, which is

$$W = m_G g,$$

where m_G is the gravitational mass and g is the acceleration of gravity. These two totally different concepts, m_I and m_G, are so closely proportional that we have come to choose our units so that they are identical. Each is measured in kilograms. This proportionality can be shown by considering two falling bodies that have different weights, for example, a lead ball and a wooden ball. (We will ignore air resistance.) Let us assume that the lead ball is one hundred times as heavy as the wood ball. If the two balls are dropped at the same time from the top of a tall building, they will fall side by side striking the ground at the same instant. At the same time that gravity is pulling down on one of the balls, its inertia is holding it back. The force of gravity is one hundred times greater on the heavy ball than on the light ball but the inertia holding the heavy ball back also is exactly one hundred times greater. The force on each ball due to the gravitational attraction between the ball and the earth can be obtained from the gravitational equation

$$F = -G\frac{M_G M_E}{R_E^2},$$

where G is the gravitational constant, M_G is the gravitational mass of the ball, M_E is the mass of the earth, and R_E is the radius of the earth. From Newton's equation of the second law, the force is

$$F = M_I a,$$

where M_I is the inertial mass of the ball and a is the acceleration the ball experiences as it falls from the building. Thus we have from these two equations an expression for the acceleration:

$$a = \frac{M_G}{M_I}\left(G\frac{M_E}{R_E^2}\right).$$

As shown by the experiment of dropping the two balls of different mass (or any number of objects with different masses), a is the same for all masses, which leads directly to

$$M_I \alpha M_G$$

and with the proper choice of units

$$M_I = M_G.$$

This fact was known for many years and various experiments have been performed to test it. Newton was able to show that the conceptually different masses differed by less than one part in one thousand. Later, Bessels, by carefully measuring the period of a pendulum, was able to show that the two masses were equivalent to within one part in 6×10^4. Eötvos in the early 1900s performed an experiment with a pendulum system supporting two masses in such a way that the forces acting on the pendulum would cause a torque if the inertial and gravitational masses differed and showed that if there were a difference it was less than two parts in 10^8. The same type of experiment was performed using modern techniques by R. H. Dicke and it showed that the difference between the inertial mass and the gravitational mass is less than one part in 10^{10}. (See the *Scientific American* article by R. H. Dicke that is referenced at the end of the chapter.) Although this equivalence was known for many years it was first interpreted by Einstein.

Let us consider a passenger in an elevator. In Fig. 5.11(a) we show an elevator being pulled through space, where we assume $g = 0$ with an acceleration of 9.8 m/sec^2. While being pulled through space, a passenger releases a ball and observes that it "falls" toward the floor with an acceleration of 9.8 m/sec^2. An outside unaccelerated observer would say that when the ball is released the floor of the elevator accelerates up toward it with an acceleration of 9.8 m/sec^2. In Fig. 5.11(b) the elevator is shown resting at the surface of the earth where the acceleration of gravity is equal to 9.8 m/sec^2. The passenger releases the ball and once again observes that the ball falls toward the floor of the elevator with an acceleration of 9.8 m/sec^2. This time an outside observer would say that when the passenger dropped the ball it accelerated toward the floor with an acceleration equal to the acceleration of gravity because of the gravitational field of the earth. For this example, the passenger in the elevator has observed exactly the same result for this simple experiment in the two situations as predicted by the principle of equivalence. There is no physical experiment that the person on the elevator can perform that would indicate whether the forces acting on the ball were forces due to inertial

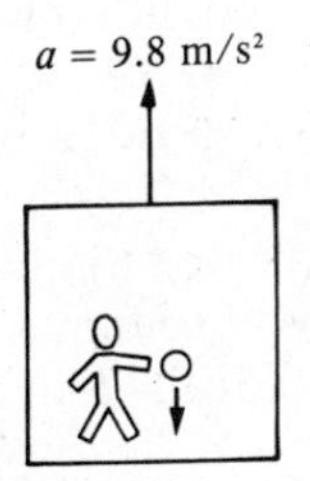

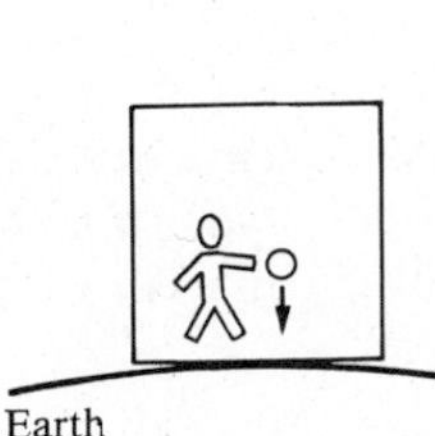

(b)

Figure 5.11 (a) Elevator being pulled with force such that it is accelerating through space at 9.8 m/s^2. (b) Elevator resting on the surface of the earth.

effects of an accelerating elevator or due to the gravitational effects of a nearby large mass.†

If we consider the elevator of Fig. 5.11(a) accelerating through the cosmos, the inertial effects can be observed inside the elevator. But Einstein asserted that we can choose the elevator to be a fixed frame of reference. The entire universe then must be considered to be accelerating past the elevator. In Einstein's view the accelerated motion of the universe generates what Newton called a gravitational field. This gravitational field then causes the objects in the elevator to press against the floor. Depending on one's viewpoint, one can say that these effects are gravitational or inertial. Which is the "proper choice" of frames of reference? According to Einstein, there is no "proper choice"; there is only relative motion of the elevator in the universe. The relative motion creates a behavior described by the equations of general relativity.

In general relativity, Einstein does away with the gravitational mass and gravitational force entirely! How then does he account for the behavior we have so long regarded as gravitational? In the discussion on special relativity, we saw that space and time are interrelated in the relativistic transformation equations. Whereas in classical physics an event is specified by its location (three space coordinates) *and* its time, in relativity an event is specified by four coordinates of space-time. (A dimensionally consistent set of space-time coordinates is x, y, z, and ct, where c is the velocity of light.) Thus Einstein arrived at the concept of four-dimensional space-time. Four-dimensional space-time cannot be visualized, but it is mathematically rational. It is a truism that $a =$ the positive root of $\sqrt{a^2}$. The two-dimensional theorem of Pythagoras states that $a = \sqrt{b^2 + c^2}$. In three dimensions, the diagonal of a rectangular parallelepiped is $a = \sqrt{b^2 + c^2 + d^2}$. It follows that in four dimensions the corresponding relationship is $a = \sqrt{b^2 + c^2 + d^2 + e^2}$.

If we accept the rationality of four-dimensional space-time, then for purposes of visualization we can adopt the "flatland" analogy. Suppose there were a two-dimensional culture with north-south and east-west but no up or down. To these people a line would be an insurmountable "fence" since for them "surmountable" would be impossible to conceive. Flatland people would have the same trouble visualizing a three-dimensional environment that we have visualizing a four-dimensional environment. We might try to make a three-dimensional environment mathematically reasonable to them by extending their understandable theorem of Pythagoras to three dimensions.

We who understand three dimensions so easily might find that flatland was truly a small portion of a very large curved surface — really three-dimensional. Similarly Einstein proposed that we live in an environment that is really four-dimensional but

†At this point we must be careful. The observer in the elevator on earth is in a spherical gravitational field. The observer in the other elevator is in a nonspherical acceleration field. The difference in these two fields can be determined by tests inside the elevator, but these tests distinguish between structures of the fields and not between inertial and gravitational fields.

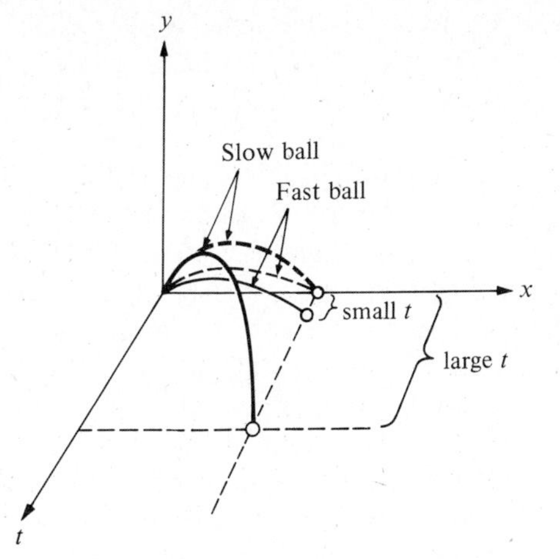

Figure 5.12 Trajectory of a baseball in x, y, t space.

that appears locally to be three-dimensional. In four-dimensional space-time, the shortest "distance" between two events is called a geodesic whereas if "flatland" is really spherical the shortest distance between two points is a great circle.

In general relativity, Einstein asserts that four-dimensional space-time is curved, just as we might find the two-dimensional space of the flatlanders is curved. In general relativity *the curvature of four-dimensional space-time is the result of mass.* Infinitely far from any mass (if that statement has meaning!), an object would move in a straight line (i.e., a geodesic and a straight line would be the same thing†), just as Newton said. Near a mass, an object moves in a curved path, not because of a gravitational force but because the mass distorts the space in its vicinity. (Four-dimensional space-time is not spherically curved. Each mass object creates a "dimple" in its curvature.)

When you throw a baseball the trajectory is curved. If you throw the ball harder, the curvature of its path is less even though the faster ball *seems* to move through the same space. But the faster ball moves through a different space-*time* than the slower. In space-time, the two balls move with the same curvature! To clarify this we note that the flight of a baseball (neglecting any wind) can be represented in two dimensions. This permits us to use the third dimension that we can visualize as time. Figure 5.12 shows the paths of the two balls in x, y, t space. The slow ball rises higher than the fast one (y-direction) but they both are chosen to have the same range (x-direction). Time advances little for the fast ball but time advances more for the slow ball. The projections of the two paths on the x-y plane are quite different. But the curvatures of the two balls in x, y, t space are the same! (This may be hard to see in the three-dimensional picture on a two-dimensional page.) The nearby earth

† The shortest distance between two "points" is called a geodesic in any space — two-dimensional, three-dimensional, etc.

has caused a curved dimple in nearby space-time. Both balls are following the shortest "distance" in space-time. Each is following a geodesic. Neither ball is experiencing gravity as a force. Each is "doing what comes naturally" in a curved space!

We usually think light travels in straight lines. According to general relativity, this effect is due only to the fact that the speed of light is very great. In Fig. 5.12, a flash of light from the origin to the landing place of the balls would be so quick that its path would be indistinguishable from the x-axis itself. The curvature of its path in space-time, however, would be the same as that of the balls.

If one can accept the idea that the paths of balls and beams of light moving near the massive earth are geodesics in space-time curved by the earth, one must still account for the "gravitational" attraction between bodies at "rest" relative to one another. Why does your body press against the floor when you are "standing still" or why do the spheres of the Cavendish experiment[†] attract one another? Newton taught that every object attracts every other object whether they move relative to one another or not — and this appears to be so!

Let two objects be at rest relative to us, each with a fixed x, y, and z. As we observe these objects time is passing. We and the objects are hurtling through space-time! Each object slightly curves the space-time of the other. Left to themselves, these objects would move on geodesics in space-time that are getting closer together in x, y, z space. Unless we hold them apart with nongravitational forces, they will accelerate toward one another. The closer they get, the greater the curvature of their space-times and they will behave as though they were experiencing an inverse square force of gravitational attraction!

An analogy for curved space that is "dangerous" because it relies on gravity to work but is useful because of its dynamics is this: Imagine a frame mounted horizontally over which a thin rubber membrane is stretched. A ball placed on the membrane will depress it, curving the space. Another ball placed on the membrane will also depress the membrane. Each ball "attracts" the other and starts to roll toward the other. If a large ball (representing the sun) is placed on the membrane and if small balls (representing the planets) are tossed onto the membrane toward but not at the large one, the small balls will "orbit" the large one. The small balls (planets) will slightly influence each other as they also depress the membrane.

But what of the situation that was definitive to Newton — the bucket of water in outer space rising up the sides of the bucket when it is rotated? This seems to prove that even if translational accelerations can be understood relativistically, rotational or centripetal accelerations cannot. To an observer who ran around the bucket with the same angular velocity as the water, the water would rise up the sides even though the water would not appear to rotate!

The water-in-a-bucket situation is poor because it requires some force to keep the water in the bucket. Suppose an astronaut experiencing "weightlessness" in space tries the experiment with a glass of water. Unless he has some skill, the water

[†]See any elementary physics text for an explanation.

will soon fly about the spaceship. Instead, let the astronaut swing an object at the end of a string. The string will become stretched. Another astronaut swinging with the object would see the string tighten, the spaceship whirling about him, and, if he could look out a window, the stars whirling about him. According to Einstein, the whirling astronaut must explain the tightness of the string using the same laws of physics as his fellow astronaut. According to Einstein, all motion is relative and all observers are entitled to think of themselves as being at rest.

The real test of this situation would be to have the first astronaut hold the object on the end of the string without twirling it. The object would float about with the string limp. Then we would cause everything *except* the object to rotate: the astronauts, the spaceship, and every star, galaxy, and nebula. What would happen? According to Einstein, the string would become tight. Conversely, if there *were* no astronauts, spaceship, stars, galaxies, or nebulas, the swinging object would not have made the string tight in the first place. The inertial mass of the universe "does something" to the space-time where the experiment is performed. Any time that an object is swung relative to the rest of the universe, the string becomes tight. Both astronauts agree that the object is rotating relative to the rest of the universe!

Now, as in special relativity, this is all very strange, awkward, and difficult. It may be nice to think that *all* motion is relative, that the laws of physics can be the same for all observers regardless of their relative motion, that the "force of gravity" is an unnecessary illusion, and that the concept of gravitational mass is an unnecessary complication. But is not curved four-dimensional space-time with all its mathematical difficulties too great a price to pay for so little gain? Newton's simple theory was so successful that even one who is at ease with general relativity will still use Newton's theory for most calculations.

The final answer to this question is not yet in, but the reason the general theory must be taken seriously is that in addition to its philosophical attractiveness it explains quantitatively a few things that Newton's theory cannot explain. It just may be correct!

As in the case of the special theory of relativity, under certain conditions Einstein's theory reduces to Newton's theory. You will recall that at velocities small compared to that of light the Einstein special-relativity equations reduced to the Newtonian equations. Also, in the case of general relativity, Einstein's equations reduce to Newton's equations in weakly curved space-time. That is, the old Newtonian laws turn out to be good approximations of the Einsteinian laws if the curvature of space-time is not too large.

5.22 TESTS OF THE GENERAL THEORY OF RELATIVITY

There have been very few tests of the general theory of relativity. Observations of the precession of the orbit of Mercury made before the publication of the general theory were explained on the basis of this new theory. Two experiments suggested by

Einstein were carried out within a few years of the presentation of the theory. A fourth test was suggested and carried out in the 1960s.

First let us discuss the precession of the orbit of Mercury. Except for minor interplanetary influences, the orbits of the planets are ellipses, the major axes of which are fixed in their relative positions. But for Mercury — the innermost planet — the major axis "precesses" 575 seconds of arc per century. According to Newton, the planets are moving in response to gravitational forces, and according to Einstein, they are following geodesics in space-time curved mainly by the sun. Newtonian forces including the effects of other planets account for all but 43 seconds of the precession of Mercury. General relativity accounts for the entire precession with no 43-second discrepancy!

Now let us consider the gravitational red shift. It is well known that the motion of a light-emitting source will cause a shift in the frequency or wavelength of the emitted light.† Galaxies that are moving rapidly away from us will have atomic spectral lines shifted toward the long wavelength or red end of the spectrum — the so-called red shift. Einstein's general theory of relativity predicts a similar shift toward the red for sources of light in regions of large space-time curvature.

Consider a tall rocket ship far out in space that at $t = 0$ starts accelerating. Assume that on the inside and at the bottom of the rocket ship is a light source producing light of a unique frequency and on the inside and at the top of the rocket ship is a second identical light source and an observer, as shown in Fig. 5.13(a). Light emitted from the light source at the bottom (light 1) at time $t = 0$, when the rocket is instantaneously at rest in the inertial reference frame, will reach clock 2 and the observer when they are moving with the velocity $v = at$. To the observer, it will appear as if the light reaching him from light source 1 was emitted by a source moving downward with the velocity v and he will observe the light to be Doppler-shifted. When he compares this light with the light from light source 2 next to him, he will observe that the frequency from light source 1 is lower. The frequency of the light from light source 2 is greater than the frequency from light source 1. The equivalence principle tells us that if we consider the rocket ship at rest near a large mass as shown in Fig. 5.13(b), the same effect should be observed. That is, light emitted from light source 1 in Fig. 5.13(b) when it reaches the observer should be shifted toward the low-frequency end of the spectrum. If h is the distance between the two light sources, then the difference in gravitational potential will be gh with light source 2 at the higher gravitational potential. The frequency of the light as it moves to the higher gravitational potential has been decreased or "shifted" toward the red end of the spectrum.

A red shift is observed for light going from the lower to the higher altitude, that is, from the lower to the higher gravitational potential in our example.‡ Alterna-

† This is the Doppler effect in light. See any elementary physics text for an explanation.

‡ The rest mass of a photon is zero. But a photon has an inertial mass $h\nu/c^2$. Without general relativity, the photon would experience a gravitational field causing a red shift. This value, however, is one half that predicted by general relativity.

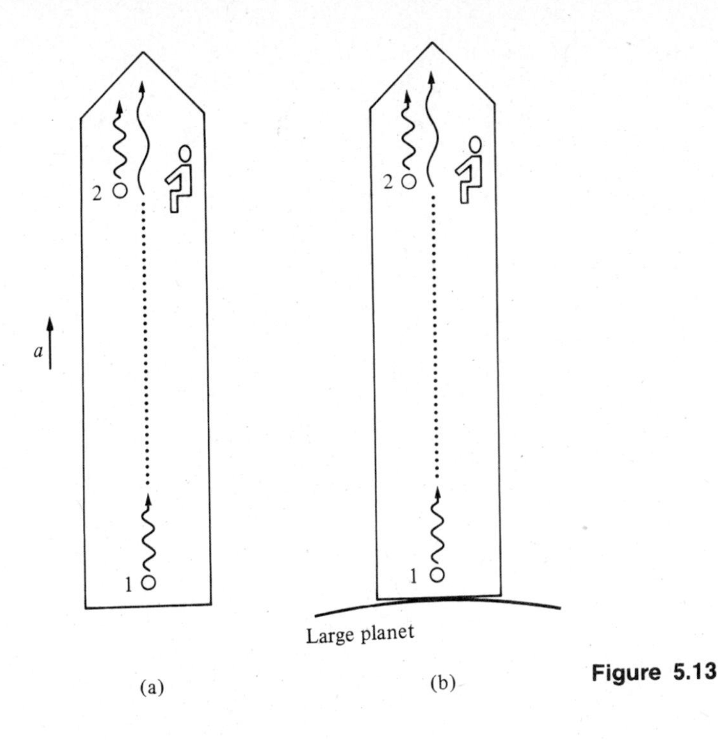

Figure 5.13

tively, the two light sources could be considered as clocks marking the passage of time. If this is the case, then our argument has shown that if the two light sources are considered as clocks with clock 1 in the position of light source 1 and clock 2 in the position of light source 2, clock 2 should register more ticks in a given interval of time and should therefore register a longer time interval than clock 1. All processes should take place faster near clock 2 on top of the stationary rocket than at the bottom of the rocket near clock 1. For example, radioactive atoms should decay at a faster rate near clock 2.

A way to observe the gravitational red shift is to look at the spectral lines of atoms on the surface of the sun (or any other star). An atom on the surface of the sun where the gravitational potential is lower than the gravitational potential far from the sun will emit light of lower frequency than that emitted by the same atom observed here on earth. This effect has been observed in the red shift of the atomic spectra from the sun and also in the atomic spectra from white dwarf stars that are more massive than the sun. The results agree with the prediction of general relativity to within an accuracy of about five percent.

With the discovery of the Mössbauer effect, it became possible to measure with extreme accuracy the gravitational red shift on the surface of the earth. (See Section 11.13.) In this experiment, the radiation was gamma radiation instead of the visible light of the experiments with atoms on the sun. The source and the absorber of the radiation were ^{57}Co samples mounted vertically one above the other with a separation of 22 m. The source and the absorber, of course, correspond to light sources 1 and 2 of Fig. 5.13. In 1960, Pound and Rebka performed the experiment and verified the gravitational red shift.

The third and most interesting prediction of the general theory of relativity is that light passing near a large mass and therefore in a large space-time curvature will be bent. Consider the spaceship now with a hole in the side allowing a ray of light to enter. If it is accelerating upward as shown in Fig. 5.14(a), an observer inside will say that the light is bent down in a parabolic path as shown. The bending of the light is due to the motion of the rocket ship during the period of time it takes the light to travel from the entrance hole to the far side. The principle of equivalence indicates that exactly the same observation should be made in a rocket ship at rest on the surface of a large planet if the space-time curvature is enough to make the effect observable. (See Fig. 5.14b.) The curvature of space-time will cause a light ray to be bent in the direction of the acceleration due to gravity. Einstein suggested a method of detecting the influence of a space curving mass on light. The position of a star relative to the other stars can be measured very accurately, say, for example, in September. Six months later when the earth is on the other side of the sun relative to that star, the light coming from the star passes near the surface of the sun. This light should be bent in the space-time curvature near the sun as shown in Fig. 5.15, causing a shift in the apparent position of the star. The general theory predicts a deflection of starlight in the vicinity of the sun of 1.75 seconds of arc.

Einstein's general theory was published in 1917 and fortunately an eclipse of the sun was predicted for 29 March 1919. (Obviously, a total eclipse of the sun is needed to observe the stars in the daytime, which of course it is if the starlight is to come

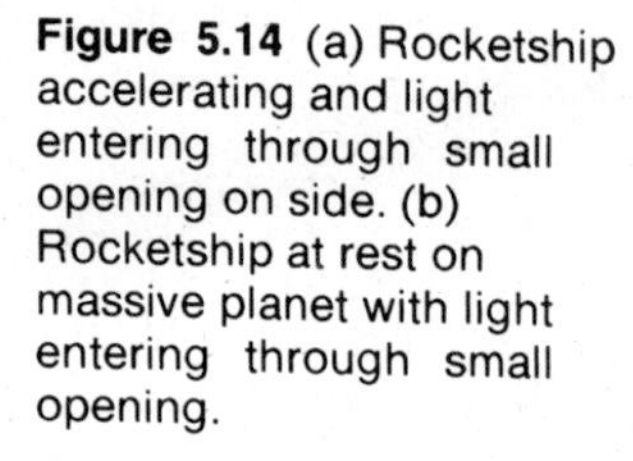

Figure 5.14 (a) Rocketship accelerating and light entering through small opening on side. (b) Rocketship at rest on massive planet with light entering through small opening.

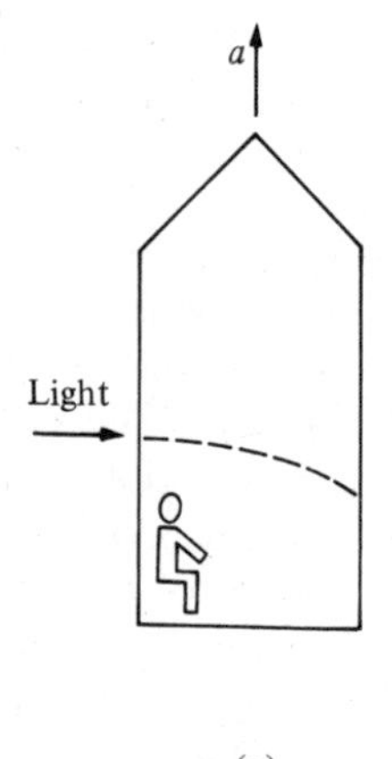

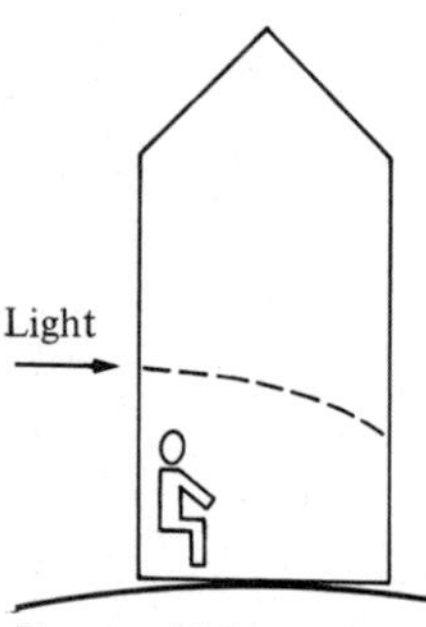

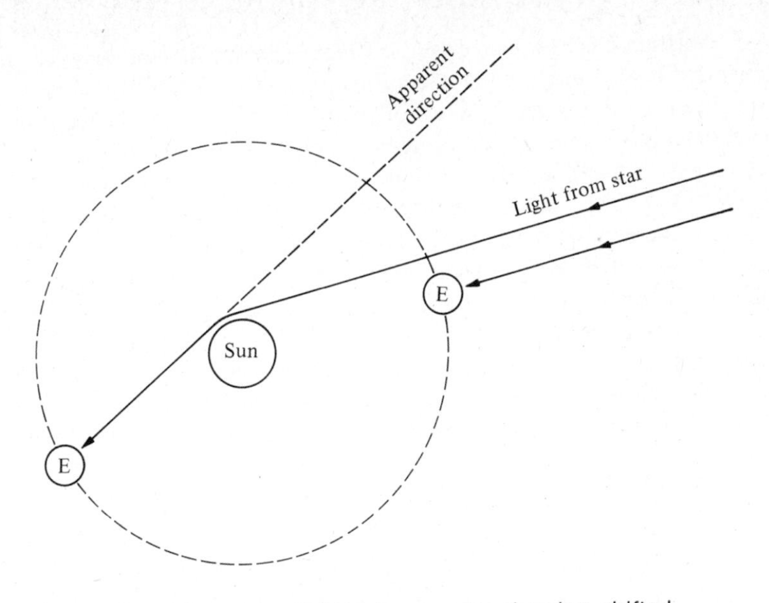

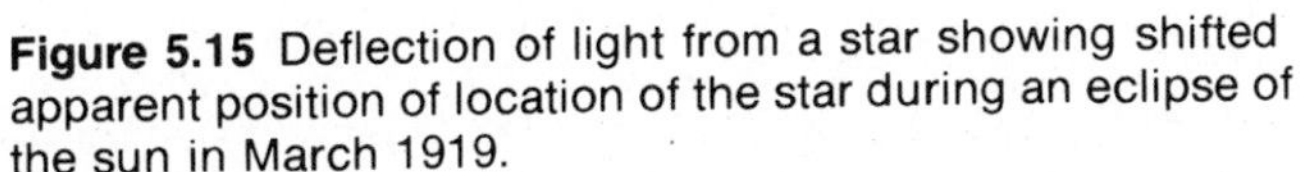
Figure 5.15 Deflection of light from a star showing shifted apparent position of location of the star during an eclipse of the sun in March 1919.

from near the direction of the sun.) Two great expeditions were organized: one to Sobral in northern Brazil and the other to the island of Principe in the Gulf of Guinea. Sir Arthur Eddington joined the expedition to Principe and described the event: "The morning of March 29th was extremely cloudy and remained so until well into the eclipse. However, towards the end of the eclipse the clouds vanished and the scientists were able to obtain clear pictures of some of the stars. Of the many plates, most were inadequate; however, one showed five stars that were in the proper positions to test the Einstein theory. After many months of careful study, the location of the stars indicated that the light had been deviated in its path by the sun 1.64 seconds of arc. This deviation compared very favorably with the 1.75 seconds predicted by Einstein two years earlier."

More recent observations have indicated the possibility of an example of light being deflected by a mass much greater than the mass of a single star. In 1979, two quasar images were discovered that were almost identical.† The images are believed to be generated from the same quasar and are a result of the light passing near a large galaxy or cluster of galaxies on its way to earth. The large mass causes the light from the quasar passing on either side of it to be bent. The slight bending causes the double image. An extensive review of gravitational lenses is given by Frederic Chaffee in the *Scientific American* article referenced at the end of this chapter.

†Quasars are objects in the cosmos that appear star-like except that they emit tremendous amounts of energy and appear to be the farthest objects from us in the universe.

The fourth consequence of the general theory of relativity was pointed out by Irwin I. Shapiro many years after Einstein's work. Professor Shapiro showed that a prediction of the general theory of relativity is that solar space-time curvature will affect the echo times of radar signals reflected from the planets. He calculated the estimated delay time for echo signals from earth to Mercury and back and from earth to Venus and back. The times of the flight were predicted to be retarded by the mass of the sun. The maximum delay would be when the radar signal passed near the sun on its trip to the planet and back, and would be expected to occur when the planet was opposite the earth. When a planet is aligned in this fashion, it is said to be in superior conjunction. Figure 5.16 shows the data from Shapiro's work with the delay in the signal plotted as a function of the time in days. Zero corresponds to the time of superior conjunction on 25 January 1970. His experimental results are in excellent agreement with the predictions of the general theory of relativity.

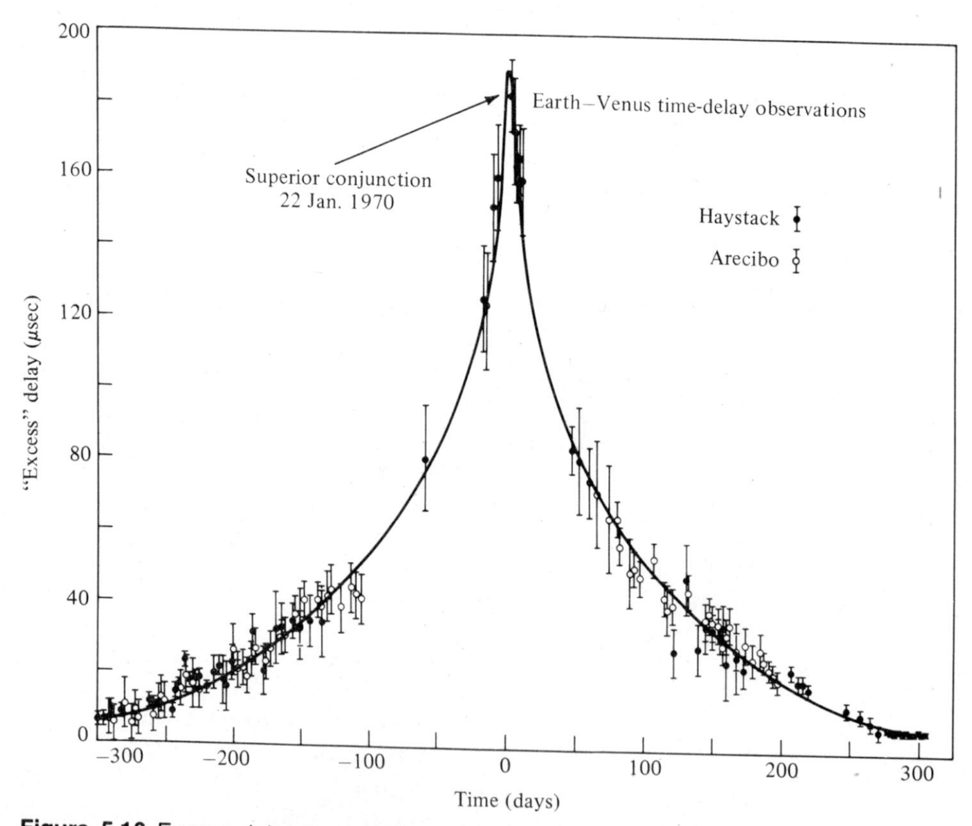

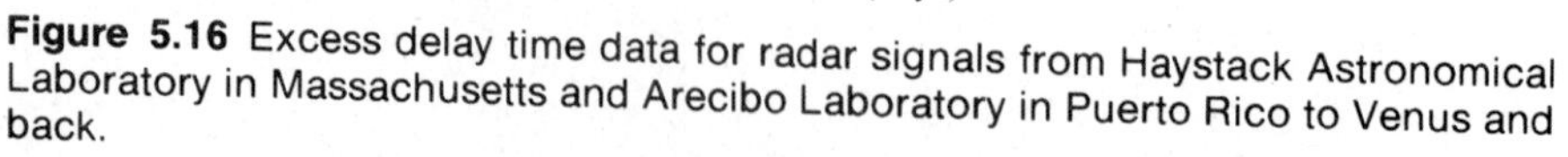
Figure 5.16 Excess delay time data for radar signals from Haystack Astronomical Laboratory in Massachusetts and Arecibo Laboratory in Puerto Rico to Venus and back.

Data courtesy of Irwin I. Shapiro of Massachusetts Institute of Technology.

Another test of the general theory of relativity has been proposed by B. Hoffman and J. P. Richard. Their calculations indicate that the frequency of the radiation coming from a pulsar will be shifted by the space curvature due to the sun. The fractional variation of the frequency arriving at the surface of the earth after passing near the sun is predicted to be on the order of 5×10^{-10}.

The results of these tests of general relativity are impressive but not conclusive. Unfortunately for the theory, the effects outlined in the tests can be explained by other theories. [For example, in his book *Electromagnetism and Relativity* (New York: Harper & Row, 1962), E. P. Ney discusses the first three basic tests using only special relativity and the principle of equivalence — but this is still relativity.]

Predictions of the general theory that have far less than conclusive experimental support include gravitational waves and black holes. First let's discuss gravitational waves. According to the theory, such gravitational waves should be produced by accelerating masses. For years, physicists have been attempting to detect directly these waves with no success. Now, however, a binary pulsar system has been discovered that gives indirect evidence that they exist.† Two masses in orbit about each other are accelerating masses and thus should radiate gravitational energy.

If the two pulsars are emitting radiation and are losing energy,‡ this loss of energy should show up as a change in the orbit of the two objects. That is, the two objects should spiral slowly toward one another causing a change in the period of rotation. Such a decrease has been observed. It should be noted that the gravitational theory of Newton applied to these masses would not predict the radiation of energy and thus the change in the period of oscillation.

Another prediction of the general theory that was pointed out in 1939 by Oppenheimer and Snyder was the possibility of the existence of what is called a black hole. To understand what is meant by a black hole, let's look at the classical quantity defined as the escape velocity. Consider a particle of mass m that is to escape from a planet with mass M. In order for the particle to escape from the gravitational pull of the larger body, the sum of its kinetic energy and potential energy must be at least zero:

$$\mathrm{KE} + \mathrm{PE} = 0.$$

Using the kinetic energy as $mv^2/2$ and the potential energy as $-GmM/r$, one obtains the escape velocity

$$v_e = \sqrt{\frac{2GM}{r}}.$$

†A pulsar is believed to be the final stage of a certain category of stars in which the mass has collapsed to a small size. The density of the material in a pulsar is equal to that of nuclear matter. A pulsar is believed to be a magnetized spinning neutron star. The spinning star sends out rotating beams of electromagnetic radiation.

‡The companion star to the pulsar in the binary system is also thought to be a neutron star, however without its beam ever pointing toward the earth.

This velocity depends on the mass and radius of the larger body. For the earth, which has a mass of 5.977×10^{24} kg and a radius of 6378 km, the escape velocity is about 111.2 km/s. For a much denser object, for example a neutron star with a mass of 1.4 times the mass of the sun and a radius of 1.5×10^4 m,† the escape velocity is 1.6×10^5 km/s. Of course, photons with velocity equal to c could escape from either the earth or a neutron star.

A question that could be asked is how large would M have to be and how small would r have to be in order that the escape velocity would be equal to c? This question cannot be answered from classical physics; however the general theory of relativity yields an answer. According to the general theory, if an object of mass M has a radius given by $R_{\text{sch}} = 2GM/c^2$, the escape velocity would be c. The subscript is for Schwarzschild, the name of the person who first correctly derived the expression. The object satisfying this equation would trap everything including photons. Nothing could escape — not even light; thus the object would be completely black and, therefore, a black hole. The Schwarzschild radius for an object with mass equal to the mass of our sun would be 3×10^3 m. In other words, for our sun to become a black hole it would have to shrink to a radius of 3000 m.

Do black holes exist or are they unfulfilled predictions of the general theory of relativity? Since a black hole cannot obviously be seen in the usual sense, evidence for its existence can only be indicated by the effect on matter in its vicinity. One possibility would be to observe a black hole as a companion to a normal star in a binary system. Another would be to observe x radiation given off by charged particles as they are accelerated toward a black hole. We shall discuss this possibility in Chapter 6.

At this point the student may wonder why there are so few tests of the general theory of relativity. The reason is the extreme difficulty of making significant measurements. The effects on physical quantities of the mass of the earth or of the mass of bodies near the earth are very small and thus the measurement of their effects is extremely difficult.

5.23 THE TWIN PARADOX

One of the most talked about features of the theory of relativity has been the so-called twin paradox or clock paradox. The paradox is usually discussed in terms of a thought experiment involving twins. One twin gets into a spaceship and makes a long trip through space at speeds approaching the speed of light. When he returns,

†A neutron star is believed to be the final stage for stars having a mass approximately 1.4 times the mass of the sun. After a star of this mass burns out its basic fuel, it is believed to collapse under the force of gravity until the gravitational force is balanced by the nuclear force between neutrons. Typical mean density for such an object is about 10^{17} kg/m^3.

he finds that his twin brother has aged far more than he has. In other words, time on the spaceship has gone at a slower rate than time on earth. If the space-traveling twin confines his trip to a short distance at relatively low speeds, the difference in ages will be very small. But over long distances at velocities close to that of light the difference can be large.

In another example, consider an astronaut traveling from the earth to the vicinity of a star 50 light years away. Assume that he makes the trip by accelerating quickly to a velocity of $0.8c$ and travels at this constant velocity to the star. Then he deaccelerates and reaccelerates back toward the earth and makes the return trip at the same velocity. We can assume that the acceleration times are negligible compared with the coasting times. If we were worried that the acceleration time was too long and would cause inaccuracies in our calculations, we could discuss a journey to a more distant star.

The distance to the star is 50 light years and the speed is $0.8c$. Therefore for a person remaining on earth

$$t = 2 \times \frac{50\,ly}{.8c} = 125 \text{ years.}$$

However, according to the astronaut's clock,

$$t = 2 \times \frac{50\,ly\sqrt{1 - \dfrac{v^2}{c^2}}}{.8c} = 75 \text{ years.}$$

The beginning of the twin paradox was the so-called clock hypothesis, which Einstein discussed in his first paper on special relativity in 1905.†

> From this there ensues the following peculiar consequence. If at the points A and B of K there are stationary clocks which, viewed in the stationary system, are synchronous; and if the clock at A is moved with the velocity v along the line AB to B, then on its arrival at B the two clocks no longer synchronize, but the clock moved from A to B lags behind the other which has remained at B by $1/2\ tv^2/c^2$ (up to magnitudes of fourth and higher order), t being the time occupied in the journey from A to B.
>
> It is at once apparent that this result still holds good if the clock moves from A to B in any polygonal line, and also when the points A and B coincide.
>
> If we assume that the result proved for a polygonal line is also valid for a continuously curved line, we arrive at this result. If one of two synchronous clocks at A is moved in a closed curve with constant velocity until it returns to A, the journey lasting t sec, then by the clock which has remained at rest the

†Excerpt from A. Einstein, "Zur Electrodynamik bewegter Körper," *Annalen der Physik* **17**, 891 (1905), translated by W. Perrett and G. B. Jeffery, in *The Principle of Relativity* by Einstein, Lorentz, Minkowski, and Weyl. New York: Dover.

> traveled clock on its arrival at A will be $1/2\ tv^2/c^2$ sec slow. Thence we conclude that a balance clock at the equator must go more slowly, by a very small amount, than a precisely similar clock situated at one of the poles under otherwise identical conditions.

The paradox part of the story arises because there is no preferred frame of reference according to the special theory. In the twin example, call the "stay-at-home" twin Al and call the astronaut twin Bill. Bill climbs aboard his rocket ship and flies off into space and returns after he has aged one year. He finds that Al has aged by many more years than this. Special relativity tells us that we can treat the problem equally by assuming that Bill climbs into his rocket ship and the earth travels in the opposite direction at a high velocity and returns to the rocket ship. For this situation, Bill is the "stay-at-home" twin and Al should come back to find that Bill is much older than he. It seems that this is a symmetrical problem. Whatever journey each has made is relative and it may therefore seem that each observer is free to say that he has not traveled at all and that all of the traveling has been done by the other. Each says the other should be older. This is the paradox. The significant point is that if the twins separate and then return, the basic postulates of the special theory of relativity cannot be satisfied. One twin must accelerate. In our example, Bill climbs into his rocket ship and travels out to some distant point and turns around and returns to earth. In his journey, he first must accelerate away from the earth and then, when he reaches the farthest point of his travel, must decelerate to zero and reaccelerate back toward the earth. At the end of his journey, he must decelerate to land on the surface of the earth. Thus from the viewpoint of special relativity, the problem is not symmetrical. One twin feels forces at the time of acceleration, and the other does not.

Although it is not necessary to invoke the general theory of relativity to solve the problem, if we use it to make the calculations of the elapsed time while the astronaut is traveling with nonuniform velocity and if we use the special theory during the uniform velocity, we would find that we would obtain the same answer that we obtained when we just considered the problem from the viewpoint of special relativity. The twin who ages least is the one who makes the journey and we know that he makes the journey because he is the one that feels the inertial forces at the time he accelerates. But something still seems wrong. Doesn't the general theory indicate that there is no absolute motion, no preferred frame of reference even for accelerating frames? The way out of this paradox is through the principle of equivalence. If we take the situation in which the earth moves away from the rocket ship, turns around at some point in space, and returns to the rocket ship, the entire universe must move with the earth. When the earth decelerates to zero and then reaccelerates back toward the rocket ship, the whole universe decelerates and then accelerates. This accelerating universe has a great slowing effect on the clock and the result is that we obtain the same answer to the problem that we would obtain when we assume it was the rocket ship traveling out and returning back to the earth.

Until recently, it was impossible to perform experiments to test the predictions of relativity on the passage of time as discussed in the twin-paradox problem. However, with the advances in atomic clocks the precision is now such that experiments can be performed. In one such experiment, very accurate clocks were flown aboard commercial airlines, first in one direction around the earth and then in the other. Comparison of the passage of time as measured by the clocks traveling in one direction and then in the other with clocks that remained fixed to the surface of the earth indicated that the predictions of relativity were correct. For a complete review of this experiment, see the articles by J. C. Hafele and R. E. Geating that are referenced at the end of the chapter.

FOR ADDITIONAL READING

Amar, Henri, "New Geometric Representation of the Lorentz Transformation," *Am. J. Phys.* **23**, 487 (1955).

Barker, E. F., "Energy Transformations and the Conservation of Mass," *Am. J. Phys.* **14**, 309 (1946). This is an excellent discussion of the mass-energy relation.

Calder, N., *Einstein's Universe*. New York: Viking, 1979.

Chaffee, F., "The Discovery of a Gravitational Lens," *Scientific American* **243**, 70 (November 1980).

Dicke, R. H., "The Eötvös Experiment," *Scientific American* **84** (December 1961).

Einstein, Albert, *Relativity, the Special and General Theory*. R. W. Lawson, translator. New York: Holt, 1920. A simple, lucid account of relativity.

Einstein, Albert, *Out of My Later Years*. New York: Philosophical Library, 1950. Essays on a number of subjects.

Einstein, Albert, "How I Created the Theory of Relativity," *Physics Today* **35**, 45 (August 1982).

French, A. P., editor, *Einstein*. Cambridge, Mass.: Harvard University Press, 1979.

Frisch, D. H., and J. H. Smith, "Measurement of the Relativistic Time Dilation Using μ-Mesons," *Am. J. Phys.* **31**, 342 (1963).

Gamow, George, *Mr. Tompkins in Wonderland*. New York: Macmillan, 1944. This delightful fantasy is an account of life in a strange, relativistic world where the velocity of light is only ten miles per hour.

Goldsmith, M., A. Mackay, and J. Woudhuyson, editors, *Einstein: The First Hundred Years*. New York: Pergamon Press, 1980.

Hafele, J. C., and R. E. Geating, "Around-the-World Atomic Clocks: Predicted Relativistic Time Gains," *Science* **177**, 166 (1972); and "Around-the-World Atomic Clocks: Observed Relativistic Time Gains," *Science* **177**, No. 168 (1972).

Holton, Gerald, *Special Theory of Relativity*, Resource Letter SRT-1 and Selected Reprints. New York: American Institute of Physics, 1963. Contains, among others, the article by Terrell and the one by Weisskopf on the invisibility of length contraction.

Holton, Gerald, "Einstein and the Crucial Experiment," *Am. J. Phys.* **37**, 968 (1969).

Katz, Robert, *An Introduction to the Special Theory of Relativity*, Momentum Book No. 9. Princeton, N.J.: Van Nostrand, 1964.

Marion, J. B., editor, *A Universe of Physics*. Chapter 4: Relativity. New York: Wiley, 1970.

Michelson, A. A., *Studies in Optics*. Chicago: University of Chicago Press, 1927. This contains descriptions of famous experiments to determine the velocity of light and to measure the ether drift.

Michelson, A. A., and E. W. Morley, "The Michelson–Morley Experiment," in W. F. Magie, *A Source Book in Physics*. Cambridge, Mass.: Harvard University Press, 1963.

Miller, A. I., *Albert Einstein's Special Theory of Relativity*. Reading, Mass.: Addison-Wesley, 1981.

Pais, Abraham, *Subtle is the Lord: The Science and the Life of Albert Einstein*. New York: Oxford University Press, 1982.

Rush, J. H., "The Speed of Light," *Scientific American* **62** (August 1955).

Scott, G. D., and M. R. Viner, "The Geometrical Appearance of Large Objects Moving at Relativistic Speeds," *Am. J. Phys*. **33**, 534 (1965).

Sears, F. W., and R. W. Brehme, *Relativity*. Reading, Mass.: Addison-Wesley, 1968.

Shankland, R. S., "Conversations with Albert Einstein," *Am. J. Phys*. **31**, 47 (1963); and "Conversations with Albert Einstein II," *Am. J. Phys*. **41**, 895 (1973).

Shankland, R. S., "Michelson–Morley Experiment," *Am. J. Phys*. **32**, 16 (1964). A historical account.

Shankland, R. S., "The Michelson–Morley Experiment," *Scientific American* **107** (November 1964).

Shapiro, I. I., "Radar Observations of the Planets," *Scientific American* **28** (July 1968).

Shapiro, I. I., et al., "Fourth Test of General Relativity: New Radar Result," *Phys. Rev. Lett*. **26**, 1132 (1971).

Shipman, Harry L., *Black Holes, Quasars, and the Universe*, 2nd ed. Boston: Houghton-Mifflin, 1980.

Steward, A. B., "The Discovery of Stellar Aberration," *Scientific American* **100** (March 1964).

Taylor, E. F., and J. A. Wheeler, *Spacetime Physics*. San Francisco: Freeman, 1966. An introductory textbook in relativity.

Terrell, James, "Invisibility of the Lorentz Contraction," *Phys. Rev*. **116**, 1041 (1959).

Webster, D. L., "Relativity and Parallel Wires," *Am. J. Phys*. **29**, 841 (1961).

Weisperg, Joel M., Joseph H. Taylor, and Lee A. Fowler, "Gravitational Waves from an Orbiting Pulsar," *Scientific American* **245**, 75 (October 1981).

Weisskopf, V. F., "The Visual Appearance of Rapidly Moving Objects," *Phys. Today* **13**, 24 (1960).

Whittaker, Edmund T., *A History of the Theories of Aether and Electricity*, revised ed. London: Thomas Nelson and Sons, Vol. 1, 1951; Vol. 2, 1953. Comprehensive treatment of the development of various concepts in these fields.

REVIEW QUESTIONS

1. What were the consequences of the null result of the Michelson–Morley experiment?
2. What were the eventual consequences of the Einstein postulates of special relativity on our views of length, time, and mass?

3. What is the significance of pair production to special relativity?
4. What is the Mach principle of equivalence? Explain.
5. What are the four tests of the general theory of relativity?
6. What is meant by simultaneous events?
7. In the twin problem, what is the paradox?

PROBLEMS

5.1 A person who knows Galilean–Newtonian mechanics leaves an earthbound laboratory and establishes an isolated one in a closed, over-the-road trailer that a truck can pull without noise or vibration along a level highway. Is is possible to perform at least one experiment in the new laboratory system to determine whether the trailer has (a) a linear acceleration, (b) a radial acceleration, and (c) a constant linear velocity, or no velocity? Describe an experiment that might be performed in each case.

5.2 An experimenter who is skilled in Newtonian mechanics moves from a laboratory in an astronomical observatory to one in a cave that is shut off from all contact with the outside world. Could any experiment be performed in this new laboratory system that would determine (a) whether the earth is rotating and (b) whether it is moving with constant linear velocity? Describe an experiment that might be performed in each case.

5.3 (a) The ends of a rectangular loop of wire are connected to a sensitive galvanometer. If the whole system (loop and galvanometer) is then moved with constant velocity in a uniform magnetic field where the lines of induction are normal to the plane of the loop, will the galvanometer deflect? Explain. (b) It is proposed to make a ground-speed indicator for an airplane in the following way: A galvanometer on the instrument panel is to be connected to the ends of a wire stretched between the wing tips (or simply to the tips themselves for a metal airplane). It is believed that there will be an induced emf in the wire that is directly proportional to the speed of "cutting" the vertical component of the magnetic field of the earth when in level flight. Will the proposed indicator work? Explain. (Assume that the magnetic field is constant.) (c) Is it possible to measure the speed of an automobile relative to the earth from the fact that it is moving in the earth's magnetic field? Explain.

5.4 A person who is unaware of Newtonian mechanics and who has always lived in a closed trailer is told that a plumb line points in the direction of the force of gravity. (a) In what direction will that person think this force acts if, as viewed by an observer on the earth, the trailer has always been moving at a constant speed around a horizontal circular track? (b) Does a plumb line on your trailer system, the earth, point in the direction of the gravitational pull of the earth? (c) Does it point in the direction of the acceleration due to gravity?

5.5 Three floats are placed as shown in Fig. 5.1 in a river that flows at 0.5 meter per second. It is 100 m from float A to C and from A to D. Assume that two swimmers, each swimming at one meter per second, start at float A and one swims to float C and back and the other swims to float D and back. Calculate the time required for each to complete the trip. What is the ratio of these times?

5.6 A satellite in an equatorial orbit carries an astronaut who with the aid of a telescope observes Polaris (the North Star). If the satellite completes a 40,000 km orbit in 90 minutes, what is the apparent angular displacement of the star due to stellar aberration?

5.7 A super-searchlight that is rotating at 120 rev/m throws its beam on a screen 50,000 mi away. What is (a) the sweep speed of the beam across the screen, (b) the speed of a photon from the searchlight to the screen, and (c) the speed with which any particular photon sweeps across the screen? (d) Do any of these results violate the second postulate of the special theory of relativity? (The value of c is 186,000 mi/s.)

5.8 Cathode-ray tubes have been made in which the sweep speed or trace speed of the electron beam across the fluorescent screen exceeds the free-space velocity of light. Does this require that the velocity of any one of the electrons exceed the limiting relativistic velocity? Explain.

5.9 Show from Eq. (5.25) that the relativistic equations to transform space and time to system S from observations on S' are

$$x = \gamma(x' + vt')$$

and

$$t = \gamma[t' + (vx'/c^2)],$$

respectively.

5.10 The coordinates of an event are x, y, z, and t in the unprimed frame and x', y', z', and t' in the primed frame. Suppose a quantity ds^2 is defined by $ds^2 = dx^2 + dy^2 + dz^2 - c^2\,dt^2$. Show that this quantity is invariant under a relativistic transformation of coordinates, that is, that $ds^2 = ds'^2$.

5.11 At what speed v will the Galilean and Lorentz expressions for length differ by (a) 0.1%, (b) 1%, and (c) 10%?

5.12 Consider two events, which when observed from the S system occur at different points x_1 and x_2 but at the same time t_0. Show that these two events are not simultaneous to an observer in the S'-system, which can move with any velocity v parallel to the x-axis, and that the time difference in S' can vary from $-\infty$ to $+\infty$.

5.13 The relative speed of S and S' is $0.98c$. The following statement is made by S': "At noon a red light flashed at my origin and a blue light flashed 10 seconds later at $x' = 9 \times 10^8$ meters, $y' = 0$, $z' = 0$." What is the temporal separation of the two flashes as measured by S?

5.14 For an observer in a rest frame S, an explosion occurs at $x_1 = 0$ at time $t_1 = 0$. A second explosion occurs at $x_2 = 500$ m at time $t_2 = 10^{-6}$ s. Calculate the velocity of a second observer relative to the first observer if the second observer is to observe the explosions simultaneously.

5.15 A rocket is traveling away from earth at a speed of $0.6c$. The crew of the rocket flashes a distress signal back to earth. Ten minutes later, as measured by someone with a clock on the rocket, the rocket explodes. How far did the earth observer calculate the rocket traveled between the time it sent the distress signal and the time it exploded?

5.16 A man in a super rocket travels between two markers on the ground placed 90 m apart in 5×10^{-7} s at constant speed. This time is measured by a ground observer. (a) How far apart does the man in the rocket judge the markers to be? (b) What time interval does his own watch show elapsed? (c) What is his estimate of the speed of the markers relative to him? (d) Compare this with the ground observer's judgment of the rocket's speed.

5.17 An observer A is on a rocket ship that passes the earth at a speed of 1.8×10^8 m/s. An observer B on earth sets her watch so that it reads zero, the same as A's watch when the ship passes. (a) If observer B looks at A's watch through a telescope, what time does B see on A's watch when B's watch reads 30 s? (b) If A observes B's watch, what time does she see when her watch reads 30 s? (Note that the transit time of the light signal must be taken into account.)

5.18 Clock C_3 in reference frame S' is moving at a speed $v = 0.6c$ along the x-axis with respect to reference frame S. Clocks C_1 and C_2, separated by a distance $d = 300$ meters on the x-axis, are synchronized and at rest in S. C_3 is synchronized with C_1 when their positions coincide. (a) When C_3 arrives at the position of C_2, by how much will he have to advance his clock in order that it read the same as C_2? (b) An observer in S accounts for the result above by the general rule that "moving clocks run slow." The purpose of the rest of this problem is to see how an observer in S' (who must think that C_1 and C_2 are slow) explains C_3 being behind C_2. Consider any two events, for example, at x_1 and x_2 (the position of clocks C_1 and C_2), that O' regards as simultaneous. By how much does O see these events differ in time? Derive an expression and substitute to get a numerical answer. Since this time difference is recorded by C_1 and C_2, what does O' conclude about the synchronization of C_1 and C_2? (c) How much time will O' say has elapsed on *one* clock, either C_1 or C_2, in the other system? Find numerical values of Δt and $\Delta t'$, where $\Delta t'$ is the time, measured by O', required for C_3 to get from C_1 to C_2. (d) According to O', how much greater will C_2 read when C_3 gets there than C_1 read when C_3 was there? Answer this by using parts (a) and (b). Compare to the result of $\Delta t = d/v$, obtained from the point of view of O. (e) How much time has elapsed on C_3? (f) So how much must C_3 be advanced if it is to be set equal to C_2? (g) In summary, O says C_3 must be set up because C_3 is running slow. Why does O' say that C_3 must be advanced (if it is to equal C_2)? This explains in detail how each observer can see the other's clock run slow, yet agree that C_3 is behind C_2.

5.19 An object A is at rest in an unprimed coordinate system and has a length l_A in that system. An object B is at rest in a primed system moving relative to A with velocity v parallel to the length l_A. The length of B in the primed system is $l_{B'}$, and $l_{B'} = l_A$. The measured length of B in the unprimed system is l_B. Assume that the length of B is measured by an observer in the unprimed frame by the method of noting the simultaneous positions of its two ends at a fixed time and measuring the distance between the noted points. (a) Prove that $l_B = (1/\gamma)l_{B'}$. (b) Prove that a similar measurement of the length of object A by an observer in the primed coordinate system yields $l_{A'} = (1/\gamma)l_A$. (c) Show that the unprimed observer sees object A to be longer than B by

$$\Delta l = l_A\left(1 - \sqrt{1 - \beta^2}\right).$$

The object of the rest of the problem is to show that it is not really paradoxical that each observer sees the other length as shorter. Parts (d) and (e) will show that the primed observer exactly accounts for the result in part (c) by observing the other to make two "errors." One of these is indicated by part (b), and the other has to do with simultaneity. (d) If l_B is the "true" length, what length would a primed observer expect the unprimed observer to see if his only error was due to their difference in simultaneity? (e) Since there are two "errors," what length does the primed observer expect the unprimed observer to see for object B? (f) Show that, according to the primed observer, the difference in the "true" length of object B and the length measured by the unprimed observer is just Δl given in part (c).

5.20 A subatomic particle with a mean proper lifetime of 3.0 μs is formed in a high-energy accelerator and projected through the laboratory with a speed of $0.8c$. (a) What is the mean lifetime of the particle in the laboratory system? (b) How far does the particle travel, on the average, from the point of formation in the laboratory before disintegrating? (c) What is distance in part (b) as observed from the reference frame in which the particle is at rest?

5.21 Spaceship A is traveling with a speed of $0.8c$ with respect to the earth. Spaceship B, traveling on a parallel course, passes A with a relative speed of $0.5c$. What is the speed of B with respect to the earth when calculated (a) classically, and (b) relativistically?

5.22 (a) With respect to the laboratory, ball A rolls eastward with a constant speed of 2 m/s and ball B rolls westward with a constant speed of 2 m/s. What is the relative velocity of A with respect to B calculated relativistically? (b) With respect to the laboratory, the electrons in accelerator A are projected toward the east with a speed of 2×10^8 m/s and electrons in accelerator B are projected toward the west with a speed of 2×10^8 m/s. What is the relative velocity of the A electrons with respect to the B electrons calculated relativistically?

5.23 Consider a source of light in the primed frame that is moving parallel to the x-axis with a velocity v relative to the unprimed frame. (a) A beam of radiation from the source is emitted in the $x'y'$-plane at an angle θ' with the x'-axis. Transform the velocity components of the beam, $u'_x = c\cos\theta'$ and $u'_y = c\sin\theta'$, to the unprimed frame and find the resultant of the transformed components. (b) Show that the angle θ made by the beam with the x-axis is given by

$$\cos\theta = \frac{\cos\theta' + (v/c)}{1 + (v/c)\cos\theta'}.$$

(c) The axis of a cone of light in the primed frame is parallel to the x-axis. Given that the half-angle of the cone is 60° and that $v = 0.8c$, find the half-angle of the cone in the unprimed frame. Why would the result in a problem like this be called the "headlight effect"?

5.24 At what speed must an observer approach earth in order that its "thickness" measured along the direction of travel is 10% of its diameter?

5.25 How fast will a ruler have to move in the direction of its length in order to appear to a stationary observer to be half its proper length?

5.26 A stick of length L_0 is at rest in the S-system and is oriented at an angle θ with respect to the x-axis. What is the length of the stick and its orientation angle as viewed by an observer in the S'-system, which is moving with the velocity v parallel to the x-axis?

5.27 In elementary optics, some derivations of the equation for the Doppler effect for light contain a step in which the velocity of the light source is added vectorially to the velocity of the light emitted. Relativistically, the velocity of light is independent of that of the source. How would you explain the existence of the Doppler effect on the basis of the special theory of relativity? [Consider time intervals (frequency), etc.]

5.28 (a) Show that the rest mass of an electron is equivalent to 0.511 MeV. (b) What is the energy equivalence of the rest mass of (1) the proton (2) the neutron?

5.29 An electron in a certain x-ray tube is accelerated from rest through a potential difference of 180,000 V in going from the cathode to the anode. When it arrives at the anode what is (a) its kinetic energy in eV, (b) its relativistic mass, (c) its relativistic velocity, and (d) the value of e/m_e? (e) What is the velocity of the electron calculated classically?

5.30 A kilogram of water is heated from 0°C to 100°C. (a) What is its increase in mass due to its increase in thermal energy? (b) What is the ratio of this increase to the initial mass? (c) Could this mass increase be measured? (d) How much mass is lost by a kilogram of water at 0°C when it freezes to ice at 0°C?

5.31 A quantity of ice at 0°C melts to become water at 0°C and gains 1 kg of mass. What was its initial mass? (The heat of fusion of water is 79.7 kcal/kg.)

5.32 To dissociate CO into C and O requires 11 eV. What is the change in mass of a CO molecule when broken into C and O?

5.33 A doubly ionized helium atom is accelerated from rest through a potential difference of 6×10^8 volts in a linear accelerator. Find (a) its kinetic energy in MeV, (b) its relativistic mass in kg, and (c) its relativistic velocity. (Assume that the rest mass of a helium atom is equal to four times that of a hydrogen atom.)

5.34 In the high energy accelerator at Fermilab (to be discussed in Chapter 11), a proton can be accelerated to an energy of 1 TeV (10^{12} eV). (a) What would be the mass of this proton? (b) What would be the speed of such a proton?

5.35 Show that the expression $E_k = \frac{1}{2}mv^2$ does *not* give the relativistic value of the kinetic energy of a body even if the relativistic mass is used. [*Hint:* Substitute m from Eq. (5.39) into the given relation for kinetic energy, expand by the binomial theorem, and compare the result with Eq. (5.48).]

5.36 Calculate, relativistically, the amount of work in MeV that must be done (a) to bring an electron from rest to a velocity of $0.4c$, and (b) to increase its velocity from $0.4c$ to $0.8c$. (c) What is the ratio of the kinetic energy of the electron at the velocity of $0.8c$ to that of $0.4c$ when computed from (1) relativistic values, and (2) classical values?

5.37 Consider an electron moving at a speed of $c/4$. (a) If the speed is doubled, by what factor will the kinetic energy be increased? (b) If its kinetic energy is increased by a factor of 100, by what factor will its speed be increased?

5.38 Plot the ratio of m/m_0 of a particle as a function of v/c for the following values of the independent variable: 0.0, 0.2, 0.4, 0.6, 0.8, 0.9, and 0.99.

5.39 An airplane has a speed of 500 mi/hr with respect to the earth. (a) For earthbound observers, what is the fractional change in the length of the airplane since take-off due to its speed? (b) What is the fractional change in mass?

5.40 What is the measured length of a meter stick moving parallel to its length when its mass is found to be twice its rest mass?

5.41 A ball has a mass of one kilogram. If you could exert a force of one newton, how far would you have to push it before it has twice that mass?

5.42 An observer at rest with respect to a cube of copper measured the density of the cube to be 8.9×10^3 kg/m^3. What would an observer moving parallel to one edge and at $0.9c$ with respect to the cube measure for the density?

5.43 A particle moves at a speed such that its kinetic energy just equals its rest mass energy. What is the speed of the particle?

5.44 At the surface of the earth the radiation intensity from the sun is

$$0.140 \text{ J/cm}^2 \cdot \text{s}.$$

The present mass of the sun is 2×10^{30} kg and the sun is 149×10^6 km from the earth. Calculate the percent loss in the mass of the sun in (a) one hour, and (b) 10^6 years.

5.45 (a) Show that the speed of a particle whose total energy is E is

$$v = c\left[1 - \left(\frac{m_0c^2}{E}\right)^2\right]^{1/2}.$$

(b) Show that the speed of a particle whose momentum is p is

$$v = \frac{pc}{\left(p^2 + m_0^2c^2\right)^{1/2}}.$$

5.46 An electron having a kinetic energy of 0.50 MeV enters a region that has a uniform magnetic induction of 5×10^{-3} T normal to the motion of the electron. Calculate the radius of the trajectory using classical physics and then repeat the calculation using relativistic physics.

5.47 Some high-energy accelerating devices produce electrons with kinetic energies as high as 10^9 eV. (a) What is the ratio of the mass of the electron at this energy to the electron rest mass? (b) What is the speed of this high-energy electron?

5.48 A certain type of charged particle accelerator fails to work when the mass of the accelerated particle becomes 25% greater than its rest mass. (a) What is the highest kinetic energy this accelerator can give to (1) a proton and (2) an electron? (b) What is the relativistic velocity of the proton and the electron in part (a)?

5.49 Assuming that calculations must be made relativistically instead of classically when the results from the two methods differ by more than 1%, find the potential difference through which a charged particle must be accelerated from rest in a vacuum so that its relativistic or moving mass exceeds its rest mass by 1% of the rest mass if the particle is (a) an electron and (b) a proton.

5.50 Assuming that calculations must be made relativistically instead of classically when the results from the two methods differ by more than 1%, find the potential difference through which a charged particle must be accelerated from rest in a vacuum so that its classical velocity is 1% more than its relativistic velocity if the particle is (a) an electron and (b) a proton.

5.51 When the radioactive isotope cobalt-60 disintegrates, it emits two gamma-ray photons and one 0.31-MeV beta particle (electron). (a) What is the relativistic velocity of the ejected beta particle? (b) If the cobalt-60 is immersed in water, is the velocity of ejection calculated in part (a) less than the velocity of light in water, which has an average index of refraction of 1.33?

5.52 What is the energy equivalent in joules and in kilowatt-hours of 1 kg of (a) uranium, (b) wood, and (c) sand?

5.53 Consider a vehicle powered by converting the mass of 1 kg of nuclear fuel completely into energy. Determine how many gallons of gasoline would be required by conventional engine in order to do the same job. A gallon of gasoline delivers approximately 10^8 joules when it is burned in the conventional way.

5.54 The fissioning of an atom of uranium-235 releases 200 MeV of energy. What percent is this fission energy of the total that would have been available if all the mass of the uranium atom had appeared as energy?

5.55 In a famous experiment on nuclear reactions performed by Cockcroft and Walton, a stationary target composed of the isotope lithium-7 was bombarded with 0.70-MeV hydrogen ions. It was found that these elements then combine and change to two alpha particles, each having a kinetic energy of 9.0 MeV. (An alpha particle is doubly ionized helium-4.) (a) Calculate and compare the total mass, in u, of the initial particles with the total mass of the final particles. (The rest masses of the neutral atoms are given in Appendix 5. Use the conversion factor 1 u/931 MeV/c^2. Assume that all data are precise enough to warrant answers to three significant figures.) (b) Calculate and compare the total energy (includes rest mass energy) in MeV of the initial particles with the total energy of the final particles. (c) Show that the difference between the sum of the kinetic energies of the final particles and the sum of the kinetic energies of the initial particles is equal to the energy equivalent of the difference between the sum of the rest masses of the initial particles and the sum of the rest masses of the final particles. (d) Using the results of the preceding parts, discuss the law of conservation of mass and the law of conservation of energy in relativity. In which part might one say that mass has been "transformed" or "changed" into energy? Why might this be said?

5.56 How much work must be done in terms of the rest mass energy, m_0c^2, to increase the velocity of a particle (a) from rest to $0.90c$, and (b) from $0.90c$ to $0.99c$? (c) What is the increase in mass in terms of m_0 for each case?

5.57 A gamma-ray photon having a wavelength of 0.0045 Å materializes into an electron-positron pair in the neighborhood of a heavy nucleus. (a) What is the total kinetic energy of the pair in MeV immediately after being produced? (b) Could two 0.75 MeV quanta that are passing through lead materialize into an electron-positron pair? Explain.

5.58 An electron and a positron that have negligible velocities combine to produce two-photon annihilation radiation. (a) What is the energy of each photon? (b) What will be the relative direction of motion of these photons? Explain.

5.59 A 2.90 MeV quantum of radiation, passing through lead, materializes into an electron-positron pair. Given that these particles have equal kinetic energies, find (a) the relativistic mass, and (b) the relativistic velocity of each. Neglect the recoil energy of the lead atom. (c) What is the direction of and the smallest value of the magnetic induction that will cause each of these particles to have a trajectory with a radius of curvature of 12 cm?

5.60 If an electron-positron pair is produced near a lead nucleus so that the positron has 11 MeV of energy and the electron has 10 MeV of energy, how far from the center of the lead nucleus was the pair produced?

5.61 Show that it is impossible for pair production to take place in free space.

5.62 Show that it is impossible for a photon to give up all of its energy and momentum to a free electron.

5.63 An electron is accelerated from rest through a potential difference of 2×10^6 volts. Calculate its momentum. What would be the energy of a photon if it were to have a momentum equal to what you calculated for the electron?

5.64 Positronium is a hydrogenlike revolving system composed of a positron and an electron. (a) Show that the energies of the Bohr orbits are one half those of hydrogen. (b) How do the frequencies of the emitted spectra compare with those of hydrogen?

6.1 DISCOVERY

In Chapter 2 we reported how the study of electric discharges through gases at low pressure led Thomson to the discovery of the electron and of isotopes. Let us now consider an earlier discovery made in 1895 in connection with electric discharges: Roentgen's discovery of x-rays. We quote from a translation of Roentgen's original paper, published in 1895:†

> If the discharge of a fairly large induction-coil be made to pass through a Hittorf vacuum-tube, or through a Lenard tube, a Crookes tube, or other similar apparatus which has been sufficiently exhausted, the tube being covered with thin, black card-board which fits it with tolerable closeness, and if the whole apparatus be placed in a completely darkened room, there is observed at each discharge a bright illumination of a paper screen covered with barium platino-cyanide, placed in the vicinity of the induction-coil, the fluorescence thus produced being entirely independent of the fact whether the coated or the plain surface is turned towards the discharge-tube...
>
> The most striking feature of this phenomenon is the fact that an active agent here passes through a black card-board envelope which is opaque to the visible

† Reprinted by permission of the publishers. From William Francis Magie, *A Source Book in Physics*. Cambridge, Mass.: Harvard University Press, 1935.

> and the ultra-violet rays of the sun or of the electric arc; an agent, too, which has the power of producing active fluorescence...
>
> We soon discover that all bodies are transparent to this agent, though in very different degrees. I proceed to give a few examples: Paper is very transparent; behind a bound book of about one thousand pages I saw the fluorescent screen light up brightly, the printers' ink offering scarcely a noticeable hindrance. In the same way the fluorescence appeared behind a double pack of cards; a single card held between the apparatus and the screen being almost unnoticeable to the eye. A single sheet of tin-foil is also scarcely perceptible; it is only after several layers have been placed over one another that their shadow is distinctly seen on the screen. Thick blocks of wood are also transparent, pine boards two or three centimeters thick absorbing only slightly. A plate of aluminum about fifteen millimetres thick, though it enfeebled the action seriously, did not cause the fluorescence to disappear entirely.

The history of science has many instances of "accidental" discovery and of these the discovery of x-rays is a prime example. Columbus's discovery of America was another instance where someone looking for one thing found another. Although accidents can happen to anyone, "accidental" discovery of truth seems to be reserved to those observers who have earned the right to be lucky. Such luck falls on the deserving few whose courage, patience, insight, and objectivity enable them to take advantage of the "breaks of the game." There is a word for this process that is better than the word "accident." It is "serendipity" — the process or art of taking advantage of the unexpected.

Of all the discoveries made by humans, there is probably none that attracted public attention more quickly than the discovery of x-rays. The fact that the rays permitted one to "see" through opaque objects was sensational "tabloid material" and there was great consternation lest, by their use, fully dressed people might be made to appear unclothed. When such speculation died down, however, there was wide appreciation of the value of x-rays in setting broken bones, and the rays were quickly put to this use.

6.2 PRODUCTION OF X-RAYS

As Roentgen said in his paper, the observed x-rays came from low-pressure gas discharge tubes like those already described in our discussion of cathode rays. We quote further from Roentgen:†

> According to experiments especially designed to test the question, it is certain that the spot on the wall of the discharge-tube which fluoresces the strongest

† Magie, *op. cit.*

is to be considered as the main centre from which the x-rays radiate in all directions. The x-rays proceed from that spot where, according to the data obtained by different investigators, the cathode rays strike the glass wall. If the cathode rays within the discharge-apparatus are deflected by means of a magnet, it is observed that the x-rays proceed from another spot — namely, from that which is the new terminus of the cathode rays.

Modern x-ray tubes still produce x-rays by causing cathode rays to strike a solid target, but the techniques have been greatly refined. In a modern Coolidge tube, the source of the cathode-ray electrons is a heated filament. There is no need for a residual gas to be ionized, and so modern tubes are evacuated to a high degree. The target of modern tubes is a metal having a high melting point and a high atomic number. In the production of x-rays, a large amount of heat is generated in the target and so it is usually made hollow to permit cooling water or oil to be circulated through it. In addition to the tube itself, the other major part of the apparatus is a source of high potential to accelerate the cathode-ray electrons. Originally, this was provided by an induction coil (spark coil), but the common technique today is to use

Figure 6.1 X-ray apparatus: Coolidge tube.

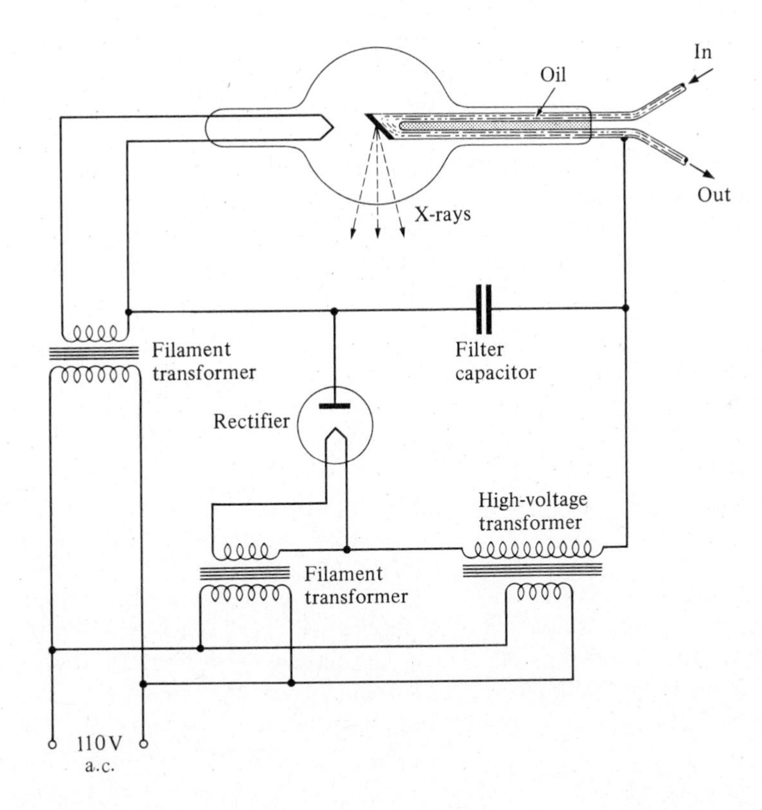

step-up transformers that operate at power-line frequency alternating current. If a.c. is applied to the tube, cathode rays and, consequently, x-rays are produced during only one-half the cycle and the potential across the tube is constantly changing. Most x-ray machines rectify and filter the high alternating potential so that there is a steady high voltage across the tube. The potential used depends on the ultimate use of the x-rays, but it usually ranges between ten thousand and a million volts. A schematic diagram of the essential features of an x-ray machine is shown in Fig. 6.1.

6.3 THE NATURE OF X-RAYS. X-RAY DIFFRACTION

Roentgen failed to determine the nature of x-rays. He eliminated some possibilities and he suggested what they might be, but his own uncertainty is shown by the fact that he chose the name x-rays.

Upon finding that x-rays are not deflected by electric or magnetic fields on the one hand or perceptibly refracted or diffracted on the other, Roentgen concluded that the rays were neither charged particles nor light of any ordinary sort. Tongue-in-cheek, he proposed that x-rays might be the "longitudinal component" of light, in analogy with the fact that acoustic vibrations in solids have both a longitudinal and a transverse part. There is some appeal in the idea that light might have a component that, being longitudinal, might be "slimmer" than the transverse part and therefore penetrate solids more easily than the "fatter" transverse waves. But Maxwell's wave theory of light, which was then at its height, accounted for light being a transverse wave motion, and an effort was made to identify x-rays with this more familiar phenomenon.

In 1912, Max von Laue conceived of a way of testing the idea that x-rays might be light of very short wavelength. Recall the diffraction grating formula,

$$n\lambda = d \sin \theta, \tag{6.1}$$

where n is the spectrum order (1, 2, 3, . . .), λ is the wavelength, d is the grating space, and θ is the angle of diffraction. We can see that if λ is very small compared with d, the diffraction angle must be very small unless the order n is large. Since the intensity in high orders is very weak, we can see that assuming λ to be very small can account for the observed failure of gratings to produce measurable diffraction for x-rays.

The obvious remedy is to make gratings with much finer rulings, although we shall see in Section 6.11 that there is another remedy. But the art of making gratings was already in a high state of mechanical perfection. To improve gratings by reducing their grating space by a few orders of magnitude would have required better materials for both the ruling machine and the grating material itself. However, the basic granularity of matter imposes limitations. This dilemma led Laue to the

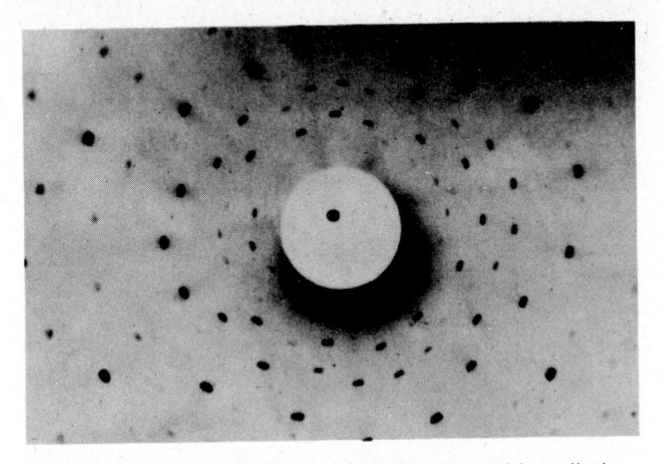

Figure 6.2 Laue diffraction of NaCl taken with radiation from a tungsten-target tube operated at 60 kV. The dark patch below the center disk was caused by scattered x-rays.

idea of taking advantage of the very granularity that stood in his way. It was felt that the regular shapes of crystals, with their plane cleavage surfaces and well-defined edges, must mean that the atoms are regularly arranged throughout their structure. Laue thought that the atoms of a single crystal might provide the grating needed for the diffraction of x-rays. At Laue's suggestion, Friedrich and Knipping directed a narrow beam (pencil) of x-rays at a crystal and set a photographic plate beyond it. The result was a picture like that shown in Fig. 6.2. (Most of the x-rays go directly through the crystal and strike the plate at its center. To prevent gross overexposure at this point it is usual to fasten a disk of lead over the center of the plate. When the photograph shown in Fig. 6.2 was taken, a lead disk masked the center for all but the last second of a 40-min exposure.) Laue knew he had met with success when he observed the complicated but symmetrical pattern on the plate. These spots could be due only to diffraction from the atoms of the crystal. [Strictly, the structural units of such crystals as NaCl and KCl are ions, not neutral atoms. These are called *ionic crystals* because they are held together by strong electrostatic forces acting between charged particles, e.g., (Na^+) (Cl^-) and (K^+) (Cl^-).]

This diffraction pattern is hardly like that produced by a manufactured grating. In optical spectroscopy, the grating consists of parallel lines in a plane. You can appreciate that its diffraction would become more complex if a grating were turned 90 degrees and then ruled again with a second set of lines perpendicular to the first. If many such gratings made of glass were stacked behind one another, the diffraction pattern would be further complicated. Since a crystal consists of a regular array of atoms in three-dimensional space, it produces an intricate pattern.

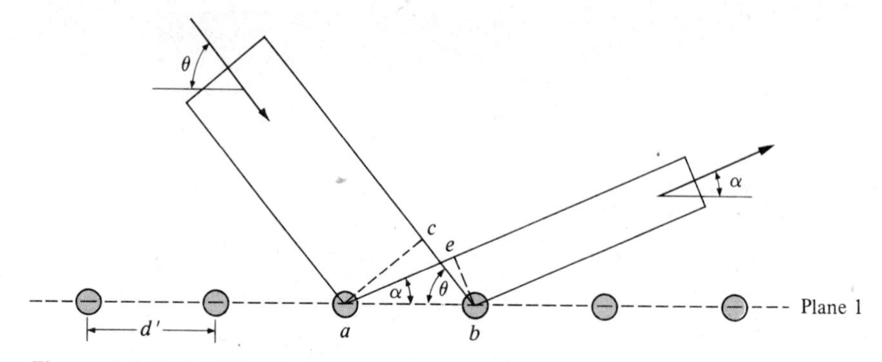

Figure 6.3 Path difference construction for waves scattered from a single plane of atoms in a crystal.

Later we shall discuss qualitatively why a Laue pattern looks as it does. Our immediate point, however, is that by diffracting x-rays Laue demonstrated that x-rays can be combined destructively — the sole, unique experimental criterion for wave motion. At the same time, Laue patterns established that x-rays have very short wavelengths, and confirmed the supposition that crystals have their atoms arranged in a regular structure.

Late in 1912, shortly after the Laue experiment, William L. Bragg devised another technique for diffracting x-rays. Instead of observing the effect created by passing the rays through a crystal, Bragg considered how x-rays are scattered by the atoms in the crystal lattice. Consider an x-ray wavefront incident on a surface row of atoms in a crystal plane, as shown in Fig. 6.3. Each atom becomes a source of scattered x-radiation. In general, the scattered x-rays from all the atoms in the crystal combine destructively as they fall on top of one another in a random manner. If certain conditions are met, however, constructive interference will occur at a few places. One of the relations that must be satisfied for reinforcement can be derived with the aid of Fig. 6.3. In this figure d' is the distance between adjacent atoms, θ is the angle between the incident rays and a row of atoms in the surface plane,† and α is the angle between scattered rays and the surface plane. To obtain the path difference between rays from adjacent atoms, we construct the lines $\overline{ac}$ and $\overline{be}$ perpendicular to the incident and scattered rays, respectively. This path difference is obviously $\overline{ae} - \overline{cb}$, and, for reinforcement, this difference must equal some integral multiple of the x-ray wavelength, or $m\lambda$. Therefore, we have $\overline{ae} - \overline{cb} = m\lambda$, which can be written as

$$d' \cos \alpha - d' \cos \theta = m\lambda. \tag{6.2}$$

The other relation that must be satisfied for maximum reinforcement is that the scattered rays from successive planes of atoms meet in phase. Referring to Fig. 6.4, in which d is the distance between successive planes, we note that rays scattered

† Note that whereas in optics it is usual to measure angles of incidence and reflection between the rays and the normal to the surface, the Bragg angle is measured between the ray and a crystal plane.

from the second plane travel a greater distance than those from the first plane. In order that the rays from successive planes reinforce one another, it is necessary that these additional distances shall be some integral multiple of the wavelength, $n\lambda$. If we draw the line $\overline{ea}$ normal to the incident rays and the $\overline{ec}$ normal to the scattered rays, we see that the length of the path for a ray from the second plane exceeds that for a ray from the first plane by the sum of the distances $\overline{ab}$ and $\overline{bc}$. Therefore the rays from the two planes will interfere constructively when $\overline{ab} + \overline{bc} = n\lambda$, which is equivalent to

$$d \sin \theta + d \sin \alpha = n\lambda. \tag{6.3}$$

In general, the conditions imposed by Eqs. (6.2) and (6.3) cannot be satisfied simultaneously without considering the scattered wavelets from atoms that are in the various layers but not in the plane of incidence. Both conditions are met, however, in the special case where $\theta = \alpha$. Then Eq. (6.2) reduces to zero and Eq. (6.3) becomes

$$n\lambda = 2d \sin \theta, \tag{6.4}$$

where n is the order of the spectrum. When $\theta = \alpha$, we have precisely the condition of regular optical reflection. Because of this, Bragg scattering is usually called Bragg "reflection." This is actually a misnomer, but we shall follow common usage. The planes of atoms in the crystal that are responsible for Bragg reflection are called *Bragg planes*. Let us now summarize the conditions for constructive interference of x-rays scattered from Bragg planes. The *first condition* is that the angle the incident beam makes with the planes must equal that made by the reflected beam; the *second condition* is that the reflections from the several Bragg planes must meet in phase, that is, they must satisfy the relation, $n\lambda = 2d \sin \theta$.

The Bragg technique of using crystals as diffraction gratings for an x-ray spectrometer is shown in Fig. 6.5. The x-rays coming from the tube at the left are

Figure 6.4 Path difference construction for waves scattered by successive planes of atoms in a crystal.

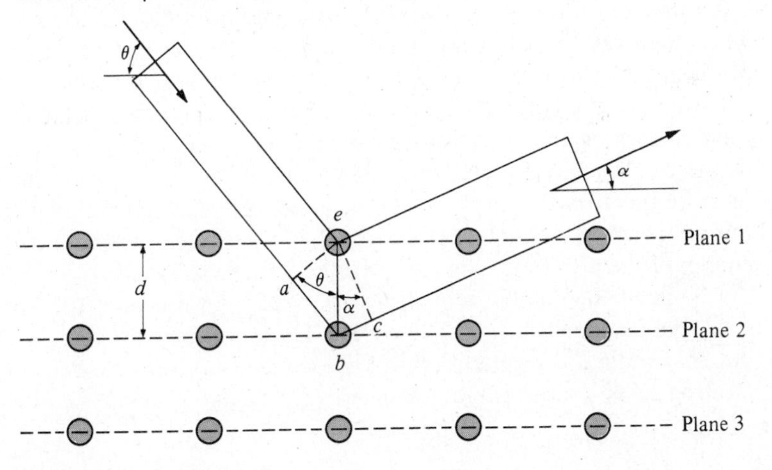

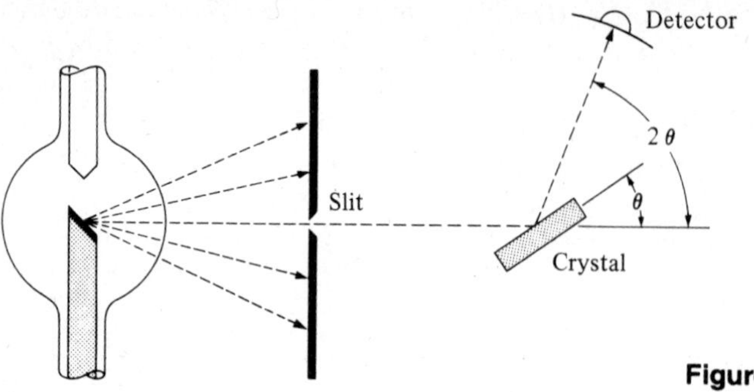

Figure 6.5 Crystal spectrometer.

restricted to a narrow beam or collimated by a lead sheet that absorbs all rays except those that pass through a small slit. These rays fall on a crystal that can be rotated about an axis parallel to the slit and perpendicular to the plane of the figure. The angle θ between the crystal planes and the original x-ray direction can be measured. Bragg reflection can take place only in the direction 2θ from the original x-ray direction. Whether or not the rays actually reinforce one another in this direction depends on whether the second Bragg condition is fulfilled. Since d is a fixed crystal property, the measurements of those angles for which reflections do occur provide a measure of the x-ray wavelengths that are present. In order that spectra measured in this way may be quantitative, the crystal spacing d must be known.

6.4 MECHANISM OF X-RAY PRODUCTION

Roentgen reported in his original paper that x-rays are produced where cathode rays strike some material object. Since we now know that cathode rays are high-velocity electrons, we can restate Roentgen's observation by saying that x-rays are produced when high-velocity electrons hit something. To explore the mechanism of x-ray production, let us ask what happens when electrons strike solid matter.

Many of the electrons that strike matter do nothing spectacular at all. Most of them undergo glancing collisions with the particles of the matter, and in the course of these collisions, the electrons lose their energy a little at a time and thus merely increase the average kinetic energy of the particles in the material. The result is that the temperature of the target material is increased. It is found that most of the energy of the electron beam goes into heating the target.

Some of the bombarding electrons make solid hits and lose most or all of their energy in just one collision. These electrons are rapidly decelerated. We have pointed out that radiation results when a charged body is accelerated. Therefore when an electron loses a large amount of energy by being decelerated, an energetic pulse of electromagnetic radiation is produced. This is an *inverse photoelectric effect* in which an electron produces a photon. In Chapter 3 we found that photons of a given energy produce photoelectrons with a certain maximum energy. Here we find that electrons of a given energy produce x-ray photons with a certain maximum energy. According to classical electromagnetic theory, there is no lower limit to the wavelength of the radiation that a moving electron can produce when it is stopped suddenly. But there is a quantum limit. Given that an electron has been accelerated through a difference of potential of V volts, we can use Eq. (3.43), which Duane and Hunt showed was valid for x-rays, to compute the wavelength of the resulting radiation in angstroms if the electron loses all its energy in a single encounter. In this case the energy loss of the electron in eV is *numerically* equal to the accelerating potential in volts. Therefore we have

$$\lambda_{\min}(\text{Å}) = \frac{12400}{V}. \tag{6.5}$$

This expression gives the minimum wavelength, since no electron can lose more energy than it has, and there will be a continuous distribution of radiation toward longer wavelengths because there are all sorts of collisions, from direct hits to glancing ones. Thus glancing collisions account for the *continuous spectrum* of x-rays from any target material and also for the inefficiency of the conversion of electron energy into x-ray energy. The Germans named this continuous radiation *bremsstrahlung*, meaning literally "braking radiation." This is a highly descriptive term, since it refers to the radiation that results from the braking or stopping of charged particles. Some continuous spectra are shown in Fig. 6.6.

Looking at the collision process more closely, however, we find there is another very important kind of collision energy exchange. The bombarding electron may also give energy to electrons bound to the target atoms. If these atomic electrons are freed from their home atoms, ions are produced. Since x-ray producing electrons have energies of the order of many thousands of electron volts, it is very easy for them to produce ions by removing outer electrons. X-ray producing electrons may also have enough energy to produce ions by removing inner electrons from the atom, even down to the innermost or K-shell. Such an ion has a low-energy hole in its electronic structure, and this vacancy is promptly filled when one of its electrons in a higher energy state falls to this low-energy level. When an electron falls into a low-energy level, it releases its energy as radiation, as happens in the Bohr theory. Although the energy required to ionize an atom by removing an outer electron is much less than 100 eV, the energy required to ionize by removing an inner electron may be as high as 120,000 eV. When an outer electron falls into such a vacancy, it will radiate a photon of this energy. Such photons are in the x-ray region and have wavelengths that are fractions of angstroms. This mechanism, which accounts for a significant part of x-ray production, produces x-rays having particular wavelengths that are *characteristic* of the target material.

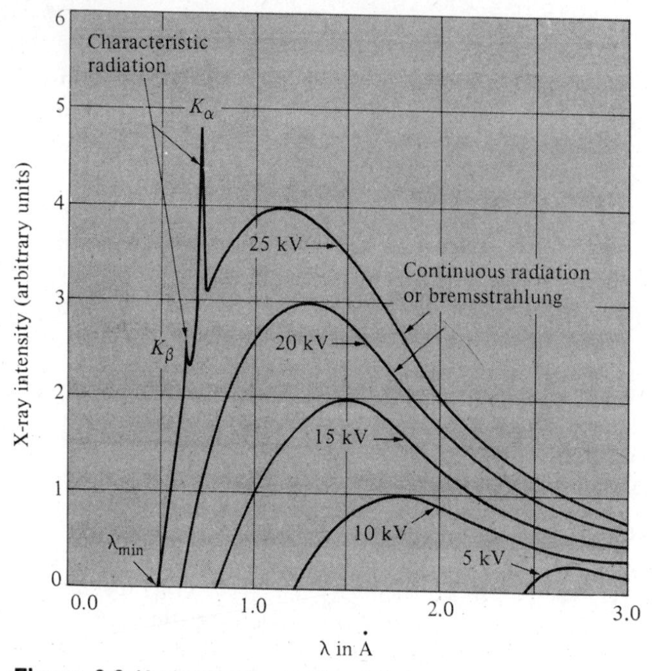

Figure 6.6 X-ray spectrum of molybdenum as a function of the applied voltage. Line widths are not to scale. The K-series excitation potential of this element is 20.1 kV. The shortest L-series wavelength is 4.9 Å.

6.5 X-RAY ENERGY LEVELS

In our discussion of the periodic table, we noted that a heavy element has two electrons with $n = 1$, eight with $n = 2$, etc. The Pauli exclusion principle requires that no two electrons can have identical quantum numbers and thus no two electrons can have identically the same energy. But n is the principal quantum number and it has the most to do with the energy of the electron. Thus the two electrons with $n = 1$ have energies that differ only slightly. The electrons for which $n = 2$ have much more energy than those for which $n = 1$, but they differ among themselves very little. In discussing x-rays, it is often convenient to ignore the minor energy differences of those electrons having the same n and concentrate on the large differences associated with different values of n. In the discussion of shell structure in Section 4.17, we said that those electrons having $n = 1$ constitute the K-shell electrons. Those electrons with $n = 2$ form the L-shell. The heaviest elements have seven populated shells, K, L, M, N, O, P, and Q. However, the N-, O-, P-, and Q-shells have energies that give rise to the optical spectrum and their energies are *negligible* as far as x-rays are concerned.

With these simplifications, we can at once make an energy-level diagram that is typical of all the heavy elements (see Fig. 6.7). It is important now to point out the similarities and differences between this energy-level diagram and that of hydrogen.

The one electron in hydrogen is normally in the level for which $n = 1$, and the higher levels in hydrogen are vacant unless the atom is excited. The x-ray diagram represents a typical heavy element having an atomic number of about 40. This element has many electrons. In general, the levels of the heavy elements are occupied by 2, 8, 18, 32, etc., electrons. A hydrogen atom may be excited from the state $n = 1$ to the state $n = 3$. But in a heavy element, a K-shell electron, $n = 1$, cannot be excited to the states $n = 2$ or $n = 3$, since these L- and M-shells are normally filled.

The electron in hydrogen is bound to its nucleus by only one proton with a charge of plus e. The inner electrons of a heavy element are bound by electrostatic forces due to a nucleus with a large charge Ze. This difference in nuclear charges leads to vast differences in the energies of the various levels. Whereas the level having $n = 1$ in hydrogen has an energy of only -13.6 eV, in a heavy element the level for which $n = 1$ has an energy of the order of minus thousands of electron volts. It is $-20{,}000$ eV in our hypothetical example.

For both light and heavy elements, each level except K is actually a narrow group of levels within which we are ignoring the energy range. Indeed, in x-ray studies, we frequently ignore the energy difference between the energy of the N-shell

Figure 6.7 Hypothetical x-ray energy-level diagram.

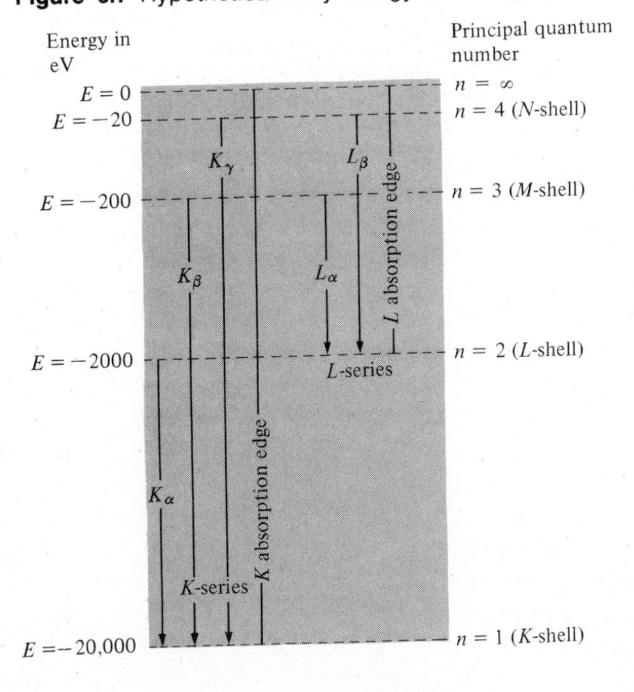

($n = 4$) and the energy of a free electron ($n = \infty$). This leads to an interesting result: All heavy elements have x-ray spectra that are similar in character.

Suppose the hypothetical substance whose energy-level diagram is shown in Fig. 6.7 were the target of an x-ray tube across which there is a potential difference of 50,000 volts; then all the cathode-ray electrons strike this target with 50,000 electron volts of energy. Most of these electrons fritter their energy into heat. Some cathode-ray electrons undergo collisions that produce radiation quanta having energy equal to the energy loss in each collision. These cause x-rays with a *continuous* spread of wavelengths down to 0.248 Å, as determined by the exciting voltage. Some cathode-ray electrons undergo collisions that cause the removal of target electrons from the various target shells. If an outer, N-shell electron is knocked out, the optical spectrum of the target material will be excited. Light will be emitted as electrons fall into the N-shell vacancy. This is a negligible effect so far as x-rays are concerned, and the light cannot be seen, since it is usually produced at a slight depth within the target, which is opaque to visible light. If an M-shell electron is removed, the same remarks apply except that the radiation is in the far ultraviolet region. If an L-shell electron is knocked out, we have a more interesting situation. An L vacancy can be filled in two important ways: An electron can fall into the L-shell from outside the atom (or from the N-shell — we ignore the difference), producing an x-ray of wavelength 6.20 Å, or the vacancy in the L-shell can be filled from the M-shell. This energy difference is 1800 eV, and causes an x-ray of wavelength 6.89 Å. (In the latter case, the vacancy in the M-shell will also be filled, by the production of a far ultraviolet photon.) These x-radiations of the target are called the L_β and L_α radiations, respectively. Since it is more probable that the L-shell will be filled from the nearer M-shell, the less energetic or longer wavelength L_α radiation is more intense than the L_β. If the bombarding electrons had less than 2000 eV of energy, they would not be able to remove electrons from the L-shell and *neither* the L_α *nor* L_β radiation could result.

We have assumed, however, that the bombarding electrons have 50,000 eV of energy, so that they can not only remove electrons from the L-shell but also from the K-shell. If a K vacancy is filled from the L-shell, the radiation would have wavelength 0.689 Å and would be called K_α radiation. (The consequent L vacancy would then result in L_α or L_β radiation.) If the K vacancy is filled from the M-shell, there will be more energetic radiation, K_β, of wavelength 0.626 Å. Or if the K vacancy is filled from the N-shell (or from infinity), there will be slightly more energetic radiation, K_γ, of wavelength 0.620 Å. These K-series radiations are listed in the order of their relative probability. The K radiations of this example are quite energetic or "hard," whereas the L-series radiations are in the "soft" x-ray region.

In this example, we have used a hypothetical substance in order to have simple numbers to work with, but what we have done is typical of all x-ray spectra. If we ignore the small energy differences of the electrons within a given shell, all elements produce x-ray spectra of similar character and with very few lines. The shortest wavelength associated with a shell is called the *absorption edge wavelength* of that shell, for reasons given later in this chapter. The corresponding energy is equal to that needed to remove an electron from the shell to infinity.

6.6 X-RAY SPECTRA OF THE ELEMENTS. ATOMIC NUMBER

In 1913, Moseley used a large number of elements as x-ray tube targets and found that the K_α radiation of each element was distinct from that of any other element. This result not only provides a unique way of identifying the various elements but also is of theoretical significance. Specifically, Moseley's law states that the frequencies of corresponding x-ray spectral lines, such as K_α's, as we go from element to element, may be represented by an equation of the form

$$\sqrt{\nu} = a(Z - b), \tag{6.6}$$

where a and b depend on the particular line and Z is the atomic number of the element.

The original ordering of the elements in the periodic table was based on their atomic weights. Initially, atomic numbers were mere ordinal numbers specifying where the element lay in a list based on these weights. These atomic numbers had no more physical significance than the house numbers that identify the houses on a given street. Moseley found that if he arranged the wavelengths of the K_α lines in the order of atomic weights, these wavelengths formed a remarkably regular progression. The sequence was not perfect, however, since he found both gaps and wavelengths out of order. He attributed the gaps in the series to undiscovered elements and proposed that there should be a unique correlation between the wavelength series and atomic number. His words were, "We have here a proof that there is in the atom a fundamental quantity, which increases by regular steps as we pass from one element to the next. This quantity can only be the charge on the central positive nucleus." In Moseley's day, nickel with an atomic weight of 58.69 was listed ahead of cobalt with an atomic weight of 58.94. There was some chemical evidence that the order of these two elements in the periodic table should be reversed, and Moseley demonstrated this by showing that the atomic number of cobalt is 27 and that of nickel is 28.

Moseley's work was announced in 1913, the same year that Bohr accounted for the hydrogen spectrum. In the case of hydrogen, Bohr came to the same conclusion that Rutherford had reached from alpha-particle scattering and that Moseley verified for all elements. Recall that in the Bohr model the charge of the nucleus was basic to his analysis and the mass of the nucleus played a minor role. Note how nicely Bohr's idea of electrical forces accounts for the Moseley observation of the regular progression of the K radiations. The electron in hydrogen is a K-shell electron held to the nucleus by the attraction of one proton. The two K electrons of helium are held by two protons and thus their binding energies are greater (more negative) than that of the electron in hydrogen. The two K electrons of a heavy element are bound by the attraction of many protons. As we move from one element to the next, the energy to remove a K electron, the K binding energy, increases in a regular way.

We can now combine the work of Bohr and Moseley in an interesting and quantitative way that strengthens both of their views. From Eq. (4.22) for the frequency of the spectral lines of hydrogen, we have

$$\nu = \left(\frac{m_e e^4}{8\varepsilon_0^2 h^3}\right) Z^2 \left(\frac{1}{n_1^2} - \frac{1}{n_2^2}\right),$$

or

$$\sqrt{\nu} = \left[\left(\frac{m_e e^4}{8\varepsilon_0^2 h^3}\right)\left(\frac{1}{n_1^2} - \frac{1}{n_2^2}\right)\right]^{1/2} Z. \tag{6.7}$$

This bears some resemblance to Moseley's law, so we are encouraged to go further. Suppose we remove one of the two K-shell electrons of an element to infinity. The remaining outer electrons will be attracted to the nucleus not by the nuclear charge Ze as Bohr assumed, but by a charge $(Ze - e)$. One may think of the remaining K electron as "screening" the outer electrons from the full nuclear attraction. Thus the b term of Moseley's law is a nuclear screening constant and it equals one for the K-series lines. Since the K_α line is a transition from $n_2 = 2$ to $n_1 = 1$, we have everything we need to attempt a Bohr theory calculation of an x-ray K_α line. Choosing molybdenum ($Z = 42$), we have

$$\nu = \frac{m_e e^4}{8\varepsilon_0^2 h^3}\left(\frac{1}{1^2} - \frac{1}{2^2}\right)(42 - 1)^2.$$

Expressed in energy units, we obtain 17.2 keV for this line, which is in remarkable agreement with the experimental value of 17.4 keV.

6.7 X-RAY ABSORPTION

The most spectacular property of x-rays is their ability to penetrate materials that are opaque to less energetic radiation. The basic mechanism of absorption of ultraviolet, visible, and infrared radiation is the transfer of photon energy to the electronic, vibrational, and rotational energy states of the material doing the absorbing. For an x-ray photon, whose energy is orders of magnitude greater than visible radiation, these mechanisms are trivial because a high-energy photon has a small probability of interacting in a comparatively low-energy process. High-energy x-ray photons are more likely to interact with electrons in the *K*- or *L*-shells. Since the probability of an x-ray interacting with the many loosely bound electrons in an absorber is small and since there are relatively few tightly bound electrons, high-energy (short-wavelength or hard) x-rays have remarkable penetrating ability.

Roentgen first detected x-rays by the fluorescence they produce in certain materials. This effect can be used for quantitative intensity measurements if it is coupled with an objective measurement of the fluorescent light produced. Roentgen also observed that x-rays blacken a photographic plate, and this may be used to measure the intensity if a densitometer is used to measure the blackening. A third method was suggested by Roentgen's experiments on the conductivity of air due to x-rays. The x-rays ionize the air, and this effect can be measured quantitatively with an ionization chamber. For a detailed description of an ionization chamber, see Chapter 10.

An ionization chamber can be used to study the penetrating ability of x-rays with apparatus like that shown in Fig. 6.8. X-rays are collimated by slits, rendered monochromatic by a Bragg reflection, and passed through a material under study; then their intensity is measured by the ionization chamber. If we vary the thickness of the absorbing material, a plot of transmitted intensity against thickness looks like that shown in Fig. 6.9. This is an exponential decay curve that can be derived by assuming that a small thickness of absorber, dx, reduces the intensity of the beam by an amount $-dI$ proportional to both the intensity I and the thickness dx. This assumption leads to the differential equation

$$dI = -\mu I \, dx, \tag{6.8}$$

Figure 6.8 Schematic diagram of the apparatus for x-ray absorption experiments.

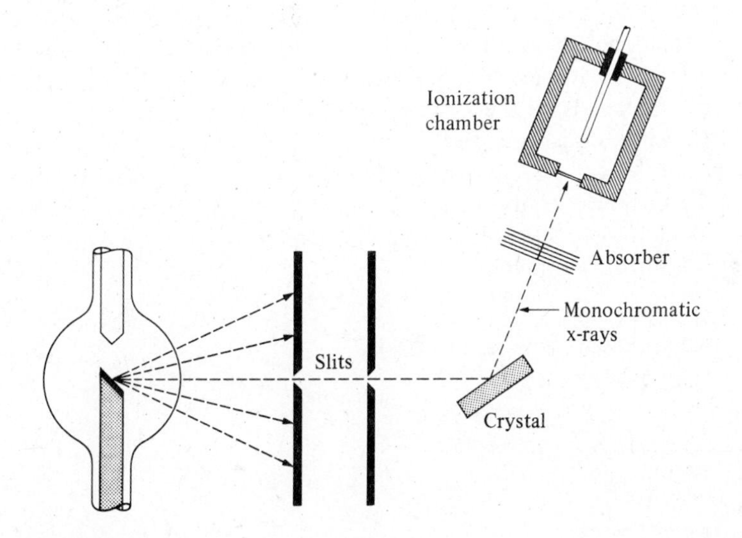

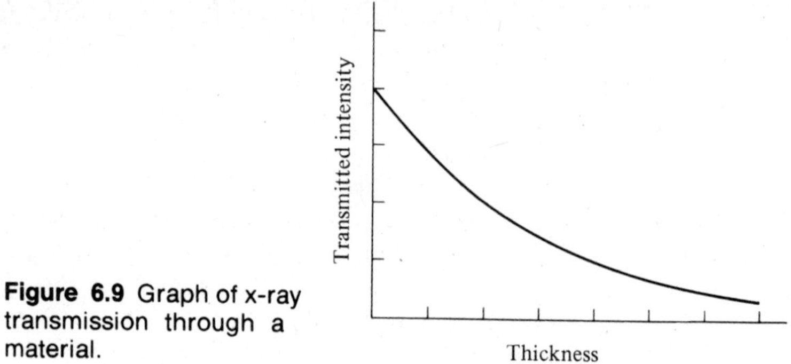

Figure 6.9 Graph of x-ray transmission through a material.

where μ is a constant of proportionality. If the intensity of the beam that is incident on a layer of material of thickness x is I_0, then I, the intensity of the *transmitted* beam, can be found by integrating Eq. (6.8). We have

$$\int_{I_0}^{I} \frac{dI}{I} = \int_0^x -\mu\, dx, \tag{6.9}$$

which gives

$$\ln I - \ln I_0 = -\mu x$$

or

$$I = I_0 e^{-\mu x}. \tag{6.10}$$

This is the equation of the curve in Fig. 6.9.

The quantity μ is called the *linear absorption coefficient*, although the terms *macroscopic absorption coefficient* and *linear attenuation coefficient* are also used. When Eq. (6.8) is solved for μ, we see that the linear absorption coefficient is equal to the fractional decrease in intensity of the radiation per unit thickness of the absorber. Thus it has the dimensions of reciprocal length. The value of this coefficient depends on the x-ray wavelength and the material used.

It is interesting to express the absorption properties of a material in another way. If we let $I = I_0/2$, we can solve for the corresponding value of x. Since this gives the thickness of absorber that reduces the intensity of the transmitted beam to one half that of the incident beam, it is called the *half-value layer*, T, or hvl. Thus, from Eq. (6.10), we have

$$\ln 2 = \mu t \tag{6.11}$$

or, if we use logarithms to the base 10,

$$2.30 \log 2 = \mu T,$$

or

$$T(\text{hvl}) = \frac{0.693}{\mu}. \tag{6.12}$$

Although the equations we have derived are correct for a *narrow* beam of x-rays, they must be modified before they are applied to the general case of shielding against dangerous radiation. The absorber in the apparatus shown in Fig. 6.8 reduces the intensity of the beam both by "soaking up" x-rays and by scattering them in a new direction. Thus when a broad beam of x-rays passes through an extended absorber, x-ray photons removed from the beam at one point may be scattered into it again at another point. When these photons have less than 200 keV of energy, a rule-of-thumb method for correcting for the scattered rays is to assume that the source is one and one-half times as intense as it is measured to be. This factor of 1.5 is called the buildup factor and has higher values for photons of higher energies.

The mechanism of absorption that we have considered is the photoelectric effect, in which the x-rays lose energy by ejecting electrons from the absorbing material. There is another effect which is weaker than the photoelectric effect in the usual range of x-ray energies but which becomes more important than the photoelectric effect at high x-ray energies (about 0.1 MeV and above). This is Compton scattering, which will be discussed in Section 6.12. In this effect an electron is ejected along with an x-ray of lower energy (see Fig. 6.10b). A scattered x-ray is not really absorbed, since it does not lose a very large fraction of its energy. But, as mentioned earlier, the apparatus of Fig. 6.8 is sensitive to scattering, since an x-ray photon deflected out of the beam will fail to reach the detector, just as if it had been absorbed completely.

Figure 6.10 Summary of x-ray and gamma-ray interaction with matter.

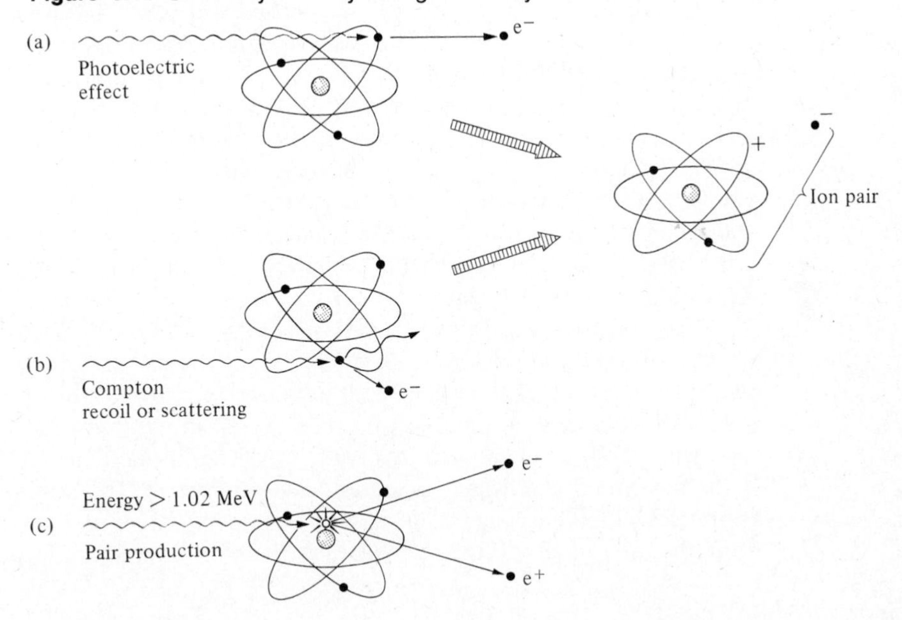

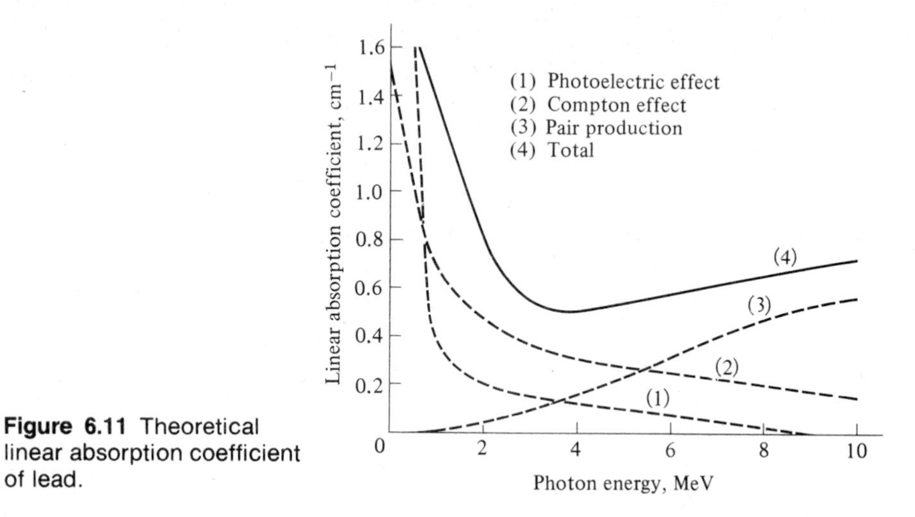

Figure 6.11 Theoretical linear absorption coefficient of lead.

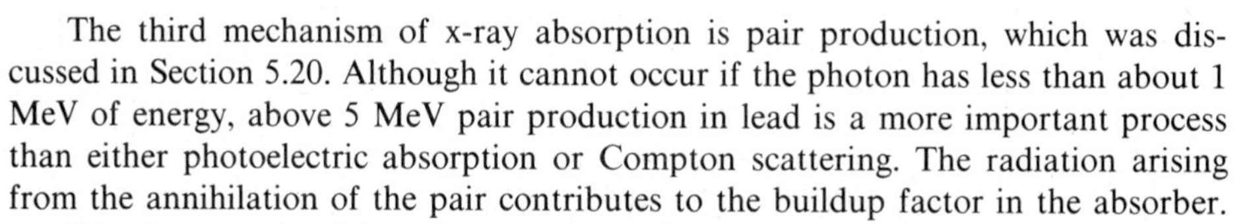

The third mechanism of x-ray absorption is pair production, which was discussed in Section 5.20. Although it cannot occur if the photon has less than about 1 MeV of energy, above 5 MeV pair production in lead is a more important process than either photoelectric absorption or Compton scattering. The radiation arising from the annihilation of the pair contributes to the buildup factor in the absorber.

The three mechanisms of x-ray absorption are summarized in Fig. 6.10. The contribution of each of these to the linear absorption coefficient in lead for photons of different energies is shown in Fig. 6.11.

The value of the linear absorption coefficient depends on the x-ray wavelength and on the absorbing material. The basic variation of μ with λ is that μ is very nearly proportional to λ^3. But this basic variation is profoundly influenced by the nature of the absorbing material. If the x-ray wavelengths are so long that they can excite only outer electrons of the absorbing material, then interaction between x-rays and the *K*, *L*, or *M* electrons of the absorber must be elastic. But if the x-rays have shorter wavelengths, they may be able to eject *M*, *L*, or even *K* electrons. The various absorption mechanisms are additive, and each varies approximately as λ^3. The variation of μ with the wavelength is shown in Fig. 6.12. In this figure, the λ's are absorption edge wavelengths.

X-rays of short wavelength are very penetrating despite the fact that they can be absorbed by ejecting electrons from any shell of the absorber, that is, by the photoelectric effect. If the wavelength is longer than the *K* absorption limit of the absorber, the x-rays are unable to ionize the absorber from the *K* level and one absorbing level is eliminated. At this wavelength, there is an abrupt change in the absorption coefficient, called an absorption edge. The absorption coefficient again rises with increasing wavelength until suddenly the x-rays are unable to ionize from the first level of the *L*-shell and there is another absorption edge. The contribution to μ for each mechanism varies approximately as λ^3 for each absorbing level and the discontinuities occur when the number of absorption mechanisms changes.

It can be seen in Fig. 6.12 that there is one K edge and there are three L edges. This fine structure of the variation of absorption coefficient with wavelength comes about because the two K electrons have practically the same energy, but the L electrons fall into three subgroups. These energy levels of a substance are more easily observed in absorption than in emission and were ignored when we discussed x-ray emission spectra. One might be tempted to assign absorption edges to electron transitions within the atom, such as from K to L. These transitions cannot occur in absorption, however, since there is normally no vacancy in the L-shell into which a K electron can go. For this reason, an atom that absorbs energy from an x-ray beam must absorb enough energy to lift some electron *entirely free* of the atom. If an absorbing atom is ionized by having one of its K electrons removed, that atom then has a K-shell vacancy and so it will emit one or more of its characteristic wavelengths as it returns to the ground state. Since this secondary emission can be radiated in any direction, the effect of the process is the removal of an x-ray photon from the direction of the original beam. The characteristic radiations emitted as a result of absorbing photons of high energy are often called *fluorescence* x-rays.

Figure 6.12 X-ray absorption spectrum for platinum.

From Harald A. Enge, M. Russell Wehr, and James A. Richards, *Introduction to Atomic Physics*, © 1972. Addison-Wesley, Reading, Mass. Reprinted with permission.

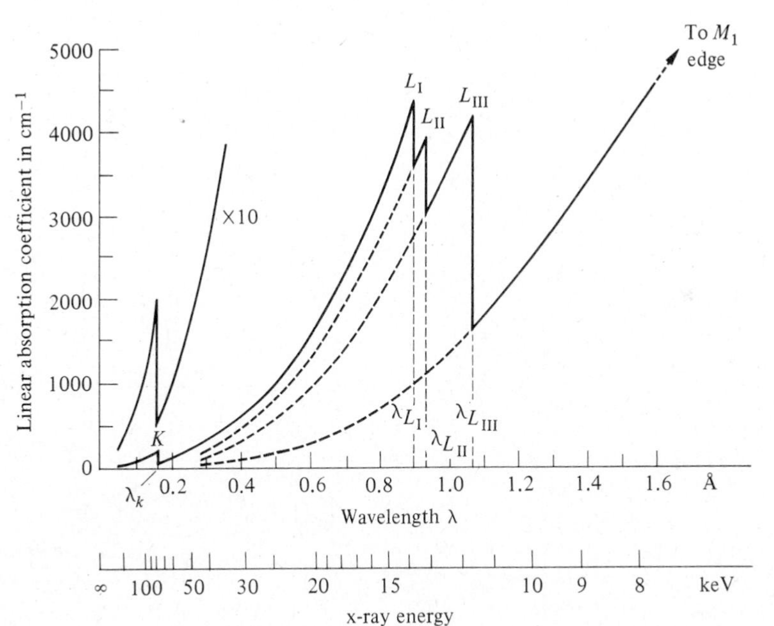

Just as there is a regular progression of x-ray emission wavelengths as we vary the atomic number of the target material, there is also a regular progression in the absorption edges of different elements. The shifting of x-ray spectra with atomic number provides an important technique for obtaining essentially monochromatic x-rays. Suppose we have an x-ray tube with a copper target. The *K*-series of the copper consists of an intense K_α line, and weak K_β and K_γ radiations, listed in the order of decreasing wavelength and of increasing energy. Nickel is a metal whose atomic number is one less than that of copper. Its energy levels are less negative than those of copper, and its emission lines and absorption edges are at slightly longer wavelengths than those of copper. The K_γ wavelength of nickel is almost identical with its λ_K absorption edge, and it falls between the strong K_α and the weaker K_β and K_γ emission lines of copper. Figure 6.13 shows the *K*-series emission spectrum of copper and the linear absorption coefficient of nickel as a function of wavelength. Since the minimum of the absorption curve is at a slightly shorter wavelength than one of the *K* lines but at a longer wavelength than the others, nickel filters out most of the radiation from the copper except its strong K_α line. There are several sets of elements such that by using one as a filter for the emission of the other, nearly monochromatic x-rays are obtained. The filtering effect of nickel for copper radiation is shown in Fig. 6.13.

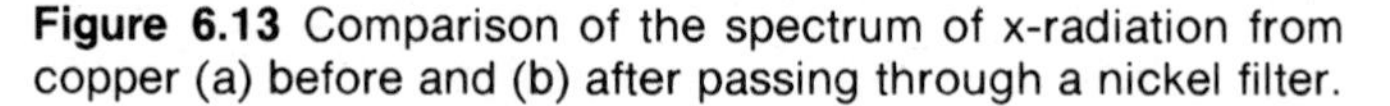

Figure 6.13 Comparison of the spectrum of x-radiation from copper (a) before and (b) after passing through a nickel filter.

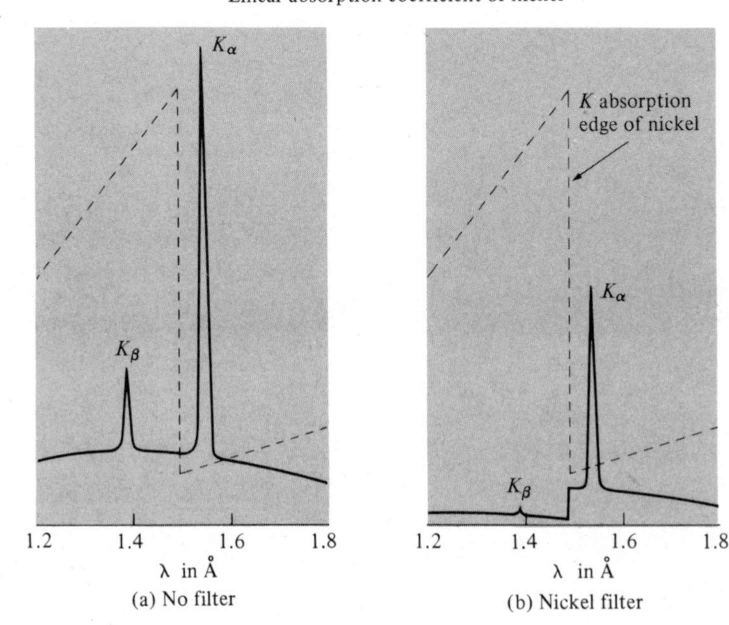

We have described how the linear absorption coefficient depends on the wavelength of the x-rays. We now consider how it depends on the absorbing material.

The absorption coefficient depends strongly on the density of the absorbing material which, of course, changes greatly if the material goes from gaseous to solid states. We can write

$$\mu x = \frac{\mu}{\rho} x\rho, \tag{6.13}$$

where ρ is the mass density of the material. The quantity μ/ρ is called the *mass absorption coefficient*, μ_m. Its units are of the form of area divided by mass. Dimensional analysis shows that $(x\rho)$ is a mass per unit area, m_a. It is the mass of a sheet or slab of the absorber that has the thickness x and a unit of surface area normal to the incident x-ray beam. In these terms, the exponent of Eq. (6.10) becomes $-\mu_m m_a$ and we have

$$I = I_0 e^{-\mu_m m_a}. \tag{6.14}$$

Although the mass absorption coefficient varies from material to material far less than the linear coefficient does, it is still far from constant. Materials with large atomic numbers absorb x-rays more readily than the lighter elements, so that a certain mass per unit area of lead is more effective than the same mass per unit area of, say, carbon. In general, μ_m varies approximately as the atomic number cubed. Combining this empirical fact with the dependence on λ stated earlier, we can say, approximately, that

$$\mu_m = k\lambda^3 Z^3, \tag{6.15}$$

where k is nearly constant as long as λ does not vary through the absorption edges of the materials being compared.

6.8 RADIATION UNITS

Some of the energy absorbed from ionizing radiation that passes through matter damages the medium by causing molecular changes or by altering the crystalline form. The amount of damage produced depends on the nature of the absorbing material, the energy of the photon, and the intensity of the radiation. The effects are large in complex organic molecules and so x-rays are injurious to living tissue. To study this quantitatively, it is necessary to define a unit for the amount of radiation absorbed.

One of the units of the amount of x-radiation is the *roentgen*, R. The roentgen is the amount of x- or gamma radiation that produces in 1 cm^3 of *air* (under standard conditions) ions carrying $1/(3 \times 10^9)$ coulomb of charge of either positive or negative sign.[†] Air was chosen for the ionized medium because its mass absorption coefficient is nearly the same as that of water and of body tissue over a considerable range of wavelengths. The *milliroentgen*, mR, is a thousandth of a roentgen. In radiobiology these units are usually called the *exposure* units for radiation delivered to a specified area or volume of the body. Note that the units of amount of radiation do not depend on time, and they are not a measure of the energy flux in a beam of radiation.

The *exposure rate* is measured by the quantity of radiation delivered per unit time. Some of the units are the *roentgen per hour*, R/h, the *milliroentgen per hour*, mR/h, and so on.

Absorbed radiation frees photoelectrons and Compton recoil electrons. These secondary particles or corpuscles will also ionize the air as they move toward the collecting electrodes in an ionization chamber. The charges carried by these ion pairs produced by the secondaries must be counted as a part of the quantity of charge to be measured. [An *ion pair*, ip, is composed of the positive ion and the negative ion (usually an electron) produced when a neutral particle is ionized. The magnitude of the charge on each member of the pair is necessarily the same. Each ion of an ion pair produced in air is singly charged.] Although all the secondary particles originate in 1 cm^3 of air where the absorption occurred, the measured charge comes from the whole volume through which these secondaries range.

When a large number of ions are formed in air, it is found that the *average* energy required to produce an ion pair is 33.7 eV. This is much greater than the minimum energy needed to ionize either oxygen or nitrogen. The value is larger because some photons eject inner electrons whose binding energies are much greater than the minimum ionization energy. If we take the product of the number of ion pairs in air corresponding to a roentgen and the average energy required to produce such a pair, we find that the roentgen is equivalent to 0.112 erg of energy absorbed per cubic centimeter of standard air or 86.9 ergs per gram of air. When a beam of x-rays enters another medium where the atoms have different atomic numbers, the absorption of energy changes. Thus it was found that exposure to a dose of 1 R of x-rays results in the absorption of *about* 97 ergs per gram of soft body tissue. The *Rep* (roentgen equivalent physical) is that dose of ionizing radiation that results in the absorption of 97 ergs per gram of body tissue. This unit is *not* very definite because of the variations in the composition of the body. However, a definite unit is the *rad* (radiation absorbed dose), which is 100 ergs (10^{-5} J) of *absorbed* energy per gram of *any* absorbing material. The *millirad*, mrad, is a thousandth of a rad. Additional units will be discussed in the chapter on natural radioactivity.

[†] The definition adopted by the International Commission on Radiological Units and Measurements is as follows: "The roentgen is that quantity of x- or gamma radiation such that the associated corpuscular emission per 0.001293 g of dry air produces, in air, ions carrying 1 esu of quantity of electricity of either sign." This mass of air has a volume of 1 cm^3 at 0°C and a pressure of 760 mm of mercury, and 1 esu of charge is equal to $1/(3 \times 10^9)$ C.

Radiation damage to the human body depends on the absorbed dose, the exposure rate, and the part of the body exposed. The safe limit for those exposed to radiation over the *whole* body during their working day is now set at 1250 mRem/calendar quarter.† Up to a few years ago, the safe tolerance level was thought to be over four times this amount. It is likely that it will be lowered again soon. Although the absorbed dose due to long exposure to low-level radiation is large, the resulting direct damage is negligible because the body has time to repair the injury. The effects of acute radiation exposure over the *whole* body are about as follows: 50–100 R, some blood changes; 100–250 R, severe illness but recovery within six months; 400 R, fatal to 50% of the persons affected (this is called the median lethal dose, MLD or LD-50); and 600 R, fatal to all.

The mechanism of tissue destruction is not completely understood. When radiation is absorbed by the various complex organic molecules in the body, an electron is either raised to a higher energy level within the molecule or removed altogether. One might expect that after the electron returns, all would be normal again, as in the case of hydrogen and other single atoms. However, during the time that the molecule is in the excited state, its constituent atoms sometimes rearrange themselves. Then, although the system is again neutral after the electron's return, it is no longer the same molecule. A reaction that occurs in some cases is that two hydroxyl radicals, OH, combine to form hydrogen peroxide, H_2O_2. The molecular situation is somewhat analogous to a high tower that a child builds with small wooden blocks. If we "ionize" the tower by pulling out a block near the base, it is evident that the tower will not rebuild itself when we return the "electron" to the heap. It is unlikely that any substantial change would have occurred if only the topmost block had been removed and then returned. Destruction of a body cell depends on which electron in the molecular structure absorbs the radiation energy.

In the human body, the hands and feet can receive a much larger dose of radiation without permanent injury than any other part. However, the genes in the cells of the body are readily damaged. Injury to those in the reproductive cells is particularly serious because it gives rise to mutations in the generations that follow. These mutations are almost always adverse and the process is irreversible. There is no safe lower limit of radiation when considering the inheritance of genetic damage. It has been said that "a little radiation is a little bad, and a lot is a lot bad." This means that the probability of absorption by any one gene or group of genes is less for a small dose than for a large one. However, the damage per quantum of radiation absorbed is the same in both cases. It is estimated that the rate of mutation will show significant increase if the exposure is more than 10 R during their reproductive lifetime — a period of about 30 years. During that length of time, one will receive about 4 R from cosmic rays and from the radioactive materials that are found in low concentrations everywhere. Any exposure to x-rays adds to the accumulated dose.

† It is recommended for a female worker of childbearing age that the dose not exceed 500 mRem in any nine month period.

Remember that the damage discussed here is transmitted to the generations to come. A whole body exposure of 10 R over a period of three decades would not harm the parent.

The things considered in this section lie in the field of radiological physics. As we advance into the nuclear age, the solution of an increasing number of problems of public health will become the responsibility of persons trained in this area of science.

6.9 CRYSTALLOGRAPHY

The regular arrangement of the space positions of the atoms in a crystal is called a *lattice array*. The basic interplanar distance or *principal grating space*, d, is shown for the cubic crystal KCl in Fig. 6.14. Note that d is the distance between *adjacent* atoms. The *unit cell* of a lattice is the smallest block or geometric figure of a crystal that is repeated again and again to form the lattice structure. The length of the side of a unit cell is the distance between atoms of the *same* kind. This is equal to $2d$ for the cubic crystal in Fig. 6.14. The length of the side of the unit cell is equal to the basic distance d only in the case of a simple cubic structure in which all of the atoms are of the same kind. Some pure metals form crystals of this type. These basic interatomic distances can be computed from knowledge of the molecular weight of the crystalline compound, the Avogadro constant, the density of the material, and its crystalline form. For the cubic type the general procedure is to calculate the number of atoms per unit volume, multiply this by the volume associated with each atom, and equate the product to unity. This procedure is illustrated in the example that follows.

EXAMPLE KCl (sylvite) is a cubic crystal having a density of 1.98 g/cm^3. (a) Find the distance between adjacent atoms in this crystal, and (b) find the distance from one atom to the next one of the same kind.

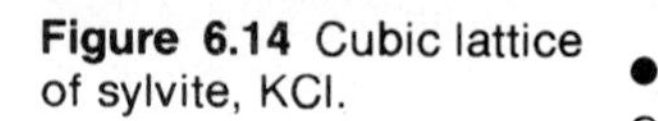
Figure 6.14 Cubic lattice of sylvite, KCl.

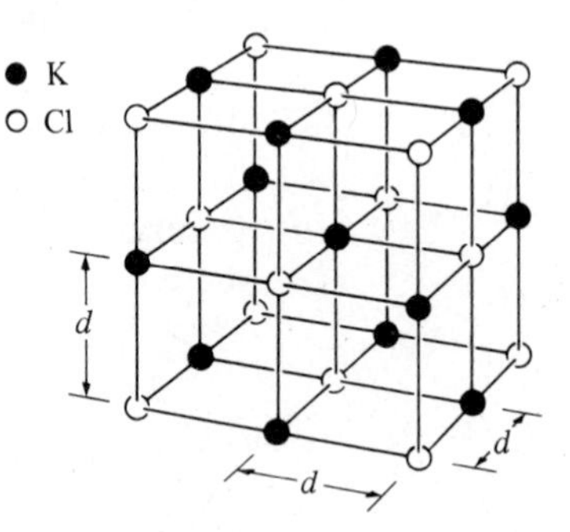

Solution (a) Molecular weight of KCl = 39.10 + 35.45 = 74.55.

$$\text{Mass of KCl molecule} = 74.6\frac{\text{g}}{\text{mole}} \times \frac{1\ \text{mole}}{6.02 \times 10^{23}\ \text{molecules}}$$
$$= 12.4 \times 10^{-23}\ \text{g};$$
$$\frac{\text{No. KCl molecules}}{\text{unit volume}} = 1.98\frac{\text{g}}{\text{cm}^3} \times \frac{\text{molecule}}{12.4 \times 10^{-23}\ \text{g}}$$
$$= 1.60 \times 10^{22}\frac{\text{molecules}}{\text{cm}^3}.$$

Since KCl is diatomic,

$$\frac{\text{No. of atoms}}{\text{unit volume}} = \frac{2\ \text{atoms}}{\text{molecule}} \times 1.60 \times 10^{22}\frac{\text{molecules}}{\text{cm}^3}$$
$$= 3.20 \times 10^{22}\frac{\text{atoms}}{\text{cm}^3}.$$

Let d be the distance, measured along the edge of the cube, between adjacent atoms in the crystal and let n be the number of atoms along the edge of a 1 cm cube. Then the length of an edge is nd, and the volume of this unit cube is n^3d^3. However, n^3 is the number of atoms in 1 cm^3; therefore

$$3.20 \times 10^{22}\frac{\text{atoms}}{\text{cm}^3} \times d^3 = 1$$

or

$$d = (3.20 \times 10^{22})^{-1/3}\ \text{cm} = 3.14 \times 10^{-8}\ \text{cm}$$
$$= 3.14\ \text{Å} = 3140\ \text{XU}.^{\dagger}$$

This is the principal grating space, or basic interplanar distance, of KCl.
(b) The distance between two atoms of the same kind is twice the above value, or 6.28 Å. This is the length of the side of a unit cell of KCl. ▬

The calculation of the grating space is simple only for a cubic crystal, and the distance thus computed is only the *basic* or *principal* interplanar distance. Within a crystal, there are many crystal planes from which Bragg reflection can result. Consider the many possibilities presented by the two-dimensional situation of a marching band on a football field (Fig. 6.15). The basic distance between any adjacent members of the band might be 5 ft. But as the band marches past, it is evident that there are many lines through the band that resemble a three-dimensional array of parallel planes. These sets of parallel lines or planes are separated

† X-ray wavelengths are commonly less than an angstrom, 10^{-10} m, and are often expressed in X-units, abbreviated XU. The XU is about 10^{-13} m. Its true definition is not based on the meter but on the interatomic distance or lattice constant of rock salt, NaCl. When the XU was first used, the lattice constant of rock salt was taken to be 2.8140 Å or 2814.0 XU. High-precision measurements have revised the rock salt lattice constant so that we now have 1000.00 XU = 1.00202 Å.

from one another by different amounts that can be computed from the basic interatomic distance. Reflections from some sets of planes are more intense than those from some other sets. Intensity variations are introduced by the differences in the nature of the planes in crystals composed of more than one type of atom. Elements with high atomic numbers scatter radiation more effectively than those with low atomic numbers. If the x-marked band members in Fig. 6.15 represent one kind of atom and the dot-marked ones another kind, we note that some planes contain only x-marked scatterers, some only the dot-marked type, and some have a mixed population. (It should be kept in mind that the lines in Fig. 6.15 are actually analogous only to the traces of the crystal planes in the plane of the paper. These planes of the crystal are not necessarily normal to the paper. It is evident from a study of Fig. 6.14, for example, that a plane containing only potassium atoms is not perpendicular to any of the faces of the cube.)

We are now in a position to appreciate both why the Bragg method of x-ray spectroscopy is simple and why Laue pictures are complex. In the Bragg method, the crystal can be set so that one set of populated planes is reflecting a beam of x-rays, and the three-dimensional crystal is used in a two-dimensional manner. In the Laue method, a pencil of x-rays is made to pierce the crystal perpendicular to a set of planes of the crystal. Most of the spots are due to diffraction from sparsely populated planes that happen to be situated so that both of the Bragg conditions for reflection are satisfied. The Bragg method is better for the study of the wavelengths of x-rays, but the Laue technique is very useful for the study of crystals. It is a tedious but rewarding task to start with a Laue pattern and work back to the geometry of the array of atoms that must have produced it. An ingenious type of microscope employing both x-rays and visible light has been devised for crystal study. With its aid an enormous, useful magnification can be obtained, so that one can "look" into the lattice structure, as shown in Fig. 6.16.

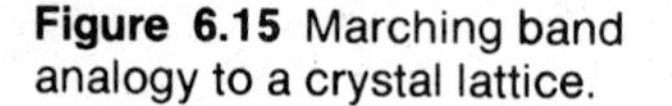
Figure 6.15 Marching band analogy to a crystal lattice.

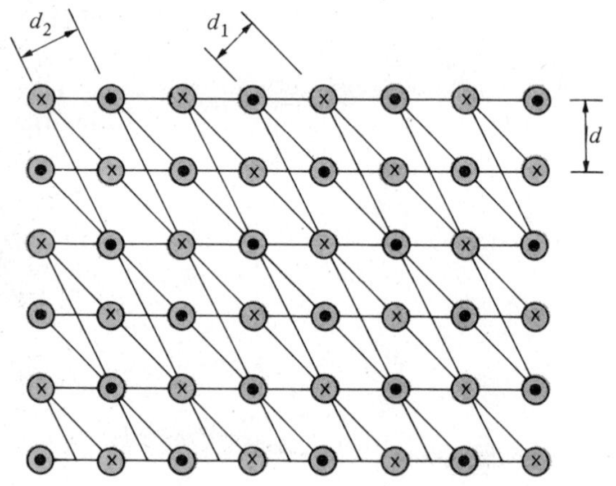

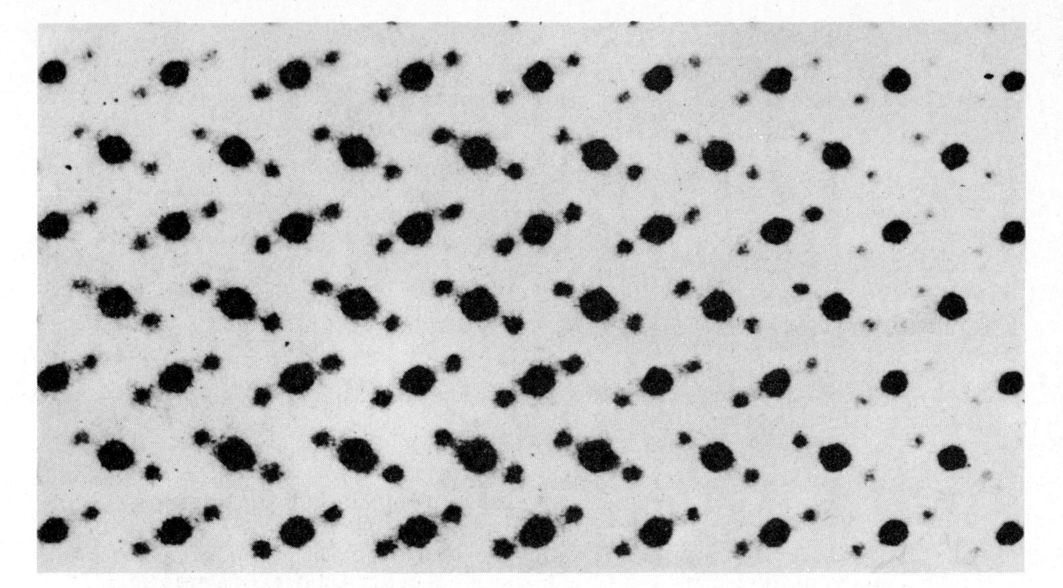

Figure 6.16 Atoms in marcasite, FeS_2, looking along the crystallographic *c*-axis magnified 4.5 million times. The larger spots are iron atoms with 26 electrons each; the smaller dots are sulfur atoms with 16 electrons each. The regular array of the atoms in the crystal is quite evident. This remarkable photograph was taken with a two-wavelength microscope.

Courtesy of M. J. Buerger, Massachusetts Institute of Technology.

We shall not dwell on the subject of crystallography, but there is another x-ray crystallographic technique we must mention in passing. The Bragg and Laue methods require single crystals large enough for study. Many materials that are basically crystalline cannot be obtained as large single crystals, but these materials may be studied by means of the *powder technique*. One form of this technique, the Debye–Scherrer method, requires that the substance be ground and powdered, so that it can be assumed that all the tiny crystals have random orientation. When a pencil of x-rays is passed through this powder, a series of rings is formed on a photographic film. Each ring is the intersection of the film plane and a cone of rays. Each cone is the locus of rays for which some set of crystal planes is so oriented that both Bragg conditions of reflection are fulfilled. The formation of a powder diffraction pattern is shown schematically in Fig. 6.17, and Fig. 6.18 is the pattern for NaCl made with the x-radiation from copper.

Another form of powder technique is used to study the orientation of crystals in an extruded or drawn material like wire. If the crystallites in the wire actually are random in their orientation, a true powder pattern of circles results. But if the crystallites are somewhat oriented, then each circle has nonuniform intensity and the

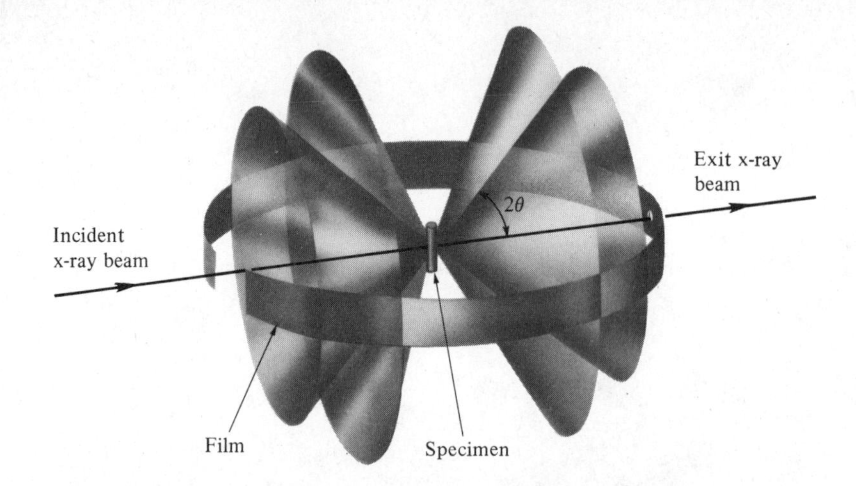

Figure 6.17 Relation of film to specimen and incident beam in the Debye-Scherrer powder diffraction method. X-rays forming the cone having a half-angle of 2θ were reflected from the same kinds of planes, all oriented at an angle θ to the incident beam. Other cones are formed in a similar way.

From B. D. Cullity, *Elements of X-ray Diffraction*, © 1956. Addison-Wesley, Reading, Mass. Reprinted with permission.

pattern on the film tends toward a Laue pattern. The degree of orientation can be determined from such pictures.

Laue patterns of metals or other materials rather opaque to x-rays are sometimes obtained by studying the x-rays scattered back from the side of the material nearer the x-ray source. This technique is used, for example, to study the crystalline structure of steel.

Figure 6.18 Powder diffraction pattern of NaCl made with the K_α and K_β wavelengths of copper. During exposure, the upper portion of the film was covered with nickel foil. This filtered out the K_β line.

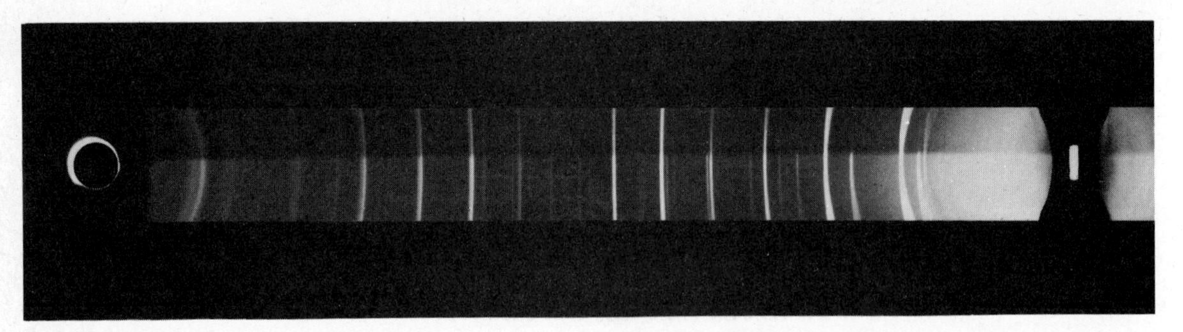

Since a single crystal is used for Laue patterns, there are only a very small number of planes that fulfill the Bragg conditions for constructive interference for any one x-ray wavelength. Therefore, to obtain many Laue spots requires the use of an incident beam containing many wavelengths. On the other hand, the minute crystals in a powder have random orientations, so that many planes are available for the production of interference maxima even when monochromatic x-rays are used.

6.10 MILLER INDICES

The orientations of the various lattice planes in a crystal are usually specified by a system that was first used in 1839 by W. H. Miller, an English mineralogist, to describe the faces of a crystal. In the general case, a face or an internal plane will be inclined to all of the crystallographic axes. These axes are a convenient coordinate frame that can be used to determine the orientation of a plane by giving the distances from the origin to its points of intersection with the three axes. All the series of planes that are parallel to this plane will have intercepts whose ratios to the axial lengths of the *unit* cell, a, b, and c, are independent of the particular axial lengths involved in the given lattice. Since we are concerned only with the orientation of a plane and not its absolute position, these ratios could serve to specify the plane. But these ratios cause difficulty when the plane is parallel to a crystallographic axis, as in Fig. 6.19(a). In this case, the planes do not intersect two of the axes at all; that is, their intercepts are said to be at infinity. To avoid the introduction of infinity into the specification of the orientation of a plane, the reciprocal of the fractional intercept is used. Thus the reciprocal will be zero when the plane and axis are parallel. This leads to the system of determining the orientation of a plane in a crystal lattice by the *Miller indices*, which are defined as *a set of integers in the ratio of the reciprocals of the fractional intercepts which the plane makes with the crystallographic axes*. It is customary to designate the Miller indices of a plane as $(h\ k\ l)$, which means that the plane has fractional intercepts of $1/h$, $1/k$, and $1/l$ with the axes and that the actual intercepts are a/h, b/k, and c/l. It is evident that the intercepts in Fig. 6.19(a) are $1, \infty, \infty$. The reciprocals of these are $1, 0, 0$, and therefore the Miller indices are (100). (This is read "one-zero-zero" or "one-aught-aught," not "one hundred.") These (100) planes are parallel to the BC-plane. A similar group of planes could be drawn parallel to the AC-plane and might be designated the (010) planes; and a set parallel to the AB-plane might be called the (001) planes. It is evident, however, that these three sets of planes in a cubic crystal are equivalent. Whether a set be called (100), (010), or (001) planes depends entirely on the selection of axes in the crystal, and hence all three of these sets of planes are named the (100) planes. The indices given for the series of planes in Figs. 6.19(b) and (c) are self-evident. This is not always so. For example, let the intercepts of the plane be 1, 2, and 3 units on the three axes, respectively. The

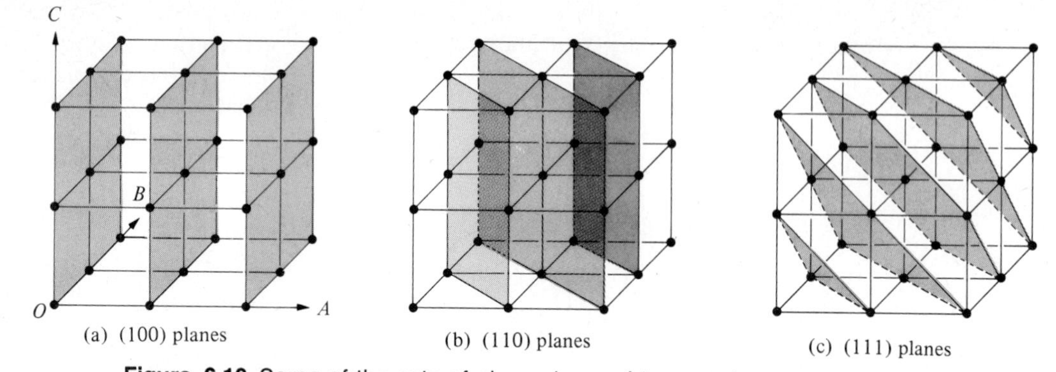

(a) (100) planes (b) (110) planes (c) (111) planes

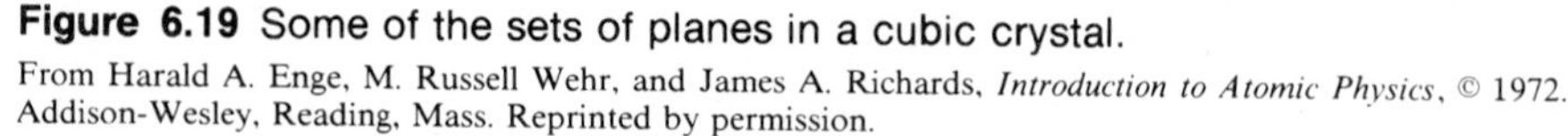

Figure 6.19 Some of the sets of planes in a cubic crystal.
From Harald A. Enge, M. Russell Wehr, and James A. Richards, *Introduction to Atomic Physics*, © 1972. Addison-Wesley, Reading, Mass. Reprinted by permission.

indices of this plane are therefore in the ratio $1/1$, $1/2$, and $1/3$. When these fractions are multiplied by their lowest common denominator, 6, we obtain (632), the Miller indices of the plane.

The distance d between successive members of a series of parallel planes involves a, b, and c, the axial lengths of the unit cell, and $(h\ k\ l)$, the Miller indices of the planes. We shall derive an expression for the interplanar distance d for the relatively simple case of an orthorhombic crystal. The three axes of such a crystal are mutually perpendicular. In Fig. 6.20, ABC is one of a series of parallel planes that has intercepts $OA = a/h$, $OB = b/k$, and $OC = c/l$. The origin of coordinates O is in the next plane of the set parallel to ABC. Therefore, ON, the length of the normal from the origin to the plane, is equal to d. Let θ_a, θ_b, and θ_c be the angles ON makes with the three crystallographic axes, respectively. Then the direction cosines of ON are $\cos\theta_a = ON/OA$, $\cos\theta_b = ON/OB$, and $\cos\theta_c = ON/OC$. Since the sum of the squares of the direction cosines of a line equals unity, upon squaring the cosine

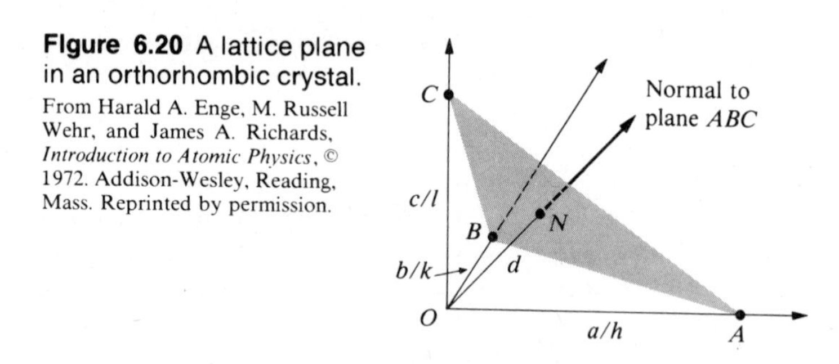

Figure 6.20 A lattice plane in an orthorhombic crystal.
From Harald A. Enge, M. Russell Wehr, and James A. Richards, *Introduction to Atomic Physics*, © 1972. Addison-Wesley, Reading, Mass. Reprinted by permission.

expressions and substituting for the intercepts OA, OB, and OC, one gets

$$\left(\frac{d}{a/h}\right)^2 + \left(\frac{d}{b/k}\right)^2 + \left(\frac{d}{c/l}\right)^2 = 1,$$

or, upon rearranging, this becomes

$$d = \frac{1}{\left(h^2/a^2 + k^2/b^2 + l^2/c^2\right)^{1/2}}.$$

If the orthorhombic crystal is cubic, the basic interplanar distances are equal. In a simple cube, these distances are the lengths of the sides of a unit cell. Then $a = b = c$, and the preceding equation reduces to

$$d = \frac{a}{\left(h^2 + k^2 + l^2\right)^{1/2}}. \tag{6.16}$$

When this relation is substituted into Eq. (6.4), we find that the *first-order* diffraction maxima in Bragg reflections are given by

$$\lambda = \frac{2a}{\left(h^2 + k^2 + l^2\right)^{1/2}} \sin\theta. \tag{6.17}$$

This equation can be used to calculate the axial length of a side of a *cubic* crystal, and the Miller indices of the set of planes involved in giving a particular maximum if the wavelength λ of the radiation and the Bragg angle θ are known. The value of λ can be found in tables of the characteristic radiation for the various x-ray tube-target materials. Referring to Fig. 6.17, it is seen that the data for finding θ can be obtained by measuring the distance along the film from the exit beam to a diffracted line and the distance halfway around the film in its circular mounting from the exit beam to the entrance of the incident beam.

To find $(h\ k\ l)$, let us first rearrange Eq. (6.17) so that we have

$$\frac{\lambda^2}{4a^2} = \frac{\sin^2\theta}{h^2 + k^2 + l^2}. \tag{6.18}$$

The sum of the squares of the Miller indices is always integral, and the left-hand side of the last equation is constant for any given diffraction pattern. Therefore, having determined θ for each line in a pattern, one can now find the Miller indices of the planes associated with each line by choosing integers h, k, and l for each line, so that the ratio $[\sin^2\theta/(h^2 + k^2 + l^2)]$ has the same value for every line.

Three types of lattice with cubic symmetry are shown in Fig. 6.21. In (a) the diffracting centers are at the corners of a simple cube; in (b), a body-centered cube, an additional lattice point is located at the center of the cube; and in (c), a face-centered cube, an additional lattice point is located at the center of each face of the cube. Both NaCl and KCl crystals have the face-centered cubic configuration. The reader should consult a book on crystallography for descriptions of configurations of crystals that are not cubic.

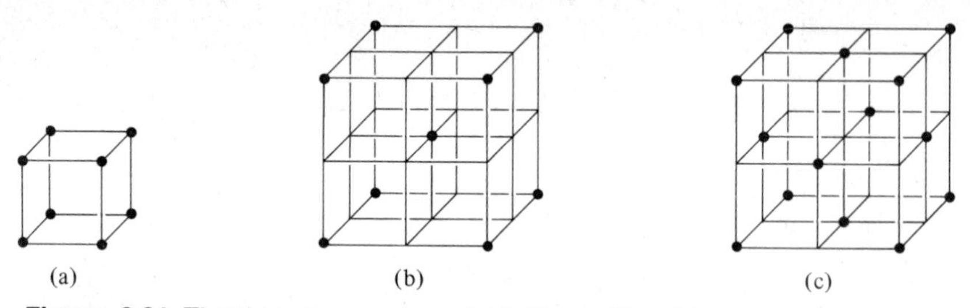

Figure 6.21 The three types of crystal lattice with cubic symmetry: (a) simple; (b) body-centered; (c) face-centered.

From Harald A. Enge, M. Russell Wehr, and James A. Richards, *Introduction to Atomic Physics*, © 1972. Addison-Wesley, Reading, Mass. Reprinted by permission.

6.11 DIFFRACTION WITH RULED GRATINGS

X-rays were first diffracted with crystals but eventually a method was also devised for diffracting them with artificially created gratings. This method is an interesting and important example of how physical research often builds crosslinks that reinforce the entire structure of our ideas.

We stated earlier that x-rays were not diffracted by gratings because the wavelength of x-rays is much less than the separation of the closest lines we can rule with the granular materials at our disposal. Compton found a way around this difficulty that was certainly ingenious. He used a grating in an unusual way. The index of refraction of glass for x-rays is slightly *less* than one. Thus when x-rays go from air or vacuum into glass, they are passing from an *optically* dense medium into one less dense, and the rays are refracted away from the normal. Recall that whenever light goes from a more dense to a less dense medium, there is a critical angle, and that if the rays approaching the interface make an angle greater than the critical angle, total reflection results. X-rays striking glass at "grazing incidence" will not enter the glass at all, but will be totally reflected. If, now, the glass surface is a ruled grating, these rulings will appear closer together from a grazing angle than when viewed from a point normal to the grating surface. Thus the very geometrical situation that makes the bands between the rulings of the grating capable of reflecting also makes the apparent grating space much less. Compton found that with his ingenious arrangement he could diffract x-rays by measurable amounts.

The important result of this work was that Compton could measure x-ray wavelengths absolutely. The grating space was determined by a screw whose pitch could be measured by counting the threads in a length measured by an ordinary length standard. Therefore the grating space was known from very direct elementary measurement. All prior x-ray wavelength measurements had been made in terms of a calculated crystal spacing that depended on the Avogadro constant, and the Avogadro constant depended in turn on the charge of the electron. When high-precision measurements were made, it was found that the wavelengths measured by a grating differed from those measured by crystal diffraction by about 0.3%. Since this was an absolute measurement, it was concluded that the crystal spacings that had been used formerly were in error. Shiba, a Japanese physicist, traced this back through the Avogadro constant to the charge of the electron and concluded that Millikan's value was in error. Shiba pointed out that the error lay in the viscosity of air used in the calculations in the oil-drop method. Once the new value of the charge was accepted, all physical constants based on that charge were revised. It is interesting that an experiment in x-ray spectroscopy should provide a technique for measuring the basic unit of electrical charge.

6.12 COMPTON SCATTERING

Planck introduced the idea that radiation must be emitted in bundles of energy, although he believed that once emitted, the energy spread in waves. Einstein extended the Planck idea to the absorption of radiation in his explanation of the photoelectric effect. He added the assumption that once a quantum of energy was radiated, it preserved its identity as a photon until it was finally absorbed. In the chapter on relativity we showed that mass and energy are identical. Since photons have energy, they must have mass. This developing concept that photons are true particles throughout their life comes to its climax in the Compton effect.

Compton assumed that the photons had the very "mechanical" property of momentum and solved the problem of impact of a photon and a material particle by means of relativistic mechanics.

Let us consider a material particle that is initially at the origin and at rest relative to the coordinate frame shown in Fig. 6.22. (The initial motion of the particle is negligible compared with the other velocities in the following analysis.) Initially, it has no momentum p_1 and its energy E_1 is only its rest-mass energy, m_0c^2. This particle is then hit by a photon moving along the x-axis. This photon brings to the impact an energy

$$E = h\nu = \frac{hc}{\lambda}. \tag{6.19}$$

The photon has no rest mass, but its moving mass is equal to its energy, divided by c^2, or

$$m_{ph} = \frac{E}{c^2} = \frac{h}{c\lambda}. \tag{6.20}$$

The momentum of the photon is its mass times its velocity, which is c. Thus its

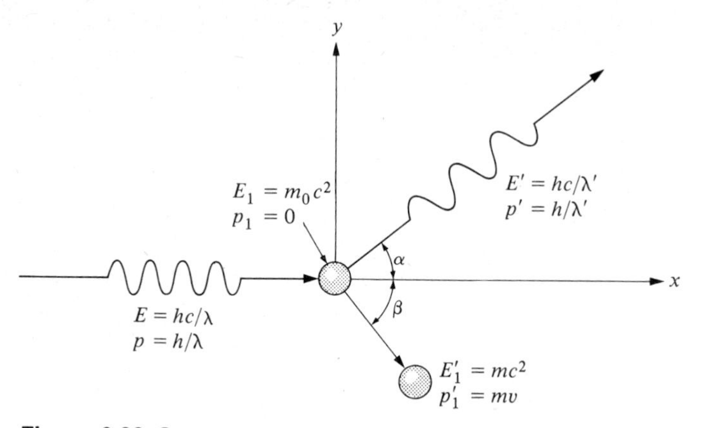

Figure 6.22 Compton scattering.

momentum p, which is all in the x-direction, is

$$p = \frac{h}{c\lambda}c = \frac{h}{\lambda}. \tag{6.21}$$

We assume that the impact is slightly glancing, so that the motion of the material particle is directed below the x-axis, while the photon is deflected above the x-axis. The impact gives the material particle kinetic energy, so that its total energy becomes mc^2. The material particle also acquires momentum, mv, which is directed at an angle β to the x-axis. The mass of the particle in these statements is the relativistic moving mass.

The photon leaves the impact with its energy changed to hc/λ' and its momentum changed in magnitude and direction to h/λ' at an angle α to the x-axis. The above statements are summarized in Fig. 6.22.

We may now use the laws of conservation of mass-energy and of momentum to describe the impact situation. If we *neglect* any binding energy that the particle may have and apply the law of conservation of mass-energy, then the kinetic energy of the ejected particle is given by

$$E_k = \frac{hc}{\lambda} - \frac{hc}{\lambda'} = mc^2 - m_0c^2. \tag{6.22}$$

For the conservation of the x-component of momentum, we may write

$$\frac{h}{\lambda} = \frac{h}{\lambda'} \cos \alpha + mv \cos \beta, \tag{6.23}$$

and for the y-component,

$$0 = \frac{h}{\lambda'} \sin \alpha - mv \sin \beta. \tag{6.24}$$

The experimental test of the theory we are developing consists in measuring the wavelength λ' of the deflected photons as a function of their angle of deflection α. The initial photon wavelength λ is known. Since the material particles are usually electrons, m_0 is also known. We find, then, that the three equations (6.22) through (6.24) involve five variables, m, v, λ', α, and β, of which three, m, v, and β, relating to the electron, will be eliminated. This requires an additional equation that is the relativistic interdependence of m and v. From Eq. (5.39), we have

$$m = \frac{m_0}{\sqrt{1 - (v^2/c^2)}},$$

or

$$m^2v^2 = c^2(m^2 - m_0^2). \tag{6.25}$$

We first eliminate β between Eqs. (6.23) and (6.24) by isolating the terms containing β and squaring both equations. We thus obtain

$$m^2v^2\cos^2\beta = h^2\left(\frac{1}{\lambda^2} - \frac{2\cos\alpha}{\lambda\lambda'} + \frac{\cos^2\alpha}{\lambda'^2}\right) \tag{6.26}$$

and

$$m^2v^2\sin^2\beta = h^2\left(\frac{\sin^2\alpha}{\lambda'^2}\right). \tag{6.27}$$

Adding these equations and using $\sin^2 x + \cos^2 x = 1$, we find that

$$m^2v^2 = h^2\left(\frac{1}{\lambda^2} - \frac{2\cos\alpha}{\lambda\lambda'} + \frac{1}{\lambda'^2}\right). \tag{6.28}$$

We next eliminate v between Eqs. (6.28) and (6.25), obtaining

$$c^2(m^2 - m_0^2) = h^2\left(\frac{1}{\lambda^2} - \frac{2\cos\alpha}{\lambda\lambda'} + \frac{1}{\lambda'^2}\right). \tag{6.29}$$

Solving Eq. (6.22) for m and squaring yields

$$m^2 = m_0^2 + \frac{2m_0h}{c}\left(\frac{1}{\lambda} - \frac{1}{\lambda'}\right) + \frac{h^2}{c^2}\left(\frac{1}{\lambda} - \frac{1}{\lambda'}\right)^2. \tag{6.30}$$

We eliminate m by substituting from Eq. (6.30) into Eq. (6.29). This gives

$$2m_0hc\left(\frac{1}{\lambda} - \frac{1}{\lambda'}\right) + h^2\left(\frac{1}{\lambda^2} - \frac{2}{\lambda\lambda'} + \frac{1}{\lambda'^2}\right) = h^2\left(\frac{1}{\lambda^2} - \frac{2\cos\alpha}{\lambda\lambda'} + \frac{1}{\lambda'^2}\right). \tag{6.31}$$

By canceling terms and multiplying both sides of the equation by $\lambda\lambda'/h$, we find that Eq. (6.31) simplifies to

$$m_0c(\lambda' - \lambda) - h = -h\cos\alpha. \tag{6.32}$$

We have finally

$$\Delta\lambda = \lambda' - \lambda = \frac{h}{m_0c}(1 - \cos\alpha). \tag{6.33}$$

When the particle involved in Compton scattering is an electron, the first term on the right-hand side of Eq. (6.33) becomes $(h/m_e c)$. This quantity is called the Compton wavelength of the electron.

The experimental test of Eq. (6.33) can be made with an apparatus such as that schematically represented in Fig. 6.23. Monochromatic x-rays are scattered through an angle α and the spectrum of the scattered radiation is measured. The measuring spectrometer must be free to swing in an arc about the scatterer, so that the scattered wavelength can be measured as a function of the scattering angle α. Compton's results for the molybdenum K_α line scattered by carbon for four values of α are shown in Fig. 6.24. The solid vertical lines correspond to the primary wavelength, λ, and the dashed ones to the calculated modified wavelength λ'. The electrons which, when hit, modify the photon wavelength must suffer a change of energy, as required by our derivation. Such electrons must be either free or loosely bound. If the photon encounters a tightly bound electron, the whole atom is involved in the collision, and the value of m_0 in Eq. (6.33) then becomes the mass of the atom rather than the mass of an electron. Since the mass of the atom is several thousand times the electronic mass, the Compton wavelength shift for it is too small to detect. These scattered photons comprise the unmodified peaks shown in Fig. 6.24.

For very high-energy photons most of the atomic electrons appear free and a large fraction suffer a wavelength shift. For low-energy x-rays most electrons appear bound unless the scattering atom has a low atomic number. For *visible photons*, all the electrons appear bound and there is *no* observed Compton shift. Although we have just seen that the likelihood of a Compton wavelength shift depends on the photon energy and the scattering material, the amount of the wavelength shift depends only on the scattering angle. This aspect of Compton scattering has been tested, and essentially the same change in wavelength was found for x-rays scattered from fifteen different elements.

In Fig. 6.24, the modified peak is broader than the unmodified one: The reason for this difference does not appear in the derivation because we assumed that the

Figure 6.23 Schematic diagram of the apparatus for measuring Compton scattering.

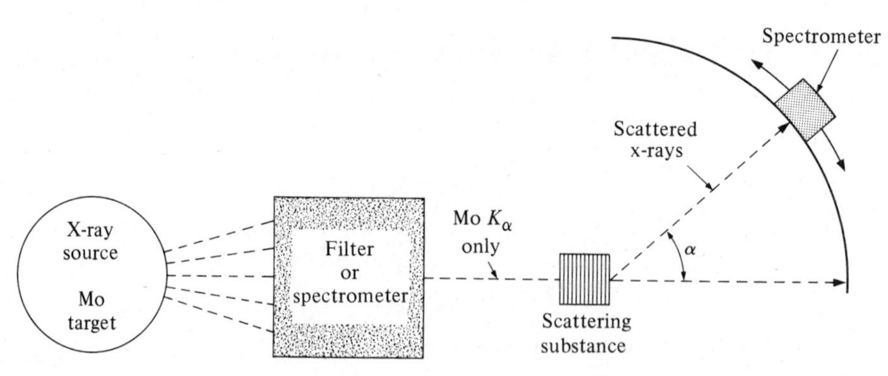

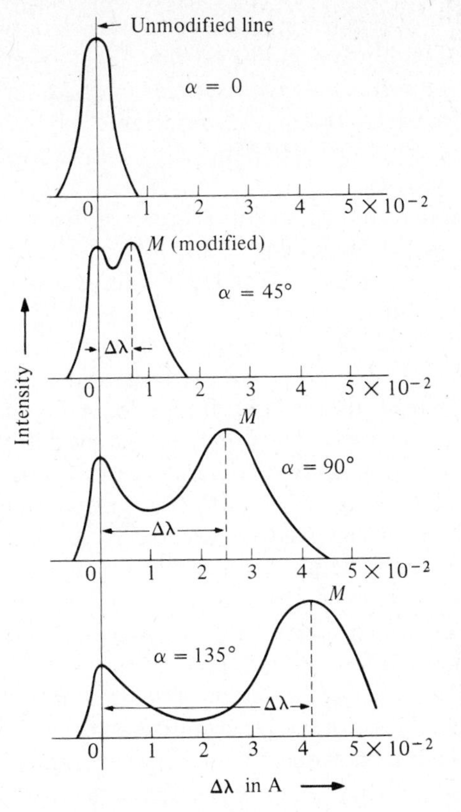

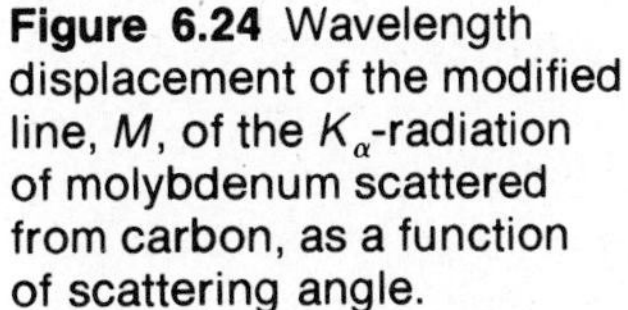
Figure 6.24 Wavelength displacement of the modified line, M, of the K_α-radiation of molybdenum scattered from carbon, as a function of scattering angle.

electron doing the scattering was initially at rest. Motion of the target electron fully accounts for the fact that, for a given angle, some photons had their energy changed a little more or less than what would be expected for stationary electrons.

Not only does the Compton effect contribute to our understanding of absorption coefficients, its theory also provides a good check of relativity and extends the particle concept of photons. With photons behaving like billiard balls, we are nearly back to Newton's particle theory of light. In the next chapter we shall synthesize the divergent wave and particle theories of radiation.

6.13 EXTRATERRESTRIAL X-RAYS

Although the earth's atmosphere is transparent for visible light, it is opaque for the entire x-ray part of the spectrum. Therefore the discovery of extraterrestrial x-ray sources had to wait until detectors could be transported above the atmosphere. The first source was found in 1964 and now more than 170 have been observed.

Extraterrestrial x-rays can be divided into two categories: soft, with wavelengths greater than 10 Å, and hard, with wavelengths less than 10 Å. There appears to be an amorphous background of soft x-rays with sources in colliding interstellar gas molecules. On very rare occasions within the observable universe, a supernova occurs, and the expanding gas from this exploding star interacts with the interstellar medium, producing soft x-rays.

There are two, and perhaps three, major sources of hard x-rays. Quasars, believed to be the farthest objects from us, emit great quantities of energy in all parts of the electromagnetic spectrum.† Included in this radiation is hard x-radiation. The mechanism for production is not understood.

A second source of hard x-radiation is believed to be neutron stars in close proximity to normal stars. As was discussed in Section 6.4, x-rays are given off when electrons are caused to decelerate as they strike solid matter. X-rays can be produced by these processes with electrons, or with any other charged particle taking the place of the electron. A binary system of a neutron star and a normal star would generate x-rays because the massive neutron star would attract large quantities of mass from the normal star. This matter would drop into the deep gravitational well resulting in ionized high-energy particles. High-energy x-rays would result from the collision of these particles with the surface of the neutron star.

A source of x-rays similar to that described in the previous paragraph has been discovered in the constellation Cygnus and is called Cygnus X-1. Many astrophysicists believe the radiation is the result of a black hole–star combination. The irregular variations in the signal is of the order of milliseconds, which indicates that the candidate for a black hole has a maximum diameter of 10^3 km.‡ The mass of the accompanying star to form the binary with the possible black hole is estimated to have a mass near 30 times the solar mass. This information along with the period of the binary orbit of 5.6 days indicates that the invisible companion is probably massive also. Therefore, since the object causing the emission of the x-rays must be small and also very massive, it is thought to be a black hole.

FOR ADDITIONAL READING

Bragg, Sir Lawrence, "X-Ray Crystallography," *Scientific American* **58** (July 1968).
Clark, George L., *Applied X-Rays*. 4th ed. New York: McGraw-Hill, 1955.
Compton, A. H., and S. K. Allison, *X-Rays in Theory and Experiment*. 2nd ed. Princeton, N.J.: Van Nostrand, 1935. A comprehensive treatise.

†See the article by P. S. Osmer referenced at the end of this chapter.

‡The maximum physical size (diameter) for an object emitting radiation is believed to be the distance that light would travel during the period of the oscillations. For this object, the distance is 10^3 km.

Cullity, B. D., *Elements of X-Ray Diffraction*. Reading, Mass.: Addison-Wesley, 1956. Good discussion of x-rays and crystal structure.

Giaconni, Riccardo, "The Einstein X-Ray Observatory," *Scientific American* **242**, 80 (February 1980).

Jauncey, G. E. M., "Birth and Early Infancy of X-Rays," *Am. J. Phys.* **13**, 362 (1945). In addition to Roentgen's account of his discovery, this article contains a number of sensational accounts of x-rays from the newspapers of 1896.

Osmer, P. S., "Quasars as Probes of Distant and Early Universe," *Scientific American* **246**, 122 (February 1982).

Richards, Horace C., "A. W. Goodspeed: A Pioneer in Radiobiology," *Am. J. Phys.* **11**, 342 (1943).

Roentgen, Wilhelm C., "The Roentgen Rays," in W. F. Magie, *A Source Book in Physics*. Cambridge, Mass.: Harvard University Press, 1935.

Stuewer, R. H., *The Compton Effect*, New York: Science History Publications, 1975.

Thumm, Walter, "Roentgen's Discovery of X-Rays," *Phys. Teacher* **13**, 207 (1975).

REVIEW QUESTIONS

1. How would you construct an apparatus for generating x-rays?
2. What are the ways that x-rays interact with matter?
3. What is the roentgen?
4. What are Miller indices?
5. What is Compton scattering? How does conservation of momentum enter into the analysis?

PROBLEMS

6.1 Potassium iodide, KI, is a cubic crystal that has a density of 3.13 g/cm^3. Find (a) the basic interplanar distance, and (b) the length of a side of the unit cell.

6.2 The spacing between the principal planes in a crystal of NaCl is 2.820 Å. It is found that a first-order Bragg reflection of a monochromatic beam of x-rays occurs at an angle of 10°. (a) What is the wavelength of the x-rays in Å and in XU, and (b) at what angle would a second-order reflection occur?

6.3 What is the shortest wavelength that can be emitted by the sudden stopping of an electron when it strikes (a) the screen of a TV tube operating at 10,000 V and (b) the plate of a high-power radio transmitter tube operating at 30,000 V? (c) Determine the spectral region in which these wavelengths lie by referring to Fig. 3.2.

6.4 (a) At what potential difference must an x-ray tube operate to produce x-rays with a minimum wavelength of 1 Å? of 0.01 Å? (b) What is the maximum frequency of the x-rays produced in a tube operating at 20 kV? at 60 kV?

6.5 An x-ray tube is operating at 150,000 V and 10 mA. (a) If only 1% of the electric power supplied is converted into x-rays, at what rate is the target being heated in calories per second? (b) If the target weighs 300 g and has a specific heat of 0.035 cal/g°C, at what average rate would its temperature rise if there were no thermal losses? (c) What must be the physical properties of a practical target material? What would be some suitable target elements?

6.6 Since the index of refraction of all materials for x-rays is very close to unity, it is impractical to make lenses for them. Therefore all radiographs (x-ray photographs) are necessarily shadowgraphs. (a) What requirement is imposed on the target area from which the radiation originates so that the pictures will be sharply defined? (b) What is the smallest size film with respect to the size of the object that can be used in radiographing it?

6.7 X-rays from a certain cobalt target tube are composed of the strong K series of cobalt and weak K lines due to impurities. The wavelengths of the K_α lines are 1.785 Å for cobalt, and 2.285 Å and 1.537 Å for the impurities. (a) Using Moseley's law, calculate the atomic number of each of the two impurities. (b) What elements are they? (For the K-series, $b = 1$ in Moseley's law.)

6.8 The K absorption edge of tungsten is 0.178 Å and the average wavelengths of the K-series lines are $K_\alpha = 0.210$ Å, $K_\beta = 0.184$ Å, and $K_\gamma = 0.179$ Å. (a) Construct the x-ray energy-level diagram of tungsten. (b) What is the least energy required to excite the L-series? (c) What is the wavelength of the L_α line? (d) If a 100 eV electron struck the tungsten target in a tube, what is the shortest x-ray wavelength it could produce, and (e) what is the shortest wavelength characteristic of tungsten that could be emitted?

6.9 The radiation from an x-ray tube operated at 80 kV is incident on a sheet of tungsten. (a) Using the energy levels characteristic of tungsten calculated in Problem 6.8, find the maximum kinetic energies of the electrons ejected from the K-, L-, and M-shells, in eV. (b) What is the range of kinetic energies of the electrons ejected by 80 keV photons from the shells between M and infinity?

6.10 A thin sheet of nickel is placed successively in a beam of x-rays from cobalt, then in one from copper, and finally in one from zinc. Discuss the effect of each of these beams on the nickel atoms and show that the filtered radiation from copper is essentially monochromatic. X-ray data for these elements are given in the accompanying table. Consider only K-shell absorption.

	Emission wavelengths, Å		
Element	K_α	K_β	K_{absorb}
Co	1.79	1.62	1.61
Ni	1.66	1.49	1.48
Cu	1.54	1.39	1.38
Zn	1.43	1.29	1.28

6.11 The table in the preceding problem gives the K_α lines for various elements. Determine the atomic number of each element from these data.

6.12 In uranium ($Z = 92$) the K absorption edge is 0.107 Å and the K_α line is 0.126 Å. Determine the wavelength of the L absorption edge.

6.13 (a) How does the linear absorption coefficient vary with the atomic number of an absorber? (b) Before an examination of the gastrointestinal tract of a patient is made with x-rays, the patient drinks a suspension of barium sulfate, $BaSO_4$. Why must this be done? (c) Some fluids in the human body can be followed by the use of x-rays if a solution of potassium iodide,

KI, is introduced into the region to be studied. Why must this be done?

6.14 How many half-value layers of a material are necessary to reduce the intensity of an x-ray beam to (a) $\frac{1}{16}$, (b) $\frac{1}{80}$, and (c) $\frac{1}{200}$ of its incident value?

6.15 The half-value layer of steel for 1.5 MeV radiation is 0.5 in. (a) Plot the curve of transmitted intensity against thickness in half-value layers for a sheet of steel 3 in. thick when the intensity of the incident beam is 200 units. (b) Using the same coordinate scales as in part (a), plot the radiation that has been absorbed against the thickness. (c) Most of the radiation absorbed appears as heat. Assuming low thermal conductivity so that the temperature gradient in the steel is approximately the slope of the absorption curve, discuss the variation of this gradient within the sheet. (d) Will the sheet have thermal stresses?

6.16 For 0.2-Å x-rays, the mass absorption coefficients in cm^2/g for several metals are: aluminum, 0.270; copper, 1.55; and lead, 4.90. (a) What is the half-value thickness of each for a narrow beam of x-rays? (b) What thickness of each is required to reduce the intensity of the transmitted beam to $\frac{1}{32}$ of its incident value? (c) If the "buildup" due to scattering and other processes for a broad beam of radiation is equivalent to making the incident beam 1.5 times its actual intensity, what thickness of each material is then needed to obtain an intensity reduction to $\frac{1}{32}$?

6.17 Show that the tenth-value layer of an absorber equals $2.30/\mu$.

6.18 The linear absorption coefficient of a certain material for x-rays having a wavelength of 1 Å is $\mu_1 = 3.0\ cm^{-1}$, and for those having a wavelength of 2 Å is $\mu_2 = 15\ cm^{-1}$. If a narrow beam of x-rays containing equal intensities of 1-Å and 2-Å radiation is incident on the absorber, for what thickness of the material will the ratio of the intensities of the transmitted rays be 4 to 3?

6.19 The linear absorption coefficients of aluminum and copper are $0.693\ cm^{-1}$ and $13.9\ cm^{-1}$, respectively, for the K_α line from tungsten. (a) What percentage of the intensity of this line will pass through a 5 mm plate of aluminum? of copper? (b) What thickness of aluminum is equivalent to the absorption of 5 mm of copper for this wavelength?

6.20 Observations made from very high altitude balloons and from rockets that go to the "top" of the earth's atmosphere show that the sun and certain stars emit x-rays (x-ray astronomy). Why does visible light from these sources reach the surface of the earth whereas the x-rays do not?

6.21 Calculate the fraction of an x-ray beam transmitted by mercury for which $\mu_m = 3.0\ cm^2/g$ when passing through (a) 0.1 mm of liquid mercury, (b) 10 cm of a solution of 0.2 mole of $Hg(NO_3)_2$ dissolved in a liter of water, and (c) an 80 cm column of mercury vapor at 500°C and 1 atmosphere pressure.

6.22 Show that when the exponent μx in $I = I_0 e^{-\mu x}$ is equal to or less than 0.1, then, with an error of less than 1%, (a) the transmitted radiation equals $I_0(1 - \mu x)$, and (b) the absorbed radiation equals $I_0 \mu x$. [*Hint:* Expand the exponential term into a series.]

6.23 For 2 MeV x- or gamma rays, the half-value layers of some commonly used shielding materials are water, 5.9 in.; ordinary concrete (relative density, 2.6), 2.3 in.; iron, 0.80 in.; and lead, 0.53 in. Find (a) the mass absorption coefficient of each in ft^2/lb, and (b) the weight per unit area in lb/ft^2 of each material for shielding walls that are 5 half-value layers thick.

6.24 Three quanta, each having 2 MeV of energy, are absorbed in a material. One of the quanta is absorbed by the photoelectric process, another is involved in a Compton scattering, and the third is involved in pair production. (a) Discuss the possible methods by which the scattered photon and each of the charged particles produced can be involved in energy interchanges until all the original photon energy is reduced to thermal energy. (b) Will all this final thermal energy be freed at the point where absorption occurred originally?

6.25 (a) Show that the mean free path of an x-ray photon in a solid is $1/\mu$. [*Hint:* Refer to the derivation of Eq. (1.37).] (b) When a part of the human body, say the arm, is overexposed to a beam of soft (long wavelength) x-rays, a "burn" appears on the skin on the side of incidence, but for rather hard (short wavelength) x-rays, the "burn" usually occurs only on the skin farthest from the incident side. Explain.

6.26 The K_α line of molybdenum (wavelength, 0.712 Å) undergoes a Compton collision in carbon. What is the wavelength change of the line scattered at 90° if the scattering particle is (a) an electron, and (b) the whole carbon atom? (c) What is the scattered wavelength in each case?

6.27 Each of three quanta of radiation undergoes a 90° Compton scattering in a block of graphite. Assuming that the ejected electron had no binding energy in carbon, compute the fractional change in wavelength $(\Delta\lambda/\lambda)$ for the scattered radiation when the incident quantum is (a) a gamma ray from cobalt, $\lambda = 1.06 \times 10^{-2}$ Å, (b) K_α x-rays from molybdenum, $\lambda = 0.712$ Å, and (c) visible light, $\lambda = 5000$ Å. (d) The photoelectric work function of carbon is actually 4.0 eV. If this is taken into account, what would be the answers to the preceding parts? (e) Comment on the feasibility of resolving the two waves in a beam composed of the incident and scattered radiation in each of the preceding cases.

6.28 An x-ray quantum having a wavelength of 0.15 Å undergoes a Compton collision and is scattered through an angle of 37°. (a) What are the energies of the incident and scattered photons and of the ejected electron? (b) What is the magnitude of the momentum of each photon? (c) Find the momentum of the electron both graphically and analytically using the values found in the preceding part.

6.29 Using the data and results of Problem 6.26 for Compton scattering of the K_α line of molybdenum by carbon, find the energy of the recoil particle given that it is (a) an electron and (b) the whole carbon atom. (c) What is the direction of motion of the recoil particle in each case with respect to the direction of incidence of the photon?

6.30 A photon of energy E undergoes a Compton collision with a free particle of rest mass m_0. (a) Show that the maximum recoil kinetic energy of the particle is

$$E_{k\,\max} = \frac{E^2}{E + m_0c^2/2}.$$

(b) What is the maximum energy that can be transferred to a free electron by Compton collision of a photon of violet light ($\lambda = 4000$ Å)? (c) Could violet light eject electrons from a metal by a Compton collision?

6.31 When a hydrogen atom goes from the state $n = 2$ to the state $n = 1$, the energy of the atom decreases by 10.2 eV. In Chapter 4, it was stated that all of this energy is radiated as a single photon. Actually, the energy of the photon must be slightly less than 10.2 eV, since a small part of the energy is needed to provide the kinetic energy of the recoiling atom. (a) What is the recoil momentum of the hydrogen atom? (b) What fraction of the 10.2 eV is taken by the recoil of the atom?

6.32 (a) Radiation of wavelength λ incident normally on a certain mirror is reflected with no loss of energy. Consider a beam of radiation to be a stream of photons. Show that the pressure exerted by a beam on the mirror is $2I/c$, where I is the incident intensity. (b) Calculate the pressure exerted by a beam having an intensity of 0.6 W/m².

6.33 A photon of wavelength λ undergoes Compton scattering from a free electron as shown in Fig. 6.22. Prove that, regardless of the energy of the incoming photon, the scattered photon cannot undergo pair production if α is greater than 60°.

6.34 The apparatus for a Compton scattering experiment is arranged so that the scattered photon and the recoil electron are detected only when their paths are at right angles to one another. Using Fig. 6.22, show that under these conditions (a) the scattered wavelength is given by $\lambda' = \lambda/\cos\alpha$ and (b) the energy of the scattered photon is m_0c^2.

6.35 A photon of energy 1.92 MeV undergoes pair production in the vicinity of a lead nucleus. The created particles have the same speed, and both travel in the direction of the original photon. Calculate the recoil momentum of the lead nucleus.

6.36 Use the conservation laws to show that it is impossible for pair production to take place in free space. Assume that the velocity v of each of the particles of the pair is parallel to the direction of motion of the incident photon.

6.37 Show that the roentgen is equivalent to (a) 2.083×10^9 ion pairs per cm^3 of standard air, and (b) 1.611×10^{12} ion pairs per g of air. The density of air is 1.293 g/liter at standard conditions.

6.38 Given that it requires 33.7 eV of energy to produce an ion pair in air, show that the roentgen is equivalent to an energy absorption of (a) 7.02×10^4 MeV per cm^3 of standard air, (b) 5.42×10^7 MeV per g of air, (c) 86.9×10^{-7} J per g of air, and (d) 2.07×10^{-6} cal per g of air, and 0.87 rad.

6.39 An x-ray beam having an exposure rate of 480 mR/h produces ions in 6 cm^3 of standard air in an ionization chamber. What is the saturated ionization current in amperes?

6.40 A common form of exposure meter or dosimeter is a cylindrical capacitor filled with a gas. One model worn by personnel working in the vicinity of x-ray equipment has an absorption equivalent to 6 cm^3 of standard air. (a) If the electrodes are charged to a difference of potential of 400 V, what must be the resistance of the electrode insulation so that the charge leakage in 8 h shall not exceed 10% of an assumed safe tolerance level of 50 mR/8 h day? (Assume that the voltage remains constant.) (b) What would be the effect on the accuracy of the dosimeter if (1) there are finger streaks on the insulator, and (2) the relative humidity is high?

6.41 The principal lines in the K x-ray spectrum of nickel have the following wavelengths: 1.656 Å, 1.497 Å, and 1.485 Å. Determine the Bragg angles at which these lines are reflected in the third order from the (111) planes in a simple cubic crystal for which the unit axial length is 6 Å.

6.42 Referring to Fig. 6.17, one finds that in a particular experiment the distance halfway around the circular film mounting was 22.53 cm, and the distance along the film from the exit beam to one of the lines was 6.34 cm, and to another one, 11.03 cm. What was the Bragg angle for each line?

6.43 When K_α radiation from copper ($\lambda = 1.54$ Å) is used to obtain a powder diffraction pattern of KCl, a face-centered cubic crystal, first-order maxima occurs at $\theta = 25.3, 29.6, 44.1, 54.5,$ and 58.4°. (a) Calculate the interplanar distance d associated with each of these Bragg angles. (b) Find the Miller indices for the set of planes for each value of d. (c) Calculate the basic interplanar distance from the data for each case. (d) What is the length of a side of a unit cube of this face-centered cubic crystal? (e) Draw three face-centered cubic crystals and sketch in the (100) planes in one, the (110) planes in another, and the (111) planes in the third.

6.44 A monochromatic K_α x-ray beam from Cu radiation falls on a crystal of sodium chloride along the x-direction. The crystal that was originally oriented with lattice directions along x, y, and z is rotated slowly about an axis in the z-direction. Certain diffracted beams will emerge in the x-y plane, when the crystal reaches suitable angles. Find the angles that these various emergent beams make with the x-axis.

6.45 Show that the relative interplanar distances $d_{100} : d_{110} : d_{111} = 1 : 1\sqrt{2} : 1/\sqrt{3}$ for a simple cube; $1 : 2/\sqrt{2} : 1/\sqrt{3}$ for a body-centered cube; and $1 : 1/\sqrt{2} : 2/\sqrt{3}$ for a face-centered cube.

7 WAVES AND PARTICLES

7.1 WAVE-PARTICLE DUALITY OF LIGHT

Electromagnetic radiation, which includes visible light, infrared and ultraviolet radiation, and x-rays, was shown to be a wave motion by interference experiments. This type of experiment, which involves constructive and destructive interference, is considered to be a test for the existence of waves since it requires the presence of two waves at the same position at the same time, whereas it is impossible to have two particles occupy the same position at the same time. On the other hand, the experimental results for blackbody radiation, photoelectric effect, and x-ray absorption can be explained by considering the radiation to appear as a stream of particles that are, for example, absorbed one at a time. It appears that light can best be considered as a wave in some experiments and as a particle in others. This does not occur randomly, however; the experiments can be sorted into the following two types. Those that can best be described by the wave nature of light are ones that may be called propagation experiments. An important part of their explanation is the consideration of the path or paths traveled by the light, as in interference experiments where a path difference is determined. The experiments that can best be described by the particle nature of light may be called interaction experiments. The radiation interacts with matter to produce a resultant absorption or scattering.

This dual nature of light was not readily accepted. The main reason for this is the apparently contradictory aspects of the two natures. A wave is specified by a frequency ν, wavelength λ, phase velocity u, amplitude A, and intensity I. These are not all independent. Thus the velocity, frequency, and wavelength are related by $u = \nu\lambda$. A wave is necessarily spread out and occupies a relatively large region of space. Actually, a sinusoidal wave with a unique wavelength would have to have infinite length; otherwise there would be a change in wavelength at the ends of the wave. A particle, on the other hand, is specified by a mass m, velocity v, momentum p, and energy E. The characteristic that seems in conflict with a wave is that a particle occupies a definite position in space. In order for a particle to be at a definite position, it must be very small. It is difficult to accept the conflicting ideas that light is a wave that is spread out over space and also a particle that is at a point in space. This acceptance is necessary, however, to explain all the results of the experiments that can be performed with light. (We use the word light to include the entire electromagnetic spectrum.)

We do have connections between the wave and particle characteristics of light. As indicated in Chapter 3, Planck related the energy of the photon, E, and the frequency of the wave, ν, by

$$E = h\nu. \tag{7.1}$$

In Chapter 6, we saw how Compton used the relation

$$p = h/\lambda, \tag{7.2}$$

between the momentum of the photon, p, and the wavelength of the wave, λ. In addition, the intensity of the wave is related to the rate at which photons pass through a unit area. Consequently, the particle characteristics of light can be found from the wave characteristics, even though the concepts of wave and particle appear to contradict each other. We shall attempt to resolve this contradiction in Section 7.5.

7.2 THE DE BROGLIE HYPOTHESIS

The dual nature of light, made necessary by experimental results, was extended by de Broglie in 1924. He felt that nature was symmetrical and the dual nature of light should be matched by a dual nature of matter. His argument was that if light can act like a wave sometimes and like a particle at other times, then things like electrons, which were considered particles, should also act like waves at times.

To specify the wave properties, de Broglie proposed that the relation between the momentum and the wavelength of a photon is a general one, applying to photons and material particles alike. Since the momentum of a material particle is its mass m times its velocity v, the de Broglie wavelength is

$$\lambda = \frac{h}{mv}. \tag{7.3}$$

These proposed waves were not electromagnetic waves but were a new kind of wave, which were called matter waves or *pilot* waves. The word *pilot* implies that these waves pilot or guide the particle. When de Broglie published his hypothesis, it was not supported by any experimental evidence. His only real argument was his intuitive feeling that nature must be symmetrical.

At this point we may consider the fact that the wavelength is not sufficient to completely specify a wave. The frequency or velocity must also be known. We choose to define the frequency of the matter wave by extending the photon analogy, $\nu = E/h$, and by using the relativistic energy expression, $E = mc^2$. Since the phase velocity of any wave motion is $u = \nu\lambda$, we find the velocity of the waves associated with a particle to be $u = \nu\lambda = (E/h)(h/mv)$. Thus we have

$$u = \frac{mc^2}{h}\frac{h}{mv} = \frac{c^2}{v}. \tag{7.4}$$

In this equation, v is the speed of the material particle, which must be *less* than the speed of light, c; thus u is *greater* than the speed of light. The speed of the mass-energy of the particle does not exceed the speed of light in free-space, but the phase velocity of its associated waves does. This result does not conflict with the concepts of relativity, since the speed of light is a limiting speed only for mass-energy. In the chapter on relativity, we warned that we would encounter situations where we would use speeds greater than c.

7.3 BOHR'S FIRST POSTULATE

The theoretical implications of the de Broglie wavelength of matter are interesting. By making a very plausible assumption, we can relate this wavelength to the Bohr model of the atom.

If a stretched string, fastened at each end, is caused to vibrate, the disturbances move along the string in both directions, are reflected at the ends, and come back on one another. In general, this disturbance of the string causes a complex motion that makes the string blurred everywhere. If the exciting frequency, the length of the string, or the string tension is varied, stationary waves can be produced. Displacement loops and nodes appear. Instead of running up and down the string, the waves appear to stand still with transverse activity at the loops. The condition for standing waves is that the length of the string be an integral number of half wavelengths of the disturbance, since there is a node at each end.

If we were to form a circular loop with the string, the disturbances would move around the loop in both directions and there would be no reflections. The condition for standing waves for this configuration is that the length of the loop (the circumference of the circle) be an integral number of whole wavelengths of the disturbance. Thus, the condition for circular standing waves is

$$2\pi r = n\lambda,$$

where r is the radius of the circle and n is an integer.

Let us now assume that the Bohr orbits correspond to standing electron waves, analogous to the circular loop of string. If we use the above condition with the de Broglie wavelength, we have

$$2\pi r = \frac{nh}{m_e v}. \tag{7.5}$$

Recalling that the angular momentum of a particle moving in a circular orbit is $m_e vr$, we find that the angular momentum of the Bohr electron is

$$m_e vr = n\frac{h}{2\pi}, \tag{7.6}$$

which is precisely Bohr's first postulate. This discussion should not be interpreted as deriving Bohr's postulate, for essentially what we have done is replace Bohr's postulate with de Broglie's postulate.

7.4 THE DAVISSON AND GERMER EXPERIMENT

Einstein pointed out that if the de Broglie hypothesis is valid, then it should be possible to diffract electrons. Schroedinger felt that if de Broglie were right, then the waves associated with matter should suffer diffraction. At that time there was no real experimental evidence that provided convincing proof of the wave nature of matter. However, in 1925 Elsasser deduced from de Broglie's theory that a beam of electrons diffracted by a crystal should show interference phenomena. This prediction eventually led to experimental verification.

In 1927 Davisson and Germer were studying the scattering of electrons by nickel. Their technique was reminiscent of both Rutherford alpha-particle scattering and Compton x-ray scattering. They directed a beam of electrons onto a block of nickel and measured the intensity of the electrons as they scattered from the nickel in different directions. In the course of the experiment, their vacuum system broke accidentally and had to be repaired.

When the vacuum system broke, the nickel target was at a high temperature, and the air caused the nickel to acquire a heavy coat of oxide. To remove the oxide from the block of nickel, Davisson and Germer reduced the oxide slowly in a high-temperature oven. When their apparatus was reassembled, they began to get very different results. Whereas the number of scattered electrons had previously become continuously less as the scattering angle increased, they now found that the number of electrons went through maxima and minima. *The electrons were being diffracted.* Using the familiar techniques of x-ray diffraction by crystals, Davisson and Germer computed the wavelength their electrons must have, and they found that this wavelength agreed with the de Broglie formula.

The prolonged heating to clean the nickel block had caused it to become a single crystal, and the electron diffraction pattern was completely analogous to x-ray

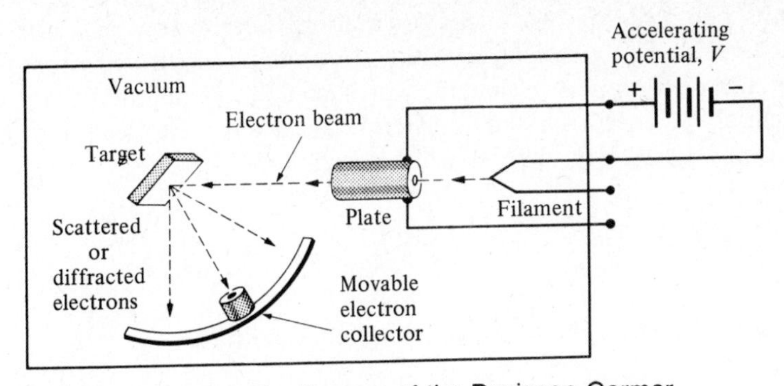

Figure 7.1 Schematic diagram of the Davisson-Germer electron diffraction apparatus.

diffraction by the Bragg technique. This experiment verified the de Broglie hypothesis and indicated that material particles have wave properties.

To consider the Davisson and Germer experiment in more detail, we show their apparatus schematically in Fig. 7.1. At the right is an electron gun that provides a collimated beam of electrons whose energy is known from the accelerating potential. These electrons were scattered by the nickel target, which could be rotated about an axis perpendicular to the page. The movable electron collector could be swung about the same axis as the target, so that it could receive the electrons coming from the target in any direction included in the plane of the diagram. Figure 7.2 shows the

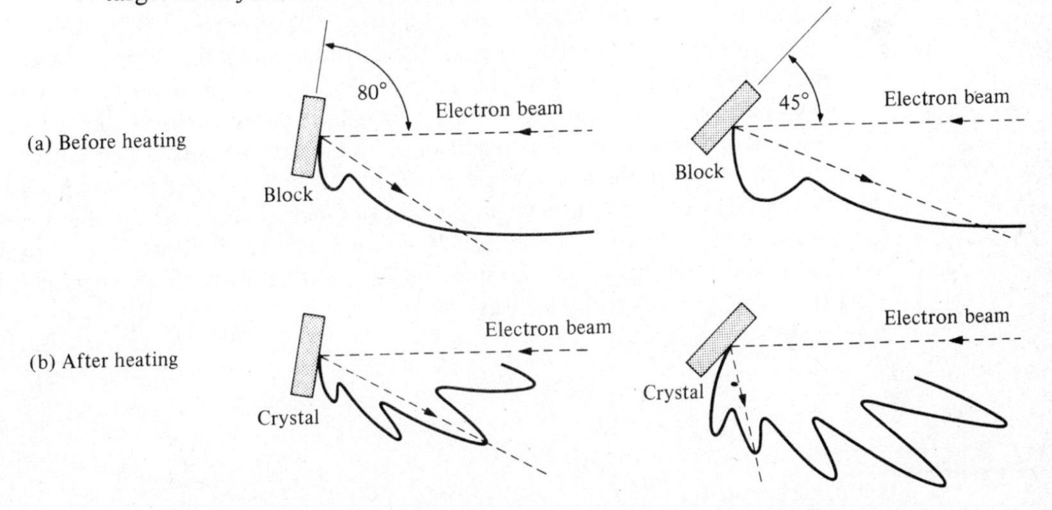

Figure 7.2 (a) Electron scattering from a block of nickel. (b) Electron diffraction from a single crystal of nickel.

From *Phys. Rev.* 30, 705 (1927). Reprinted by permission of the American Physical Society.

results obtained for two target orientations both before and after the target was heat-treated. The wavelength of the electrons observed experimentally agreed well with the de Broglie wavelength, which could be computed from the accelerating potential, V. Since Davisson and Germer used 75-eV electrons, they could obtain the electron velocity from the classical expression

$$E_k = Ve = \frac{m_e v^2}{2}, \tag{7.7}$$

and substitute into the de Broglie relation, obtaining

$$\lambda = \frac{h}{m_e v} = \frac{h}{\sqrt{2Vem_e}}. \tag{7.8}$$

The de Broglie hypothesis was further verified in Germany when Estermann and Stern diffracted helium atoms from a lithium fluoride crystal, and in the United States when Johnson diffracted hydrogen from the same kind of crystal. G. P. Thomson, son of J. J. Thomson, obtained excellent powder diffraction patterns by sending a collimated beam of electrons through very thin sheets of various metals. Figure 7.3(a) shows an electron diffraction pattern of aluminum. For comparison an x-ray diffraction pattern of aluminum is shown in part (b) of the figure. The effective wavelengths for the two patterns are different but the similarities are evident. Davisson and Germer, using electrons instead of x-rays, repeated Compton's experiment of diffracting soft x-rays from a grating at grazing incidence.

With the complete verification of the de Broglie hypothesis, we have arrived at a point where our atomic world has strange aspects. Compton scattering showed that waves have particle aspects and the de Broglie hypothesis shows that particles have wave characteristics. Another result of all these experiments is contained in Bohr's principle of *complementarity*, which states that in no single experiment does a photon show both wave and particle properties, nor does a particle simultaneously show both particle and wave properties. In Section 7.5 we shall attempt to resolve this duality more seriously than Eddington did when he humorously suggested that the primordial entity is really a "wavicle."

7.5 WAVE GROUPS

Although the interference effects compel us to accept the fact that, like light, material particles have the dual nature of waves and particles, it is still difficult to see how the wave extended over a region of space does not conflict with the small particle located at a specific position. This apparent conflict can be resolved by the

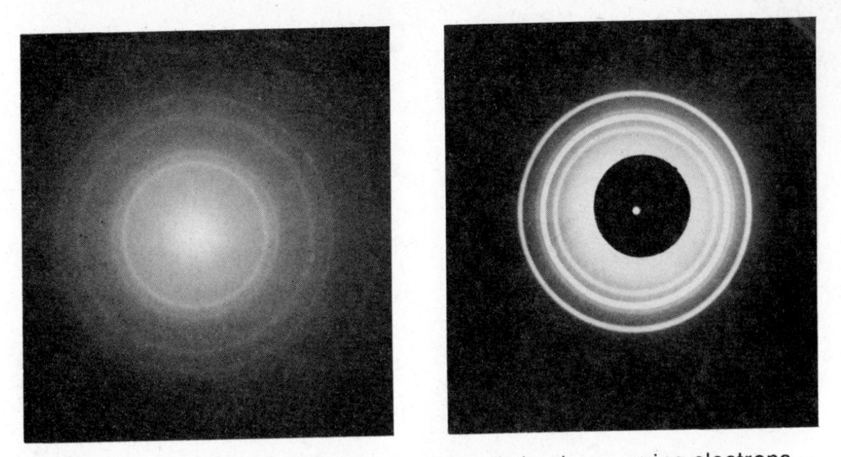

Figure 7.3 (a) Electron diffraction pattern of aluminum, using electrons accelerated through 8000 volts. (b) X-ray diffraction pattern of aluminum, using *K*-radiation from copper. The wavelengths and geometry are different for the two photographs. A lead disk was placed over the center of the x-ray film for most of the exposure.

formation of what is called a wave group or wave packet. In essence this is a wave that extends over a very limited region of space. It is formed by applying the principle of superposition to many waves. When many waves travel through the same medium, the resultant wave is the summation (in our case it will be algebraic) of all the waves. In acoustics, such a summation of two waves of slightly different frequencies produces beats, which are fluctuations in the amplitude of the resultant wave. We shall see that when more and more frequencies are added, the regions where the amplitude is large get smaller.

To simplify the discussion of the superposition we will use a one-dimensional wave; that is, a wave whose displacement, y, depends on the position along a line, x, and the time, t. A sinusoidal wave with amplitude A may be written as

$$y = A \sin 2\pi\left(\frac{x}{\lambda} - \nu t\right) = A \sin(kx - \omega t), \tag{7.9}$$

where we have introduced k, the propagation number which equals $2\pi/\lambda$, and the angular frequency $\omega = 2\pi\nu$. Even with one-dimensional conditions it is difficult to draw the wave, since there are two independent variables. What we usually do is hold one fixed and vary the other. Thus we either draw the displacement as a function of position at some fixed time (equivalent to a snapshot) or the displacement as a function of time at some fixed position.

Let us now try to form a wave group. Consider two waves, one with wave number $\bar{\nu}(\bar{\nu} = 1/\lambda)$ equal to 5 and the other with wave number equal to 6. Such waves are shown in Fig. 7.4. The superposition of these two waves yields the envelopes of waves shown in the lower part of Fig. 7.4. For the superposition of the two waves, we would obtain an infinite number of wave groups extending from

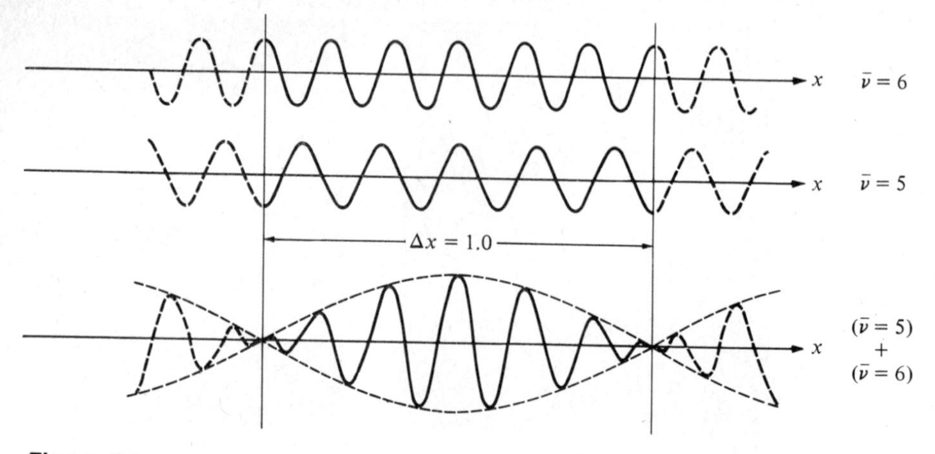

Figure 7.4

minus infinity to plus infinity along the x-axis. Each of these groups would be moving in the plus-x direction with the velocity we will call the group velocity v_g.

Figure 7.5 shows a number of waves with wave numbers from $\bar{\nu} = 7$ to $\bar{\nu} = 13$. Also the amplitude of each is different by a small amount. A superposition of these waves yields the wave group shown at the bottom of that figure. This wave group moves in the plus-x direction with a velocity of v_g. (Even with the superposition of this many waves, an infinite number of wave groups would be obtained. However, they are farther apart than the wave groups for the addition of the two waves.)

Now we shall take the step that will give us a single wave packet. The superposition of an infinite number of periodic waves differing infinitesimally in wave number and frequency will yield a single wave packet. Such a wave packet is shown in Fig. 7.6. The contributing waves cancel each other everywhere except in the one localized region. At a given moment in time at this one point the positive displacements of all of the contributing periodic waves add up to a maximum total displacement. At no other place in space at that time do the maxima all coincide. At any other place in space it is equally as probable to have a negative displacement as a positive displacement, so the total displacement approaches zero.

Let us consider two of the phase waves, which can be written as

$$y_1 = A\cos(k_1 x - \omega_1 t)$$

and

$$y_2 = A\cos(k_2 x - \omega_2 t),$$

where $u_1 = \omega_1/k_1$ and $u_2 = \omega_2/k_2$. The functions have been chosen such that the two waves are in phase and maximum at $x = 0$ and $t = 0$. The addition of these two waves can be done by using the trigonometric identity

$$\cos a + \cos b = 2\cos\tfrac{1}{2}(a - b)\cos\tfrac{1}{2}(a + b).$$

The resultant wave becomes

$$\begin{aligned} y &= y_1 + y_2 \\ &= 2A \cos \tfrac{1}{2}[(k_2 - k_1)x - (\omega_2 - \omega_1)t] \cos \tfrac{1}{2}[(k_1 + k_2)x - (\omega_1 + \omega_2)t]. \end{aligned}$$

If the propagation numbers and frequencies differ only by differential amounts, we have

$$\begin{aligned} k_2 - k_1 &= dk, \\ \omega_2 - \omega_1 &= d\omega, \\ \tfrac{1}{2}(k_1 + k_2) &\doteq k_1, \end{aligned}$$

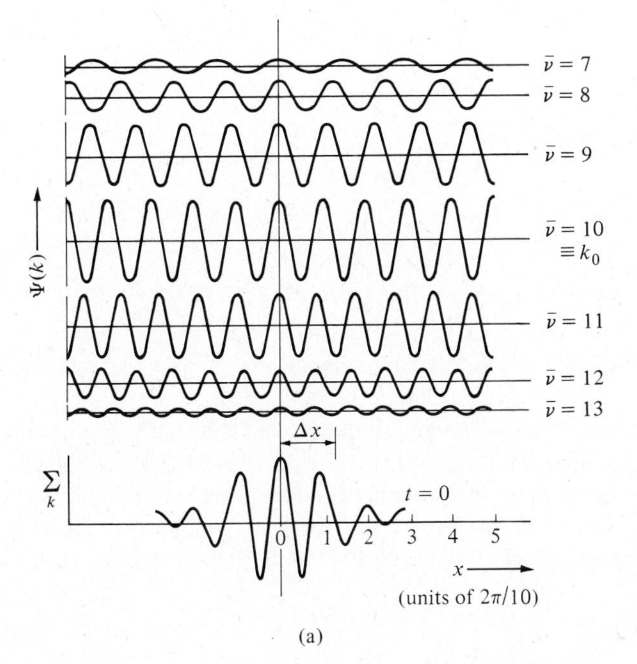

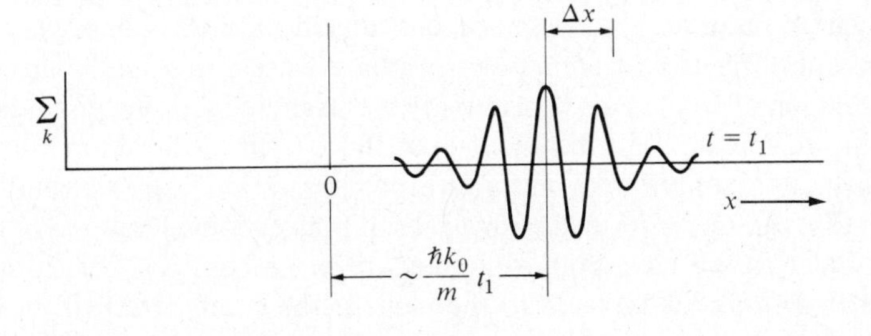

Figure 7.5 Superposition of waves.

Adapted from *Introduction to Quantum Mechanics* by Charles M. Sherwin. Copyright © 1959 by Holt, Rinehart and Winston, Inc. Re - printed by permission of Holt, Rinehart and Winston, CBS College Publishing.

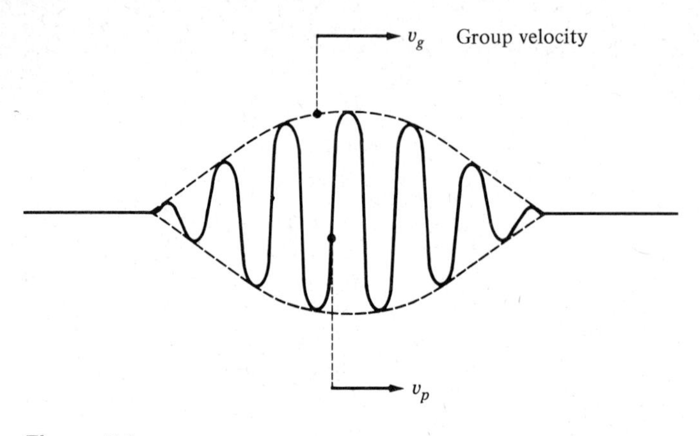

Figure 7.6

and

$$\tfrac{1}{2}(\omega_1 + \omega_2) \doteq \omega_1.$$

The resultant wave is then

$$y = 2A \cos \tfrac{1}{2}[dk\, x - d\omega\, t] \cos(k_1 x - \omega_1 t). \tag{7.10}$$

The second cosine function is the original wave. The coefficient of this cosine can be considered to be an amplitude that varies with x and t. This variation of amplitude is called the *modulation* of the wave and has a long wavelength $4\pi/dk$ and a low frequency $d\omega/4\pi$. From the product of these we can get the velocity of propagation of the modulation, which is the group velocity v_g:

$$v_g = \left(\frac{4\pi}{dk}\right)\left(\frac{d\omega}{4\pi}\right) = \frac{d\omega}{dk}. \tag{7.11}$$

The group velocity depends on the way in which the frequency varies with the propagation number. Since we have considered only two phase waves, it is possible that v_g will differ for different pairs of phase waves; that is, the group velocity may be a function of frequency. When many waves are added, the derivative of Eq. (7.11) is evaluated for the central frequency of those used in the summation.

There is a simple and instructive demonstration experiment illustrating these results that can be performed with a pair of pocket combs that have slightly different tooth spacings. We have found that a set of black combs is best — one with 12 and the other with about 14 teeth to the inch. If the combs are held in front of a light surface (or on an overhead projector) and superimposed with their teeth parallel, light and dark regions can be observed. These bands are like acoustical "beats" and the number of bands per inch is the difference in the number of teeth per inch in the combs. The teeth of the two combs represent the phase waves and the light bands of constructive interference are the wave groups. By adding many more combs with different tooth spacings one behind the other, we could form an arrangement such that only one constructive region remained.

If we now slide one comb over the other in a direction perpendicular to the teeth, we observe that the bands move with a velocity different from that of the comb motion. Groups move faster or slower than the phase teeth and have a direction with or against the comb motion, depending on which comb is moved. It is much better to experience this than to read a description.

If the wave group is to represent a particle, then in some way it is necessary that the speed of the group and the speed of the particle be the same. If these speeds differed, the particle would soon be in a region where the amplitude of the wave is negligible and the wave would not give a useful indication of the position of the particle. Using the mass-energy relation from relativity, we can find the particle velocity in terms of its momentum and energy:

$$v = p/m = pc^2/mc^2 = pc^2/E.$$

To get the group velocity in the same terms we use $p = h/\lambda$ and $E = h\nu$, which are written as

$$k = \frac{2\pi p}{h} \quad \text{and} \quad \omega = \frac{2\pi E}{h}. \tag{7.12}$$

The group velocity can then be written as

$$v_g = \frac{d\omega}{dk} = \frac{dE}{dp}. \tag{7.13}$$

To evaluate this, we need the energy-momentum relationship from relativity, Eq. (5.50):

$$E^2 = p^2c^2 + m_0^2c^4.$$

Differentiating this, we obtain

$$2E\,dE = 2pc^2\,dp.$$

The group velocity becomes

$$v_g = \frac{dE}{dp} = \frac{pc^2}{E}. \tag{7.14}$$

This is the same as the above expression for the particle velocity and we see that the choices that were made for the frequency and wavelength to be associated with the particle also lead to the satisfying result that the wave group and the particle have the same speed. Thus it is possible to have a wave motion that has the particle characteristic of being in a small region of space and that will move with the particle's speed.

7.6 WAVE-PARTICLE DUALITY

The wave-group idea is an attempt to resolve the conflict contained in the fact that some explanations of experiments require the use of waves while others require the use of particles. This interpretation means that an electron is a group of matter waves and a photon is a group of electromagnetic waves; that is, there are only waves in nature but they form wave packets that we call particles. Macroscopic objects consist of many of these wave groups and the interactions between objects are simply interactions of waves.

Although this "reality" of the waves seems reasonable, there are some difficulties associated with it. When a wave is incident on a boundary between two media, it generally splits into a reflected wave and a refracted wave. The incident wave group becomes two groups. If the wave group is to be an electron, where the boundary would be produced by a change in electric potential energy, we know that the electron does not split. It is difficult to see how the two wave groups, one reflected and one refracted, can be the single electron. In addition, the Coulomb force, which is stated for point charges and acts between electrons, cannot be handled conveniently with this interpretation.

Another possible interpretation was presented by de Broglie. In this the particles are "real" and the associated waves are pilot waves that guide the particle. These waves are abstract quantities that are to be regarded as probability waves. The amplitude of these probability waves at a certain position is a measure of the probability of the particle being at that position. The pilot waves are abstract quantities and the word "wave" is used as it is, for instance, in the phrases "wave of enthusiasm" and "crime wave." Since the waves are not directly observable, there is no necessity for a medium. This interpretation removes the difficulty of the wave group concept when the wave is partially reflected and partially refracted. For the probability waves, the amplitude of the reflected wave determines the probability that the particle reflects from the boundary and the amplitude of the refracted wave determines the probability that the particle penetrates the boundary. A single particle will do *only* one or the other, but a large number of particles will divide according to the probabilities.

This pilot wave interpretation, which means that a light beam is a stream of photons, also runs into difficulty. Suppose that we pass a light beam through a pair of slits separated by a small distance, as in the interference experiments of Young.

The interference pattern produced on a screen consists of alternate light and dark bands. The positions of these bands are determined by the differences of the paths of the two waves proceeding from the slits to the screen. The alternate light and dark bands represent alternate large and small probabilities of a photon arriving at the respective positions on the screen. If one of the slits is covered, the interference pattern changes, which means the probabilities of where the photon will arrive at the screen changes. For example, the photon may now have a large probability of arriving at a position where the previous probability was very small. If the photons are real, however, they *must* go through one slit or the other. Consequently, as the photon goes through one slit, its motion is influenced by the other slit and the photon is able to know whether the slit through which it did not pass is open or not. This gives some "intelligence" to the particles that does not seem reasonable. We will return to this problem in Section 7.8.

The resolution of the conflict between waves and particles lies in an appraisal of what we mean by a wave and a particle. Both of these terms, when applied to the fundamental entities, are abstractions of the human mind that are arrived at by extrapolation from the macroscopic world of grains of sand and waves on strings. The following is another very clever and useful trick of the human mind. A hollow rubber ball has its center of gravity at its center. Discussion of most, but not all, of the motion of the ball can be greatly facilitated by regarding the ball as a point mass with all its mass at the center of gravity. The center of gravity has no objective reality, and if someone cuts the ball open, points to the center and says, "Ha! You see, there is no mass there," we reply that the center of gravity makes a poor description of what is at the center of the ball, but that it continues to be useful in describing the motion of the ball. No one description of the ball can ever completely represent what the reality of the ball is. In the same way the particle description and the wave description are each incomplete in attempting to describe physical reality.

The mistake of those who say that interference shows that light *is* a wave phenomenon is a verbal mistake that is made every day. We point to a map and say this *is* the United States. What we mean is that this diagram on a piece of paper is a scale representation of many of the physical and political features of the United States. We know that the real United States cannot be folded, rolled up, or burned. We know that the states are not different in color, only a few square inches in area, and completely flat. The map is a clever, useful, elegant model invented by the human mind. More may be learned about some aspects of the United States in one hour of map study than in a lifetime of looking at the real United States. We do not scoff at maps because they are unreal; we admire them as useful descriptions.

Both the wave and the particle are models we have constructed in attempting to describe matter. Quite naturally, we do not expect either model to give a complete description. Some properties, such as interference, are contained in one model, the wave, while other properties, such as mass, are contained in the other model, the particle. The two models complement each other in that together they give a description of matter. Thus we should say that the electrons are waves *and* particles, not waves *or* particles. The same statement can be made about electromagnetic radiation. During an experiment, the particular model that is used is determined by the apparatus used.

Even though we admit that waves and particles are not "real," it is very awkward to talk about experimental procedures and results in such a way as to indicate this. Consequently, we will still make statements that seem to imply that particles exist; for example, we say that the intensity of the wave is a measure of the probability of the location of the particle. This is for convenience only. Both the wave and the particle are incomplete models and both are necessary for a description of all the properties of matter that are experimentally determined. With this interpretation, there is no conflict in the dual nature of matter or electromagnetic radiation.

7.7 THE HEISENBERG UNCERTAINTY PRINCIPLE

An important consequence of the wave-particle duality can be developed from the wave group analysis of Section 7.5. In an attempt to get a wave that had a limited extent, we added many waves to form a wave group. If we correlate the wave model to the particle model by assuming that the amplitude of the group measures the probability of the particle being at that position, we see that there is still no certainty in knowing the location of the particle. It could be anywhere in the group.

To decrease the uncertainty in location of the particle, we have to reduce the size of the group, Δx (see Fig. 7.7). From the theory of Fourier analysis, it can be shown that a combination of waves with wave numbers covering a range Δk will yield a wave group spread out in space Δx, where the product is of the order of unity.[†] Thus

$$\Delta x\, \Delta k \approx 1. \tag{7.15}$$

The size of the group can be reduced by increasing the spread of propagation numbers, Δk. It appears that we may eliminate the uncertainty in position of the particle by using an infinite spread of propagation numbers. We see from $k = 2\pi p/h$ that the momentum of the particle is determined by the propagation number of the wave. If we use an infinite spread of propagation numbers, we will have an infinite spread in the momentum of the particle. When we decrease the uncertainty of the particle's position, we increase the uncertainty of the particle's momentum. If we put Eq. (7.15) in terms of momentum from Eq. (7.12), we have

$$\Delta x\, \Delta p_x \approx \frac{h}{2\pi} = \hbar. \tag{7.16}$$

The subscript is added to the momentum to indicate that it is the momentum associated with the x-displacement. Equation (7.16) is interpreted as indicating that the uncertainty in position of the particle times the uncertainty in the associated

† The exact value of the product will depend on the amplitude of the waves.

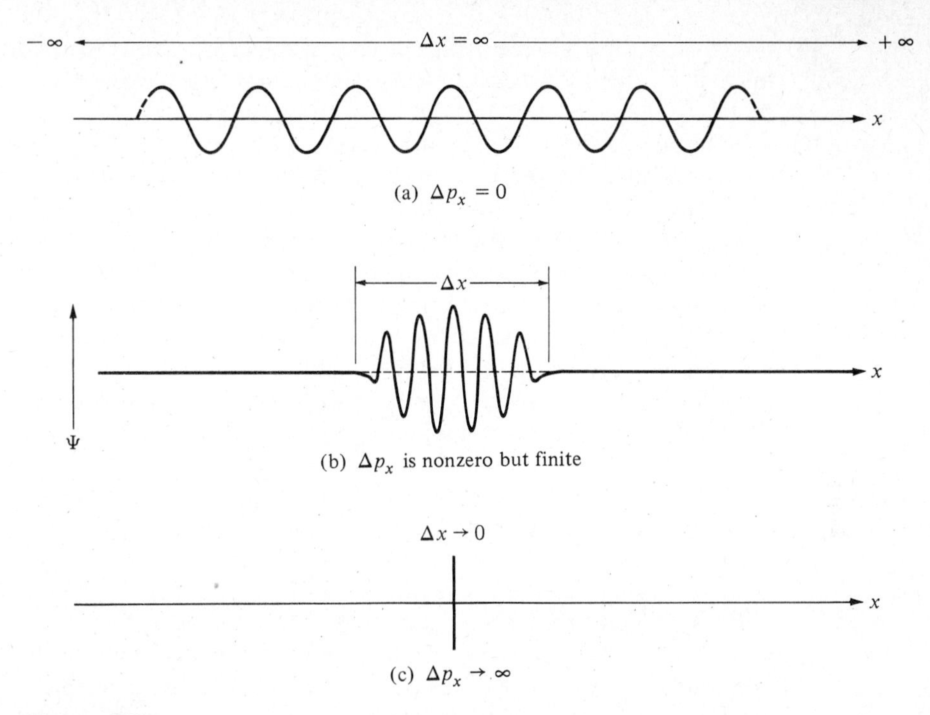

Figure 7.7

momentum is approximately the Planck constant divided by 2π. The words *associated momentum* are used because in the three-dimensional case there are also momentum components in the y- and z-directions and with the x-position we use the x-component of momentum. There are equivalent equations for the other directions:

$$\Delta y \, \Delta p_y \simeq \hbar, \tag{7.17}$$

$$\Delta z \, \Delta p_z \simeq \hbar. \tag{7.18}$$

These uncertainties are involved with the nature of matter and are not the same as the uncertainties introduced by the limited precision of some measuring device. In a practical experiment the uncertainties introduced by the equipment will usually be much larger than the ones associated with the wave-particle duality.

We can perform a similar analysis on the waves for time to obtain

$$\Delta t \, \Delta E \simeq \hbar. \tag{7.19}$$

This relationship can be interpreted as meaning that the uncertainty of the energy of the particle is dependent on the time interval used, since the particle must be observed for a time Δt to be certain that the particle has passed the point of observation. In a broader sense, if an energy measurement is performed in a time interval Δt, there will be a corresponding minimum uncertainty in the energy given by Eq. (7.19).

This result throws light on a question not considered in the Bohr theory of atomic energy states. An electron in an atom spends most of its time in its unexcited state, and since in this state Δt is large, then ΔE is small and the energy of the electron is well determined. Although the time electrons spend in excited states is usually rather short, it is highly variable. We said in Chapter 4 that phosphorescence is due to some energy states being metastable, in which case Δt may be hours. The longer an electron remains in a particular state, the better is its energy in that state determined. This means that the energy difference between the excited state and the ground state has less uncertainty and there is little spread in the radiated frequency; that is, the spectral line is sharper. Experimentally, the transitions from short-lived excited states to the ground state result in broad intense spectral lines, while the transitions from metastable states result in sharp weak spectral lines. The fact that weak spectral lines are sharper than intense lines agrees with these uncertainties in energies.

These equations concerning uncertainties, that is, Eqs. (7.16), (7.17), (7.18), and (7.19) are variant statements of the *Heisenberg uncertainty principle*. We look upon these statements as indicative of the inherent nature of the physical world. Thus it is impossible for us to know an exact position and an exact momentum of a photon or an electron. A precise knowledge of one can be obtained only at the expense of the precision of the other. Frequently, the uncertainty principle is introduced as being due to the act of measurement. For example, when the position of an electron is determined, the measurement introduces the uncertainty in position and momentum because of an interaction between the apparatus and the electron. This implies that the electron had an exact position and momentum before the measurement. We prefer the interpretation that the uncertainties are inherent in the nature of the things we do experiments with. Admittedly, this choice cannot be made on the basis of any physical result. Physics deals with the results of measurements and either interpretation leads to the same uncertainties in measured quantities.

As anticipated in Chapter 1, another philosophical idea affected by the uncertainty principle was that of causality. According to classical theory, the path of a particle is determined by its initial position and momentum and by the forces acting on it. If the forces between the particles of the universe were known and if it were possible to measure, at a given time, the exact position and momentum of every particle, then the past and future positions and momenta of every particle could be calculated. The past and future are completely determined by the information known at an instant of time. The uncertainty principle indicates that we cannot know an exact position and momentum for each particle; we can only determine what the particles will probably do. For macroscopic objects the uncertainties are so small that the probable motion does not differ significantly from classical motion. However, for photons and electrons, the classical prediction of their motion gives a poor idea of the experimental results.

7.8 THE DOUBLE-SLIT EXPERIMENT

To crystallize the ideas we have been discussing about the wave-particle duality and to emphasize the limits on our ability to describe the behavior of the basic entities of the physical world, we take the double-slit diffraction experiment mentioned in Section 7.6 as an example. A schematic diagram of the experiment is shown in Fig. 7.8. The source emits a beam that is incident on two slits separated by a distance D. The beam could be electromagnetic, that is, a stream of photons, or a beam of cathode rays, that is, a stream of electrons. The following analysis applies to either, but for convenience we shall consider the stream to be photons. At a distance from the slits, which is large compared to D, we have a detector, which could be a photographic film or a series of photoelectric cells.

If the detector is a film, it will show, after exposure, the characteristic interference pattern indicated by the curve labeled intensity on the figure. The positions of the maxima and minima are found by considering the superposition of the wave from one slit with the other; that is, the interference pattern is a prediction of the wave model. The angular separation of the central maximum and the adjacent minimum is determined from

$$D \sin \theta = \lambda/2 = h/2p_x,$$

where λ is the wavelength of the light and p_x is the associated momentum of the photon.

When angle θ is small, it becomes

$$\theta = h/2p_x D, \tag{7.20}$$

and θ is the angle between any adjacent maximum and minimum.

If the detector is a series of photoelectric cells, the light will release photoelectrons and the maxima will indicate those cells where many electrons are emitted. Each photoelectron represents the arrival at the detector of a photon and thus more photons strike the detectors around the maxima in the pattern than around the

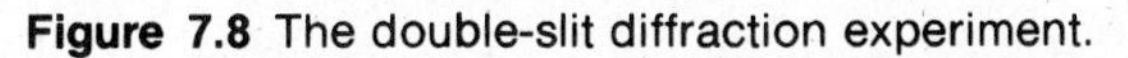

Figure 7.8 The double-slit diffraction experiment.

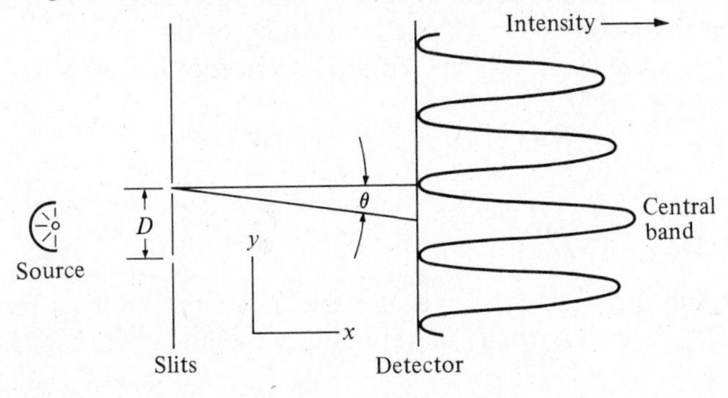

minima. The pattern gives very little information pertaining to a single photon but only the effect produced by many photons. The correlation between the wave model and the particle model is statistical. The maxima of the wave pattern represent regions of greatest probability for the arrival of a photon.

If one of the slits is covered during the experiment, the resulting single-slit pattern will differ from the preceding one. If a cover is placed over the bottom slit for the first half of the experiment and then the cover is placed over the top slit for the second half, the total exposure at the detector is different than when both slits are uncovered. The positions on the screen, where the photon will most likely strike, change. If we consider the light beam to be a stream of photons, a single photon must pass through either one slit or the other. The fact that the pattern is different for two slits than one slit means that the motion of the photon is influenced by the other slit. It appears that the photon, as it passes through one slit, "knows" whether the other slit is open. Thus the particle model seems to lead to an absurd result.

An analysis of the experiment shows that no information was obtained experimentally about the passage of a photon through either slit. Consequently a statement about such passage is meaningless. This emphasizes that physics deals with predictions for experimental results. The definition of every physical quantity *must* contain, explicitly or implicitly, the operation necessary to measure the physical quantity. It makes *no sense* to talk about the results that cannot be observed. If we want to know through which slit the photon passes, we must revise the experiment so that we can determine this. To detect the passage of a photon we place many small particles just to the right of the slits. After passing through a slit, the photon will collide with one of these particles. Observation of the recoil of the small particle will allow us to determine the slit used by the photon. In this ideal experiment the only uncertainties introduced will be those predicted by the uncertainty principle; we will assume that the precision achieved in constructing the apparatus was so high that there are no uncertainties in dimensions, such as the separation of the slits.

To be certain of the slit from which the small particle recoiled, the uncertainty in its y-position must be much smaller than the separation of the slits; we must have

$$\Delta y \ll D.$$

There will be an exchange of momentum during the collision, but there is an uncertainty in the amount since we don't know the details of the collision. The uncertainty in the y-component of the momentum of the photon, Δp_y, must not be great enough to cause it to deviate from the interference pattern. This means that we must have

$$\Delta p_y / p_x \ll \theta = h/2p_x D,$$

or

$$\Delta p_y \ll h/2D.$$

Since momentum is conserved, Δp_y is also the uncertainty in the y-component of the momentum of the small particle. If we multiply the uncertainty in the y-position of the small particle with the uncertainty in its y-component of the momentum, we obtain

$$\Delta y \, \Delta p_y \ll Dh/2D = h/2. \tag{7.21}$$

If we compare this with the prediction of the uncertainty principle, $\Delta y \, \Delta p_y \simeq \hbar$, we see that the small uncertainties necessary in Eq. (7.21) are not possible. If Δy is small enough that we can determine through which slit the photon passes, Δp_y is so large that we do not get the interference pattern at the detector. *We cannot detect both the particle nature, as indicated by the collision with the small particle, and the wave nature, as indicated by the interference pattern.* This example also illustrates how the uncertainty principle is used to resolve apparent conflicts between the wave model and the particle model.

7.9 SUMMARY

The wave-particle duality of matter is substantiated by experiment. Neither the wave model nor the particle model is sufficient alone to include all of the properties of the physical world. The two models are synthesized to complement each other in providing a description for all experimental results. This synthesis is contained in the relations between the wave properties of wavelength and frequency and the particle properties of momentum and energy. The predictions of the two models seem to be correlated statistically; a large wave amplitude means a large probability of the particle being at that position. For material particles like electrons, the particle model has been developed so that, for example, we can calculate electron trajectories in magnetic fields. The corresponding wave model for material particles requires further elaboration. Although we can find the associated wavelength and frequency, we have no method for calculating the amplitude of the "electron wave." Our experience with previous wave motions, such as sound waves, indicates that a wave equation is needed. The solution of this equation would give the amplitude of the wave, which we can correlate with the particle model. A *complete* acceptance of the dual nature of matter requires a further development of the wave model.

FOR ADDITIONAL READING

Born, Max, *Physics in My Generation*. London and New York: Pergamon Press, 1956. Thought-provoking essays on relativity, wave mechanics, and other fields of modern physics.

Carruthers, P., *Quantum and Statistical Aspects of Light*, Resource Letter QSL-1 and Selected Reprints. New York: American Institute of Physics, 1963.

de Broglie, Louis, *Matter and Light*. W. H. Johnston, translator. New York: Dover Publications, 1946.

Gamow, George, *Mr. Tompkins Explores the Atom*. New York: Macmillan, 1944. A fantasy describing life in a wave-mechanical atomic world.

Gamow, George, "The Principle of Uncertainty," *Scientific American* **198**, 50 (January 1958).

Hund, F., "Paths to Quantum Theory Historically Viewed," *Physics Today* **23** (August 1966).

Margenau, Henry, *The Nature of Physical Reality*. New York: McGraw-Hill, 1950. Comprehensive and thorough.

Margenau, H., and G. M. Murphy, *The Mathematics of Physics and Chemistry*. Princeton, N.J.: Van Nostrand, 1943.

Price, William C., and Seymour S. Chissick, editors, *The Uncertainty Principle and Foundations of Quantum Mechanics*. New York: Wiley, 1977.

Squire, Charles F., *Waves in Physical Systems*. Englewood Cliffs, N.J.: Prentice-Hall, 1971.

REVIEW QUESTIONS

1. What is the de Broglie hypothesis?
2. What experiment confirmed the work of de Broglie?
3. Why is a wave packet needed in wave mechanics?
4. What are the significant consequences of the Heisenberg uncertainty principle?
5. What phenomena can be best explained on the basis of the wave nature of light?
6. What phenomena can be best explained by the particle nature of light?

PROBLEMS

7.1 Show that relativistically the wavelength associated with a particle having a rest mass m_0 and a velocity v is

$$\lambda = \frac{h(1 - v^2/c^2)^{1/2}}{m_0 v}.$$

7.2 Show that the de Broglie wavelength in angstroms for an electron accelerated from rest through a potential difference V in volts is (a) classically, $\lambda = 12.27/V^{1/2}$ and (b) relativistically,

$$\lambda = \frac{12.27}{V^{1/2}}\left(\frac{Ve}{2m_e c^2} + 1\right)^{-1/2}.$$

7.3 What is the de Broglie wavelength of the waves associated with an electron that has been accelerated from rest through a potential difference of (a) 100 V and (b) 8000 V?

7.4 Show that the de Broglie wavelength for a particle of mass m moving with the rms speed of a Maxwellian distribution at temperature T is $\lambda = h/(3mkT)^{1/2}$, where k is Boltzmann's constant.

7.5 Calculate the de Broglie wavelength for a neutron moving with the rms speed of a Maxwellian distribution at (a) 20°C and (b) 100°C. (c) At what angle will these neutrons undergo a first-order Bragg reflection from the principal planes of MgO, which is a cubic crystal with a principal grating space of 2.101 Å?

7.6 The spacing between the nuclei of a certain crystal is 1.2 Å. At what angle will first-order Bragg reflection occur for thermal neutrons (kinetic energy of 0.025 eV)?

7.7 Compare velocities and kinetic energies of an electron and of a neutron for which the de Broglie wavelengths are equal to 1 Å.

7.8 (a) What is the de Broglie wavelength associated with an electron in the $n = 1$ orbit of the Bohr atom? (b) What is the circumference of the first Bohr orbit? (c) Make the same calculations for the $n = 2$ orbit.

7.9 What is the de Broglie wavelength (a) for a 22-caliber rifle bullet having a mass of 1.1 g and a speed of 3×10^4 cm/s and (b) for a 75 kg (165 lb) student ambling along at 0.5 m/s? (c) What would have to be the grating space of a plane grating that would diffract the "particles" in parts (a) and (b) through an angle of 30° in the first order assuming normal incidence? (d) Discuss the feasibility of such diffraction.

7.10 Compare the wavelength of an electron with kinetic energy of 1 MeV, a neutron with kinetic energy of 1 MeV, and a 1 MeV photon.

7.11 Electrons from a heated filament are accelerated by a difference of potential between the filament and the anode of 3 kV. A narrow stream of electrons coming through a hole in the anode is transmitted through a thin sheet of aluminum. Using Bragg's law, calculate the angle of deviation of the first-order diffraction pattern.

7.12 An alpha particle (doubly ionized helium) is ejected from the nucleus of a radium atom with 5.78 MeV of kinetic energy. (a) What is the de Broglie wavelength of this particle? (b) How does this wavelength compare with the nuclear diameter, which is about 2×10^{-14} m?

7.13 (a) By actual experiment with two combs, determine whether the group velocity of the waves is greater or less than their phase velocity when the "long-wavelength" comb is moved faster than the "short-wavelength" one. (b) Which is the greater, the group velocity or the phase velocity, (1) in a beam of white light in glass and (2) in a beam of sound waves having different frequencies in air? (c) Except in free space, the group and the phase velocity of light are identical only for ideally monochromatic light. Is it possible to obtain such light? (Consider the Doppler effect in the source of light, filters, finite slit widths of monochromators, etc.)

7.14 A wave group is formed by the addition of an infinite number of waves. The ratio of the angular frequency and propagation number is the phase velocity. (a) Given that the phase velocity is constant, show that the group velocity is equal to the phase velocity. (b) Given that the phase velocity is proportional to the propagation number, show that the group velocity is twice the phase velocity.

7.15 Show that the group velocity v_g can be obtained from the phase velocity u by the relation

$$v_g = u - \lambda \frac{du}{d\lambda}.$$

7.16 Under proper conditions, the velocity of water waves having the wavelength λ is given by $u = (g\lambda/2\pi)^{1/2}$, where g is the acceleration due to gravity. Show that the group velocity for such waves is one-half the phase velocity.

7.17 (a) Show from Eqs. (7.3) and (7.4) that the phase velocity in free space of the de Broglie waves associated with a moving particle having a rest mass m_0 is given by

$$u = c\sqrt{1 + \left(\frac{m_0 c}{h}\lambda\right)^2}.$$

(b) According to this equation, which wavelengths will have the greater phase velocity, the long ones or the short ones? Will each exceed c? (c) By referring to the experiment in Problem 7.13, show that the answer to part (b) of this problem involves no contradiction in saying that the velocity of a particle is less than c and that the phase velocity of its associated waves is greater than c.

7.18 (a) What is the momentum of a photon of wavelength 0.02 Å? (b) What is the momentum of an electron that has the same total energy as a 0.02 Å photon? (c) What is the de Broglie wavelength of the electron in part (b)?

7.19 Compute the minimum uncertainty in the location of a 2-g mass moving with a speed of 1.5 m/s and the minimum uncertainty in the location of an electron moving with a speed of 0.5×10^8 m/s, given that the uncertainty in the momentum is $\Delta p = 10^{-3} p$ for both.

7.20 Assume that the uncertainty in the location of a particle is equal to its de Broglie wavelength. Show that the uncertainty in its velocity is equal to its velocity.

7.21 A 20 gram bullet and an electron each are measured with an uncertainty of 0.01% to have a speed of 100 meters per second. With what fundamental accuracy can the location of each be obtained if the position is measured simultaneously with the speed in the same experiment? Compare the relevance of the uncertainty principle to electrons and bullets.

7.22 Compute from the uncertainty principle the minimum possible energy of an electron localized inside a nucleus of radius 10^{-13} cm. Does this value of the energy seem reasonable? (The rest mass of an electron converts to 0.511×10^6 eV.) What problem do you see for the old model of a nucleus which was made up of only protons and electrons?

7.23 The electron in a hydrogen atom moves into the excited state $n = 2$ and remains there for 10^{-8} s before making a downward transition to the ground state. Calculate the uncertainty of the energy in the state $n = 2$. Is this a significant correction to the Bohr theory prediction of -3.39 eV?

7.24 A hydrogen spectrum has one spectral line of wavelength 4340.5 Å with a measured width of 10^{-11} Å. What is the average time that the atomic system remains in the corresponding upper energy state?

7.25 We wish to simultaneously measure the wavelength and position of a photon. Assume that our measurement of wavelength gives $\lambda = 6000$ Å and that our equipment allows an accuracy of one part in a million in the measurement of λ. What is the minimum uncertainty in the position of the photon?

7.26 What is the uncertainty in the location of a photon of wavelength 3000 Å if this wavelength is known to an accuracy of one part in a million?

8 QUANTUM MECHANICS

8.1 THE CLASSICAL WAVE EQUATION

The subject of quantum mechanics cannot be treated meaningfully without employing far greater mathematical sophistication than has been necessary heretofore. The reader will encounter functional notation, operators, partial differential equations, probability, and the so-called complex algebra. Although we attempt to explain each mathematical step, our purpose here is to use mathematics rather than to teach it. We hope that if some readers of this book cannot follow every step in the argument, the flavor and elegance of the wave mechanics will, nevertheless, be conveyed.

A development of the wave nature of matter requires that a wave equation be used to find the displacement of the matter wave, which will be a function of position and time. Before the equation for matter waves is developed, let us consider the classical wave equation. In particular, we shall consider the equation for transverse waves in a string. This means that there will be one position coordinate, x. Since the displacement, y, is a function of two independent variables, x and t, a differentiation of y with respect to x must be a partial differentiation; that is, one for which t is held constant. When y is differentiated with respect to t, x is held constant; thus we expect a partial differential equation for y.

The phase velocity of a traveling wave in the string is determined by the elastic properties of the string. If the string is uniform, this velocity, u, is constant and the wave travels along the string without change of shape. An observer moving parallel to the string with velocity u will see this stationary shape. If the position coordinate used by the moving observer is x', he or she will say that the displacement of the string is a function of x' but not of t, or that

$$y = f(x').$$

The coordinate x' is related to the original coordinate x by the Galilean–Newtonian transformation from Chapter 5: $x' = x - ut$. Consequently, for an observer at rest with respect to the string, the displacement is the same function of $(x - ut)$ or

$$y = f(x - ut).$$

The sinusoidal wave of Eq. (7.9) was simply a particular function of $(x - ut)$. Since there can also be a traveling wave moving in the opposite direction, this functional relation can be generalized to

$$y = f(x \pm ut). \tag{8.1}$$

(The use of the relativistic transformation would not change this result.)

We can use Eq. (8.1) to obtain the partial differential equation for y. Some partial derivatives of y with respect to x are

$$\frac{\partial y}{\partial x} = \frac{\partial f}{\partial x} = \frac{\partial f}{\partial (x \pm ut)} \frac{\partial (x \pm ut)}{\partial x} = \frac{\partial f}{\partial (x \pm ut)}$$

and

$$\frac{\partial^2 y}{\partial x^2} = \frac{\partial}{\partial x}\left[\frac{\partial f}{\partial (x \pm ut)}\right] = \frac{\partial^2 f}{\partial (x \pm ut)^2} \frac{\partial (x \pm ut)}{\partial x} = \frac{\partial^2 f}{\partial (x \pm ut)^2}.$$

Differentiating with respect to t, we have

$$\frac{\partial y}{\partial t} = \frac{\partial f}{\partial t} = \frac{\partial f}{\partial (x \pm ut)} \frac{\partial (x \pm ut)}{\partial t} = \frac{\partial f}{\partial (x \pm ut)}(\pm u) = \pm u \frac{\partial f}{\partial (x \pm ut)}$$

and

$$\frac{\partial^2 y}{\partial t^2} = \pm u \frac{\partial^2 f}{\partial (x \pm ut)^2} \frac{\partial (x \pm ut)}{\partial t} = u^2 \frac{\partial^2 f}{\partial (x \pm ut)^2}.$$

A second-order partial differential equation, which includes both directions of the velocity, can be formed from the two second-order derivatives obtained above:

$$\frac{\partial^2 y}{\partial x^2} = \frac{1}{u^2} \frac{\partial^2 y}{\partial t^2}. \tag{8.2}$$

This wave equation can also be derived by considering the forces acting on a differential element of the string and its resulting acceleration. The solutions of Eq. (8.2) are the possible displacements of the string. A *particular solution* is determined by the initial displacement of the string in conjunction with any imposed boundary conditions. Thus, if the string is fixed at two points, the displacements at these two points must always be zero. The particular solution is then the usual standing wave. Even though we obtained Eq. (8.2) from a traveling wave, we see there are other types of waves possible. Since the equation is linear, any linear combination of solutions will also be a solution. This implies the *principle of superposition* and gives the usual method of forming a standing wave by a linear combination of traveling waves moving in opposite directions.

A review of the development of Eq. (8.2) will show that use of a string was not vital to the argument. The equation is called the classical wave equation and applies to the one-dimensional wave motion of strings, air columns, etc. Many of the properties that are associated with such waves are a result of the form of Eq. (8.2). We seek the corresponding equation for the waves that are associated with material particles.

8.2 THE SCHROEDINGER EQUATION

Since the concept of matter waves is not a result of previous physical theories, it is *impossible* to derive the corresponding wave equation for a particle. We are dealing with a new field of physics and cannot expect the basis for this field to be dependent on the basis of other fields. We are using the wave model, however, and the equation is developed in a manner analogous to that used for other waves.

The dependent variable of a matter wave is called the wave function and is denoted by Ψ (called capital psi). For one-dimensional problems the wave function is a function of the coordinates x and t. Analogous to the classical wave, we may expect that for a traveling matter wave, Ψ will be a function of $(x - ut)$. Since the velocity of the wave, u, is the ratio of the circular frequency and the propagation number, $u = \omega/k$, the wave function can be written as a function of $(kx - \omega t)$. From Eq. (7.25) we recall that the propagation number of the wave is proportional to the momentum of the particle and that the angular frequency is proportional to the energy; therefore we can write

$$\Psi = f\left[\frac{2\pi}{h}(px - Et)\right] = f\left(\frac{px - Et}{\hbar}\right),$$

where $\hbar$ (called h-bar) stands for $h/2\pi$. We may select as our function a sine or cosine wave. A more general wave would be a sum of a sine and cosine wave. By taking a certain combination of sine and cosine, we can use Euler's identity to express Ψ in exponential form:†

$$\Psi = A\exp\left[\frac{i}{\hbar}(px - Et)\right]. \tag{8.3}$$

This exponential form is very convenient when derivatives are to be found.

†Euler's identity is $e^{i\theta} = \cos\theta + i\sin\theta$, where i is the imaginary number $\sqrt{-1}$. This has the properties $i^2 = -1$ and $i^{-1} = -i$.

To obtain a differential equation for Ψ, we seek a combination of partial derivatives for which Ψ is a solution. We assume that the energy and momentum of the particle are constant. If Eq. (8.3) is differentiated with respect to x, we obtain

$$\frac{\partial \Psi}{\partial x} = A\frac{ip}{\hbar}\exp\left[\frac{i}{\hbar}(px - Et)\right]$$

or

$$\frac{\partial \Psi}{\partial x} = \frac{ip}{\hbar}\Psi,$$

which can be rearranged as

$$p\Psi = \frac{\hbar}{i}\frac{\partial \Psi}{\partial x} = -i\hbar\frac{\partial \Psi}{\partial x}. \tag{8.4}$$

This relationship is interpreted as meaning that the momentum p may be replaced by the differential operator $-i\hbar(\partial/\partial x)$, which operates on the wave function. Differentiation of Eq. (8.3) with respect to t gives

$$\frac{\partial \Psi}{\partial t} = -\frac{iE}{\hbar}A\exp\left[\frac{i}{\hbar}(px - Et)\right] = -\frac{iE}{\hbar}\Psi,$$

which is rearranged to give

$$E\Psi = i\hbar\frac{\partial \Psi}{\partial t}. \tag{8.5}$$

Thus the differential operator $i\hbar(\partial/\partial t)$, when operating on Ψ, is equivalent to the energy. Although Eqs. (8.4) and (8.5) were obtained for constant momentum and energy, we assume that they are valid even when p and E are not constant.

The partial derivatives with respect to x and t are connected by means of the relation between the energy and the momentum. The classical expression for the kinetic energy, in terms of the momentum, is

$$E_k = \frac{mv^2}{2} = \frac{(mv)^2}{2m} = \frac{p^2}{2m}, \tag{8.6}$$

where m is the rest mass of the particle. The total energy and momentum are related by the expression for energy conservation of a particle, which is

$$\frac{p^2}{2m} + V = E, \tag{8.7}$$

where V is the potential energy, which is a function of x. This step restricts us to a nonrelativistic wave equation. To obtain the wave equation, Eq. (8.7) is multiplied by Ψ and the differential operators are substituted for p and E. When this is done, we obtain

$$-\frac{\hbar^2}{2m}\frac{\partial^2 \Psi}{\partial x^2} + V\Psi = i\hbar\frac{\partial \Psi}{\partial t}. \tag{8.8}$$

This is the Schroedinger equation for one-dimensional matter waves. It was obtained with the use of analogies to other wave equations and solutions. Since the wave aspects of particles are not a result of previous properties, this equation cannot be derived. The test of the Schroedinger equation is whether its predictions agree with experimental results. To apply this equation to a particular problem, we need to know the particle mass and the potential function.

Let us compare the Schroedinger equation with the classical wave equation,

$$\frac{\partial^2 y}{\partial x^2} = \frac{1}{u^2}\frac{\partial^2 y}{\partial t^2}.$$

Both equations are linear partial differential equations. This means that the principle of superposition is valid and therefore that a sum of solutions is also a solution. This is necessary in order to have interference phenomena. Equation (8.8) differs from the classical wave equation by having a term with no derivative, $V\Psi$, and having a first-order time derivative. In addition, the Schroedinger equation contains i, which makes it a complex equation. This means that Ψ is a complex function but of real variables x and t. Because the two wave equations have these differences, we cannot expect the analogy between them to be maintained. In particular, the complex nature of Ψ indicates that Ψ cannot be observed like the displacement of a string. It also means that we should not look for a medium that transmits our matter waves. There is no "ether" for matter waves, as opposed to the existence of, say, water for water waves.

8.3 INTERPRETATION OF THE WAVE FUNCTION

As with other wave motions, the wave equation we have obtained must be supplemented by additional conditions in order that the solutions have physical significance. These conditions are that Ψ and $\partial\Psi/\partial x$ must be single-valued, finite, and continuous. For brevity we will call a function that satisfies these conditions a *well-behaved function*. A comparison with the string shows that these conditions are reasonable. At some specific values of x and t, there must be only one value of the displacement, and it cannot be infinite. Because the energy of the string is proportional to the square of the amplitude, an infinite amplitude would require infinite energy. The need for continuity of the displacement of the string is obvious. The fact that similar conditions are placed on $\partial\Psi/\partial x$ is also reasonable and eliminates the possibility of a kink in the string.

Now that we have a wave equation with its associated conditions requiring it to be well-behaved, we must decide on an interpretation of Ψ. The discussion in Chapter 7 indicates that the intensity of the wave should provide a prediction of the probable location of the particle. Since the position coordinate x varies continuously, we will have a probability density as we did in the Maxwellian speed distribution in

Chapter 1. We shall consider the probability of the particle being located in an interval between x and $x + dx$. In general, the probability is time-dependent and the probability for this location is written as $P(x, t)\,dx$, where $P(x, t)$ is the *probability distribution function*. This probability must be real, whereas the solutions to the Schroedinger equation are complex; consequently Ψ cannot directly be used to give the probability. To obtain a real quantity we postulate that the probability that a particle is located within the interval between x and $x + dx$ is

$$P(x, t)\,dx = \Psi^*(x, t)\Psi(x, t)\,dx, \tag{8.9}$$

where Ψ^* (called psi-star) is the *complex conjugate* of Ψ. For all functions, the complex conjugate can be obtained by reversing the sign of any imaginary, i, that may be in the function. The product $\Psi^*\Psi$ is real; it is also written as $|\Psi|^2$, where the bars mean absolute value. This is similar to other waves where the intensity is proportional to the square of the amplitude. The probability that the particle is in a finite interval between x_1 and x_2 is found by summing the probabilities, as indicated by the integral:

$$\text{probability between } x_1 \text{ and } x_2 = \int_{x_1}^{x_2} \Psi^*\Psi\,dx. \tag{8.10}$$

This probability may be time dependent, since the wave function is time dependent. The interpretation of $\Psi^*\Psi$ as a probability density leads to another restriction on the wave function. The particle must be *somewhere* in the entire range of x; therefore the probability of the particle being in this range must be unity, which means that

$$\int_{-\infty}^{\infty} \Psi^*\Psi\,dx = 1. \tag{8.11}$$

When this restriction has been satisfied, the solution is said to be *normalized*.

A well-behaved and normalized solution of the Schroedinger equation for a specific system determines the probability distribution function for the system. This probability density is not directly observed experimentally for a single particle. To show this and to introduce some of the uses of probabilities, we consider a discrete system; that is, a system with a discrete number of possible results, each with a certain probability. Suppose that we use a loaded coin for which the probability of obtaining a head is twice that of obtaining a tail when the coin is tossed. If this probability distribution is normalized, the probability of a head is $\frac{2}{3}$ and the probability of a tail is $\frac{1}{3}$. The problem is to determine how these probabilities will affect experiments performed with the coin. We certainly do not expect to obtain a tail on every third toss. We would not be surprised if we didn't obtain four tails in twelve tosses. As the number of tosses increases, however, we expect that approximately one-third of them will be tails, and that the approximation will improve with the number of tosses. A quantitative test of the approximation is to give each possibility a numerical value and then to calculate the average number obtained in a series of tosses. If we let x represent the value of a toss and we give a tail the value of one and a head the value of two, the average of N tosses is

$$\bar{x} = \frac{\sum_{i=1}^{2} x_i N_i}{\sum_{i=1}^{2} N_i} = \frac{(1) N_1 + (2) N_2}{N_1 + N_2},$$

where N_1 is the number of tails obtained and N_2 is the number of heads, so that $N_1 + N_2 = N$. As N increases, we expect this average value to approach the average value calculated on the assumption that the results coincide with the prediction of the probabilities. We shall call this the *expectation value* and denote it by $\langle x \rangle$. The expectation value is

$$\langle x \rangle = \frac{\sum_{i=1}^{2} x_i P_i}{\sum_{i=1}^{2} P_i} = \sum_{i=1}^{2} x_i P_i,$$

where P_i is the probability of obtaining the value x_i and $\sum_{i=1}^{2} P_i = 1$ when the probabilities are normalized. For our coin $P_1 = \frac{1}{3}$ and $P_2 = \frac{2}{3}$, so that we have

$$\langle x \rangle = (1)(\tfrac{1}{3}) + (2)(\tfrac{2}{3}) = \tfrac{5}{3}.$$

We expect the average value of a large number of tosses to be very close to $\frac{5}{3}$. (Note that the expectation value cannot be observed experimentally in a single toss of the coin.) A knowledge of P_i also allows us to calculate the expectation value of any function of x, in the same manner:

$$\langle f(x) \rangle = \sum_{i=1}^{2} f(x_i) P_i.$$

As an example, let us calculate the expectation value of x^2 for the coin. We have

$$\langle x^2 \rangle = \sum_{i=1}^{2} (x_i)^2 P_i = (1)^2(\tfrac{1}{3}) + (2)^2(\tfrac{2}{3}) = 3.$$

Note that $\langle x^2 \rangle$ for this system is not equal to the square of $\langle x \rangle$.

It is entirely possible to encounter probability distributions that have the same $\langle x \rangle$ but different $\langle x^2 \rangle$. The value of $\langle x^2 \rangle - \langle x \rangle^2$ indicates the shape of the distribution — low values indicate a sharp distribution and large values indicate a broad one. A distribution that has no spread is one for which the probability of a certain value is one and the probabilities of all other values are zero. If the value x_j has $P_j = 1$, we have

$$\langle x \rangle = x_j$$

and

$$\langle x^2 \rangle = x_j^2.$$

Thus for a distribution that permits only one value to be observed, we see that

$$\langle x^2 \rangle = \langle x \rangle^2. \tag{8.12}$$

This equation is usually used as a test to see if the probability distribution corresponds to a system where only one value of the variable is observed.

When there is a continuous range of possible values, we must use a probability distribution function, and the probability of obtaining a value between x and $x + dx$ is given by $P(x)\,dx$. In the expressions for expectation values, the sums are replaced by integrals and we have

$$\langle x \rangle = \int_{-\infty}^{\infty} xP(x)\,dx$$

and

$$\langle f(x) \rangle = \int_{-\infty}^{\infty} f(x)P(x)\,dx.$$

This applies to the situation where we have a solution to the Schroedinger equation with $P(x, t) = \Psi^*\Psi$. When calculating expectation values with this probability distribution, we choose to write the function between Ψ^* and Ψ. The expressions for expectation values become

$$\langle x \rangle = \int_{-\infty}^{\infty} \Psi^* x \Psi\,dx \tag{8.13}$$

and

$$\langle f(x) \rangle = \int_{-\infty}^{\infty} \Psi^* f(x) \Psi\,dx. \tag{8.14}$$

The wave function does not give the position of the particle but only permits us to compute the expectation value of the position. For the wave group constructed in Chapter 7, this corresponds to the center of the group and is the average value of the position after many measurements have been taken. When we say that the wave function gives a complete description of the particle, we mean that we can calculate the expectation value of any physical quantity that can be measured. Such quantities are called *physical observables*. A necessary consequence of this is that expectation values of physical observables must be real, since physical measurements give real numbers.

Two physical observables of interest, in addition to position, are momentum and energy. The expectation value of the momentum of a particle is given by

$$\langle p \rangle = \int_{-\infty}^{\infty} \Psi^* p \Psi\,dx.$$

This integration cannot be performed unless the momentum is expressed as a function of x. We already used a function of x to replace the momentum in our development of the Schroedinger equation. According to Eq. (8.4) we replace p with the operator $-i\hbar(\partial/\partial x)$. Our decision to place the variable between Ψ^* and Ψ is due to the fact that we must use operators. For the momentum, the partial differentiation is performed on Ψ only. The expectation value of momentum is

$$\langle p \rangle = \int_{-\infty}^{\infty} \Psi^* \left(-i\hbar \frac{\partial}{\partial x} \right) \Psi \, dx. \tag{8.15}$$

For the energy, we have the operator from Eq. (8.5); thus the expectation value of the energy is

$$\langle E \rangle = \int_{-\infty}^{\infty} \Psi^* \left(i\hbar \frac{\partial}{\partial t} \right) \Psi \, dx. \tag{8.16}$$

If we were interested in other physical observables, such as angular momentum, we would have to select equivalent operators for them, either by analogy or as some combination of previous operators. All of these expectation values may be dependent on time but they are not dependent on position.

Let us review the use of the Schroedinger equation. The system in which the particle is contained is described by a potential function $V(x, t)$. For this potential a well-behaved solution of the partial differential equation is found. This solution Ψ, when normalized, gives the probability distribution function $\Psi^*\Psi$ from which the expectation value of any physical observable can be determined. The expectation values are related to the average values obtained from experiments on the system. Since we can predict all experimental results from Ψ, we say that it provides a complete description of the motion.† If the expectation value of the square of a physical observable is equal to the square of the expectation value of that observable, a measurement of the observable will always yield the expectation value.‡ A wave function that yields this result represents a state of the system that can be said to have a definite value of the observable. If this is not true, there is a spread in the distribution and a definite value of the observable has not been found.

8.4 THE TIME-INDEPENDENT EQUATION

Before we consider a specific potential function, we shall try a standard technique for solving partial differential equations. The technique is to see if there is a solution $\Psi(x, t)$ that is a product of a function of x, say $\psi(x)$ (called small psi), and a function of t, say $\phi(t)$. We assume that the solution can be written as

$$\Psi(x, t) = \psi(x)\phi(t) \tag{8.17}$$

† It should be noted that only statistical experimental results can be so predicted.

‡ Strictly, the condition is that $\langle x^n \rangle$ must equal $\langle x \rangle^n$ for all values of n. In practice, the value $n = 2$ is used.

and insert this in Eq. (8.8) to obtain

$$-\frac{\hbar^2}{2m}\frac{\partial^2}{\partial x^2}(\psi\phi) + V\psi\phi = i\hbar\frac{\partial}{\partial t}(\psi\phi).$$

In each differentiation, the function of the other variable can be placed in front of the derivative and then the partial derivative can be replaced by an ordinary derivative, since it operates on a function of the variable only. This gives

$$-\frac{\hbar^2}{2m}\phi\frac{d^2\psi}{dx^2} + V\psi\phi = i\hbar\psi\frac{d\phi}{dt}.$$

This equation is divided by $\Psi = \psi\phi$ and we have

$$-\frac{\hbar^2}{2m}\frac{1}{\psi}\frac{d^2\psi}{dx^2} + V = i\hbar\frac{1}{\phi}\frac{d\phi}{dt}. \tag{8.18}$$

We now impose the restriction that the potential energy V is a function of x only. This restriction means that the entire left-hand side of Eq. (8.18) is a function of x only while the right-hand side is a function of t only. Since x and t are independent variables, the only way a function of x can always equal a function of t is to have both functions equal to a constant. Otherwise, the two functions would vary independently and even if they were equal for certain values of x and t, they would not be equal for all values of x and t. The constant that each side of Eq. (8.18) must equal is called the *separation constant* E and we obtain two equations:

$$-\frac{\hbar^2}{2m}\frac{1}{\psi}\frac{d^2\psi}{dx^2} + V = E$$

and

$$i\hbar\frac{1}{\phi}\frac{d\phi}{dt} = E.$$

We have separated the partial differential equation into two ordinary differential equations which can be written as

$$-\frac{\hbar^2}{2m}\frac{d^2\psi}{dx^2} + V\psi = E\psi, \tag{8.19}$$

called the *time-independent* Schroedinger equation, and

$$i\hbar\frac{d\phi}{dt} = E\phi. \tag{8.20}$$

Thus letting $\Psi = \psi\phi$ enables us to obtain two ordinary differential equations to be solved.

Since we restricted the potential to be a function of x, it does not appear in Eq. (8.20); therefore we can solve for the time dependence of the product solution without specifying the potential. If Eq. (8.20) is written as

$$\frac{d\phi}{dt} = -\frac{iE}{\hbar}\phi,$$

it is similar to Eq. (6.8) for the absorption of x-rays. The solution is

$$\phi = e^{-iEt/\hbar}. \tag{8.21}$$

Ordinarily, there would be a constant before the exponential. Since ϕ is multiplied by ψ, we can absorb this constant into ψ. Equation (8.21) gives the time dependence for all product solutions, $\psi\phi$, and we have

$$\Psi = \psi e^{-iEt/\hbar}. \tag{8.22}$$

Because of the similarity to sinusoidal waves, ψ is called the amplitude and the differential equation for ψ, Eq. (8.19), is also called the *amplitude equation*.

Since ϕ must be well behaved, the separation constant E is restricted to a real number. If E were complex, the exponential would have a real part and would approach infinity as t approached infinity. To see what the separation constant represents, we calculate the expectation value of the energy from Eq. (8.16) and obtain

$$\langle E \rangle = \int_{-\infty}^{\infty} \Psi^* \left(i\hbar \frac{\partial}{\partial t} \right) \Psi \, dx$$
$$= \int_{-\infty}^{\infty} \psi^* e^{iEt/\hbar} \left(i\hbar \frac{\partial}{\partial t} \right) \psi e^{-iEt/\hbar} \, dx.$$

Performing the differentiation, we have

$$\langle E \rangle = \int_{-\infty}^{\infty} \psi^* e^{iEt/\hbar} E \psi e^{-iEt/\hbar} \, dx = E \int_{-\infty}^{\infty} \Psi^* \Psi \, dx.$$

Since the wave function is normalized, the last integral is unity and we have

$$\langle E \rangle = E.$$

Using the fact that the square of an operator means the operator applied successively, we can also obtain

$$\langle E^2 \rangle = E^2,$$

which means that $\langle E^2 \rangle = \langle E \rangle^2$. Since this is the requirement for a probability distribution to yield a single value of the observable, the value E will always be obtained from a measurement of the energy of the system.† (This statement cannot be made just from the fact that $\langle E \rangle = E$.) *A product solution corresponds to a state of the system with a definite energy E.* Values of the separation constant for which the solutions are well behaved are the possible energies of the system. The probability distribution function for these solutions is independent of time, since we have

$$P = \Psi^* \Psi = \psi^* e^{iEt/\hbar} \psi e^{-iEt/\hbar}$$
$$= \psi^* \psi.$$

† This energy will be the kinetic energy plus the potential energy of the particle.

This means that the expectation value of any function of x will be constant. States that have this time-independent probability distribution are called *stationary states.* The wave function of these stationary states has the exponential time dependence of Eq. (8.21). The position-dependent part of the wave function must be found from the amplitude equation, which requires knowledge of the potential distribution. Since the amplitude equation has only real terms in it, we shall be able to find real solutions for ψ. This is convenient because it will permit us to graph the solutions and thus obtain a picture of the amplitude as well as the probability distribution.

8.5 THE INFINITE SQUARE WELL

In order to show some of the characteristic features of the solutions to the amplitude equation without the complication of a difficult differential equation, we choose a simple potential distribution that is an idealization similar to the rigid walls of mechanics. The particle is restricted to a region ranging from $x = 0$ to $x = L$, where the potential is constant and most conveniently taken to be zero. Outside this region the potential is very large and we shall simplify our problem by taking it to be infinite. This potential distribution is shown in Fig. 8.1(a). It is called a square well because the potential rises vertically from the horizontal potential inside the well. This is an approximation for a more realistic potential, such as that shown in part (b) of the figure. Our task is to solve the time-independent Schroedinger equation for the amplitude function ψ for the infinite square well.

The potential walls of the well divide the problem into three parts: two outside the well and one inside the well. For the regions outside the well the infinite potential causes the second term of the amplitude equation,

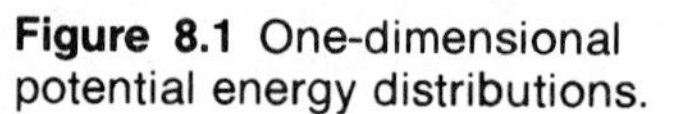

$$-\frac{\hbar^2}{2m}\frac{d^2\psi}{dx^2} + V\psi = E\psi,$$

to be infinite unless $\psi = 0$. If this term is infinite, the equation can be satisfied in these regions only if the first term is also infinite. A function with an infinite second derivative is not a well-behaved function; therefore the solution for $x < 0$ and $x > L$ must be $\psi = 0$. This zero probability of being outside the well corresponds to the

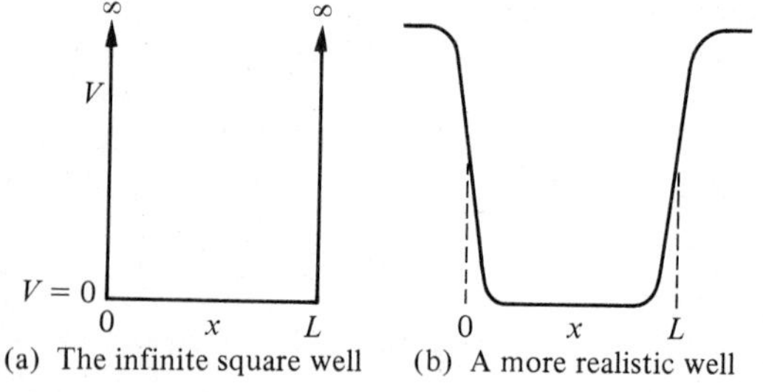

(a) The infinite square well (b) A more realistic well

Figure 8.1 One-dimensional potential energy distributions.

classical rigid wall where the particle cannot penetrate the wall. The continuity condition requires that the solution inside the well must be zero at the walls. There will be discontinuity of the slope at the walls, but this is due to the practical impossibility of the potential we are using.

The amplitude equation inside the well, where the potential is zero, is

$$-\frac{\hbar^2}{2m}\frac{d^2\psi}{dx^2} = E\psi. \tag{8.23}$$

The boundary conditions at the walls are

$$\psi = 0 \quad \text{at} \quad x = 0, L. \tag{8.24}$$

The differential equation can be written as

$$\frac{d^2\psi}{dx^2} = -\text{k}^2\psi, \tag{8.25}$$

where

$$\text{k}^2 = \frac{2mE}{\hbar^2}. \tag{8.26}$$

The general solution of Eq. (8.25) is

$$\psi = A \sin \text{k}x + B \cos \text{k}x.$$

The boundary condition at $x = 0$ gives

$$0 = A \sin(0) + B \cos(0) \quad \text{or} \quad B = 0.$$

At the other wall, $x = L$, we have

$$0 = A \sin \text{k}L.$$

Except for the trivial case when $A = 0$ (and thus $\psi = 0$), this equation is satisfied for the values of k given by

$$\text{k}L = \pi n, \qquad \pm n = 0, 1, 2, 3, \ldots,$$

or

$$\text{k} = \pi n/L, \qquad \pm n = 0, 1, 2, 3, \ldots. \tag{8.27}$$

From Eq. (8.26) these discrete values of k lead to discrete values of the energy:

$$E_n = \frac{\hbar^2\pi^2 n^2}{2mL^2}, \qquad n = 1, 2, 3, \ldots, \tag{8.28}$$

where we have used the integer n, called the quantum number, as a subscript on E to indicate this discrete character. The only constant to be determined is A, for which we use the normalization condition, Eq. (8.11):

$$\int_0^L A^2 \sin^2\left(\frac{n\pi x}{L}\right) dx = 1.$$

For all values of n, the value of A is $(2/L)^{1/2}$. The final solutions are

$$\psi_n = \sqrt{2/L}\,\sin(n\pi x/L), \qquad n = 1, 2, 3, \ldots. \tag{8.29}$$

The first three energy levels and amplitude functions are shown in Fig. 8.2.

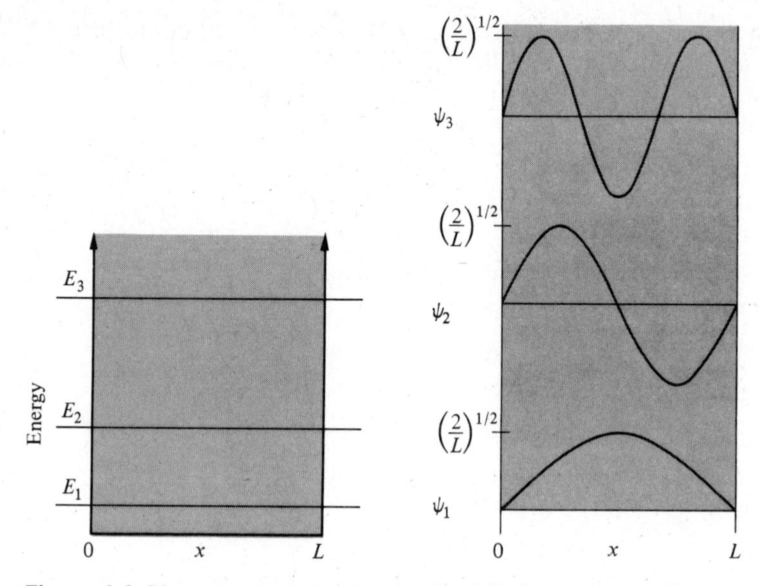

Figure 8.2 Discrete energy states for the infinite square well.

These results for the infinite square well show one of the most significant features of wave-mechanical solutions. When a particle is in a bounded system, there are discrete energy levels. These quantized energy levels enter the problem naturally and are a result of the application of the condition that the solutions of the Schroedinger equation be well behaved. Classically, a particle inside a one-dimensional box with rigid walls could have any positive energy. This disagreement

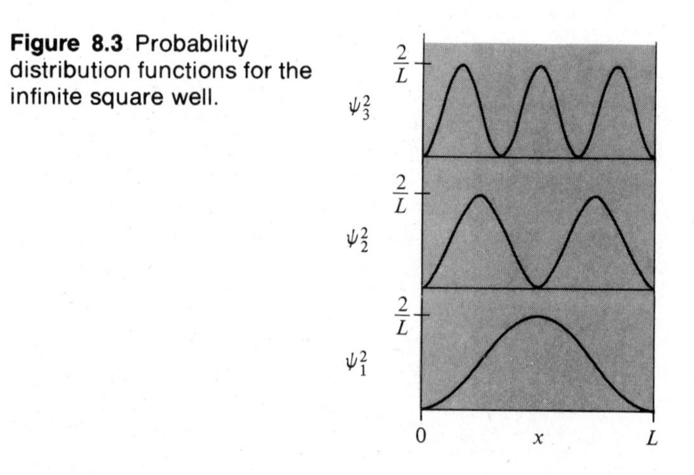

Figure 8.3 Probability distribution functions for the infinite square well.

between wave and classical mechanisms is also evident in the probability distribution functions. These are shown in Fig. 8.3 for the lowest three states.

Classically, the particle passes back and forth between the walls and it has an equal probability of being found anywhere between $x = 0$ and $x = L$. The classical probability function is constant with a value of $1/L$. The quantum functions show peaks of magnitude $2/L$ and valleys where the probability is very small. The number of peaks is equal to the quantum number, n. As n gets very large, the number of peaks becomes large and they get closer together. The average of the distribution is $1/L$ and when n is very large, the distribution approaches the classical distribution. This is an example of the correspondence principle introduced in Problem 4.32. The discrete character of the energy levels disappears also as n gets large, since the energy difference when n changes by one becomes very small compared to the energy.

The expectation value of x for all states of the infinite well is $L/2$, which can be seen from the symmetry of the probability distributions about $x = L/2$. The expectation value of x^2 is *not* $L^2/4$. This is just an indication of the many experimental values possible in a measurement of the position of the particle. The expectation value of the momentum p is zero for all states. The expectation value of p^2 is not zero and again this indicates that there is more than one possible value for a measurement of the momentum of the particle. These results show that the solutions of the Schroedinger equation are not in conflict with the Heisenberg uncertainty principle. The fact that these states correspond to a definite energy does not conflict with $\Delta E \, \Delta t \simeq h$, since the wave functions represent stationary states; that is, $\Delta E = 0$ and $\Delta t = \infty$.

If the square well has a finite potential at the walls, the qualitative features are the same, except that ψ is not zero outside the walls but decreases exponentially as it "penetrates" the walls. Discrete energy levels are still obtained, although there are a finite number of them. This limit on the number occurs because eventually the energy becomes equal to the value of the potential at the walls. For energies higher than this, the wave function does not decrease outside the walls and the particle can have any energy. We have what is called a *continuum* of energy states.

8.6 THE BARRIER POTENTIAL

In the last section, we investigated the properties of a particle in a well with infinite walls on each side. We now want to look at the properties of a particle incident on a barrier with finite height. Our reason for wishing to investigate this potential is its significance to alpha decay, which we shall discuss in Chapter 11.

Figure 8.4 show this potential and divides it into three regions: I, II and III. In regions I and III, the potential is zero and in region II the potential is V_0. The

Schroedinger equation for each region will be

$$\frac{d^2\psi_{\text{I}}}{dx^2} + \frac{2mE}{\hbar^2}\psi_{\text{I}} = 0, \qquad \text{Region I} \tag{8.30}$$

$$\frac{d^2\psi_{\text{II}}}{dx^2} + \frac{2m(E - V_0)}{\hbar^2}\psi_{\text{II}} = 0, \qquad \text{Region II} \tag{8.31}$$

and

$$\frac{d^2\psi_{\text{III}}}{dx^2} + \frac{2mE}{\hbar^2}\psi_{\text{III}} = 0. \qquad \text{Region III} \tag{8.32}$$

Let $\alpha^2 = 2mE/\hbar^2$ and $\beta^2 = 2m/\hbar^2(V_0 - E)$ and solutions to the Schroedinger equation for the three regions become

$$\psi_{\text{I}} = Ae^{i\alpha x} + Be^{-i\alpha x}, \qquad \text{Region I} \tag{8.33}$$

$$\psi_{\text{II}} = Ce^{\beta x} + De^{-\beta x}, \qquad \text{Region II} \tag{8.34}$$

and

$$\psi_{\text{III}} = Fe^{i\alpha x} + Ge^{-i\alpha x}. \qquad \text{Region III} \tag{8.35}$$

In region I, we have the possibility of a particle moving to the right ($e^{i\alpha x}$) or a particle moving to the left ($e^{-i\alpha x}$). In region III, we have the same possibility. However, we choose $G = 0$ since we have no source for particles or point for reflection in region III. Our problem is reduced to solving for the remaining five constants, which can be done by applying the appropriate boundary conditions. The wave function and derivative of the wave function must be continuous at $x = 0$ and $x = a$. In other words, at $x = 0$,

$$\psi_1 = \psi_2, \tag{8.36}$$

$$\frac{d\psi_1}{dx} = \frac{d\psi_2}{dx}, \tag{8.37}$$

and at $x = a$,

$$\psi_2 = \psi_3, \tag{8.38}$$

$$\frac{d\psi_2}{dx} = \frac{d\psi_3}{dx}. \tag{8.39}$$

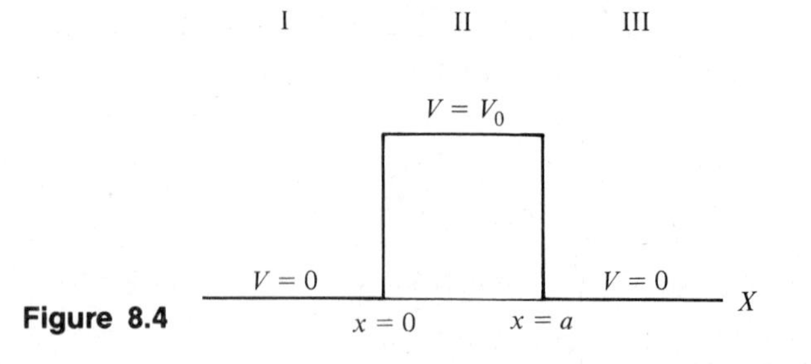

Figure 8.4

Substituting ψ_1, ψ_2, and ψ_3 and their derivatives from Eqs. (8.33), (8.34), and (8.35) will yield the following:

$$\begin{aligned} A + B &= C + D, \\ i\alpha(A - B) &= \beta(C - D), \\ Ce^{\beta a} + De^{-\beta a} &= Fe^{i\alpha a}, \\ \beta Ce^{\beta a} - \beta De^{-\beta a} &= i\alpha Fe^{i\alpha a}. \end{aligned} \tag{8.40}$$

These four equations contain five unknowns. Thus we cannot directly solve for each. We can solve for any of the unknowns in terms of any of the others. After that we can use the normalization condition to evaluate the final constant and solve our problem. However, we are interested only in solving the part of the problem that will yield information about penetrating the barrier. We would like to have an expression for the transmission coefficient for the particle through the potential barrier between $x = 0$ and $x = a$. This transmission coefficient is the ratio of the amplitudes of the wave function in regions III and I. That is,

$$T = \frac{|F|^2}{|A|^2}. \tag{8.41}$$

Using Eq. (8.40), we can solve for A in terms of F to obtain

$$|A|^2 = |F|^2\left[\frac{(e^{\beta a} + e^{-\beta a})^2}{4} + \frac{(\alpha^2 - \beta^2)^2}{16\alpha^2\beta^2}(e^{\beta a} - e^{-\beta a})^2\right]. \tag{8.42}$$

This equation shows that the transmission coefficient goes to zero only if βa is infinite. But let's make some simplifying assumptions. If the energy E is of the same order of magnitude as the potential energy V_0, then α and β are of the same magnitude. Thus α^2 minus β^2 would be small, and the second term of Eq. (8.42) could be neglected. Also, since β has $\hbar^2$ in the denominator and the barrier width a would not be infinitesimal, the product βa for some practical problems could be considerably greater than one. Therefore $e^{-\beta a}$ would be small compared to $e^{\beta a}$ in the first term and Eq. (8.42) could be simplified. With these assumptions, the transmission coefficient becomes

$$T = 4e^{-2\beta a}. \tag{8.43}$$

Substituting for β yields for the transmission coefficient

$$T = 4e^{(-2a/\hbar)\sqrt{2m(V_0 - E)}}. \tag{8.44}$$

Note that the transmission coefficient will not be zero unless V_0, a, or m is infinite. Also, the transmission coefficient increases as m decreases or as a or V_0 decreases.

8.7

THE HARMONIC OSCILLATOR

When other potential distributions are considered, there are very few for which exact analytic solutions of the Schroedinger equation can be found. Two of these are the harmonic oscillator and the Coulomb potential of electrostatics. We shall not solve the amplitude equation for these but we shall discuss the wave-mechanical solutions and compare them with older theories.

The simple harmonic oscillator is a very important system in classical mechanics. When there are oscillations about an equilibrium position, many systems at least approximate a simple harmonic oscillator. The potential for the harmonic oscillator is $V = Kx^2/2$, where K is the force constant of the system. The amplitude equation is

$$-\frac{\hbar^2}{2m}\frac{d^2\psi}{dx^2} + \frac{1}{2}Kx^2\psi = E\psi.$$

As with the square well, this equation has well-behaved solutions only for discrete values of E and they can be denoted by a quantum number, $n = 0, 1, 2, \ldots$. The potential distribution is shown with the three lowest energy levels in Fig. 8.5(a). The solution for $n = 1$ is shown in part (b) of the figure. The displacements where the energy E_n is equal to the potential energy V are labeled a_n and are the classical turning points. According to classical mechanics, this is where the particle stops and reverses its motion; these points also indicate the amplitude of the sinusoidal motion. The solution for ψ_1 (which must be squared to give the probability distribution) shows that there is a probability of the particle being *beyond* the classical turning point.

The classical treatment permits the oscillator to have any energy. The wave-mechanical treatment quantizes the energy. In terms of the quantum number n, the energies are given by

$$E_n = \left(n + \tfrac{1}{2}\right)h\nu, \tag{8.45}$$

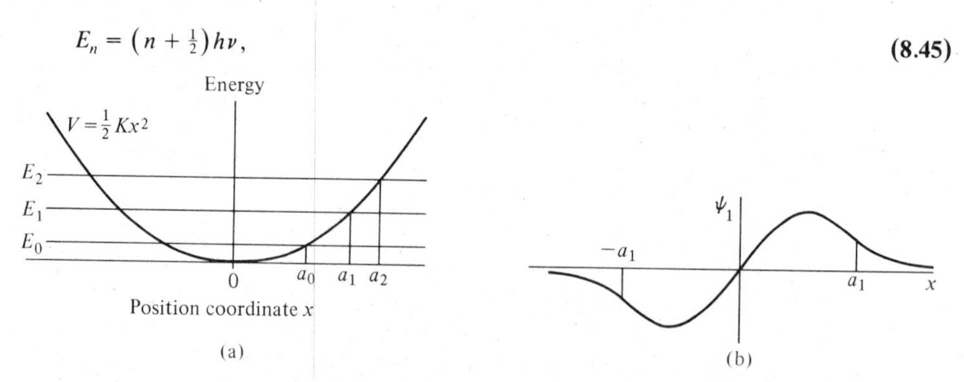

Figure 8.5 (a) Potential distribution and energy levels for the harmonic oscillator. (b) The solution of the amplitude equation for $n = 1$.

with ν being the frequency of oscillation:

$$\nu = (1/2\pi)\sqrt{K/m}\,.$$

For a harmonic oscillator the energy levels are uniformly spaced and separated by $h\nu$. This is the smallest permissible energy change, which Planck introduced to explain blackbody radiation, as was discussed in Chapter 3. The only difference between the energy levels predicted by the Schroedinger equation and those of Planck is the term $\frac{1}{2}$ in Eq. (8.45). This does not affect the energy spacing but it does mean that the lowest energy level is $hv/2$ instead of zero. This is necessary in order that the Heisenberg uncertainty principle not be violated. If the oscillator had zero energy, it would have zero position and zero momentum, which are not simultaneously possible.

If, for the sake of comparison, we choose to give the classical solution only those energies permitted by the quantum solution, we get a set of amplitudes $a_0, a_1, a_2, \ldots,$ where the subscripts are the quantum numbers. These amplitudes are shown as dashed vertical lines in Fig. 8.6. The curves that are drawn concave upward show the relative probability, on a classical basis, of finding the moving mass at any distance from its equilibrium position. These curves are concave upward because the mass moves through the equilibrium position with maximum velocity and spends most of its time near one end of the path or the other.

The shaded areas represent the wave-mechanical solution for the probable location of the mass for several energies. In the lowest energy state, with $n = 0$, the classical and quantum solutions violently disagree. Instead of the equilibrium position being least likely, it turns out to be the most likely in the new mechanics.

As the energy becomes greater, the classical probability retains its form, but the number of peaks in the quantum solution increases. This number is always $(n + 1)$. As the number of these peaks increases, their maxima begin to resemble the classical curve. Projecting this tendency to large values of n, we see that the classical curve and the average of the quantum curve merge and finally become experimentally indistinguishable. The probability of finding the particle outside the classical turning point, as represented by the shaded area outside a_n in Fig. 8.6, decreases as n increases and becomes negligibly small for very large n. This agreement between the classical theory and the quantum theory for large n is in accordance with the correspondence principle.

8.8 QUANTUM MECHANICAL MODEL OF ATOM

Another system that can be solved analytically is that of a charge moving in the Coulomb potential of another charge. This corresponds to the hydrogen atom, where the electron has a potential energy due to the presence of the proton, which we assume remains fixed. This is a three-dimensional problem, however, and we must

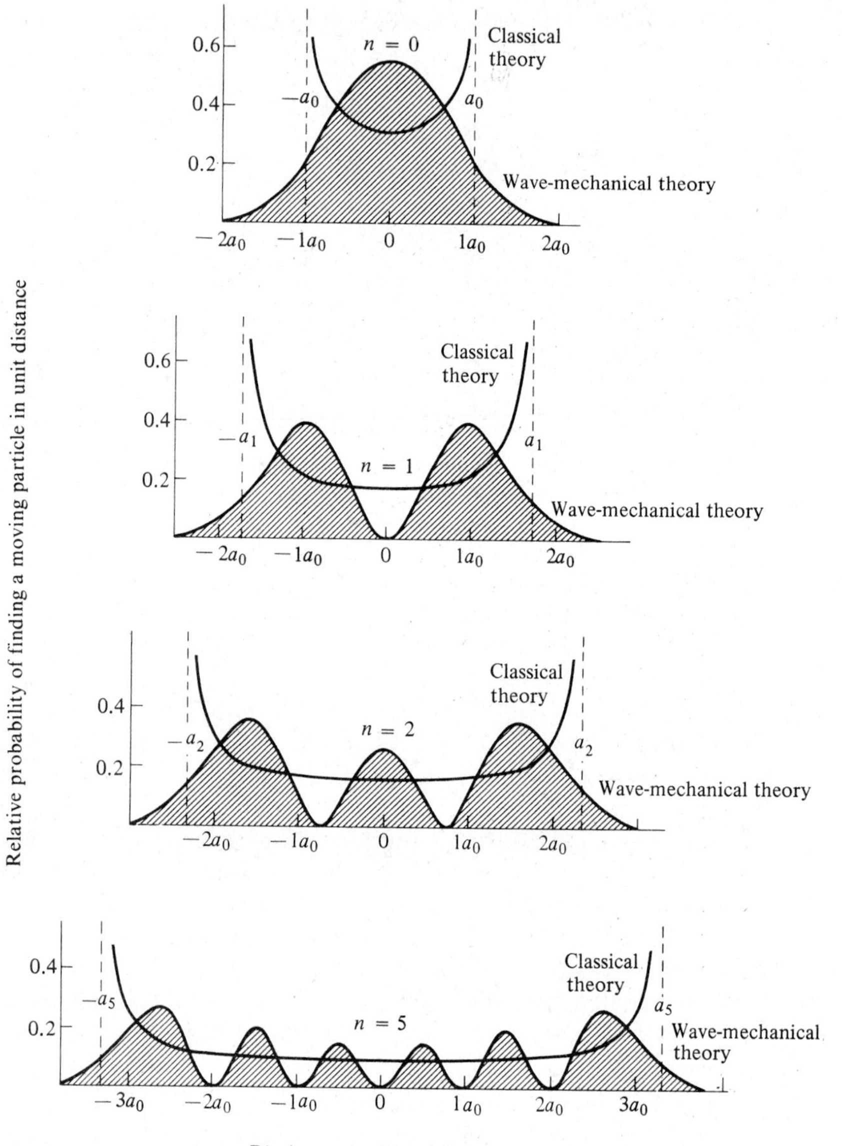

Figure 8.6 Each diagram shows the relative probability of finding a harmonic oscillator at various displacements both classically and wave-mechanically. Four different energies are shown, corresponding to quantum numbers $n = 0, 1, 2$, and 5. The classical amplitudes are a_2 for $n = 2$, etc. Thus $2a_0$ is twice the classical amplitude for $n = 0$.

Reprinted with permission from Blackwood, Osgood, and Ruark, *An Outline of Atomic Physics*, 3rd ed. New York: J. Wiley, 1955.

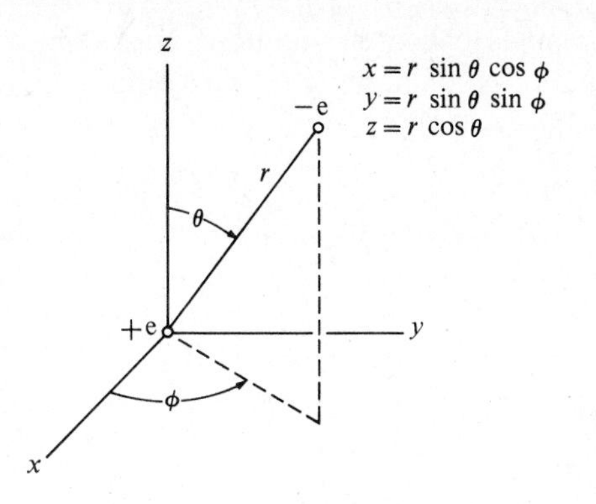

Figure 8.7 The spherical polar coordinate system.

extend the theory. By analogy with the classical wave equation, the amplitude equation becomes†

$$-\frac{\hbar^2}{2m_e}\left(\frac{\partial^2\psi}{\partial x^2}+\frac{\partial^2\psi}{\partial y^2}+\frac{\partial^2\psi}{\partial z^2}\right)+V\psi=E\psi. \tag{8.46}$$

The partial derivatives are used in Eq. (8.46) because ψ is a function of x, y, and z. This combination of partial derivatives is also written in operator form as

$$\frac{\partial^2}{\partial x^2}+\frac{\partial^2}{\partial y^2}+\frac{\partial^2}{\partial z^2}+\nabla^2,$$

where ∇^2 is called the Laplacian and is read "del-squared." The normalization condition becomes a volume integral:

$$\int\psi^*\psi\,dv=1, \tag{8.47}$$

where the integration is performed over all space.

The potential energy for the hydrogen atom is

$$V(r)=-\frac{1}{4\pi\varepsilon_0}\frac{e^2}{r}.$$

For this potential the problem is simplified by using spherical polar coordinates r, θ, and ϕ. This coordinate system and its relation to the xyz-system are shown in Fig. 8.7.

The Schroedinger equation for this potential in spherical polar coordinates becomes

$$\frac{1}{r^2}\frac{\partial}{\partial r}\left(r^2\frac{\partial\psi}{\partial r}\right)+\frac{1}{r^2\sin\theta}\frac{\partial}{\partial\theta}\left(\sin\theta\frac{\partial\psi}{\partial\theta}\right)+\frac{1}{r^2\sin^2\theta}\frac{\partial^2\psi}{\partial\phi^2}+\frac{2\mu}{\hbar^2}[E-V(r)]\psi=0, \tag{8.48}$$

†Since the potential is independent of time, the Schroedinger equation will still separate and the time dependence is given by $e^{-iEt/\hbar}$.

where $\mu = m_e m_p / m_e + m_p$ and is called the reduced mass. The resulting partial differential equation is separated by assuming a product solution,

$$\psi(r,\theta,\phi) = R(r)\Theta(\theta)\Phi(\phi), \tag{8.49}$$

where the functional dependencies are indicated. This separation of the equation results in three ordinary differential equations, one for each coordinate. The three equations are

$$\frac{d^2\Phi}{d\phi^2} = A\Phi, \tag{8.50}$$

$$\frac{1}{\sin\theta}\frac{d}{d\theta}\left(\sin\theta\frac{d\Theta}{d\theta}\right) + \frac{A\Theta}{\sin^2\theta} = -B\Theta, \tag{8.51}$$

and

$$\frac{1}{r^2}\frac{d}{dr}\left(r^2\frac{dR}{dr}\right) + \frac{2\mu}{\hbar^2}\left[E - V(r) - \frac{h^2}{2\mu}\frac{B}{r^2}\right]R = 0. \tag{8.52}$$

The partial differential equation with three variables has been separated into three equations each involving only one of the coordinates. Now the problem is to solve each of these equations separately.

Single-valued solutions of Eq. (8.50) can be found if we let $A = -m_l^2$, where $m_l = 0, \pm 1, \pm 2, \pm 3, \ldots$. Appropriate solutions to Eq. (8.51) can be found if $B = l(l+1)$, where l is an integer. Also, it is found in solving Eq. (8.51) that a further restriction is placed on m_l, that is $|m_l| \leq l$. Finally, solution of Eq. (8.52) can be found only if l is a positive integer and we introduce another integer n related to l by

$$l = 0, 1, 2, \ldots (n-1).$$

In summary, the solution of the Schroedinger equation, Eq. (8.48), is reduced to solving three ordinary differential equations, Eqs. (8.50), (8.51), and (8.52). The solutions then yield the quantum numbers n, l, and m_l with the following restrictions:

$$\begin{aligned} n &= 1, 2, 3, \ldots, \\ l &= 0, 1, 2, 3, \ldots (n-1), \\ m_l &= -l, (-l+1), \ldots (l-1), l. \end{aligned} \tag{8.53}$$

These quantum numbers and restrictions are the same as those shown in Table 4.2.

Consideration of the radial wave equation, Eq. (8.52), yields interesting results. Let's consider the ground state of the hydrogen atom which would be where the quantum number n equals 1. Because of the conditions on l, it would have to equal zero. For $l = 0$, the constant B would be

$$B = l(l+1) = 0.$$

Then using

$$V(r) = -\frac{e^2}{4\pi\varepsilon_0 r}$$

gives

$$\frac{d^2R}{dr^2} + \frac{2}{r}\frac{dR}{dr} + \frac{2\mu}{\hbar^2}\left[E + \frac{e^2}{4\pi\varepsilon_0}\frac{1}{r}\right]R = 0. \tag{8.54}$$

A solution to this equation is

$$R(r) = Ce^{-r/a_0}.$$

Taking the derivatives with respect to r yields

$$\frac{dR(r)}{dr} = -\frac{C}{a_0}e^{-r/a_0} = -\frac{1}{a_0}R(r)$$

and

$$\frac{d^2R(r)}{dr^2} = \frac{C}{a_0^2}e^{-r/a_0} = \frac{1}{a_0^2}R(r).$$

Substituting these values back into Eq. (8.54) will yield

$$\frac{1}{a_0^2}R(r) - \frac{2}{ra_0}R(r) + \frac{2\mu}{\hbar^2}\left[E + \frac{e^2}{4\pi\varepsilon_0}\frac{1}{r}\right]R(r) = 0 \tag{8.55}$$

or finally

$$\left(\frac{1}{a_0^2} + \frac{2\mu}{\hbar^2}E\right) + \left(\frac{2\mu e^2}{4\pi\varepsilon_0\hbar^2} - \frac{2}{a_0}\right)\frac{1}{r} = 0. \tag{8.56}$$

For this equation to hold for all values of r, it is necessary that the terms enclosed in parentheses must both be equal to zero. Equating the term that multiplies $1/r$ to zero yields

$$a_0 = \frac{4\pi\varepsilon_0\hbar^2}{\mu e^2}. \tag{8.57}$$

Now if we set the other term of Eq. (8.56) equal to zero, we obtain

$$E = -\frac{\mu e^4}{32\pi^2\varepsilon_0^2\hbar^2}. \tag{8.58}$$

Equation (8.57) is the expression we found for the radius of the first Bohr orbit back in Chapter 4 when we were discussing the Bohr model of the atom. Equation (8.58) is the expression for the energy obtained from the Bohr model with $n = 1$.[†]

The probability of finding the electron in a spherical shell a distance r from the nucleus will be the square of the radial wave function times the volume of the spherical shell. Thus the probability $P(r)$ of finding the electron in the spherical shell between r and $r + dr$ is given by

$$P(r) = |R(r)|^2 4\pi r^2\, dr = C^2 e^{-2r/a_0} r^2\, dr. \tag{8.59}$$

[†] These are the same equations we obtained in Chapter 4 with $h/2\pi$ replaced by $\hbar$ and the mass of the electron replaced by the reduced mass μ.

We can obtain the value of r where the probability of finding the particle is the greatest by differentiating Eq. (8.59) with respect to r and setting it equal to zero. Thus,

$$\frac{d}{dr}\left(C^2 e^{-2r/a_0} r^2\right) = 0$$

and

$$C^2 e^{-2r/a_0}(2r) + C^2\left(-\frac{2}{a_0}\right)e^{-2r/a_0} r^2 = 0.$$

Solving this equation for r yields

$$r_{\max} = a_0.$$

We have shown here that a_0 is the radius of the first Bohr orbit. The Bohr theory dictated that the electron had to be located in an orbit with this radius. Quantum mechanics allows the electron to be anywhere between the nucleus and $r = \infty$, but the most probable location is a distance a_0 from the nucleus.

We have seen for the special case where $n = 1$ the energy agrees with the energy found for the ground state of the Bohr atom. The general solution of the radial wave equation, Eq. (8.52), also yields quantized energies of the hydrogen atom. These energies depend only on the quantum number n and are given by

$$E_n = -\frac{\mu e^4}{32\pi^2\varepsilon_0^2\hbar^2 n^2}, \qquad n = 1, 2, 3, \ldots . \tag{8.60}$$

These are the energies that were derived from Bohr's postulates in Chapter 4.[†] Instead of Bohr's circular orbits, we have a stationary probability distribution that is a function of r, θ, and ϕ and is characterized by three quantum numbers. In general, we see that for every independent coordinate of a bound system the well behaved conditions introduce a quantum number. These quantum numbers specify the wave function of the system and one of the permissible discrete energy levels.

The spin quantum number, s, is not obtained from the Schroedinger equation, which is applied only to nonrelativistic systems, since we used the classical expression for the kinetic energy. The spin quantum number does appear in the relativistic quantum mechanics, which was developed by Dirac in 1928.

8.9 SUMMARY

With the help of an analogy to the classical wave equation we have developed a matter wave equation. This equation, with the associated boundary conditions, has solutions that yield probability distribution functions and quantized energy levels. It

[†]Schroedinger's derivation of Eq. (8.60) in 1926 was the first verification of the validity of the new quantum mechanics.

is the agreement between these predictions and experimental results that is the sole justification for the Schroedinger equation. For instance, we assumed a product solution of the time-dependent equation in order to separate the equation. At that point, we were not assured that these mathematical solutions were also physical solutions. The fact that the ensuing energy levels agree with experimental results shows that these are physical solutions. In particular, this wave theory for matter predicts the correct energy levels for the harmonic oscillator and the hydrogen atom. These two systems were originally quantized in two entirely different ways. The Schroedinger theory is seen to be applicable to any system that can be specified by a potential energy. Although there are only a few systems that can be solved exactly, approximation techniques have been developed that are used to find the solutions for a great many problems.

There is a major flaw in the development of the wave theory of matter presented here: There must be time-dependent probability distributions. If there were only time-independent probabilities, there would be no transitions from one state to another. Generally, every experiment is concerned with causing a transition between two states. The radiation from an excited atom originates from a transition to a lower energy level. Even though the stationary state solutions that we have found are physical solutions, they are not the only solutions. We will not go into the details but merely state that the theory has been developed to include the time-dependent features necessary for an understanding of transition probabilities. We recall that these transition probabilities, which determine the intensities of the discrete wavelengths emitted by atoms, were lacking in the Bohr theory of the atom.

There is some conflict between previous classical ideas and those of the theory of matter waves. For example, the concept of a classical trajectory of a particle must be discarded and replaced by a probability distribution spread over a large region of space. It may be useful at this point to consider the purpose and limitations of a physical theory. A theory is developed to answer questions. In physics the questions for which answers are to be provided by the theory must be able to be answered also by experiment. It is the agreement between the theoretical prediction and the experimental result which determines the status of the theory. (The experimental result, properly interpreted, is always correct.) A question that cannot be tested experimentally has no meaning. As an illustration of this, the wave theory of matter predicts the probability of finding a particle in a certain region. If an experiment is performed to determine whether the particle is there, the result will be that it is or it isn't; only the average result of many such experiments must agree with the predicted probability. Nothing is said about where the particle was before the experiment was performed and no questions concerning a previous position can be answered. In order to make predictions, theories use models (such as the particle and the wave that we have been using) and abstract concepts that are not themselves directly measured (such as Ψ). It is the correctness of the predictions that justifies the models and the abstract concepts.

It is also not disturbing when theories developed for one field of physics do not apply in other fields. We expect that, as physics is extended to new regions, new theories will be developed. At the present time the quantum theory, by including the previous classical theory through the correspondence principle, has successfully predicted the experimental results for a wider range of problems than any other theory.

FOR ADDITIONAL READING

Bockhoff, Frank J., *Elements of Quantum Theory*. Reading, Mass.: Addison-Wesley, 1976.

Bridgman, P. W., *The Logic of Modern Physics*. New York: Macmillan, 1949. This classic discusses the significance of measurable and unmeasurable quantities; it also discusses meaningless questions.

Eisberg, Robert M., *Fundamentals of Modern Physics*. New York: Wiley, 1961.

Eisberg, R., and R. Resnick, *Quantum Physics*. New York: Wiley, 1974.

Park, David, *Introduction to the Quantum Theory*. New York: McGraw-Hill, 1964.

REVIEW QUESTIONS

1. What are the similarities and differences between the Schroedinger wave equation, Eq. (8.8), and the classical wave equation, Eq. (8.2)?
2. How would you obtain the probability distribution function from the solutions to the Schroedinger wave equation?
3. How would you find the expectation value for energy from the wave function for a particle?
4. What are the steps that must be taken to solve the problem of a particle trapped in an infinite square well?
5. How does energy quantization arise in the quantum mechanical treatment of the infinite square well problem?
6. What are the steps that must be taken in order to solve with quantum mechanics the problem of an electron in the vicinity of a proton (the hydrogen atom)?
7. What are the restrictions on the quantum numbers n, l, and m_l? What are the origins of these restrictions?

PROBLEMS

8.1 The possible values obtained from a measurement of a discrete variable, called x, are 1, 2, 3, and 4. (a) Given that the respective probabilities are $\frac{1}{4}$, $\frac{1}{4}$, $\frac{1}{4}$, and $\frac{1}{4}$, calculate the expectation values of x and x^2. (b) Given that the respective probabilities are $\frac{1}{12}$, $\frac{5}{12}$, $\frac{5}{12}$, and $\frac{1}{12}$, calculate the expectation values of x and x^2. (c) Compare the results for parts (a) and (b).

8.2 Show that the normalizing constant A in the solutions for the infinite square well of width L is $(2/L)^{1/2}$ for all values of the quantum number n.

8.3 An electron is bounded by a potential that is approximated by an infinite square well with a width of 2×10^{-8} cm. Calculate the lowest three permissible energies of the electron.

8.4 An infinite square well has a width of 2.00 Å. What is the fractional change in the lowest two permissible energies of an electron in this well if the width is increased by (a) 0.02 Å and (b) 2.00 Å? (c) What is the fractional change in the two lowest energy levels if a proton replaces the electron in this well?

8.5 Show that the fractional difference in energy between two adjacent energy levels of a particle in an infinite square well is

$$\frac{\Delta E}{E} = \frac{2n+1}{n^2}.$$

What is the classical limit of this fraction ($n \to \infty$)?

8.6 If Ψ_1 and Ψ_2 are solutions to Eq. (8.8), show that $\Psi_1 + \Psi_2$ is also a solution.

8.7 Show that if Ψ_1 is a solution to Eq. (8.8), a constant times Ψ_1 is also a solution.

8.8 Show that the amplitude functions

$$\psi_0 = \left(\frac{a}{\sqrt{\pi}}\right)^{1/2} e^{-a^2x^2/2}$$

and

$$\psi_1 = \left(\frac{a}{2\sqrt{\pi}}\right)^{1/2} 2axe^{-a^2x^2/2}$$

are both normalized.

8.9 A particle of mass m is in an infinite square well of width L in a state specified by the quantum number n. Use Eq. (8.29) to show (a) that the expectation value of x is $L/2$ and that the expectation value of x^2 is

$$\frac{L^2}{3}\left(1 - \frac{3}{2\pi^2 n^2}\right)$$

and (b) that the expectation value of p is zero and the expectation value of p^2 is $(\hbar\pi n/L)^2$.

8.10 Show that $\psi = A \sin kx + B \cos kx$ is a solution to Eq. (8.25).

8.11 A steel ball with a mass of 30 g is bouncing back and forth between rigid walls. The walls are 2 m apart and the ball has a velocity of 0.3 m/s. (a) Find the kinetic energy of the ball. (b) Assuming that this system corresponds to a particle in an infinite square well, find the quantum number of the state of the steel ball. (c) How much additional energy must the ball acquire to reach the next higher state?

8.12 A particle in a box is in its lowest state. What is the probability of finding the particle between $x = 0$, and $x = L/3$?

8.13 The particle in a box is an overly simple model for an electron bound to an atom by coulombic potential. Although one may roughly estimate the energy for a transition in an atom with a square well model by using the lowest levels, the estimate becomes worse for higher levels. To show this, calculate by what margin the transition from the first to the 100th level using a square well model exceeds the experimental energy for complete ionization of a typical heavy atom (5 – 15 eV). Assume a square well for an atom of 1 Å radius.

8.14 The classical probability distribution function for a harmonic oscillator with amplitude a is $P(x) = A/(a^2 - x^2)^{1/2}$, where x lies between $-a$ and $+a$. (a) Determine the constant A by normalizing the distribution. (b) Calculate the expectation values of x and x^2.

8.15 The wave function that is the solution of the amplitude equation for the lowest energy state of a harmonic oscillator is $\psi(x) = Be^{-a^2x^2/2}$. (a) Determine the constant B by normalizing the wave function. (b) Calculate the expectation values of x and x^2. (c) Calculate the expectation values of p and p^2.

8.16 The wave function for the harmonic oscillator in the lowest state is

$$\psi_0 = \left(\frac{a}{\sqrt{\pi}}\right)^{1/2} e^{-a^2x^2/2}$$

where $a^2 = \omega m/\hbar$. Find the expectation value of potential energy for this lowest state. Express the answer in terms of $h\nu$. (See Table 1.2.)

8.17 Find the expectation value of potential energy for the lowest state of the harmonic oscillator. Express your answer in terms of $h\nu$.

8.18 Assume that the uncertainty in position for a harmonic oscillator is $\Delta x = (\langle x^2\rangle - \langle x\rangle^2)^{1/2}$ and that the uncertainty in momentum is

$$\Delta p = (\langle p^2\rangle - \langle p\rangle^2)^{1/2}.$$

Use the results of Problem 8.15 to show that these uncertainties for the lowest state are consistent with the Heisenberg uncertainty principle.

8.19 (a) Show that the function given below is a wave function of the amplitude equation for the harmonic oscillator. [*Hint:* This means that when the kinetic and potential energy operators are applied to this function, the result is the original function multiplied by a constant.] (b) What is the energy E of the system described by this wave function? The function is

$$\psi(x) = Axe^{-\sqrt{mK}x^2/2\hbar},$$

where A is a constant, x is the displacement, m is the mass of the oscillator, and K is the force constant.

8.20 A 40 g mass suspended from a helical spring is oscillating vertically with an amplitude of 3 cm and a period of $\pi/5$ s. (a) Find the energy of the oscillating mass. (b) Assuming that mechanical energy can be quantized according to $E = h\nu$, how many quanta of energy does this mass have? (c) What is the distance between the peaks of the probability curve for locating the oscillating mass along its path? (d) Could this distance be observed?

8.21 Discuss why the probability distribution function $\psi^*\psi$ must be finite.

9 THE ATOMIC VIEW OF SOLIDS

9.1 INTRODUCTION

The solid state of matter has always been of interest to physicists, but there have been important advances in this field since quantum mechanics has been applied to it. In the last decade or so, solid-state physics has come to the forefront as one of the most fruitful fields of research.

Whereas the classical study of the solid state was primarily concerned with the measurement of the macroscopic properties of solids:— mechanical, optical, thermal, and electrical:— the modern study of the solid state is primarily concerned with the microscopic properties.

We have already discussed some solid-state physics. In Chapter 1, we presented the law of Dulong and Petit and showed how the specific heat capacities of solids assisted in the determination of the atomic masses. In Section 4.17, we discussed the interatomic forces that account for molecular structure and we suggested that a projection of these arguments could account for the structure of crystals. In this chapter, we shall show how some of the properties of solids were "explained" by classical theory and then we shall go on to see how the introduction of quantum mechanics led to better explanations. Transistors are among the practical devices that have come from our deepening understanding of the solid state.

9.2

CLASSICAL ATOMISTIC APPROACH TO MOLAR HEAT CAPACITY OF SOLIDS

In Sections 1.7 and 1.8, we developed the classical kinetic theory of gases. In particular, we found that the molar heat capacity of a gas could be expressed as

$$C_v = \frac{fR}{2}, \tag{9.1}$$

where f is the number of degrees of freedom:— three for a monatomic gas, five for a diatomic gas, etc.

We can extend the kinetic theory of matter to account for the molar heat capacities of solids by generalizing our definition of the number of degrees of freedom over that given in Section 1.8. In vibrating systems the number of degrees of freedom is *not* equal to the number of coordinates necessary to specify the positions of the particles.

Consider a mass m that is acted on by a system of springs so that it moves linearly with simple periodic motion of amplitude A and angular frequency ω. If the position varies as $\cos \omega t$, the velocity varies as $\sin \omega t$. The expression for the potential energy is $E_p = \frac{1}{2}mA^2\omega^2 \cos^2 \omega t$, and the kinetic energy is $E_k = \frac{1}{2}mA^2\omega^2 \sin^2 \omega t$. The total energy of the oscillator is the sum of these two expressions. It is constant and equal to the common coefficient of the squared trigonometric functions. This constant value could be expected from the law of conservation of energy. The graphs of the sine and cosine functions have the same shape; they differ only in phase. Thus the time average of the sine squared over a cycle equals the time average of the cosine squared. From this we may conclude that the average value of the kinetic energy of the oscillating body equals the average value of its potential energy, and the total energy is "partitioned" equally between these two types.

Since the average energy of a body or a mechanical system is distributed equally among the various ways in which it may have energy, it remains only to state the new rule for determining the number of these ways. Although in mechanics the number of degrees of freedom is taken to be the number of space coordinates necessary to completely specify the positions of the parts of a system, a more general rule used in statistical mechanics is the following: Write the complete energy expression for the system in terms of position coordinates and velocities, using any coordinate system that may be convenient. Count the number of variables in this expression that appear to the *second power*. This number is the number of statistical degrees of freedom. This count will include all kinetic energy terms and all elastic potential energy terms, but it will exclude such terms as gravitational potential energy, since this kind of energy depends on the height to the first power. We now apply this new rule to a solid.

A solid consists of a collection of atoms each of which is more or less bound, but which are, nevertheless, able to vibrate. Although each atom has its position completely specified by three coordinates, application of the rule for determining the number of degrees of freedom shows that there are six for each. The total energy of an atom in a solid has the form

$$E = \tfrac{1}{2}m\left(v_x^2 + v_y^2 + v_z^2\right) + \tfrac{1}{2}k\left(x^2 + y^2 + z^2\right) + mgy, \tag{9.2}$$

where k is the force constant. The velocity terms represent the translational kinetic energy and have the same form regardless of the state of matter. The middle terms represent the elastic potential energy of an oscillator. These depend on the displacement of the atom from its equilibrium position. The last term represents the gravitational potential energy. This energy expression contains six variables that appear to the second power.

Since the energy is shared equally by these six degrees of freedom, we should expect the molar heat capacity at constant volume to be

$$C_v = 6R/2 = 3R = 5.97 \text{ cal/mole} \cdot \text{K}, \tag{9.3}$$

where $R = 1.99$ cal/mole · K. According to this theory, the value of C_v should be the same for all solids. This is a derivation of the law of Dulong and Petit, which was rather well verified in Problem 1.3. Although it is pleasing to have extended the kinetic theory to solids with apparent success, we have led ourselves into a trap. According to the argument above, the specific heat of a solid should be independent of temperature, whereas in reality the specific heats decrease when the temperature is lowered. Furthermore, the theory above is inconsistent with the classical view of the theory of electrical conductivity.

Einstein was the first to show a way out of the trap mentioned above. In the classical approach, each atom of the solid has three modes of vibration. The energy associated with each mode is considered to be continuous, which results in an average energy per mode of $h\nu$. These considerations led to the Dulong–Petit value of specific heat. Einstein also assumed that each atom could vibrate with three modes. However, each mode in his model had one frequency associated with it and each mode had only certain specific values of energy,

$$E = nh\nu \qquad n = 0, 1, 2, 3, \ldots,$$

(which is similar to the assumption made by Planck for blackbody radiation). Einstein obtained an expression for the average energy for each mode, which was

$$E_{\text{ave}} = \frac{h\nu}{e^{h\nu/kT} - 1}. \tag{9.4}$$

(This equation is the same as Eq. 3.24.) Using this value for the average energy per mode yields an expression for the specific heat more complicated than the expression of Eq. (9.3). Einstein's complicated expression is

$$C_v = 3R\left(\frac{h\nu}{kT}\right)^2 \frac{e^{h\nu/kT}}{\left(e^{h\nu/kT} - 1\right)^2}. \tag{9.5}$$

Figure 9.1 illustrates the classical and Einstein curves for the specific heat for a

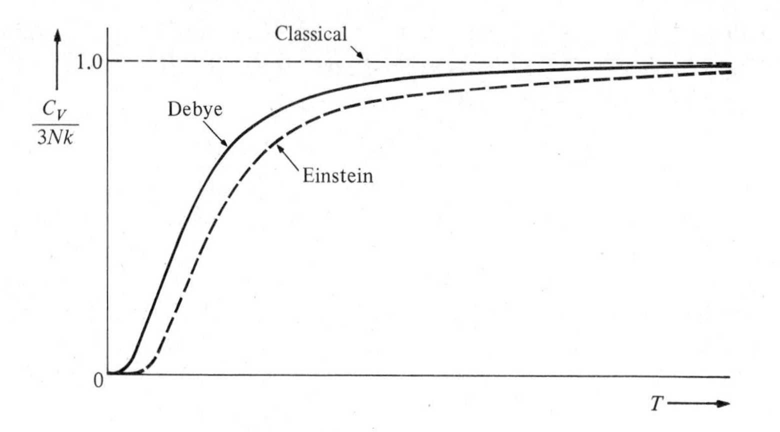

Figure 9.1 Molar heat capacity.

typical material. As the temperature increases, the Einstein curve approaches the classical curve. However, as the temperature is lowered, the Einstein curve decreases, which is known to be the case for all materials. The fit of the Einstein curve with the data is not perfect, but it was a remarkable improvement over the classical model. The main discrepancy between the curve and the data is in the low-temperature region. In practice, it is found that C_v depends on T^3 at low temperature rather than on the exponential behavior predicted by Einstein.

Debye did as Einstein did and postulated that a solid of N atoms would have $3N$ vibrational modes. Debye's model, however, was more realistic in that it did not assume that all vibrational modes were at the same frequency. He also assumed that there was a maximum frequency corresponding to a minimum wavelength due to the spacing of the atoms on lattice sites. This minimum wavelength is dictated because the wavelength cannot be shorter than twice the spacing of the atoms. Debye's model also gave an expression for the specific heat that approached the classical Dulong–Petit value at high temperatures and decreased as the temperature was decreased. At low temperatures, the Debye model yielded the accurate T^3 dependence of specific heat.

9.3 CLASSICAL THEORY OF ELECTRON GAS IN SOLIDS

The classical theory of insulators is that in these materials the electrons are bound to their molecules, which are in turn tied to fixed locations in the solid. Conductors, on the other hand, are assumed to have atoms whose outer electrons are free to migrate from atom to atom throughout the solid crystalline structure. These electrons are assumed to behave within a conductor like an ideal gas. This assumption leads to the conclusion that the number of degrees of freedom in a conductor should be nine: three for the kinetic energy of the molecules, three for the elastic potential energy of the molecules, and three for the kinetic energy of the electrons. We shall explain the inconsistency between the nine degrees of freedom that every conductor should have classically and the experimentally verified six degrees of freedom of the law of Dulong and Petit after seeing where we are led by the "electron gas" theory as it applies to conduction.

In order to calculate the resistivity of the metal, we assume that the electron gas is thermally agitated and has a root-mean-square velocity, which may be obtained by solving Eq. (1.43) for v. We get

$$v_{\text{rms}} = \sqrt{\frac{3kT}{m_{\text{e}}}}. \tag{9.6}$$

If the electrons have a mean free path L, the the average time $\bar{t}$ between collisions is

$$\bar{t} = \frac{L}{v_{\text{rms}}} = L\sqrt{\frac{m_{\text{e}}}{3kT}}. \tag{9.7}$$

Suppose that the electrons are in a bar of cross-sectional area A and length l, across which there is a potential difference V. This potential difference causes an average electric field V/l which exerts a force on each electron. The force on the electron is

$$F = eE = \frac{eV}{l}. \tag{9.8}$$

This force acceleratates the electrons an amount

$$a = \frac{e}{m_{\text{e}}}\frac{V}{l}. \tag{9.9}$$

They accelerate for an average time $\bar{t}$ and acquire an average drift velocity v_d given by

$$\overline{v_d} = \frac{eV}{2m_{\text{e}}l}L\sqrt{\frac{m_{\text{e}}}{3kT}}. \tag{9.10}$$

This drift velocity is along the bar and is very small compared with the random v_{rms}. Indeed, $\overline{v_d}$ is so small that each interval between thermal collisions begins a new acceleration "from rest" and the transport of electrons proceeds at an average velocity of $\overline{v_d}$.

If the concentration of free valence electrons is n per unit volume, then the number that pass across a plane perpendicular to the axis of the bar per unit time is $nA\overline{v_d}$, and the electrons constitute a current given by

$$I = enA\overline{v_d} = \frac{e^2nLA}{2m_{\text{e}}l}\sqrt{\frac{m_{\text{e}}}{3kT}}V. \tag{9.11}$$

Comparison with Ohm's law shows that this development gives the resistance of the bar as

$$R = \frac{2l}{e^2nLA}\sqrt{3kTm_e}. \tag{9.12}$$

Since in terms of the specific resistance we may write $R = \rho(l/A)$, we see that

$$\rho = \frac{2\sqrt{3kTm_e}}{e^2nL}, \tag{9.13}$$

and the electrical conductivity is

$$\sigma = \frac{1}{\rho} = \frac{e^2nL}{2\sqrt{3kTm_e}}. \tag{9.14}$$

It is interesting to find that for some pure metals, assuming that n is the number of valence electrons per unit volume and that L is the interatomic distance, quite good agreement is reached at room temperature. Unfortunately for this theory, it is well known that the resistivity of most metals is proportional to the absolute temperature over a wide range, so that this agreement is largely coincidental. Still worse, we have already pointed out that this theory would assign a molar heat capacity of $3R/2$ to the electron gas, while in fact the molar heat capacity of metals is quite readily explainable on the basis of translational and vibrational energy of the molecules alone. Good conductors of heat are also good conductors of electricity. The proportionality of these two kinds of conduction is called the *Wiedemann-Franz* relationship, and it strongly implies that the two types of conduction have the same mechanism. If the motion of electron gas accounts for electrical conductivity and therefore thermal conductivity, it is paradoxical that the thermal motions of the electrons do not contribute to the specific heat of the material.

9.4 QUANTUM THEORY OF ELECTRONS IN SOLIDS

In our efforts to account for the thermal and electrical properties of conducting solids by assuming an "ideal electron gas," we have implied that the character of the gas was in every respect like an ideal molecular gas except that the "molecules" are electrons. In the kinetic theory of gases, we assumed that the molecules were not subject to any forces except during collisions. This is not true of an "ideal electron gas" within a solid. Whenever an electron passes near a positive nucleus it is electrostatically attracted to it, and the electrons are always in this electric field. Even though the electrons may be "free" to migrate, they are still associated with atoms. In Chapter 4, we found that the electrons associated with atoms can exist only in certain permitted energy levels. Just as the electrons in an atom may have only the energies permitted by that atom, so the electrons in a solid may have only the energies permitted by that solid. Whereas the molecules in a gas may have *any* energy, the electrons in a solid have their energies quantized.

In Chapter 4, we found that if an atom has more than one electron, no two of these electrons may have identical quantum numbers. This is the Pauli exclusion principle, and it imposes a restriction on the number of electrons that can occupy any one energy state. Thus we saw that even in an unexcited atom most of the electrons are forced to have energies well above that of the lowest energy state.

Sommerfeld, Fermi, and Dirac assumed that the electrons in a solid must have their energies quantized and that the electrons in a solid must obey the Pauli exclusion principle. Whereas the maximum number of electrons in any atom is about 10^2, the number of valence electrons per cubic centimeter of solid metal exceeds 10^{20}. If the Pauli principle applies to a solid over any appreciable region, it implies that although a few of the valence electrons may have very small energies, most of the electrons must occupy energy states extending to high energies. Following these assumptions, Fermi and Dirac derived an expression for the distribution of the energies and the speeds of the electrons in the electron gas in a solid. This Fermi–Dirac distribution is in marked contrast to the Maxwell distribution of gas molecules because the energies of the gas molecules are not quantized. Compare Fig. 9.2 with Fig. 1.2. In Fig. 9.2, we note that there are few electrons in the region of low speed and low energy, and that in accordance with the Pauli principle most of them are forced to have high speed and high energy. A quantitative comparison of the Maxwell and Fermi–Dirac distributions may be made by stating that those electrons at the high-speed edge or maximum of the Fermi–Dirac distribution have speeds that would correspond to the most probable speed they would have at about 30,000 K in a Maxwellian gas if the electron energies were not quantized. Contrary to the classical view, this statement applies even when the solid is at absolute zero.

Another important difference between these two distributions is their sensitivity to temperature. Whereas the Maxwell distribution retains its shape but is "stretched" toward higher speeds at higher temperatures, the Fermi–Dirac distribution is nearly independent of the temperature of the metal. Figure 9.2 suggests how it changes from 0 K to 1000 K.

We are now in a position to resolve two of the difficulties we have encountered in classical theory. First, when a conductor is heated the electrons contribute a

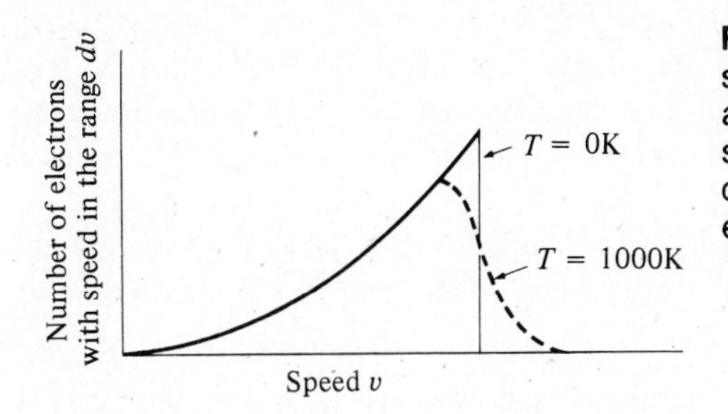

Figure 9.2 Distribution of speeds in an electron gas according to Fermi-Dirac statistics. (The spread of the dashed-line portion is exaggerated in this graph.)

negligible amount to the molar heat capacity. The electrons in the "Fermi gas" may indeed absorb a little energy when a conductor is heated, but those that are free to absorb energy are so few that the amount they absorb is a negligible fraction of the energy absorbed by the molecules of the crystal. Second, the high energy of the electrons at the maximum of the Fermi–Dirac distribution does not prevent them from acquiring a low-energy drift velocity in the presence of either an electric field or a thermal gradient. Thus the electrons contribute insignificantly to the specific heat, but supply a common mechanism for electrical and thermal conductivity.

9.5 WAVE-MECHANICAL TREATMENT OF ELECTRICAL CONDUCTIVITY

We have used the Fermi–Dirac distribution of electron speeds to resolve two of the classical difficulties, but we still must account for the temperature coefficient of resistance.

The electrons that contribute to electrical conductivity are those at the high-speed edge of the Fermi–Dirac distribution. These have much higher root-mean-square velocities than would be implied by Eq. (9.6). We can use this equation if, instead of the absolute temperature T of the conductor, we use the "equivalent" temperature of the maximum of the Fermi distribution. We have already mentioned that this "equivalent" temperature is very high and almost independent of the physical temperature of the conductor. Thus "T" in our earlier derivation takes on a new meaning and is no longer a significant variable.†

Since we have argued that the equivalent T is essentially constant, it is difficult to see how the resistivity and conductivity, Eqs. (9.13) and (9.14), can be temperature-dependent. The least well defined quantity in these equations is now L, the mean free path, which we supposed earlier to be the interatomic distance. The question is whether we can redefine L so as to restore to the resistivity the magnitude and temperature dependence we know it has experimentally.

Since the value of the equivalent T is much greater than room temperature, the required value of L must be greater than the interatomic distance if ρ and σ are to be consistent with the experiment. Furthermore, if L were inversely proportional to the actual absolute temperature of the sample, the resistivity in Eq. (9.13) would vary with temperature, as it should. A proper proof that L varies inversely with temperature is complicated and not particularly enlightening. We therefore substitute a qualitative argument.

A vibrating atom presents a larger cross section than one at rest. The area increase is proportional to the square of the amplitude, just as the area of a circle is

† This will be discussed further under "kinetic temperature" in Section 12.16.

proportional to the square of its radius. The amplitude of an oscillator varies as the square root of its energy, and the average energy of an atomic oscillator is proportional to the temperature. Thus the increment in cross-sectional area is proportional to the temperature. If the atoms in a crystal had a negligible area when at rest, this argument could explain the temperature coefficient of resistance. At absolute zero the atom would have no significant cross section and electrons could streak through a conductor without encountering any resistance at all. Since, experimentally, the resistance of a conductor does tend to zero as the temperature approaches absolute zero, we must now justify the statement that nonvibrating atoms have a cross section for electron collisions that really is negligible.

This justification comes about most easily by considering the wave nature of the electrons and by regarding the scattering of electrons to be analogous to the scattering of light. We are all familiar with the scattering of a beam of light by dust particles in air or by colloidal suspensions in a liquid. Even in the absence of particles, molecules of air will themselves contribute to the scattering, thus producing the blue color of the sky and the reddish tint of the sun when low on the horizon. If the widely dispersed molecules of the gaseous air or water vapor molecules can cause such appreciable scattering, what would be expected of the same substance, say water, in the liquid form? Since the density of the liquid is about 1000 times greater than that of the gas, it is surprising to find the actual scattering of the liquid is less than 50 times as great.† As a matter of fact, it is not the particles themselves, but the variation in density of groups of particles that contributes to the scattered light. Thus regularly arranged crystalline solids, whose appearance from a macroscopic view approaches a continuous medium, are strikingly free of scattering. If sunlight is focused first on a clean cover glass, such as is used for microscope slides, and then on a freshly cleaved flake of mica, the strong scattering by the first, in contrast to the lack of scattering by the second, can be easily observed if both are viewed against a very dark background.

We are thus able to support our original contention that it is not the cross-sectional area of the undisturbed atoms that must be considered in calculating the collision frequency and the mean free path of the electron in the metal, but rather that area which exhibits the characteristic random thermal fluctuations associated with the vibrational motion of the crystal lattice.

An alternative explanation of this interesting lack of scattering of electrons by regular arrays of atoms may be seen by considering the Bragg reflection conditions that we introduced in our discussion of x-rays:

$$n\lambda = 2d\sin\theta. \tag{9.15}$$

If we apply this formula to the waves associated with electrons within a crystal, λ is the de Broglie wavelength of the electrons, n is an integer, d is the basic crystal spacing, and θ is the reflection angle (see Fig. 6.4). The de Broglie wavelengths of

† R. W. Wood, *Physical Optics*. 3rd ed., p. 428. New York: Macmillan, 1936.

electrons at the high-speed edge of the Fermi–Dirac distribution is large compared with the crystal spacing. Since $\sin\theta$ cannot exceed unity, the only possible value for n is zero and no lateral scattering by diffraction can take place. Thus the electrons should move undeflected through the crystal, and the electrical resistance should be zero.

This conclusion is justified if the crystal is absolutely pure and its atoms are in perfect order, as at absolute zero. If the crystal contains impurities or structural imperfections, these "flaws" will be separated by some multiple of the crystal spacing. If d in the Bragg formula is interpreted as the distance between imperfections, then it may become greater than λ, and n and θ may become greater than zero. Since these spacings are irregular, incoherent scattering will be observed, rather than ordinary interference diffraction patterns. This scattering of electrons hinders their progress through the crystal and causes resistance. In this interpretation, the temperature coefficient of resistance arises from the fact that thermal agitation introduces random structural irregularities.

In summary, we have shown that the linear relationship between the resistance of a conductor and its absolute temperature can be justified. The classical derivation contained a temperature to the wrong power. We reinterpreted T on the basis of the Fermi–Dirac distribution of electron velocities and found it to be a large constant rather than the absolute temperature. Having removed the temperature dependence altogether, we next reintroduced it by showing that the electron mean free path, L, varies inversely with temperature.

9.6 ELECTRIC POTENTIALS AT A CRYSTAL BOUNDARY

Ordinary gas molecules are usually confined by some sort of container, but the atmosphere is confined to the earth by the earth's gravitational field, which in the large scale is by no means negligible. The most probable reason the moon has no atmosphere is that its gravitational field is too weak to hold one. What is the container that tends to confine electrons within a metal? For a metal, the field that contains the electrons is electrical. A piece of metal is ordinarily uncharged, but if an electron escapes, the piece becomes positive. The positive piece then has an electric field that attracts the electron back to itself. Electric fields are so strong that electrons which would otherwise get away "gravitate" back to the metal. Thus if an electron is near the surface and its motion is directed so as to escape, it encounters a *potential barrier* that ordinarily prevents the escape. Of course, if the electron has sufficient energy, it may surmount the potential barrier and become free of the metal. We have discussed both the photoelectric effect, in which electrons are given excess energy by photons, and thermoelectric emission, where the excess energy comes from thermal excitation. In both cases we found that there was a definite threshold energy that electrons must be given to enable them to surmount the potential barrier, and we called this energy the work function.

9.7 ENERGY BANDS IN SOLIDS

If we turn our attention from the boundary to the body of a metal, we find that the electric potential is not uniform from a microscopic point of view. Although the valence electrons in which we are interested are shielded from the positive nuclei by inner, bound electrons, nevertheless the valence electrons are attracted to the nuclei. These attractions cause potential variations. Figure 9.3 shows the nature of the potential variation in a crystal along a line that intersects the surface of the crystal.

While our discussion so far has apparently been able to explain several properties of metals on the basis that the electrons have a Fermi–Dirac distribution, we still have no theoretical basis for distinguishing the metals from other classes of solids. In this section, we shall introduce refinements in the theory to account for the

Figure 9.3 Schematic diagram of the variation of the electron potential energy along a row of atoms within a crystal and the potential barrier at the edge.

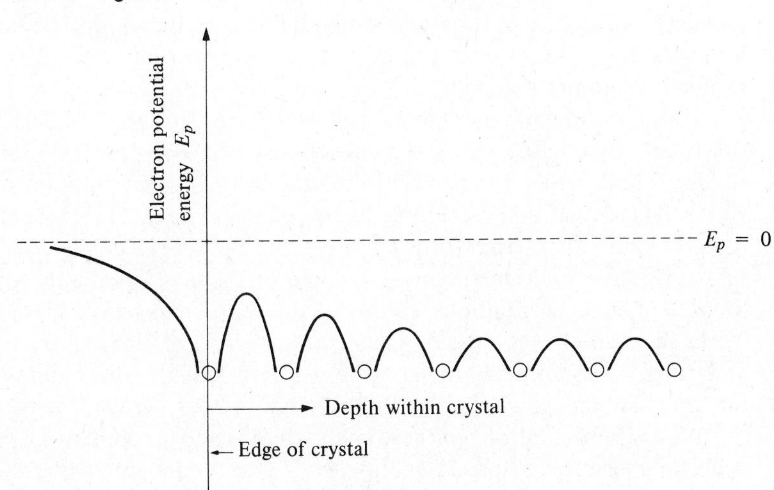

variation in electrical conductivity that distinguishes insulators and semiconductors from conductors.

The permissible energy levels of the electrons in a crystal can be found with the Schroedinger equation in Chapter 8. The potential distribution is put in the equation and solutions are determined that are well behaved. The crystal is three-dimensional and the actual distribution of the electric potential depends on the specific crystal. If a representative simplified potential is used, the form of the solutions to the Schroedinger equation and the energy levels can be found. When the well-behaved conditions are used, the permissible energy levels are found to be collected in what are called *energy bands*. Within these energy bands, the discrete energy levels are so close that they appear to be continuous. These bands are separated by regions, called *forbidden bands*, where there are no permissible energies.

Since these energy levels are for the entire crystal, they are to be occupied by a large number of electrons. When the Pauli principle is applied to these electrons, the limit on the number of electrons that may occupy a single energy level means that a large number of these levels must be filled. The implications of this result will be discussed in Section 9.8.

There is an analogy to this situation that could be stated in either electrical or mechanical terms. We choose the mechanical case and consider a group of identical simple pendulums, each of which will oscillate with the same frequency ν. If two of these pendulums are hung from a flexible common support, they become coupled oscillators and the system has *two* natural frequencies, $\nu + \delta$ and $\nu - \delta$, a little above and below the natural frequencies of the isolated pendulums. If the entire group of pendulums is suspended from the same flexible support, the system has as many *different* frequencies as there are pendulums. Thus the system has a *band* of frequencies. The identical electron energies of isolated atoms correspond to the identical frequencies of the isolated pendulums. As the atoms become coupled in the solid, the band of electron energies corresponds to the band of frequencies of the coupled pendulum system.

The following argument is more sophisticated, in that it is closer to the rigorous treatment. It is based on Bragg reflection of de Broglie waves, first discussed in Section 9.5. Whether an electron is moving through the crystal or not, its de Broglie waves extend over a considerable region of the crystal. Therefore we must consider the possibility of internal diffraction effects. Because of many partial reflections in successive layers within the crystal, some of these reflections become "total," and because of angular variations, the wavelengths and energies of these back "reflected" waves have an energy spread. Thus certain bands of electron energy are permitted and other bands are forbidden. Again we have a qualitative argument that accounts for the possibility that some electron energies can occur while others cannot.

We repeat that although these energy bands appear continuous, they are actually collections of closely spaced energy levels. The arguments we are about to present depend on the fact that each permitted band can hold only a certain number of electrons.

9.8 ELECTRICAL CLASSIFICATION OF SOLIDS

The energy bands we have just discussed provide an especially simple and satisfying explanation of the electrical classification of solids. If the highest energy band that has some electrons is unfilled, as in Fig. 9.4(a), these electrons may be excited from a lower to a higher energy level within the band. The separation of the levels within the band is so slight that the electrons are readily able to be accelerated to higher energies by weak electric fields. A material with such an electronic band structure is a conductor. All metals fall into this class, and it is this class that we have been discussing in this chapter.

Most nonmetals are insulators. For insulators, the highest occupied band is always completely filled. Although one might suppose that this case should be rare, it is actually the most common. Not only is the highest occupied band filled, as shown in Fig. 9.4(b), but also the unoccupied forbidden band just above the filled band is wide on the energy scale. Since thermal agitation is unable to lift electrons

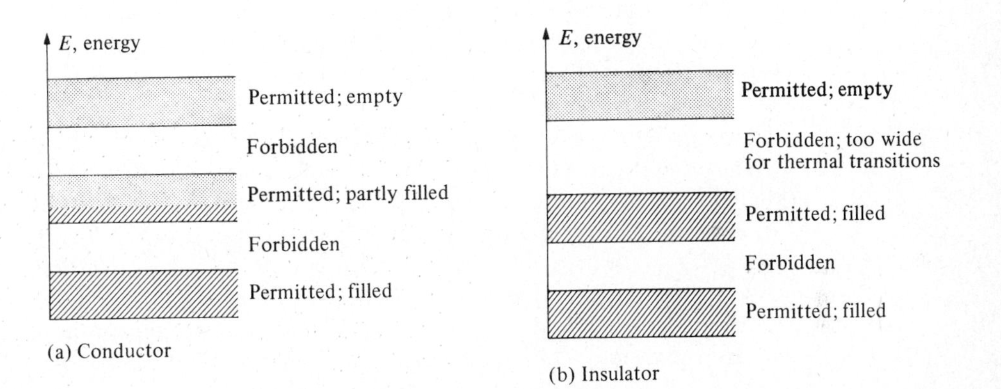

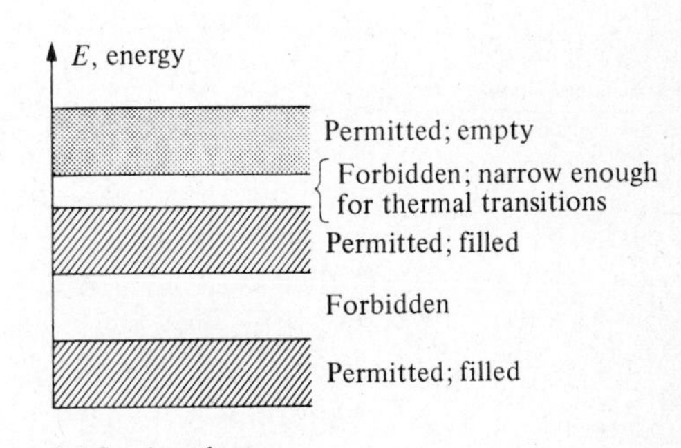

Figure 9.4 Schematic energy bands of (a) a conductor, (b) an insulator, and (c) a pure intrinsic semiconductor at absolute zero.

from the filled band to the next higher permitted band, the electrons cannot absorb small amounts of energy due to electric fields or thermal agitation. Therefore these electrons are "fixed," and the material must be both a thermal and electrical insulator. In an intense electric field, of course, some electrons may be pulled into the higher permitted band, where they may migrate conductively. This is dielectric breakdown and results in permanent damage to a solid dielectric.

Semiconductors comprise a third class of materials. The difference between semiconductors and insulators is that the gap between the band that is filled at absolute zero, called the *valence band*, and the higher permitted band is narrow in a semiconductor, as shown in Fig. 9.4(c). In this case, thermal energy at room temperature is sufficient to raise some electrons from the valence band to the higher permitted band, called the *conduction band*. Electrons in the conduction band are free to transport electrical charge. Their number per unit volume is the n in Eqs. (9.11) through (9.14). For a semiconductor, n increases with temperature, and it is easy to see that if n increases with temperature faster than L decreases, the electrical conductivity of a semiconductor will increase with temperature instead of decreasing as it does for conductors.

When electrons are transferred to the conduction band from the valence band, the valence band is no longer filled, and there are then energy states within the valence band available to the valence electrons. Valence electrons can then move successively, like automobiles in a traffic jam. When an opening appears, one car moves to fill it, leaving an opening behind. This is repeated over and over, so that the net effect is that the physical motion of the cars is forward while the gap between them moves backward. In the electrical case, the vacancy created by the removal of an electron is a small positively charged region called a *hole*. When a nearby electron moves into a positive hole, it leaves another positive hole at its original location, and so on. Thus there is a net transport of positive charge that is called *hole migration*. Holes are less mobile than electrons, and therefore hole currents are usually less than electron currents. Collectively, electrons and holes are called *carriers*. There is a dynamic equilibrium between mutual cancellation and thermal creation of conduction electrons and holes.

While amorphous materials have no proper place in our discussion, there is nevertheless an interesting demonstration that may be performed with a glass rod to illustrate the possibility of an insulator becoming a conductor at a high temperature. If two leads are connected to the ends of a thick glass rod, with suitable resistance in series with the rod, and connection is made to an ordinary 110-volt line, no charge will flow, of course. But application of a Bunsen burner flame to the midpoint of the rod will soon heat the glass to the point where it becomes conducting. If the burner is then shut off, the I^2R heating alone will sustain the temperature of the glass, so that it finally glows brightly and begins to melt.

In actual insulators, there are usually sufficient imperfections and chemical impurities in the lattice so that additional intermediate conducting levels exist. One would certainly expect, then, that such an irregularly arranged solid as glass would possess a large proportion of such localized conducting regions and hence exhibit the behavior described above.

Detailed examination of the known crystalline structure of various solids, plus study of the electronic structure of individual atoms in the solid, have made it possible to predict fairly well what the band structure of various materials should be, and hence which particular arrangements should be conductors, insulators, or semiconductors. Details of these calculations can be found in some of the references given at the end of this chapter.

We conclude this portion of our discussion with a summary of the differences between conductors, insulators, and semiconductors. A *conductor* is a solid with a large number of current carriers, a number independent of temperature. An *insulator* contains very few carriers at ordinary temperatures. A *semiconductor* contains relatively few carriers at low temperatures but a rather large number at higher temperatures. For semiconductors, the actual dependence of resistance on temperature is a result of two opposite effects. First, the increase in scattering of the electron wave with temperature tends to diminish the conductivity. This tends to cancel the second effect, which is due to the increase in the number of carriers. At room temperature semiconductors may have either positive or negative resistance-temperature slopes.

9.9 IMPURITY SEMICONDUCTORS

The characteristics of *pure* semiconductors, called *intrinsic* semiconductors, can be changed in very important ways by the introduction of trace amounts of impurities. Intrinsic semiconductors to which impurities have been added are said to be *doped*. Consider the intrinsic semiconductor, germanium, shown in Fig. 9.5(a). Each germanium atom has four valence electrons, and therefore each atom has four neighboring atoms bonded to it. Figure 9.5(b) shows the germanium doped with atoms of antimony. The antimony atoms will fit into this structure, but they have five valence electrons of which four participate in bonding to neighbor atoms. The fifth electron is superfluous to the structure and is therefore loosely bound to the antimony atom. Since thermal energy is sufficient to cause some of the germanium electrons to leave their valence bonds and jump to the conduction band, it is easy to see that the fifth antimony electron is even more easily excited into this conduction band. Thus practically every antimony atom introduced into the germanium lattice contributes a conduction electron without creating a positive hole. Of course, each antimony atom has become a positive ion, but this ion is tied into the lattice structure so that it cannot contribute to conduction. Thus in addition to the electrons and holes intrinsically available in germanium, the addition of antimony greatly increases the number of conduction electrons. In this case antimony is called

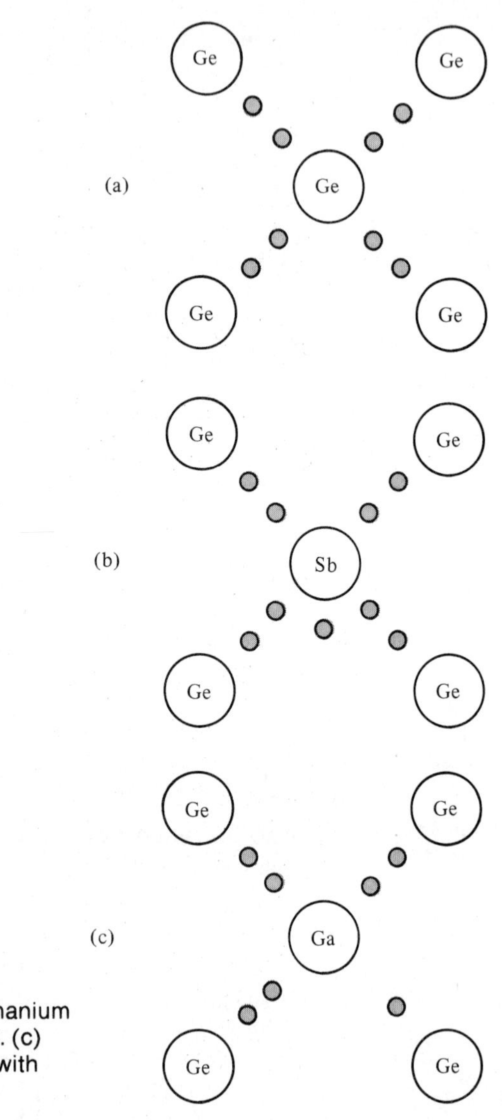

Figure 9.5 (a) Pure germanium. (b) Germanium doped with antimony. (c) Germanium doped with gallium.

a *donor* impurity and it makes the germanium an *n-type* (*n* is for negative) semiconductor.

Gallium, on the other hand, has three valence electrons. If it is introduced into germanium, it can supply only three of the four electrons necessary to fit into the germanium lattice, shown in Fig. 9.5(c). Since thermal energy is sufficient to excite some bonded germanium electrons into the conduction band, it is easy to see that thermal excitation is sufficient to cause germanium valence electrons to complete the lattice structure by attaching themselves to the gallium without leaving the valence band. This causes the gallium to become a fixed negative ion and it leaves an electron hole in the valence band. Thus gallium is an *acceptor* which, at room temperature, causes as many positive holes as there are gallium impurity atoms. The acceptor, gallium, makes germanium a *p-type* (*p* for positive) semiconductor.

When an electric field is placed across the pure germanium sample of Fig. 9.5(a), almost no electrons are able to move and therefore there is no current. However, when an electric field is placed across the antimony-doped germanium, the extra electron is able to move and thus a current exists. When an electric field is applied to the gallium-doped germanium, the weakly bound electrons associated with the germanium atoms can move to fill the vacancy. The electron then will leave a vacancy at the site of the germanium atom. An electron from a neighboring germanium atom can move to fill this vacancy, leaving a vacancy with the next germanium atom. The process is repeated with the result that a vacancy or hole moves through the crystal in the direction of the electric field just as though it were a positive charge.

An interesting effect takes place when an *n*-type semiconductor is placed next to a *p*-type semiconductor. Consider our germanium crystal grown so the antimony-doped part adjoins the gallium-doped part, as shown in Fig. 9.6(a). If a battery is connected across this crystal with the positive terminal attached to the *p*-type and the negative terminal to the *n*-type, as shown in Fig. 9.6(b), the electric field will

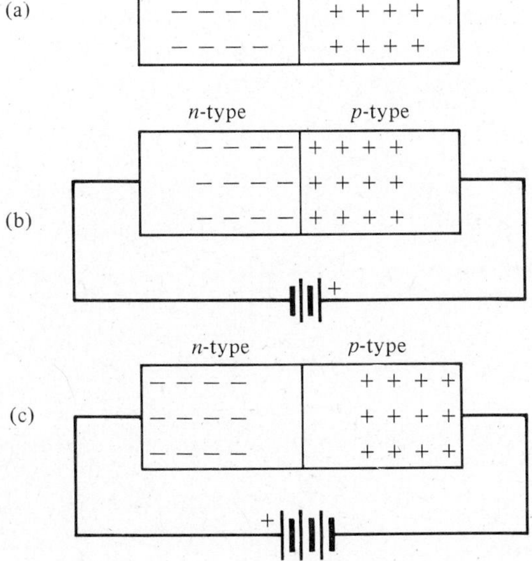

Figure 9.6

push the mobile positive holes toward the mobile negative electrons. The electrons and holes will annihilate at the junction of the n-type and the p-type material. Since the battery will continuously furnish electrons and holes, a current will be present in the circuit. The device is said to be forward biased. If on the other hand the battery is attached to the crystal with the positive terminal to the n-type and the negative terminal to the p-type, as shown in Fig. 9.6(c), there will be no current. The battery will provide an electric field in the crystal, which will tend to push the holes away from the electrons. The device is said to be reverse biased. The characteristics of this device, called a *junction diode*, will be discussed from the basis of Fermi levels in the next section.

9.10 FERMI LEVELS

The concept of the *Fermi level* is very useful in describing what happens when n- and p-type semiconductors are brought together. The Fermi level is a characteristic energy of a material.

We shall define the Fermi level in a rather qualitative way since a proper definition requires concepts beyond the level of this discussion. Our qualitative definition is that the Fermi level is the energy that corresponds to the "center of gravity" of the conduction electrons and holes "weighted" according to their energies. In an intrinsic semiconductor, there is an equal number of electrons and holes. Furthermore, both have energy distributions that fall off almost exponentially from the band edges, in close analogy to the Boltzmann distribution given by Eq. (1.45). Thus there is a decreasing concentration of electrons at energies farther above the bottom of the conduction band and there is a decreasing concentration of holes at energies farther below the top of the valence band. Figure 9.7 is a schematic representation of this situation, which is presented without proof. An examination of

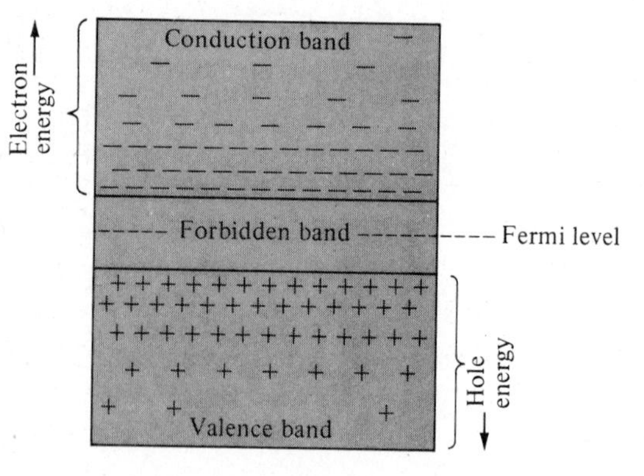

Figure 9.7 Schematic energy-level diagram of an intrinsic semiconductor at room temperature.

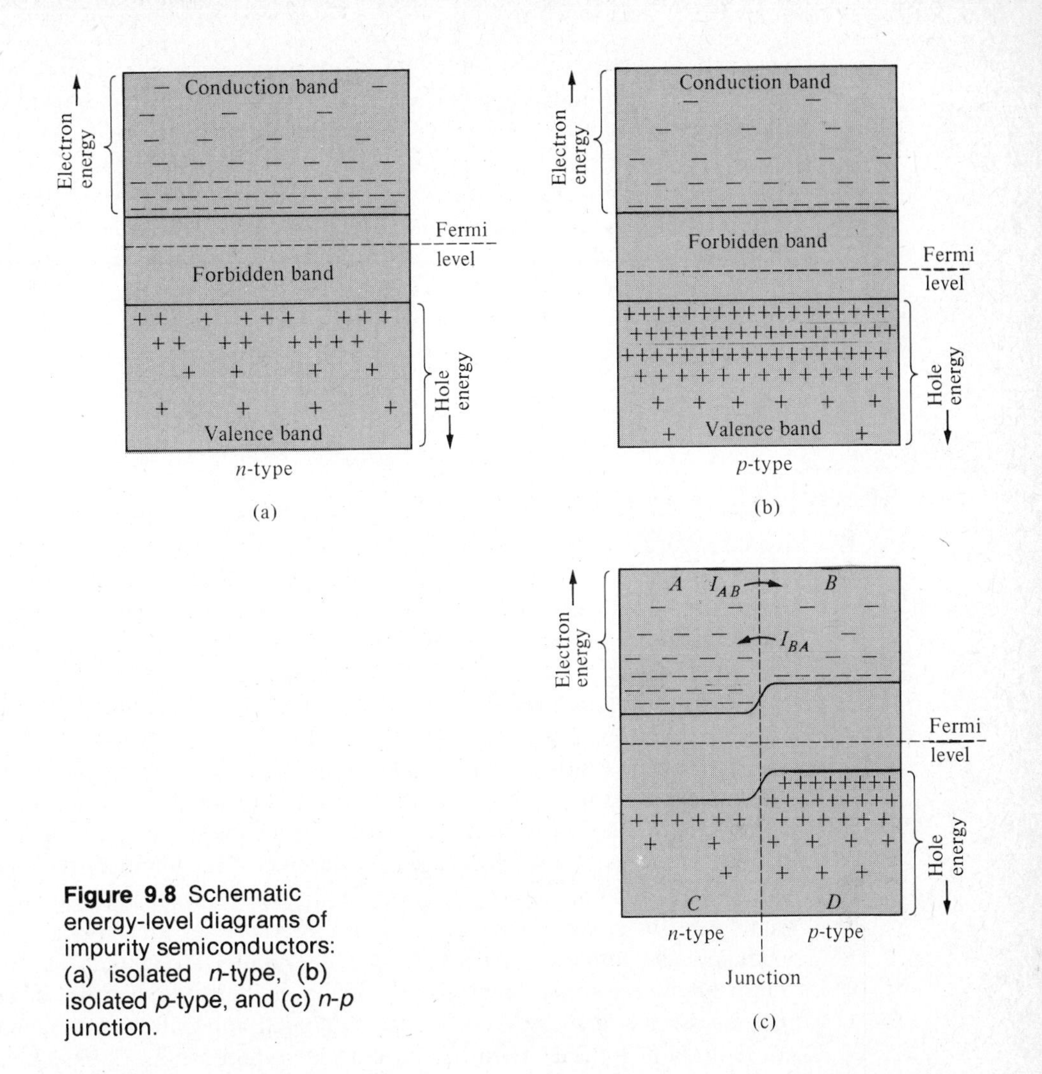

Figure 9.8 Schematic energy-level diagrams of impurity semiconductors: (a) isolated *n*-type, (b) isolated *p*-type, and (c) *n-p* junction.

the figure shows that the "center of gravity" of the pluses and minuses lies in the exact middle of the forbidden gap, where the Fermi level is drawn. Note that the Fermi level is an abstraction. Just as a hollow body can have a center of gravity where there is no matter, so a material can have a Fermi level at an energy forbidden to all electrons.

Fermi levels have two important characteristics. First, if an intrinsic semiconductor has a donor impurity added to it, it becomes *n*-type and has more conduction electrons than holes, as in Fig. 9.8(a). This moves the "center of gravity" up so that the Fermi level is above the middle of the forbidden band. Similarly, the addition of acceptors converts the intrinsic semiconductors into a *p*-type, with more holes than conducting electrons, as in Fig. 9.8(b). In this case, the Fermi level falls below the middle of the forbidden band. The second characteristic enables us to describe how the energy levels of the two types fit together at a *junction* between types. *If a crystal is part n-type and part p-type and if the crystal is isolated from any circuit, the various band levels shift so that the Fermi level is common to both types*, as shown in Fig. 9.8(c). If the crystal is in a circuit, so that one type is held at a different potential than the other, then there is a discontinuity in their Fermi levels equal to this potential difference. These facts will help us explain semiconductor rectifiers and transistors.

9.11 SEMICONDUCTOR RECTIFIERS

Rectifiers are electrical components that have the property of passing an electric current in one direction much more easily than in the other. They are useful for the conversion of alternating currents into direct currents. We propose to discuss diffusion and then use the diffusion concept to show why the junction between n- and p-type semiconductors has rectifying properties.

The principal mechanism of conduction within a semiconductor is diffusion. An example of diffusion is the distribution of gas throughout a room even though there may be no convection or other gross motion of the air in the room. If a pungent substance like ammonia is spilled on the floor at one end of a room, the smell slowly diffuses across the room until the distribution of the gas is uniform. This motion is not due to any systematic force on the molecules. The ammonia NH_3 molecules are lighter than either the N_2 or O_2 molecules of the air, so that the gravitational force would tend to cause a layer of ammonia molecules at the ceiling. In spite of this tendency, the ammonia molecules will be found throughout the room because thermal agitation causes microscopic mixing of the gases. Because of diffusion, the molecules tend to move from a region of high concentration to a region of low concentration. Indeed, the diffusion current is proportional to the change in concentration per unit distance, or the *concentration gradient*. The diffusion current across any layer is equal to a constant times the concentration gradient, and this proportionality constant is called the *diffusion coefficient*.

Both the electrons and the holes in semiconductors move about because of thermal diffusion, but we shall fix our attention on the electrons, which are more mobile than the holes. If we refer to Fig. 9.8(c), we see that there are two electron currents between regions A and B. The one to the right, I_{AB}, results from the fact that region A is n-type whereas region B is p-type. The concentration of conduction electrons is far greater in A and they extend to higher energies in spite of the fact that the bottom of the conduction band in B is higher than that in A. The magnitude of this current is sensitive to the relative concentrations of electrons at the top of the conduction bands. The current to the left, I_{BA}, is due to electrons near the junction and at the bottom of the B conduction band "falling over" the potential "hill" into region A. This current is called a *saturation current* because its magnitude depends on the concentration of conduction electrons in region B and on *the slow rate at which they can diffuse toward the junction*. The saturation current is very small because the concentration of electrons in B, the p-type region, is small. *The saturation current does not depend on the height of the potential "hill."* In the situation pictured, these two opposing electron currents are both small and equal, since the saturation current *must* be small and since the two currents are in equilibrium. The exchange of positive holes has a similar explanation.

The equilibrium situation just described is upset if electrodes are connected to the right and left sides of the crystal and if these electrodes are maintained at different potentials. Figure 9.9(a) depicts the situation when the n-type end is made negative and the p-type end is made positive. This is called biasing in the forward direction. This bias increases the band height and consequently the energy of the electrons in A relative to those in B. The saturation electron current, I_{BA}, has the same magnitude as before, but now the more energetic electrons in A can surmount the junction barrier far more easily and the net electron current is overwhelmingly to the right. Of course, the transport of electrons to the right and holes to the left constitutes a conventional current to the left.

If the forward bias situation has been understood, there is hardly a need to describe the reverse bias situation depicted in Fig. 9.9(b). The fact that the back current is small and independent of the backward potential is clear, since it consists

Figure 9.9 Schematic energy-level diagrams of biased semiconductor junctions with (a) forward bias, ΔV, and (b) reverse bias, ΔV.

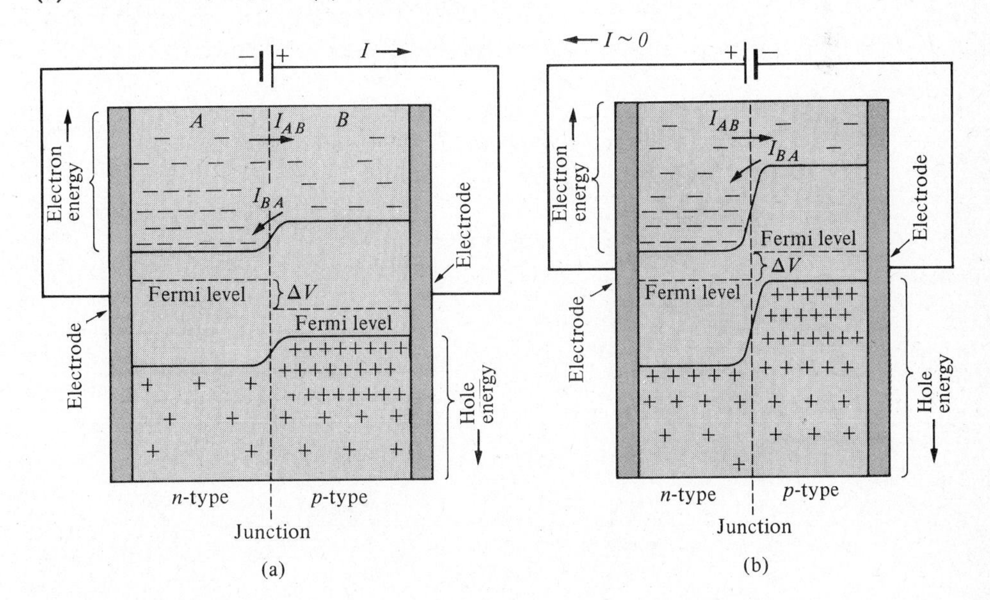

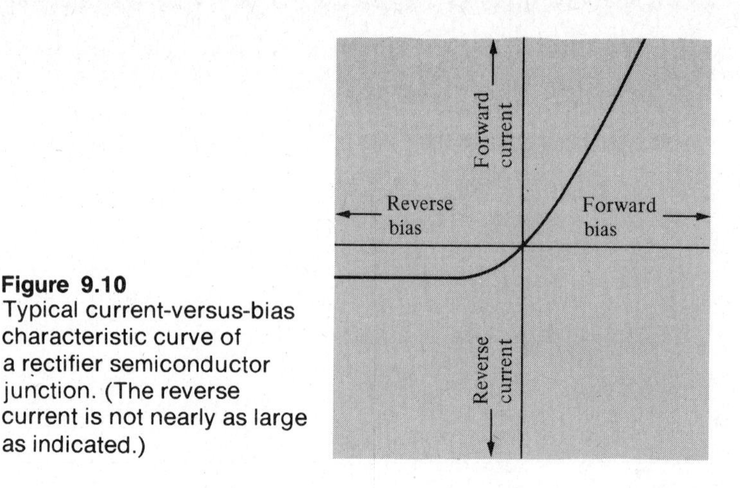

Figure 9.10
Typical current-versus-bias characteristic curve of a rectifier semiconductor junction. (The reverse current is not nearly as large as indicated.)

only of the saturation current. The characteristics of an *n-p* rectifier junction are shown in Fig. 9.10. Such a junction is often called a *crystal diode* because it performs the rectifying function of a two-electrode vacuum tube.

9.12 TRANSISTORS

We have discussed the diode rectifier rather fully because all of the ideas will be useful in describing a much more important device, the *transistor*. A transistor can transform a small electrical charge into a large electrical charge, that is, amplify. We shall consider the case of a junction transistor that is used so that a small amount of power can control a relatively large amount of power.

Figure 9.11 Schematic energy-level diagram of an isolated *n-p-n* transistor.

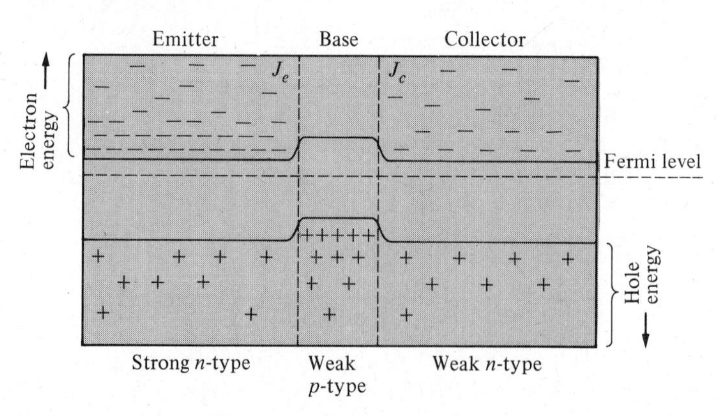

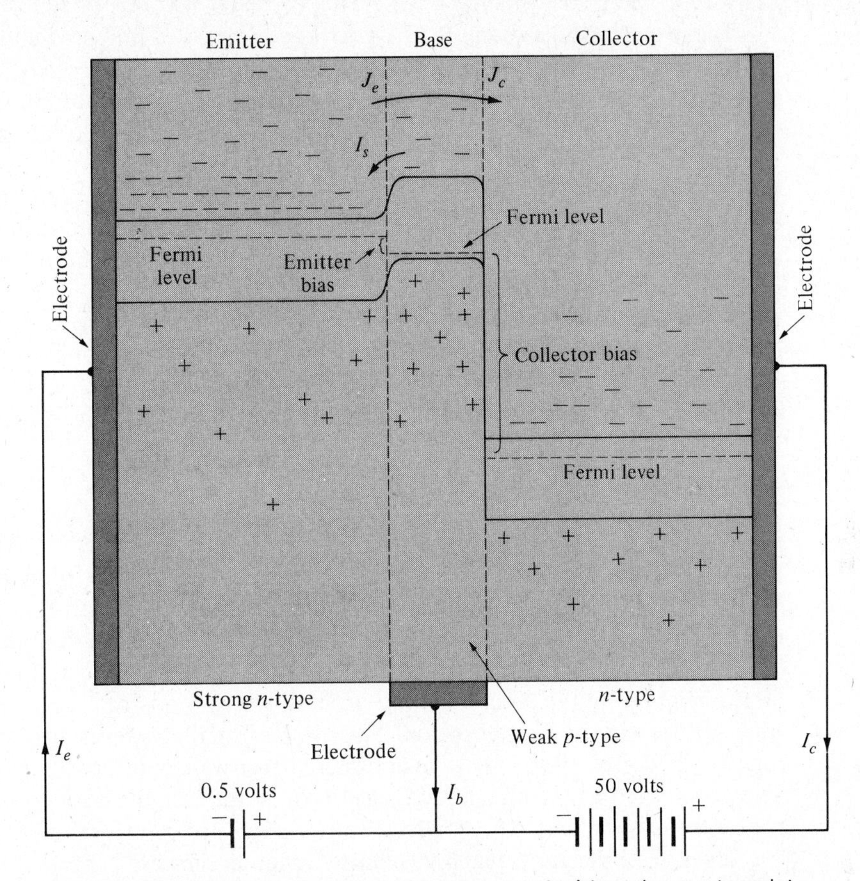

Figure 9.12 Schematic energy-level diagram of a biased *n-p-n* transistor. (Arrows indicate the directions of electron flow.)

A schematic energy-level diagram of an isolated *n-p-n* junction transistor is shown in Fig. 9.11. A comparison with earlier figures shows that it may be regarded as two rectifiers back-to-back. An *n-p-n* junction transistor can be used as an amplifier when it is biased, as shown in Fig. 9.12, where the left-hand *n*-type region, called the *emitter*, is heavily doped, so that it has a large electron conductivity. The central *p*-type region is called the *base*, and the junction between these regions, J_e, has a small forward bias of a few tenths of a volt. Considering this junction as a rectifier leads to the conclusion that the emitter-to-base resistance is small and the electron current is large.

The right-hand region is called the *collector*. It is made *n*-type and the junction between the base and the collector, J_c, has a strong reverse bias :— up to fifty volts. Considered as a rectifier, we would conclude that the base-to-collector resistance is large and the electron current is small.

The next considerations lead us to alter what we can predict from rectifier behavior. The essential new fact is that the base region is *very thin*:— about one thousandth of an inch. This distance is small compared with the mean distance an electron moves before recombining with a hole. Almost all the electrons leaving the emitter and entering the base pass right through the base into the collector. The thinness of the base leads to an artificially increased "saturation current," I, to the collector. After the electrons penetrate the collector junction, they are "over the wall" and cannot diffuse back into the base. Since the number of electrons that penetrate the base from emitter to collector includes almost all of the electrons in the large emitter current I_e, the current I to the collector is large also. In fact, the collector current is less than the emitter current by only the amount of the very small base current. Since this result is very different from what we would conclude from considering the transistor as two back-to-back rectifiers, we now review what happens in more detail.

The applied bias potentials appear mainly at the junctions, since each region of the transistor is a rather good conductor. Away from the junctions, the more important forces on electrons and holes are due to thermal agitation. When electrons pass into the emitter from the electrode, an excess of electrons is produced at the left side of this region. This concentration of electrons has two effects. First, it upsets the equilibrium between electrons and holes, which are constantly recombining due to electrostatic attraction and re-forming because of thermal agitation. The excess electrons promote the recombination process, which tends to reduce the number of useful carriers in the emitter region. This effect is small, however, because this region is a heavily doped *n*-type material and there never were many positive holes anyway. The second effect of the added electrons is the establishment of an electron concentration gradient. This gradient causes electrons to diffuse toward the base.

When these electrons reach the emitter junction they are "emitted" into the base. We have already seen that electrons in *n*-type material have an energy distribution high enough to enable them to surmount an *n-p* junction barrier that is somewhat reduced, in this case by the small emitter bias. If the base were thick, most of these electrons would either migrate to the base electrode or combine with holes in the *p*-type base region. Such undesirable behavior is minimized in two ways. First, the base is a weak *p*-type material, so that the number of positive holes is not large. Second, the base is made very thin, so that the probability of an electron from the emitter encountering a hole in the base is very small. Once these electrons have been "collected" into the collector, they suffer collisions that reduce the kinetic energy of most of them below the relatively high collector bias potential barrier, so that very few of them can diffuse back into the emitter. Thus the current to the electrode of the thin, weak, *p*-type base is kept small. This second discussion of transistor behavior leads to the same conclusion: Almost all the current entering the transistor in the low-resistance emitter region finally emerges from the high-resistance collector region with only a small loss to the base electrode.

We must next show that these transistor characteristics can be used to accomplish amplification. There are many different circuits in which transistors are employed. A full analysis of any of these circuits is outside our present interest, but each analysis involves its own set of simplifying assumptions as different aspects of the transistor are emphasized. In what follows we will assume that the emitter-to-base resistance, R_e, is constant and low, that the base-to-collector resistance, R_c, is constant and large compared with R_e, and that the base current is zero, so that I_e equals I_c. If the emitter current is changed in some way by an amount ΔI_e, the power fed to the transistor is $2I_eR_e\,\Delta I_e$. This causes a change in the power developed in the collector circuit, equal to $2I_cR_c\,\Delta I_c$. Since we are assuming that I_e equals I_c, we see that the output collector power is greater than the input emitter power by the factor R_c/R_e. This is but one of several ways of explaining amplification by transistors. Of course, the output power really comes from the batteries; the small input power enables the transistor to liberate the greater power available from the batteries.

Transistors have difficulties when operated at a high temperature. We saw that the addition of an impurity into an intrinsic semiconductor shifts the Fermi level from the middle of the forbidden band. Both rectification and amplification depend on this fact. Raising the temperature of either an *n*- or *p*-type material greatly increases the number of conduction electrons and valence holes. Since the number of impurity conductors is largely independent of the temperature, the increased conductivity caused by a rise in temperature tends to swamp the impurity conductivity and the semiconductor loses its *n*- or *p*-type characteristics. As the temperature rises, the conductivity becomes more nearly intrinsic, the Fermi levels tend toward the middle of the forbidden band, and the junctions between types lose their significance. The narrow temperature range of transistor operation is an important problem that must be overcome for circuits to be used in high-temperature situations.

9.13 INTEGRATED CIRCUITS

One characteristic common to the solid-state electronic devices that we have been discussing is their discreteness, or singleness. Each diode or transistor is a single package containing a single component. These components can be assembled by soldering to other components such as resistors, capacitors, and inductors to form useful circuits. Using modern techniques, however, it is possible to produce an entire electronic circuit on a single semiconductor chip and package it in a container roughly the same size as the container for a single transistor. These *integrated circuits*, as they are called, are combinations of interconnected circuit elements mounted on single substrates. An integrated circuit may be the equivalent of a relatively simple circuit containing a few transistors, resistors, etc., as shown in Fig. 9.13, or it may be the equivalent of a very complex circuit composed of many thousands of components as shown in Fig. 9.14. Figure 9.13 is an experimental circuit that is being used to test manufacturing techniques for the construction of high-density integrated circuits. Individual components can be distinguished. In the circle labeled A are four transistors and in the circle labeled B is one transistor. Resistors are shown at points C and D. The region marked E and similar regions are aluminum contacts for the circuit. This photograph shows about 40 percent of the circuit and is 0.06 inch across the top. On a silicon wafer two inches in diameter there are approximately 300 such circuits. Figure 9.14 is a photograph of the microprocessor unit of the Intel Microcomputer. This whole circuit is on a single crystal silicon chip approximately two tenths of an inch square. The advancement in integrated circuit technology has been rapid. In the early 1960s, perhaps a dozen transistors or other devices could be mounted on a plastic circuit board about 3″ × 5″ in size. Today a thousand similar circuits can be put on a single chip of silicon one tenth of an inch square. To illustrate further the rapid change, it can be pointed out that in 1969 a single transistor was on one chip. Now over 65,000 components can be placed on a single chip of the same size (see Fig. 9.15).

The importance and desirability of the integrated circuits are the result of a number of characteristics. The small size allows large, complicated systems to be of manageable physical dimensions, thus saving money on hardware, cabinets, wire, and building space. Another advantage is the ease of mass fabrication of larger arrays of components. Furthermore, integrated circuits are more reliable because there are fewer points of likely difficulty such as solder joints and mechanical connections. Finally, the most important advantage of integrated circuits may be the

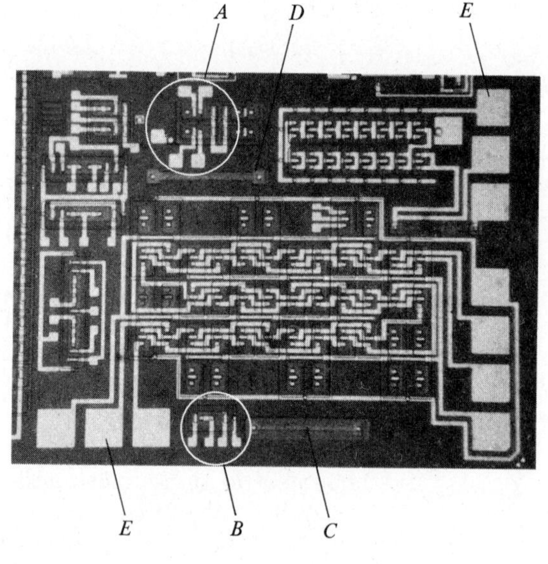

Figure 9.13 Integrated circuit: (A) four transistors, (B) one transistor, (C) and (D) resistors, and (E) aluminum contacts.

Photograph courtesy of J. L. Stone, Texas A & M University.

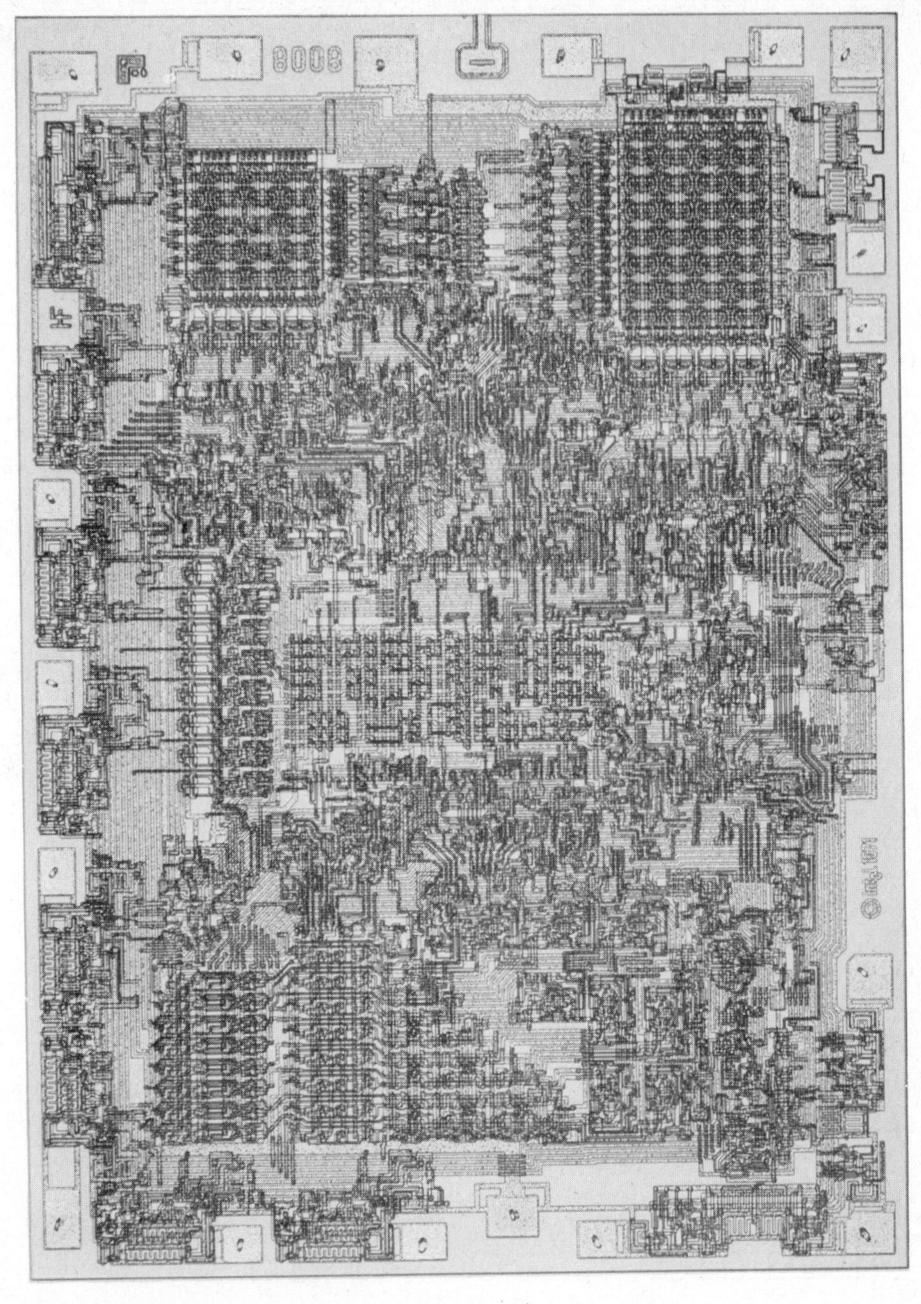

Figure 9.14 Intel's 8008 control processor unit.

Courtesy Intel Corporation.

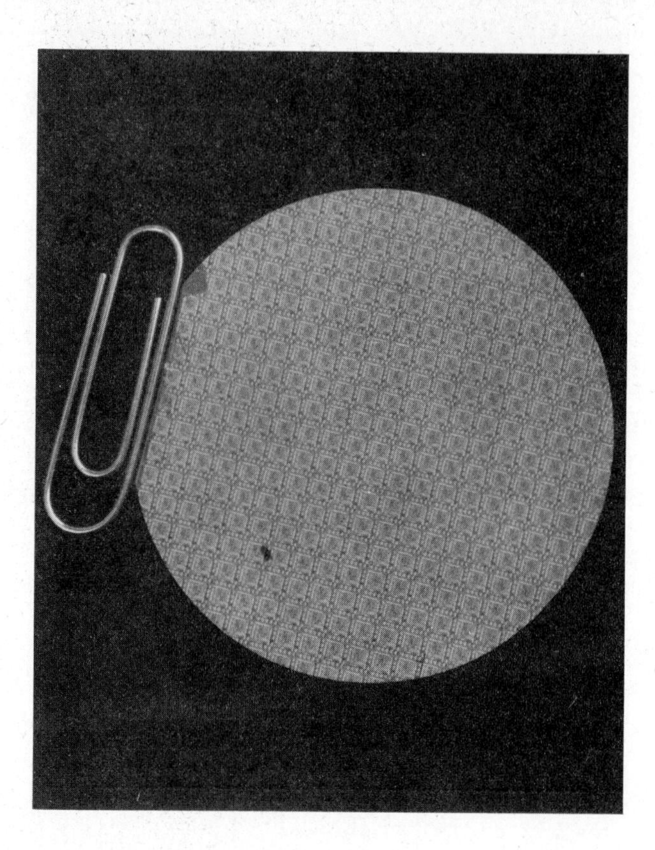

Figure 9.15 Silicon wafer containing 320 of the integrated circuits shown in Fig. 9.13.

Photograph courtesy of J. L. Stone, Texas A & M University.

low cost. The expense of a single integrated circuit containing dozens of transistors and other components is not appreciably different from the cost of a single transistor. The cost per component in ten years has dropped by a factor of more than 100. The size and cost of computer systems due to the incorporation of integrated circuits has decreased at a rate averaging 30 percent per year. The reduction in cost is most visibly demonstrated by the numerous low-priced, pocket-sized calculators now on the market. Some contain more than 10,000 transistors in a single chip.

9.14 SOLAR CELLS

Another use of semiconductor material is in the construction of the solar cell, a device to generate electrical current. At a typical *n-p* junction like the one shown in Fig. 9.16, the migration of negative charges into the *p*-type material with the accompanying migration of holes into the n-type material will cause an electric field across the junction in the direction from n-type to p-type. Thus if an electron is transferred into the conduction band at the junction, this field will cause it to move toward the n-type material.

A solar cell can be constructed of a relatively thick p-type crystal with a thin n-type crystal covering it (or a thick n-type with a thin p-type). Figure 9.16 shows a typical device with thin n-type material (about 0.5 μ) so that photons may penetrate to the junction. When a photon arrives at the junction, it interacts with an atom and causes an electron to be transferred to the conduction band. These electrons then drift through the n-type material to the current collector and through the external circuit.

Common materials for solar cells include silicon, indium-phosphorus, galium-arsenide, and cadmium-tellurium. The most common is silicon and much research has been done on it. For silicon, the band gap :— the energy necessary to transfer an electron from the upper valence level to the conduction band :— is 1.12 eV. The maximum theoretical efficiency of a solar cell depends on this band gap. For silicon, the maximum efficiency is 22%. Band gaps for indium-phosphorus, galium-arsenide, and cadmium-tellurium are 1.25, 1.35, and 1.47 eV, respectively. These materials thus have the potential to produce more efficient solar cells than silicon and research efforts are underway to accomplish this goal.

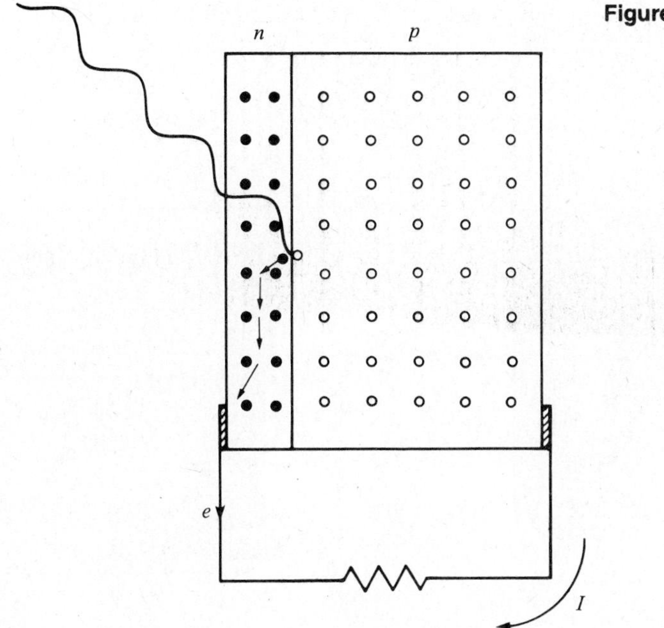

Figure 9.16

9.15

LIGHT-EMITTING DIODES AND SOLID STATE LASERS

In Section 9.10, we saw that in an intrinsic semiconductor the Fermi level was located halfway between the valence band and the conduction band (Fig. 9.7). In an n-type material, the doping caused the Fermi level to shift toward the conduction band, as in Fig. 9.8(a), while in a p-type material the doping caused the Fermi level to shift toward the valence band, as in Fig. 9.8(b). Then we saw that if we put an n-type and a p-type material together, the levels shifted to the situation shown in Fig. 9.8(c). If the two materials are heavily doped such that we have a large number of excess electrons in the n-type and a large number of excess holes in the p-type, the respective Fermi levels will be shifted such that they are located in the conduction band for the n-type and the valence band for the p-type, as shown in Fig. 9.17(a). The physical situation represented by this diagram is that in the vicinity of the junction many electrons and holes are present. If we now forward bias the junction, we shift the Fermi levels by an amount equal to the bias voltage as shown in Fig. 9.17(b). In this situation, there will be in the region near the junction a very large number of electrons in the conduction band and holes in the valence band. The relative number will depend on the bias voltage as well as the doping.

An electron in the conduction band can combine with a hole in the valence band, resulting in the emission of a photon. The energy of the photon will be equal to the energy difference between the two states. This type of radiation is called recombination radiation and a device operating in this way is a light-emitting diode (LED).

Figure 9.17

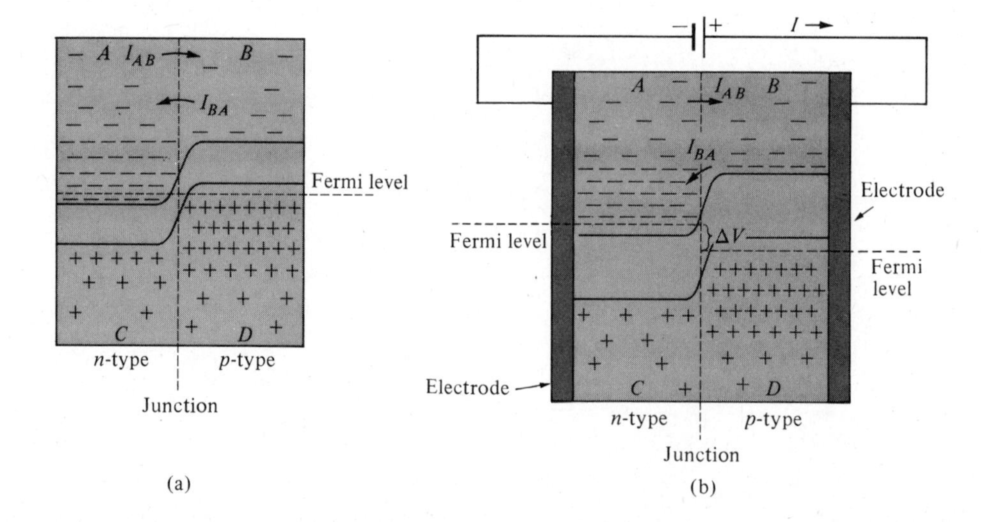

If the forward biasing is great enough, there will be a large number of electrons in the conduction band in the vicinity of a large number of holes in the valence band. This situation is analogous to the population inversion that is obtained between the metastable state and the lower state in a laser. Thus the diode is set for lasing. All that is necessary is to provide a mechanism for reflecting the radiation back and forth through the region near the junction. This radiation then will stimulate the transition of electrons from the conduction band to the valence band. A laser can be built by taking an appropriately doped *n-p* junction, polishing the ends to yield reflective surfaces, and then forward biasing the junction enough to give the proper population inversion. In this type of laser, no external mirrors are necessary because of the high reflectivity due to the difference in index of refractions between the semiconductor material and air. The semiconductors are cleaved along microcrystalline planes thus ensuring parallel ends and good reflecting surfaces.

Semiconductor lasers are usually pulsed lasers. The bias voltage is obtained by charging a large capacitor and then discharging it across the semiconductor. When the capacitor is discharged across the *n-p* material, the large forward bias is obtained and the material lases.

9.16 OPTICAL PROPERTIES

It is interesting to note that the same quantum-mechanical explanation of the electrical properties of solids can be used to gain an understanding of their optical properties. In particular, when a beam of light containing photons of low energy strikes a metal in which there are many electrons in the conduction band with empty energy levels above them, the light will be absorbed. Thus a good conductor should be expected to have the opacity to visible light that experimentally it proves to have. On the other hand, since these low-energy photons cannot excite the electrons in the filled bands of an insulator to the next higher unfilled band, the light must pass through unabsorbed. Thus, many good insulators are also transparent to visible light. Experimentally, we know that as the wavelength of the electromagnetic radiation shortens toward the ultraviolet, these transparent solids become strongly absorbing. This is what one would expect for those insulators in which the energy gap in the forbidden region is just that corresponding to energies of ultraviolet photons. Because semiconductors have very narrow forbidden zones at room temperature, they are opaque to visible light but transparent in the far infrared.

9.17 DISLOCATIONS

While it is possible by the above methods to reach a good understanding of the thermal, electrical, and optical properties of solids, we must at the same time be aware of the fact that actual solids cannot truly be perfectly regular in atomic arrangement. Even the most carefully grown crystal has some imperfections in its structure. It has become increasingly evident over the past few years that imperfections play a vital role in the behavior of solids. In particular, it has long been known that the theoretical breaking strength of a solid is about 1000 times the actual maximum breaking strength attainable. Similarly, design engineers have been aware of the tendency of matter under continued stress to deform plastically, that is, to flow, even though the total stress was maintained well below the elastic limit. This *creep*, as it is called, can now be understood on the basis of a new theory of solids dealing with the imperfections alone. A complete comprehension of the mechanical properties of solids has by no means been reached, and such problems as the fatigue of metals under continued cyclic stressing have come increasingly to the fore in modern aircraft design.

9.18 SUPERCONDUCTIVITY

H. Kamerlingh Onnes took a giant step forward for low-temperature physics when in 1908 he successfully liquefied helium.† Helium boils at 4.2 K and therefore it is possible to study materials at extremely low temperatures by immersing them in the cold bath. Kamerlingh Onnes used liquid helium for a variety of studies of metals at low temperature and in 1911 observed that the electrical resistivity of pure mercury dropped abruptly to zero at about the boiling point of helium. He concluded that mercury had passed into a new state, which, because of its remarkable electrical properties, he called the superconducting state. The temperature at which the resistance disappears is called the transition temperature. Data for the resistance of mercury are shown in Fig. 9.18 and transition temperatures for several superconducting elements and compounds are in Table 9.1.

The term that has been chosen to describe a material that exhibits these phenomena is a superconductor, not perfect conductor. There is some significance in this distinction because superconductors have an additional property that a resistanceless conductor would not have. A material in the superconducting state has the peculiar feature that it does not permit any magnetic flux to exist within the body of the material. When a superconducting sample is placed in a magnetic field, a current is set up on the surface of the material creating a magnetic field everywhere within the material just canceling the applied external field. Figure 9.19 illustrates the difference between a superconductor and a hypothetical perfect conductor. Since most metals are nonmagnetic the magnetic flux passes through, as shown in Fig. 9.19(a). If the material is cooled to a low temperature such that it is below its transition temperature and becomes a perfect conductor, the resistivity goes to zero but the flux distribution remains unaltered, as shown in Fig. 9.19(b). A superconductor behaves quite differently. If the temperature is lowered below the transition

† We shall discuss the unusual properties of liquid helium in Section 9.21.

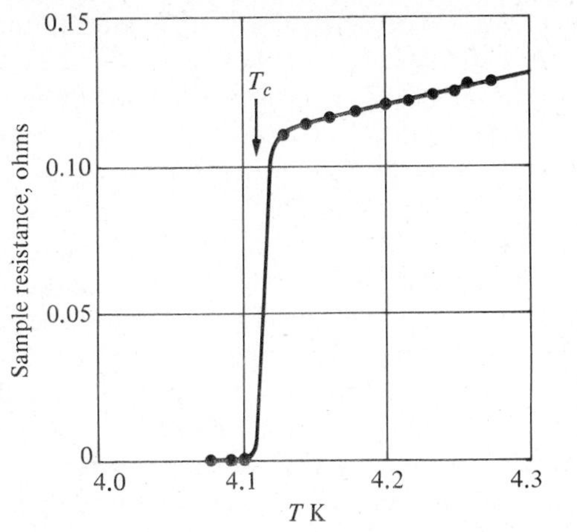

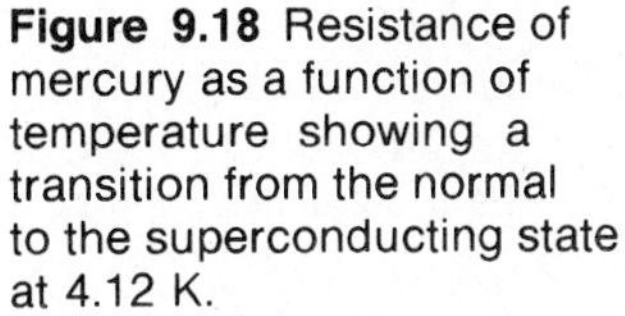

Figure 9.18 Resistance of mercury as a function of temperature showing a transition from the normal to the superconducting state at 4.12 K.

temperature in the presence of a weak magnetic field, at the transition temperature all flux is expelled from the material, as shown in Fig. 9.19(c). This behavior, which is very different from that of a perfect conductor, was first observed by Meissner in 1933.

The current in the superconductor is, of course, carried by the electrons as it is in normal metals. If the current is increased so that the electrons have more than a certain allowed momentum, the superconductivity is destroyed and the superconductor changes to a normal conductor. The maximum current that a superconductor can carry, called its critical current, decreases as the temperature is raised and falls to zero at the transition temperature of the metal. This maximum current leads to a maximum applied magnetic field. As the field strength is increased, the circulating surface currents that cancel the magnetic flux in the material must also increase. Eventually, however, the critical current is reached, and when this happens superconductivity is destroyed. The magnetic field strength at which superconductivity is destroyed is called the critical magnetic field, H_c. Since the critical current of a superconductor falls as its temperature is raised toward the transition temperature, the critical magnetic field also will decrease as the transition temperature is ap-

Table 9.1 Transition temperatures of some superconducting elements and compounds

Element	T_c, K	Compound	T_c, K
Tungsten	0.01	$ZrAl_2$	0.30
Cadmium	0.56	AuBe	2.64
Aluminum	1.19	NiBi	4.25
Tin	3.72	Nb_3Al	17.5
Mercury	4.15	Nb_3	18.05
Lead	7.19	$Nb_3Al_{0.8}Ge_{0.2}$	20.05
Niobium	9.46		

proached. The variation of critical magnetic field with temperature for a typical superconductor is shown in Fig. 9.20. At a point such as P, where the temperature and magnetic field are within the shaded region, the metal will be in the superconducting state. It can be driven into the normal state by increasing either the temperature or the magnetic field or both. We see, therefore, that a superconductor has two possible states: the superconducting state, which is resistanceless and perfectly diamagnetic, and the normal state, which is the same as a normal metal.

The feature that we have been describing, that is the destruction of superconductivity by relatively small magnetic fields, prevents the superconductor from being used in a solenoid to produce extremely large magnetic fields. However, there are superconductors that behave differently from what we have described. Those that we have been describing are called type-I superconductors and they have two states: superconducting and normal. The other type of superconductor called type-II exists in three states: superconducting, mixed, and normal. The mixed state is resistanceless but unlike the superconducting state flux from an applied magnetic field penetrates through it. That is, the Meissner effect does not occur in these type-II superconductors. In consequence, the mixed state persists up to high magnetic fields. Type-II superconductors can be used to wind solenoids capable of producing magnetic fields above 10 T.

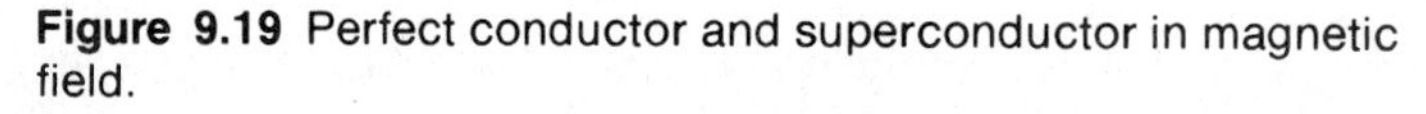

Figure 9.19 Perfect conductor and superconductor in magnetic field.

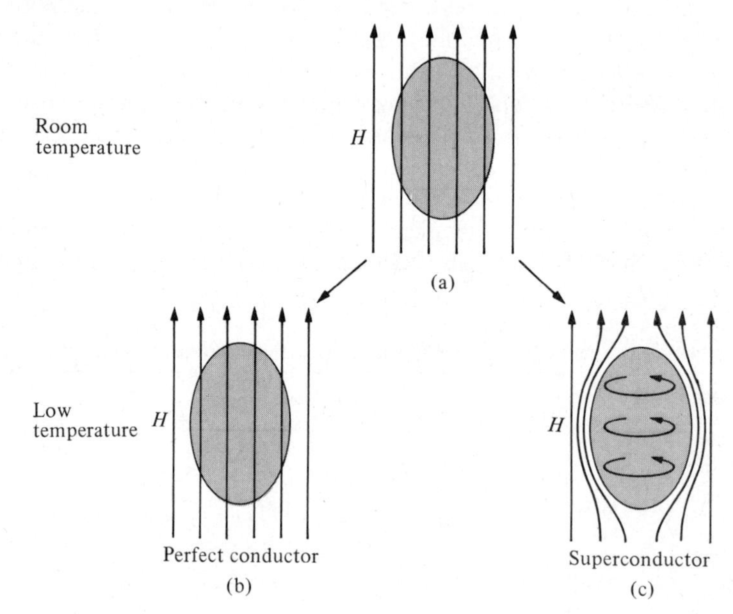

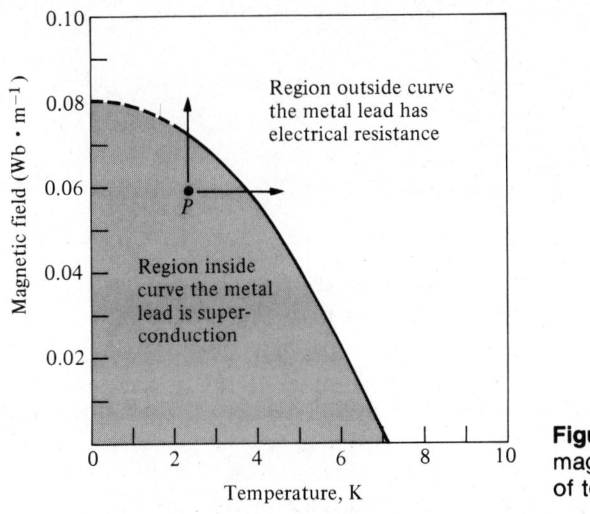

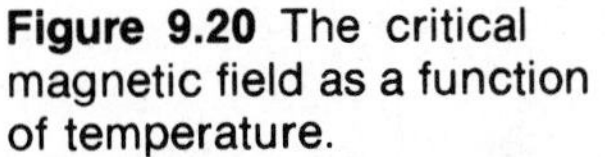
Figure 9.20 The critical magnetic field as a function of temperature.

The first successful theory for superconductivity, published in 1957 by Bardeen, Cooper, and Schrieffer, had as its major feature the pairing of electrons. There is an electron-lattice-electron interaction that results in an overall attraction between two electrons that at low temperature overcomes the Coulomb repulsion.

9.19 JOSEPHSON EFFECT

The introduction of quantum mechanics in 1926 and its subsequent development, particularly in the following decade, turned physicists from the deterministic or classical point of view of natural phenomena to a probabilistic one. There was often a marked difference between the results calculated from these two views when considering events in the microscopic world. For example, one might predict with quantum mechanics that under some circumstances a particle could be at a place that was absolutely unavailable to it classically. Naturally, most of the surprising results came during the early years when the new theory was being applied to various problems in atomic physics for the first time. But another unexpected effect was forecast in 1962 by Brian Josephson, then a graduate student at the University of Cambridge.

He studied theoretically a very low-temperature junction-sandwich system composed of superconductor-insulator-superconductor, as shown in Fig. 9.21. Such a junction can be made by first evaporating a strip of tin or lead on to a glass substrate at low pressures. The surface of the strip is then oxidized to produce a very

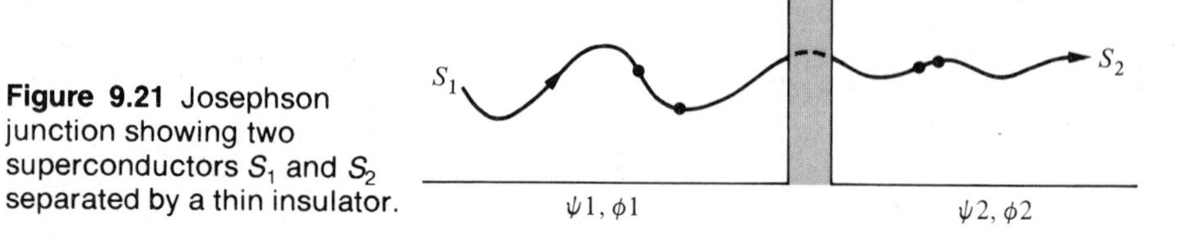

Figure 9.21 Josephson junction showing two superconductors S_1 and S_2 separated by a thin insulator.

thin layer of oxide, and finally, a second strip of superconductor is evaporated over the first and at right angles to it. At that time, it was known that electrons can go through an insulating barrier separating one superconductor from another. In the structures they investigated, the insulator was about 50 Å thick, and no current was observed to flow until an applied threshold voltage, determined by the energy gap of the superconductor, was exceeded. But Josephson analyzed the behavior of a similar junction when the insulating layer was only 10 Å or so thick. When this is the situation, one no longer has two independent bulk superconductors separated by an insulator, but one has at least weak coupling through the thin insulator between the two parts. Consider a thin junction sandwich, now called a Josephson junction, in the superconducting state. The electrons are paired in each superconductor — the Cooper pairs. As the first case, Josephson solved for current when the superconductors S_1 and S_2 are each of the same material and they are each at zero potential. The general procedure of solving involved letting ψ_1 be the probability amplitude of the electron pair on one side of the junction and ϕ_1 their phase with respect to an arbitrary zero level; letting ψ_2 and ϕ_2, respectively, be the values on the other side; then applying the time-dependent Schrödinger equation containing a term representing the effect of the electron-pair coupling or transfer interaction through the insulator to the two amplitudes. The coupling term is a measure of the leakage of ψ_1 into the region S_2, and of the leakage of ψ_2 into the region S_1. When S_1 and S_2 are at the same potential, the solution of the equations predicts that there will be a direct current flow caused by the tunneling of electron pairs through the thin insulator. This is the dc Josephson effect. The derived value of the current density J in it was shown to be equal to some maximum current density J_0 multiplied by the sine of $(\phi_2 - \phi_1)$, the phase difference of the probability amplitudes of the ψ waves. This relation clearly shows that a superconducting current of limited amount will flow even though there is no barrier voltage difference. The existence of this effect was first verified experimentally by Anderson and Rowell of the Bell Telephone Laboratories in 1963. A typical current-voltage curve is shown in Fig. 9.22 where the dc Josephson current between the two points labeled "critical current" is due to pair tunneling at zero voltage across the barrier. The continuous line at finite voltage represents a single-particle tunneling curve.

Josephson also considered the effect of placing the thin film in a magnetic field that was normal to the direction of current flow. The conclusion reached was that the dc current would vary periodically as the field is increased with minima at flux values of $n(h/2e)$, where n is some integer, and h/e is the ratio of the Planck constant to the electronic charge. This too was verified in 1963 by Rowell of the Bell Telephone Laboratories with the results shown in Fig. 9.23. Indeed this was considered a crucial experiment because the dc Josephson current described previously could conceivably be due to the presence of small "bridges" of superconductor or shorts across the junction. Although such a "bridge" current can be modified by a rather intense magnetic field, it will not, however, show the observed periodic variations in small magnetic fields.

The third situation analyzed by Josephson was that in which a potential difference was applied across the junction. In this case, the phase difference of the probability amplitudes on the two sides of the insulating barrier is time-dependent so that the supercurrent is oscillatory. This is called the *ac Josephson effect*. The alternating supercurrents are accompanied by the emission or absorption of electromagnetic radiation (see Fig. 9.24). If there is a finite potential difference $\Delta\mu$ between the conductors, the electron pairs on opposite sides of the barrier differ in energy by an amount $2\Delta\mu = 2$ eV. Therefore, the frequency ν of the associated photon will be given by $h\nu = 2$ eV, or $\nu = 2(e/h)$ V. From this relation one finds that the frequency is high even at very low voltages, 483 MHz/μ V. The existence of such oscillating supercurrents was first verified in 1963.

Perhaps the most famous use of the ac effect is the measurement of h/e that was suggested very early by Josephson himself and that was carried out between 1967 and 1968. Determining the value of h/e from the ac Josephson relation after measuring the applied voltage and the frequency of the emitted radiation is perhaps the simplest method available to determine a fundamental constant. However, it required taking extraordinary precautions, and the possession of great experimental skill to carry out the measurements to a precision of a few ppm so that the final

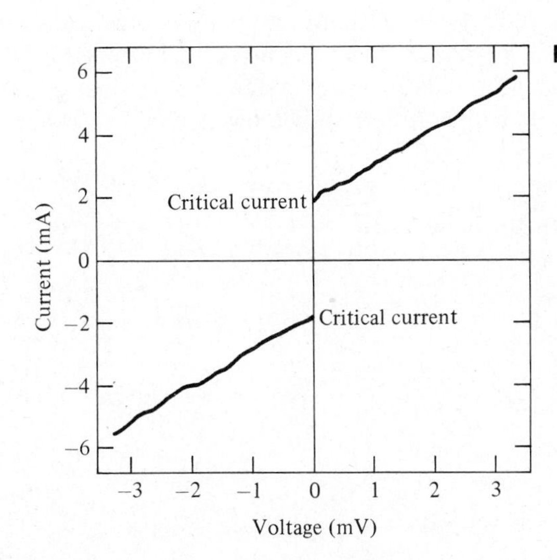

Figure 9.22 Dc Josephson current.

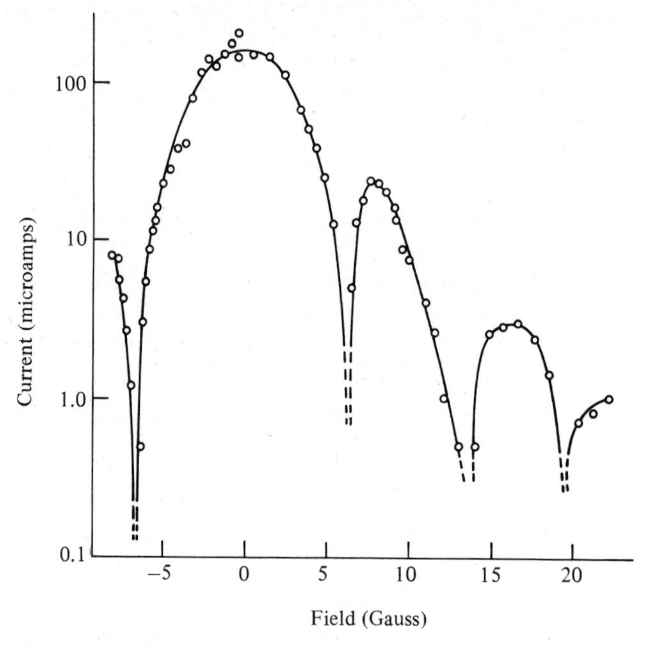

Figure 9.23 Dependence of dc Josephson current on magnetic field.

value of h/e had an uncertainty of only ± 2.2 ppm. With this constant so precisely known, it is evident that a very accurate standard of voltage measurement can be made through the observations of frequency and the use of the Josephson equation.

If monochromatic radiation is incident on the junction, the Josephson oscillator will lock onto it so that the phase of the junction is determined by the phase of the applied radiation. Such an oscillator-mixer system can serve as a sensitive radiation detector that, because its temperature is only a few kelvin, has a very low noise level.

Figure 9.24 Josephson currents.

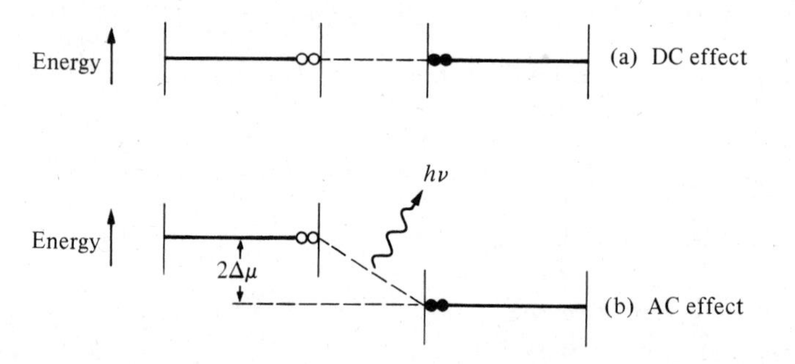

In 1970, a Josephson junction was employed in making a scan of the sun at 1-mm wavelength. This detecting system was found to be very stable, and also far more sensitive than the best bolometer previously used for such study.

9.20 THE SQUID

In recent years, Josephson junctions have been used to make very useful devices. By placing one or more of these junctions in a circuit, a sensitive magnetometer can be constructed. Such a device is called a SQUID, which is an acronym for superconducting quantum interference device.

A SQUID can be built using a Josephson junction like the one described in the previous section. Such a junction in the circuits is referred to as a weak link. Two other types of weak links are found in SQUID devices. They are two superconducting films separated by a very thin superconducting strip and a superconducting screw with an oxidized point that presses against a superconducting film. All that is necessary to have a weak link is to have the junction between the two superconducting films such that a current above a certain critical value will cause the section to go resistive. Figure 9.25 shows the different types of weak links used for SQUIDS.

There are two types of SQUID circuits — dc and ac. Usually a dc SQUID has two weak links in a superconducting ring. An ac device, which is the more common commercially available SQUID, employs one Josephson junction in a ring, as shown in Fig. 9.26. In each case, the circulating current in the superconducting ring provides a measure of the magnetic flux through the ring. In the single junction SQUID, an ac current is induced in the ring through the inductive coupling with a coil in a tank circuit, as shown in Fig. 9.26(a). The induced current is set such that it just exceeds the critical current of the weak link. If there has already been a current induced in the superconducting coil due to the presence of an external magnetic flux, the weak links will switch to a resistive state at different applied currents. The timing of the switching will give a measure of the magnitude of the circulating current and will allow for an extremely accurate measurement of the magnetic flux density.

Figure 9.25

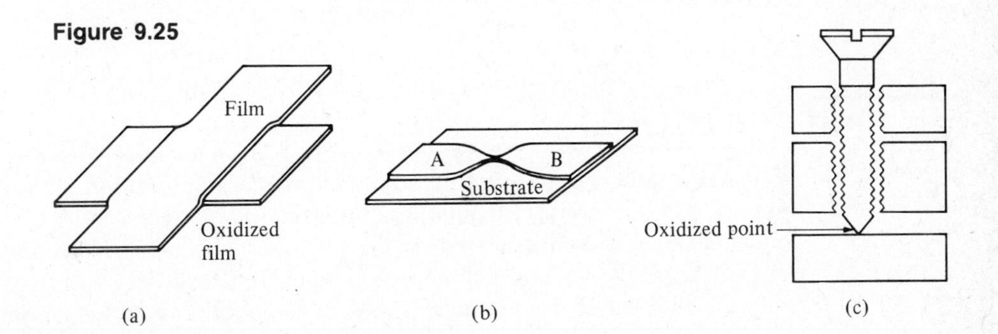

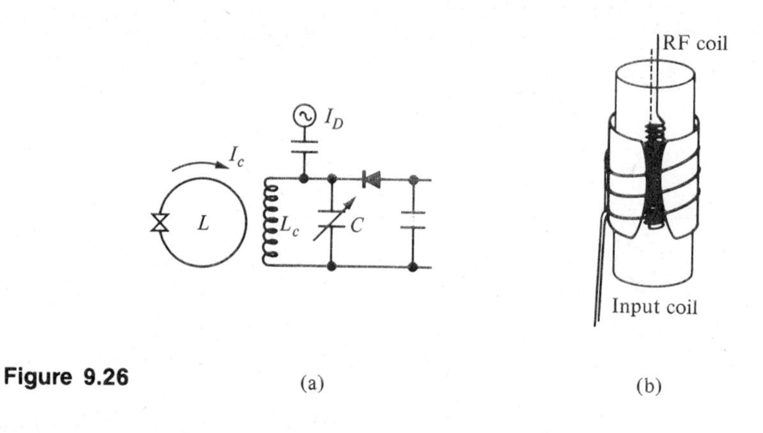

Figure 9.26

Let's consider a type-I superconducting loop without a weak link. If we were to introduce an external magnetic flux, a current would be induced in the loop in order to keep the net flux through the ring at zero. Now, if we insert a Josephson junction in the ring, it is possible that the induced current would exceed the critical current for this weak link. At that point, the circulating current would give an additional contribution to the flux through the loop. This condition is what makes it possible to correlate the output of the SQUID circuit with the applied external field. Magnetic fields of the order of 10^{-14} tesla can be detected. A typical SQUID device is shown in Fig. 9.26(b).

With these SQUID devices it is possible to make magnetocardiographs and magnetoencephalographs capable of providing rapid and detailed medical data through the detection of the varying magnetic fields produced by the heart and the brain; to provide magnetic detection of geomagnetic anomalies with the possibility of contributing to earthquake prediction; and to locate ore bodies and underwater vehicles. As previously described, the junction can be used as a voltage standard, and can be used further as a voltmeter with a sensitivity of about 10^{-16} volts. It can be a parametric amplifier with a noise temperature of less than 10^{-2} K. It can be a detector of infrared and millimeter radiation with a sensitivity that is considerably greater than that of any other type of detector. This list of uses could be extended but enough possibilities have been given to show that the applications are indeed diverse.

9.21 LIQUID HELIUM

For many years, the study of liquid helium has been an important part of modern physics. The discovery of helium was discussed in Chapter 4, where it was pointed out that the gas was discovered on the surface of the sun in 1868 and on earth in 1895. The study of liquid helium was initiated when the gas was first liquefied on July 10, 1908 by H. Kamerlingh Onnes at his laboratory in Leiden in the Netherlands. He was successful because of his extensive experience over many years with liquid air and liquid hydrogen. These two liquids are used for precooling the helium gas before its final expansion and cooling to its liquefaction temperature of 4.2 K. Helium liquefaction initiated a long and extremely interesting study of the unusual properties of this liquid.

We shall start our discussion of liquid helium with Fig. 9.27, the phase diagram showing the regions of gas, liquid, and solid. There are two outstanding features about this pressure temperature diagram. First, at one atmosphere, cooling all the way to absolute zero will not yield the solid state. It can be seen that in order to obtain the solid, the pressure over the liquid bath must be raised to at least 25 atmospheres. Helium is the only substance that under its own vapor pressure will not solidify. Second, at a pressure of one atmosphere, cooling the temperature of the liquid to 2.19 K changes the liquid helium to a new liquid state. Liquid helium above 2.19 K is called helium–I and that below is called helium–II. Helium–I is not too

Figure 9.27 Phase diagram for helium.

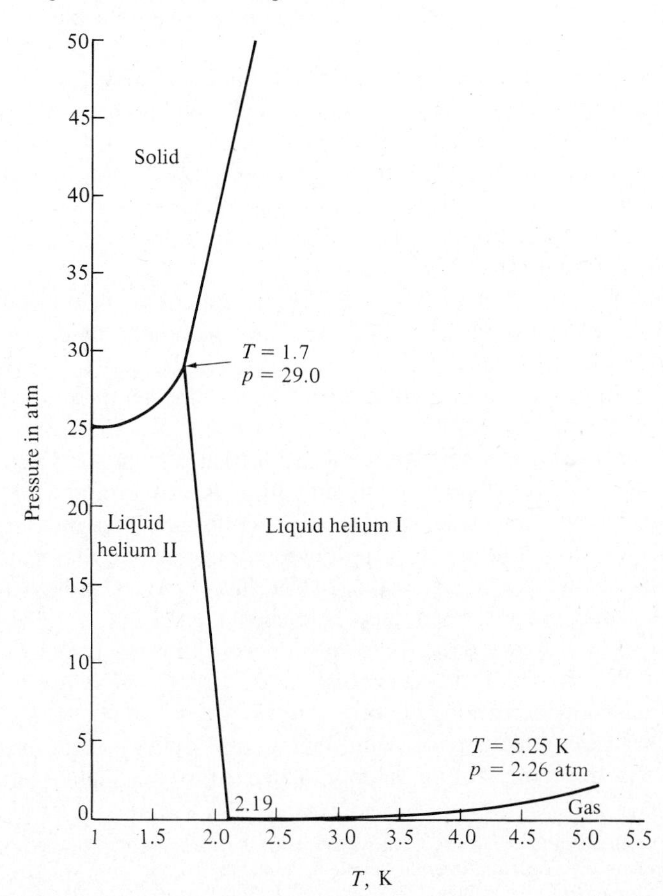

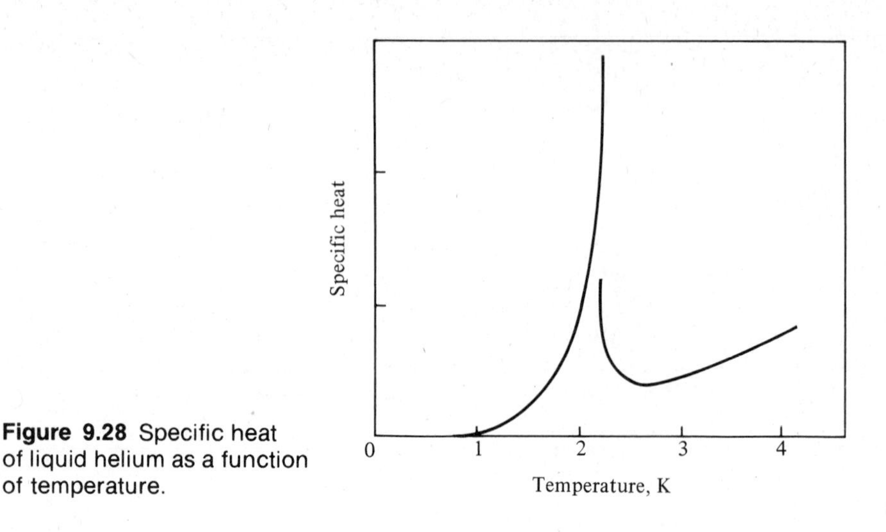

Figure 9.28 Specific heat of liquid helium as a function of temperature.

unusual; however, helium–II exhibits very strange phenomena, some of which we shall discuss.

The first evidence of strange behavior in liquid helium was observed soon after it was liquefied by Kamerlingh Onnes himself. The density of helium is very small — approximately 12% that of water. In 1911, Kamerlingh Onnes found much to his surprise that not only did the helium have a low density but its density passed through a maximum at about 2.2 K. That the density decreases at the low temperature is a very unusual and significant fact.† Something truly unusual must be going on in helium below 2.2 K.

Measurements of the specific heat show conclusively that indeed something unusual does take place at 2.2 K and that the liquid must be considered as having two different phases. Figure 9.28 shows the specific heat as a function of temperature from just above zero up to above 4 K. As the temperature is lowered, the specific heat decreases slowly until just above 2.2 K, where it starts increasing rapidly. Then below 2.2 K, the specific heat decreases from an extremely large value and continues to decrease as the temperature is lowered toward absolute zero. Kamerlingh Onnes and Dana, an American colleague, first measured the specific heat in this region and found the rapidly increasing specific heat as the temperature was lowered toward 2.2 K. The value of specific heat was so high that they felt it must be a mistake. Thus when they published the results in 1926, they failed to include the extraordinary data. These results were first published by Keesom, who because of the shape of the curve called the transition between helium–I and helium–II a λ transition and the temperature T_λ.

The thermal conductivity of liquid helium also undergoes an unusual transition at 2.2 K. As this temperature is approached from the higher temperatures, the

† Water behaves in much the same way (density decreases below about 4°C). However, the water molecule has a more complicated structure than the simple helium atom.

thermal conductivity increases by a factor of 10^6. The thermal conductivity of liquid helium–II is enormous — hundreds of times larger than that of pure copper, gold, or silver. This large value shows up experimentally in a way that must have been observed by Kamerlingh Onnes in 1908 but was not mentioned in print for thirty years. As the liquid helium is cooled by reducing the pressure above the bath toward 2.2 K, the bath is boiling vigorously at all points throughout. After passing through the λ point, the bath becomes extremely quiet. Evaporation occurs only at the surface and energy is distributed quickly throughout the liquid due to the high thermal conductivity.

One of the most interesting properties of liquid helium–II, and the property that gives it the name "superfluid," is its viscosity. There are two ways of measuring the viscosity of a liquid. One is to allow the liquid to flow through a small opening and measure the time necessary for a known quantity to pass through. The other is to move an object through the liquid and measure the drag exerted on it — for example, moving a steel ball through molasses. The velocity will give a measure of the value of viscosity. In an experiment using the first technique, the viscosity of helium was measured to be extremely small, about 10^{-10} poise. (The viscosity of air is 2×10^{-5} poise.) Because of this extremely small viscosity, the term superfluid has been applied to helium–II. Measurements of the viscosity using the second technique, drag on an object moving through the liquid, produced quite different results. The object was a cylinder rotating in the liquid as a torsion pendulum. These measurements were found to be several orders of magnitude greater than the viscosity measured with the flow technique and only three times smaller than that of helium–I. This paradox was not resolved until the theoretical work of Landau and Tisza, which will be discussed later in this section.

An unusual experiment that gained the most attention was the film flow work. As early as 1922, it was observed that two concentric vessels each containing liquid helium–II would adjust themselves automatically so that the levels of the liquid would be the same. Later experiments showed the levels were adjusted by a film of liquid helium–II moving between vessels. Consider the small beaker in Fig. 9.29. If the beaker is lowered halfway into the bath and held in that position, liquid helium will crawl up the outer wall and into the beaker until the level of helium in the beaker is equal to the level of the outer bath. The film will move in the opposite direction if the beaker is then pulled up above the bath. Helium–II will crawl up the inside of the beaker and over the edge and drip into the main bath until the beaker is empty.

Two related phenomena are the thermo-mechanical and the mechano-caloric effects. Figure 9.30(a) shows the apparatus for observing the thermo-mechanical effect. A small inner chamber with a superleak is filled with liquid helium–II and immersed in a larger bath of helium–II.† When heat is supplied by sending a current through the resistor, the level of the helium in the inner bath will rise above the outer

†A superleak is usually constructed of extremely fine powder packed so tightly that a fluid with nonzero viscosity would not pass through it.

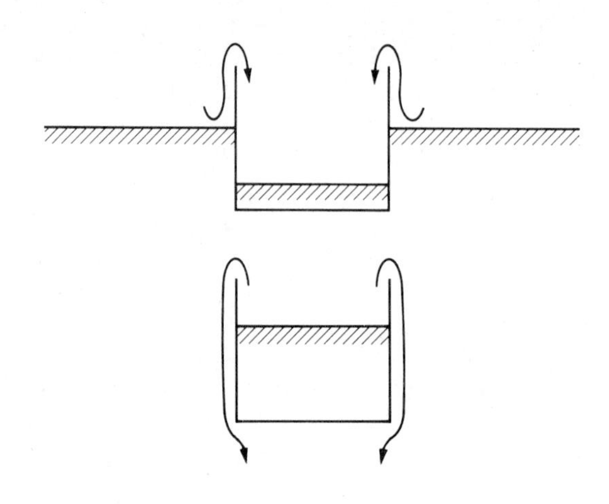

Figure 9.29

level. This is accomplished by the flow of fluid through the superleak from the outer bath into the inner chamber. The reverse effect can be demonstrated in the apparatus shown in Fig. 9.30(b) where a small chamber filled with helium–II with a superleak is suspended above a larger bath of helium–II. A thermometer in the innerbath will indicate that the temperature of this bath has increased when fluid flows from this chamber into the main bath.

An exciting variation of the thermo-mechanical effect is called the fountain-effect. To observe this effect, the apparatus of Fig. 9.30(a) is modified so that the top joins with a capillary. Now if heat is supplied to the inner chamber by shining a light on the fine powder, helium–II will rush in the bottom and out the top of the small capillary yielding the fountain-effect. These effects can be understood from the two-fluid theory of Landau and Tisza.

The two-fluid theory developed by Tisza predicted the existence of a thermal energy wave propagating through helium–II with a velocity different from that of conventional sound. This thermal wave was first detected by Kirskov in Russia and was measured in detail by Squire and Pellum in the United States. Whereas first sound is a variation in density as energy propagates through a medium, the thermal energy wave is a transfer of energy where the density remains constant. The temperature wave propagates at approximately 20 m/s while standard sound travels at above 200 m/s. This thermal wave has been named second sound.[†]

[†] First sound is an oscillation of normal fluid and superfluid together giving a density variation. In second sound, the normal fluid moves opposite to the direction of superfluid flow, keeping the density constant but giving fluctuations in temperature. There are two modes of transport in helium–II, where the normal fluid is held in place by its viscosity while the superfluid oscillates back and forth. Third sound occurs in a thin film of helium–II, where the viscosity holds the normal fluid fixed. Fourth sound occurs in bulk helium–II that is contained in a cavity filled with fine powder. The viscosity of the normal component holds the normal component fixed in the fine powder. Still another mode is fifth sound, which is a temperature wave that propagates in a superleak. It is analogous to the second sound mode but has a different velocity because the normal fluid is clamped by the superleak.

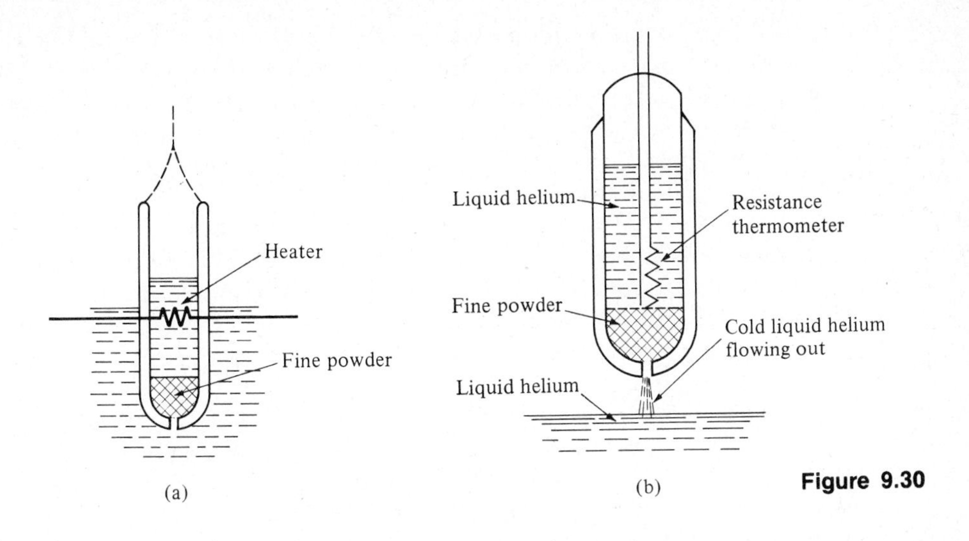

Figure 9.30

Most of the experiments we have discussed can be understood using a theoretical model proposed independently by Landau and Tisza. In this model, all of the helium atoms are condensed at absolute zero into the lowest energy state and are characterized by fluid flow without friction and atoms carrying no entropy or thermal energy. As the temperature is increased, some of the atoms are excited into the upper energy state where they are characterized by viscous flow and carry all of the thermal energy for the liquid. Helium between zero and 2.2 K is thought of as being made up of these two interpenetrating components, one called superfluid and the other normal fluid. The superfluid has a density of ρ_s and flows with velocity v_s while the normal fluid has density ρ_n and flows with velocity v_n. The total density of the liquid then is $\rho_s + \rho_n$. As the temperature is increased, the proportion of the liquid that is normal increases while the proportion of the liquid in the superfluid state decreases until at T_λ all of the atoms are in the normal or excited state.

The two-fluid theory allows us to understand the discrepancy in the values obtained for the viscosity obtained by flow through a capillary and by drag on a torsion pendulum. In the experiment where liquid helium is allowed to flow through a capillary, the normal component is held back in the reservoir and the superfluid with zero viscosity flows rapidly through the capillary. The flow is large through the small capillary and a small value of viscosity is obtained. However, in the experiment of a torsion pendulum with a cylinder in the liquid, the cylinder is influenced by the presence of the normal fluid. The normal fluid causes drag on the pendulum and the value obtained for the viscosity is due to the normal fluid.[†]

[†] If the fluid has zero viscosity and if it can be set in motion within a doughnut-shaped vessel, for example, then it should remain in motion indefinitely. Experiments of this type have been carried out and the superfluid component has been found to remain circulating for many hours.

The mechano-caloric effect is the result of superfluid helium flowing out of the chamber and carrying no entropy or thermal energy. This process causes the fluid remaining in the reservoir to be more concentrated with normal fluid, which thus increases the energy density and the temperature. The superfluid helium coming out of the upper chamber dilutes the liquid in the lower chamber with superfluid and thus decreases its temperature.

A similar explanation of the thermo-mechanical effect comes from the two-fluid model. When heat is added to the inner chamber, the superfluid is excited to the normal fluid state, which thus increases the amount of normal fluid and superfluid rushes in through the superleak to bring the concentration back to equilibrium for that temperature. The level in the inner chamber thus increases above the level in the outer chamber.

Understanding the thermo-mechanical effect now shows us that when heat is added at a point in the liquid helium, superfluid helium is excited into normal fluid helium. To reestablish equilibrium, a counter flow of helium is produced. Superfluid helium moves toward the heat source and normal fluid moves away. The normal fluid would not merely carry thermal energy due to the specific heat of the liquid but would also carry that due to the very high excitation energy necessary to get the atoms from the superfluid state to the normal state. This phenomenon is an extremely efficient transport process for thermal energy. It is much more efficient than normal diffusion or convection. Thus the two-fluid model accounts nicely for the extremely large thermal conductivity observed in helium–II.

Further developments of the theory of rotating liquid helium have predicted that the flow of superfluid is constrained by a condition analogous to the Bohr orbit

Figure 9.31

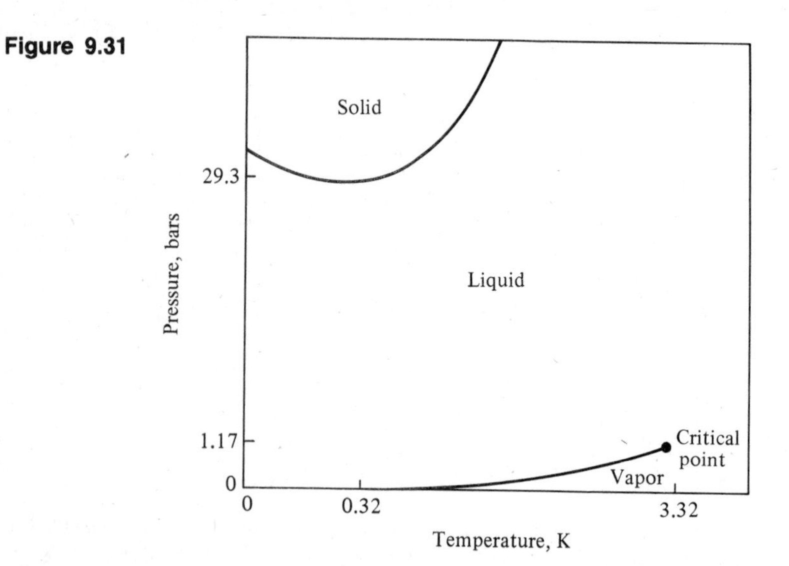

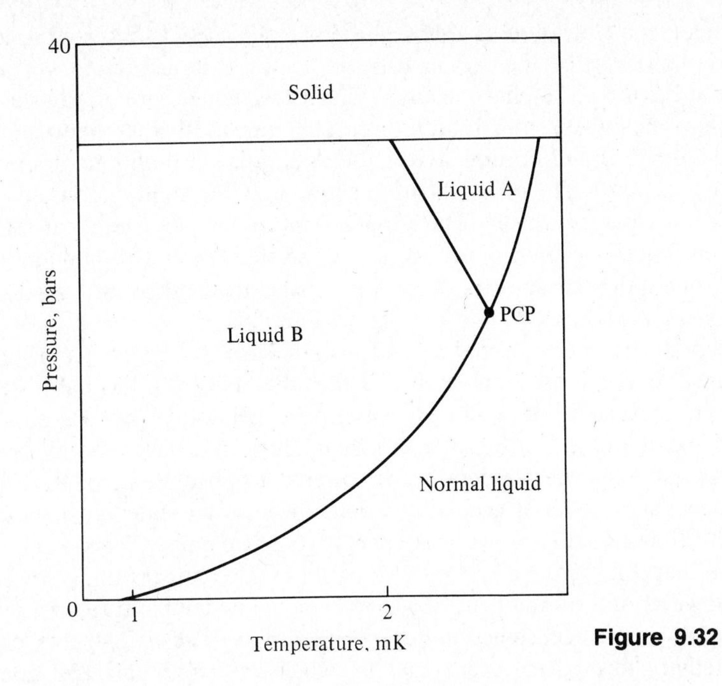

Figure 9.32

conditions for electrons in an atom. The circulation of the superfluid around any closed path in the liquid is quantized in integral multiples of h/m where h is Planck's constant and m is the mass of the helium atom. Thus in any chamber of rotating helium, one would expect to find a system of parallel vortex lines. Such vortices have been observed with the technique of introducing ions that interact with the vortices and then show up as images on a photographic plate or television screen. Photographs have been produced that indicate the existence of these vortices in the helium.

In 1972, the excitement in the study of liquid helium shifted dramatically to the other helium isotope, ^{3}He. Measurements on liquid ^{3}He by Osheroff, Richardson, and Lee at Cornell University showed that when this isotope is cooled to about 2.7 mK (0.0027 K), a new phase is observed. Liquid ^{3}He, like liquid ^{4}He, will not solidify under its own vapor pressure. Figure 9.31 shows the phase diagram for ^{3}He from zero kelvin up to 4 K. The critical point is at a temperature of 3.32 K and a pressure of 1.17 bar. A very interesting feature of the P-T diagram is the negative slope of the solid-liquid interface below 0.32 K.

The truly exceptional features of liquid ^{3}He were not observed until 1972 because they occur in the millikelvin temperature range — a range inaccessible for continued study until the 1970s. Figure 9.32 shows the phase diagram from about 3 mK to below 1 mK. As can be seen, if the temperature is lowered along the solid liquid curve, a transition to a new phase of liquid ^{3}He is observed at about 2.7 mK. Further lowering of the temperature indicates a transition to yet another liquid phase at 2.2 mK. The phase from 2.7 down to 2.2 mK is referred to as liquid A and the phase below 2.2 mK is liquid B. The transition lines between liquid A and normal, liquid B and normal, and liquid A and liquid B all join at a point called the polycritical point. The phase diagram shows how the solid and liquids appear in the absence of a magnetic field. The application of a magnetic field causes the transition between liquid A and liquid B to be shifted to lower temperatures and the polycritical point disappears. Also, in a magnetic field an intermediate phase, A_1, appears between liquid A and the normal liquid.

Liquid ^{3}He was expected to and does behave differently from liquid ^{4}He. The primary difference between the two is that the ^{3}He atom has half-integral spin and therefore must obey the Pauli exclusion principle, while the ^{4}He atom has integral spin. Thus it would not be expected that ^{3}He would have a condensation into one quantum state as the temperature is lowered to absolute zero. Rather it would be expected that the liquid gradually would fill allowed states up to a certain energy and the allowed states above that energy would be empty. This would be analogous to what happens to the electrons in a metal as the temperature is lowered.

However, interesting behavior at extremely low temperatures was expected to be analogous to the occurrence of superconductivity. The BCS theory explains superconductivity through a mechanism by which electrons pair and condense into a quantum state causing the superconductivity. Physicists thought that a similar pairing might occur in liquid ^{3}He and for many years looked for evidence of this. The search was concluded in 1972 with the discovery of the new phases.

Theory requires that the nuclear spins in liquid ^{3}He be parallel instead of antiparallel as they are with the electrons in superconductivity. The pair of ^{3}He atoms thus have one quantum unit of orbital angular momentum and in addition they have one unit of spin angular momentum. The coupling of these units of angular momentum are what give rise to the different phases of liquid ^{3}He. The two phases A and B differ in the way the different units of angular momentum couple.

As would be expected, mixtures of liquid ^{3}He and ^{4}He are also interesting. Because of the cooperative nature required between the ^{4}He atoms for the transition at T_λ, the dilution of the ^{4}He with ^{3}He should lower the temperature at which the transition occurs. This effect is analogous to the lowering of the freezing point of a liquid by the addition of an impurity. Figure 9.33 shows the temperature as a function of the fraction of ^{3}He mixed with ^{4}He. The λ transition decreases from a temperature of about 2.2 K as the molar fraction of ^{3}He is increased.

The third law of thermodynamics requires that a mixture of this type should separate and give a lower entropy (an increase in the degree of order or a decrease in the amount of disorder). This is what occurs as the temperature of the mixture is decreased to about 0.87 K. The mixture separates into a nonsuperfluid, ^{3}He-rich phase that floats on top of the superfluid, ^{4}He-rich phase. As the temperature is continuously lowered, the ^{3}He-rich phase increases in ^{3}He content to become pure ^{3}He. However, in the ^{4}He-rich phase, a small quantity of ^{3}He (about 6%) remains soluble all the way down to the absolute zero. This is a very important phenomenon for the production of low temperatures.

One of the primary ways to obtain a constant low temperature for the study of various materials is to put the sample in thermal contact with a liquid bath held at a temperature of interest. If one is interested in measuring a particular characteristic of a sample at 100°C, an easy way to do this is to bring a bath of water to the boiling point and have the sample in thermal contact with this bath. For many years, the standard method for measuring properties of materials at low temperatures was the same with the water replaced by such cryogenic liquids as oxygen, nitrogen, hydrogen, and helium.

In the example of the water, if a measurement is to be obtained at a temperature below 100°C, the pressure above the boiling water bath can be reduced to produce a lower temperature. If you want to hard-boil an egg, it is well known that it takes longer in a high-altitude city such as Denver, Colorado, than in a coastal city such as Houston, Texas. The reason for this is that the atmospheric pressure at the high altitude is lower than that at sea level, and thus the boiling temperature is lower. For convenience, pressure over the boiling water bath can be controlled not by change in altitude but by the change in pressure using a vacuum pump.

The boiling point at one atmosphere for liquid nitrogen is 77 K. Vigorously pumping on the liquid nitrogen will reduce its temperature to 63 K where it freezes.

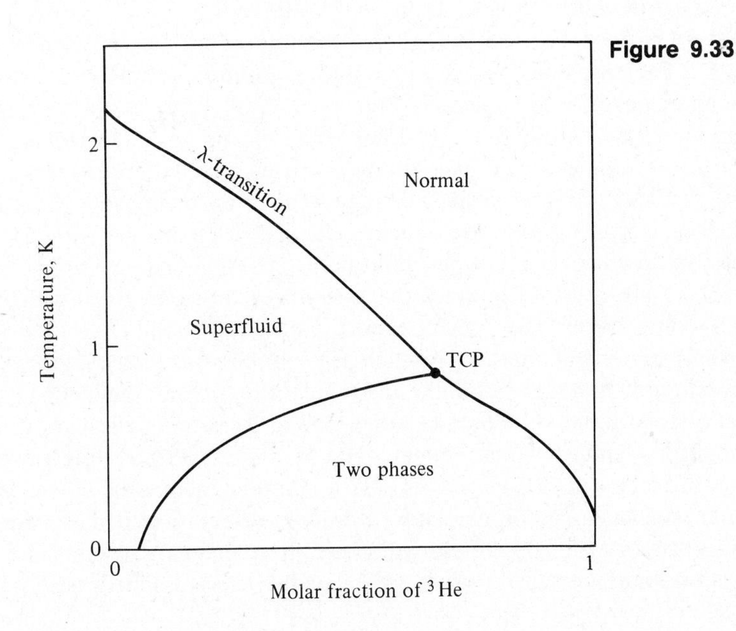

Figure 9.33

Thus measurements can be made on a sample in thermal contact with a nitrogen bath over this temperature range. Replacing the liquid nitrogen with liquid hydrogen and using the same procedure will yield a workable temperature range from about 12 K up to above 20 K. For this procedure at the lowest temperatures liquid helium is the cryogenic liquid. The liquid ^{4}He boils at 4.2 K and vigorous pumping of the bath can reduce this temperature to about 0.9 K.

One remaining step can be taken using this procedure if ^{4}He is replaced by ^{3}He. The boiling point of ^{3}He is about 1 K lower than ^{4}He. Also, the vapor pressure for ^{3}He is considerably greater than that for ^{4}He at a specific temperature. Therefore by vigorously pumping on the ^{3}He, a temperature lower than that obtained with ^{4}He can be reached. If ^{3}He is used, a temperature range from as low as 0.3 K up to a few degrees kelvin is possible.

The interesting studies in liquid ^{3}He are found at temperatures below those accessible with this technique. Thus it is not surprising that the discovery of superfluidity in ^{3}He had to wait for the development of another technique for maintaining the extremely low temperature. This technique takes advantage of the interesting characteristics of the ^{3}He and ^{4}He mixtures shown in Fig. 9.33. When the mixture is cooled to below 0.87 K, the liquid separates into the two phases. The superfluid phase acts as a background that the ^{3}He can expand into. This "evaporation" of ^{3}He into the background phase is accompanied by an absorption of latent heat. The ^{3}He and ^{4}He dilution refrigerator is a device that continually circulates ^{3}He, allowing it to be dissolved in the ^{4}He background, and thus produces temperatures in the millikelvin range. Temperatures of the order of a few millikelvin can be maintained with a dilution refrigerator.

As has been pointed out, the interesting properties of ^{3}He occur at around 2 mK. To reach this temperature and maintain it, one may make use of the characteristic of ^{3}He shown in Fig. 9.31. The negative slope of the liquid-solid interface below about 0.32 K (320 mK) means that if ^{3}He is cooled to this temperature, for example by a dilution refrigerator, compression of the liquid will cause further cooling. A device making use of this principle is called a Pomeranchunk cell. The Cornell group used such a cell to reach the temperatures of 1–2 mK necessary to discover the superfluid phases of ^{3}He.

There are other techniques that use magnetic fields to obtain these extremely small temperatures. These techniques primarily allow one to reach low temperatures for short periods of time after which the samples start warming back to the higher temperatures. In order to obtain low temperatures using magnetic cooling, a system must be found that still contains a measure of disorder (significant value of entropy) at about 1 K and on which we may operate in a way with the magnetic field to bring about order (lower entropy). A number of paramagnetic salts have this characteristic in that their magnetic dipoles show a random orientation at these temperatures. The application of a magnetic field will cause these magnetic spins to be aligned in the direction of applied field. When the magnetic field is applied, heat is given off from

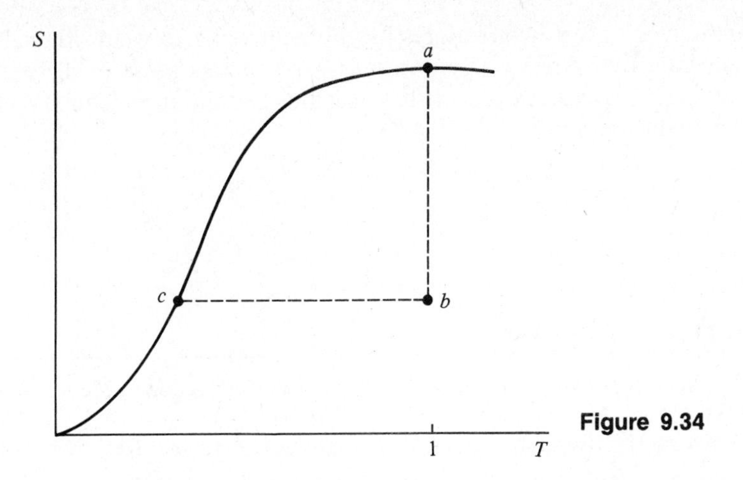

Figure 9.34

the salt to the surroundings. If the salt is held at 1 K by keeping it in contact with a vigorously pumped liquid helium bath, the thermal energy will be transported to the bath. At this point, thermal contact is broken between the salt and the bath. This may be achieved by pumping a chamber between the bath and the sample to hard vacuum. Then when the magnetic field is removed the temperature of the sample will decrease considerably.

Figure 9.34 diagrams the entropy versus temperature for a typical paramagnetic salt. The entropy approaches zero as the temperature approaches zero according to the third law of thermodynamics. Point a corresponds to the sample at 1 K in contact with the helium bath in zero magnetic field. Applying the magnetic field corresponds to moving from point a to point b on the curve lowering the entropy. When thermal contact with the bath is broken and the magnetic field is removed, the sample returns to the entropy curve along the path of bc, which carries the temperature of the sample down to a temperature of the order of hundredths of a degree.

Much lower temperatures may be obtained by using a similar technique where the remaining high entropy is in the nuclear spin system. As an example, copper that has a nuclear spin can be cooled to a few thousandths kelvin.

9.22 CONCLUSION

It is remarkable that quantum theory, originally designed to explain the optical spectra of excited atoms, should, with suitable modifications, prove so successful in explaining such diverse problems as thermal, electrical, mechanical, and optical properties of solids. The application of the quantum theory of metals to their specific heat at low temperatures has brought order to what originally appeared to be an unrelated series of observations. And in this same field, too, a satisfactory theory has finally been found that accounts for the extraordinary phenomenon of superconductivity.

FOR ADDITIONAL READING

Allen, Philip B., and William H. Butler, "Electrical Conduction in Metals," *Physics Today* **31**, No. 12, 44 (December 1978).

Brattain, W. H., "Developments of Concepts in Semiconductor Research," *Am. J. Physics* **24**, 421 (1956).

Broers, A. N., and M. Hatzakis, "Micro Circuit by Electron Beam," *Scientific American* **227**, 34 (November 1972).

Buchold, T. W., "Applications of Superconductivity," *Scientific American* **74** (March 1960).

Chalmers, Bruce, "The Photovoltaic Generation of Electricity," *Scientific American* **235**, 34 (October 1976).

Geballe, T. H., "New Superconductors," *Scientific American* **22** (November 1971).

Ginsburg, D. M., *Superconductivity*, Resource Letter Scy-1 and Selected Reprints. New York: American Institute of Physics, 1964.

Guyer, R. A., R. C. Richardson, and L. I. Zane, "Excitations in Quantum Crystals (A Survey of NMR Experiments in Solid Helium)," *Rev. Modrn. Phys.* **43**, 532 (1971).

Hovel, H. J., "Solar Cells," *Semiconductors and Semi-Metals*, **11**, Academic Press, 1975.

Keller, William E., *Helium–3 and Helium–4*. New York: Plenum Press, 1969.

Kittel, Charles, *Introduction to Solid State Physics*. 5th ed. New York: Wiley, 1976.

Komerek, P., "An Introduction to Lasers and Their Applications," *Contemporary Physics* **17**, No. 4, 355 (1976).

Landenburg, Donald, Douglas Scalapino, and Barry Taylor, "The Josephson Effects," *Scientific American* **214**, 30 (May 1966).

Lane, C., *Superfluid Physics*. New York: McGraw-Hill, 1962.

Le Croissette, Dennis, *Transistors*. Englewood Cliffs, N.J.: Prentice-Hall, 1963.

London, F., *Superfluids, Volume II, Macroscopic Theory of Superfluid Helium*. New York: Dover, 1954.

Matthias, B. T., "Superconductivity," *Scientific American* **197**, 92 (November 1957).

McWhorter, E. W., "The Small Electronic Calculator," *Scientific American* **234**, 88 (March 1976).

Smith, C. S., "The Prehistory of Solid State Physics," *Physics Today* **17**, No. 12, 18 (December 1964).

Swithenby, Stephen J., "SQUIDS and Their Applications in the Measurement of Weak Magnetic Fields," *Journal of Physics E: Scientific Instruments* **13**, 801 (1980).

Vacroux, A. G., "Microcomputers," *Scientific American* **232**, 32 (May 1975).

Wilks, J., *The Properties of Liquid and Solid Helium*. Oxford: Clarendon Press, 1967.

REVIEW QUESTIONS

1. What was the major success of the Einstein theory of specific heats?
2. What are the differences in the band structure of a conductor, an insulator, and a semiconductor?
3. In a metal, what is the Fermi energy?
4. What is occurring with electrons and holes at a *p-n* junction that leads to the rectifier effect?
5. What is the difference between the resistivity of a normal metal like copper and that for a superconducting sample like lead?
6. What is the significant feature of the Meissner effect?

PROBLEMS

9.1 Verify the statement that at room temperature the order of magnitude of the electrical conductivity of copper is given by Eq. (9.14). The interatomic spacing for copper is 3.6 Å.

9.2 In a metal, all of the electronic energy levels may be filled to an energy of the order of several eV. This means that the free electrons contain a great deal of energy. Can this energy be obtained from the metal by electron transition to a lower state? Why?

9.3 The specific heat of a sample will decrease to a very small quantity as the temperature falls below 1 K. Discuss the problems that this would cause in trying to study a sample at these very low temperatures.

9.4 Figure 6.3 indicates the position of six atoms in one plane in a single crystal. Diagram the position of these six atoms for a lattice wave propagating through this crystal with wavelength (a) $10 \times d'$, (b) $6 \times d'$, (c) $4 \times d'$, and (d) $2 \times d'$. (e) Can you indicate the positions of these atoms for a wave with wavelength shorter than $2 \times d'$?

9.5 The photoelectric threshold for a certain metal is 2900 Å. (a) Assuming a Maxwellian distribution of speeds, what temperature would be assigned to the "electron gas" if the average kinetic energy of translation of the electrons in this metal is 70 percent of the work function? (b) What is the ratio of the root-mean-square speed of these electrons to that of light? Should relativistic equations be used for these electrons?

9.6 Consider a one-dimensional stretched spring of force constant k with a mass m attached at the end performing simple periodic motion. Show that the time average of the kinetic energy of the mass over a whole cycle is equal to the time average of its potential energy, and hence that the total energy, a constant, is twice the average kinetic energy.

9.7 Assuming that the interatomic distance in a metal crystal is 2 Å and that the conduction electrons have the root-mean-square speed of a Maxwellian distribution at 30,000 K, determine the ratio of the de Broglie wavelength of the electrons at this speed to the interatomic spacing.

9.8 (a) How could one determine experimentally whether electrical conduction is by holes or by electrons? [*Hint:* Consider the effect of a transverse magnetic field on an electric current.] (b) The ratio of the electric field per unit current density per unit magnetic induction, where all three are mutually perpendicular, is known as the Hall coefficient. Show that the Hall coefficient should be a sensitive measure of the number of conducting holes or electrons per unit volume. (c) Show that this coefficient has opposite signs for conduction by electrons and by holes.

9.9 The "mobility," μ, of an electric charge is defined as the velocity increment per unit accelerating electric field. If n is the number of conducting electrons or holes per unit volume of a solid, show that the electrical conductivity is given by

$$\sigma = ne\mu.$$

9.10 Discuss why the work function of a metal is approximately independent of temperature.

9.11 The band structures of silicon and diamond are quite similar, but silicon has a metallic appearance while diamond is transparent. Explain the difference, keeping in mind that the energy gap between the valence and the conduction bands in silicon is 1.14 eV and in diamond 5.33 eV.

9.12 An n-type semiconductor has an optical absorption band whose long wavelength limit is at 36,000 Å. What is the energy gap between the donor level and the bottom of the first empty band?

9.13 The energy gap in Ge is about 0.75 eV. At what wavelength would Ge begin to absorb electromagnetic radiation?

9.14 NaCl does not form a perfect crystal. A special type defect called an F-center is present where a missing ion leaves a positive vacancy that traps an electron. This trapped electron is in a level 2.65 eV below the conduction band. What wavelength does it absorb? What color would a NaCl crystal be that contained a large number of these F-centers?

9.15 Describe an experiment that could be performed to prove that the resistance of a superconducting material is very near or equal to zero.

9.16 Describe a simple demonstration experiment that could be performed to show the Meissner effect.

10 NATURAL RADIOACTIVITY

10.1 DISCOVERY OF RADIOACTIVITY

We now go back to March 1896, the year after Roentgen discovered x-rays, when Becquerel announced the discovery of radioactivity. Although Roentgen rays had been discovered less than four months earlier, it was already known that x-rays came from the fluorescent walls of the discharge tube, and thus it was thought that fluorescence and phosphorescence might be responsible for them. Becquerel knew that uranium salts became luminescent when exposed to bright sunlight, and he had heard that the phosphorescent radiations from these activated salts could penetrate opaque bodies. Upon studying these effects, he found that the radiations from light-activated uranium did cast shadows of metallic objects on photographic plates that were wrapped in black paper. But the particularly *new* thing Becquerel found was that the radiations came from uranium salts whether those salts had been excited by light or not. He found that uranium salts that had been protected from all known exciting radiations for months still emitted penetrating radiations without any noticeable weakening. He recognized the parallel between his discovery and the discovery of x-rays, and he found that the new radiations could discharge electrified bodies as x-rays do. He realized that these radiations were not due to fluorescence, but that the uranium metal was their source. This property of uranium, that it spontaneously emits radiation, is called *radioactivity*.

10.2 THE SEAT OF RADIOACTIVITY

In showing that radioactive radiations came from uranium metal, Becquerel worked with many uranium salts and the metal itself. He used these materials crystallized, cast, and in solution. In every case it appeared that the radiations were proportional to the concentration of the uranium. It has been found that this proportionality between radiation intensity and uranium concentration continues unchanged through variations of temperature, electric and magnetic fields, pressure, and chemical composition. Since the radioactive behavior of uranium is independent of the environment of the uranium atom or its electronic structure, which changes from compound to compound, the radioactive properties of uranium were attributed to its nucleus.

10.3 RADIUM

Becquerel's discovery of the radioactivity of uranium immediately raised the question whether other elements are radioactive. Pierre and Marie Curie investigated a uranium ore called pitchblende, which contains uranium, bismuth, barium, and lead. Upon chemical separation, the uranium showed the expected activity, and the bismuth and barium fractions also showed activity. Since neither bismuth nor barium shows activity when pure, the Curies assumed that each fraction contained a new element, one chemically like bismuth and the other chemically like barium. They called these new elements polonium and radium and set out to isolate each. The skill, enthusiasm, and patience that went into this task has been beautifully told by her daughter in Madame Curie's biography.

Radium had to be separated from barium on the basis of its slightly different physical properties by the technique of fractional crystallization. The magnitude of the task may be seen from the fact that the Curies separated about one fifth of a gram of a radium salt from a ton of pitchblende. They could trace their progress by noting the increased activity of the samples as they became more and more concentrated. Weight for weight, polonium is about 10 billion times more active than uranium, and radium is 20 million times more active than uranium.

10.4 THE RADIATIONS

Although the penetrating radiations from radioactive substances were immediately likened to x-rays, Rutherford found, in 1897, that the radiations were of more than one kind, some rays being more penetrating than others. He called the less

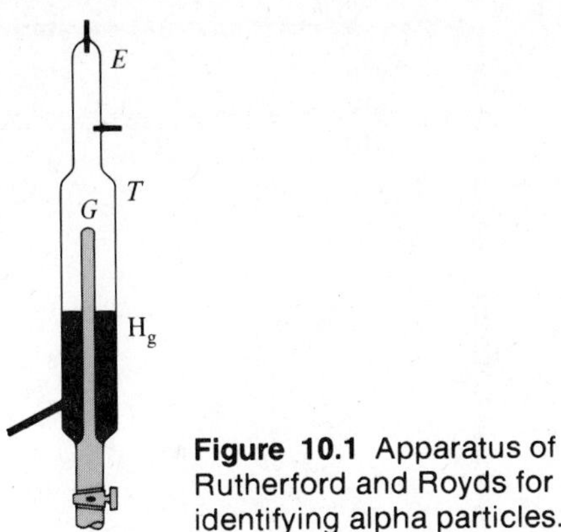

Figure 10.1 Apparatus of Rutherford and Royds for identifying alpha particles.

penetrating rays alpha (α) rays and the more penetrating ones beta (β) rays. In 1899, several investigators found that the beta component of the radiation could be deflected by a magnetic field, and that it had about the same charge-to-mass ratio (e/m_e) as the cathode corpuscles that had been discovered by Thomson just two years earlier. We now know that beta rays are electrons.†

Madame Curie deduced from their absorption properties that alpha rays were material particles and, in 1903, Rutherford succeeded in deflecting alpha particles with a magnetic field, where the deflection direction showed them to be positive. By causing alpha particles to discharge an electrometer and by simultaneously counting their scintillations, Rutherford determined that the alpha-particle charge was about twice the electronic charge. Since the charge was larger than that of electrons and the magnetic deflection much less, it was obvious that alpha particles are much more massive than electrons.

Conclusive proof that the alpha particles are helium nuclei was given by Rutherford and Royds in 1909. Their apparatus is shown in Fig. 10.1. The radioactive sample was placed inside a glass tube G which was so thin that alpha particles could pass through it. The tube G was inside an evacuated tube T which had a narrow end E with electrodes sealed into it. By raising the level of the mercury, any gas in T could be compressed into E. An electric discharge through this gas produced its spectrum and permitted the positive identification of the gas. With the tube G empty no helium was found, but with radioactive material in G, helium was found in the apparatus after two days. The alpha particles were trapped in the tube T, where they picked up electrons and became helium atoms.

† During the 1930s it was discovered that positrons are emitted by some radioactive isotopes, and the term "beta ray" or "beta particle" has now come to mean an electron or a positron of nuclear origin. Their symbols, β^- and β^+, respectively, are used only in connection with nuclear reactions. Because of earlier custom, "beta ray" still usually means an electron when the sign of the charge is not stated.

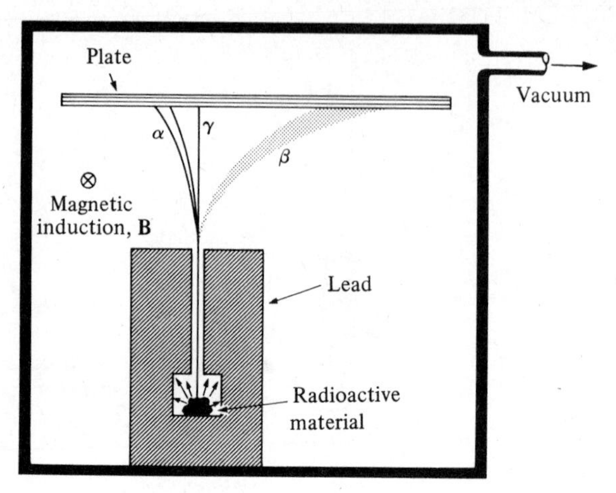

Figure 10.2 Magnetic separation of the radiations from radioactive materials.

The identification of alpha particles as helium nuclei makes it clear that radioactivity is a drastic process in which elements change in kind. If radium emits helium, it can no longer be radium. The *parent* or original substance becomes a new element called the *daughter* or *decay product* substance. Although the element lead cannot be changed into gold (the alchemists' dream of centuries before), the experiment by Rutherford and Royds leaves no doubt that one element may be transmuted into another element.

In 1900, Villard found a third kind of radiation from radioactive materials that is even more penetrating than either alpha or beta rays. These rays, called gamma (γ) rays, are not influenced by magnetic fields and therefore carry no charge. Their energy can be measured by measuring the energy of the photoelectrons they produce. They can be diffracted by crystals. Their wavelengths are found to range from about 0.5 to 0.005 Å. Crystal diffraction of these very short wavelengths is difficult and special techniques must be used, but we know that these rays are photons whose energy range overlaps that of x-rays and extends to several MeV. The term "gamma ray" is restricted to short-wavelength radiation from the nucleus, and "x-ray" is used only for similar radiation from the extranuclear part of the atom.

Most pure radioactive substances emit gamma rays accompanied by either alpha or beta rays, but since samples are seldom pure, one usually finds all three types of rays present. The three types of radiation can be characterized by the famous diagram of Fig. 10.2, which shows how the rays are deflected by a magnetic field.

These radiations from the nucleus are absorbed as they pass through matter. Gamma rays are absorbed exponentially. The linear absorption equation for x-radiation, Eq. (6.10), is valid for gamma rays, although the absorption coefficients for gamma rays are usually much less than they are for x-rays. The absorption of beta rays is more complex. It is approximately exponential up to a certain thickness of the absorber and then absorption becomes complete. The exponential relation is entirely fortuitous in this case. As beta particles are stopped in an absorber some of their energy is converted into bremsstrahlung. The absorption coefficients for beta rays are less well defined than for x- and gamma rays, but in one respect the absorption of beta rays is simpler than photon absorption. Because the mass absorption coefficient is nearly the same for all absorbing materials, the mass per unit area, m_a, required to produce a given beta absorption is independent of the material. The value of m_a for an absorber of either alpha or beta rays is often called the "thickness" of the material. The absorption of alpha particles will be discussed in Chapter 11.

10.5 GAS-FILLED DETECTORS

Before considering the laws of radioactive disintegration, we digress to describe the techniques used to observe the radiation from radioactive elements. The four basic types of detectors, which we shall describe, are (1) gas-filled, (2) scintillation, (3) visual, and (4) semiconductor.

The gas-filled detectors are of three general types: ionization chambers, proportional counters, and Geiger–Mueller counters. A typical gas-filled chamber is diagramed in Fig. 10.3. An outer cylinder encloses an inert gas and acts as the outer electrode. A thin wire that is insulated from the outer electrode is located at the center and coaxial with the outer cylinder. Incoming radiation ionizes some of the atoms of the gas freeing electrons to move toward the positively charged electrode. The positive ions drift slowly toward the outer electrode. This movement of the electrons and ions causes a current in the outer circuit and a voltage drop across the resistor R. The type of detector (ionization, proportional, or Geiger) is determined by the gas, gas pressure, and voltage. The number of ions collected versus the voltage between the electrodes is shown in Fig. 10.4. In region A, the applied voltage is small and the ions created recombine before they reach the electrodes. In region B, the region for the ionization counter, most ions created are collected by the electrode

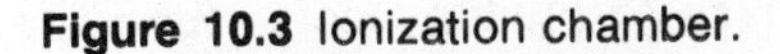
Figure 10.3 Ionization chamber.

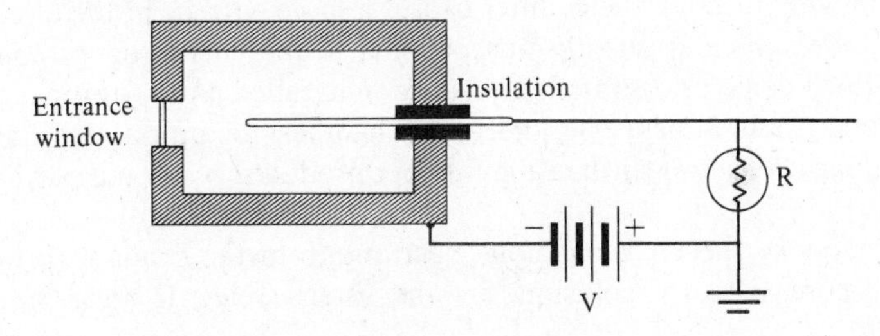

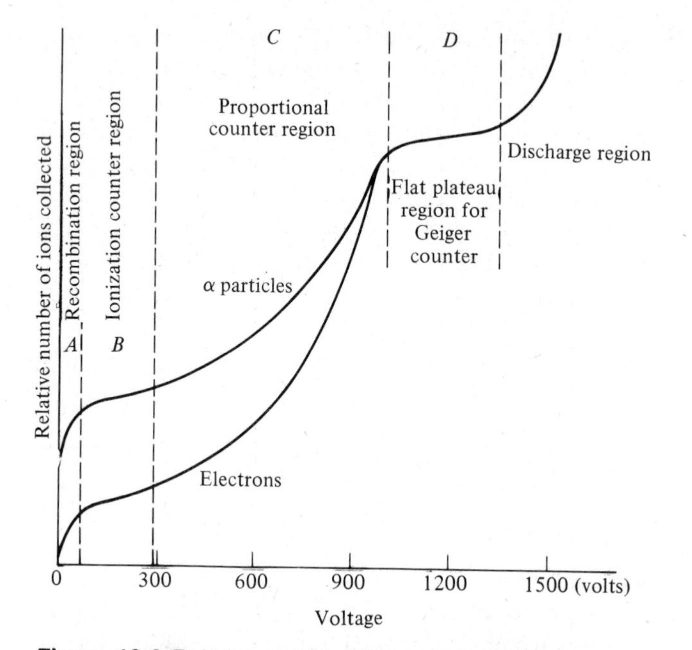

Figure 10.4 Relative number of ions collected versus voltage V for a gas-filled detector of incident radiation consisting of α particles or electrons. Note the almost flat Geiger-counter region, which is independent of the type of radiation.

From Atam P. Arya, *Elementary Modern Physics*, © 1974. Addison-Wesley, Reading, Mass. Reprinted with permission.

before recombination takes place. The ionization chamber is too slow to detect single particles, but the combined ionization effects of many nuclear particles causes a constant current and voltage drop across the resistor R. This current and resulting voltage drop is a direct measure of the intensity of the ionizing radiation. Note that in this region there is a difference between the number of ions due to alpha particles and those due to beta particles.

In region C, the voltage is large enough so that when an ion pair is created, the electrons rapidly gain enough energy to produce more ions. Therefore, a single particle moving through the counter causes a large current in the circuit and a large voltage pulse, which is directly proportional to the energy of the ionizing particle. The gas-filled detector operated in this region is called a proportional counter. In the proportional counter, as in the ionization chamber, the number of ions produced by an alpha particle is greater than the number produced by a beta particle of the same energy.

In region D, the Geiger region, electrons from the original (primary) ion pair produce more ions by collision. In the intense electric field near the central electrode, the secondary ions produce still more ions, in a geometric progression. Thus a single electron may initiate a cascade of a million or more electrons. When this avalanche strikes the central electrode, the drop in potential is easily detected electronically. Meanwhile, the relatively massive positive ions migrate toward the outer cylinder, where they pick up electrons and become excited atoms or molecules. If these atoms emit photons that can liberate photoelectrons from the outer cylinder, these photoelectrons may initiate other avalanches that tend to keep the tube in continuous discharge. But if the gas of the tube is an organic gas that dissociates easily, the excitation energy may cause the molecule to come apart rather than to emit a photon. In the first case, the electric discharge must be stopped by reducing the potential across the tube to nearly zero. This available electrode potential difference does become very small each time a pulse of current resulting from ionization flows around the circuit containing the high resistance R shown in the figure. In the second case, the life of the tube is reduced as the chemical composition of the organic gas is changed. Thus in either type of Geiger counter the discharge must be "quenched" between counts. Since a finite time is required for quenching and for starting the voltage rise across the electrodes, there is always a dead time between possible counts. This is of the order of 10^{-4} s. Geiger counters readily respond to low-intensity radiations which have enough energy to enter their sensitive region. If, however, the intensity of the radiation is too high, a Geiger counter "freezes" or "jams" because the interval between successive primary ionizing events is less than the dead time.

Modern G-M counters can have no open window, since the gas is not air and the pressure is below atmospheric. High-energy particles can penetrate into the sensitive region within the metal cylinder from any direction, but lower energy particles enter more readily through the end. Since particles of very low energy are stopped by the walls, the absorption of the wall is sometimes reduced by making an extremely thin window at one end. These windows are almost invisible and are very fragile. Their "thickness" is usually given in terms of their mass per unit area. Values of the order of 3 mg/cm^2 are common. Such windows are so thin that they easily transmit short-wavelength ultraviolet light. These photons have enough energy to "trigger" the tube by causing photoemission from the electrodes, or they even ionize the gas in a few cases. A match flame furnishes ample ultraviolet light to operate a thin-window, clear-glass tube.

10.6 SCINTILLATION DETECTORS

We have discussed fluorescence, which is sensitive to radiations with energies equal to or greater than those of visible photons. Fluorescence gives visible evidence of the radiation, but the radiation must be intense for this to occur.

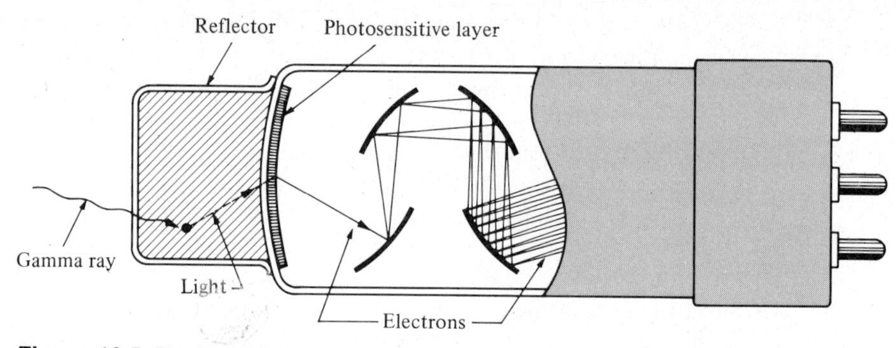

Figure 10.5 Photomultiplier tube.

We have discussed ionization chambers, which are used to measure the ionization currents radiation produces. This technique is quite sensitive in the same energy range as fluorescence, and it is quantitative.

We now turn to *scintillation*, which is really fluorescence examined closely. Because it can detect individual particles, this technique is extremely sensitive. If the scintillations are to be visually observed, the particles of radiation must have high energy, that is, they must be capable of producing a large number of visible photons when they strike the fluorescent material. However, very sensitive scintillation techniques developed during the past few years have eliminated the necessity for visual observation. Instead, the light is allowed to fall on a very sensitive photoelectric cell called a *photomultiplier tube* (Fig. 10.5), in which a single photoelectron is accelerated until it can knock "secondary" electrons from an anode. This anode is the cathode relative to another anode, where more secondary electrons are liberated. This process continues through several stages until the burst of electrons becomes great enough to constitute a disturbance that can be amplified by an electronic amplifier. Pulses from this amplifier are then electronically counted. By this technique, individual radiation particles of rather low energy can be counted.

A remarkable property of a scintillation detector and photomultiplier tube combination is that the electric pulses produced are usually proportional to the energy of the incident gamma rays. This permits not only gamma ray detection but a measurement of their energy. A sophisticated electronic device (really a specialized computer) called a *multichannel analyzer* can accept the electrical pulses, sort them into size intervals, and keep track of the number of pulses of each size entering the instrument. The resulting data constitute an energy spectrum of the gamma rays entering the scintillation detector. (Energy spectra are discussed in Section 10.9.) Although this method of measuring gamma-ray energies is relative — the instrument must be calibrated with a source of gamma rays of known energy — and although the precision is low compared to diffraction from crystals, the method is quick and has many applications.

10.7

TRACK DETECTORS

The counters we have been discussing use a sensitive volume of gas, liquid, or solid with well-defined dimensions to detect the presence of a particle. Now we shall discuss four detectors that not only record the presence of a particle but indicate the path that the particle traveled as well. We call these detectors "track" detectors. The four we shall discuss are the Wilson cloud chamber, the bubble chamber, photographic emulsions, and the spark chamber.

Certainly the most dramatic radiation detectors are cloud and bubble chambers. The principle of their operation is basically the same, although cloud chambers contain a supercooled vapor, while a bubble chamber contains a superheated liquid.

A supercooled vapor is most likely to condense on some discontinuity, and once condensation begins it continues readily by condensing onto droplets already formed. Supercooled water vapor may remain in the vapor phase and not condense, but in dusty and smoky regions there are discontinuities on which condensation can begin and therefore fog or smog form readily.

The cloud chamber illustrates the fact that charged particles are discontinuities on which condensation can be initiated. The schematic diagram of a Wilson cloud chamber is shown in Fig. 10.6. It consists of a gastight chamber having a large glass window, so that its volume can be illuminated and the events within seen or photographed. One wall of the chamber is a movable piston. The chamber contains a saturated vapor from an excess amount of some liquid, usually a mixture of alcohol and water. If the volume of the chamber is suddenly increased by moving the piston, the adiabatic expansion causes cooling that renders the vapor supersaturated, unstable, and likely to condense. If there are ions within the chamber, condensation occurs preferentially on them.

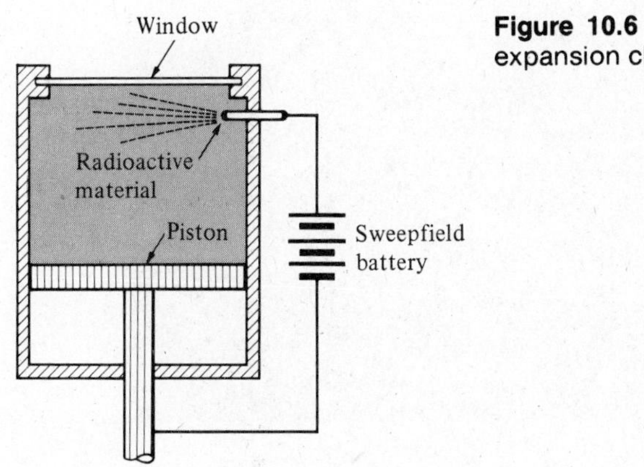

Figure 10.6 A simple expansion cloud chamber.

Since we have no interest in stray ions that may happen to be about, the cloud chamber is provided with electrodes that permit maintaining a weak electric "sweep field."

If a high-energy particle enters the chamber just before the piston expansion makes the chamber sensitive, the ions will not have had time to be swept away (the sweep field is sometimes cut off just before the expansion), and the droplets that form make the location of the ions visible. The droplets tend to move because of gravity and gas turbulence, and since they also tend to evaporate as a new equilibrium condition is established, the "picture" soon spoils. If the piston is then returned to its original position for about a minute the chamber can again be expanded.

If a high-energy electron passes through this chamber, the picture formed (Fig. 10.7) is a thin, beady line or "track" that shows very beautifully where the electron went. The droplet on each ion in its wake becomes visible. If a magnetic field is established in the space of the cloud chamber, the velocities of the electrons can be

Figure 10.7 Beady track produced by a high-energy electron. The broader tracks are due to low-energy photoelectrons.

Courtesy of C. T. R. Wilson and the Royal Society, London.

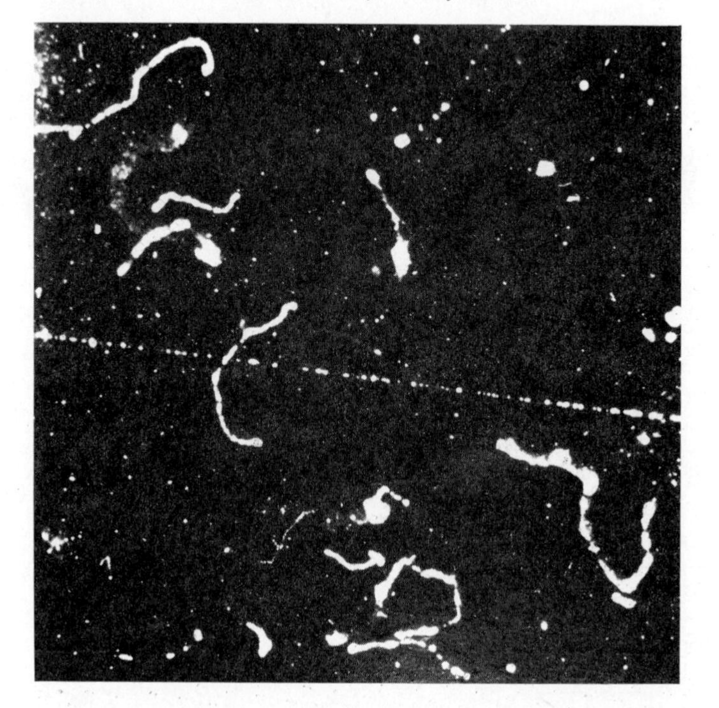

Figure 10.8 Alpha-particle tracks. Note that there are two groups, which differ in range.
(Courtesy of P. M. S. Blackett and D. S. Lees, Imperial College of Science and Technology, London.)

measured. Cloud-chamber photographs, such as Figs. 5.9 and 5.10, which accompanied the discussion of pair production, permit the measurement of the radius of curvature of the path of a charged particle, and since m_e, e, and B are known, the velocity can be computed from Eq. (2.5).

We first considered alpha particles in the Rutherford scattering experiment. Figure 10.8 shows that these doubly ionized helium nuclei produce very dense cloud-chamber tracks that resemble the vapor trails of high-flying airplanes. Alpha trails are easily distinguished from those made by electrons. The former are more dense, primarily because alpha particles carry twice as much charge as do electrons. But since alpha particles are much more massive than electrons, they are very difficult to deviate in a magnetic field. It turns out, however, that alpha-particle energies are a function of the lengths of their tracks, or ranges. This energy-measuring stick has been calibrated in the following ingenious way. If the relative proportions of vapor and gas in the chamber are known, the average ionization energy of the chamber atoms can be computed. If there is a momentary lapse between the chamber expansion and the photographic exposure, the track becomes diffuse and it is then possible to count the individual drops. Determining the number of ions produced by the alpha particle permits calculating the energy loss along its path.

Alpha particles produce an average of about 50,000 ion pairs per centimeter in air at atmospheric pressure, while beta particles produce only about 50 ion pairs per centimeter. Since both kinds of particles lose energy by giving up about 34 eV per ion pair produced, alpha tracks are short and thick, and beta tracks are long and thin. If the two kinds of particles have the same initial energies, each produces about the same total number of ions. The slower the particle moves, the more ions it produces per unit length of path. (This will be discussed in Section 11.1.) Thus tracks are more dense near their ends, and the range-energy relationship is not linear. Only charged particles produce cloud-chamber tracks, although photoelectrons ejected by photons do leave short, feathery trails that give some idea of the photon path. To the practiced eye, cloud-chamber tracks due to different particles are as dissimilar as the tracks that different animals leave in the snow.

Cloud-chamber studies have been exceedingly revealing. These chambers are much more than radiation detectors. They almost enable us to see atomic processes, and sometimes the analysis of a single picture has led to a basic discovery.

Extremely high-energy particles, such as those found in cosmic rays or produced by the most modern nuclear accelerators, produce cloud-chamber tracks that are so long the chamber shows only a small fraction of the event. Such events can be studied better with bubble chambers, where the instrument is filled with a dense liquid instead of a gas.

Consider for a moment a soft drink. Before it is opened, the carbon dioxide over the drink and that dissolved in it are in equilibrium. When the bottle is opened, the pressure is reduced and the solution of gas in the liquid is supersaturated. Bubbling results from the unstable condition and we have effervescence. That the bubbles form most readily on discontinuities is best demonstrated by pouring the drink into a glass and noting that often there are streams of bubbles rising from some speck inside the glass. These bubbles are formed because the same discontinuity is a bubble nucleus over and over again. Adding sugar, salt, cracked ice, or ice cream provides many discontinuities, so that most of the carbon dioxide effervesces almost at once. Similarly, when ions are produced in a bubble chamber they act as nuclei for the formation of small bubbles.

In actual bubble chambers, a liquid such as isopentane or liquid hydrogen is maintained at a temperature above its normal boiling point but is prevented from boiling by the application of pressure. The chamber is made sensitive by suddenly reducing the pressure. Since boiling starts preferentially on the ions in the liquid, the bubbles formed along the path of an ionizing particle are visible, much as are the droplets in a cloud chamber. Such a trail of bubbles is shown in Fig. 10.9. A bubble track can be observed only during the brief period (a few milliseconds) before general boiling begins throughout the liquid.

It is difficult to study penetrating radiations with Wilson cloud chambers because the absorption of the gas within the chamber is so small. We have already pointed out that bubble chambers containing dense material are better in this respect, but both these devices are sensitive only intermittently and neither is portable. These considerations led to the development of special photographic emulsions. Ordinary photographic materials are unsuitable because they contain too small a fraction of sensitive silver halides and because they are too thin. C. F. Powell

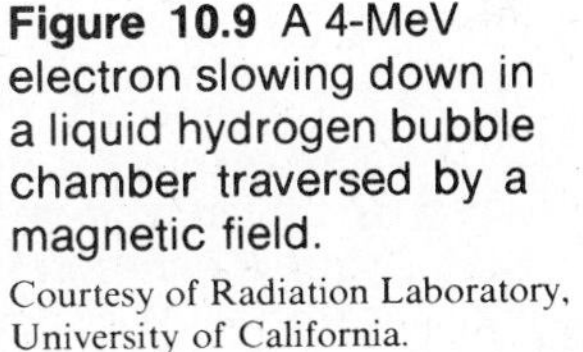

Figure 10.9 A 4-MeV electron slowing down in a liquid hydrogen bubble chamber traversed by a magnetic field.
Courtesy of Radiation Laboratory, University of California.

and others developed satisfactory emulsions that contain about 80% silver bromide by dry weight and are about 1 mm thick. These can be stacked to make a sensitive region of practically any size. Such blocks of emulsion are continuously sensitive and are portable enough to be carried by balloons and other high-altitude craft. It requires elaborate techniques to develop, piece together, and read nuclear emulsions, but the resultant "tracks" closely resemble those formed in cloud and in bubble chambers. The important characteristics of a track are its density, range, direction, and deviation in direction, and these characteristics often permit the positive identification of the particle causing the track. The emulsion record of a nuclear disintegration is shown in Fig. 10.10.

The spark chamber is a detector that can be easily activated at the appropriate time to investigate selected events. A typical spark chamber consists of a stack of metal plates separated by a couple of millimeters and surrounded by an inert gas. When a charged particle passes through the region it leaves a trail of ions and thus a conducting path between the plates. When a high voltage is applied to alternate plates, arcing will occur along the path giving a visible indication of where the particle has been. Figure 10.11 is a diagram of a typical spark chamber. The spark chamber is a relatively new detector that has become very popular recently.

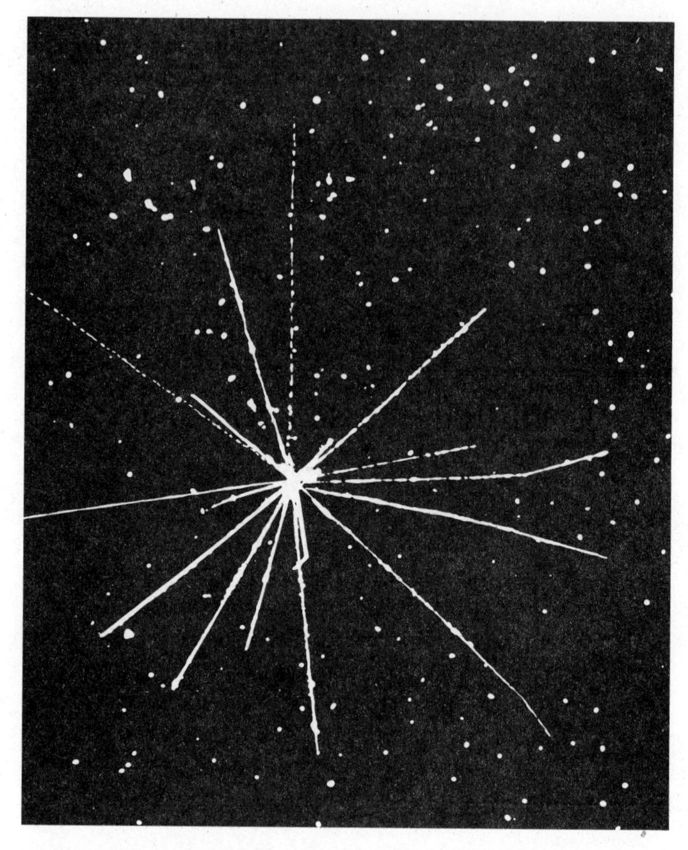

Figure 10.10 Photomicrograph, enlarged 100 times, of photographic emulsion film exposed to 2 GeV nuclear particles in the cosmotron. The incoming particle, presumably a neutron, which is invisible, hit the nucleus of an atom of the emulsion and exploded it into 17 visible particles that formed tracks in a star-shaped pattern. In general, the broad tracks are made by slow particles (protons) and the narrow tracks by fast particles. The spots in the background are grains in the emulsion.
Courtesy of Brookhaven National Laboratory.

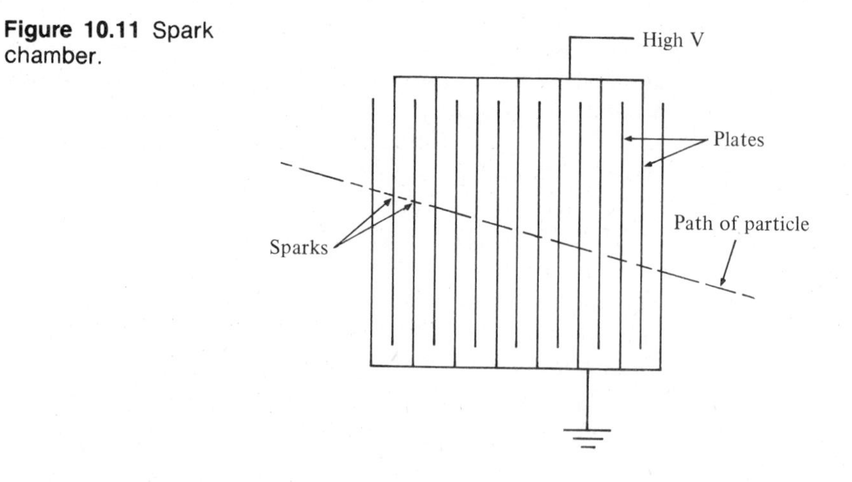

Figure 10.11 Spark chamber.

However, it still has two significant disadvantages when compared to the bubble chamber. In the spark chamber, the incoming particle can interact with many different molecules, whereas in a hydrogen bubble chamber there is only a possibility of an interaction between the particles being studied and protons (or electrons). A second disadvantage is that the trail created in the spark chamber by a particle is much larger and the uncertainty in the location of the particle is greater.

10.8 SEMICONDUCTOR DETECTORS

In Chapter 9, we indicated that the energy to create conduction electrons and holes could be supplied from thermal energy. Another way to supply this energy is from high-energy radiation such as γ-rays or x-rays. A photoelectron produced within the solid becomes a conduction electron and a conduction hole is left in the source atom. If the semiconductor is in an electric field, the electrons and holes can migrate, which constitutes a measurable current. This behavior of the semiconductor material is in close analogy with the gas in an ionization chamber. But since the concentration of atoms in a solid semiconductor radiation detector is much greater than in a gas, the semiconductor detector can be much smaller than an ionization detector.

10.9 ENERGIES OF THE RADIATIONS. NUCLEAR SPECTRA

One gram of radium gives off nearly two gram-calories of heat per second. This may not seem like much energy, since the oxidation of a gram of carbon to carbon dioxide liberates about 8000 gram-calories. But whereas the carbon might be oxidized in one second and be consumed, the gram of radium will continue to liberate two calories per second for years and the rate will only be reduced to one calorie per second in 1600 years. If we allow infinite time for the gram of radium to disintegrate, more than 6×10^8 calories would be liberated, and this calculation does not take into account the energy liberated by the daughter products of the radium. If radium were cheap and plentiful, it would be a useful source of energy.

Since the energies of alpha, beta, and gamma rays from radioactive substances can be measured, we speak of the "spectra" of various radioactive substances. The nucleus is not nearly so well understood as the atom's electronic structure. There is no nice theory of the nucleus to compare with the Bohr analysis of the hydrogen atom. But the evidence from nuclear spectroscopy indicates that nuclei do have energy levels. Thus Pa^{231} (protactinium) emits alpha particles with nine distinct energies and also thirteen different gamma-ray wavelengths.

The existence of energy levels makes it clear that nuclear phenomena are quantum-mechanical. The spectra of beta particles, however, presented an interesting problem. Although a given substance emits beta particles with a definite upper limit of energy, a continuous range of energies below that limit is found. The reason for an upper limit is evident from Fig. 10.12(a), which shows the energy-level diagram for a typical beta-ray emitter. In this case, a nucleus having an atomic number Z ejects an electron and thus changes to an excited state of a new nucleus, $Z + 1$. This then goes to the stable state by emitting two gamma-ray photons in cascade. The single energy transition of the electron accounts for the maximum beta-particle energy, but fails to account for the continuous range of smaller beta energies found when a large number of atoms of a substance are observed. It appeared, at first, that the law of conservation of energy was violated. Of all the generalizations of physics, conservation of energy is the most basic. We would like to think that this principle holds exactly and without exception.

The difficulty was resolved by supposing that there is a new particle, called a *neutrino* (ν), which carries away the energy difference (refer to Fig. 10.12(b)). Conservation of energy and of momentum required that this new particle have much less mass than the electron and that it be uncharged. This particle, proposed by Pauli in 1931, could carry away the energy difference undetected because of its small mass and zero charge. In addition to preserving the exactness of conservation of energy, the neutrino fitted into other theoretical considerations in the theory of beta decay developed by Fermi in 1934. The existence of this new particle was generally accepted long before it was experimentally observed in 1956 by Reines and Cowan.

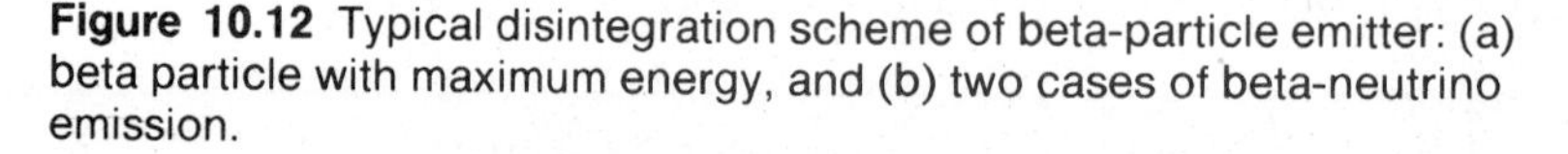
Figure 10.12 Typical disintegration scheme of beta-particle emitter: (a) beta particle with maximum energy, and (b) two cases of beta-neutrino emission.

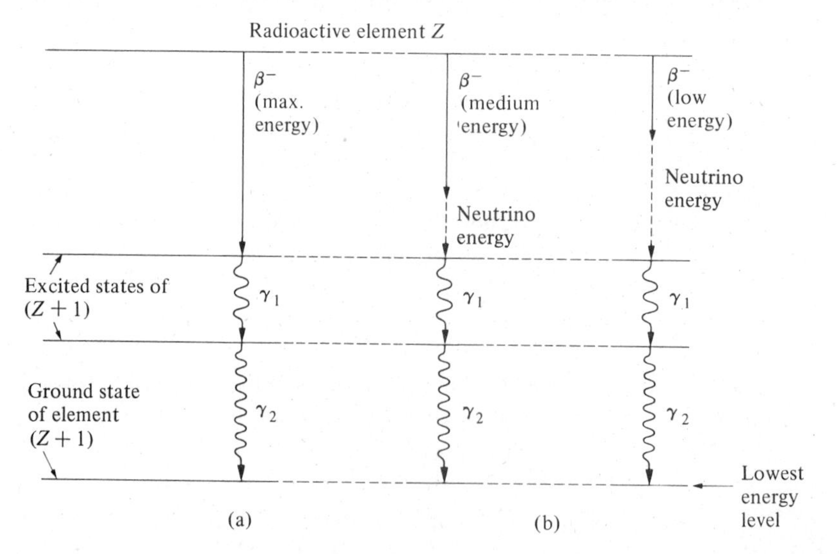

Although nuclear energy states, unlike electronic energy states, cannot be computed theoretically, we apply the electronic language to the nucleus and speak of excited states, ground states, etc.

10.10 LAW OF RADIOACTIVE DISINTEGRATION

Every radioactive disintegration involves the emission of either an electron or a helium nucleus from the nucleus of the disintegrating atom. This leaves the original nucleus changed, so that the number of atoms of the original kind is reduced. Furthermore, scintillation observations show that these disintegrations occur in a random manner that can be discussed only statistically. The idea of treating the disintegrations statistically may have been originally a matter of convenience, but we have seen that wave mechanics makes the statistical analysis of such quantum phenomena a necessity.

The discovery of radioactivity antedated Planck's quantum idea, Einstein's quantum interpretation of the photoelectric effect, Bohr theory, and wave mechanics. The unpredictability of individual radioactive disintegrations was really the *first* experimental indication that the mechanistic determinism in the physical sciences of the nineteenth century was due for revision. Thus radioactivity had two shocking aspects: It showed that the elements were not inviolate but could change from one kind to another, and that radioactivity was not subject to causal analysis. This latter aspect was assimilated as an integral and useful part of physics, particularly after the advent of wave mechanics in 1925.

To present the quantitative description of radioactivity, let us begin at time $t = 0$ with a large number of radioactive atoms N_0. Experimentation shows that the number of disintegrations per unit time $(-dN/dt)$, or the *activity*, is proportional to the number of atoms N of the original kind present. The constant of proportionality λ is called the *disintegration* or *decay constant*. Therefore we have

$$\frac{-dN}{dt} = \lambda N. \tag{10.1}$$

The quantity dN must be an integer, but if it is a relatively small integer compared with N, then N will vary in an approximately continuous way and we may treat dN as a mathematical differential. Upon integrating and using the initial condition, we have for the *number of atoms of the original kind still present* at time t

$$N = N_0 e^{-\lambda t}. \tag{10.2}$$

The statistical nature of the exponential law of radioactive decay, Eq. (10.2), is evident from the following derivation, in which the law is obtained without any special hypothesis about the structure of the radioactive atoms or about the mechanism of disintegration. Let us assume that the disintegration of an atom depends on the laws of chance, and that the probability P for an atom to disintegrate in a short time interval Δt is the same for all the atoms of the same kind and that it is independent of the past history or age of the atom. (Note that assuming that a quantity such as P is independent of age would not hold for life-experience tables.) Thus the probability of disintegration depends only on the time interval, and, for very short intervals, is proportional to the interval. Therefore $P = \lambda\,\Delta t$, where λ, the constant of proportionality, is the decay constant. The probability Q_1 that a given atom will *not* disintegrate during the interval Δt is $Q_1 = 1 - P = 1 - \lambda\,\Delta t$. The probability of the atom's not disintegrating in time $2\,\Delta t$, assuming that P is independent of the age of the atom, is

$$Q_2 = (1 - P)(1 - P) = (1 - \lambda\,\Delta t)^2.$$

Therefore, in general, the probability of the atom's surviving n intervals is $Q_n = (1 - \lambda\,\Delta t)^n$. Considering a finite time t made up of a large number n of intervals Δt, we find that $t = n\,\Delta t$ or $\Delta t = t/n$. Therefore the probability of the atom's surviving or remaining unchanged for a time t is $Q_n = (1 - \lambda t/n)^n$. But the limit of this quantity as n becomes very large is N/N_0, which is the ratio of the number of atoms that remain unchanged at the end of time t to the total number of atoms originally present. To find this limit, we first expand the expression for Q_n by the binomial theorem. This gives

$$Q_n = \left(1 - \frac{\lambda t}{n}\right)^n = 1 - n\frac{\lambda t}{n} + \frac{n(n-1)}{2!}\frac{\lambda^2 t^2}{n^2} - \frac{n(n-1)(n-2)}{3!}\frac{\lambda^3 t^3}{n^3} + \cdots$$

$$= 1 - \lambda t + \left(1 - \frac{1}{n}\right)\frac{\lambda^2 t^2}{2!} - \left(1 - \frac{1}{n}\right)\left(1 - \frac{2}{n}\right)\frac{\lambda^3 t^3}{3!} + \cdots. \tag{10.3}$$

If n is very large, the preceding relation reduces to

$$Q_n = \frac{N}{N_0} = 1 - \lambda t + \frac{\lambda^2 t^2}{2!} - \frac{\lambda^3 t^3}{3!} + \cdots = e^{-\lambda t} \tag{10.4}$$

or

$$N = N_0 e^{-\lambda t}.$$

This decay law is thus the result of a large number of events subject to the laws of chance. Therefore the activity λN is an average value. The actual values during successive short time intervals fluctuate around this average. These variations must be taken into account when measurements are made with weak radioactive sources.

Each nuclear disintegration makes an equal contribution to the beam of radiation from a radioactive material. Thus the intensity I of the beam is directly proportional to $-dN/dt$. If we call I_0 the intensity when $t = 0$, then Eqs. (10.1) and (10.2) lead to

$$I = I_0 e^{-\lambda t}. \tag{10.5}$$

If we plot ln N or ln I against t, both Eqs. (10.2) and (10.5) become straight lines whose slopes are $-\lambda$. This sometimes is a convenient way of obtaining λ from experimental data. These equations are identical in form to that which describes the intensity of monochromatic x-rays as a function of absorber thickness. They indicate that the number of atoms and the intensity of radiation require infinite time to disappear completely. (It must be remembered that differential calculus breaks down and dN becomes a step function when N is small.)

Although we cannot get a meaningful answer to the question "How long will a given radioactive sample last?" we can compute how long it will be before the sample is reduced to some fraction of its initial amount. Just as we computed the thickness of absorber that will reduce the intensity of an x-ray beam to one-half its initial value, we can compute the *half-life*, $t_{1/2}$ or T, of a radioactive sample. If we let $N = N_0/2$ in Eq. (10.2), then $t = T$ and we have

$$T = \frac{1}{\lambda}\ln 2 = \frac{0.693}{\lambda}, \tag{10.6}$$

which is similar to Eq. (6.12), the half-value layer for x-rays. Figure 10.13 is a typical graph of radioactive decay as a function of time in half-lives.

Figure 10.13 Typical radioactive decay curve.

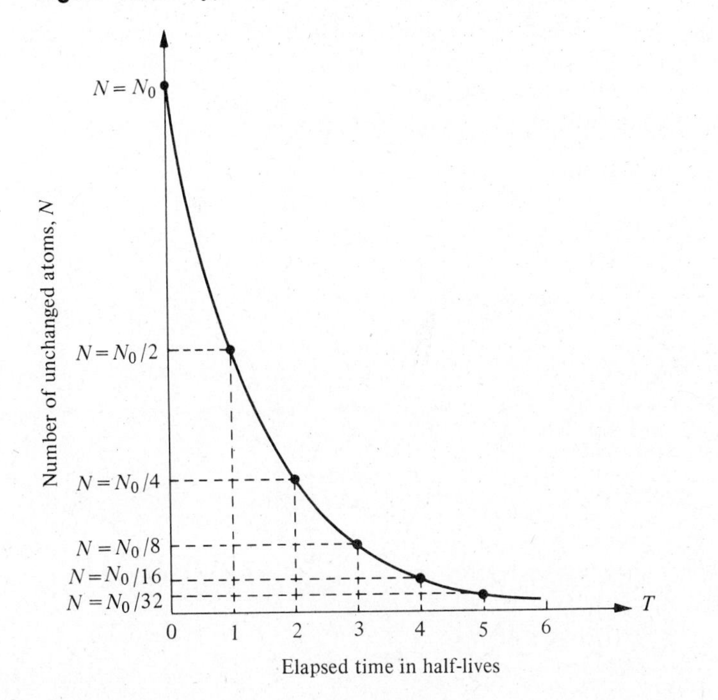

Before we can use any of these equations, the disintegration (decay) constant λ must be known. It depends on the radioactive material and varies widely from material to material, but this constant is completely independent of the environment, whether it is chemical combination, pressure, temperature, or whatever.

If λ is very small, the half-life is very large (1620 years for radium) and the reduction in N during an experiment is not significant. By measuring the disintegration rate dN/dt (with a scintillation counter, for example) from a sample of N atoms, λ can be computed from Eq. (10.1).

If λ is large enough so that the decay of the sample during an experiment is significant, then the reduction in intensity may be observed as a function of time and fitted to Eq. (10.5). A simple way to do this is to measure the time it takes for the intensity to be reduced to one half (T) and solve for λ by means of Eq. (10.6).

If λ cannot be measured by either of the two methods described, then the decay constant must be determined from radioactive equilibrium, which will be discussed in Section 10.14.

We can make another calculation that is conceptually important to the statistical view we are taking. Let us find the *mean* or *average life*, $\bar{t}$, of a radioactive atom. Some atoms survive much longer than others, so what we seek is the numerical average of the ages of the atoms as the number of atoms decreases from N_0 to 0. The expression for determining this can be found by considering the radioactive decay curve of Fig. 10.14. Let N be the number of atoms that have survived for a time t, and $N - dN$ those still existing at the time $t + dt$; then dN is the number of atoms that disintegrated during the time dt. Therefore the combined ages of the atoms in this group at the time of disintegration is $t\,dN$. The average life of an atom will be the sum of the combined ages of all the age groups of atoms from N_0 to 0 divided by the total number of atoms. Stated in mathematical terms, this is

$$\bar{t} = \frac{\int_{N_0}^{0} t\,dN}{\int_{N_0}^{0} dN}. \tag{10.7}$$

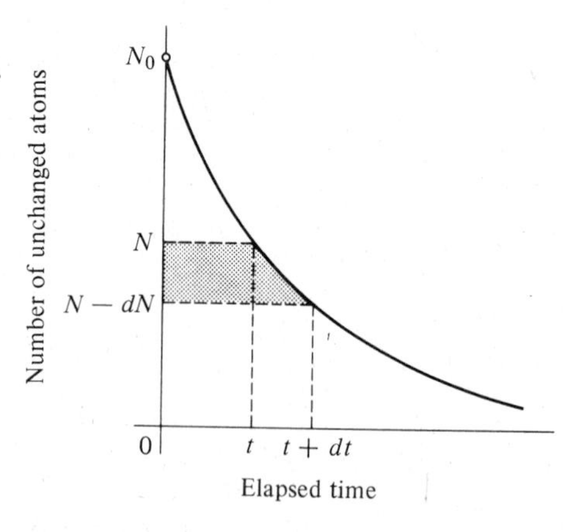

Figure 10.14 Typical radioactive decay curve.

The integrals in this equation may be evaluated in the same way as in the derivation of Eq. (1.37). The result is

$$\bar{t} = \frac{1}{\lambda}. \tag{10.8}$$

Thus $\bar{t} = 1/\lambda$ is the mean life or life expectancy of an individual radioactive atom.

EXAMPLE (a) If a radioactive material initially contains 3.00 mg of ^{234}U, how much will remain unchanged after 62,000 y? (b) What will be its ^{234}U activity at the end of that time? ($T = 2.48 \times 10^5$ y, $\lambda = 8.88 \times 10^{-14}$ s^{-1}.)

Solution (a) From Eq. (10.2), we have

$$N = N_0 e^{-\lambda t}.$$

The N's are computed from the Avogadro constant and the mass number (strictly isotopic mass) of the material. In this example, it will be convenient to express λ in terms of T. Let m be the mass of uranium remaining; then the quantities to be substituted for the terms in the equation are

$$N_0 = 3.00 \times 10^{-3}\text{ g} \times \frac{6.02 \times 10^{23}\text{ atoms}}{234\text{ g}},$$

$$N = m \times \frac{6.02 \times 10^{23}\text{ atoms}}{234\text{ g}},$$

and

$$\lambda t = \frac{0.693}{T} t = \frac{0.693}{2.48 \times 10^5\text{ y}} \times 6.2 \times 10^4\text{ y} = 0.173.$$

Substituting these quantities in the radioactive decay equation, we get

$$m \times \frac{6.02 \times 10^{23}\text{ atoms}}{234\text{ g}} = 3.00 \times 10^{-3}\text{ g} \times \frac{6.02 \times 10^{23}\text{ atoms}}{234\text{ g}} \times e^{-0.173},$$

or

$$m = 3.00 \times 10^{-3} \times e^{-0.173}\text{ g}.$$

Therefore

$$\ln \frac{3.00 \times 10^{-3}}{m} = 0.173,$$

or

$$m = 2.52 \times 10^{-3}\text{ g}.$$

(b) From Eq. (10.1) we have

$$\frac{dN}{dt} = \lambda N$$

$$= \frac{8.88 \times 10^{-14}}{1\text{ s}} \times 2.52 \times 10^{-3}\text{ g} \times \frac{6.02 \times 10^{23}\text{ atoms}}{234\text{ g}}$$

$$= 5.76 \times 10^5\text{ dis/s.}$$ ▬

Table 10.1 The uranium series

Radioactive Species — Historic Names	Nuclide	Type of Disintegration	Half-life T	Disintegration Constant λ, s^{-1}	Principal Particle Energy, MeV
Uranium I (UI)	$^{238}_{92}U$	α	4.51×10^9 y	4.87×10^{-18}	α, 4.195; γ, 0.048
Uranium X_1 (UX_1)	$^{234}_{90}Th$	β	24.1 d	3.33×10^{-7}	β, 0.191; γ, 0.03–0.09
Uranium X_2 (UX_2)	$^{234}_{91}Pa$	β	1.17 m	9.79×10^{-3}	β, 2.29; γ, 0.23–1.8
⋮					
Uranium Z (UZ)	$^{234}_{91}Pa$	β	6.75 h	2.852×10^{-5}	β, 0.51; γ, 0.04–1.7
Uranium II (UII)	$^{234}_{92}U$	α	2.47×10^5 y	8.899×10^{-14}	α, 4.773; γ, 0.5–0.12
Ionium (Io)	$^{230}_{90}Th$	α	8.0×10^4 y	2.747×10^{-13}	α, 4.684; γ, 0.07–0.25
Radium (Ra)	$^{226}_{88}Ra$	α	1600 y	1.374×10^{-11}	α, 4.781; γ, 0.19
Ra emanation (Radon)	$^{222}_{86}Rn$	α	3.823 d	2.098×10^{-6}	α, 5.486; γ, 0.51
*Radium A (RaA)	$^{218}_{84}Po$	β, α	3.05 m	3.78×10^{-3}	α, 6.002; β...
Astatine-218 (0.02)	$^{218}_{85}At$	α	1.3 s	0.53	α, 6.697
Radium B (RaB) (99^+)	$^{214}_{82}Pb$	β	26.8 m	4.31×10^{-4}	β, 0.65; γ, 0.05–0.35
*Radium C (RaC)	$^{214}_{83}Bi$	β, α	19.7 m	5.86×10^{-4}	α, 5.616; β, 0.40–3.2 γ, 0.61–2.43
Radium C' (RaC′) (99^+)	$^{214}_{84}Po$	α	1.64×10^{-4} s	4.23×10^3	α, 7.687
Radium C'' (RaC″) (0.02)	$^{210}_{81}Tl$	β	1.30 m	8.75×10^{-3}	β, 1.9; γ, 0.10–2.43
Radium D (RaD)	$^{210}_{82}Pb$	β	21 y	1.047×10^{-9}	β, 0.015; γ, 0.05
*Radium E (RaE)	$^{210}_{83}Bi$	β, α	5.01 d	1.60×10^{-6}	α, 4.7; β, 1.16
Radium F (RaF) (99^+)	$^{210}_{84}Po$	α	138.4 d	5.80×10^{-8}	α, 5.305; γ, 0.80
Thallium-206 (2×10^{-4})	$^{206}_{81}Tl$	β	4.19 m	2.757×10^{-3}	β, 1.524
Radium G (RaG)	$^{206}_{82}Pb$	Stable			

Note: In Tables 10.1, 10.2, and 10.3 space limitations preclude listing more than the principal energy of the type of particle ejected and the energy range of the gamma-ray spectrum. A branch-point element is marked with an asterisk. The percentage proportion of each of the two daughter products of branch-point atoms is given in parentheses. Nuclear isomers are connected by a dotted line. The abbreviations under half-life are: s, second; m, minute; d, day; and y, year. (Data for these tables were obtained from the source listed in Appendix 5.)

10.11

RADIOACTIVE SERIES

We have stated that when a radioactive disintegration occurs with the emission of an alpha or beta particle, the original atom, called the parent, changes into something else, called the daughter. In 1903 Rutherford and Soddy proposed that the nature of the daughter could be inferred from the nature of the parent and the particle emitted. Since we have already established the concepts of the atomic nucleus, atomic number, and atomic mass numbers, we can state the Rutherford–Soddy rules for balancing nuclear reaction equations in modern terms. They are (1) *the total electric charge (atomic number) or algebraic sum of the charges before the disintegration must equal the total electric charge after the disintegration*, and (2) *the sum of the mass numbers of the initial particles must equal the sum of the mass numbers of the final particles*. Thus if uranium with atomic number 92 emits an alpha particle with atomic number 2, the daughter must have an atomic number 90. This element is thorium. Similarly, since uranium has a mass number of 238 and the alpha particle a mass number of 4, the thorium must have a mass number of 234. Both rules are

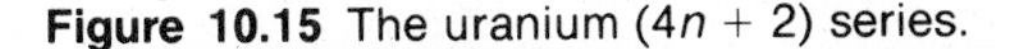

Figure 10.15 The uranium ($4n + 2$) series.

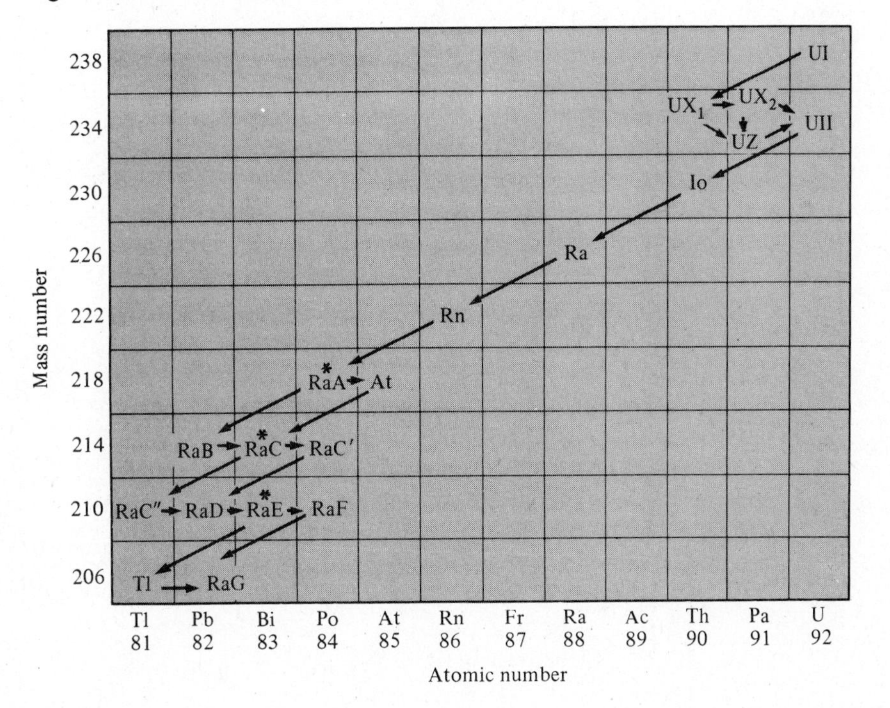

Table 10.2 The actinium series

Radioactive Species — Historic Names	Nuclide	Type of Disintegration	Half-life T	Disintegration Constant λ, s^{-1}	Principal Particle Energy, MeV
Actinouranium (AcU)	$^{235}_{92}U$	α	7.13×10^{8} y	3.08×10^{-17}	α, 4.18; γ, 0.19
Uranium Y (UY)	$^{231}_{90}Th$	β	25.5 h	8.12×10^{-6}	β, 0.299; γ, 0.08–0.31
Protoactinium (Pa)	$^{231}_{91}Pa$	α	3.25×10^{4} y	6.75×10^{-13}	α, 5.00; γ, 0.03–0.36
*Actinium (Ac)	$^{227}_{89}Ac$	β, α	21.6 y	1.08×10^{-9}	α, 4.950; β, 0.04
Radioactinium (RdAc) (1.2)	$^{227}_{90}Th$	α	18.5 d	4.40×10^{-7}	α, 5.976; γ, 0.05–0.34
Actinium K (AcK) (98.8)	$^{223}_{87}Fr$	β	22 m	5.25×10^{-4}	β, 1.15; γ, 0.05–0.08
Actinium X (AcX)	$^{223}_{88}Ra$	α	11.43 d	6.86×10^{-7}	α, 5.714; γ, 0.03–0.45
Ac emanation (Actinon)	$^{219}_{86}Rn$	α	4.0 s	0.174	α, 6.817; γ, 0.27–0.40
*Actinium A (AcA)	$^{215}_{84}Po$	β, α	1.78×10^{-3} s	3.86×10^{2}	α, 7.38
Astatine-215 (5×10^{-4})	$^{215}_{85}At$	α	10^{-4} s	7×10^{3}	α, 8.01
Actinium B (AcB) (99^{+})	$^{211}_{82}Pb$	β	36.1 m	3.20×10^{-4}	β, 1.36; γ, 0.06–1.1
*Actinium C (AcC)	$^{211}_{83}Bi$	β, α	2.15 m	5.26×10^{-3}	α, 6.622; γ, 0.35
Actinium C′ (AcC′) (0.3)	$^{211}_{84}Po$	α	0.52 s	1.33	α, 7.448; γ, 0.89–1.06
Actinium C″ (AcC″) (99^{+})	$^{207}_{81}Tl$	β	4.78 m	2.41×10^{-3}	β, 1.44; γ, 0.89
Actinium D (AcD)	$^{207}_{82}Pb$	Stable			

summarized in the equation

$$^{238}_{92}U \rightarrow ^{234}_{90}Th + ^{4}_{2}He.$$

If the parent is a beta emitter, the atomic number of the daughter must be one higher than that of the parent, since the beta particle is a negative electron. Furthermore, the electron is so light that the atomic mass numbers of the parent and daughter are the same. For example, ^{234}Th, daughter of ^{238}U, is radioactive and is a beta emitter. The daughter of ^{234}Th must have the atomic number 91 (protactinium, Pa) and the mass number 234, as shown in the equation

$$^{234}_{90}Th \rightarrow ^{234}_{91}Pa + _{-1}e^{0}.$$

The Rutherford–Soddy rules came before the establishment of the physical atomic masses or the discovery of isotopes by Thomson and Aston, described in Chapter 2. These rules predicted daughter atoms with atomic masses quite different from the then-accepted chemical atomic masses. Rutherford and Soddy had more faith in their rules (based on conservation of mass and charge) than they had in the chemical atomic masses. Their rules implied that the same chemical element could exist in forms having different masses and they predicted the discovery of isotopes, which we have already discussed.

We now know that there are three series of naturally radioactive elements that form a sequence of parent-daughter relationships: the uranium, actinium, and

Figure 10.16 The actinium ($4n + 3$) series.

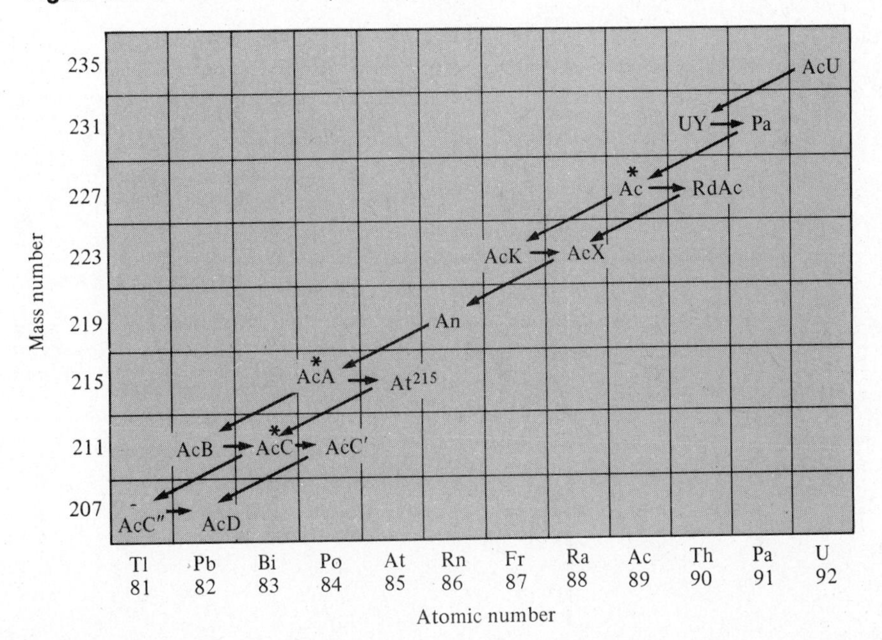

Table 10.3 The thorium series

Radioactive Species — Historic Names	Nuclide	Type of Disintegration	Half-life T	Disintegration Constant λ, s^{-1}	Principal Particle Energy, MeV
Thorium (Th)	$^{232}_{90}Th$	α	1.41×10^{10} y	1.56×10^{-18}	α, 3.994; γ, 0.06
Mesothorium 1 (MsTh 1)	$^{228}_{88}Ra$	β	5.77 y	3.81×10^{-9}	β, 0.048
Mesothorium 2 (MsTh 2)	$^{228}_{89}Ac$	β	6.13 h	3.14×10^{-5}	β, 1.18; γ, 0.06–1.64
Radiothorium (RdTh)	$^{228}_{90}Th$	α	1.913 y	1.149×10^{-8}	α, 5.424; γ, 0.08–0.21
Thorium X (ThX)	$^{224}_{88}Ra$	α	3.64 d	2.20×10^{-6}	α, 5.684; γ, 0.24
Th emanation (Thoron)	$^{220}_{86}Rn$	α	55 s	1.26×10^{-2}	α, 6.287; γ, 0.54
Thorium A (ThA)	$^{216}_{84}Po$	α	0.15 s	4.62	α, 6.777
Thorium B (ThB)	$^{212}_{82}Pb$	β	10.64 h	1.81×10^{-5}	β, 0.35; γ, 0.11–0.41
*Thorium C (ThC)	$^{212}_{83}Bi$	β, α	60.6 m	1.91×10^{-4}	α, 6.051; β, 2.25 γ, 0.04–2.2
Thorium C′ (ThC′) (66.3)	$^{212}_{84}Po$	α	3.04×10^{-7} s	2.28×10^{6}	α, 8.785
Thorium C″ (ThC″) (33.7)	$^{208}_{81}Tl$	β	3.10 m	2.52×10^{-3}	β, 1.792; γ, 0.23–2.61
Thorium D (ThD)	$^{208}_{82}Pb$	Stable			

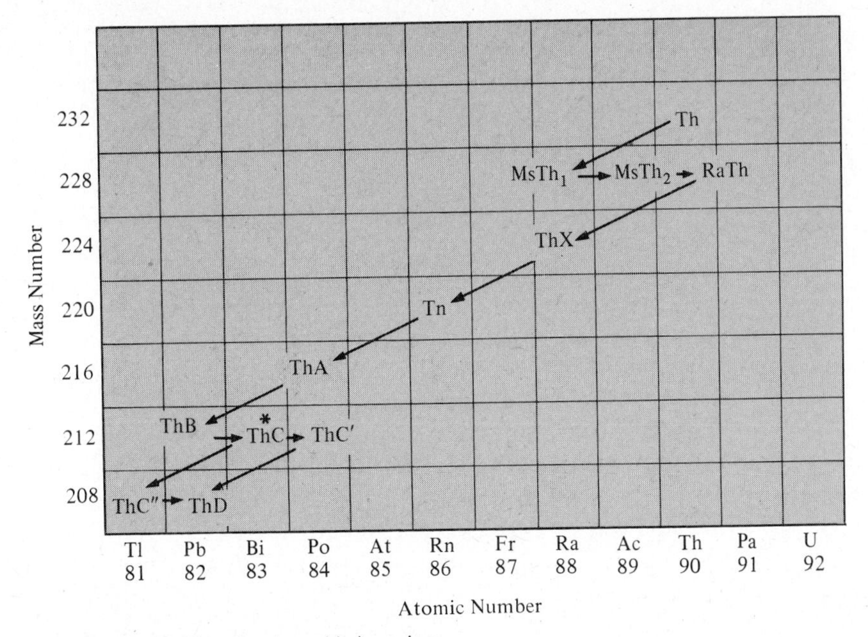

Figure 10.17 The thorium ($4n$) series.

thorium series.† They are shown graphically in Figs. 10.15, 10.16, and 10.17, respectively, and they also are tabulated in Tables 10.1, 10.2, and 10.3. Each series ends with a stable isotope of lead. Note that some elements can emit either an alpha or a beta particle. Although no one atom can go both ways, some atoms can go either of two ways and cause *branches* in the series. No matter which way the parent goes, the daughter goes the other way, so that even though the series branches, they always come together again. The two types of disintegration of a given branch-point element always occur in the same proportion. This proportion is independent of the amount of the element.

The time required for the nuclear transitions that produce gamma rays is usually of the order of 10^{-12} s. However, some *delayed* transitions with half-lives as long as several years have been observed. The delayed transitions occur in *nuclear isomers*, which are nuclides that have the same atomic number and same mass number, but exist for measurable times in different excited nuclear states with different energies and radioactive properties. In an *isomeric transition*, IT, an excited nucleus goes from a higher energy metastable state either to a lower energy metastable level or

† The uranium series is also called the $4n + 2$ series because the mass numbers of the atoms in it are given by this expression. The quantity n is an integer that decreases by unity in going from any radioelement to the next one below it. The actinium series can be represented by $4n + 3$, and that of thorium by $4n$. The value of n is not the same for the first element in each series.

directly to the ground state with the emission of gamma rays. The isotope of Pa (protactinium) in the ^{238}U series shows nuclear isomerism. Some of the nuclei of $^{234}_{91}Pa$ (UX_2) decay with a half-life of 1.18 m directly to the ground state of its isomer $^{234}_{91}Pa$ (UZ) with the emission of an energetic gamma ray. The remaining UX_2 nuclei disintegrate with the same half-life directly to $^{234}_{92}U$ (UII) by beta-particle and gamma-ray emission. On the other hand, all the nuclei of UZ decay with a half-life of 6.7 hours to $^{234}_{92}U$ accompanied by the emission of beta particles and rather low-energy gamma rays.

The tables of the radioactive series are redundant in the naming of the various isotopes. The names at the left antedate our full understanding of isotopes; they are names used in early discussion of the series. Even the more compact modern notation, like $^{238}_{92}U$, is redundant, since element 92 must be uranium.

The distribution of the natural radioactive materials is nonuniform over the surface of the earth. It has been estimated that, on the average over the earth's crust, each square mile has, within one foot of the surface, 8 tons of uranium and 12 tons of thorium.

To complete the account of radioactive series, we now consider one that has been obtained only from an artificially produced radioactive material. The first element in this series is ^{241}Pu and the stable end product is ^{209}Bi. This series is called the neptunium or $4n + 1$ series because the longest-lived element in it is ^{237}Np, which has a half-life of 2.14×10^6 years. If this element existed when the earth was formed about 4.4×10^9 years ago, the amount now remaining must be infinitesimal, because about 2000 half-lives of ^{237}Np have elapsed since that time.

It has been traditional to describe as naturally occurring only those elements ranging from hydrogen with atomic number 1 to uranium with atomic number 92. This tradition simply reflects the fact that all nuclei with atomic number greater than 92 had natural lifetimes that were short compared to the lifetime of the earth. It is believed that many nuclei with atomic number greater than 92 were present in the early stages of our part of the universe but that by now the short lifetime nuclei have completely decayed. Evidence that all of them have not disappeared has been presented recently. Darlene Hoffman of the Los Alamos Scientific Laboratory and her collaborators in 1972 detected ^{244}Pu in a rare earth ore bastnasite from the Mountain Pass Mine in California.

10.12 RADIOACTIVE GROWTH AND DECAY

These series of radioisotopes present interesting problems in relative abundances. Instead of considering a whole series, let us consider what happens to the abundance of substance B if A decays to B and B decays to C.

Let us call the number of A atoms at any instant N_1 and the initial number N_0. We can call the number of B atoms N_2 and assume that the initial number of B atoms is zero.

Every time an A atom disintegrates, it increases the number of B atoms. But every B disintegration reduces the number of B atoms. From Eq. (10.1) we have

$$\text{Number entering the } B \text{ category} = \frac{-dN}{dt} = \lambda_1 N_1 \tag{10.9}$$

and

$$\text{Number leaving the } B \text{ category} = \lambda_2 N_2, \tag{10.10}$$

where λ_1 is the decay constant for element A and λ_2 is that for element B. The net change in N_2 is the difference between Eqs. (10.9) and (10.10), or

$$\frac{dN_2}{dt}(\text{net}) = \lambda_1 N_1 - \lambda_2 N_2. \tag{10.11}$$

From Eq. (10.2) we can write

$$N_1 = N_0 e^{-\lambda_1 t}$$

and use this to eliminate N_1 from Eq. (10.11). This gives

$$\frac{dN_2}{dt} = \lambda_1 N_0 e^{-\lambda_1 t} - \lambda_2 N_2. \tag{10.12}$$

If we transpose $\lambda_2 N_2$ and multiply by the integrating factor $e^{\lambda_2 t}\,dt$, we have

$$e^{\lambda_2 t}\,dN_2 + \lambda_2 N_2 e^{\lambda_2 t}\,dt = \lambda_1 N_0 e^{(\lambda_2 - \lambda_1)t}\,dt. \tag{10.13}$$

This may be integrated at once to yield

$$N_2 e^{\lambda_2 t} = \frac{\lambda_1}{\lambda_2 - \lambda_1} N_0 e^{(\lambda_2 - \lambda_1)t} + C. \tag{10.14}$$

Since we are assuming that $N_2 = 0$ when $t = 0$, we have

$$0 = \frac{\lambda_1 N_0}{\lambda_2 - \lambda_1} + C. \tag{10.15}$$

Eliminating C from Eq. (10.14), we obtain

$$N_2 e^{\lambda_2 t} = \frac{N_0 \lambda_1}{\lambda_2 - \lambda_1}\left(e^{(\lambda_2 - \lambda_1)t} - 1\right), \tag{10.16}$$

and finally

$$N_2 = \frac{N_0\lambda_1}{\lambda_2 - \lambda_1}(e^{-\lambda_1 t} - e^{-\lambda_2 t}). \tag{10.17}$$

The decay of A and the growth and decay of B are shown in Fig. 10.18.

We have considered the mathematics of but one link in the chain of events that constitute a radioactive series. The complete solution requires the simultaneous solution of as many differential equations as there are elements in the series. The general solution gives the number of daughter atoms as a function of time under the assumptions that at time zero there were no daughter atoms, and that there were a known number of atoms of the parent. These equations are called the Bateman equations, after the man who derived them.

It is possible to make a hydrodynamic model to demonstrate the growth and decay of elements in a radioactive series. Let us construct a number of tanks that can drain water from one to another in succession (Fig. 10.19). The tanks are shaped so that the velocity of efflux is proportional to the amount of water within the tank and the nozzles have openings proportional to λ. If the first tank is filled and the water is released at time zero, then the water in each tank is the same function of time as the concentration of radioactive atoms in a radioactive series. Since each radioactive series ends on a stable isotope, the bottom tank should have no outlet.

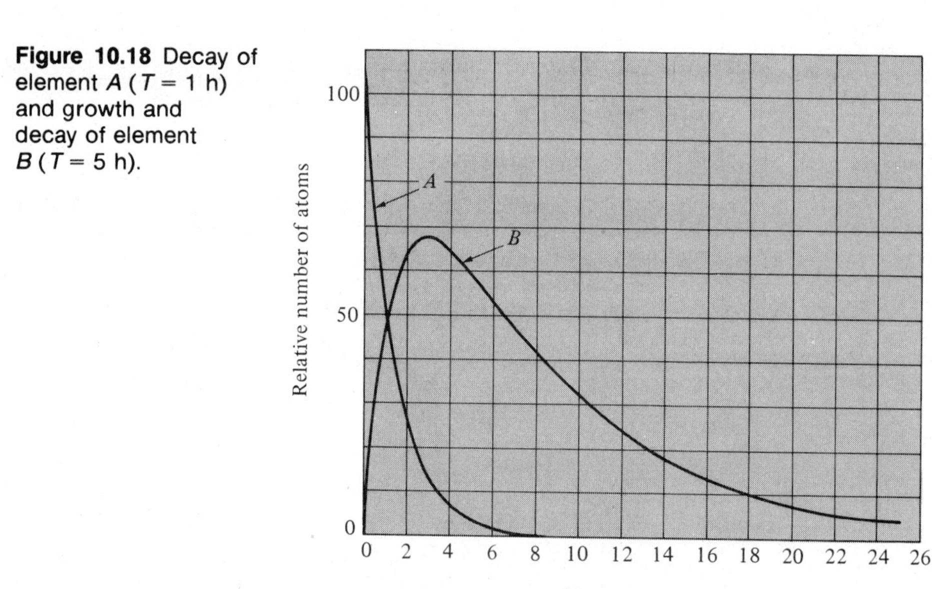

Figure 10.18 Decay of element A ($T = 1$ h) and growth and decay of element B ($T = 5$ h).

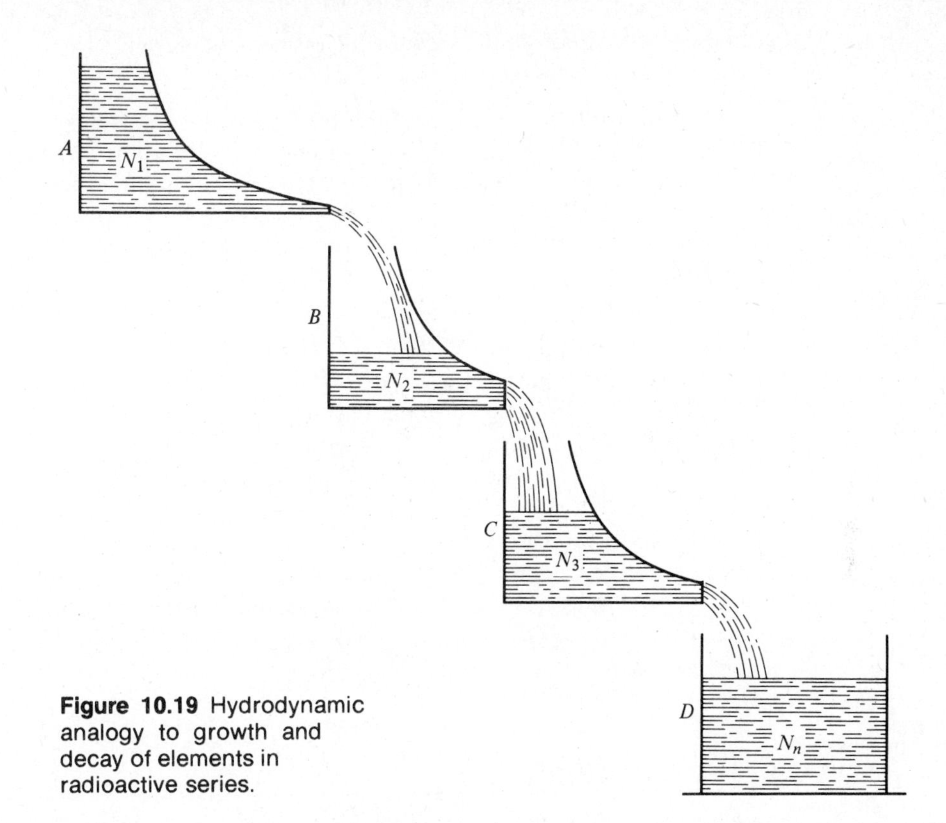

Figure 10.19 Hydrodynamic analogy to growth and decay of elements in radioactive series.

10.13 THE AGE OF THE EARTH

Another way of avoiding a detailed treatment of the involved Bateman equations is to take advantage of the fact that in any radioactive series the decay constants vary widely. This permits many reasonable approximations. The ^{238}U series, for example, begins with uranium whose half-life is 4.5 billion years and whose decay constant is less than 1/1000 that of any other element in the series, until we come to ^{206}Pb, an isotope of lead that is uniquely associated with ^{238}U. This lead is stable, has an infinite half-life, and has a decay constant of zero. This means that after a billion years or so the only elements present in any appreciable concentration will be uranium and lead. In terms of the water analogy, the intervening tanks are practically empty and serve only as a pipe through which the water in the first tank drains into the last. Thus it becomes reasonable to apply our Eq. (10.17) not only to the first and second elements of the series but also to the first and last. Since λ_2

becomes the decay constant of lead, which is zero, we have

$$N_{\text{Pb}} = N_{0\text{U}}(1 - e^{-\lambda_{\text{U}} t}). \tag{10.18}$$

^{238}U ore always contains ^{206}Pb because this lead is the end element of the uranium radioactive series. It is reasonable to assume that this is the only reason lead is present, since there is no other compelling reason why uranium and lead should be found together. It then follows that the present number of lead atoms plus the present number of uranium atoms must equal the number of uranium atoms originally present, or

$$N_{\text{Pb}} + N_{\text{U}} = N_{0\text{U}}. \tag{10.19}$$

The present concentrations of ^{206}Pb and ^{238}U can be measured experimentally. With these data and the value of λ_{U}, we can find t after eliminating $N_{0\text{U}}$ between Eqs. (10.18) and (10.19). This t is the time that has passed since the earth solidified and the original pocket of uranium was sealed in the rock. There have been many assumptions in our argument, but they are justified by the fact that t has been computed for many ore samples from different parts of the earth and the results have consistently shown its age to be about four billion years. Since radioactive processes are independent of conditions of temperature and pressure, pockets of radioactive material constitute reliable clocks that have been running throughout geological history.

10.14 RADIOACTIVE EQUILIBRIUM

Another useful approximation in discussing the radioactive series is the concept of *radioactive* or *secular equilibrium*. We have seen that decay constants vary widely and it frequently happens that the parent substance has a much longer half-life than the daughter. In this case, if only the parent is present initially, the daughter product will grow and there will be a corresponding increase in the activity of this product. The more daughter atoms there are, the more daughters disintegrate, until the rates of production and disintegration of daughter atoms become equal. When these rates are equal, the decay product is in radioactive equilibrium with its parent.

The expression for this equality can be obtained from Eq. (10.17). This equation may be rewritten as

$$N_2 = \frac{N_0\lambda_1 e^{-\lambda_1 t}}{\lambda_2 - \lambda_1}[1 - e^{-(\lambda_2-\lambda_1)t}],$$

or, since $N_1 = N_0 e^{-\lambda_1 t}$, we have

$$N_2 = \frac{N_1\lambda_1}{\lambda_2 - \lambda_1}[1 - e^{-(\lambda_2-\lambda_1)t}]. \tag{10.20}$$

If $\lambda_2 \gg \lambda_1$, then to a good approximation

$$N_2 = \frac{N_1\lambda_1}{\lambda_2}(1 - e^{-\lambda_2 t}) = \frac{N_1\lambda_1}{\lambda_2}[1 - e^{-(0.693/T_2)t}]. \tag{10.21}$$

After several half-lives of the daughter product have elapsed, the value of the exponential term is negligible and Eq. (10.21) reduces to

$$N_1\lambda_1 = N_2\lambda_2. \qquad \textbf{(10.22)}$$

If, instead of considering just one parent-daughter relationship, we treat a whole radioactive series in which the first element has a much longer half-life than any of the disintegration products, then the preceding analysis can be applied to each daughter product in the series. The resulting equilibrium relation is

$$N_1\lambda_1 = N_2\lambda_2 = N_3\lambda_3 = \text{etc.}^\dagger \qquad \textbf{(10.23)}$$

Each N in Eq. (10.23) decreases slowly and exponentially according to the small decay constant of the long-lived parent substance.

In terms of our water analog, if some tank leaks so slowly that the water level will not change noticeably while we watch it, then we can regard its leakage rate as constant. That tank will begin to fill the one below it. Presently the second tank will leak to a third, etc. Once equilibrium is established, the level in each tank will become constant at a level determined by its nozzle size. Tanks with large nozzles (λ's) will contain few water molecules (N's). The tank levels will not now be time-dependent.

Furthermore, if we leave the system running for a long time, so that the level in the original tank is noticeably decreased, then we will find the level in all the succeeding tanks decreased in the same proportion.

It must be appreciated, however, that the extent to which equilibrium is achieved depends on the relations among the half-lives and on the time, since all the atoms were of the original kind. The age of the earth is about one ^{238}U half-life, which is much greater than any half-lives of uranium daughters except ^{206}Pb. Thus for the past three billion years there has been very precise equilibrium between ^{238}U and most of its daughters. The age of the earth can be computed only because ^{206}Pb has an infinite half-life ($\lambda_{\text{Pb}} = 0$) and is *never* in equilibrium with uranium.

At the other extreme, equilibrium is very nearly established in a few seconds between such pairs as $^{214}_{83}$Bi and $^{214}_{84}$Po. The bismuth half-life is 19.7 m, while that of the polonium is 1.64×10^{-4} s. Equilibrium between these two permits the evaluation of the very short polonium half-life. The bismuth decay constant can be found by methods already described in Section 10.10. If the relative concentrations of bismuth and polonium can be measured after equilibrium has been established, then the very large polonium decay constant can be found from the equilibrium equation, Eq. (10.23).

10.15 SECONDARY RADIATIONS

Extensive study of the radiations from disintegrating nuclei have revealed two additional processes that result in the emission of electromagnetic radiation.

† Note that the N's in Eq. (10.23) refer to different kinds of atoms and so their isotopic masses will *not* divide out of the equation as they did in the example at the end of Section 10.10.

In some cases, the nucleus will absorb one of the orbital electrons of the atom. Since this probability is greatest for the adjacent K-shell electron, the process is often called *K-electron capture*. When such electron capture occurs, the product atom will have an atomic number that is one less than that of the original nucleus. The remaining orbital electrons rearrange themselves to correspond to the structure of this new atom. X-rays characteristic of the product atom are emitted during the formation of the stable state of the new atomic system. Orbital electron capture is shown schematically in one part of Fig. 10.20.

The emission of a charged particle from a parent nucleus often leaves the daughter or product nucleus in an excited state. The latter usually goes to the ground state by emitting one or more gamma photons. In some elements, however, the transition energy is given to an orbital electron by direct interaction with the nucleus. This process is called *internal conversion*, IC. It causes a photoelectron to be ejected from a shell with a kinetic energy equal to the difference between the nuclear transition energy and the binding energy of the emitted electron. Thus internal conversion electrons give a line spectrum of beta particles. Note that these beta particles come from the extranuclear structure. Those from the nucleus always have a continuous energy distribution up to some maximum value. X-rays characteristic of the daughter element from which an electron was ejected are emitted when this element returns to the normal state.

The data obtained from the study of internal conversion show that the gamma rays in radioactive decay are often emitted *after* particle disintegration. Therefore they come from the product nucleus instead of the parent nucleus. This was

Figure 10.20 Mechanisms of radiation in radioactivity.

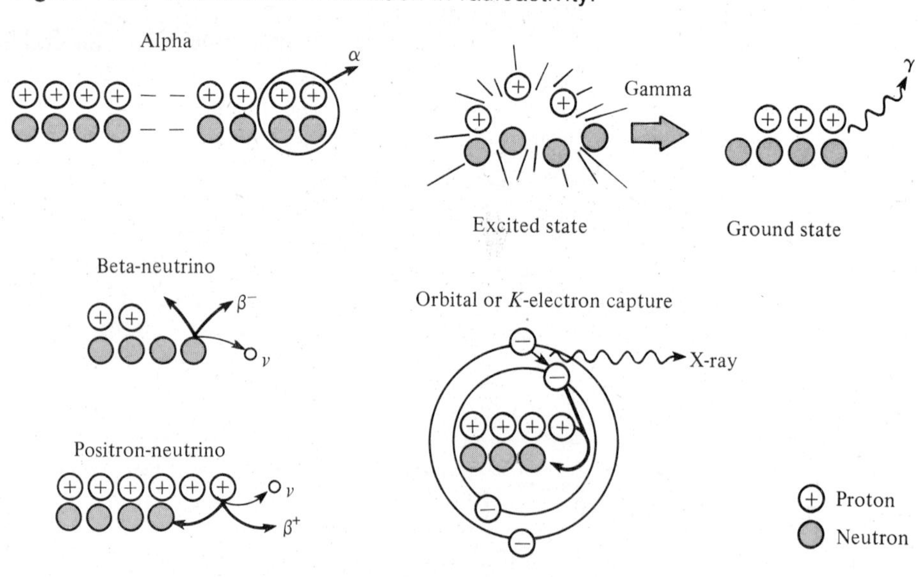

represented in Fig. 10.12. Since the half-life of the parent element determines the rate of formation of the daughter element, and since gamma rays are emitted as the product "settles down," the gamma decay rate in this case follows the decay of the parent. This is the reason for a confusion in terminology. For example, when RaB, which has a half-life of 26.8 m, disintegrates to RaC, the associated gamma activity also decays with a half-life of 26.8 m. Thus it has become customary to call the radiation gamma rays from RaB even though they actually come from RaC.

10.16 RADIATION HAZARDS

All high-energy rays are hazardous to living tissue. They have enough energy to dissociate the complicated molecules of living tissue and to kill cells. We saw in Section 10.5 how the instability of organic molecules is utilized to quench Geiger counters. Thus high-energy radiation is able to convert the molecules of living structures into alien forms, or produce "burns."

Beta particles are not particularly hazardous because they are charged and therefore are not very penetrating. Clothing and skin provide adequate shielding for our vital organs.

Alpha particles are still less penetrating and on this basis are less dangerous than beta particles, although the specific ionization of the former greatly exceeds that of other radiations. There is a gaseous alpha-particle emitter in each one of the three natural radioactive series. Each of these gases is an isotope of the element having the atomic number 86. This element is usually called radon (sometimes emanation), and the gases are ^{219}Rn (actinon), ^{220}Rn (thoron), and ^{222}Rn (radium emanation). Radon is chemically inert. These gases are liberated whenever radioactive ore is dug or crushed. Since they are inert, no gas mask can separate them from the air chemically. Those who work with radioactive ores must necessarily breathe these gases unless they are flushed away by very active ventilation.

In the lungs, these gases are very destructive, since they are literally within a vital organ. The maximum permissible safe concentration of radon in the atmosphere is about 10^{-15} percent by volume. It is said that every miner who has worked in the Joachimstal uranium mines in Czechoslovakia for more than ten years has died of lung cancer. It may be that Pierre and Marie Curie were very fortunate that they had to perform their separation of radium from uranium ore in a dilapidated shed. The then-unwanted ventilation blew the radon away.

Because beams of different radiations cause very different biological damage even when the body absorbs the same amount of energy from each type, it is necessary to define certain units in addition to the roentgen and the rad, which were discussed in Section 6.8. The absorbed dose of any ionizing radiation is the amount of energy imparted to matter by ionizing particles per unit mass of irradiated

Table 10.4

Radiation	RBE
X-ray, gamma ray, beta ray	1
Thermal neutrons	2–5
Fast neutrons	10
Alpha particles and high-energy ions of O, N, etc.	10–20

material at the place of interest. This leads to defining the *relative biological effectiveness*, RBE, of an ionizing radiation as the ratio of the absorbed dose in rads of x-rays of about 200 keV of energy that produces a specific biologic effect, to the absorbed dose in rads of the ionizing radiation that produces the same effect. A few RBE factors are listed in Table 10.4.

From the values listed we see that if the absorption in body tissue of 1 rad from a 200 keV x-ray beam produces a particular injury, then the absorption of only 0.1 rad from a beam of fast neutrons will produce the same injury. The unit of biological dose is the *roentgen equivalent man*, rem. The biological dose in rems is equal to the absorbed dose in rads times the RBE factor. The *millirem*, mrem, is a thousandth of a rem. The variations in body sensitivity to radiation are evident from the recommended maximum permissible doses. Some of these are appendages of the body, 1500 mrem per week; the lens of the eye, the gonads, and the blood-forming organs, each 100 mrem per week.

The radiation to which a large group in the general population is exposed is given in *person rems*. This quantity is calculated by multiplying the total number of people exposed by their average individual doses in rems. The Environmental Protection Agency estimates that in 1978 the radiation exposure to the general population in the United States due to various sources was as follows (measured in 1,000 person rems per year): natural background, 20,000; healing arts, 17,000; technologically enhanced — mining and processing radioactive ores — 1,000; nuclear weapons — fallout, 1,300, and development, testing, and production, 0.165; nuclear energy, 56; and consumer products, 6. Since the population of the U.S. was then 220 million, the average yearly dose per individual was about 179 millirems. The annual permissible dose in a nuclear power plant is 5 rems.

In general, the most hazardous radiations are the penetrating radiations. These include x-rays and gamma rays, together with neutral particles and the extremely high-energy particles produced by cosmic rays, which are penetrating despite their charge.

Not only are these penetrating radiations the ones that are destructively hazardous to vital organs, they are also the kind used purposefully when cancerous tissues are to be killed.

10.17
THE "RADIUM RADIATIONS" IN MEDICINE

The first radioactive material used in medical treatments was radium, ^{226}Ra. But radium itself is primarily an alpha emitter. It was neither radium nor its daughter radon, ^{222}Rn, that was effective in destroying cancerous tissue. Indeed, one must go to the fourth radioactive generation below radium before coming to a strong gamma emitter. Even when radium was administered to a patient, it was the ^{214}Bi or RaC in equilibrium with the radium that was the important gamma-ray emitter.

Radium treatment required careful attention in the hospital. Radium has a half-life of 1620 years, which is certainly long compared with the duration of the treatment. In order to specify a given dose, the radiologist had to specify both the activity of the radium sample and the length of time it should be administered.

The radium daughter (radon or radium emanation, ^{222}Rn) has a much shorter half-life, 3.82 days. In administering radon to a patient, the length of time of the application was often less critical or even inconsequential. A given amount of radon would give half its dose in 3.82 days and practically all its dose in 10 days or so. Thus the amount of radon to be administered could be specified with less concern for the removal of the radioactive material.

Hospitals usually kept radium as a solution of radium chloride. Periodically, the radon accumulated over the solution was chemically separated from the other gases over the solution and sealed into small glass or gold tubes called "radon seeds." These seeds grew in gamma activity as the radon descendants ^{218}Po, ^{214}Pb, and ^{214}Bi came to equilibrium with the radon. Maximum gamma activity was reached in about 4 hr, since these products have half-lives of 3.05, 26.8, and 19.7 m, respectively. Once equilibrium was established, the number of ^{214}Bi or RaC atoms was proportional to the number of the longer-lived radon atoms that were present. When the gamma activity began to decline, the seed could be calibrated and its future activity calculated from the decay constant of radon.

Today, medical treatments are made with radioisotopes manufactured in the laboratory, especially ^{60}Co, a gamma-ray emitter. We shall discuss these materials in Chapter 11.

10.18
UNITS OF RADIOACTIVITY AND EXPOSURE

The procedure we have just described led to the historical unit of radioactivity, the curie. Originally, it was the activity of 1 g of radium or the activity of the amount of radon in equilibrium with 1 g of radium. The original definition of the curie was replaced, in 1950, by a more general one that is closely equivalent to the old definition. By international agreement, *the curie*, Ci, is 3.7000×10^{10} disintegrations per second.† The *millicurie*, mCi, and *microcurie*, μCi, are also frequently encountered. Since the definition of the curie was peculiarly associated with radium, another unit of activity was proposed for all decaying elements. It is the *rutherford*, rd, which is 10^6 dis/s.

Medical doses are expressed in millicurie-hours. If a radon seed had reached its maximum activity and were found to have an activity of I_0 mCi at a time which we call zero, and if the seed were administered to a patient during the period from t_1 to t_2 (hours) later, the dose was

$$\text{dose in mCi-h} = I_0 \int_{t_1}^{t_2} e^{-\lambda t}\, dt, \tag{10.24}$$

where λ is the radon decay constant in reciprocal hours.

There is no one conversion factor between the activity of a radioactive material and the exposure rate in the surrounding medium. This is obvious from the fact that the ionization produced depends on the activity of the material and on the *kind* and the *energy* of the emitted radiation. It can be shown (Problem 10.48) that at a point P in air the *gamma-ray* exposure rate due to a *point* source of radioactive material is given approximately by

$$I = \frac{5.3 \times 10^3 CE}{d^2}. \tag{10.25}$$

In this expression, I is the exposure rate in roentgens per hour, C is the activity of the source in curies, E is the energy of the gamma photon in MeV, and d is the distance from the source to the point P, in centimeters. If more than one photon is emitted per disintegration, the total exposure rate will be the sum of the individual exposure rates. Since Eq. (10.25) is a point source law, it is essentially a differential expression that must be integrated subject to boundary conditions for an actual distribution of radioactive material.

10.19 CONCLUSION

In this chapter, we have described some of the most important properties of natural radioactivity, together with some of the uses that have been found for it. Most of what was said also applies to artificial radioactivity, which is discussed in Chapter

† The value given for the number of disintegrations per second of one gram of radium had to be changed several times because of the increasing precision in the determinations of its half-life. To avoid further changes, the international curie was arbitrarily assigned a fixed value. According to current data, 1 g of ^{226}Ra undergoes 3.61×10^{10} dis/s — about 2.5% less than the adopted curie.

11. In Chapter 11, we shall also discuss the reasons why some nuclei are radioactive and others are not.

FOR ADDITIONAL READING

Anderson, Carl D., "The Positive Electron," in R. T. Beyer, *Foundations of Nuclear Physics*, pp. 1–4. New York: Dover, 1949.

Beers, Yardley, *Introduction to the Theory of Error*. Reading, Mass.: Addison-Wesley, 1953. Chapter 7 is a concise account of the statistics employed in measurements of radioactivity.

Bohn, J. L., and F. H. Nadig, "Hydrodynamic Model for Demonstrations in Radioactivity," *The American Physics Teacher* (*Am. J. Phys.*) **6**, 320 (1938). Gives theory and directions for making a very instructive model of the decay and growth of the elements in a radioactive series.

Bowler, M. G., *Nuclear Physics*. New York: Pergamon Press, 1973.

Curie, Eve, *Madam Curie*. Vincent Sheean, translator. Garden City, N.Y.: Doubleday, 1937. An excellent biography.

Eisenbud, Merril, *Environment, Technology, and Health*; *Human Ecology in Historical Perspective*. New York: New York University Press, 1978.

Eisenbud, Merril, *Environmental Radioactivity*. 2nd ed. New York: Academic Press, 1973.

England, J. B. A., *Techniques in Nuclear Structure Physics, Part I*. New York: Wiley, 1974.

Environmental Protection Agency, Document EPA 520/1/77-009, *Radiological Quality of the Environment in the United States*, 1977.

Friedlander, G., and J. W. Kennedy, *Nuclear and Radiochemistry*. New York: Wiley, 1955. Discussion and derivation of the statistical equations used in radioactivity measurements given in Chapter 10.

Gentner, W., H. Maier-Leibnitz, and H. Bothe, *An Atlas of Typical Expansion Chamber Photographs*. London: Pergamon, 1954. An excellent collection of photographs of cloud chamber events with detailed description and interpretation of each. This pictorial history of nuclear physics should be examined by every student.

Glaser, Donald A., "The Bubble Chamber," *Scientific American* **46** (February 1955).

Magie, W. F., *A Source Book in Physics*. Cambridge, Mass.: Harvard University Press, 1963. Contains the following papers: "The Radiation from Uranium," by Henri Becquerel, and "Polonium" and "Radium," by Marie S. Curie and Pierre Curie.

Mann, W. B., and S. B. Garfinkel, *Radioactivity and Its Measurement*, Momentum Book No. 10. Princeton, N.J.: Van Nostrand, 1966.

Morgan, K. Z., and J. E. Turner, *Principles of Radiation Protection*. New York: Wiley, 1967.

O'Neill, Gerard K., "The Spark Chamber," *Scientific American* **46** (February 1955).

Reines, F., and C. L. Cowan, Jr., "The Neutrino," *Nature* **178**, 446 (1958). A good account of the work of these authors in verifying the existence of this elusive particle.

Rutherford, E., J. Chadwick, and C. D. Ellis, *Radiations from Radioactive Substances*. New York: Macmillan, 1930. A complete account of the work in radioactivity during the first three decades after its discovery.

REVIEW QUESTIONS

1. What are the three basic radiations that come from radioactive materials? How would you identify each?
2. What are the differences between ionization, proportional, and Geiger counters?
3. What is the basic physical phenomenon that allows a Wilson cloud chamber to detect particles? What is it for the bubble chamber?
4. What evidence led Pauli to postulate the existence of the neutrino?
5. What is the basic relationship between the rate of decay and the number of radioactive nuclei present?
6. What are the basic radioactive series?
7. What technique uses the uranium series to determine the age of the earth and the moon?

PROBLEMS

(The constants of various radioactive elements needed for solving problems can be found in Tables 10.1, 10.2, and 10.3, and in Appendixes 5 and 6.)

10.1 A certain radioactive element has a half-life of 20 days. (a) How long will it take for $\frac{3}{4}$ of the atoms originally present to disintegrate? (b) How long will it take until only $\frac{1}{8}$ of the atoms originally present remain unchanged? (c) What are the disintegration constant and the average life of this element?

10.2 The original activity of a certain radioactive element is I_0. (a) Plot the subsequent activity, I, as ordinate against half-life as abscissa for a period of six half-lives. (b) Mark the average life, $\bar{t}$, on the abscissa and estimate the activity then in terms of I_0. (c) Assuming that the daughter product of the element is stable, plot the growth of this product on the graph for part (a).

10.3 (a) How many alpha particles are emitted per second by 1 microgram of ^{226}Ra? (b) How many alpha particles will be emitted per second after the original microgram has disintegrated for 500 years? Neglect the contributions of daughter products. (c) What would be the answer to part (b) if the contributions of the daughter products were included? (Radioactive equilibrium will have been established in 500 years.)

10.4 What fraction of a given sample of radium emanation, ^{222}Rn, will disintegrate in two days?

10.5 The activity of a certain radioactive sample decreases by a factor of 8 in a time interval of 30 minutes. What are its half-life, disintegration constant, and mean life?

10.6 What is the activity of one gram of $^{226}_{88}$Ra whose half-life is 1620 years?

10.7 How much time is required for 5 mg of ^{22}Na ($T_{1/2} = 2.60\ y$) to reduce to 1 mg?

10.8 Particles emitted by a certain radioactive substance are suspected of being electrons. Describe an experiment that would prove that they are.

10.9 What are the lowest value and the highest value of the integer n for (a) the uranium $4n + 2$, (b) the actinium $4n + 3$, and (c) the thorium $4n$ series?

10.10 Radium C, ^{214}Bi, is a branch point in the uranium series. Some of its atoms disintegrate by alpha emission and some by beta emission. Write the nuclear reaction equation (a) for each case, and (b) for the disintegration of each daughter product. (c) Identify the elements resulting from the decays in part (b).

10.11 A hypothetical radioactive series starts with element ${}^{130}_{50}A$. Element ${}^{130}A$ is an alpha emitter, its "daughter" is a beta emitter, and its "granddaughter" is a branch point in the series. Find the atomic number and mass number of E, the "great-great granddaughter" of element A.

10.12 Discuss the changes in the amount of water in tank B after opening A in the hydrodynamic analog shown in Fig. 10.19. Correlate the growth and decay of a radioactive element with your discussion for the following sets of relative sizes of the nozzles: (a) A small, B large; (b) both equal; (c) A large, B small; (d) A any size, B zero (closed); and (e) both zero (closed).

10.13 How much radium B, ${}^{214}Pb$, will be present after 1 mg of radium A, ${}^{218}Po$, has been disintegrating for 20 m? Assume that all of RaA decays into RaB.

10.14 How much radium emanation, ${}^{222}Rn$, will be present after 1 g of radium, ${}^{226}Ra$, has disintegrated for one day?

10.15 Show from Eq. (10.17) that if there are initially N_0 radioactive atoms of the parent present, the time at which the number of radioactive daughter atoms is maximum is

$$t_{max} = \frac{\ln(\lambda_2/\lambda_1)}{\lambda_2 - \lambda_1}.$$

10.16 (a) How long after obtaining a pure sample of radium A, ${}^{218}Po$, will the amount of radium B, ${}^{214}Pb$, be maximum? (b) How long after obtaining a pure sample of radium will the amount of radium emanation be maximum?

10.17 If a radioactive element disintegrates for a period of time equal to its average life, (a) what fraction of the original amount remains, and (b) what fraction will have disintegrated?

10.18 Using particle accelerators, it is possible to produce radioactive nuclei. Assume that a particular type of nucleus whose decay constant is λ is produced at a steady rate of p nuclei per second. Show that the number of nuclei N present t s after the production starts is

$$N = \frac{p}{\lambda}(1 - e^{-\lambda t}).$$

10.19 How many alpha particles and how many beta particles are emitted in the actinium series, where ${}^{235}_{92}U$ decays to ${}^{207}_{82}Pb$?

10.20 (a) If a piece of uranium ore contains 2 g of ${}^{238}U$ after disintegrating for a period equal to its average life, how much ${}^{238}U$ did it contain originally? (b) If all the helium released during the disintegration of the original amount of uranium for one average life were collected, what would be its mass and its volume under standard conditions? (Note that substantially all the disintegrated uranium will have become stable lead in this period of time.)

10.21 Show that the slope of the curve of the logarithm to the base 10 of the activity of a radioactive substance plotted against time is equal to $(-\lambda/2.30)$. Start with Eq. (10.5).

10.22 The activity of a certain radioactive sample is measured every half hour and the following activities in millicuries are measured: 78, 49, 32, 20, 13, 8, 5. Plot the activity against time and determine the half-life from the graph.

10.23 A certain radioactive element is a beta-particle emitter. When the beta activity of a freshly prepared sample of this element is measured as a function of time, the following data are obtained:

Time, m	Activity, units	Time, m	Activity, units
0	1080	100	173
20	730	130	106
40	504	160	63
70	303	190	35

(a) Plot the logarithm of the activity against time. (b) Determine the disintegration constant from the slope of the curve. (c) Find the half-life of the element.

10.24 When the activity of a mixture of radioactive elements is measured as a function of time, the following data are obtained:

Time, m	Activity, units	Time, m	Activity, units
10	252	110	11.5
30	105	130	9.1
50	46.8	160	6.8
70	25.0	190	5.2
90	15.5	220	4.0
		280	2.4

(a) Plot the logarithm of the activity against time. (b) How many radioactive elements are in the mixture? (c) Find the disintegration constant of each from the slopes of the corrected curve, and (d) the half-life of each. [*Hint:* To obtain the activity of each element from the combined curve, project the curve for the longest-lived element back to $t = 0$. The actual activity ordinates (not log activity) of this projected curve must be subtracted from those of the total activity curve to obtain the activity due solely to the other element present. The use of semilog paper will simplify plotting and solving.]

10.25 Lead-206 is found in a certain uranium ore. This indicates that the lead is of radioactive origin. What is the age of the uranium ore if it now contains 0.80 g of ^{206}Pb for each g of ^{238}U?

10.26 What is the mass of one curie of (a) ^{238}U, (b) ^{210}Pb (RaD), and (c) ^{214}Po (RaC′)?

10.27 What is the mass and the volume under standard conditions of one curie of radium emanation, ^{222}Rn?

10.28 A container holds 2 Ci of ^{210}Po, which emits alpha particles with an energy of 5.30 MeV. If all of the alpha particles are stopped in the container, calculate the rate at which heat is evolved.

10.29 Calculate the wavelength associated with an alpha particle emitted by a nucleus of an atom of ^{210}Po. Compare this wavelength with the diameter of a nucleus.

10.30 Free neutrons disintegrate into photons and electrons. Calculate the kinetic energy of the electron measured relative to a frame of reference in which the original neutron was at rest.

10.31 The activity of 20 g of ^{232}Th is 2.18 μCi. Calculate the disintegration constant and the half-life of ^{232}Th.

10.32 Natural uranium is that which is found on or near the surface of the earth. It is a mixture of three isotopes, ^{234}U, ^{235}U, and ^{238}U. What is the activity of each contained in 0.5 kg of natural uranium?

10.33 If 3×10^{-9} kg of $^{200}_{79}$Au has an activity of 58.9 Ci, what is its half-life?

10.34 A hypothetical beta emitter ^{220}X has a half-life of 6.93 days. Assume that a 20 mCi sample of this element is present at $t = 0$. (a) At what time will the activity be 5 mCi? (b) How many beta particles are emitted in the first second? (c) How many beta particles are emitted between $t = 10$ days and $t = 20$ days?

10.35 (a) Assuming that radioactive equilibrium has been established, how many grams of radium, ^{226}Ra, and of radium emanation, ^{222}Rn, are contained in a 2-lb piece of ^{238}U ore that is 40% uranium oxide, U_3O_8? (b) What are the activities of these amounts of uranium, radium, and radium emanation in millicuries and in rutherfords?

10.36 One cubic centimeter of a solution containing the artificial radioactive isotope ^{24}Na with an activity of 2000 particles per second was injected into the blood of a man. The activity of one cubic centimeter of the blood, taken after a time $t = 5$ hours, proved to be 16 particles per minute. Assuming the half-life of ^{24}Na to be 15 hours, determine the volume of blood in the man's body.

10.37 Given two radioactive elements, a parent having a half-life T_1 and a daughter product having a half-life T_2. If $T_1 \gg T_2$ and if initially there is only the parent element, how long will it take, in terms of T_2, until these elements are (a) within 1 percent of their equilibrium value, and (b) within 0.1 percent of that value?

10.38 For health protection, the maximum permissible concentration of radium emanation in air for continuous exposure is 10^{-8} microcuries per milliliter of air. For this concentration, what is the ^{222}Rn content of 1 ml of standard air (a) in percent by mass and (b) in percent by volume?

10.39 If the piece of uranium ore in Problem 10.35 is ground to a fine powder so that the entrapped radium emanation, ^{222}Rn, is released, with how many cubic feet of air must this radon be mixed to reduce the concentration to the safe tolerance level of 10^{-8} microcuries per milliliter?

10.40 A coal-fired generating plant burns 7000 short tons of bituminous coal to produce electrical energy at the rate of 1000 megawatts for a 24-hour day. Bituminous coal has an average of 2 ppm (parts per million) by weight of natural uranium entrapped in it. (a) How many pounds of natural uranium are released during the burning of the coal required to provide a megawatt day of electrical energy? (b) What is the activity of that amount of uranium, assuming all of it is ^{238}U? (c) How many grams of ^{226}Ra were in radioactive equilibrium with that amount of uranium? (d) What is the volume of the associated ^{222}Rn measured under standard conditions released per megawatt day? (e) Where will the nonvolatile radioactive elements found with the uranium go as the coal is burned? (f) What is the daily average of the activity of the ^{222}Rn released in the U.S. where 600 million short tons of coal are burned annually?

10.41 The human body contains 0.20% natural potassium, and 0.0118% of that is the naturally radioactive ^{40}K. What is the radioactivity due to potassium in a person weighing 70 kg (154 lbs)?

10.42 The analysis of a certain garden fertilizer is given as 10–6–4, which means that, by weight, it contains 10% nitrogen (N), 6% phosphorus (calculated as P_2O_5), and 4% potash (calculated as K_2O). Natural potassium contains 0.0118% of the radioactive ^{40}K. (a) What is the activity of 1.0 lb of potash? (b) What is the radioactivity of the potash in a 50-lb bag of a 10–6–4 fertilizer?

10.43 The amount of a radioactive element within the human body decreases with time due to disintegration of some of the element and due to excretion of some before disintegration.

Thus λ, the effective disintegration constant of the element in the body, is equal to λ_r, the radioactive disintegration constant, plus λ_b, the biological "disintegration" constant or fractional decrease per unit time due to excretion. Assuming that each of these λ's is mathematically related to its half-life, T, in the same way that λ_r is to T_r, show that the effective half-life is $T = T_r T_b / T_r + T_b$.

10.44 The radioactive ^{14}C is found in the human body and ^{11}C is used as a tracer in some physiological studies. The value of the biological half-life of each is 10 d, and the radioactive half-life of ^{14}C is 5740 y and of ^{11}C is 20.4 m. Find the effective half-life of each.

10.45 Uranium ore contains not only the three natural isotopes of that element but also the disintegration products in radioactive equilibrium with them in each series. (a) Considering only ^{238}U, what fraction of the initial activity of the ore will remain in the mine tailings, the wastes, after extracting the ^{238}U? (Count the two $^{234}_{91}Pa$ as one, and also each branch pair as one.) (b) What element or elements in the tailings are gaseous? (c) What fraction of the activity would remain after the gas has blown away?

10.46 (a) Assuming that there are 8 tons of ^{238}U and 12 tons of ^{232}Th uniformly distributed in the first foot of depth under each square mile of the surface of the earth, find the combined activity of these two radioelements in microcuries in the cubic foot of soil under each square foot of surface. (b) Assuming radioactive equilibrium, find the mass of ^{226}Ra in one cubic foot of soil. (c) What would be the kinds and energies of the radiations from this soil?

10.47 If either radium or radioactive strontium were swallowed, where would these elements concentrate in the body? To answer this, consider the chemical properties of the elements in the groups or columns in the periodic table.

10.48 (a) In the medical use of radon, ^{222}Rn, it is actually the combined gamma radiation from ^{214}Pb and especially from ^{214}Bi that irradiates the part of the body treated. However, the half-life of radium emanation is used to calculate the exposure. Why? (b) After reaching its maximum gamma activity, calibration shows that a tube containing radium emanation, ^{222}Rn, has an activity of 2 mCi. Assuming that it is placed in a tumor 24 h later, what is the dose in millicurie-hours if it remains there (1) 2 days, (2) 4 days, and (3) forever?

10.49 A point-source radioactive element has an activity of C curies and emits one gamma-ray photon having E MeV of energy per disintegration. (a) Neglecting the rather small absorption of the air, show that the gamma-ray energy flux at d cm from the source is $3.7 \times 10^{10}\ CE/4\pi d^2$ MeV/(cm^2/s). (b) The linear absorption coefficient of air for gamma rays in the 0.5 to 2 MeV range is about 3.4×10^{-5}/cm. Show, to two significant figures, that the energy absorbed per cm^3 of air in the region between d and $(d + 1)$ cm is 13×10^5 $CE/4\pi d^2$ MeV/s, and (c) that the exposure rate in R/h at the distance d is 5.3×10^3 CE/d^2 when d is in centimeters and $5.7\ CE/d^2$ when d is in feet.

10.50 What is the gamma-ray exposure in mR/hr in air at a distance of 10 ft from an unshielded 50-curie point source of Co^{60}? Each cobalt atom emits a 1.1 and a 1.3 MeV gamma-ray photon per disintegration.

11 NUCLEAR REACTIONS AND ARTIFICIAL RADIOACTIVITY

11.1 PROTONS FROM NITROGEN

We have seen how Rutherford and his associates discovered the atomic nucleus in 1911 by analyzing the way in which alpha particles are scattered by thin metal foils. Their continued study of the interaction between alpha particles and matter led to another striking discovery in 1919. As before, Rutherford used alpha particles from a radioactive source and scintillation as the detection technique, but in this work attention was directed to the absorption of the alpha particles.

In the case of x-ray absorption, we found that the intensity of the rays penetrating an absorber depends exponentially on the thickness of the absorber, as in Eq. (6.10). This type of absorption equation applies whenever the absorption is due to a process or processes in which the incident rays are removed from a beam in one catastrophic event. We saw that for x-rays these processes could be the production of photoelectrons, Compton scattering, or pair production. As we mentioned in our discussion of the Wilson cloud chamber, the process of absorption of alpha particles is quite different. The energy of the alpha particle is usually lost by producing a very large number of low-energy ion pairs. Thus the alpha particle loses energy more or less continuously rather than in one catastrophic event. Instead of being absorbed according to an exponential law, the absorption is slight until the particles are stopped, as shown in Fig. 11.1. Alpha particles of a given energy are completely stopped by traversing a certain thickness of absorber and, for a particular kind of absorber, they have a rather definite range that is a function of their energy. Alpha particles lose more energy per unit length of path near the end of their

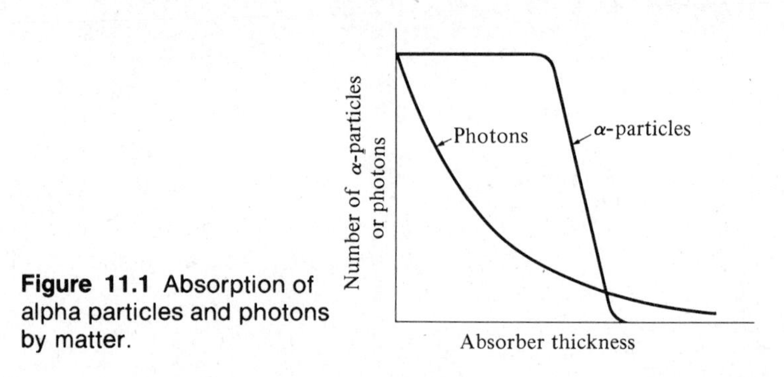

Figure 11.1 Absorption of alpha particles and photons by matter.

range, since they are moving more slowly and have more time to interact with atoms near which they pass. This leads us to assume that the energy lost per unit length of path is inversely proportional to the velocity, and therefore we have

$$-\frac{dE_k}{dx} = \frac{k}{v}, \tag{11.1}$$

where k is a constant that depends on the absorbing material. Since the velocity of alpha particles is considerably less than the velocity of light, we can eliminate dE_k by substituting the derivative of the classical expression for kinetic energy, $E_k = \frac{1}{2}mv^2$. We obtain

$$-mv\frac{dv}{dx} = \frac{k}{v}. \tag{11.2}$$

Since the particle is stopped in a distance R, its range, we have $v = V_0$ for $x = 0$ and $v = 0$ for $x = R$, and we can separate the variables in Eq. (11.2) and write

$$\int_{v_0}^{0} -mv^2\,dv = \int_0^R k\,dx. \tag{11.3}$$

Upon integration, this becomes

$$\frac{mV_0^3}{3} = kR$$

or

$$V_0^3 = k'R, \tag{11.4}$$

where k' is another constant depending on the absorbing material. This is *Geiger's rule*, which is approximately correct for any charged particles in any medium if the velocities of the particles are small compared with the velocity of light.

If we use alpha particles from the same source, V_0 is then a constant, and variations in k' can be studied by observing the range in different materials. The interaction between alpha particles and the absorbing medium is electrical, and it is not surprising that k' depends both on the atomic number of the absorber and its concentration. Indeed, if the absorber is a gas, it is convenient to study alpha-particle velocities by having a chamber in which the source and detector are separated a fixed distance and the "stopping power" of the absorber is varied by changing the pressure. Since k' is proportional to the pressure P and since R here is constant, Eq. (11.4) may be written

$$V_0^3 = k''P, \tag{11.5}$$

where k'' depends on the kind of gas and the fixed distance, and where P is the gas pressure.

Rutherford found that his variable-pressure gas chamber worked very nicely with dried oxygen or carbon dioxide as the gas. But with dried air in the chamber, he found *more scintillations than when the chamber was evacuated.* He was using alpha particles from radium-C′, whose range in air at atmospheric pressure is about 7 cm. When Rutherford increased the pressure in the chamber until the absorption of the gas was equivalent to 19 cm of atmospheric air, the number of scintillations was about twice that observed when the chamber was evacuated. The brightness of the scintillations suggested that the long-range scintillations might be due to hydrogen nuclei (protons) rather than to alpha particles.

Here was an effect obviously related to the particular gas in the chamber. Since the scintillations appeared to be hydrogen nuclei, Rutherford first suspected that the alpha particles were knocking protons off water-vapor molecules in the air. But since the effect was greatest in nitrogen even when the nitrogen was prepared so as to be scrupulously free of any hydrogen, Rutherford decided the new particles came from the nitrogen itself.

It remained to be shown whether the scintillations that appeared to be due to protons were actually protons or were nitrogen atoms. Rutherford attempted to make this distinction by deflecting the particles in a magnetic field. The geometry of his beam was poorly defined because narrowing the beam further reduced the already small number of scintillations. He felt reasonably sure, however, that the mass of the particles was two or less — certainly not 14.

Thus Rutherford became sure that either the alpha particle was breaking a proton fragment off the nitrogen nucleus or the nitrogen was breaking a fragment off the alpha particle. He sensed that the alpha particle is particularly stable, since he regarded the nitrogen nucleus as consisting of three alpha particles and two hydrogen nuclei.† He felt that the alpha particles preserved their identity and that the fragment he was observing was one or both of the hydrogens that he called outriders of the main system of mass 12.

Although Rutherford was uncertain about the details of what he observed, he was aware that he had produced the first man-controlled nuclear rearrangement and that by similar techniques one might expect to break down the nuclear structure of many of the lighter elements.

† Note that the masses balance but the charges don't. The rules for balancing nuclear equations were just being developed and were not completely understood by Rutherford or anyone else.

Figure 11.2 Cloud-chamber stereograph of the ejection of a proton from a nitrogen nucleus. The slight curvature of the tracks is due to mass motion of the gas caused by a small leak in the expansion chamber.
Courtesy of P. M. S. Blackett.

Rutherford could speculate about the kind of nucleus remaining after the alpha particle knocked a proton from nitrogen. The alpha particle might simply knock the proton off and continue with reduced energy according to the equation

$$^{4}_{2}\mathrm{He} + {}^{14}_{7}\mathrm{N} \rightarrow {}^{1}_{1}\mathrm{H} + {}^{4}_{2}\mathrm{He} + {}^{13}_{6}\mathrm{C}, \qquad \textbf{(11.6)}$$

or the alpha particle could penetrate the nucleus and remain with it, as in the equation

$$^{4}_{2}\mathrm{He} + {}^{14}_{7}\mathrm{N} \rightarrow {}^{18}_{9}\mathrm{F} \rightarrow {}^{1}_{1}\mathrm{H} + {}^{17}_{8}\mathrm{O}. \qquad \textbf{(11.7)}$$

Figure 11.3 Elastic collision of an alpha particle with hydrogen — the long, thin track.
Courtesy of P. M. S. Blackett and D. S. Lees, Imperial College of Science and Technology, and the Royal Society, London.

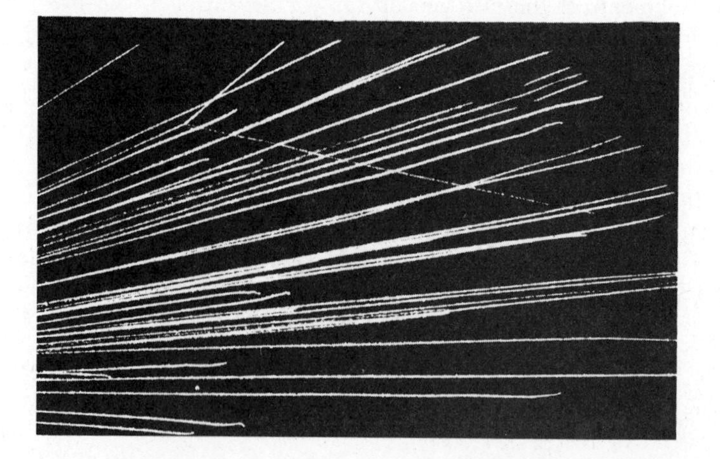

A choice between these possibilities could have been made if carbon, fluorine, or oxygen could be found in the gas by chemical or spectrographic means. But the amount of any of these elements formed was far too small to identify.

In 1925, Blackett reported that out of more than 20,000 cloud chamber photographs, including 400,000 alpha-particle tracks in nitrogen, eight were like Fig. 11.2, which shows the nuclear disintegration taking place. The two parts of the figure are two pictures of the same event taken from different angles — a stereograph. As in all Blackett's pictures, most of the tracks are straight alpha-particle tracks. But in Fig. 11.2 one alpha track ends abruptly. At the end of the alpha track may be seen two tracks, a thin one caused by a proton and a thick one due to a heavy nucleus. There appear to be two rather than three products of the nuclear reaction.

Furthermore, the stereographic picture permits a complete analysis of the three-dimensional situation. The fact that the two product-particle tracks determine a plane that includes the original alpha-particle track makes it highly unlikely that a third particle escaped undetected. Thus we can be sure that Eq. (11.7) is the correct one.

Blackett's photographs (Figs. 11.3 and 11.4) also show that the probability of having an alpha particle collide with another atom while traveling several centimeters in a gas is only about 1 in 50,000. This verifies the concept that most of the volume occupied by a gas is free space and that the atomic nucleus is actually a very minute target.

Rutherford and Chadwick bombarded all the light elements with alpha particles. They found about ten cases where protons were produced. By studying the range of these protons, they determined that in some cases the proton has more energy than

Figure 11.4 Elastic collision of an alpha particle with helium.

Courtesy of P. M. S. Blackett.

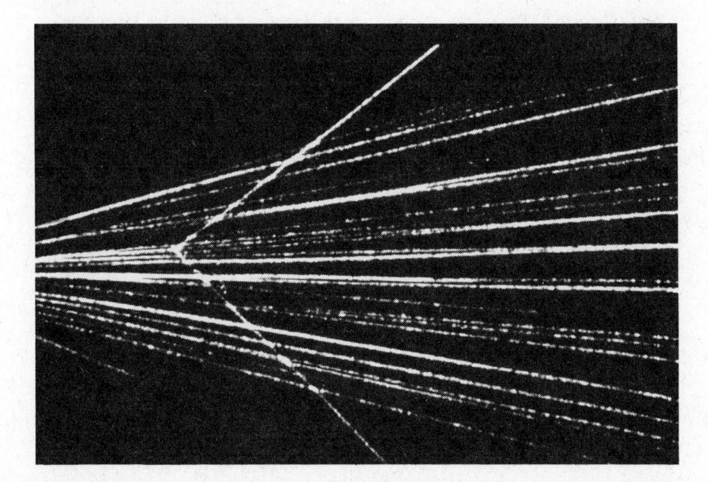

the original alpha particle. This further refutes the idea that the proton is knocked off the bombarded nucleus and supports the idea that the alpha particle is first absorbed by the nucleus, which then ejects a proton whose energy is largely determined by the instability of the intermediate nucleus. These experiments also indicated that the proton is a fundamental nuclear particle.

11.2 PENETRATING RADIATION PUZZLE

One of the light elements that failed to give protons under alpha-particle bombardment was beryllium. In 1930, Bothe and Becker in Germany reported that beryllium bombarded with alpha particles from polonium produced a very penetrating radiation that they assumed to be gamma rays. Others showed that boron also had this property, and the energy of the photons was estimated to be about 10 MeV — greater than that of any gamma rays from radioactive sources.

In 1932, the daughter and son-in-law of Madame Curie, Mme. Curie-Joliot and M. Joliot, announced that these radiations could eject rather high-energy protons from matter containing hydrogen. When paraffin wax was placed in front of their ionization chamber, the ionization was nearly doubled. They *assumed* (incorrectly as it turned out) that this process was analogous to the Compton effect. Instead of an x-ray knocking out an electron, they thought a gamma ray was striking a proton. They could compute the energy the gamma-ray photon must have in order to give the proton its observed energy. Assuming a direct hit in which the gamma ray scatters back on itself, they computed the gamma-ray energy to be about 50 MeV, a tremendous amount.

It is interesting to see how this type of calculation is made. Repeating the Compton equation, Eq. (6.33), which gives the change in wavelength of a photon upon being deflected through an angle α by a particle of mass m_0, we have

$$\Delta\lambda = \frac{h}{m_0 c}(1 - \cos\alpha). \tag{11.8}$$

The photon suffers the largest change of wavelength and transfers the greatest amount of energy when the angle of scattering is 180°, in which case Eq. (11.8) becomes

$$\Delta\lambda = \frac{2h}{m_0 c}. \tag{11.9}$$

Since we shall show that only a small fraction of the photon energy E_f is transferred to the proton, the changes in the wavelength, frequency, and energy of the photon are slight, and these quantities may be treated as differentials. Thus the maximum change in the energy of the photon is

$$dE_\nu = h\,d\nu = h\,d\left(\frac{c}{\lambda}\right) = -\frac{hc\,d\lambda}{\lambda^2}. \tag{11.10}$$

Using $\lambda = c/\nu$ and Eq. (11.9) to eliminate λ and $d\lambda$, we have

$$dE_\nu = -\frac{2(h\nu)^2}{m_p c^2} = -\frac{2E_\nu^2}{m_p c^2}. \tag{11.11}$$

Now, the energy lost by the photon is just that gained by the proton, dE_p. Furthermore, m_0c^2 is the rest mass-energy of the proton being hit, which is about 1 u. Thus we have

$$dE_p = 2E_\nu^2 \text{ in u}. \tag{11.12}$$

The observed proton ranges were up to about 26 cm in air, which corresponds to an initial proton velocity of about 3×10^7 m/s. This velocity permits a nonrelativistic calculation of the kinetic energy, which is 7.5×10^{-13} joule or 4.7 MeV. Since Eq. (11.12) is in u, we must convert the observed proton energy of 4.7 MeV to 0.005 u. This is the change in the proton energy, dE_p; therefore we can compute the photon energy, E_ν:

$$E_\nu = \sqrt{dE_p/2} = \sqrt{0.005/2} = 0.05 \text{ u} = 47 \text{ MeV}. \tag{11.13}$$

Thus it would take about a 50 MeV gamma-ray quantum to give a proton having about 5 MeV of kinetic energy in an ideal Compton-type collision.

It was found that the penetrating radiations could also expel nitrogen nuclei from nitrogenous compounds. The kinetic energy of the nitrogen was 1.2 MeV, and a calculation such as we have just made indicated that the necessary photon energy was about 95 MeV. It appears not only that these gamma-ray energies are very great, but that they are not unique.

By 1932, the isotopic masses of the elements had been measured fairly accurately. All these masses were very nearly integers, and the minor differences were attributed to the relativistic equivalence of mass and energy, $E = mc^2$. Thus if the mass of an atom was found to be less than the sum of the masses of its parts (then assumed to be alpha particles, protons, and electrons), the difference was assigned to the mass-energy that would be released if the nucleus were assembled from those parts.

We have seen that the equivalence of mass and energy permits the use of mass units for the measurement of energy and that one atomic mass unit is equivalent to 931 MeV. Using the available data for the energies of the particles, one might, in the early thirties, have made the following calculation. The alpha particle from polonium-210 has a kinetic energy of 5.26 MeV, which may be expressed as 0.00565 u. The rest energy or isotopic mass of an alpha particle was then thought to be 4.00106 u. Let this alpha particle strike a boron atom assumed to be at rest and to have a mass-energy of 11.00825 u.† There would then be a total mass-energy of 15.01496 u going into the reaction. If the products of the reaction are assumed to be a gamma ray and a new nucleus, the new nucleus would have to be $^{15}_{7}N$, whose rest mass-energy was thought to exceed 14.999. Assuming that the nitrogen nucleus is left with no kinetic energy and that the gamma ray carries away all the energy difference, the gamma ray would have an energy of $15.015 - 14.999 = 0.016$ u. Thus the maximum energy that the gamma ray could have would be $0.016 \times 932 =$ 14.9 MeV, far less than 50 or 90 MeV.

Similar results were obtained from the alpha-particle bombardment of beryllium.

11.3 DISCOVERY OF THE NEUTRON

Ever since 1924, Rutherford and Chadwick had believed that there was an electrically neutral particle and Chadwick had been looking for experimental proof of its existence. He found that proof in the data we are discussing. Chadwick assumed that when boron is bombarded with alpha particles, the products are not $^{15}_{7}N$ and a gamma ray, but $^{14}_{7}N$ and $^{1}_{0}n$, a *neutron*. Such a particle would have tremendous penetrating ability, since it has no charge. Whereas charged particles interact with the electric fields of the nuclei in the matter they penetrate, an uncharged particle would interact with nuclei only when within the influence of the very short-range nuclear forces. Most important, however, Chadwick assumed that the mass of the neutron was nearly equal to the mass of the proton, so that if it hit a proton, the transfer of its energy to the proton could be complete. To find this energy transfer let us consider in detail an *elastic* collision between two bodies. We assume a direct (head-on) hit so that both bodies move on along the extension of the line of incidence. Let the first body have a mass m_1 and velocity v_1. Then its kinetic energy, E_{k_1}, is $\frac{1}{2}m_1v_1^2$. Its momentum p_1 is m_1v_1, which may be conveniently expressed as $\sqrt{2m_1E_{k_1}}$. After the collision, v_1, E_{k_1}, and p_1 change to v_1', E_{k_1}', and p_1'. Let the target particle have a mass m_2, which is initially at rest, so that

$$v_2 = E_{k_2} = p_2 = 0.$$

After the collision, these become v_2', E_{k_2}', and p_2', respectively. Since the collision is elastic, kinetic energy is conserved, and

$$E_{k_1} = E_{k_1}' + E_{k_2}' \qquad \text{or} \qquad E_{k_1}' = E_{k_1} - E_{k_2}'. \tag{11.14}$$

Because the collision is linear, the momenta add algebraically and we have

$$p_1 = p_1' + p_2'$$

†Since it was known that alpha particles produced protons from $^{10}_{5}B$, it was logical to assume that the penetrating radiation came from $^{11}_{5}B$.

or

$$\sqrt{2m_1E_{k_1}} = \sqrt{2m_1E'_{k_1}} + \sqrt{2m_2E'_{k_2}}. \tag{11.15}$$

We can eliminate one quantity between these two equations, and we choose to eliminate E'_{k_1} since it is difficult to measure. Then, substituting E'_{k_1} from Eq. (11.14) into Eq. (11.15), we have

$$\sqrt{2m_1E_{k_1}} = \sqrt{2m_1\left(E_{k_1} - E'_{k_2}\right)} + \sqrt{2m_2E'_{k_2}}. \tag{11.16}$$

Squaring and dividing by 2, we obtain

$$m_1E_{k_1} = m_1E_{k_1} - m_1E'_{k_2} + m_2E'_{k_2} + 2\sqrt{m_1m_2\left(E_{k_1} - E'_{k_2}\right)E'_{k_2}}. \tag{11.17}$$

Canceling, transposing, squaring again, and dividing by E'_{k_2}, we get

$$\left(m_1^2 - 2m_1m_2 + m_2^2\right)E'_{k_2} = 4m_1m_2E_{k_1} - 4m_1m_2E'_{k_2}. \tag{11.18}$$

Upon rearranging, we obtain

$$(m_1 + m_2)^2E'_{k_2} = 4m_1m_2E_{k_1},$$

or

$$E'_{k_2} = \frac{4m_1m_2E_{k_1}}{(m_1 + m_2)^2}. \tag{11.19}$$

By putting E'_{k_2} from Eq. (11.14) into Eq. (11.19), we can get E'_{k_1} in terms of E_{k_1} and thus

$$E'_{k_1} = E_{k_1} - \frac{4m_1m_2E_{k_1}}{(m_1 + m_2)^2} = \left(\frac{m_1 - m_2}{m_1 + m_2}\right)^2 E_{k_1}, \tag{11.20}$$

which shows that E'_{k_1} is zero from $m_1 = m_2$ so that in this circumstance $E_{k_1} = E'_{k_2}$, and all the incident energy is transferred.

We now follow Chadwick's procedure and apply these equations to collisions between particles of unknown mass and energy and particles of known mass and energy, namely hydrogen and nitrogen nuclei. (These may be considered at rest because the thermal kinetic energies are relatively insignificant.) If m_2 is a proton, $m_2 = 1$ u, $E'_{k_2} = 4.7$ MeV and Eq. (11.19) becomes

$$4.7 = \frac{4m_1E_{k_1}}{(m_1 + 1)^2}, \tag{11.21}$$

whereas if m_2 is a nitrogen nucleus, $m_2 = 14$ u, $E'_{k_2} = 1.2$ MeV, and we have

$$1.2 = \frac{4m_1 \times 14E_{k_1}}{(m_1 + 14)^2}. \tag{11.22}$$

Solving Eqs. (11.21) and (11.22) simultaneously, we find that $m_1 = 1.03$ u and $E_{k_1} = 4.7$ MeV.

Thus Chadwick proposed that the penetrating radiation was not gamma radiation but neutral particles, each having a mass of about 1 u. Such a particle would transfer energy to other nuclei more efficiently than photons, and both recoil protons and nitrogen nuclei can be attributed to a particle having a single reasonable energy, instead of the two large contradictory values required by the puzzling gamma-photon proposal.

Our equations have been applied to the collision between neutrons and other observable nuclei. Chadwick went back and analyzed the nuclear reaction itself to obtain a further check of his hypothesis and a better estimate of the neutron's mass.

For the case of boron, the reaction equation assumed was

$$^{4}_{2}\mathrm{He} + ^{11}_{5}\mathrm{B} \rightarrow ^{15}_{7}\mathrm{N} \rightarrow ^{14}_{7}\mathrm{N} + ^{1}_{0}\mathrm{n}. \qquad \textbf{(11.23)}$$

To apply the law of conservation of mass-energy, he had to know the various particle energies. The alpha particle was known to have a kinetic energy equivalent to 0.00565 u. The boron was assumed to be at rest. The neutrons were detected by letting them collide with protons. The proton range indicated their energy to be 0.0035 u, and since neutrons and protons have nearly the same mass, Chadwick could take the measured proton energy to be the neutron energy. The nucleus of $^{15}_{7}\mathrm{N}$ recoils as it emits the neutron and becomes $^{14}_{7}\mathrm{N}$. From the law of conservation of momentum we have

$$\sqrt{2m_\alpha E_{k_\alpha}} = \sqrt{2m_N E_{k_N}} + \sqrt{2m_n E_{k_n}}. \qquad \textbf{(11.24)}$$

Here everything is known except E_{kN}, which turns out to be 0.00061 u. Aston had measured the isotopic masses of $^{4}_{2}\mathrm{He}$, $^{11}_{5}\mathrm{B}$, and $^{14}_{7}\mathrm{N}$ to be 4.00106, 11.00825, and 14.0042, respectively. Chadwick wrote the complete mass-energy equation:

$$^{4}_{2}\mathrm{He} + E_{k_\alpha} + ^{11}_{5}\mathrm{B} = ^{14}_{7}\mathrm{N} + E_{k_N} + ^{1}_{0}\mathrm{n} + E_{k_n}, \qquad \textbf{(11.25)}$$

where every quantity was known except the mass of $^{1}_{0}\mathrm{n}$. Substituting, we have

$$4.00106 + 0.00565 + 11.00825 = 14.0042 + 0.00061 + ^{1}_{0}\mathrm{n} + 0.0035.$$

Solving for the mass of the neutron, Chadwick obtained 1.0067 u. This is surprisingly near the best modern value, which is 1.008665 u on the $^{12}\mathrm{C}$ scale.

We have treated the discovery of the neutron in great detail for several reasons. The neutron is a new fundamental particle so important that we shall devote Chapter 12 to its effects. The discovery is an excellent example of discovery by inference. [Neutrons do not cause fluorescence, do not cause photographic images, do not make cloud-chamber tracks, and do not trip Geiger counters. Neutrons cause weak ionization by colliding with electrons (about one ion per meter of path) and neutrons interact with nuclei.] Furthermore, the methods used in this discovery, the mechanics of impact and conservation of mass-energy, will clarify material that still lies before us.

In 1950, it was found that a free neutron, one outside the nucleus, is radioactive. It has a half-life of 12 m and disintegrates into a proton, an electron, and a neutrino.

11.4

NEUTRON DIFFRACTION

The study of the structure of solids has developed through several stages: gross visual appearance, microscopic observation with visible light, x-ray diffraction, electron microscopy, and neutron diffraction. All of these procedures are still useful and complement rather than supplement one another in the information they supply.

Unquestionably, the first great step forward, beginning in 1912, was the use of x-rays to determine crystal structure, which we discussed in Chapter 6. Although much is revealed by x-ray diffraction techniques, the method has certain limitations. The amount of scattering of these electromagnetic waves by atoms is directly dependent on the total number of electrons per atom. Since the scattering cross section of an atom for x-rays is proportional to the atomic number of the atom, the effect of the light element such as hydrogen cannot be discerned in the much more intense diffraction effect of the heavy element with which it is associated — for example, in the compound potassium hydride, KH.

In 1936, W. M. Elsasser suggested that moving neutrons should have de Broglie waves associated with them and that they could, therefore, be diffracted. The de Broglie wavelength of a neutron moving with the most probable speed in a Maxwellian distribution at 20°C is 1.80 Å. This is of the order of the interplanar spacing in crystals and thus is suitable for use in investigating their structure. Shortly after this suggestion, it was experimentally demonstrated that when a beam of neutrons from a radium-beryllium source was diffracted by MgO crystal, a maximum occurred where predicted by the Bragg "reflection" relation. Subsequent investigation showed that, generally, the coherent scattering cross sections of nuclei for thermal neutrons does not depend on the atomic number of the element, as it does for x-rays. However, it does sometimes show a marked difference between the isotopes of an element because of the change in the nucleon population for the particular element considered.

Also in 1936, F. Bloch of Stanford University predicted that the neutron should have spin as well as a magnetic moment.† He predicted that it should have a magnetic moment of the same order of magnitude as the measured moment of the proton but opposite direction with respect to the angular momentum. This was soon verified by producing a beam of polarized neutrons from an incident unpolarized beam by passing it through or reflecting it from magnetized iron. Further, as expected, the amplitude for magnetic scattering was of the same order as that for nuclear scattering, and the cross section for scattering from magnetized iron was significantly greater when the magnetic moment vector of the neutron was parallel to the magnetic induction in the iron than when it was antiparallel.

†Scattering experiments indicate that the neutron is composed of a positive charge and an equal negative one, and that the distribution of these is not uniform in this particle. The neutron spins and because of the way the charges are distributed, their magnetic effects, although opposite, do not cancel. The resultant magnetic moment vector of a neutron is antiparallel to its spin vector.

Together, these experiments showed that a low-energy neutron is a versatile probe that can be used to investigate the structure of matter. It is unique in its use to locate low atomic number elements, such as hydrogen, in a crystal; and in its use to sense the magnetic condition of a material. Eventually it was utilized to study the magnetic order in a crystal, even when it is antiferromagnetic, that is, a material in which the magnetic moments of the atoms in adjacent planes are opposite in direction so that there is no resultant magnetization from a macroscopic point of view.

Another type of experiment using neutrons was pioneered by E. O. Wollan. He obtained a beautiful photographic record of the Laue diffraction pattern produced when a narrow beam of neutrons composed of many wavelengths was incident on a single crystal of sodium chloride. Figure 11.5 is the Laue photograph from his first paper on the subject and shows neutron diffraction by NaCl. The Laue technique with x-rays was discussed in Section 6.3.

Figure 11.5 Laue photograph showing neutron diffraction by NaCl.
Courtesy of E. O. Wollan and C. G. Shull.

11.5 ACCELERATORS

Although important discoveries came from nuclear reactions produced by alpha-particle bombardment, the only elements disintegrated by alpha particles were the light elements. In our discussion of the Rutherford alpha-particle scattering, we obtained an equation for the distance of closest approach when an alpha particle is directly aimed at a nucleus, Eq. (4.3). In words, the equation states that the minimum distance is proportional to the charge of the alpha particle times the charge of the target nucleus divided by twice the kinetic energy of the alpha particle. The reason it is easy to disintegrate light elements is that these elements have small nuclear charges. If the heavier elements repel alpha particles and therefore resist disintegration, the remedy is to use bullets with more energy, less charge, or both.

Protons are not ejected by radioactive atoms, but they are easily formed by ionizing hydrogen. They have half the charge of alpha particles and should be good bullet particles if given enough energy. Because of their charge, they can be accelerated by causing them to fall through a potential difference.

The *Cockcroft-Walton accelerator* consisted of an ion source, an evacuated tube in which the ions could be accelerated, and a source of high potential to do the accelerating. The high voltage was furnished by a voltage multiplier circuit† which produced a high potential difference between the ends of an evacuated tube. The charged particles traveled through this tube and arrived at the target with energies of from one to two MeV.

Another high-voltage system is the *Van de Graaff electrostatic generator*, in which charges are transported to bodies by means of moving belts to supply the required large potentials. Figure 11.6 is a schematic diagram of a typical Van de Graaff accelerator. A high voltage at spray point S causes the gas in this region to break down and the resulting positive ions are attached to belt B. These charges are transported to the charge remover at point C. Since this point is inside a conducting sphere, there is no electric field due to previously collected charges. This sphere is the high-voltage terminal. Negative charges are applied to the belt at spray point S' and the belt carries these charges back down to ground. After a period of time the sphere is charged to a high voltage (about eight million volts) limited only by the breakdown voltage of the surrounding gas. Hydrogen and deuterium stored in tanks ^1H and ^2H are leaked into the positive-ion source area and ionized. These protons and deuterons are accelerated down and arrive at the target with an energy of 8 MeV. A tandem Van de Graaff accelerator makes use of two generators of this type in a back-to-back position. The target is in contact with a second sphere

†A voltage multiplier circuit effectively charges a set of capacitors in parallel and discharges them in series.

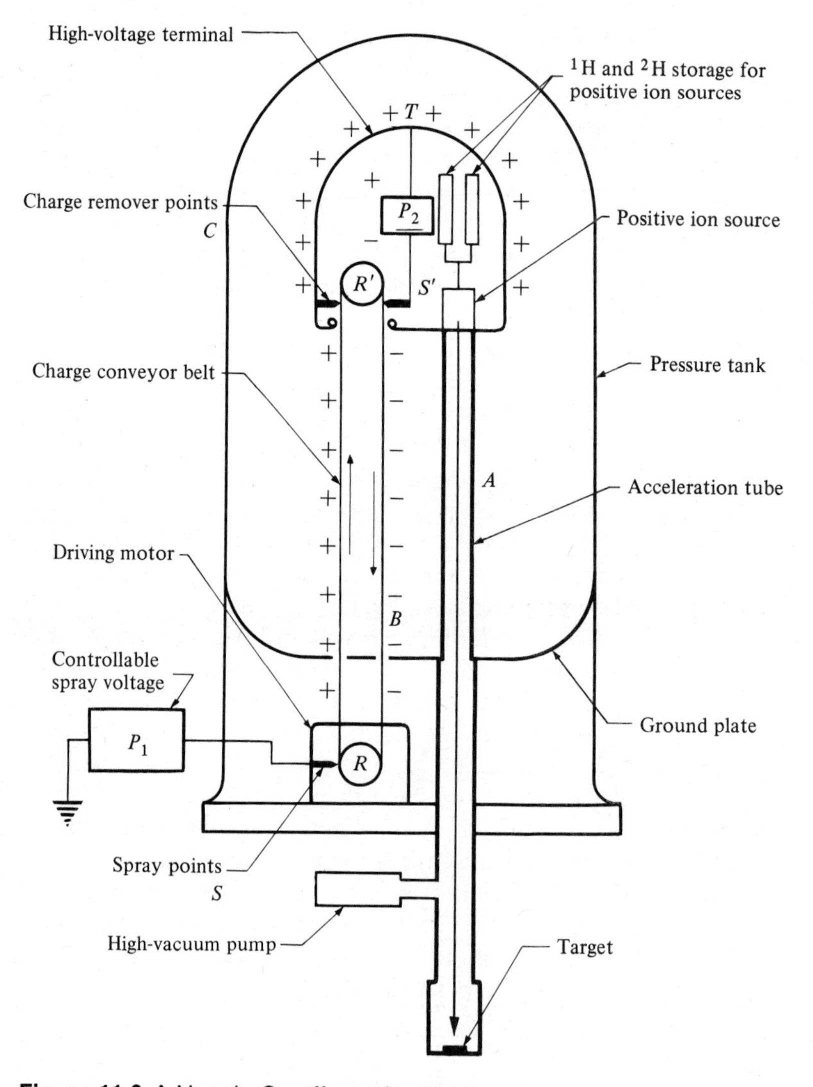

Figure 11.6 A Van de Graaff accelerator.

From Atam P. Arya, *Elementary Modern Physics*, © 1974. Addison-Wesley, Reading, Mass. Reprinted with permission.

charged in the opposite way to the source sphere. Tandem accelerators capable of accelerating protons to 20 MeV with current of about 1.5 microamperes are available commercially. Van de Graaff and tandem Van de Graaff accelerators provide a constant potential that accelerates the ions to a well-defined and measurable energy.

The Cockcroft-Walton and the Van de Graaff accelerators depend on an extremely high voltage to accelerate charged particles. The maximum energy of the particles is limited by the maximum voltage that one can obtain. This maximum voltage is limited by the problem of providing adequate insulation. A *linear accelerator* accelerates charged particles to high energies without the need for very high voltages. A charged particle will accelerate when placed between two electrodes; however, if it is inside a conductor its motion is unaffected by changes of potential on the conductor. In a typical linear accelerator, Fig. 11.7, there is a succession of electrodes. A proton leaves the ion source and is accelerated toward electrode 1, which is charged negative. When it is inside electrode 1 the polarity of all of the electrodes is switched and the particle, upon entering the gap, is presented with a second accelerating voltage. Once again, after having been accelerated through the gap between electrodes 1 and 2 the proton drifts with a uniform velocity through electrode 2. While it is inside electrode 2, the polarity is once again switched so that electrode 3 will accelerate the proton across the next gap. This process can be repeated indefinitely with the energy of the particle being built up in small steps. Since the frequency of the oscillator providing the voltage difference between the electrodes is constant and since the protons gain speed as they go through gap after gap, the drift tubes further from the ion source are longer than those closer. With the technique described above, the linear accelerator at Brookhaven National Laboratory can accelerate protons to 50 MeV and the largest proton linear accelerator, which is at the University of Minnesota, can accelerate protons to 68 MeV.

Linear accelerators for electrons are simpler than those for protons because the electrons at comparatively low energies have speeds almost equal to the speed of

Figure 11.7 A proton linear accelerator.

From Atam P. Arya, *Elementary Modern Physics*, © 1974. Addison-Wesley, Reading, Mass. Reprinted with permission.

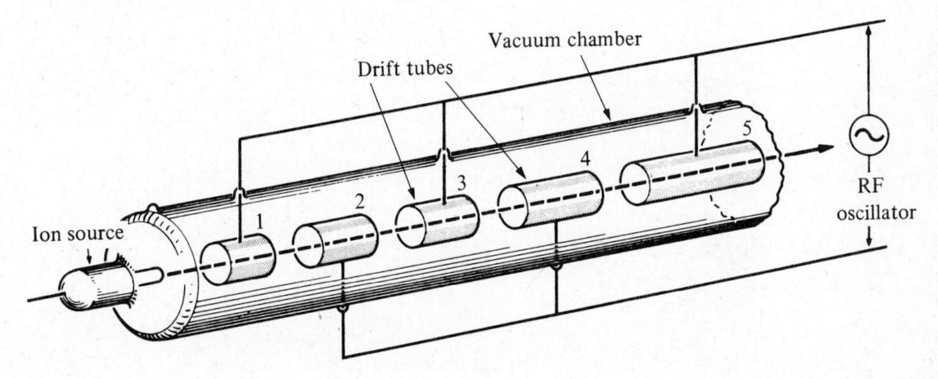

light. Thus, the speed will not increase significantly as the energy is increased further. If electrons are supplied to the linear accelerator from an outside source with an energy of 3 or 4 MeV, and in the correct phase with the electric field traveling down the tube, they will remain in phase and acquire additional energy. The largest linear electron accelerator, shown in Fig. 11.8, has been built at the Stanford Linear Accelerator Center (SLAC). This 3000-meter-long accelerator is contained inside two parallel tunnels. One, the lower one, is 3 meters wide and 4 meters high and contains the accelerator. The upper one is 7.2 meters wide and 6.6 meters high and contains the associated electronic equipment. The SLAC accelerator produces 20 GeV electrons with a beam current of about 12 microamperes. Linear accelerators are very useful as preaccelerators or injectors for the large cyclic accelerators, which will be discussed later.

Another type of accelerator is the *cyclotron* developed by Lawrence and Livingston in 1932. It is like a linear accelerator in that a rather low potential is used successively. In the cyclotron, however, the beam is not linear but is "curled up" by a magnetic field.

Figure 11.8 The Stanford two-mile linear accelerator.
Stanford Linear Accelerator Center, Stanford University.

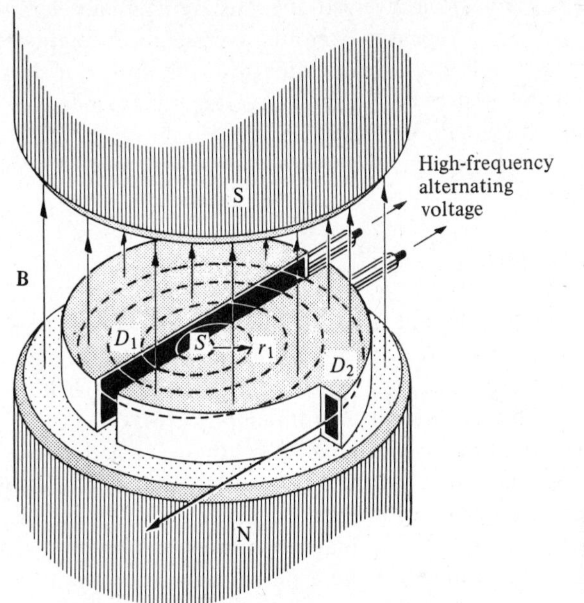

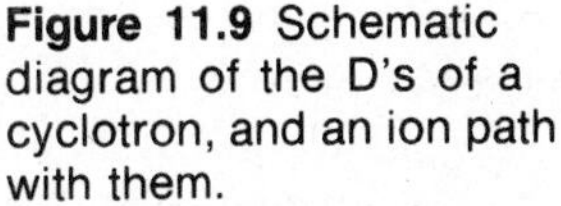
Figure 11.9 Schematic diagram of the D's of a cyclotron, and an ion path with them.

We have seen (Eq. 2.5) that a charged particle moving with a velocity perpendicular to a magnetic field moves in a circular path. The time it takes for the particle to move around the circle once is the length of the path divided by the particle's tangential velocity, or $T = 2\pi r/v$. Using Eq. (2.5), $mv = qBr$, to eliminate the velocity, we have

$$T = \frac{2\pi m}{qB}.$$

Since ω, the angular velocity of the particle, is equal to $2\pi/T$, then

$$\omega = \frac{qB}{m}. \tag{11.26}$$

Thus the angular velocity of the charged particles is independent of their translational velocity. Lawrence utilized this property of moving charges in a magnetic field with the device shown schematically in Fig. 11.9. The heart of the instrument consists of two chambers called "D's" (because of their shape) which have roughly the size and shape of a circular metal cake box sawed in two. The ions to be accelerated are introduced near the center of these D's at S. If D_2 is negative relative to D_1, the positive ion is attracted into D_2 where, because it is within a conducting chamber, it experiences no electric field. A magnetic field perpendicular to the plane of the D's forces the ions into a path with the angular velocity given by Eq. (11.26). In a time $T/2$, the ions return to the gap between the D's, and if by this time the potentials of the D's have been reversed, the ions are again accelerated in going from D_2 into D_1. This process is repeated over and over again. It might appear difficult to switch the potentials back and forth in time with the ions, but this is not so. If the D's are connected to a source of alternating voltage of frequency $\nu = qB/2\pi m$, the potential and the ions remain in phase. The tangential velocity v of the ion is equal to $r\omega$. Therefore the kinetic energy of an ion emerging from the cyclotron along a path of radius r is

$$E_k = \tfrac{1}{2}mv^2 = \tfrac{1}{2}mr^2\omega^2 = \frac{q^2B^2r^2}{2m}. \tag{11.27}$$

The magnitude of the potential used does not appear in the equation. The main problem in building a cyclotron is to produce a very intense magnetic field that is uniform over a large area. Large cyclotrons (Fig. 11.10) can produce 10 MeV protons, 20 MeV deuterons, and 40 MeV alpha particles.

The real upper limit of cyclotrons is set by the relativistic mass variation of the ions accelerated. As the velocity of the ions approaches the velocity of light, the mass increases and the angular velocity is no longer constant. Some instrument designs correct for this effect by making the magnetic induction nonuniform or by frequency modulating the exciting voltage. If the exciting voltage is frequency modulated, the machine is called a *synchrocyclotron* or *frequency modulated (FM) cyclotron*. Although the synchrocyclotron can accelerate particles to a much higher energy than is possible with a cyclotron, its output beam current is much less. There is no theoretical limit to the size of a synchrocyclotron and, therefore, to the energy of the accelerated particles. A practical limit, however, is dictated by the magnet. It is not economically feasible to build magnets for machines of this design for accelerating particles to energies of more than about one GeV.

The development of the synchrotron solved the problem of obtaining higher and higher energy particles without the need for a giant magnet. A synchrotron, rather than having a single large magnet, has many relatively small magnets located along only one orbital radius. Protons are accelerated to a fairly high kinetic energy by a Van de Graaff generator and then injected into the orbit of fixed radius. There they are accelerated once on each revolution by an R.F. accelerator that provides an oscillating electric field. The frequency of the field is synchronized with the orbiting frequencies so that on each turn the protons gain an increment of energy. After making several million revolutions, protons exit the machine with energies of about 3 GeV. To keep the protons in the fixed orbit as they gain energy, the magnetic field must be increased. In a typical machine (for example, the Brookhaven machine shown in Fig. 11.11), the magnetic field changes from 300 to 14,000 gauss while the radio-frequency of the accelerator changes from 0.37 to 4.20 MHz.

Synchrotrons use variable magnetic fields with constant field gradients. An *alternating gradient synchrotron*, on the other hand, uses a fixed field but alternating field gradient. The alternating gradient (AG) technique results in strong focusing of the particles and the variation in their orbit is relatively small. For this reason, the

(a)

(b)

Figure 11.10 (a) The M.I.T. cyclotron. (b) The D's removed from the gap between the poles of the electromagnet.

Courtesy of the MIT Museum.

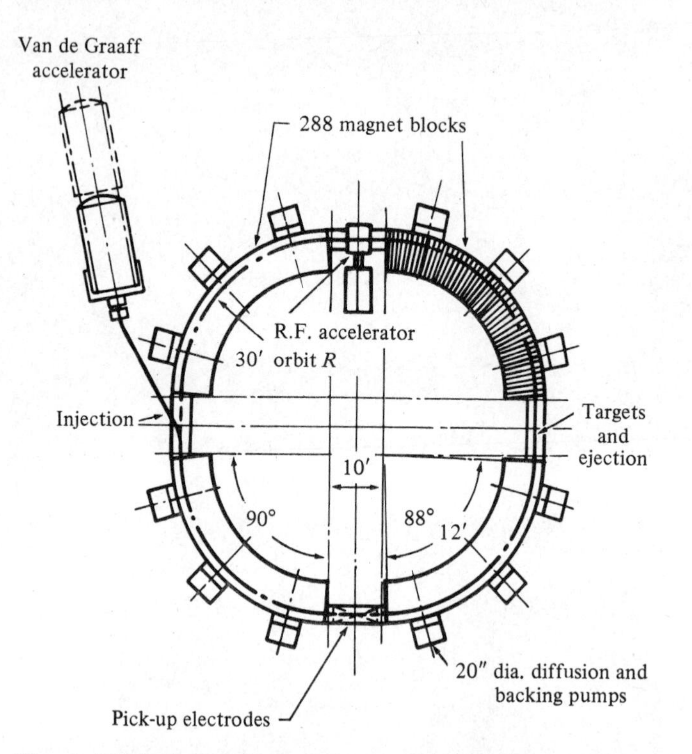

Figure 11.11 The proton synchrotron at Brookhaven National Laboratory.

M. S. Livingston, J. P. Blewett, G. K. Green, L. J. Haworth, *Rev. Sci. Int.* **21**, 7 (1950). Reprinted by permission.

vacuum chamber that contains the beam can be much smaller than for a regular synchrotron; hence, the magnets can be smaller. The Brookhaven AG synchrotron, which was completed in 1960, is a typical machine. Ions are accelerated first by a Cockcroft-Walton accelerator, then by a linear accelerator before being injected into the main ring. Around the center line of the orbit, which is 842 feet in diameter, there are 240 magnets and 12 R.F. acceleration stations.

The largest and most energetic AG synchrotron, which is shown in Fig. 11.12, is at the National Accelerator Laboratory in Batavia, Illinois. Protons are accelerated to 0.75 MeV by a Cockcroft-Walton generator. A linear accelerator further accelerates the protons to an energy of about 200 MeV. After leaving the linear accelerator, the protons enter a booster ring similar to the Brookhaven synchrotron, which raises the energy to 8 GeV. The protons next are injected into the main accelerator ring, which is 6.3 kilometers in circumference. The main ring is divided into six identical sections with each section containing radio frequency accelerators, bending magnets, and focusing magnets. Up to the time of the design of the Batavia accelerator all accelerators used the same magnets to focus the beam as well as to bend it. This arrangement was not suitable in this case and thus the bending magnets are separate from the focusing magnets. Each of the 774 bending magnets is 20 feet long, one foot high, and $2\frac{1}{2}$ feet wide, and weighs 11 tons. Focusing magnets are 7 feet long and weigh 5 tons. These focusing magnets are called quadrapole magnets because they have four poles. The magnetic field is zero at the center of the four poles and increases in every direction away from the center. A proton that moves out from the center of the path has a force exerted on it forcing it back toward the middle.

Protons after 200,000 revolutions in the main ring have about 400 GeV of energy, which is limited by the size of the magnetic field necessary to keep the particles in orbit. For higher energies, larger magnetic fields are needed. Plans for the expansion of the facility include a ring of superconducting magnets that will generate much larger magnetic fields and allow acceleration to energies of about 1000 GeV (one tera-electron volt or TeV). When this new phase is completed, the

Figure 11.12 Aerial view of the Fermi National Accelerator Laboratory, Batavia, Illinois. The large circle is the main accelerator chamber containing conventional electromagnet ring and superconducting magnet ring.
Photograph courtesy of Fermi lab.

present ring of conventional magnets will serve as a booster accelerator from which the protons will be transferred to the new superconducting ring. The present electricity bill for the National Accelerator Laboratory is several million dollars per year. A substantial savings can be had if the protons are accelerated only to 100 GeV in the present ring and then transferred to the superconducting ring. Therefore, the motivation to build the superconducting facility is one of economics as well as one of increased energies.

One technique for greatly increasing the effective energy of an accelerator is to store the particles in intersecting storage rings. Figure 11.13 is a diagram of the intersecting storage rings at CERN near Geneva, Switzerland. Protons are accelerated to 28 GeV in the proton synchrotron on the left of the diagram. From there they are deflected by large magnets alternately to the two intersecting storage rings shown on the right of the figure. The two beams of 28 GeV protons travel in opposite directions intersecting each other in the two regions marked "experimental areas." When two 28 GeV protons collide, the energy available to create new particles is equivalent to that available in the collision of a 1500 GeV proton with a stationary proton.

To illustrate this last point, we will consider the situation of two automobiles colliding. In case one, an automobile collides with an identical stationary automobile (analogous to the situation in a conventional accelerator where a proton hits a

Figure 11.13 Intersecting storage rings at CERN.
From Atam P. Arya, *Elementary Modern Physics*, © 1974. Addison-Wesley, Reading, Mass. Reprinted with permission.

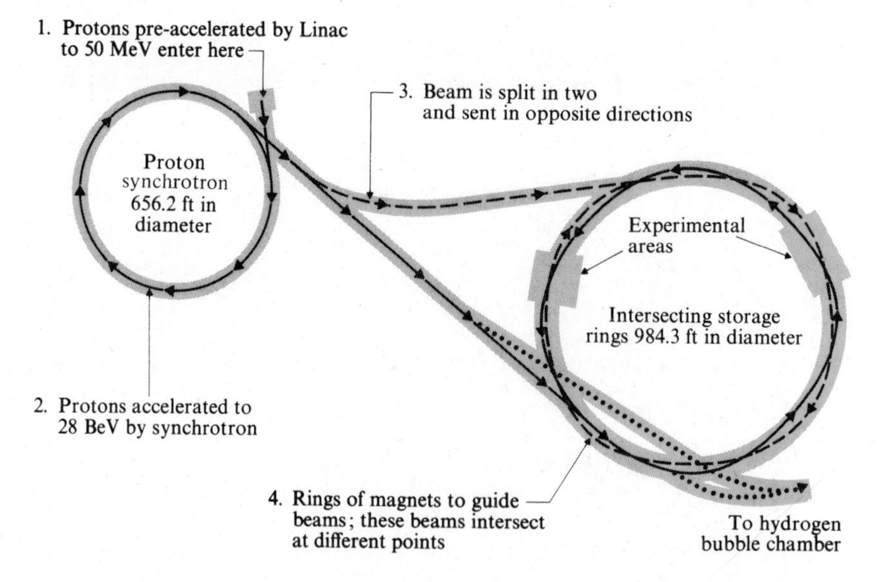

stationary target). Case two involves a head-on collision between two automobiles moving toward each other at the same speed (analogous to the situation using intersecting storage rings where both particles are moving with maximum energy). Simple considerations show that if an automobile hits a stationary automobile of equal mass, the subsequent velocity of the wreckage (both vehicles locked together) is just one half of the velocity of the first automobile before collision. From this we can calculate the amount of energy available for the bending or heating of the metal. This energy will be the difference between the initial kinetic energy and the final kinetic energy, or $\frac{1}{4}mv^2$ (m is the mass of each auto and v is the initial velocity of the first). If we were discussing particles, this so-called collision energy would be the amount of energy available for creating new particles according to $E = mc^2$.

Now consider the second case, which would leave the wreckage motionless. The total kinetic energy before collision is $\frac{1}{2}mv^2 + \frac{1}{2}mv^2$ and the kinetic energy after collision is zero. The available energy to go into the damage of the automobile again will be the difference of the kinetic energy before and after collision. For this case the total is mv^2. Thus four times more energy is available in this case than in the case involving the stationary automobile. (See Problem 11.21.)

The situation with colliding particles that are moving at relativistic speeds is not this simple. However, the qualitative result is the same — much more energy is available for the creation of new particles in the case where the colliding particles are both moving.[†] Intersecting storage rings are planned for the Batavia accelerator discussed above, which if filled with 100 GeV protons from the superconducting ring will allow interactions equivalent to 2.2 million GeV protons striking stationary targets.

Our discussion of the synchrotron has been primarily about accelerating massive particles such as protons. Synchrotrons can be designed to accelerate electrons to a few GeV. However, the acceleration of electrons to higher energies is inefficient because of the radiation losses arising from the acceleration. For this reason, the primary electron accelerators are the linear accelerators, such as the SLAC accelerator in California.

11.6 THE COCKCROFT–WALTON EXPERIMENT

The first nuclear disintegrations induced without the help of alpha particles from radioactive substances were made by Cockcroft and Walton in 1932, the same year that Chadwick discovered the neutron. They used protons as bombarding particles. The protons were accelerated by a high voltage that was achieved by charging

[†] For a discussion of the energy available in collision processes, see the article by G. K. O'Neill referenced at the end of the chapter.

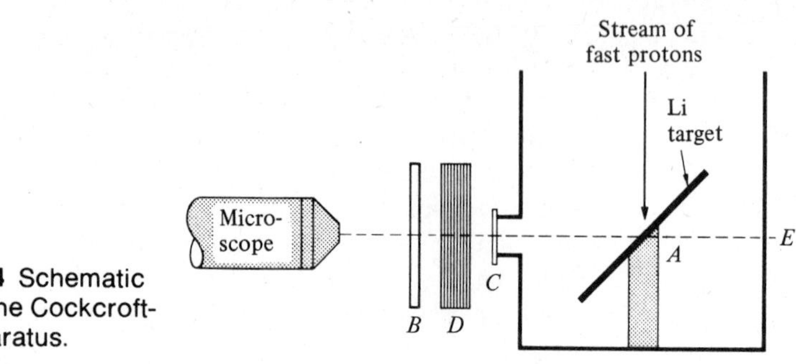

Figure 11.14 Schematic diagram of the Cockcroft-Walton apparatus.

capacitors in parallel and discharging them in series, a technique in use at that time in x-ray technology. Proton energies up to 700,000 eV were obtained in this way.

Cockcroft and Walton found reactions that were often the converse of reactions already known. It was known that alpha bombardment often produced protons, but they also found that proton bombardment often produced alpha particles. Although they bombarded 16 different elements, the quantitative data they reported for lithium and fluorine are of particular interest.

The target region of their apparatus is shown in Fig. 11.14. The fast protons struck the target material at A. A zinc-sulphide fluorescent screen was placed at B, and any scintillations produced could be observed with the microscope at the left. The mica sheet at C held the vacuum in the accelerating tube and it had sufficient stopping power to prevent scintillations due to scattered protons. With lithium as the target, scintillations were observed that appeared to be alpha particles. When additional mica sheets, D, of known stopping power were introduced, the rays were found to have a range equivalent to about 8 cm of air. An ionization chamber and a cloud chamber were also used to positively identify the alpha particles and to determine that their range was 8.4 cm (Fig. 11.15). This range indicated an energy of 8.5 MeV for each particle. Evidently the reaction taking place was

$${}_1^1\mathrm{H} + {}_3^7\mathrm{Li} \rightarrow {}_4^8\mathrm{Be} \rightarrow {}_2^4\mathrm{He} + {}_2^4\mathrm{He}. \tag{11.28}$$

Since the lithium atom had only thermal motion, it could be regarded as being at rest, and since the proton velocity was small compared with that of the observed alpha particle, Cockcroft and Walton assumed that the two product alpha particles must acquire equal and opposite velocities (conservation of energy and of momentum require this if the initial energy and momentum may be neglected). Cockcroft and Walton tested this assumption in the following way. They made the target very thin and added another scintillation detector at E, opposite the original one, as shown in Fig. 11.14. They could then observe *simultaneous* scintillations, which indicated that both alpha particles came from a single disintegration.

They also checked their assumption by setting up the mass-energy equation of the reaction, as Chadwick did:

Proton mass + lithium mass + input energy in mass units

= 2(alpha-particle mass + alpha-particle energy in mass units).

When they substituted numerical values in this equation, they disregarded the kinetic mass-energy of the proton because this was less than the uncertainty in the isotopic mass of lithium known at that time. Using values from *their* data, we have, in u,

$$1.0072 + 7.0134 = 2(4.0011 + E_{k\alpha})$$

or

$$E_{k\alpha} = 0.0092 \text{ u} = 8.6 \text{ MeV per particle.} \qquad (11.29)$$

This is in good agreement with the energy determined from the range, 8.5 MeV.

In the case of fluorine, the reaction was

$${}^{1}_{1}\text{H} + {}^{19}_{9}\text{F} \rightarrow (\text{excited})\,{}^{20}_{10}\text{Ne} \rightarrow {}^{16}_{8}\text{O} + {}^{4}_{2}\text{He}. \qquad (11.30)$$

They again wrote the mass-energy equation to find the energy liberated. By dividing the energy between the oxygen and the alpha particle according to the laws of

Figure 11.15 Cloud-chamber photograph of pairs of alpha particles from proton disintegration of lithium. The lithium target is close to a thin mica window at the center of the chamber.
Courtesy of P. I. Dee, University of Glasgow.

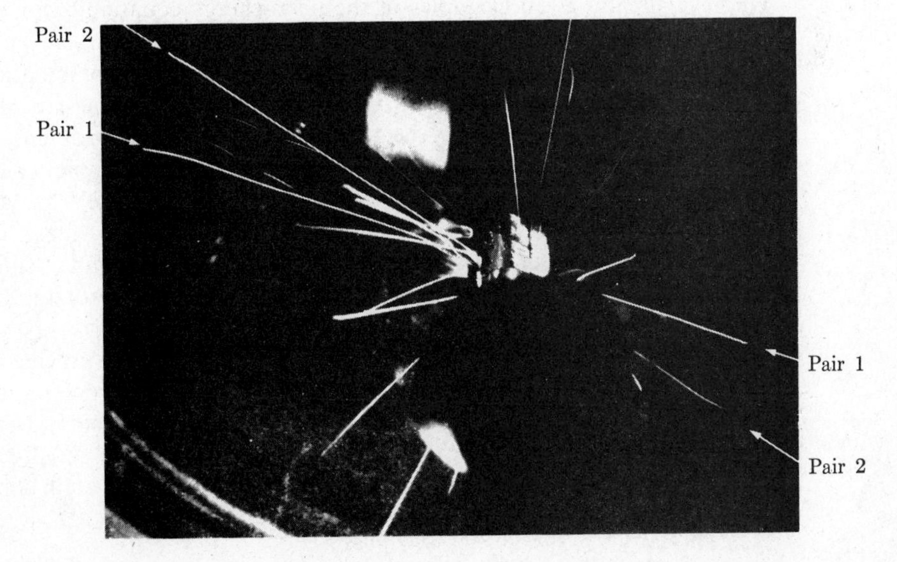

conservation of energy and of momentum, they again obtained good agreement between the measured and computed alpha-particle energies.

The significance of the Cockcroft–Walton experiments can hardly be overemphasized. They were the first to use an ion accelerator to produce nuclear disintegrations, and they proved, beyond any doubt, the equivalence of mass and energy. Together with Chadwick, their effective use of the mass-energy equation to account for the mass-energy balance of nuclear reactions provided a powerful tool for nuclear physics.

It is appropriate that we pause here and reflect on the power of abstract reasoning. When the search for the ether drift failed, Einstein decided to make physics valid without an ether medium. His theory of relativity gave a unique significance to the speed of light. He proposed that mass and energy are different manifestations of the same thing, with the speed of light squared as the constant of proportionality. This consequence of relativity was in violent disagreement with the highly successful Newtonian mechanics, and Einstein had not a shred of experimental evidence to support him. In the Cockcroft–Walton experiment, we find not only that Einstein is justified, but also that his concept of mass-energy equivalence provides the basic key to the application of conservation of energy to nuclear processes. Einstein, who helped Perrin establish the atomic view of matter, had never heard of the nucleus when he proposed the theory of relativity.

11.7 NUCLEAR MASS-ENERGY EQUATIONS. *Q*-VALUE

We have already given examples of the mass-energy equation; we now consider its use in detail. This equation is the relativistic combination of the two classical principles of conservation of mass and of energy. The combination is effected through the use of the relation $E = mc^2$, which expresses the equivalence of mass and energy. Mass-energy may be measured in either mass or energy units. The conversion factor is c^2. Using units convenient to nuclear physics, we may say that 1 u has an energy of 931 MeV or that an energy of 931 MeV has a mass of 1 u. Nuclear mass-energy equations are usually written in atomic mass units, although any consistent units of mass or energy can be used. The fundamental equation is simply that the total mass-energy before a reaction equals the total mass-energy after the reaction.

Let us consider a general kind of reaction involving moving particles and photons. Let m_1 and m_2 be the rest masses of the initial particles and E_{k1} and E_{k2} their respective kinetic energies; let m_3 and m_4 be the rest masses of the final particles and E_{k3} and E_{k4}, their respective kinetic energies; finally, let ν_0 be the frequency of an initial photon and ν be the frequency of a final photon. Upon equating the total mass-energy before the reaction to that afterwards, we obtain

$$m_1c^2 + E_{k_1} + m_2c^2 + E_{k_2} + h\nu_0 = m_3c^2 + E_{k_3} + m_4c^2 + E_{k_4} + h\nu \tag{11.31}$$

or, transposing and collecting similar terms,

$$[(m_1 + m_2) - (m_3 + m_4)]c^2 = [(E_{k_3} + E_{k_4}) - (E_{k_1} + E_{k_2})] + h(\nu - \nu_0). \tag{11.32}$$

The right-hand side of this equation is the change in energy resulting from the reaction. The energy that is released or absorbed in a nuclear reaction is called the *Q-value* or *disintegration energy* of the reaction. Thus from Eq. (11.32) we have

$$\begin{aligned} Q &= [(m_1 + m_2) - (m_3 + m_4)]c^2 \\ &= [(E_{k_3} + E_{k_4}) - (E_{k_1} + E_{k_2})] + h(\nu - \nu_0). \end{aligned} \tag{11.33}$$

If the rest masses are in u, then the Q-value in u is

$$Q = (m_1 + m_2) - (m_3 + m_4). \tag{11.34}$$

The Q-value may be either positive or negative. If it is positive, then rest mass-energy is converted to kinetic mass-energy, radiation mass-energy, or both and the reaction is said to be exoergic or exothermic. If it is negative, the reaction is endoergic or endothermic and then either kinetic mass-energy or radiation mass-energy must be supplied if the reaction is to take place.

We are particularly interested in applying the equations we have just obtained to nuclear reactions. When we calculate the Q-value of a nuclear reaction, the masses used in either Eq. (11.33) or Eq. (11.34) must be the rest masses of the nuclei. However, the isotopic masses given in tables such as that in Appendix 5 are the rest masses of the neutral atoms. To obtain the rest mass of the nucleus of an atom, we must subtract the rest masses of all the orbital electrons from the isotopic mass. Are there any circumstances in which the rest masses of the neutral atoms may be used instead of those of the bare nuclei to calculate the disintegration energy?

Consider the reaction

$${}^{11}_{5}\mathrm{B} + {}^{1}_{1}\mathrm{H} \rightarrow {}^{8}_{4}\mathrm{Be} + {}^{4}_{2}\mathrm{He}. \tag{11.35}$$

To obtain the Q-value of this reaction, we will substitute the nuclear masses into Eq. (11.34). Let the chemical symbols represent the isotopic masses of the atoms in u and let e represent the rest mass of an electron in the same units; then we have

$$\begin{aligned} Q &= [(\mathrm{B} - 5\mathrm{e}) + (\mathrm{H} - \mathrm{e})] - [(\mathrm{Be} - 4\mathrm{e}) + (\mathrm{He} - 2\mathrm{e})] \\ &= (\mathrm{B} + \mathrm{H} - 6\mathrm{e}) - (\mathrm{Be} + \mathrm{He} - 6\mathrm{e}) \\ &= (\mathrm{B} + \mathrm{H}) - (\mathrm{Be} + \mathrm{He}). \end{aligned}$$

This result shows that the *isotopic masses of the atoms can be used* to compute the Q-value in this case.

Next we consider a radioactive substance undergoing beta decay, that is, emitting an electron. An example is

$${}^{106}_{47}\mathrm{Ag} \rightarrow {}^{106}_{48}\mathrm{Cd} + {}^{0}_{-1}\mathrm{e} \ (\text{or } \beta^-). \tag{11.36}$$

Following the same procedure as before, we find

$$\begin{aligned} Q &= [(\text{Ag} - 47\text{e})] - [(\text{Cd} - 48\text{e}) + \text{e}] \\ &= (\text{Ag} - 47\text{e}) - (\text{Cd} - 47\text{e}) \\ &= \text{Ag} - \text{Cd}. \end{aligned}$$

Therefore, *to find the disintegration energy in reactions involving electron emission, the isotopic masses may be used if the mass of the emitted electron is disregarded.* It should be remembered that the kinetic energy of the beta particle calculated from this and similar reactions is the maximum value it can have under the circumstances. In general, this amount of energy is actually shared between the emitted electron and the neutrino. Since the latter has negligible rest mass, it does not enter into the calculation of the Q-value.

In some cases positrons are emitted from nuclei. A reaction of this type is

$$^{13}_{7}\text{N} \rightarrow {}^{13}_{6}\text{C} + {}^{0}_{+1}r \text{ (or } \beta^{+}). \tag{11.37}$$

In this case, we have

$$\begin{aligned} Q &= [(\text{N} - 7\text{e})] - [(\text{C} - 6\text{e}) + \text{e}] \\ &= (\text{N} - 7\text{e}) - (\text{C} - 7\text{e} + 2\text{e}) \\ &= \text{N} - (\text{C} + 2\text{e}). \end{aligned}$$

Therefore, to compute the energy change *in reactions involving positron emission, the isotopic masses may be used if two electron masses are included with those of the product particles.* In this case too, in general, the positron shares kinetic energy with the neutrino.

In following these rules for reactions involving alpha particles, α, or protons, p, we use the isotopic masses of $^{4}_{2}$He and $^{1}_{1}$H.

In obtaining data to determine isotopic masses, we find that the results of mass spectroscopy provide only a starting point. We saw how Chadwick used the mass-energy equation to compute the mass of the neutron. In Eq. (11.25), the rest-mass terms are much larger than the kinetic energy-mass terms; therefore Chadwick could obtain the mass of the neutron to five significant figures. The repeated application of the mass-energy equation to a large number of nuclear reactions has enabled the computation of isotopic masses with great precision. Many are now known to either seven or eight significant figures.

In nuclear collisions having a Q-value, the treatment is like that given in Eqs. (11.14) through (11.20), except that the sum of the kinetic energies of the final particles is equal to $E_{k1} + Q$ instead of E_{k1}. Also, if the collision is not head-on, the law of conservation of momentum must be written in component form.

11.8

CENTER-OF-MASS COORDINATE SYSTEM. THRESHOLD ENERGY

When the law of conservation of energy and the law of conservation of momentum were written to solve the nuclear collision discussed in Section 11.3, it was *implicit* that the velocities in the various expressions were the values in the *laboratory coordinate system*. In this *laboratory*, or *L-system*, it was assumed that the target nucleus was at rest before the collision and that only the bombarding particle was in motion. It turns out that the treatment of nuclear collisions is easier if the center of mass of the particles is taken as the reference system. In the *center-of-mass* or *C-system*, the center of mass of the incident particle and the target nucleus is considered to be at rest and both particles are approaching it. The equations obtained by an observer moving with the center of mass are relatively simple and may be readily solved. The results can easily be transformed back to the *L*-system.

The relationship between the *L*- and *C*-systems is shown in Fig. 11.16. According to the law of conservation of momentum, the momentum of the center of mass in the *L*-system must equal the sum of the momenta of the particles in the same system. Therefore, assuming low velocities so that classical mechanics may be used, we have

$$mv = (m + M)V_c$$

or

$$V_c = \frac{m}{m + M}v. \tag{11.38}$$

In the *C*-system, before collision, the incident particle has a velocity of $v - V_c$ toward the right and the target has a velocity of V_c toward the left. Therefore the

Figure 11.16 An elastic collision between two particles as described in the laboratory system and the center-of-mass system.

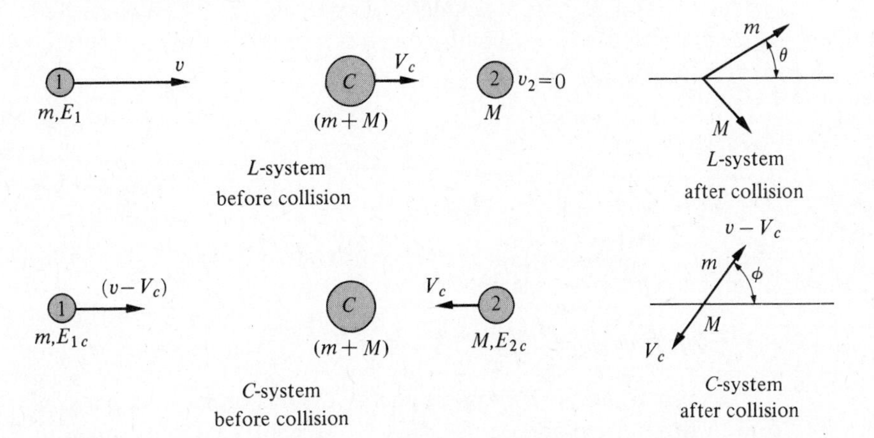

total momentum p_c in this reference frame is

$$p_c = m(v - V_c) - MV_c.$$

Substituting V_c from Eq. (11.38) into the last equation, we get

$$p_c = m\left(v - \frac{mv}{m + M}\right) - M\frac{mv}{m + M}$$
$$= m\frac{Mv}{m + M} - M\frac{mv}{m + M} = 0.$$

Because momentum is conserved, it follows that after collision the magnitude of the momentum of each particle will be the same but oppositely directed, as seen from the C-system. If the collision is *elastic*, kinetic energy is also conserved and therefore the speeds of the particles in the C-system will be unchanged by the collision. Thus an observer in the C-system will describe an elastic collision as one in which the magnitude of each velocity is unchanged but in which the direction of each changes by the same amount, ϕ in Fig. 11.16. The velocity of each particle in the L-system is obtained by adding V_c vectorially to the velocity of each in the C-system.

Let us now consider an *inelastic* collision, particularly an endoergic nuclear reaction. As before, the total momentum in the C-system before collision is zero and so is the momentum afterward. In the L-system, the initial momentum of the system is not zero and therefore the momentum of the final particles in it cannot be zero. Thus these final particles must have kinetic energy. Because of this, not all of the energy of the bombarding particle in a nuclear reaction is available to provide the Q-value required to cause the reaction. The minimum amount of energy that a bombarding particle must have in the L-system in order to initiate an endoergic reaction is called the *threshold energy*, E_{th}. The expression for this energy can be found from the Q-value relation, Eq. (11.33), which is valid in all inertial reference frames. Writing this for the C-system, we have

$$Q = (E_{3c} + E_{4c}) - (E_{1c} + E_{2c}), \tag{11.39}$$

where E_{3c} and E_{4c} are the kinetic energies of the product particles. At threshold, these particles are formed with no velocity in the C-system. Therefore $(E_{3c} + E_{4c}) = 0$, and Eq. (11.39) becomes

$$Q = -(E_{1c} + E_{2c})$$

or

$$-Q = \tfrac{1}{2}m(v - V_c)^2 + \tfrac{1}{2}MV_c^2. \tag{11.40}$$

When V_c in this equation is replaced by its equivalent from Eq. (11.38) and the resulting expression is simplified, we find that

$$-Q = \frac{1}{2}mv^2\left(\frac{M}{m + M}\right). \tag{11.41}$$

The quantity $\frac{1}{2}mv^2$ is the kinetic energy of the bombarding particle in the L-system, and in this particular case it equals E_{th}. When E_{th} is substituted in Eq. (11.41), the equation becomes

$$-Q = E_{\text{th}}\left(\frac{M}{m+M}\right)$$

or

$$E_{\text{th}} = (-Q)\left(1 + \frac{m}{M}\right). \tag{11.42}$$

Thus the kinetic energy of the incident particle in the laboratory system must exceed the Q-value to cause an endoergic reaction.

11.9 ARTIFICIAL (INDUCED) RADIOACTIVITY

Although the work of Cockcroft and Walton stimulated the investigation of many nuclear reactions with better and better ion accelerators, we have not come to the end of important discoveries made with alpha particles from radioactive sources. In 1934, while I. Curie-Joliot and F. Joliot were bombarding aluminum with alpha particles from polonium, they observed neutrons, protons, and positrons coming off the aluminum. We have seen that alpha bombardment of light elements often produces protons and sometimes neutrons. Even the presence of positrons was not too surprising, since, on the basis of isotopic masses known in 1934, protons could be thought of as a neutron and positron combined. Their new discovery was made when they noted that the emission of positrons continued *after* the alpha bombardment was stopped. The positron activity decreased with time according to an exponential law and the phenomenon was clearly just like natural radioactivity. They assumed that the artificially radioactive element was ${}^{30}_{15}\text{P}$, formed according to the reaction

$${}^{27}_{13}\text{Al} + {}^{4}_{2}\text{He} \rightarrow {}^{30}_{15}\text{P} + {}^{1}_{0}\text{n}. \tag{11.43}$$

The phosphorus decays with a half-life of 2.55 m into silicon and a positron:

$${}^{30}_{15}\text{P} \rightarrow {}^{30}_{14}\text{Si} + {}^{0}_{+1}\text{e}. \tag{11.44}$$

To justify their assumption, the Joliots irradiated aluminum for a long period and then separated the phosphorus chemically. Since the radioactivity went with the phosphorus fraction, their assumption was verified.

In their original paper, the Joliots also reported the formation of radioactive nitrogen and silicon isotopes by bombardment of boron and magnesium, respectively. These new isotopes have half-lives of 10.1 m and 4.9 s. We can call these isotopes "new" since, if they were ever abundant in nature, they are now present in undetectable amounts.

In the years since the original discovery of artificial radioactivity, it has been found possible to form radioactive isotopes of all the elements, and for most elements there are many radioactive isotopes. Thus many more than half the known isotopes are radioactive.

When Madame Curie separated radium, she could follow the progress of the separation by observing the increased activity of the radium fraction. The radium was radioactively "tagged." In the discovery of artificial radioactivity, a crucial point was that in the separation of the aluminum and phosphorus the radioactivity went with the phosphorus. The actual amount of phosphorus was exceedingly small, but radioactivity is easy to detect and small amounts can be traced through successive chemical processes. This property of radioisotopes has been developed into a fine art called "tracer technique." We give a few examples of its applicability.

When iodine is taken into the body, it tends to collect in the thyroid gland. Patients who have cancer of the thyroid may be given radioactive iodine. The radioactivity goes just where it should to fight the malignancy.

One pipeline may be used to transport, successively, several kinds of petroleum products with surprisingly little mixing at the boundaries. At the source end of the line, a radioactive material is introduced between products. At the receiving end of the line, Geiger counters outside the pipe announce the arrival of the new product.

If a small amount of a radioelement is mixed with a large amount of the same stable element, the relative proportions of these remain fixed except for the predictable decay of the radioelement. Thus the concentration of that element can be followed through all kinds of chemical and physical processing by measuring the intensity of the activity. It may well be that the new knowledge of phenomena in the basic sciences, obtained by the use of radioactive tracers, will be more important than power from nuclear reactors.

11.10 AGE DETERMINATION OF SAMPLES YOUNGER THAN THE EARTH

One of the radioactive isotopes that is midway between natural radioactivity and artificial radioactivity is ^{14}C, which disintegrates by beta decay into stable ^{14}N. It is natural radioactivity in the sense that it occurs in nature, and it is artificial in the sense that it would not occur in nature if it were not constantly being re-formed. Carbon-14 is formed in the atmosphere by high-energy particles from outer space called cosmic rays (refer to Problem 11.41). Carbon-14 has a half-life of 5730 years. In the atmosphere there is a kind of "radioactive equilibrium" between the production of ^{14}C by cosmic rays and its diminution by radioactive disintegration. Fortunately for us, the concentration of ^{14}C in the air we breathe and the food we eat is very small. In the body the concentration is only about $1.8 \times 10^{-10}\%$ of the ^{12}C in living tissue.

But we and every other living thing contain carbon, and all living things have an amount of ^{14}C that is in equilibrium with the ^{14}C in the atmosphere. When death comes, living things stop breathing and stop eating food. The intake of ^{14}C stops. From the time of death, the ^{14}C disintegrates without further replacement. Thus, at death, equilibrium is ended and exponential radioactive decay is the only process remaining.

Old wood contains less ^{14}C than new wood (Problem 11.44). Old bones have less ^{14}C than new bones. By measuring the concentration of ^{14}C, the time since death occurred can be computed. Thus ^{14}C provides a radioactive clock for anthropologists just as uranium provides a radioactive clock for geologists. The ^{14}C half-life is suitable for the dating of cultural history, up to an age of several tens of thousands of years, just as the half-life of uranium is suitable for the dating of the history of the earth.

A technique for measuring the age of samples of the order of millions of years old uses the radioactive isotope ^{40}K. This isotope of potassium decays into ^{40}Ar with a half-life of 1.28×10^9 years. This process takes advantage of the fact that volcanic material is ejected from beneath the surface of the earth at high pressure and high temperature. When it begins to cool in the atmosphere it coalesces. The solidification takes place very quickly and thus the samples are completely free of argon. From that time on, ^{40}K contained in the sample changes into ^{40}Ar, which is trapped inside the crystal where it remains until the sample is recovered for analysis. The analysis consists of melting the sample to free the argon gas in the chamber of a mass spectrometer. With the known half-life, this measurement of the amount of ^{40}Ar yields the age of the sample. Paleoanthropologists knowing the age of the volcanic rock surrounding various fossils thus know the age of the fossils themselves.

11.11 NUCLEAR BINDING ENERGY

In Chapter 10, we said nothing about the *cause* of radioactivity. Why are some isotopes stable while others are not? Why do those that are unstable postpone their disintegration and have probable lives ranging from microseconds to billions of years? We can say something definite about the first question and something plausible about the second.

We shall limit our discussion to radioactive alpha emitters, although a similar argument can be made for the other cases. Let us inquire whether ordinary aluminum, $^{27}_{13}Al$, is alpha radioactive on the basis of mass-energy. The assumed reaction would be

$$^{27}_{13}Al \rightarrow ^{23}_{11}Na + ^{4}_{2}He + Q. \tag{11.45}$$

From the table of isotopic masses, we find for this case that

$$26.98153 = 22.98977 + 4.00260 + Q$$

or

$$Q = -0.01084 \text{ u} = -10.05 \text{ MeV}. \tag{11.46}$$

The negative Q-value indicates that instead of taking place spontaneously, this reaction cannot proceed unless we supply 10.05 MeV of energy per disintegration. This isotope of aluminum is very stable against alpha decay. Now consider radium,

$$^{226}_{88}Ra \rightarrow ^{222}_{86}Rn + ^{4}_{2}He + Q. \tag{11.47}$$

This gives

$$226.0254 = 222.0175 + 4.0026 + Q$$

or

$$Q = +0.0053 \text{ u} \qquad \text{or} \qquad +4.93 \text{ MeV}. \tag{11.48}$$

The positive Q-value indicates that radium is unstable and can emit an alpha particle, giving off 4.93 MeV of energy in the process. This result is the combined kinetic energy of the alpha particle and the radon nucleus. (The combined kinetic energy actually is not quite as high as 4.93 MeV because this Q-value includes the energy of the gamma-ray photon emitted during the disintegration.)

Rather than carry out in detail energy calculations like those of Eqs. (11.46) and (11.48) for every reaction in which we might be interested, it is very helpful to view the matter graphically. Every energy calculation involves an energy difference and requires that the energy reference level be specified. The total *binding energy* of a nucleus is defined as the energy that would be required to separate the nucleus into isolated particles. Before the discovery of the neutron, nuclei were thought to be composed of protons and electrons. This view presented certain theoretical difficulties that were resolved by regarding the nucleus as composed of protons and neutrons — collectively called *nucleons*. When an atom is discussed with particular emphasis on its nuclear composition, it is called a *nuclide*.

The total binding energy is equal to the difference between the isotopic mass of the atom and the sum of the masses of its neutrons, protons, and extranuclear electrons. (The mass of a proton and an extranuclear electron is equal to that of hydrogen.) It is obvious that the total binding energy of heavy nuclei is greater than that of light nuclei because they have more nucleons to separate. But one might expect the binding energy per nucleon to be about constant, and indeed it is about constant for heavy elements. However, the variations are significant, as is shown in Fig. 11.17, which is based on data for stable and nearly stable nuclides. This figure is *usually* presented with the curve concave downward, and is then called the binding-energy curve. Our figure is consistent with the scheme we used in discussing the Bohr energy levels in hydrogen. The distance from the curve up to the zero energy level is proportional to the energy per nucleon that would be required to take the nucleus apart, just as the distance from a Bohr energy level to the top is proportional to the energy required to free an electron from an atom. The important characteristic of this curve is that it is concave upward with a minimum around mass number 56. From this curve we can see, qualitatively, the results we obtained for the alpha emission of aluminum and radium.

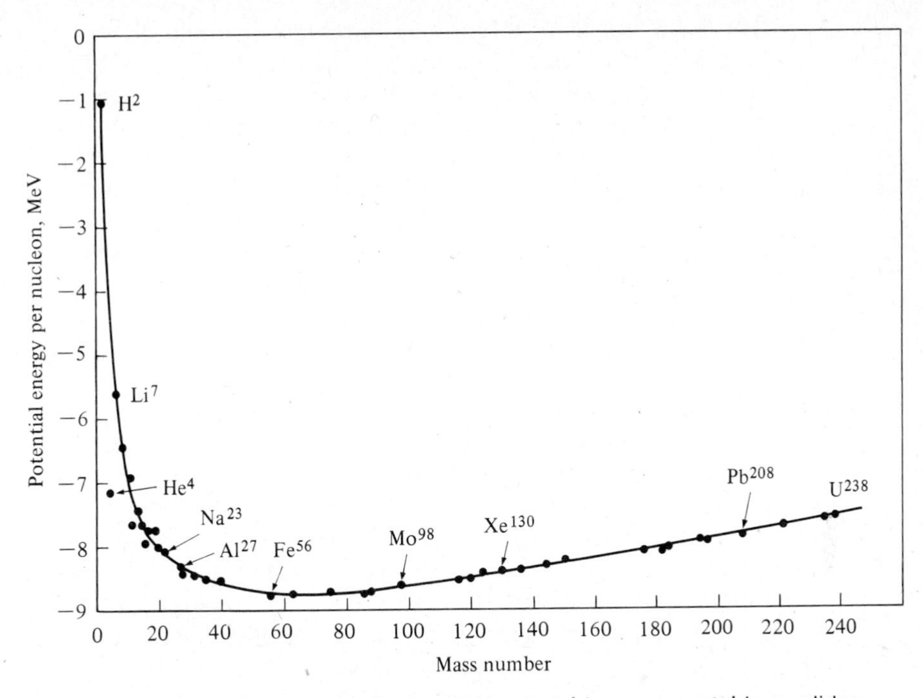

Figure 11.17 Potential energy per nucleon of stable or near-stable nuclides relative to the state of complete separation.

Alpha-particle emission always produces a new nucleus with a mass number that is four less than that of the original nucleus. If we find aluminum with mass number 27 on the curve and move to the left to sodium with mass number 23, the ordinate becomes less negative. Therefore, in going from aluminum to sodium, we must have done some of the work necessary to take aluminum apart. Since we must do work to make the reaction proceed, the reaction is endothermic and cannot take place spontaneously. In going from radium to radon, however, we go to a new element that is harder to take apart than the element with which we started. Since the new element is more stable than the original element, the reaction is exothermic and can take place spontaneously. This graph makes it evident that those reactions which produce new nuclei with mass numbers nearer the minimum of the curve are the ones that are likely to be exothermic.

In general, to release energy a nuclide having a mass number greater than about 56 has to disintegrate, whereas one below about 56 will have to combine with a particle. The fission of uranium is an outstanding example of the release of energy by disintegration; the fusion of hydrogen is a case of releasing energy by nuclear combination. From this curve we can understand too why there is a limit to the number of naturally occurring elements. Since the naturally radioactive series of elements are all at the right where the curve bends upward, it is logical to assume that if any heavier elements were formed, they would be very unstable and would have long since ceased to exist. We shall see in Chapter 12 that several of these elements have been artificially created and have been found to be unstable, with short half-lives compared with the age of the earth.

In addition to the mass-energy equation, there are subtler factors that influence the stability of a nuclide. For example, a nuclide of atomic number Z and mass number A has A nucleons, of which Z are protons and $A - Z$ are neutrons. Nuclides having the same value of Z are *isotopes*; those having the same value of A are called *isobars*; those having the same number of neutrons are called *isotones*; and those having the same excess of neutrons over protons, $A - 2Z$, are *isodiapheres*. Thus $^{37}_{17}Cl$ is an isotope of $^{35}_{17}Cl$, an isobar of $^{37}_{16}S$, an isotone of $^{39}_{19}K$, and an isodiaphere of $^{39}_{18}Ar$.

A survey of 281 stable nuclides shows that 165 of them have an even number of protons and an even number of neutrons; 53 have an odd number of protons and an even number of neutrons; 57 have the number of neutrons odd and the number of protons even; and only 6 have odd numbers of both kinds of nucleon. All of these odd-odd nuclides except $^{14}_{7}N$ are rare. The even-even type are more stable and are, therefore, the least likely to be radioactive. Six nuclides of this type together comprise about 80% of the earth's crust: $^{16}_{8}O$, $^{24}_{12}Mg$, $^{28}_{14}Si$, $^{40}_{20}Ca$, $^{48}_{22}Ti$, and $^{56}_{26}Fe$. Further light on the question of nuclear stability comes from quantum mechanics.

11.12 RADIOACTIVITY AND QUANTUM MECHANICS

No elements except hydrogen could be stable were it not for very strong, short-range attractive forces between nucleons. Without such forces, the protons would fly apart by coulomb repulsion. Figure 11.18 is an approximate potential energy diagram for an alpha particle in or near a nucleus having an effective radius b. When the alpha particle is far from the nucleus $r > b$, it is repelled according to Coulomb's law, and its electric potential energy is then given by $E_{p\alpha} = (1/4\pi\varepsilon_0) \times (2Ze^2/r)$. In the figure, the solid curve sloping away from the nucleus represents $E_{p\alpha}$ as a function of r. The Coulomb force on the alpha particle is equal to the slope of the potential energy curve, $F = -dE_{p\alpha}/dr$. The central part of the curve is shown dashed because we do not know its shape, but we do know that the curve must dip as the strong attractive short-range nuclear forces overcome the Coulomb repulsion. The distance b is equal to the nuclear radius introduced in our discussion of Rutherford's alpha-particle scattering. This potential energy wall around the nucleus tends to prevent the entrance or exit of alpha particles whose energies are less than the height h of the potential barrier. The barrier is higher for nuclei of larger atomic number and would be half as high for protons as it is for alpha particles. Thus this barrier illustrates the difficulty of disintegrating heavy elements with alpha particles and

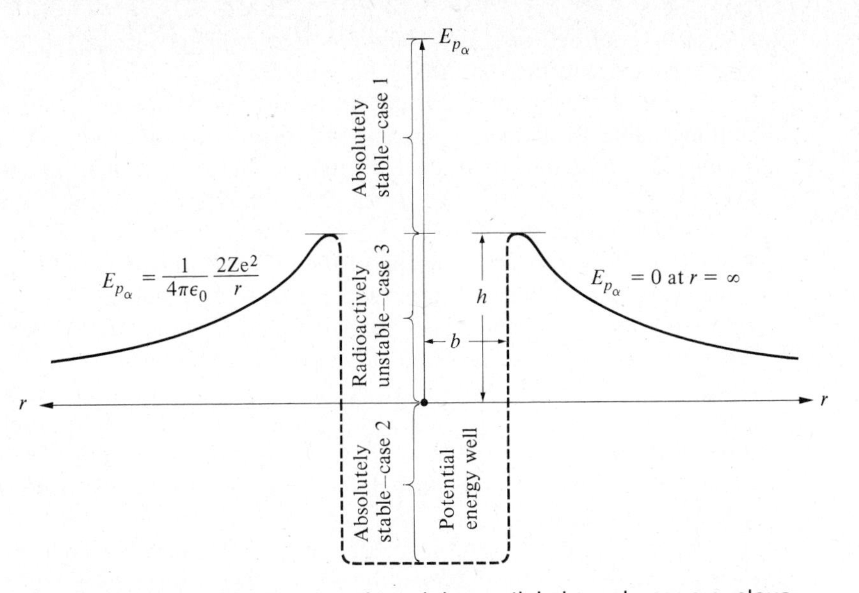

Figure 11.18 Potential energy of an alpha particle in and near a nucleus.

shows why protons with a given energy can disintegrate more nuclides than can alpha particles. The situation is much like the problem of rolling a ball up a volcanic cone into the crater.

Although the alpha particle emitted by a radioactive substance like radium probably forms from two protons and two neutrons at the moment of ejection, we can think of the alpha particle as an entity within the radium nucleus. There are three cases we must consider. If this alpha particle had more energy than the height of the potential barrier, it could easily overcome the attractive force and thus nuclear disintegration would occur at once. If the alpha particle in the nucleus has a negative energy, the barrier is insurmountable and the particle can never get out. If the alpha particle has positive energy but it is less than h, the nucleus is radioactive.

An example of the first case is the intermediate nuclide, (excited) ^{20}Ne, Eq. (11.30). The neon nucleus immediately emits an alpha particle. An example of the second case is the fluorine, ^{19}F, of this same reaction, which is completely stable. Samples of the third case include all the radioactive nuclides, both natural and artificial. Radium C′, ^{214}Po, is very unstable, with a half-life of 1.64×10^{-4} s, and ^{238}U is nearly stable, with a half-life of 4.5×10^9 yr.

Cases one and two are perfectly understandable from the classical energy viewpoint. Particles with more energy than the barrier are not hindered by it. Particles with energy far below the barrier are completely stopped by it. But, according to the classical energy viewpoint, the nuclides of case three should also be stable. If the nuclear alpha particle has less energy than the height of the barrier, it should be confined just as certainly as if its energy were negative. If this classical view were correct, there would be no radioactivity.

In our discussion of quantum mechanics, we compared the classical and quantum mechanical treatments of a harmonic oscillator such as a mass oscillating on a spring. The mass has a fixed energy that is alternately potential and kinetic, and it moves subject to an attractive force determined by the spring. This force may be described as producing a potential barrier that prevents the mass from getting away. According to the classical solution of the problem, the mass never gets farther from its central position than its amplitude, which is the distance to the potential barrier. But in the wave-mechanical solution, we found that there is a finite probability that the mass can have a displacement from the center greater than the amplitude. We also looked at the situation of a particle in a well with a barrier potential on one side. We found that there was a finite probability that the particle would tunnel through the barrier and exit the well. Because of the wave nature of matter, there is a finite probability that the particle will escape. Every radioactive disintegration demonstrates this.

The energy of the emitted alpha particle (and the recoil nucleus) does not depend on the height of the potential barrier. The kinetic energy after the disintegration is the same as the energy of the nucleus and its alpha particle before the disintegration. But this energy does have a bearing on the decay process. If the energy is nearly equal to the energy height of the potential barrier, then disintegration is likely, the decay constant is large, and the half-life is small. If the energy is small compared with the barrier height, then the radioactive nucleus is relatively stable and has a long half-life. This point is not covered by our analogy between an alpha particle in a nucleus and a mass on a spring. The reason is that whereas the spring exerts an attractive force on the mass everywhere the mass moves, the alpha particle experiences a repulsive force after r exceeds b. If the potential energy well for the oscillating mass had had a shape like that in which the alpha particle moves, the probability of finding the mass beyond its amplitude would increase as its energy increased.

One might try to explain radioactivity classically by imagining that the nucleons in a nuclide are like gas molecules having a Maxwellian energy distribution. Every once in a while some nucleons "gang up" on one or more of their companions and give them enough energy to eject them from the nuclide. If this were so, the ejected particle would have energy at least as great as the potential barrier. This is not observed experimentally.

We may use quantum-mechanical ideas to resolve another dilemma. If radium disintegrates into radon and an alpha particle with the release of 4.78 MeV of energy, why should not radium be formed when radon is bombarded with alpha particles of about this energy? No doubt this reverse reaction can and does occur, but consider how unlikely it is. If you could place a radium atom on the table and watch it disintegrate, you might be rewarded in a few seconds. On the other hand, you might have to watch for millions of years. If you want a 50–50 chance of seeing the disintegration, you must be prepared to sit with rapt attention for about 2340 years, the average life. The probability of finding the radium alpha particle outside the nucleus in a short time is exceedingly small. Conversely, suppose you were to try to introduce a 4.78 MeV alpha particle into radon to form radium. To have a 50–50 chance of succeeding, you would have to keep the alpha particle near the radon nucleus for about 2340 years. This would be even more difficult than waiting for radium to disintegrate. It would be necessary to bombard the radon nucleus an untold number of times. If one must penetrate the radon nucleus with an alpha particle, it is far more promising to bombard it with particles whose energy exceeds the radon potential barrier. Recall that alpha particles from radioactive materials were used successfully in disintegrating light nuclei whose potential barriers are relatively low.

11.13 MÖSSBAUER EFFECT

The Mössbauer effect provides a means for producing and studying gamma rays whose energies are extremely well defined. We will begin this discussion by reviewing the factors that cause a gamma ray to be emitted with a range of energies.

Cobalt-57 decays to an excited state of iron-57 by *K* electron capture. Nine percent of the excited iron-57 nuclei go directly to the ground state with the emission of a 137 keV gamma-ray photon, but 91 percent of the transitions involve two steps. After emitting a 123 keV gamma ray, the nucleus is in an intermediate excited state that has a half-life of 10^{-7} s. In the transition from this state to the ground state a gamma ray of low energy, 14.4 keV, is emitted. What factors limit the precision of the energy value of this transition?

Knowing the half-life of the excited state, we can use Heisenberg's uncertainty principle to estimate the "inherent uncertainty" of the energy. This turns out to be about 4×10^{-8} eV, and comparing this to the energy of the gamma ray, we find an uncertainty of about three parts in 10^{12}, a very small fraction.

Another consideration in discussing the energy of the emitted gamma ray is the energy imparted to the recoiling nucleus. This always reduces the energy of the gamma ray. The data for computing the recoil energy can be obtained from momentum considerations. If a gamma-ray photon has a nominal energy E (14.4 keV for iron-57), then its momentum is given by the relativistic expression $p = h/\lambda = E/c$. Because momentum is conserved, the emitting system must also have E/c units of momentum imparted to it in a direction opposite to that of the emitted quantum of radiation. If this recoil momentum is given only to the emitting nucleus, we can calculate its recoil energy. For iron-57, which has a mass of about 10^{-25} kg, we find the recoil velocity to be about 80 m/s. This is low enough to justify a classical calculation. Thus the kinetic energy of the recoiling nucleus is about

2×10^{-3} eV, and the energy of the emitted gamma ray is *less* than the transition energy by this amount. It is to be noted that this recoil energy is about 5×10^4 times greater than the uncertainty in the gamma-ray energy calculated from Heisenberg's principle.

A third influence on the gamma-ray energy would seem to be the Doppler effect of the source nucleus. This effect depends on the velocity of the source. The thermal velocity of a *free* atom may be computed by the classical methods discussed in Chapter 1. The root-mean-square speed of an atom that has a mass of 57 u and that is in the gaseous state at room temperature is about 350 m/s. But an atom that is *in a crystal* cannot be treated classically. Since it is bound in a crystal, the atom has only quantized vibrational energy states. These quantized states are variously populated and widely separated compared to the low recoil energy of our example. At reduced temperature the occupancy of the lowest energy state is greatly increased. In the discussion that follows we shall assume that the temperature is so low that the effects of thermal motion and crystal lattice vibrations are negligible.

Returning to gamma-ray emission, if the energy of the photon is low enough, as in iron-57, then the recoil energy is insufficient to separate the atom from the crystal lattice or to change the vibrational state of the emitting atom. In this case the recoil energy is not transferred to *one* emitting atom but to a *group* of atoms. This means that the recoiling mass is relatively tremendous, the recoil velocity is practically zero, and the gamma ray carries away substantially all the energy of the transition. As a consequence, the remaining uncertainty of the energy of the gamma ray is almost completely determined by the Heisenberg principle alone.

This is the *Mössbauer effect*, which is also called *recoilless emission*. Mössbauer emitters (there are others in addition to iron-57) must emit gamma rays of low enough energy so that the emitting atom is neither freed from the crystal lattice nor excited to a high vibrational energy state.

The Mössbauer effect is important in realizing the necessary conditions for the absorption of gamma rays by nuclei that are identical to those of the emitter. A nucleus can absorb only that gamma-ray photon whose energy is equal, within the limits set by the uncertainty principle, to the difference in energy between an excited state and the ground state of the nucleus. This is called *resonance absorption*. It will occur only under Mössbauer conditions because, unless the gamma-ray emission is recoilless, the gamma ray has its energy reduced so much that it cannot resonate with a nucleus of the same kind. A Mössbauer emitter, however, produces gamma rays with an extremely narrow spread and these gamma rays can be detected with equal energy precision by means of resonance absorption.

To emphasize the degree of precision obtainable and the importance of the Mössbauer effect, we cite two of its many applications.

In discussing radioactivity, we stated that magnetic fields and other environmental influences had no detectable effect on nuclear events. However, infinitesimal effects have actually been observed by using a Mössbauer emitter whose radiation is being detected by a Mössbauer absorber. Placing either the emitter or the absorber in a magnetic field will put the system out of resonance because of slight changes of the nuclear energy levels. Indeed if one is in a different chemical state from the other, the difference can be detected. The change is usually measured by means of the Doppler effect. Resonance may be restored by moving the emitter relative to the absorber with almost unbelievable slowness, of the order of a few centimeters per minute. By varying this motion in a carefully controlled way, the system can be "tuned" over a very narrow energy spectrum.

One of the phenomena predicted by the general theory of relativity is the gravitational red shift, which may be considered as the change in energy of a photon as it travels from one place to another of different gravitational potential. For a photon of energy E, traveling between two points that are separated by a height h near the earth where the acceleration due to gravity is g, the change in energy of the photon is given by $(E/c^2)gh$. Obviously, the corresponding fractional change of energy, gh/c^2, is minute. Nevertheless this change has been verified by placing an absorber about 75 ft above a Mössbauer emitter, and then determining the energy shift from the rate at which one or the other must be moved to achieve resonance absorption. This experiment provides a terrestrial test of the general theory of relativity.

11.14 THE BOMBARDING PARTICLES

Thus far we have discussed the use of alpha particles and protons to bombard nuclei. With the advent of particle accelerators the study of nuclear physics became very popular and a great number of reactions were studied. Many of these transitions were excited by high-energy electrons and high-energy photons or gamma rays. Many new isotopes were discovered. One of these, which deserves special mention, is deuterium.

As early as 1920, Harkins and Rutherford predicted that there should be a heavy isotope of hydrogen having mass number 2. It was not found in mass spectroscopy because it was masked by the hydrogen molecule H_2. In 1931, Urey, Brickwedde, and Murphy separated heavy hydrogen from ordinary hydrogen by evaporating liquid hydrogen. Since light hydrogen is more volatile, it evaporated a little more readily than the heavy hydrogen. This isotope was positively identified spectroscopically. The atomic spectrum of heavy hydrogen is shifted from that of ordinary hydrogen because its greater mass causes a small but measurable difference in the Rydberg constant. Soon after its discovery, heavy hydrogen was separated in considerable quantities by the electrolysis of water, where again the molecules formed with light hydrogen were more mobile and electrolyzed more readily. Since heavy hydrogen has twice the mass of its common isotope, it has many striking properties. Water formed from heavy hydrogen, called "heavy water," has a specific gravity of 1.108, which is quite different from that of ordinary water. Because of its special importance, heavy hydrogen has been given a special name, *deuterium*, and the symbol D. Thus ${}_1^2H$ and ${}_1^2D$ have identical meaning. (Ordinary hydrogen, ${}_1^1H$, also has a special name, *protium*, but it is seldom used.) Just as ionized hydrogen is called a proton, p, ionized deuterium is called a *deuteron*, d. A third hydrogen isotope, ${}_1^3H$, is called *tritium* and is sometimes represented by the symbol T. The tritium nucleus is called *triton* and is indicated by t. Tritium is radioactive, with a half-life of 12.26 years.

One of the important uses of deuterium was as a bombarding particle. It has unit charge like the proton, but it has twice the mass of a proton. Deuterium bombardment led to still more nuclear reactions, such as

$$ {}_1^2D + {}_1^2D \rightarrow {}_2^3He + {}_0^1n + Q. \qquad \textbf{(11.49)} $$

We choose this particular reaction as an example because it is one of many reactions that produce neutrons.

It is easy to see why neutrons were also used as bombarding particles. Because they have no charge, they are not subject to Coulomb repulsion. With no potential barrier to penetrate, neutrons can enter almost any nucleus with ease. The number of nuclear transitions that can be produced with neutrons is greater than the number induced by all other particles combined.

11.15 NEUTRON REACTIONS. MODES OF NUCLIDE DECAY

Nuclear reactions are often represented in a shorthand notation in which the participating particles are identified in this order: target particle, bombarding particle, ejected particle, and product particle. The symbols for the bombarding and ejected particles are enclosed in parentheses. Thus Eq. (11.49) may be written ${}^2D(d, n){}^3He$. Equation (11.43) becomes ${}^{27}Al(\alpha, n){}^{30}P$. Any reaction in which a deuteron produces a neutron is called a (d, n) reaction, and the second example is called an (α, n) reaction.

The first type of neutron reaction observed was of the (n, α) type, where the target was, successively, nitrogen, oxygen, fluorine, and neon. A second type is called simple *radiative capture*, (n, γ). The first observed reaction of this type was the formation of deuterium from hydrogen and a neutron with the emission of a gamma photon. Nearly all elements undergo radiative capture of neutrons and it is probably the most common nuclear process. It is particularly important that the probability of this type of reaction taking place is greater for slow neutrons than for fast ones.

A third type of neutron reaction is the (n, p) type. Except for the lighter elements, the potential barrier inhibiting the escape of the proton causes these reactions to have negative Q-values. Therefore the incident neutron must be fast, with an energy of 1 MeV or more. Although they are less probable than (n, p) reactions, (n, d) and (n, t) reactions are also found.

The second type of reaction, (n, γ), is especially interesting. It adds a neutron to the bombarded nucleus and frequently makes it radioactive. An examination of the periodic table shows that atomic weight is not proportional to atomic number. The higher the atomic number of a nuclide, the more the number of neutrons exceeds the number of protons. If we plot the number of neutrons, $A - Z$, against the number of protons, Z, for the stable nuclides, we obtain the result shown in Fig. 11.19.

Figure 11.19 Modes of decay of nuclides.

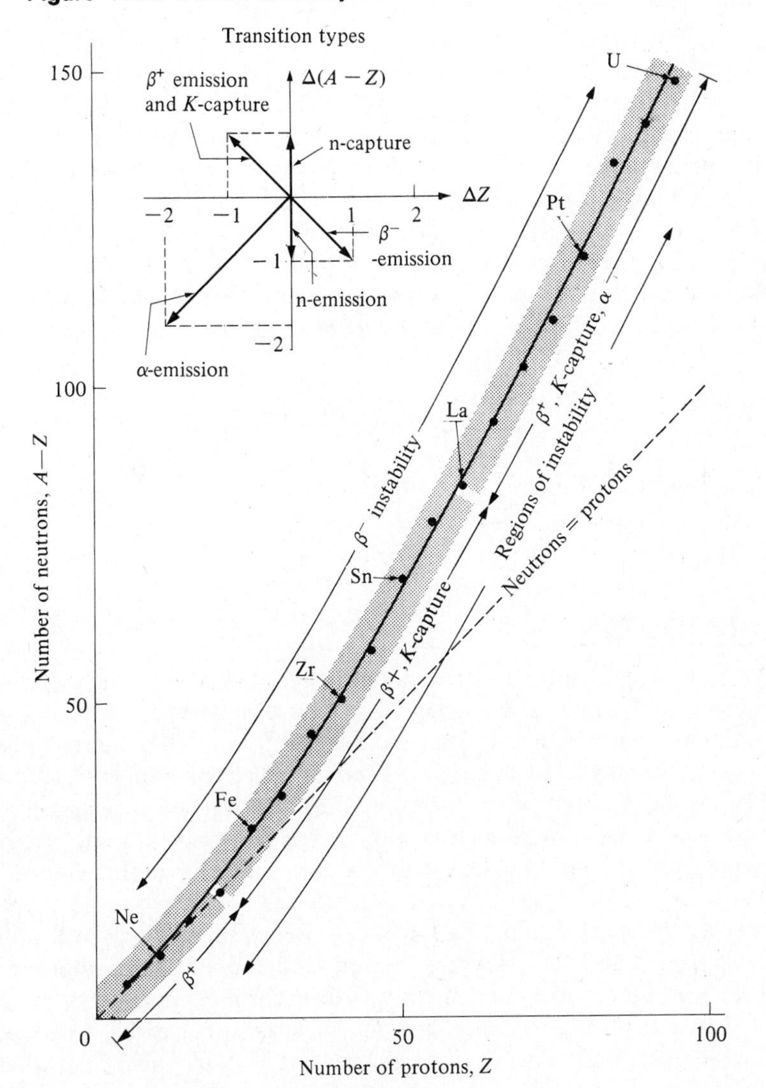

In the inset of this figure, the abscissa is the change ΔZ in atomic number due to a nuclear reaction, and the ordinate is the corresponding change $\Delta(A - Z)$ in the neutron content of the nucleus. Five arrows are shown in this inset. One, directed upward, represents the change in a nuclide for simple neutron radiative capture, (n, γ). Another, downward and to the right, represents beta decay in which the emission of a negative electron is accompanied by the conversion of a neutron into a proton. The one pointing directly downward represents neutron emission. The fourth arrow represents the emission of an alpha particle in which two neutrons and two protons are removed. The remaining arrow, upward and to the left, represents both positron emission and orbital electron or *K*-electron capture, where a proton is converted to a neutron. Nuclides off the full-line curve tend to be unstable, and the particle each emits is likely to be one which brings the new nuclide nearer the curve. Thus the region above the curve is the region of likely beta decay. A nuclide in the region below the curve disintegrates in such a way that its atomic number decreases. In general, elements of low mass number disintegrate by positron emission; those in the middle range, either by positron emission or by orbital electron capture; and the heavy elements, by positron emission or by orbital electron capture or by ejecting an alpha particle. The natural radioactive series involve alpha and beta decay in such a way that the series of nuclides tend to follow the curve of stability. Since neutron capture makes a new nuclide above the original one on the chart of Fig. 11.19, the new nuclide tends to be beta-unstable. As a typical example, we have

$$^{103}_{45}\mathrm{Rh} + ^{1}_{0}\mathrm{n} \rightarrow ^{104}_{45}\mathrm{Rh} + \gamma. \qquad \mathbf{(11.50)}$$

The ^{104}Rh is beta-unstable, with a half-life of 42 s, and disintegrates according to the equation

$$^{104}_{45}\mathrm{Rh} \rightarrow ^{104}_{46}\mathrm{Pd} + ^{\;0}_{-1}\mathrm{e}. \qquad \mathbf{(11.51)}$$

11.16 NUCLEAR MODELS

As we saw in Chapter 4, atoms could be visualized as consisting of a nucleus with charge $+Z$ located at the center of Z electrons. Also in Chapter 4, we introduced quantum numbers and the Pauli exclusion principle. The electrons filled the allowed electronic energy levels in a way that no two electrons were in exactly the same state. Also, as the electrons filled shells, the spins were aligned antiparallel. In this scheme, hydrogen would have a nucleus with charge plus one with one electron orbiting it. Helium would have a nucleus with $+2$ electronic charge surrounded by 2 electrons that filled the outer shell. As we proceed up the list of atoms, the next to have a closed outer shell would be neon. Then, following suit, the next atoms with filled shells would be argon, krypton, xenon, and radon. The significant evidence that these atoms had filled outer shells was their chemical inactivity.

There is strong evidence that the nucleus also can be described by a model consisting of shells. Part of that evidence is shown in Table 11.1, where the number

Table 11.1

Z-N	Number of Stable Nuclei	Nuclear Spin
even-even	155	0
even-odd	53	1/2, 3/2, 5/2, 7/2...
odd-even	50	1/2, 3/2, 5/2, 7/2...
odd-odd	4	1, 3

of stable nuclei for each nuclear type — even Z even N, even Z odd N, odd Z even N, odd Z odd N — is given. The large number of stable nuclei with even Z and even N is evidence for a shell model. Further evidence is given by the experimentally measured nuclear spin for these nuclei also shown in Table 11.1. The fact that each even-even nucleus has zero total nuclear spin is evidence that the protons fill their levels in pairs and the neutrons fill their levels in pairs with each pair having antiparallel spins yielding zero net spin. Even-odd and odd-even nuclei have half integral nuclear spin. The half integral spin is attributed to leftover protons or neutrons. The interpretation of the odd-odd nuclear type is that the one proton left over after the other protons are paired off has its one-half spin added to the one-half spin of the leftover neutron resulting in integral nuclear spin.

Other evidence for the shell structure of the nucleus can be found in the binding energy of the nucleons. Consider the binding energy of the last neutron placed in a nucleus. Since the binding energy can be obtained from the difference between the mass of the nucleus and the mass of the individual particles of the nucleus, the calculated binding energy is given by the following:

$$BE = M(A-1, Z)c^2 + m_n c^2 - M(A, Z)c^2.$$

Figure 11.20 shows the difference between the experimentally determined binding energy of the last neutron and that calculated from the above equation. Note the large differences in the binding energy in the regions of $N = 28$, 50, 82, and 126. These are called magic numbers.

Each nucleon within a nucleus is assumed to have total angular momentum made up of the angular momentum due to its own spin, $\hbar/2$, and that due to its motion in the nuclear field of all other nucleons, $l\hbar$. In analogy to the notation for electron configuration, values for the nuclear orbital angular momentum quantum number l have letters associated with them

$$\begin{aligned} l &= 0\ 1\ 2\ 3\ 4 \ldots \\ &\ \ \ \ \mathrm{s\ p\ d\ f\ g} \ldots \end{aligned}$$

The total angular momentum for a single nucleon is thus $J = l + 1/2$ or $J = l - 1/2$.

The shell model yields a system of energy levels (shown in Fig. 11.21) for the nucleons in a nucleus. We can see from the figure that the first available state for the

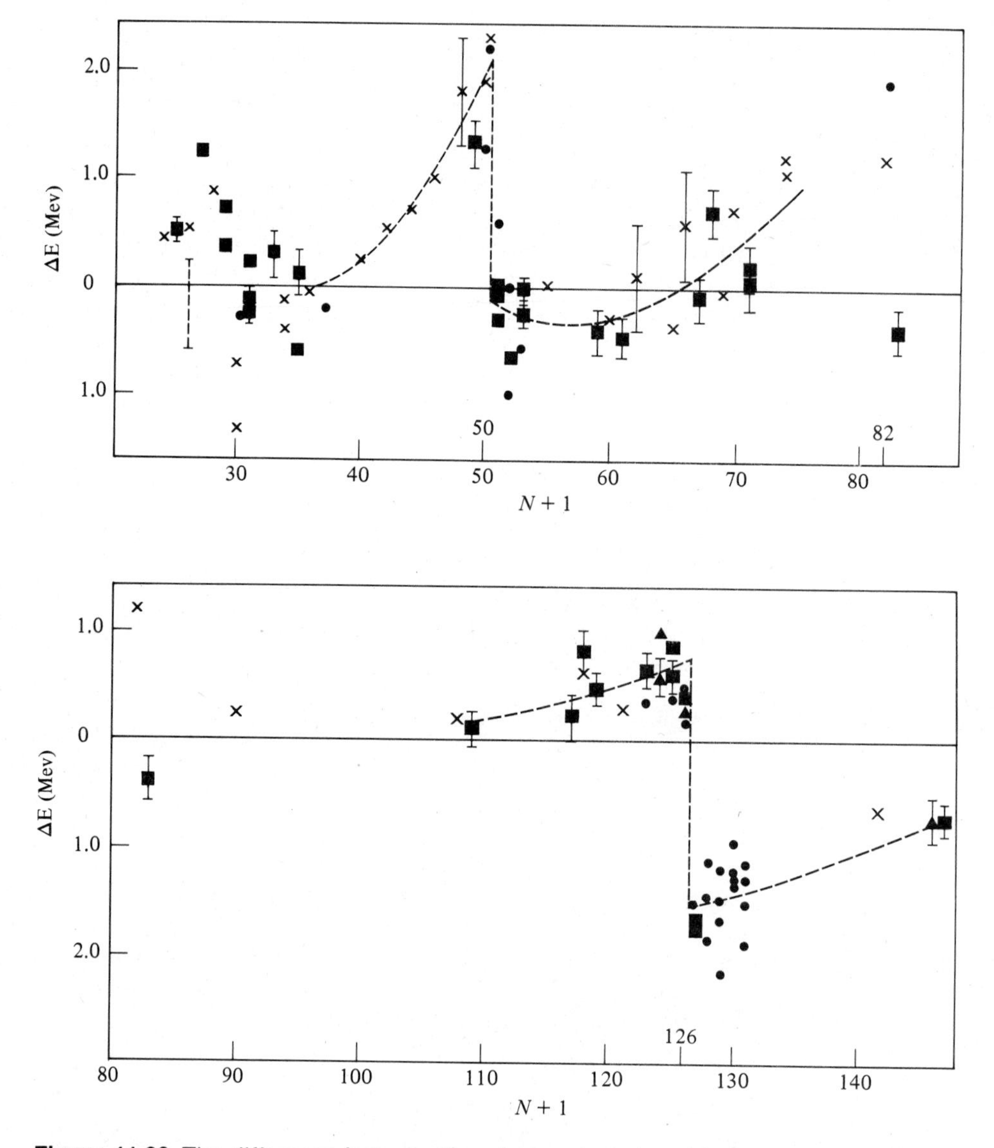

Figure 11.20 The difference between the observed neutron binding energy for a nucleus with $N + 1$ neutrons and the neutron binding energy calculated from the mass formula, plotted as a function of the number of neutrons $N + 1$.

From J. Harvey, *Phys. Rev.* **81**, 353 (1951). Reprinted by permission of the American Physical Society.

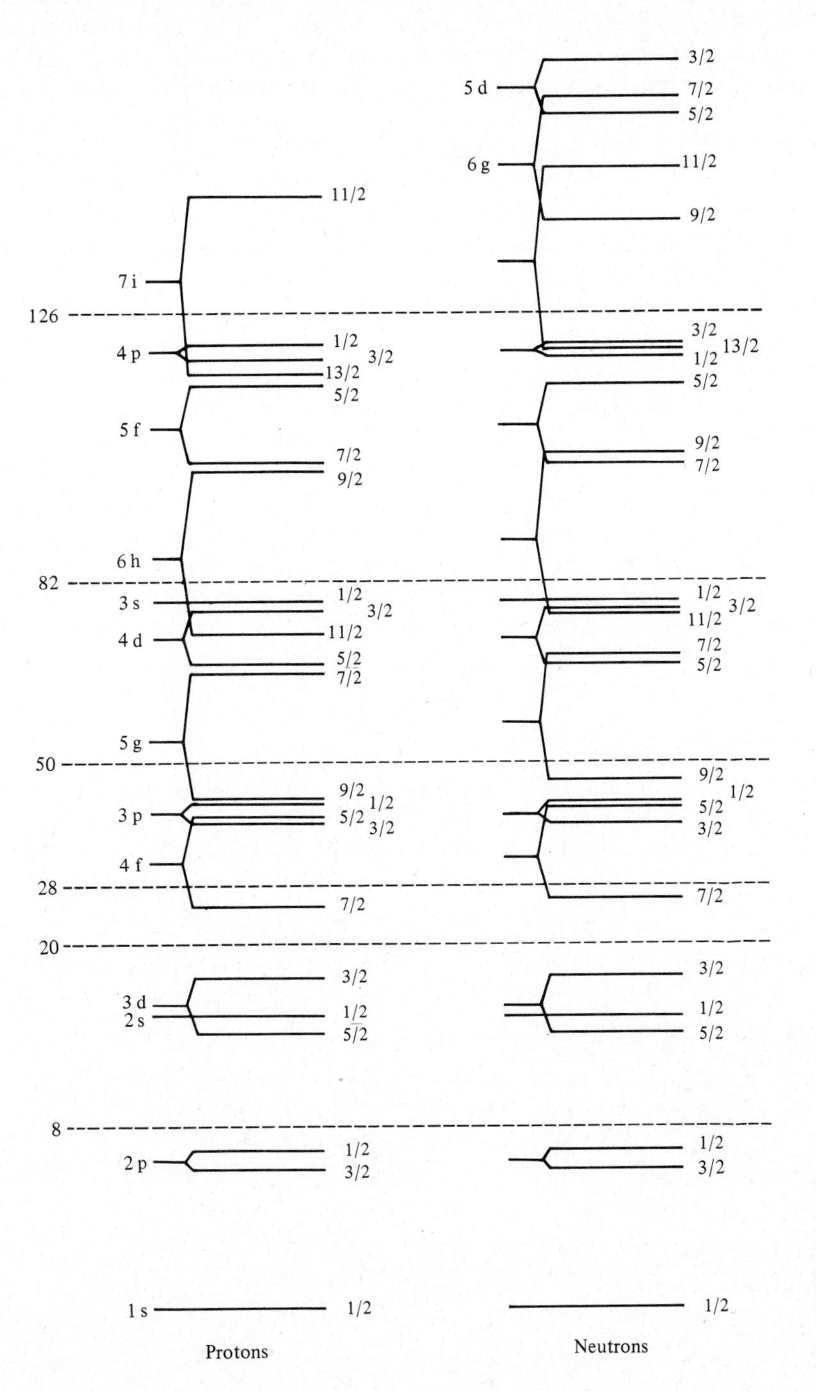

Figure 11.21 Energy levels as predicted by the shell model.

From Klinkenbers, *Rev. Mod. Phys.* **24**, 63 (1952). Reprinted by permission of the American Physical Society.

neutron is the $1s_{1/2}$ state and the first state for the proton is the $1s_{1/2}$ state. Each will hold two nucleons. Therefore, if we consider helium with its two neutrons and two protons, we would expect the two neutrons to fill the neutron $1s_{1/2}$ state and the two protons to fill the proton $1s_{1/2}$ state. Hydrogen has only one proton but an isotope of hydrogen, deuterium, has 1 proton and 1 neutron. For deuterium, the proton would be in the $1s_{1/2}$ state and the neutron would be in the $1s_{1/2}$ state, each bringing the states to half full. The experimental value for the total spin of deuterium indicates that the value is 1, with this being the evidence that the spin of the neutron and the spin of the proton are parallel. The assignment of the nucleons can be continued using the shell model.

Models of the nucleus are broken into two major groups: the individual particle models, which include the shell model we just discussed, and the collective models, which include the liquid-drop model. This nuclear model is analogous to a liquid drop. In a liquid, the forces are short-range forces in effect between a given molecule and its nearest neighbors. The nuclear force is known to be extremely short-range and would primarily act between a nucleon and its nearest neighbors. In a drop of liquid, the forces act uniformly on a molecule because it is surrounded by other molecules (see Fig. 11.22). At the surface of the liquid, however, the molecules are not surrounded and the force is uneven. This arrangement is what yields surface tension, an effect that causes a droplet to become spherical. The same is assumed to be true in the nucleus. There is an energy associated with the attractive force between the nucleons and this energy is proportional to the number of nucleons present:

$$\text{attractive energy} = a \cdot \text{A},$$

where a is a constant. Since the nucleons at the surface will be less tightly bound than those in the interior of the nucleus, the energy given by this relation is greater than the actual energy and the excess depends on the amount of surface. The larger the surface area, the larger the number of nucleons there will be that are not

Figure 11.22

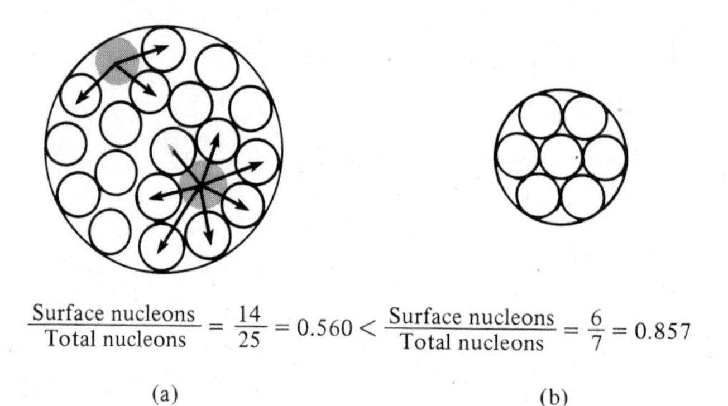

surrounded by the nucleons (Fig. 11.22). This is referred to as the surface tension effect and is analogous to the effect in the liquid droplets. Surface tension energy is given by

$$STE = -bR^2,$$

where b is a constant. As was true for a liquid drop, the most stable shape for the nucleus is spherical. The liquid-drop model predicts that if the sphere is large enough, there will be a critical size where the force of Coulomb repulsion will be greater than the surface tension force holding the nucleus together. The calculated critical size is approximately a sphere large enough to hold a couple of hundred nucleons in good agreement with the observed number of stable or near stable nuclei. For large nuclei near this critical size, the surface tension energy is just in balance with the Coulomb energy of repulsion for the large number of protons in the nucleus. A deformation of this spherical shape will result in a decrease in the surface tension energy and may allow the Coulomb energy to become greater than the surface tension energy. If this occurs, fission is possible.

FOR ADDITIONAL READING

Bacon, G. E., "The Applications of Neutron Diffraction," *Endeavour* **25**, 129 (September 1966).

Beyer, Robert T., *Foundations of Nuclear Physics*. New York: Dover, 1949. Everyone interested in the scientific method and how it revealed properties of the nucleus should read this book. It contains facsimiles of the fundamental studies as originally reported in the scientific journals of Europe and the United States. The articles include "The Positive Electron," by Carl D. Anderson, pp. 1–4; "Experiments with High Velocity Positive Ions," by J. D. Cockcroft and E. T. S. Walton, pp. 23–38; "Un Nouveau Type de Radioactivité," by Mme. Irene Curie and M. F. Joliot, pp. 39–41; "Possible Production of Elements of Atomic Number Higher than 92," by Enrico Fermi, pp. 43–44; "Uber den Nachweis und das Verhalten der bei der Bestrahlung des Urans mittels Neutronen entstehenden Erdalkalimetalle," by O. Hahn and F. Strassman, pp. 87–91; "The Production of High Speed Light Ions without the Use of High Voltages," by E. O. Lawrence and M. S. Livingston, pp. 93–109; "Collision of α Particles with Light Atoms. An Anomalous Effect in Nitrogen," by E. Rutherford, pp. 111–137.

Cohen, B. L., *Concepts of Nuclear Physics*. New York: McGraw-Hill, 1971.

Curtiss, Leon F., *Introduction to Neutron Physics*. Princeton, N.J.: Van Nostrand, 1959. Broad coverage.

Libby, Willard F., *Radiocarbon Dating*. 2nd ed. Chicago: University of Chicago Press, 1955. A very interesting book.

Libby, Willard F., "Chemistry and the Atomic Nucleus," *Am. J. Phys.* **26**, 524 (1958). An informal account of radiochemistry and carbon dating.

Livingston, M. Stanley, *Particle Accelerators: A Brief History*. Cambridge, Mass.: Harvard University Press, 1969.

Livingston, M. Stanley, and John P. Blewett, *Particle Accelerators*. New York: McGraw-Hill, 1962.
Muller, R. A. "Radio Isotope Dating with Accelerators," *Physics Today* **32**, 23 (February 1979).
Murphy, George M., editor, *Production of Heavy Water*. New York: McGraw-Hill, 1955.
O'Neill, G. K., "Particle Storage Rings," *Scientific American* **107** (November 1966).
Reid, J. M., *The Atomic Nucleus*. New York: Penguin, 1972.
Wertheim, G. K., *Mössbauer Effect*, Resource Letter ME-1 and Selected Reprints. New York: American Institute of Physics, 1963.
Wertheim, G. K., *Mössbauer Effect: Principles and Applications*. New York: Academic Press, 1964.
Wilson, R. R., "The Batavia Accelerator," *Scientific American* **72** (February 1974).
Wilson, R. R., "The Next Generation of Particle Accelerators," *Scientific American* **242**, 42 (January 1980).

REVIEW QUESTIONS

1. In the alpha particle bombardment of beryllium experiment, what was the evidence that led Chadwick to believe that the product particles were neutrons?
2. What are the differences between particle accelerators that cause charged particles to fall through one large potential difference and those that cause particles to fall through small potential differences many times?
3. What is meant by the term Q-value?
4. How would you determine the ages of two bones, one on the order of thousands of years old and the other on the order of millions of years old?
5. What are the similarities and differences between the shell model of the nucleus and the Bohr model of the atom?

PROBLEMS

11.1 Show that the range of a charged particle in a material is proportional to the $3/2$ power of its kinetic energy at the beginning of its range.

11.2 Radium-C′, ^{214}Po, emits some of the most energetic alpha particles observed from naturally radioactive elements. The energy of these particles is 7.68 MeV and their range is 6.90 cm in air at 15°C and 1 atm pressure. (a) Show that the moving mass and the relativistic velocity of an alpha particle just ejected from RaC′ do not differ from the rest mass and the classical velocity, respectively, by more than 1%. (b) Repeat part (a) for a 7.68 MeV proton. (c) Do you think that the differences between the classical and relativistic results are so significant that relativistic mechanics must be used in calculations involving high-energy particles such as those in this problem?

11.3 Using the data of Problem 11.2, calculate (a) the classical velocity of an alpha particle from RaC′, ^{214}Po, and (b) the total number of ion pairs it produces over its range in air. (It requires 34 eV to produce an ion pair in air.) (c) If all the ion pairs produced by the alpha particles from a microcurie of RaC′ constitute the current in an ionization chamber, what would be the current in amperes?

11.4 An alpha particle just ejected from RaC′, ^{214}Po, undergoes an elastic collision with one of the nuclei of molecular hydrogen in a region filled with this gas. If both particles continue in the original direction of motion of the alpha particle, find classically (a) the ratio of the velocity of the proton after collision to that of the incident particle, and (b) the kinetic energy of the proton. (It requires about 4 eV of energy to dissociate the hydrogen molecule. The data for ^{214}Po are given in Problem 11.2.)

11.5 An alpha particle going through a cloud chamber undergoes an elastic collision with a nucleus of unknown mass number initially at rest. A photograph of the event shows a forked track. It is seen from measurements of the track that the collision deviated the alpha particle 60°, and the struck nucleus went off at an angle of 30° with the direction of motion of the incident particle. What is the mass number of the unknown nucleus? (Since this is not a head-on collision and since momentum is a vector quantity, the law of conservation of momentum must be applied in component form.)

11.6 A 5.15 MeV alpha particle has an elastic collision with a hydrogen nucleus. If the path of the alpha particle is deflected upward 14° from its original direction and this particle then has 2.57 MeV of energy, what is the velocity and direction of motion of the proton?

11.7 (a) Show from Eq. (11.20) that when a particle of mass m_1 has a head-on elastic collision with a particle of mass m_2 which is initially at rest, then the fraction of the energy lost by m_1 is $4m_1m_2/(m_1 + m_2)^2$. (b) Considering m_1 constant and m_2 variable, show that this loss is maximum for $m_1 = m_2$. (c) How much is this maximum loss in percent?

11.8 Calculate the density of the ^{12}C nucleus. (See Eq. 4.4.)

11.9 Thermal neutrons have an average kinetic energy $\frac{3}{2}kT$, where T is room temperature, 300 K. (a) What is the average energy in eV of a thermal neutron? (b) What is the corresponding de Broglie wavelength?

11.10 How many head-on, elastic collisions must a neutron have with other particles to reduce its energy from 1 MeV to 0.025 eV if the particles are atoms of (a) deuterium, (b) carbon, and (c) lead?

11.11 A nucleus of ^{60}Ni in an excited state decays to its ground state by emitting a 1.33 MeV photon. What is the recoil energy and recoil speed of the ^{60}Ni nucleus?

11.12 Before the discovery of the neutron, it was thought that the nucleus consisted of protons and electrons. The wave nature of the electron requires that a large energy is necessary to confine an electron in a region as small as the nucleus. If an electron has a wavelength of 10^{-13} cm, calculate its kinetic energy. (Relativistic expressions must be used.)

11.13 The beam in a cyclotron has a maximum diameter of 1.6 m. Given that the magnetic induction is 0.75 T, calculate the kinetic energies of the particles if they are (a) protons and (b) deuterons.

11.14 Deuterons in a cyclotron describe a circle of radius 32.0 cm just before emerging from the D's. The frequency of the applied alternating voltage is 10 MHz. Neglecting relativistic effects, find (a) the flux density of the magnetic field, and (b) the energy and speed of the deuterons upon emergence.

11.15 Design a cyclotron that will accelerate protons to 10 MeV.

11.16 The new superconducting ring in the Fermi Lab accelerator is 2 km in diameter and can increase the energy of a proton to 1 TeV. What must be the magnetic field of the magnets?

11.17 What is the length of the Stanford two-mile accelerator as viewed by a 20 GeV electron?

11.18 The two-mile linear accelerator at Stanford is aligned with the aid of a laser. How many wavelengths of the red laser light are there in two miles? (He–Ne laser with $\lambda = 6328$ Å.)

11.19 The energy per unit time radiated by a particle with charge e moving with relativistic velocity in a circular orbit is

$$P = \frac{1}{6\pi\varepsilon_0}\omega\frac{e^2}{R}\left(\frac{v}{c}\right)^3\left(\frac{E}{m_0c^2}\right)^4,$$

where R is the radius of the trajectory, ω is the angular velocity of the particle, m_0 is the rest mass, and E is the total energy. (a) Calculate the power radiated from an electron synchrotron operated at 300 MeV when the electron orbit radius is 100 cm. (b) Calculate the energy loss by radiation during a single revolution of an electron. (c) Determine the ratio of the radiated energy per cycle to the total energy of the electron.

11.20 (a) Calculate the power radiated from a proton synchrotron operated at 350 MeV when the orbit radius is 200 cm. (b) Calculate the energy loss by radiation during a single revolution of the proton. (c) Determine the linear and angular velocity of the protons in this orbit.

11.21 (a) A 2000 kg automobile traveling at its maximum speed of 200 km per hour crashes head-on into an identical automobile that is at rest. The speed of the wreckage (both automobiles locked together) immediately after collision is 100 km per hour. How much energy has gone into the bending and heating of metal? (b) Two automobiles identical to the two in part (a) move toward each other at their maximum speed of 200 km per hour. After they collide the wreckage is at rest. How much energy has gone into the bending and heating of metal?

11.22 Sulfur-32 is bombarded with neutrons. Write the following reactions: (n, p), (n, α), (n, d), (n, γ).

11.23 Complete each of the following by writing the nuclear reaction equation:

(a) $^{10}B(n, \alpha)?$ (g) $^{31}P(d, p)?$
(b) $^{14}N(\alpha, n)?$ (h) $^{12}C(\gamma, \alpha)?$
(c) $^{31}P(\gamma, n)?$ (i) $^{14}N(?, p)^{14}C$
(d) $^{7}Li(p, n)?$ (j) $^{39}Ca \rightarrow {}^{39}K + ?$
(e) $^{29}Al \rightarrow {}^{29}Si + ?$ (k) $^{120}I \rightarrow ? + {}^{0}_{+1}e$ (or β^+)
(f) $^{27}Al(n, \gamma)?$ (l) $^{59}Co(n, ?)^{60}Co$

11.24 A certain photonuclear reaction is $^{24}Mg(\gamma, n)^{23}Mg$. What is the least energy the photon must have to produce this reaction, *assuming* that the products have negligible kinetic energy? The isotopic mass of the product nuclide is 23.001453. (To conserve momentum, the product particles must actually have velocities and therefore some kinetic energy.)

11.25 The Q-value of the disintegration of ^{226}Ra by alpha-particle emission to form ^{222}Rn was found, from Eq. (11.48), to be 4.93 MeV. Using this result and the laws of conservation of energy and of momentum, find (a) the recoil energy of the radon nucleus, and (b) the kinetic energy of the alpha particle. Assume that all of the disintegration energy appears as kinetic energy of the particles.

11.26 (a) Calculate the Q-value of the reaction in the Cockcroft–Walton experiment in which ^{7}Li, when bombarded with 0.7 MeV protons, produced two alpha particles having equal kinetic energies. (b) What is the kinetic energy of each alpha particle? (c) What is the range of each of these alpha particles in air at 15°C and 1 atm pressure? [Use either Eq. (11.1) or the result of Problem 11.1. Comparative data for alpha particles from ^{214}Po are given in Problem 11.2.]

11.27 (a) Calculate the Q-value for the decay of a free neutron into a proton and an electron. (b) Calculate the Q-value for the decay of a proton into a neutron and a positron.

11.28. For lead with $Z = 82$, the number of neutrons ranges from 112 to 132. For what values is alpha decay possible?

11.29 The maximum beta-particle energy from the decay of ^{10}Be to ^{10}B is 0.56 MeV. Calculate the isotopic mass of the ^{10}Be atom.

11.30 Suppose that a thermal neutron with an energy of 0.025 eV is absorbed by a ^{32}S nucleus to produce ^{33}S. What is the energy of the photon that is radiated if the ^{33}S nucleus goes to the ground state? The isotopic mass of ^{33}S is 32.97146.

11.31 One student claims that he has observed ^{16}N to decay by alpha-particle emission. Another student claims that ^{16}N decays by beta-particle emission. Which student is correct?

11.32 The Q-value of the reaction ^{6}Li(p, α)^{3}He is found experimentally to be 3.945 MeV. Calculate the isotopic mass of ^{3}He from this information.

11.33 What target isotope must be used to form the compound nucleus $^{24}_{11}$Na when the projectile is (a) a neutron, (b) a proton, and (c) an alpha particle?

11.34 A bombarding particle of mass m_1 and kinetic energy E_{k1} strikes a nucleus at rest having a mass m_2. The resulting nuclear reaction produces two particles having masses m_3 and m_4 and kinetic energies E_{k3} and E_{k4}, respectively. The directions of their motions are shown in Fig. 11.23. Using classical mechanics, show that

$$Q = \left(1 + \frac{m_3}{m_4}\right) E_{k_3} - \left(1 - \frac{m_1}{m_4}\right) E_{k_1} - \frac{2\sqrt{m_1 m_3 E_{k_1} E_{k_3}}}{m_4} \cos\theta_3 .$$

11.35 A neutron beam is incident on a stationary target of ^{19}F atoms. The reaction ^{19}F(n, p)^{19}O has a Q-value of -3.9 MeV. What is the lowest neutron energy that will cause this reaction to take place?

11.36 (a) Compute the Q-value of the reaction ^{14}N(α, p)^{17}O, which occurred in Rutherford's alpha-range-in-nitrogen experiment. (b) If a 7.68 MeV alpha particle causes the reaction in part (a) in a nitrogen atom that is initially at rest, what is the kinetic energy in the laboratory system of the proton produced, assuming that it and the product nucleus continue in the direction of the motion of the incident alpha particle? (c) What is the threshold energy

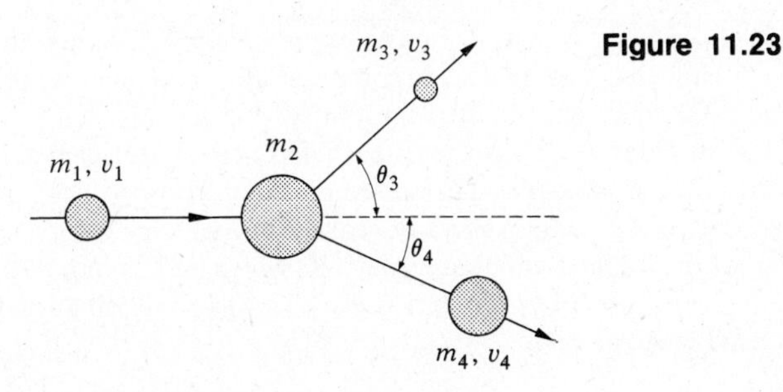

Figure 11.23

of the alpha particle for the reaction in part (a), assuming that the nitrogen target atom is at rest? (d) If the nitrogen atom were moving with the most probable speed in a Maxwellian distribution at 20°C, what would be its kinetic energy? Would this be significant with respect to the threshold energy of the alpha particle?

11.37 Under the proper conditions two alpha particles can combine to produce a proton and a nucleus of ^{7}Li. (a) Write the equation for this reaction. (b) What minimum kinetic energy must one of the alpha particles have in the laboratory system in order to make the reaction proceed if the other alpha particle is at rest? (c) Now assume that both alpha particles are moving, each directly toward the other, with identical speeds. What is the minimum energy of the alpha particles that will allow the reaction to proceed? (Note that laboratory system and the center-of-mass system are the same in this part.)

11.38 A common means of obtaining neutrons in the laboratory is to bombard tritium with deuterons. Assume that the incident deuterons have an energy of 1 MeV. (a) Write the equation for this reaction. (b) Compute the Q-value. (c) Calculate the energy of the neutrons that are emitted at an angle of 90° with respect to the beam of incident deuterons. (Mass of ^{3}H = 3.01605 u.)

11.39 A neutron of zero kinetic energy interacts with lithium at rest and the reaction $^6_3$Li(n, α)$^3_1$H occurs. (a) Write the equation for this reaction. (b) Calculate the Q-value. (c) Calculate the kinetic energy of the ejected alpha particle. (Isotopic mass of ^{3}H = 3.01605 u.)

11.40 A sample of ^{31}P weighing 25 g is bombarded with neutrons until the activity of the ^{32}P produced is 3mCi. What is the ratio of the number of radioactive phosphorus atoms to the number of stable atoms?

11.41 Two reactions that occur in the upper atmosphere due to the effects of cosmic "rays" are ^{14}N(n, p)?; and ^{14}N(n, ?)^{12}C. (a) Complete the nuclear reaction equation for each and calculate the Q-value of each. (b) Is the nuclide formed in each case stable or radioactive? (See Appendixes 5 and 6. Isotopic mass of ^{14}C = 14.003233.)

11.42 Natural carbon is 18% of human body weight. The activity of ^{14}C in a person weighing 70 kg is 0.1 μCi. What fraction of the carbon in the body is ^{14}C?

11.43 Some of the carbon atoms in trees and lumber are ^{14}C. Why are there none among the carbon atoms in petroleum products?

11.44 Due to ^{14}C, the charcoal from the fire pit in an ancient Indian camp site has an average beta activity of 12.9 counts or disintegrations per minute per gram of carbon in the sample (cpm/g). The absolute specific activity of the ^{14}C in the wood from living trees is independent of the species of the tree and its averages 15.3 cpm/g. What is the age of the charcoal sample?

11.45 If each of the average counts in Problem 11.44 has a numerical uncertainty of ± 0.1 cpm/g and if the half-life of ^{14}C is 5730 $\pm$ 30 years, what is the numerical uncertainty in the determination of the age of the charcoal sample?

11.46 The tritium content of the water from certain deep wells is only 33% of that in the water from recent rains. (a) How long has it been since the water in the well came down as rain? (b) What assumption did you make about tritium in solving this problem?

11.47 In November 1974, Don Johanson discovered a partial skeleton (later named Lucy) that changed the thinking of anthropologists and caused great controversy. The bones were found in rock containing trapped potassium with a typical sample yielding a total of 0.1 g. A fraction (0.0000118) of the trapped potassium was ^{40}K, which decays into ^{40}Ar. This sample contained 2.39×10^{-9} g of ^{40}Ar, of which 7.25×10^{-10} g was contamination from the air. What is the approximate age of Lucy?

11.48 What must be the activity of a radiosodium (^{24}Na) compound when it is shipped from Oak Ridge National Laboratory so that upon arrival at a hospital 24 h later its activity will be 100 millicuries?

11.49 The steel compression ring for the piston of an automobile engine has a mass of 30 g. The ring is irradiated with neutrons until it has an activity of 10 microcuries due to the formation of ^{59}Fe ($T = 45.0$ days). Nine days later the ring is installed in an engine. After being used for 30 days, the crankcase oil has an average activity due to ^{59}Fe of 12.6 disintegrations per minute per 100 cm^3. What was the mass of iron worn off this piston ring if the total volume of the crankcase oil is 6 qt?

11.50 (a) What energy must be supplied to overcome the electrostatic repulsion in order to bring two widely separated protons together until they are separated by 2×10^{-13} cm, which is of the order of nuclear dimensions? (b) What electrostatic force acts on two protons separated by 2×10^{-13} cm?

11.51 By referring to Fig. 11.17, determine whether energy must be supplied or will be released if we assume that the following transmutations can be accomplished: Fe $\rightarrow$ Xe, Li $\rightarrow$ Na, and Pb $\rightarrow$ Al.

11.52 (a) Find the total binding energy and the binding energy per nucleon of ^{2}H, ^{4}He, and ^{56}Fe. (b) How much energy would be involved in removing the nucleons in He to infinity? Would energy be released or have to be supplied for this dispersion?

11.53 Assuming that the lifetime of the excited state involved in the 14.4 keV transition in the nucleus of ^{57}Fe is 10^{-7} s, calculate (a) the associated uncertainty in the energy using Heisenberg's principle, Eq. (7.19), (b) the corresponding fractional uncertainty in the energy of the emitted gamma ray, and (c) the fractional uncertainty in its frequency. (d) What is the absolute value of the uncertainty in its frequency and in its wavelength?

11.54 What is the recoil velocity and the recoil energy of a free ^{57}Fe nucleus when its emits a 14.4 keV gamma-ray photon?

11.55 When the nucleus of a Mössbauer emitter is placed in a magnetic field, there is fractional increase in the energy of the emitted gamma-ray photon of 10^{-12}. What must be the magnitude and the direction of the velocity, with respect to the emitter, of an absorber of the same material so that resonance absorption will occur? (The fractional change in observed frequency in the Doppler effect when a receiver moves toward a wave source is v/c, where v is the velocity of the receiver with respect to the source and c is the speed of propagation of the waves.)

12 NUCLEAR ENERGY

12.1 THE DISCOVERY OF FISSION

Since neutron capture followed by beta decay produces a new nuclide of higher atomic number, the process suggested to Fermi an intriguing possibility. Why not cause neutron capture in uranium, the last (then known) element of the periodic system? If uranium formed a beta-unstable isotope, the disintegration product would be element number 93, a "transuranic" element. In 1934, Fermi and his associates attempted the experiment. Apparently the uranium did undergo neutron capture. Apparently the product was beta-unstable. But the beta activity included no less than four different half-lives. As in the discoveries of natural and artificial radioactivity, an attempt was made to identify the product nuclide by chemical separation. Early in 1938, Hahn and Strassmann showed that the beta-active product was chemically like radium and concluded that the new nuclides were new isotopes of radium. In order to get from uranium to radium, whose atomic number is four less, the uranium would have to emit two alpha particles. These alpha particles were not found.

The chemical analysis of such tiny amounts of material is very difficult, but in September 1938 Mme. Joliot-Curie and P. Savitch published a report of their work that identified one activity as due, apparently, to *lanthanum*, a rare-earth element that has a much smaller mass number than uranium. This led Hahn and Strassmann to repeat their own earlier work very carefully. In December 1938 they concluded that chemists should replace the symbols Ra, Ac, and Th with Ba, La, and Ce. They as nuclear chemists, closely associated with physics, could not decide to take this step in contradiction to all previous experience in nuclear physics.

The correct interpretation of the puzzling data came in January 1939 when Meitner and Frisch wrote, "It seems possible that the uranium nucleus has only small stability of form, and may, after neutron capture, divide itself into two nuclei of roughly equal size." Because of its resemblance to the splitting of one living cell into two of equal size, this nuclear process was named after the biological process, *fission*.

All previous nuclear reactions had had as their products no particles more massive than alpha particles and no nuclei farther removed from the original nucleus than two atomic numbers. Even in this decade of rapid advance, it took five years to break away from the limitation of the old conceptions and break through to the conception of this strikingly new process.

Once the fission process was regarded as plausible, the chemical evidence for its correctness was almost immediately reinforced by physical evidence. There were literally scores of laboratories around the world that were already equipped to verify fission. If one uranium atom splits into two atoms near the middle of the periodic series, consideration of the energy per nucleon by either a glance at Fig. 11.17 or a few moments of calculation will show that the Q-value of the reaction is tremendous. It is 200 MeV, which is about ten times that of the most energetic reaction previously known. Within a short time after hearing the news, many laboratories confirmed this energetic reaction.

12.2 NUCLEAR ENERGY

Although one wink of your eye requires the expenditure of many billions of electron volts of chemical energy, nuclear processes are potentially far more energetic than chemical processes. The Q-values we have been discussing are energies per single disintegration, whereas the energy to wink an eye comes from the chemical conversion of many billions of molecules. The liberation of chemical energy is usually expressed on a per-gram, per-pound, or per-kmole basis, and the conversion factor between per kmole and per molecule basis is the Avogadro constant, about 6×10^{26}. In Section 10.9 we compared the energy from oxidizing carbon and from radioactive disintegration on a per-gram basis. Table 12.1 lists some energies on a per-particle basis. The first and third processes of Table 12.1 account for most of the energy utilized in industrial and life processes.

We have seen that radioactive disintegrations are nuclear processes in which an appreciable amount of rest mass-energy is converted to kinetic mass-energy. The mass-energy equation that we have repeatedly applied to nuclear reactions is perfectly general and can be applied to chemical reactions as well. We can write

$$C + 2(O) = CO_2 + Q, \qquad (12.1)$$

but nothing useful can be learned from applying the equivalence of mass-energy to a

Table 12.1 Energies per particle

Kinetic energy of one water molecule, 450-ft waterfall	0.00025	eV
Average kinetic energy of a gas molecule at room temperature	0.025	eV
Carbon atom oxidized to CO_2	4.0	eV
Visible photon	2.0	eV
Ultraviolet photon	3.0 to 100.0	eV
Hard x-ray photon	0.1 to 1.0	MeV
Gamma-ray photon	1.0 to 3.0	MeV
Radium disintegration	4.8	MeV
Fission disintegration, uranium	200.0	MeV
Cosmic-ray particle	1.0 to 10.0	GeV

chemical equation. The Q-value in u is $4.0/(931 \times 10^6)$ or 4.3×10^{-9} u. In the chemical reaction, the carbon and oxygen would be mixtures of isotopes, but even if pure isotopes were used, the uncertainty of the isotopic masses far exceeds the Q-value. Thus in chemical reactions, it is proper to assume that mass and energy are conserved separately. We write Eq. (12.1) in order to emphasize the contrast between chemical and nuclear reactions.

12.3 CHAIN REACTION

Highly energetic radioactive disintegrations are not very promising as practical sources of energy. Although radioactive nuclides made in the laboratory are available in far greater quantities than the naturally radioactive nuclides, there is none whose abundance would justify fleeting consideration as a prime source of energy in a class with coal or oil. Nevertheless, radioactivity may be utilized to make electric batteries of exceptionally long life. Thus energy from radioactivity will find limited use in special applications.

Some nuclear transmutations provide more energy per event than the natural radioactive disintegrations. The $^6Li(d, \alpha)^4He$ process has a Q-value of 22.4 MeV. But this reaction requires accelerated deuterons, and the energy required to operate the apparatus is millions of times greater than the energy derived from the process.

Although fission reactions release more energy per event than any type of reaction known earlier, it is another property of fission that makes it a practical source of energy.

The principal isotope of uranium that undergoes fission is ^{235}U. This nuclide, which has about a 0.7% concentration in natural uranium, fissions in many ways. The *fission yield* of a fission product is the percentage of the fissions that lead to the

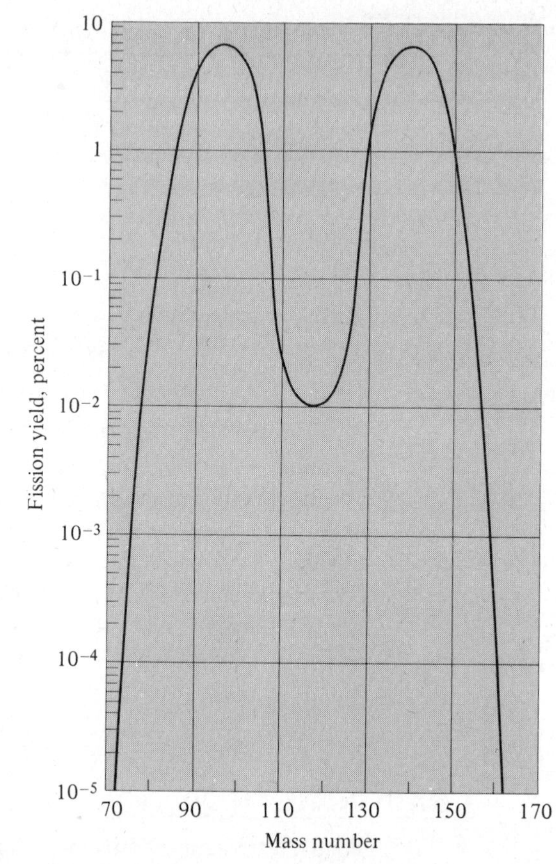

Figure 12.1 Fission yield from ^{235}U.

formation of a particular nuclide or a group of isobars. Since there are two nuclides produced per fission, the total of all the fission yields for a given process is 200%. The fission yield from ^{235}U of nuclides of different mass numbers is shown in Fig. 12.1 (note that the ordinate scale is logarithmic). Obviously, we can write no unique reaction, but a typical one is

$$^{235}_{92}U + ^{1}_{0}n \rightarrow ^{236}_{92}U \rightarrow ^{140}_{54}Xe + ^{94}_{38}Sr + 2\,^{1}_{0}n + \gamma + 200 \text{ MeV}. \qquad \textbf{(12.2)}$$

The fission energies and the fission-yield curves for ^{238}U and ^{239}Pu do not differ appreciably from ^{235}U. Some heavy nuclei undergo *spontaneous fission* in which the nucleus divides in the ground state without being bombarded by neutrons or other particles. The half-life of ^{238}U and certain other heavy nuclei for this process is of the order of 10^{16} years. Therefore, only about 20 nuclei per gram of these elements undergo fission every hour. Their contribution to the operation of a reactor is insignificant and will be neglected in further discussion.

Figure 11.19 is a graph of the number of nuclide neutrons as a function of nuclide protons for the stable elements. This figure shows that the heavier the element, the greater its neutron excess. When a fission occurs, the product nuclides are lighter than the parent element and, to be stable, they must have a smaller excess of neutrons. The nuclides first formed have neutron excesses that make them either absolutely or radioactively unstable against neutron emission (see the first and third cases of Section 11.12). Both ^{140}Xe and ^{94}Sr are radioactive beta emitters; therefore Eq. (12.2) does not take us to the end of the disintegration. Before these two nuclides were formed, they had a parent nuclide that was a neutron emitter. Its two neutrons appear on the right-hand side of the equation along with the large Q-value.

The *average* number of neutrons from ^{235}U is 2.5 per fission. They provided the real potentiality of fission as a power source because they made a chain reaction appear feasible. Here was a reaction that had among its products the same kind of particle that initiated it. If, on the average, at least one neutron from the first fission could produce another fission, and at least one neutron from the second could produce a third, etc., then a self-sustaining chain of events would take place with no necessity for accelerators or other devices to keep the process going. The first fission reactions were obviously not self-sustaining chain reactions, since the fission activity stopped when the neutron source was removed. Although some of the neutrons from the first fission reactions undoubtedly caused other fission reactions, it was evident that too many of the neutrons produced were being absorbed in some nonfission-producing manner. The problem that faced nuclear physicists can be stated simply: Under what experimental conditions — if any — could the fission reaction be made self-sustaining?

Soon after the fission of ^{235}U was accomplished, two more reactions were found. These produced the transuranic nuclides Fermi set out to produce in his original experiments in 1934. Just as he expected, simple neutron capture is followed by beta activity and the formation of the transuranic element neptunium, Np:

$$^{238}_{92}\text{U} + ^{1}_{0}\text{n} \rightarrow ^{239}_{92}\text{U} \rightarrow ^{239}_{93}\text{Np} + ^{0}_{-1}\text{e}. \tag{12.3}$$

The neptunium, which is radioactive with a half-life of 2.35 days, becomes plutonium, Pu:

$$^{239}_{93}\text{Np} \rightarrow ^{239}_{94}\text{Pu} + ^{0}_{-1}\text{e}. \tag{12.4}$$

The plutonium is also radioactive, but its half-life is 24,360 yr. It, like ^{235}U, is fissionable, and we shall see that it is a very important nuclide.

Enough has now been said to account for the fact that a chain reaction did not occur when the first uranium sample was bombarded with neutrons. In the first place, we know that neutrons are very penetrating. Thus the most likely fate of neutrons from the first fission was escape from the uranium sample. The original uranium samples were mixtures of ^{235}U and ^{238}U. Since ^{238}U is much more abundant than ^{235}U, the neutrons that did not escape from the first fission reactions had an excellent chance of nonfission capture by ^{238}U. It was anticipated that if ^{235}U could be separated from ^{238}U and other neutron-capturing nuclides, and if enough of this pure ^{235}U could be prepared so that neutrons had a good chance of causing fission before they could escape, then a chain reaction would probably occur. If the chain reaction grew within a large body of ^{235}U, energy should be liberated in explosive amounts.

12.4 NEUTRON CROSS SECTIONS

In our previous discussion of nuclear reactions, we have been concerned with "what happens" rather than with "how much happens." Although the reaction products have been so minute that identification was difficult, the minuteness did not detract from interest in "what happened." In discussing a chain reaction where one event must successfully cause another, "how much" is just as important as "what."

In Chapter 1, we introduced the concept of mutual collision cross section as a way of describing the probability of one particle hitting another. We showed that this cross section was a property of both the bullet and the target particles, since it is defined as

$$\sigma = \pi(r_{\text{bullet}} + r_{\text{target}})^2. \tag{12.5}$$

By assigning this area to the target particle, we could regard the bullet particle as a point and still have a measure of the likelihood of a collision. When we introduced this concept, we were discussing the kinetic theory of gases. The particles were whole, uncharged atoms, and we were thinking of the collisions as being pure mechanical hits, like those between billiard balls. If we considered collisions between positive ions, the cross section for deflection would be bigger. The coulomb force of repulsion would extend beyond the physical limits of the particles, so that the bullet particle would be deflected without coming so near the target particle. Although the ion-deflection cross section would be bigger than if the same atoms were not ionized, the cross section for chemical interaction would be less. The mutual repulsion that increases the probability of deflection diminishes the probability of near approach. Thus the concept of cross section becomes a general and useful measure of the probability of many classes of "collisions." If two particles can interact in more than one way, the probabilities of the various interactions can be measured in terms of the cross section of each.

If, instead of thinking of a gas molecule moving through a gas, we think of a neutron moving through matter, we find that a variety of kinds of collision interactions are possible. Each has its own probability, which is directly proportional to its cross section. Some of the possibilities are listed in Table 12.2.

In Chapter 1, we showed that the probability of a particle making a simple collision in going through a gas for a distance dx was $n\sigma\,dx$, where σ is the mutual collision cross section, and n is the number of particles per unit volume (Eq. 1.29). In complete analogy, we may now say that the probability of a neutron causing fission

Table 12.2

Process name	Reaction type	Cross-section symbol
Radiative capture	(n, γ)	σ_r
Fission	(n, f)	σ_f
Elastic scattering	(n, n)	σ_s
Proton capture, etc. (absorption)	(n, p), (n, d), etc.	σ_a
Total	$\sum \sigma$'s	σ_t

in moving a distance dx through matter composed of an element whose atoms are at rest is

$$P_f = \sigma_f N\, dx, \tag{12.6}$$

where the nuclear concentration, N, is the number of nuclei per unit volume. The *microscopic cross sections*, σ's, become the *macroscopic cross sections*, $N\sigma$'s, when multiplied by N. The total probability that some type of interaction will occur is

$$P_t = (\sigma_f + \sigma_s + \sigma_a + \text{etc.})\,N\,dx. \tag{12.7}$$

If the material is composed of more than one element, then the neutron may collide with any nuclide present, depending on the concentration of the nuclide and its cross sections. Thus if there are two nuclides present in concentration N and N', the *total probability* that the neutron will undergo some interaction in going a distance dx is

$$P_t = (N\sigma_t + N'\sigma_t')\,dx. \tag{12.8}$$

Before these equations can be used quantitatively, the values of the σ's must be known. Although many σ's were known before the period of secrecy about the fission process, one of the first tasks in calculating the feasibility of a chain reaction was the determination of many more cross sections. Mathematically, this process is very much like the determination of the probability of radioactive decay, the decay constants.

If I_0 incident neutrons per second produce I_f fissions per second, then the probability of fission is

$$P_f = \frac{I_f}{I_0}. \tag{12.9}$$

But P_f is also given by Eq. (12.6), so we may write

$$\frac{I_f}{I_0} = N\sigma_f\, dx$$

or

$$\sigma_f = \frac{I_f}{I_0 N\, dx}. \tag{12.10}$$

Thus if we know the number of fissions produced per unit time by a known number of neutrons per unit time in a material of known concentration and thickness, the fission cross section can be determined. By measuring the number of absorptions per unit time or the number of scatters per unit time, etc., the cross sections can be determined.

In this differential method, it is assumed that dx is so small that the intensity of the neutron beam can be considered constant throughout the thickness of material, and that the nuclei in the material cannot hide behind one another.

If the slab of material has finite thickness, then the neutron intensity is a variable, I, and we must integrate to determine its value. In penetrating a slab of the material of thickness dx, the intensity of the neutron beam is changed an amount $(-dI)$ by any interactions that remove neutrons from the beam. The total probability of interaction is then

$$P_t = -\frac{dI}{I} = N\sigma_t\, dx. \tag{12.11}$$

Upon integrating, and noting that for $x = 0$ we have $I = I_0$, we obtain

$$I = I_0 e^{-N\sigma_t x}, \tag{12.12}$$

where I is the intensity of the *transmitted* beam. This equation permits the evaluation of the total cross section from data such as those given in Fig. 12.2 and provides a check on the more difficult differential method through the equation

$$\sigma_t = \sigma_f + \sigma_s + \sigma_a + \text{etc.} \tag{12.13}$$

As in Eq. (1.33), we can now speak of the neutron mean free path:

$$L_t = \frac{1}{N\sigma_t}. \tag{12.14}$$

Or for the mean neutron path per fission we may write

$$L_f = \frac{1}{N\sigma_f}. \tag{12.15}$$

If the material is a mixture of two fissionable materials having different fission cross sections and different nuclear concentrations, the mean neutron path per fission becomes

$$L_f = \frac{1}{N\sigma_f + N'\sigma_f'}. \tag{12.16}$$

The measurement of cross sections is complicated by the fact that not only does each nuclide have several cross sections but also cross sections are often complicated functions of the neutron energy. The total cross sections of three nuclides are shown in Fig. 12.3. Since the microscopic cross sections are usually very small quantities, it is convenient to have a special unit for them, the *barn*. One barn is an area equal to

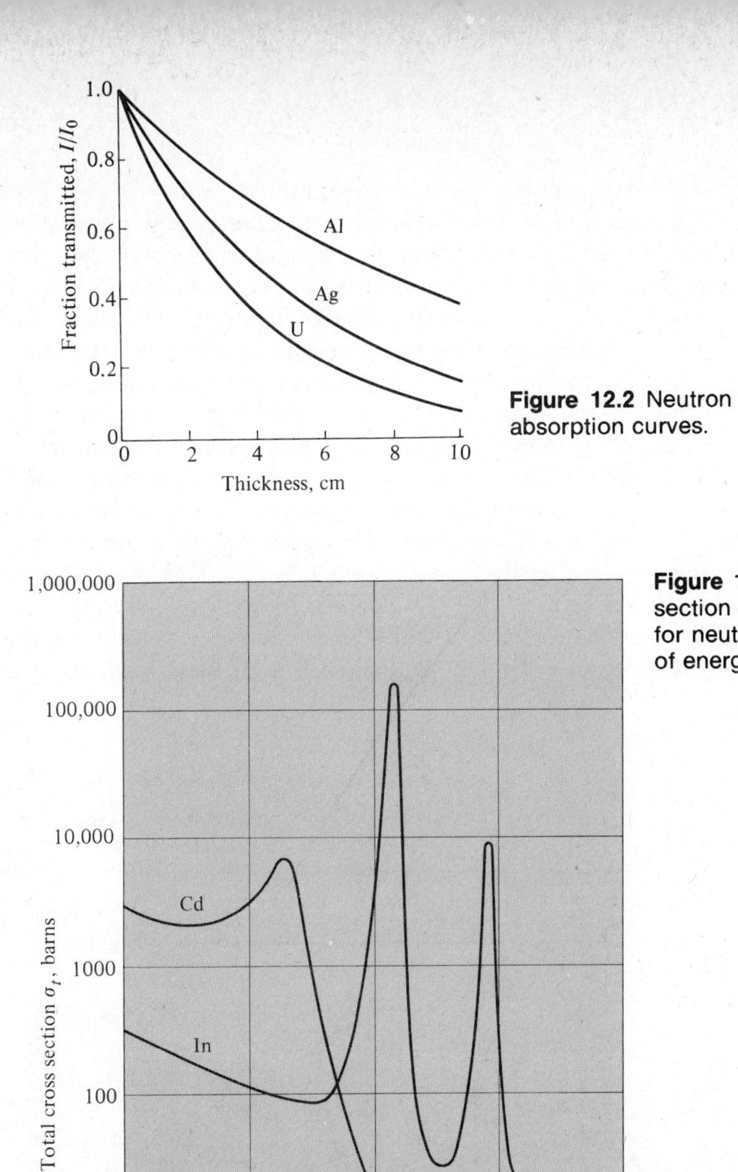

Figure 12.2 Neutron absorption curves.

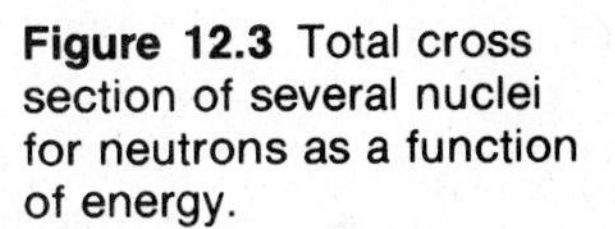

Figure 12.3 Total cross section of several nuclei for neutrons as a function of energy.

10^{-24} cm^2. Cross sections are proportional to the probabilities of neutron reactions. Although cross sections have area units, they are not the physical cross sections of the nuclei. Nevertheless, it is interesting to note that most nuclides have "diameters" of about 10^{-12} cm and consequently their physical cross sections are about 1 barn.

Although every cross section should be expressed as a function of energy, as in Fig. 12.3, many elements have cross sections that can be represented by a simple equation. These nuclides have cross sections that vary inversely as the velocity of the neutron. Such materials are called *$1/v$-absorbers*. The most important neutron energy distinction is that between fast neutrons and thermal neutrons. Neutrons from a fission reaction are fast and have an average energy of about 2 MeV. Neutrons that have traversed enough matter to be in thermal equilibrium with the molecular motions of the material are called *slow* or *thermal* neutrons, and these have energies distributed about the value of 0.025 eV. Table 12.3 gives values of the cross sections of several important nuclides for these two energies.

Obtaining neutron cross sections as a function of velocity requires methods for measuring neutron velocities. We cannot discuss these methods in detail, but it is

Table 12.3 Neutron cross sections in barns

	σ_t (slow)	σ_t (fast)	σ_s (slow)	σ_a (slow)	σ_f (slow)	σ_f (fast)
^{1}H	(38)	(4.3)	38	0.33	0	0
H_2O	(110)					
^{2}D	(7)	(3.4)	7	0.0005	0	0
D_2O	(14.5)				0	0
^{3}T	5400	(1.9)	1	5400	0	0
Li	(72)	(1.7)	1.4	71	0	0
^{6}Li	(951)	(0.26)	(6)	945	0	0
^{7}Li	(1.4)		(1.4)	(0)	0	0
B	(760)		4	755	0	0
C	(4.8)		4.8	0.0032	0	0
O	(4.2)		4.2	0.0002	0	0
Zr	(8)		8	0.18	0	0
Cd	(2560)		7	2550	0	0
Hf	(113)		8	105	0	0
U (natural)		(7)	4.7 (fast)	(2) (fast)		0.5
^{235}U	697	6.5	10	107	580	1.3
^{238}U	2.8		0	2.8	0.0005	0.5
^{239}Pu	1075		9.6	315	750	

Values in parentheses are less certain than the others. Data taken or deduced from *Neutron Cross Sections*, Brookhaven National Laboratory.

easy to see that for uncharged particles many familiar methods must be excluded. We can diffract neutrons by crystals and thus obtain the de Broglie wavelength, from which the velocity can be computed. There are also mechanical devices that can "chop" a slow neutron beam so as to eliminate all neutrons outside a narrow energy range.

12.5 REACTOR CRITICALITY

We are now in a position to make simple calculations that give some idea of what we may expect in various circumstances. Consider first that we have a very large body of pure ^{238}U. It is known that each fission produces an average of 2.5 neutrons having energies of about 2 MeV. Since the body of metal is very large, we can neglect the loss of neutrons by escape and consider the probable fate of neutrons within the metal. We consider three possibilities:

1. Neutrons may scatter elastically within the metal. These collisions have a cross section of 4.7 barns for fast neutrons, so that scattering collisions are rather common. That scattering collisions deflect the neutrons is of no consequence. As long as the neutrons remain within the metal, their direction of motion makes no difference. Scattering collisions slow the neutrons but, since the uranium nuclei are 238 times as massive as the neutrons, it takes many collisions to slow the neutrons appreciably. We therefore ignore the influence of scattering.
2. The neutrons may be absorbed in some nonfissioning manner, such as that described in Eq. (12.3). These absorptions have a cross section of 2 barns for high-energy (fast) neutrons and represent neutron loss so far as the possibility of a chain reaction is concerned.
3. The neutrons may produce fission reactions for which the cross section is 0.5 barn. These neutrons are productively absorbed and tend to maintain the chain reaction. Under our assumptions, the probability that one neutron will produce another fission is the microscopic fission cross section divided by the total cross section, or

$$P_f = \frac{\sigma_f}{\sigma_a + \sigma_f} = \frac{0.5}{2 + 0.5} = 0.2.$$

Thus about one out of five neutrons produces a fission. Since each fission produces 2.5 neutrons on the average, each neutron probably produces $0.2 \times 2.5 = 0.5$ neutron. We can summarize this result in the following way. If there were ten neutrons in one neutron generation, eight of these would be unproductively absorbed and two would produce fission. Each fission would produce 2.5 neutrons, so that in the next generation there would be five neutrons. Thus each generation has half as many neutrons as the preceding one and no chain reaction can occur. The ratio of the number of neutrons in one generation to the number in the preceding generation is called the production factor or *multiplication constant*, k. When the body of material is so large that leakage may be neglected, the multiplication constant is represented by the symbol k_∞, called *k-infinity* or *infinite* (k). Since we have found $k_\infty = 0.5$ for a body of ^{238}U, we know that any amount of ^{238}U will be *subcritical*. This means that k is less than unity and that any fission process that may be initiated will quickly die out.

If we repeat these calculations for ^{235}U under the same assumptions, we find a very different result. Every ^{235}U nucleus that absorbs a high-energy neutron produces a fission. Since every absorbed neutron produces another fission, the multiplication constant is simply the number of neutrons per fission, or $k_\infty = 2.5$, and it is evident that each neutron generation is much larger than the preceding one. A large mass of ^{235}U would explode. The only way to prevent a body of pure ^{235}U from exploding would be to break it into pieces so small that 60% of the neutrons would escape without producing fission. These pieces would each be under the critical size for ^{235}U.

The concept of critical size results from the fact that fissions occur throughout the volume of a reacting body and neutrons escape through the surface of that body. As the size of a body is changed, the volume changes according to the cube of its dimensions, while its area changes as the square of its dimensions. Thus the ratio of neutron production to neutron leakage varies as the first power of the dimensions. If the configuration of the reactor is such that k_∞ is greater than unity, there is always a smaller size for which the material will be just critical, with k equal to unity. Below this size k is less than unity. In summary: above the critical size, k is greater than unity and the neutron population increases exponentially with time; *at the critical size, k is equal to one* and the neutron population is constant; below the critical size, k is less than unity and the neutron population falls off exponentially with time.

The calculations we have made are sufficient to show that natural uranium, which is about 99.3% ^{238}U and 0.7% ^{235}U, cannot chain react and that the concentration of the ^{235}U must be increased. In order to estimate the degree of enrichment required, we make a third calculation under the same assumptions as before except that we now have a metal composed of a mixture of ^{235}U and ^{238}U in which the number of nuclei per unit volume is designated by N_5 and N_8, respectively. The calculation is made just as before, but now we must use the fast-neutron cross sections of both nuclides weighted according to their concentrations. Since $\sigma_a = 0$ for fast neutrons in ^{235}U, we have

$$k_\infty = 2.5\frac{N_5\sigma_{f5} + N_8\sigma_{f8}}{N_5(\sigma_{a5} + \sigma_{f5}) + N_8(\sigma_{a8} + \sigma_{f8})} \tag{12.17}$$

$$= 2.5\frac{1.3N_5 + 0.5N_8}{(0 + 1.3)N_5 + (2 + 0.5)N_8}. \tag{12.18}$$

Letting $N_5 = 0$ gives the result of our first calculation and letting $N_8 = 0$ gives our second result. Letting $k_\infty = 1$, we can find the concentration ratio that will make a large mass of uranium critical. Solving for N_8/N_5, we obtain 1.5, which tells us that for a large body of uranium metal to be critical for fast neutrons, it must be about 40% ^{235}U. For a body of finite size, the enrichment of ^{235}U would have to be greater than 40%. Calculations such as these led to the decision to attempt the enrichment of ^{235}U in uranium metal.

The separation of isotopes is always difficult, but since uranium is the heaviest natural element, the fractional mass difference between ^{235}U and ^{238}U is especially small.

The most effective method of separation found was the gaseous diffusion of uranium hexafluoride, UF_6. Although the enrichment of ^{235}U to about 99% purity requires about 4000 diffusion stages, a tremendous plant based on this principle was built at Oak Ridge, Tennessee, which has been in production since 1945. A measure of the difficulty of this process is contained in the fact that in 1955 the AEC evaluated natural uranium at $40 per kg and 95% ^{235}U at $16,258 per kg. With fuel enriched with ^{235}U, the construction of a fast-neutron bomb reactor is, in principle, simple. The uranium parts must each be less than critical size lest they explode spontaneously from stray neutrons or an occasional spontaneous fission. Detonation is accomplished by bringing the subcritical parts together into a *supercritical* whole. The critical size of the ^{235}U enriched uranium is about the size of a grapefruit.

Weapons grade uranium contains at least 50% ^{235}U. The uranium in a nuclear power plant is enriched to about 3.6% ^{235}U. Obviously, that is much too lean a mixture to produce a nuclear explosion.

12.6 MODERATORS

A glance at Table 12.3 shows that the uranium cross sections for slow neutrons are very different from those for fast neutrons. Whereas most nuclear reactions are best induced with high-energy particles, neutron-induced reactions usually have cross sections that increase as the neutron energy is decreased. This may be phrased in a somewhat naive way by saying that since there is no coulomb repulsion, slow neutrons spend more time near the nuclei they pass and therefore the short-range nuclear attractive forces have a better chance to take effect. This is why many cross sections vary as $1/v$.

The nonfission capture cross section of ^{238}U increases as the neutron energy is reduced, but the fission cross section of ^{235}U increases even more. Thus the relative probability of fission in natural uranium is much greater for low-energy neutrons than for high-energy neutrons. To show how marked this difference is, we next compute the production factor that a large body of natural uranium *would* have *if* each fission produced 2.5 *thermal* neutrons. We again use Eq. (12.17), except that we

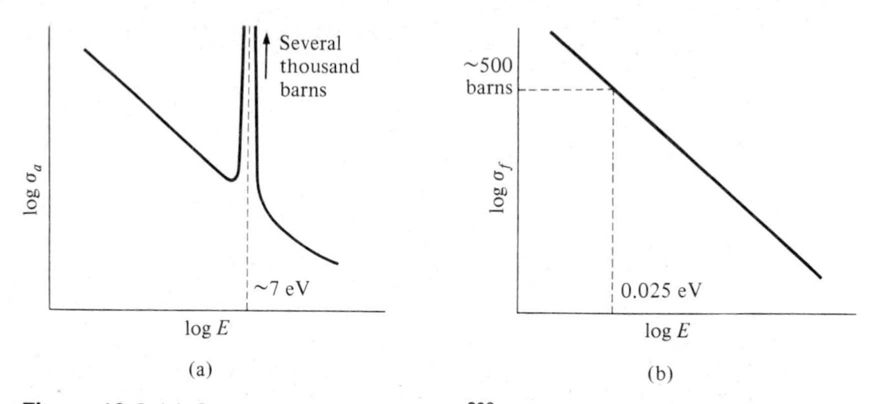

Figure 12.4 (a) Capture cross section of ^{238}U for slow neutrons, and (b) fission cross section of ^{235}U for slow neutrons, assuming a $1/v$ relationship.

substitute the thermal cross sections and take $N_8/N_5 = 1/0.007 = 143$, which yields

$$k_\infty = 2.5\frac{1 \times 580 + 143 \times 0.0005}{1 \times (107 + 580) + 143(2.8 + 0.0005)} = 1.33. \tag{12.19}$$

Since k_∞ is considerably greater than unity, it appears that even a finite size of natural uranium would chain react if the fission neutrons were slow. Since fission neutrons are fast rather than slow, the calculation we have just made is purely academic unless neutrons can be slowed. Even before the period of secrecy, it was proposed that if the fast-fission neutrons could be slowed (*moderated*), critical conditions might be achieved in natural uranium.

Another important consideration is brought out qualitatively in Fig. 12.4(a). For neutrons with an energy of about 7 eV, ^{238}U has a neutron resonance capture in which the ^{238}U nucleus has a very large probability of going into an excited state. Thus, in a natural mixture of ^{235}U and ^{238}U, there is very little chance that a neutron will be slowed below 7 eV, to the energy for which the probability of ^{235}U fission capture is very great.

The idea of using a moderator to increase the probability of a chain reaction was based on these variations of cross section with neutron velocity. If the fast-fission neutrons could escape from a lump of uranium and be slowed to very low energies, and were then allowed to fall on natural uranium, the relative probability of fission might be increased enough so that a chain reaction would result.

There are reactions in which a neutron is absorbed and then one, two, or even three neutrons of lower energy are emitted. Such reactions have the effect of slowing neutrons and even increasing their number, but these reactions are most easily produced in heavy nuclei with high-energy neutrons. For the slowing of neutrons with the average energy of fission neutrons, about 2 MeV, elastic scattering is much more suitable.

In our discussion of the discovery of the neutron, we worked out the mechanics of head-on collisions between moving particles and particles at rest (Section 11.3). We found that the energy transfer depends on the relative masses of the bodies involved in the collision and, in particular, that if a body strikes another body having the same mass, all the energy of the first body is transferred to the second one. If the collision is not head-on, the fraction of the energy transferred is certainly less. It is not too difficult to compute the *average fractional energy loss* between two bodies whose relative masses are known when the effect of *glancing collisions* is taken into account. This fraction is greatest when the two bodies have the same mass.

If a neutron enters a material, it loses a certain average fraction of its energy with each collision as long as the material nuclei can be considered at rest. If the neutron survives capture until its energy is reduced nearly to that due to thermal motion of the material molecules, then its energy loss per collision is much less than before. Indeed, at these energies there is a chance that the neutron may acquire energy from the molecules. Thus thermal agitation sets a lower limit to the energy a neutron acquires by the collision process. In Chapter 1, we showed that the average kinetic energy of a gas molecule depends only on its temperature, and we can consider neutrons in equilibrium with a moderator to be a gas at the same temperature as the moderator. Thus no moderator can reduce the energies of neutrons below the energies of the moderator molecules. The energies of the molecules of a gas at room temperature, 20°C, are distributed about a value of 0.025 eV. Neutrons in thermal equilibrium with matter at 20°C will have the same energy distribution as such gas molecules. These neutrons are called thermal neutrons.

Assuming that fission neutrons are emitted with an average energy of 2 MeV and have an energy of 0.025 eV when they have been "thermalized," we find the total fractional energy remaining to be $0.025/(2 \times 10^6) = 1.3 \times 10^{-8}$. The fractional energy remaining after a single collision can be calculated from Eq. (11.20). This value, raised to the nth power, where n is the number of collisions required to produce thermalization, is equal to the total fractional energy remaining. Since the absorption cross sections are functions of the neutron energy, it is awkward to compute the chance a neutron has to survive the thermalization process, but if criticality is to be achieved this chance times the number of neutrons per fission must be at least one.

Table 12.4 shows some of the characteristics of light elements as neutron moderators. Mechanically, hydrogen is the best moderating nuclide, since it requires, on the average, only 18 collisions to achieve thermalization. But since hydrogen can absorb neutrons to form deuterium, the chance of a neutron surviving the 18 collisions is not too good. We shall see, however, that hydrogen in ordinary water can be used as a moderator.

Deuterium requires 25 collisions on the average, but it has more promise as a moderator because its capture cross section is small. Oxygen also has a small capture cross section. Since a gas moderator has a very small number of nuclei per unit volume, both hydrogen and deuterium are used as moderators in the form of water.

Table 12.4

Element	σ (absorption), barns	σ (scatter), barns	Average fractional energy loss per collision	Average number of collisions to thermalize, n
H	0.33	38	0.63	18
D	0.0005	7	0.52	25
He	0.000	1	0.35	42
Li	71	1.4	0.27	62
Be	0.01	7	0.18	90
B	755	4	0.17	98
C	0.0032	4.8	0.14	114
N	1.7	10	0.12	132
O	0.0002	4.2	0.11	150

Helium would be a superior moderator except that its inertness prevents its concentration in a chemical compound.

Both lithium and boron have absorption cross sections that are too high. Thus beryllium and carbon are the lowest atomic weight elements that are solid at room temperature and whose neutron absorption cross sections are low enough for consideration as moderators. Of these, carbon is by far the more abundant and thus used more frequently.

After neutrons have undergone a number of random glancing collisions in a material, the motions of some of them will have been deviated so much that they return to the region from which they came. This "back scattering" is equivalent to diffuse reflection. Any material that has a low neutron absorption cross section can be used as a *neutron reflector*.

12.7 THE FIRST REACTOR

In 1942, many physicists believed that chain reaction would occur in pure ^{235}U and possibly in ordinary unenriched uranium in a properly moderated reactor. The first alternative required the very difficult separation of a considerable amount of ^{235}U from its isotope, ^{238}U. The second alternative appeared to be the faster way of testing the theoretical calculations.

Graphite for the moderator had to be prepared in an especially pure form, since any impurities such as boron and cadmium (neutron absorbers) would reduce the number of neutrons that survived the slowing-down process.

Uranium had never been used commercially, and in 1941 there were only a few grams of the pure metal in the United States. By the fall of 1942 about six tons of uranium metal had been prepared. It, too, had to be free of neutron-absorbing impurities.

Before the reactor (pile) could be built, there had to be instrumentation for its control. In the first pile and in all controlled reactors that have followed it, the basis for control is the presence of materials that are good neutron absorbers. We have seen that boron has a large absorption cross section and cadmium is another good neutron absorber. These elements, alloyed with steel in the form of rods, can be pushed into the pile to make it subcritical and to adjust the multiplication constant k.

Before the pile could be built, it was important to know how rapidly it would respond to control-rod manipulation. Fortunately, not all the neutrons emitted by fission appear at once. Although most neutrons are produced without delay, i.e., *prompt neutrons*, some of the fission products are radioactive neutron emitters that produce *delayed neutrons*, as shown in Fig. 12.5. Measurements showed that 0.4% of the neutrons are delayed at least 0.1 s and that 0.01% are delayed about a minute.

Figure 12.5 Uranium fission with delayed neutron emission in the bromine fission fragment decay chain. The six fission neutrons shown in this example are many more than the 2.5 average.

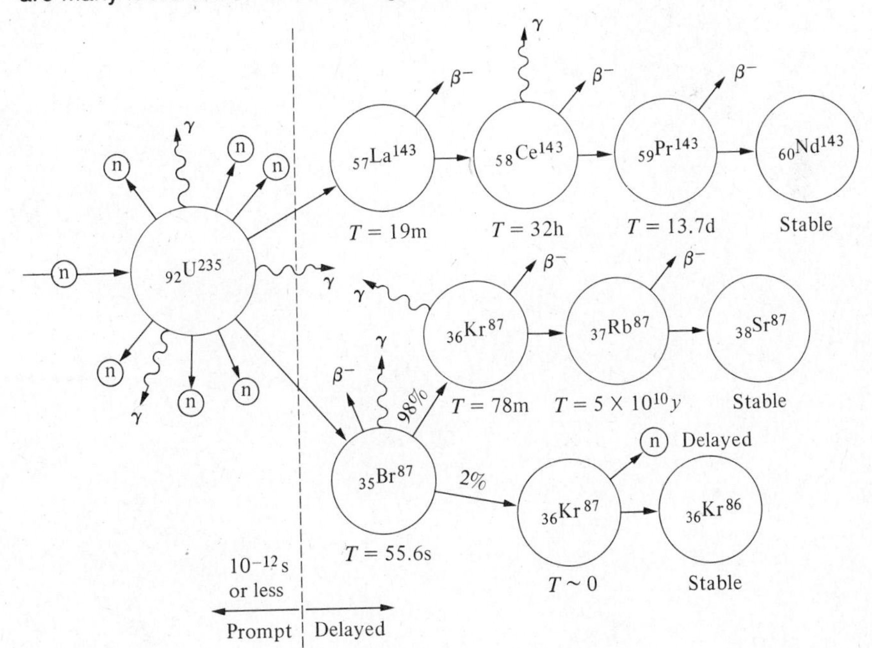

Thus a reaction that is subcritical for prompt neutrons can be critical for prompt and delayed neutrons together. (This latter state is sometimes called *delayed critical.*) The idea was that by putting the reactor together with the control rods in place, the rods could be carefully withdrawn at frequent intervals during construction. By seeing to it that the reactor never went critical for prompt neutrons, that is, *prompt critical*, it was assumed that the reactor would be sufficiently sluggish so that human responses could prevent the reactor from "running away" destructively. (See Fig. 12.6.)

Figure 12.6 The neutron balance in a natural uranium reactor that is just critical. The sketch shows what might happen to 1000 representative neutrons out of the many millions present within a reactor.
Courtesy of Westinghouse Electric Corporation.

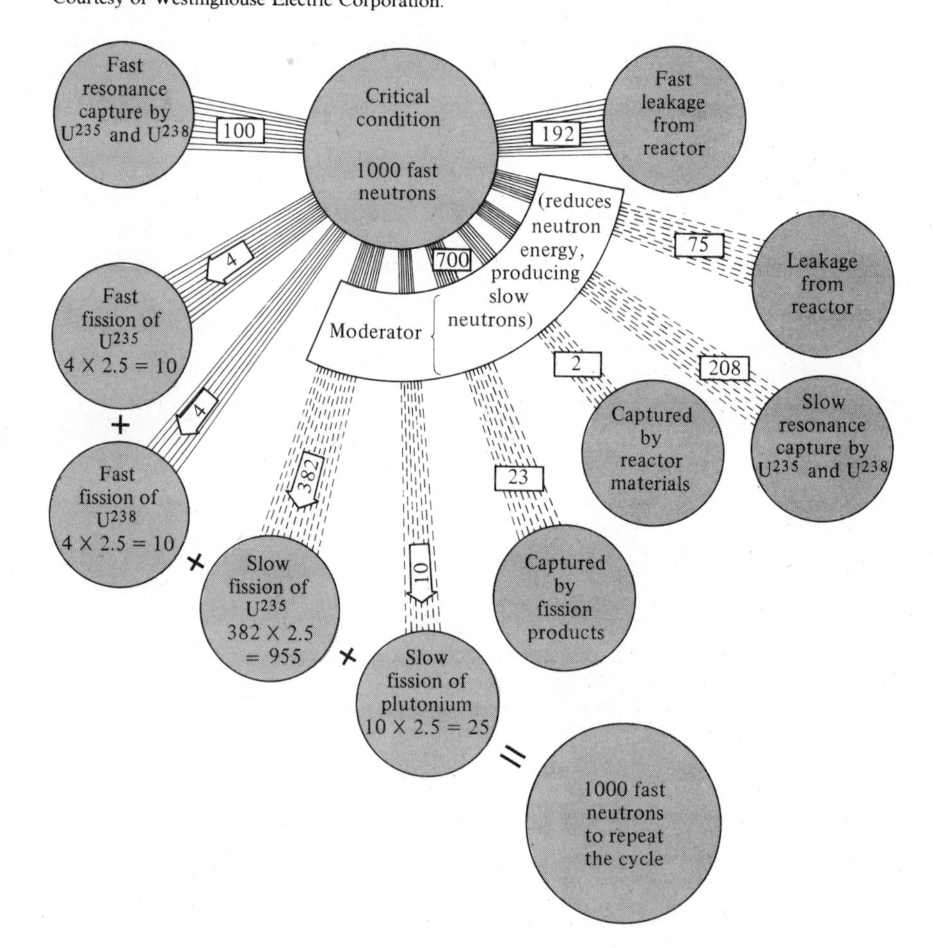

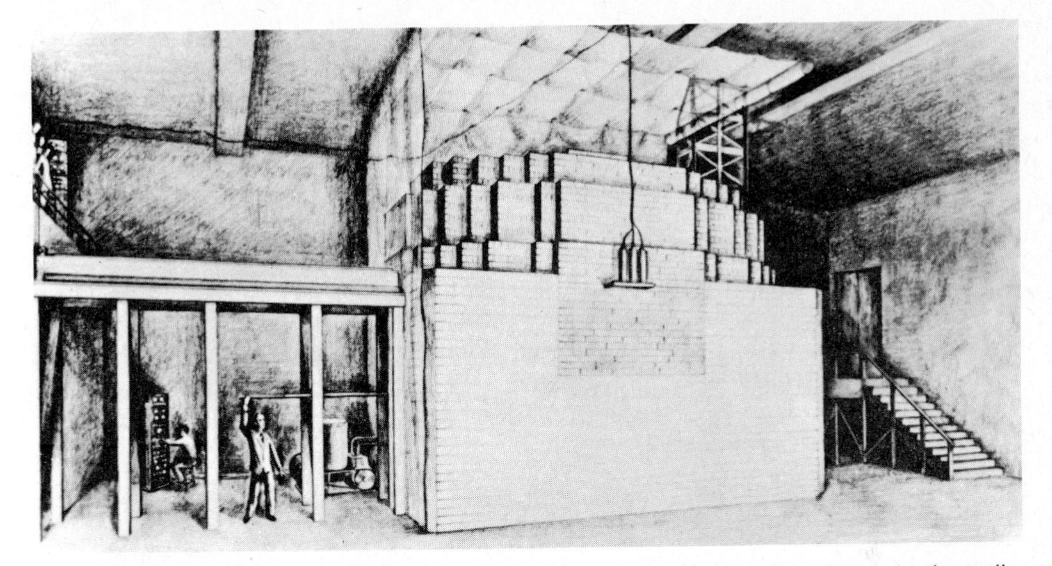

Figure 12.7 Sketch of the first nuclear reactor. An inscription on a plaque on the wall of the building (now torn down) read: "On December 2, 1942, man achieved here the first self-sustaining chain reaction and thereby initiated the controlled release of nuclear energy."
Courtesy of General Electric Company.

This first pile (Fig. 12.7) was assembled by Fermi and his associates in a squash court under the west stand of Stagg Field, the stadium of the University of Chicago. It was a cubic lattice of lumps of uranium in a *pile* of graphite blocks. The lump size was a compromise. They were made large, so as to include as many ^{235}U nuclei as possible, and yet small enough so that most of the fast-fission neutrons escaped without capture by ^{238}U. They were about 10 cm in diameter. The graphite space between the lumps was chosen so that most of the neutrons from one lump would be thermalized by the time they reached the next lump about 30 cm away. On the average, then, fission neutrons left the lumps at high energy, were in the graphite when their energy was 7 eV, and reentered the uranium at 0.025 eV. Since the neutrons were not in the uranium when their energy was at the resonance absorption of ^{238}U, the loss of neutrons to ^{238}U was small.

The reactor was built around a neutron source using alpha particles from a radioactive source to excite an (α, n) reaction, as in beryllium. As the blocks of graphite and the lumps of uranium were assembled, some fissions occurred and the number of neutrons increased. The neutrons that were not absorbed in the pile material escaped. As the size of the structure grew, the chance that a neutron from one lump would find another lump became greater and the neutron intensity increased. By observing the neutron intensity as the pile grew, Fermi and his associates could anticipate at what stage of the construction the pile would become self-sustaining. The pile first became critical on the afternoon of 2 December 1942, and operated at a power level of $\frac{1}{2}$ watt. It was later increased to 200 watts — the maximum considered safe in that location.

Fermi was concerned that nitrogen of the air in the gaps of the pile would absorb enough neutrons to prevent criticality. He was prepared to encase the entire pile in balloon cloth and flush out the nitrogen with helium or some other gas that has a low absorption cross section. This cloth can be seen in Fig. 12.7.

This first reactor was of tremendous significance. It proved that a chain reaction could be produced and controlled. It is the prototype of all the controlled power-production reactors that have followed it. But in 1942 the motivation was to produce a destructive reactor, and the Chicago pile facilitated this end by making the conversion process practical.

12.8 THE CONVERSION PROCESS

We have noted that the ^{238}U, which absorbs neutrons and hinders the operation of a thermal pile, is ultimately converted to plutonium according to Eqs. (12.3) and (12.4). Plutonium is fissionable and, since it is a different chemical element, it can be separated from the uranium by chemical methods. Once the Chicago reactor proved that a chain reaction could be maintained, reactors specifically designed for the manufacture of plutonium were built at Oak Ridge, Tennessee, and at Hanford, Washington. In these cases, the reactors served only as prolific neutron sources for the bombardment of ^{238}U. Although these reactors were designed to permit easy insertion and removal of fuel elements and the removal of heat energy, they used natural uranium and a graphite moderator, just as in the first reactor. A fuel element is left in the reactor until a point of diminishing returns is reached in the conversion of ^{238}U. This point comes well before all the ^{235}U is used up because the fission products have large neutron capture cross sections and begin *poisoning* the reactor. The impurities absorb an appreciable fraction of the neutrons that are needed to maintain the reaction and to convert ^{238}U into plutonium. Once a fuel rod is spent, it is dissolved and the plutonium is separated by chemical processes. This seemingly straightforward operation is complicated by the fact that tremendous quantities of radioactively dangerous material must be handled. This radioactivity is due principally to the fission products that have poisoned the fuel element. The mere disposal of these radioactive fission products becomes a problem of mammoth proportions.

A large amount of this transuranic, artificially produced element was prepared by this method during World War II. The first atomic bomb was composed of uranium, but the second contained plutonium.

12.9 RESEARCH REACTORS

Chain reactors may be used for explosions, for research, and for power. The primary value of reactors for research is that they are prolific sources of neutrons. Neutrons are produced from reactions excited by particles from radioactive elements and by those produced with accelerators, but the neutrons from chain reactions are many orders of magnitude more numerous than from these other sources. Of all the bombarding particles, neutrons are the best able to produce nuclear rearrangements, and an intense source of neutrons is a powerful tool for studying nuclear reactions.

In some reactors, a window in the protective shielding permits the escape of an intense beam of neutrons that can be used in many ways to learn more about nuclear physics. Another technique is to insert a slug of material into the reactor, where it is subjected to intense neutron irradiation. Practically every element thus irradiated forms radioactive isotopes in significant amounts. Such artificially created radioisotopes are used in medical and tracer applications, where they have virtually replaced the naturally radioactive elements. These artificial radionuclides are available to qualified users at very reasonable prices.

The availability of intense radioactive sources permits the study of the effects of their radiations. Cobalt-60 is an example of an artificially created isotope that is an intense gamma-ray source. One effect of gamma-ray irradiation under active study is the sterilization of food. Beta or electron irradiation produces promising structural changes in molecules, particularly in high polymers. By this means rubber can be vulcanized, and the melting temperature of polyethylene can be raised.

Since ^{60}Co is an intense source of gamma radiation, it is dangerous to be near this material. In laboratories equipped with "gamma facilities," the ^{60}Co is immersed to a depth of about ten feet in a deep tank filled with water, to shield the personnel from its radiation. Materials to be irradiated are either lowered into the water near the gamma source or, more commonly, placed near the top of the tank, and the ^{60}Co is raised out of the water by remote control. Upon disintegrating, ^{60}Co nuclei emit beta particles whose maximum energy is 0.31 MeV and gamma rays having energies of 1.17 MeV and 1.33 MeV.

In this section, we have seen that a research reactor provides a series of research possibilities. One may study the characteristics of the reactor itself, the nuclear effects produced by excess neutrons, and the properties and behavior of materials made by the reactor.

12.10 POWER REACTORS

Any chain reactor can be adjusted to produce tremendous amounts of energy. It is only necessary to let the chain reaction become supercritical and wait for it to grow. The Chicago reactor was operated first at one-half watt and then at 200 watts, but the power level could have been run much higher. The main reason that the first reactor was operated at so low a level was that it was inadequately shielded and the neutron leakage might have endangered the personnel. The second reason was that there was no provision for cooling the reactor, and if it became too hot, it would destroy itself. (There was little chance that the reactor would explode violently. The entrapped gases could have become hot enough to push the pile apart, but this would have made the reactor subcritical and stopped the reaction.) Thus the main physical limitation on the power a reactor can develop is the provision for the removal of heat.

Research reactors can be aircooled because no attempt is made to utilize the heat energy developed, but for power reactors a more dense material is indicated. We naturally think of water as a cooling material, especially since we shall see that water can also serve as a moderator. Boiling-water reactors will be discussed in Section 12.11. An efficient coolant should have a large specific heat capacity (specific heat times density). Since it must flow through the reactor, the use of dense liquid is indicated; and since it should have good thermal conductivity, metals appear attractive. Mercury, a dense liquid metal, has been used for the transfer of thermal energy and has replaced water in some boiler-turbine power systems. But mercury has a slow-neutron absorption cross section of 380 barns for the natural mixture of its isotopes, and this is too great. Not only would mercury absorb

Figure 12.8 Schematic representation of a liquid-metal coolant reactor.

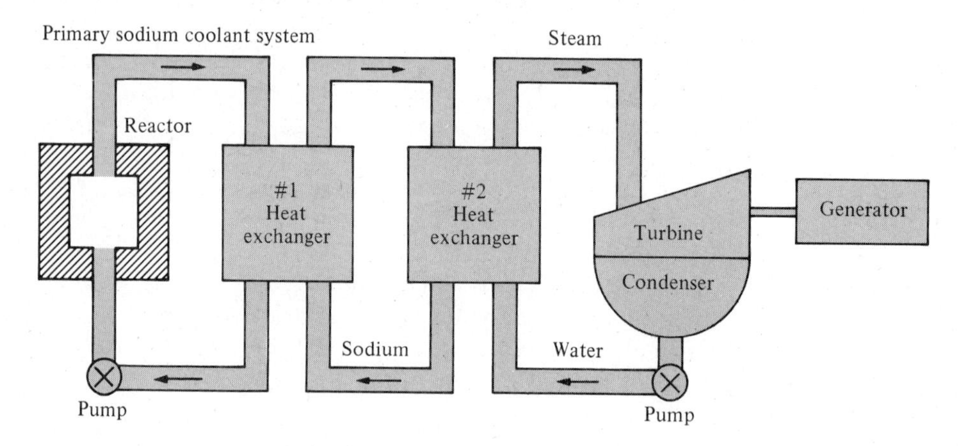

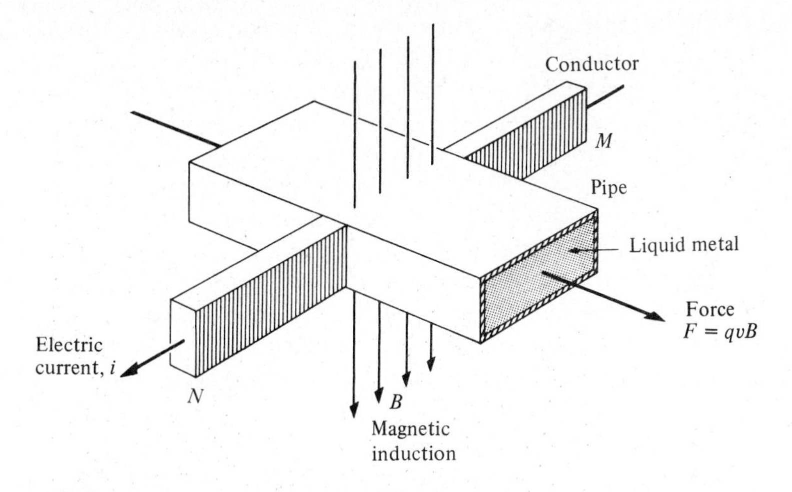

Figure 12.9 Schematic diagram of an electromagnetic pump for liquid metals.

neutrons from the chain reactor, but it would also form radioactive isotopes that would render all the apparatus near which it passed radioactive. It turns out that despite its chemical activity, especially when molten, liquid sodium is a good reactor coolant. It is a rather dense material with good thermal conductivity and with a thermal neutron absorption cross section of only 0.5 barn. The sodium is usually alloyed with potassium to lower the melting temperature. The essentials of a sodium-cooled reactor power system are sketched in Fig. 12.8.

The reactor is cooled by the molten metal in the primary sodium system and the energy transferred to the sodium in the secondary system by heat exchanger 1. The thermal energy then is transferred to water in heat exchanger 2. This two-stage system is necessary because sodium near the reactor core becomes radioactive. The steam that is generated operates a conventional turbine and electric generator.

The way in which the molten metal is pumped is an ingenious application of basic electrical theory. It is a well-known fact that good thermal conductors are good electrical conductors, and molten sodium is a good electrical conductor. Electric charges moving perpendicular to a magnetic field experience a force perpendicular to both the current and the field, as expressed in Eq. (2.3). Figure 12.9 is a sketch of a liquid-metal pump based on this law. The electric current from M to N flows partly through the pipe but largely through the conducting liquid metal. Since there is a magnetic field directed toward the bottom of the page, the moving charges experience a force parallel to the pipe. By varying the current or the field or both, the force and pumping action can be controlled. There is obviously no need for sealed bearings, since there are no moving parts.

On the average, the fission energy of either uranium or plutonium is 200 MeV. This is distributed approximately as follows: kinetic energy of the fission fragments, 167 MeV; kinetic energy of the fission neutrons, 5 MeV; prompt gamma rays and delayed gamma rays from the fission fragments, each 6 MeV; and the beta particles and neutrinos from the fission products, 5 MeV and 11 MeV, respectively. All of the neutrino energy and some of the gamma-ray and neutron energy escape through the walls of the reactor. This loss is just about compensated by the heat produced in moderating the neutrons plus the energy released in various parts of the reactor by the radioactive decay of atoms that were activated by gamma rays and neutrons. Because the fission fragments and the beta particles are easily absorbed, their energies are converted into heat close to their points of origin. Therefore, about 75% of the heat appears in the fuel rods and their cladding.

12.11 THE BOILING-WATER REACTOR

We have reserved the boiling-water reactor for more detailed discussion because the first large all-nuclear-powered generating station in the United States is of this type.

Although it is impossible to achieve a chain reaction with unenriched uranium and ordinary water as the moderator, ordinary water can moderate a reactor with enriched fissionable fuel. The boiling-water reactor is particularly attractive for two reasons. The moderating water can be converted to steam within the reactor and fed to a turbine without an intermediate heat exchanger, and a boiling-water reactor has an inherent stability that contributes to safety.

If the control rods in a reactor having a solid moderator are set so that the multiplication factor is greater than one, the chain reaction grows without practical limit until the reactor destroys itself. It is not too difficult to prevent this from happening by having safety control rods that can be thrust into the reactor quickly in the event anything goes amiss. (Such a shutdown is called a *scram*.) But a reactor having a solid moderator is basically unstable and always presents the possibility of a minor explosion that might break the shielding walls and spread dangerous radioactive materials in its vicinity.

In a water-moderated reactor even this unlikely hazard is eliminated. If the reactor gets out of hand, the water moderator will boil excessively, and since the bubbles of vapor are not dense enough to moderate the neutrons effectively, the reactor automatically goes *subcritical*. To test this point, the USAEC built a boiling-water reactor and tried to blow it up. They were able to produce a moderate explosion only when all the control rods were jerked out of the reactor simultaneously.

Indeed, a simple boiling-water reactor is too stable. If it is made supercritical for a short time in order to increase the level of power production, the water vaporizes and the reactor goes subcritical. Thus a simple boiling-water reactor cannot meet a sudden demand for increased power.

The Dresden reactor is designed to meet this problem in an ingenious way that utilizes the thermodynamic properties of water. It is a dual-cycle system, as shown in Fig. 12.10. The reactor proper, at the left, operates normally at a pressure of 600 psi (lb/in^2). Half the thermal energy leaves the reactor as steam, which is applied to the high-pressure blades of the turbine. The other half of the thermal energy leaves the reactor as water, which is removed from below the water level of the reactor. This water is "throttled" down to a pressure of 350 psi in what is called a "flash tank." At the lower pressure, much of the water vaporizes into steam, which is applied to the low-pressure blades of the turbine. But in converting water to steam, the heat of vaporization of the steam is provided at the expense of the temperature of the water not converted to steam. This low-temperature water is then mixed with the condensate from the turbine and pumped back into the reactor. The cool water injected into the reactor prevents boiling and keeps the moderator liquid.

The entire system is really two systems in parallel. The high-pressure system is inherently stable, as we have described. The low-pressure system is inherently unstable, since it tends to maintain the moderator as a liquid. The combination is

Figure 12.10 Schematic diagram of a dual-cycle, boiling-water reactor.
Courtesy of General Electric Company.

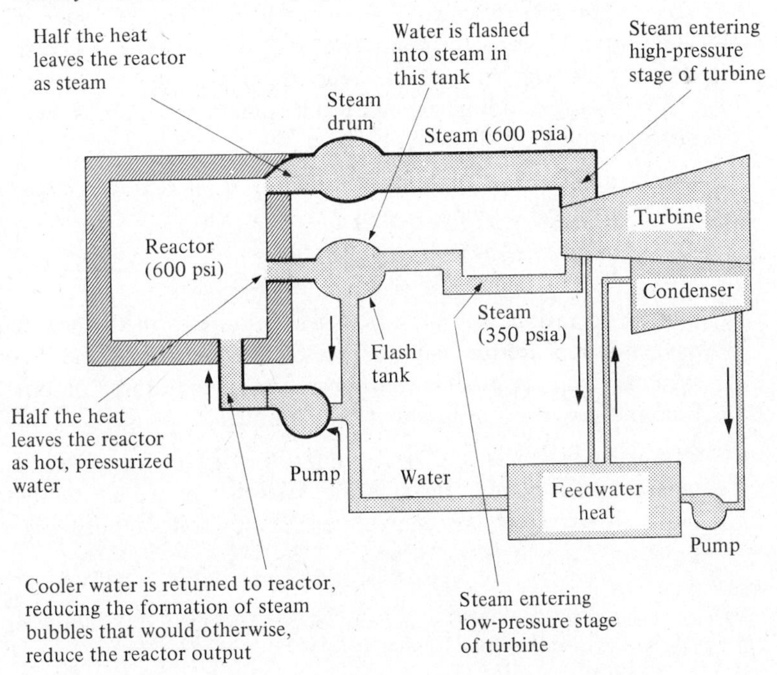

still stable, but enough "positive feedback" is provided to give the reactor a reasonable "response time."

The nuclear reactor supplies only thermal energy, and this must be converted to mechanical energy in the turbine and electrical energy in the generator. As with conventional power plants, the overall utilization of the prime thermal energy is low, about 33%. If the plant is to deliver 180,000 kW to the electrical lines, the nuclear reactor must supply 545,500 kW of thermal power, of which 365,500 kW is extracted from the condenser by water flowing over it. If available, this water is taken from a large river, lake, or bay in the vicinity. Otherwise it is used repeatedly in a large cooling tower where air currents and evaporation combine to carry the discarded heat away. Such a heat sink is required by the heat engine, a turbine in this case, no matter what fuel is used — coal, gas, nuclear, oil, or solar. This cooling water is not radioactive nor is the fog, or "steam," formed in the tower and at the top by the condensing of the vaporized coolant.

The following excerpt describes the nuclear-powered plant at Dresden, Illinois:[†]

> The principal station components of Commonwealth Edison's nuclear power plant are housed in an airtight sphere 200 feet in diameter [Fig. 12.11]. This assures safety to the surrounding area in case of an incident because the sphere is designed to contain the internal pressure that would result if all the water in the reactor were to escape in the form of steam. The sphere is ventilated via a stack that can be blocked off during an emergency.
>
> The reactor and associated equipment are surrounded by a thick concrete shield. The control room — the plant's nerve center — is located on the upper level to give a good top view of the reactor during maintenance and reloading. (Throughout reloading, enough water is maintained above the reactor core to provide biological shielding.)
>
> During a power failure, heat from the reactor is removed by condensing steam and returning the condensate by gravity to the reactor vessel. Evaporating the water at atmospheric pressure cools the condenser. An overhead tank stores water for cooling the shut-down condenser.
>
> During operation, isotopes of oxygen that are formed when neutrons strike oxygen atoms are the major source of radioactivity in the steam-and-water system. When the plant is shut down, the value of these isotopes in the water becomes insignificant in about five minutes.
>
> Corrosion products and other impurities in the water that have become radioactive in passing through the reactor core are an annoying source of radioactivity. Tests have shown that the only solid impurities are those that are

[†]Quoted by permission from the *General Electric Review*, November, 1955. (The design of the Dresden power station has been changed somewhat since 1955.)

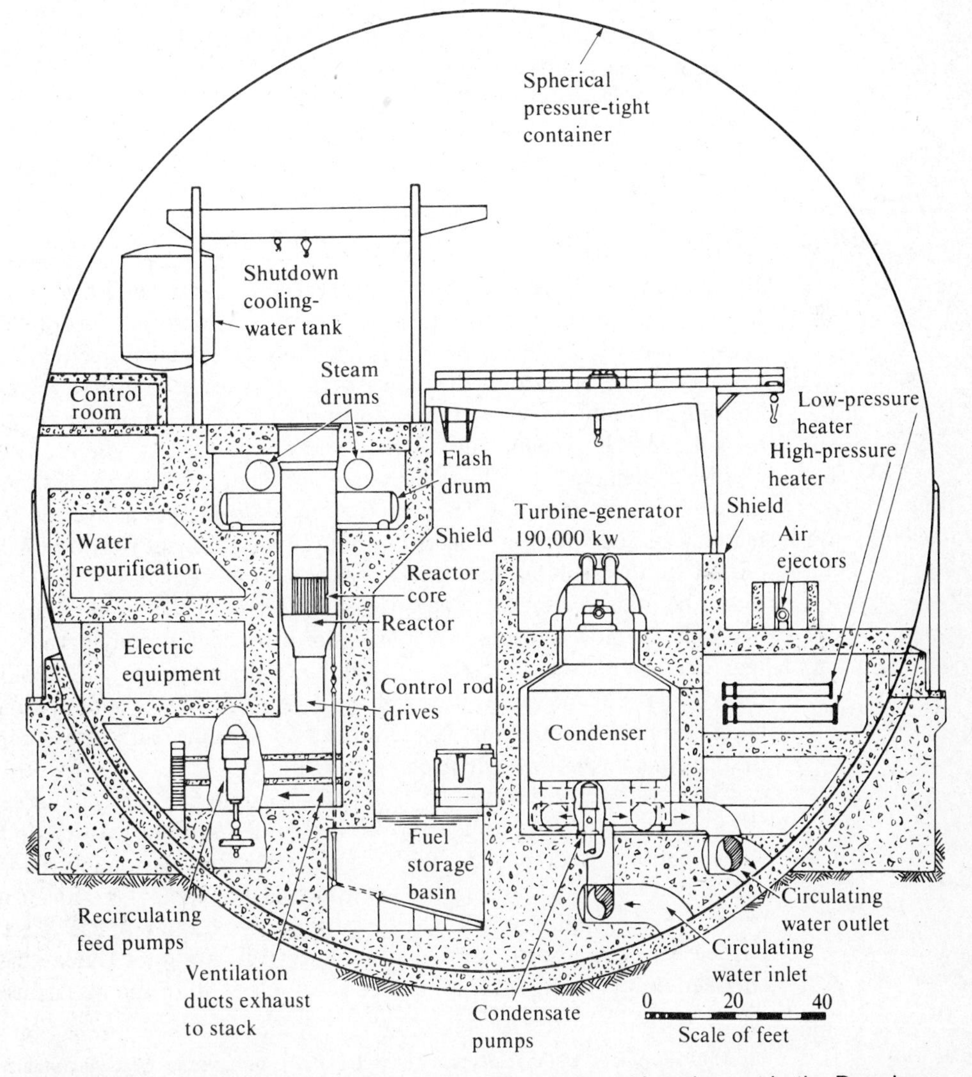

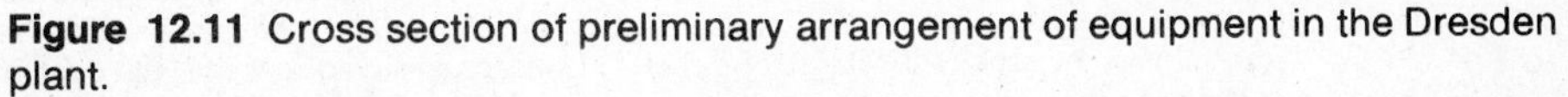

Figure 12.11 Cross section of preliminary arrangement of equipment in the Dresden plant.

Courtesy of General Electric Company.

entrained by minute droplets of water. But with efficient moisture separators, the steam will contain only about one ten-thousandth of the concentration of radioactive solids present in the water.

To reduce this residual radioactivity, the concentration of impurities in the reactor water is maintained at a low level by continuously bypassing a portion of the water through a cleanup filter and demineralizer.

Only a small fraction of the impurities in the steam will adhere to the turbine. Thus any difficulty appears unlikely in maintaining the turbine and associated equipment under normal conditions.

12.12 BREEDER REACTORS

It has been estimated that during the next 30 years we will need more energy than has been consumed throughout all of history. A large portion of this energy must come from nuclear reactors.† Although there is enough fissionable material in the earth's crust to operate regular nuclear reactors for a long time to come, there is an exciting possibility to extend that source manyfold. It is possible to operate a reactor on ^{235}U or ^{239}Pu, which will not only produce power but will also convert ^{238}U into ^{239}Pu. An element like ^{238}U, which can be transformed into a fissionable substance, is called a fertile material. The new fuel that is formed in a layer or blanket of fertile material placed within a reactor can supply a significant portion of the total energy output of the system. Eventually, however, the new fissionable material can no longer be utilized effectively because of poisoning of the blanket, that is, because the neutrons are now captured by the fission products already formed. This material then must be removed and reprocessed. Nevertheless, since ^{238}U is otherwise useless, this conversion operation may greatly reduce the net cost of fissionable material. Nonfissioning thorium is more plentiful than uranium, and it also can be converted into fissionable nuclei by neutron bombardment.

The cycle for converting ^{238}U into fissionable ^{239}Pu is shown in Fig. 12.12(a). ^{238}U absorbs a neutron to become ^{239}U. ^{239}U emits a beta particle to become ^{239}Np, which emits a beta particle to become ^{239}Pu, the fissionable end product. The conversion cycle for thorium is shown in Fig. 12.12(b). ^{232}Th absorbs a neutron to become ^{233}Th, which emits a beta particle to become ^{233}Pa. ^{233}Pa emits a beta particle to become the fissionable nuclide ^{233}U. To some extent the conventional enriched uranium reactor is a breeder because not all of the fast neutrons emitted in

† It should be pointed out that *all* energy conversion systems have significant environmental impact. We do not have space to discuss this topic; however, there are many interesting articles on the subject. See, for example, the articles by Hammond, Bethe, Cohen, and Vook as well as the book by Teller referenced at the end of this chapter.

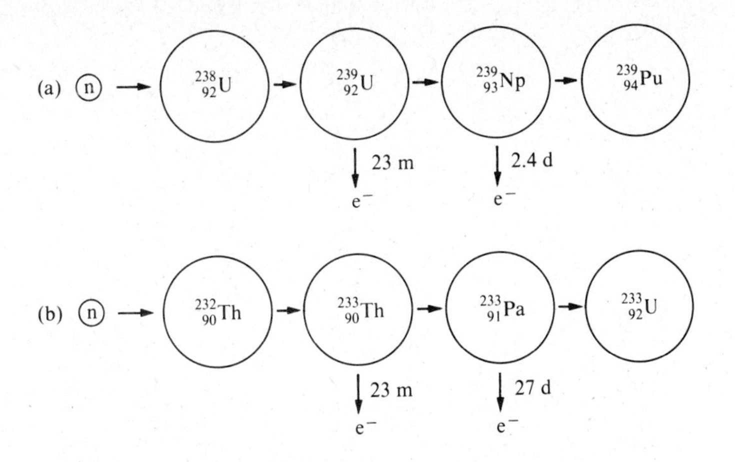

Figure 12.12 Conversion of fertile nucleus into fissionable nucleus.

fission are thermalized. When these bombard ^{238}U atoms, several isotopes of plutonium are formed according to the uranium-neptunium conversion chain. The approximate yield is 2% ^{238}Pu, 60% ^{239}Pu, 24% ^{240}Pu, 14% ^{241}Pu, which disintegrates into americium, $^{241}_{95}Am$. The fissioning of these products, especially ^{239}Pu, by thermal neutrons contributes to the energy output of the reactor.

If, on the average, one fissionable atom is produced to replace each atom that is destroyed in the original fission reaction, the reactor will operate with no net loss of fissionable material. A measure of its performance is the breeding ratio or the ratio of fissionable atoms produced to fissionable atoms destroyed. A breeder reactor is a reactor that has a breeding ratio greater than one and, therefore, is capable of producing more nuclear fuel than it consumes (but, of course, not forever).† A quantitative measure of the efficiency of this production is the doubling time: the time required to produce as much net additional fissionable material as was originally present in the reactor. At the end of the doubling time, the reactor will have produced enough fissionable material to refuel itself and one other identical reactor. A reasonable doubling time is in the range of 7 to 10 years.

If in the fission process two neutrons are produced, then theoretically the breeding ratio can be one. One of the neutrons goes to produce the next fission and continue the chain reaction while the second is absorbed by the fertile nucleus to produce a new fissionable nuclei. In practice, however, more than two neutrons are

†A reactor that converts fertile atoms into fuel by neutron capture is called a "converter" reactor. A converter reactor that produces more fissionable atoms than are consumed is called a "breeder" reactor. Sometimes "converter" is used to refer to a reactor that uses one kind of fuel and produces another (for example, ^{235}U produces ^{238}Pu). It usually requires careful reading to determine which meaning is being used in the various articles on reactors.

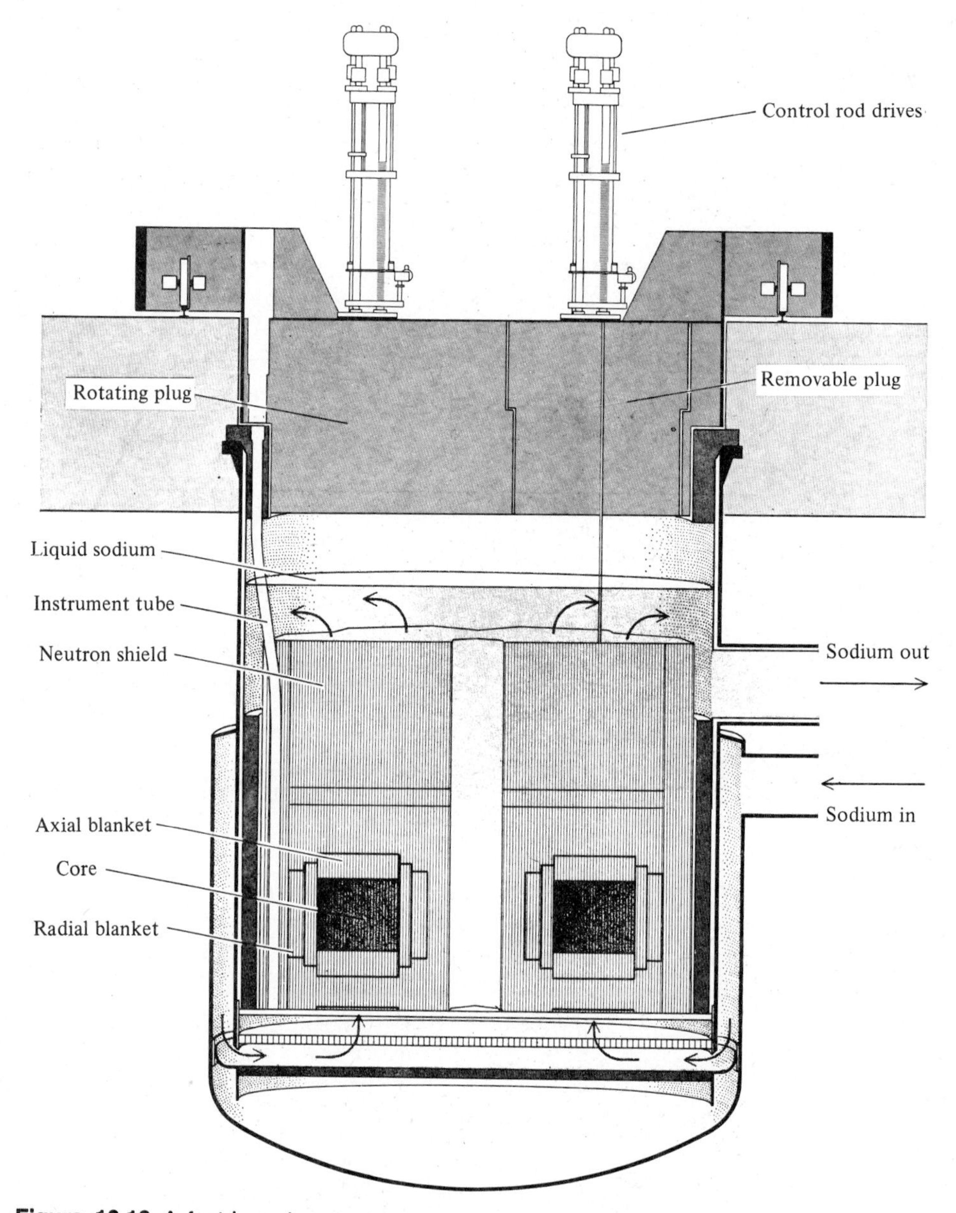

Figure 12.13 A fast breeder reactor.

From "A Third Generation of Breeder Reactors" by T. R. Bump. Copyright © May 1967 by Scientific American, Inc. All rights reserved.

needed because some are lost by absorption in the control rods and structural materials of the reactor as well as in the fission products. Because of these losses, the reactor must produce at least 2.2 neutrons per fissionable atom to achieve a breeding ratio of one. When a nucleus of ^{239}Pu absorbs a thermal neutron, on the average 2.09 neutrons are liberated. When it absorbs a fast neutron, 2.7 neutrons are released. On the other hand, when ^{233}U absorbs a thermal neutron, approximately 2.3 neutrons are released and for a fast neutron 2.4 are released. Reactors designed to breed with thermal neutrons are called thermal breeders, while reactors designed to breed with fast- or high-energy neutrons are called fast breeders. (Do not get confused and think that the fast or slow has to do with doubling time.)

The primary effort in recent years has been in fast breeder reactors. The fast breeder shown in Fig. 12.13 has two cores. Each is a cylinder approximately one meter in diameter and one meter long surrounded by a blanket of fertile fuel. Liquid sodium is used for cooling and for the transfer of usable energy. Sodium at about 400°C is circulated in a heat exchange system similar to the one for the regular reactor system shown in Fig. 12.8. The rotating plug and removable plug at the top of the reactor are situated such that the excess fuel may be removed for preparation and transfer to another reactor. These plugs also furnish access to the core vicinity for maintenance.

Research is also progressing on fast breeder reactors using high-pressure helium gas as the cooling medium. There are three advantages to this high-pressure technique. The helium gas does not become radioactive and there is no need for an intermediate heat exchanger. The helium does not react chemically with the water in the system. Finally, the helium is transparent, which provides visibility during refueling and maintenance of the core.

12.13 CHERENKOV RADIATION

When an intense radioactive source ejects high-energy charged particles into a transparent material such as water, plastic, or glass, a ghostly bluish glow extending some distance into the medium can be seen. This phenomenon is easily observed when a room containing a swimming pool reactor or a gamma facility is darkened. This *Cherenkov radiation* has an interesting explanation and use. Let us consider a particular case, ^{60}Co immersed in water. The most energetic beta particles emitted by this radioelement have a relativistic velocity of about $0.8c$. But these electrons move through water, where the velocity of light is about $0.75c$ (computed from c/n, where n is the index of refraction of water). Thus the electrons in the water travel faster than the phase velocity of light *in that medium*. The situation is much like that of a boat going through water faster than the velocity of water waves, or of a jet plane going through air faster than the speed of sound in air. The electron produces a "bow-wave" of light. Let a projectile (boat, plane, or electron) move with a velocity v from A to B in time t, as shown in Fig. 12.14. This projectile causes a sequence of disturbances in the medium that combine constructively to form the wavefronts represented by the lines from B to C and from B to D. These wavefronts move with a phase velocity v_p, and while the projectile goes from A to B the Huygens' wavelet (disturbance) from A goes from A to C, a distance $v_p t$. Thus the angle θ between the direction of motion of the projectile and the wavefront is given by

$$\sin \theta = \frac{v_p t}{vt} = \frac{c}{nv}. \tag{12.20}$$

It can be shown that the number of quanta emitted as Cherenkov radiation in a wavelength interval $\Delta\lambda$ is proportional to $[1 - (c^2/n^2v^2)]/\lambda^2$. It is evident from this expression that the short wavelengths will be predominant, and therefore that the color will be bluish-white.

Cherenkov radiation can be used to measure electron velocities. The radiation is most intense in the direction of advance of the wavefront, as shown in Fig. 12.14. If the electron beam is collimated so that its direction is known and if the direction of maximum radiation intensity is measured, then θ is determined. Since c and n are known, v can be calculated. The value of n depends on the wavelength of the light. For precision measurements, the direction of Cherenkov radiation should be measured for a particular color.

Figure 12.14 Bow wave in Cherenkov radiation.

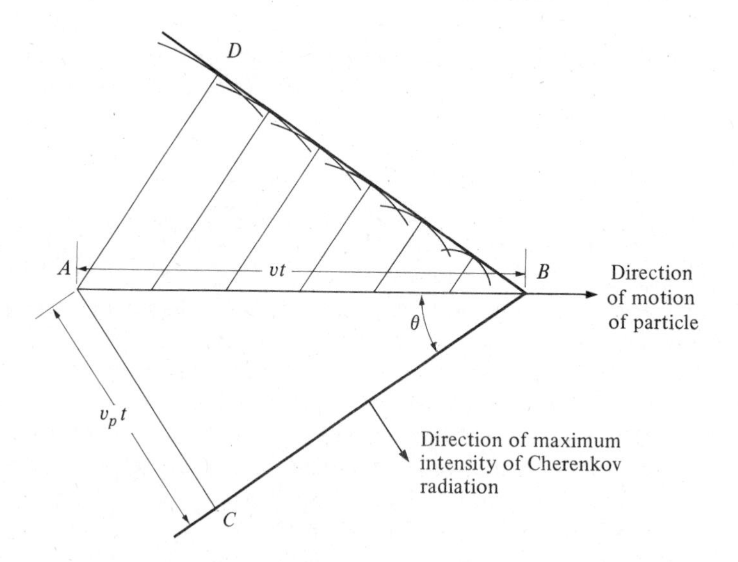

12.14 ACTIVATION ANALYSIS

Since the addition of a neutron to a nucleus converts nearly every element to a radioactive form and since radioactivity can be readily detected, a method that supplements chemical analysis is possible.

Suppose that one has a sample of material that might contain a trace of, say, indium. The problem is to determine whether indium is present and, if it is present, its concentration. We assume that the amount present is too small to detect by normal chemical methods. Indeed, there is a possibility that reagents used in the chemical analysis might introduce more indium than exists in the given sample.

Now let this sample be bombarded with neutrons. Then some of the atoms of each of the elements present will be converted to a radioactive form. In particular if any indium is present, some of it will become radioactive. Next a solution containing a high concentration of stable indium is added to the unknown. We now know that the sample contains a lot of indium of which most is not radioactive. But if indium were present originally, then some of the indium in the sample is radioactive and this portion of it will, of course, emit the gamma-ray spectrum characteristic of that element. Through a chemical procedure, all of the indium is next separated by standard techniques. This is comparatively easy because now indium is abundantly present. If the separated indium is found to be radioactive, then the original sample must have contained indium. Since the intensities of the lines in the gamma-ray spectrum are directly proportional to the number of radioactive atoms present and since this spectrum can readily be obtained with a multi-channel analyzer, the amount of indium in the original sample is easily determined.

This is but one example of neutron activation analysis, which is a whole new technique that can be used to determine, both qualitatively and quantitatively, trace elements of chemical elements present in very minute samples of a substance.

12.15 NATURAL FUSION

In 1928, Professor A. A. Knowlton wrote a physics text in which a unifying theme was the question, where does the sun's energy come from? He considered the possibility that the sun was leaking energy stored within it during some stage of astronomical evolution. The sun was treated as a coal pile deriving its energy from chemical reaction. Dr. Knowlton estimated the energy brought to the sun by falling meteors. He also assumed the possibility of the sun contracting, with the potential energy of its parts becoming kinetic and thermal. His conclusion was as follows:†

† From A. A. Knowlton, *Physics for College Students*. New York: McGraw-Hill, 1928.

The English astronomer Eddington has calculated that if $\frac{1}{10}$ of the hydrogen now present in the sun were to be converted into helium, the reaction would liberate a supply of heat enough to keep the sun radiating at its present rate for a billion years. If such a process is going on anywhere, it must be at some place where atomic nuclei are forced to approach one another either as in the impact of swiftly moving atoms or by enormous external pressure. A few years ago, we had not the slightest ground for believing that either atomic disintegration or atom building was possible. Perhaps there are other things happening in the great solar crucible of which we have as yet no suspicion. At any rate we have here a plausible and fascinating hypothesis as to the source of solar energy which, unlike any of those previously considered, is quantitatively adequate.

It is now believed that the primary energy source in the sun is the *proton cycle*, the fusing of hydrogen into helium in the following reactions:

$$^{1}_{1}H + ^{1}_{1}H \rightarrow ^{2}_{1}H + ^{0}_{+1}e + \nu + \gamma$$
$$^{2}_{1}H + ^{1}_{1}H \rightarrow ^{3}_{2}He + \gamma$$
$$^{3}_{2}He + ^{3}_{2}He \rightarrow ^{4}_{2}He + ^{1}_{1}H + ^{1}_{1}H.$$

Note that the first two interactions must be carried out twice to provide the two ^{3}He nuclei for the third interaction. The net result of this process is the fusing of four hydrogens to form one helium with a total mass energy conversion of 26.7 MeV.

By 1939, enough was known about nuclear reactions for Bethe to propose a second set of reactions by which hydrogen might be converted to helium in the sun. It is called the *carbon* or *CNO cycle*, since carbon, nitrogen, and oxygen serve as nuclear catalysts. The steps are as follows:

Reaction	Product half-life
$^{12}_{6}C + ^{1}_{1}H \rightarrow ^{13}_{7}N + \gamma$	9.96 m
$^{13}_{7}N \rightarrow ^{13}_{6}C + ^{0}_{+1}e + \nu$	
$^{13}_{6}C + ^{1}_{1}H \rightarrow ^{14}_{7}N + \gamma$	
$^{14}_{7}N + ^{1}_{1}H \rightarrow ^{15}_{8}O + \gamma$	124 s
$^{15}_{8}O \rightarrow ^{15}_{7}N + ^{0}_{+1}e + \nu$	
$^{15}_{7}N + ^{1}_{1}H \rightarrow ^{12}_{6}C + ^{4}_{2}He$	

(12.21)

If we add these equations, cancel out nuclides that appear on both sides, and calculate the total Q-value, we have

$$4(^{1}_{1}H) \rightarrow ^{4}_{2}He + 2(^{0}_{+1}e) + 24.7 \text{ MeV}. \quad \textbf{(12.22)}$$

The annihilation of the positrons supplies an additional 2 MeV of energy, so that the total released is actually 26.7 MeV. The proton cycle is believed to be the primary cycle in our sun while the carbon cycle is thought to be the primary process in hotter stars.

To produce the reactions of the CNO cycle on earth, the carbon and nitrogen nuclides must be bombarded with accelerated protons. But the temperature at the center of a star is so great that some thermal protons at the high-energy end of their Maxwellian velocity distribution are able to react. The fraction of high-energy thermal protons is so small that it may take a million years to go through this cycle once with particular atoms, but the total number of protons able to produce the reactions is so great that it accounts for the enormous solar energy. Measurements indicate that there is enough hydrogen in the sun to maintain this carbon cycle for about 30 billion years.

The sun as well as any other star is in equilibrium because of two opposing effects. If the temperature were to rise, the fraction of the protons able to cause the cycle would increase. This would increase the rate of energy production and further increase the solar temperature. On the other hand, if the temperature increases, the sun tends to expand and the concentration of the nuclides tends to decrease, making reactions less likely. These two effects balance each other.

There is every reason to believe that the sun is not unique in this respect, and that all stars are nuclear furnaces of one sort or another. Sometimes weak or unknown stars grow to great brilliance quite suddenly. These *new stars* (*nova*) are unquestionably unstable stars whose nuclear reactions have gone wild.

Since all animals derive their energy ultimately from plants, which get their energy from photosynthesis of the sun's light, since coal and oil were formed from plant life, and since water power is derived from solar heat, we can trace almost all the energy we use to the sun. Thus we live and breathe and have our being because solar hydrogen is converted to helium.

Radioactivity and nuclear fission are processes in which matter is taken apart. A solar cycle is a "putting-together process" in which matter is fused, and the process is called *fusion*.

The discussion of nuclear binding energy in Chapter 11 made it clear that the binding energy per nucleon was greater for helium than for hydrogen. In general, this increase in binding energy per nucleon for heavier nuclei increases up to $^{56}_{26}\text{Fe}$. From $^{56}_{26}\text{Fe}$ up to the heaviest naturally occurring nuclei, $^{238}_{92}\text{U}$, the binding energy per nucleon decreases. Thus for the lighter elements, it is in general advantageous for the nuclei to combine to become heavier nuclei. That is, it is advantageous for hydrogen to become helium and for helium to become beryllium and so forth. In a star then, it is generally advantageous for the star in its fusion process to create heavier and heavier elements. In our star, the sun, we have seen that hydrogen is fused to form helium. In other types of stars, other fusion processes take place. If the conditions are right (high temperature and high pressure), beryllium can be formed in the following process:

$$^{4}_{2}\text{He} + {}^{4}_{2}\text{He} \rightarrow {}^{8}_{4}\text{Be} + \gamma.$$

There is a problem for the fusion chain in getting past beryllium. Two helium nuclei contain more binding energy than one beryllium nucleus. This results in ^8_4Be being unstable and decaying into alpha particles (helium nuclei). Astrophysicists Salpeter and Hoyle showed that under the proper conditions, a third helium nucleus could fuse with beryllium to form $^{12}_6C$. The process would be

$$^8_4\text{Be} + {}^4_2\text{He} \rightarrow {}^{12}_6\text{C} + \gamma.$$

The smooth decrease in potential energy shown in Fig. 11.17 and the resulting smooth increase in binding energy per nucleon for these same nuclei indicate that once past carbon the fusion process easily results in nuclei up to and including $^{56}_{26}Fe$. Two of the steps that would follow would be

$$^4_2\text{He} + {}^{12}_6\text{C} \rightarrow {}^{16}_8\text{O} + \gamma \quad \text{and}$$
$$^{12}_6\text{C} + {}^{12}_6\text{C} \rightarrow {}^{24}_{12}\text{Mg} + \gamma.$$

The conditions are right in the sun for the fusing of hydrogen. Once the hydrogen fuel is consumed, the conditions quite likely will become favorable for the fusion of helium. The charge on helium is twice that for hydrogen and the coulomb repulsion between two helium nuclei thus would be much greater, which means that the nuclei being fused would have to be traveling toward each other at much higher velocities. To obtain these velocities, the temperature within the star would have to be much greater. In general, to fuse heavier elements requires higher temperatures.

At this point, a good question might be, where do the heavier elements on earth come from?† The above discussion would lead one to believe that the most likely source of these elements would be the fusion process in the center of stars. Helium and other light nuclei would be formed in rather young stars like our own. Heavier elements would be formed in older and hotter stars. The older stars, after forming the heavy nuclei and consuming their basic fuel, would subsequently spit out these heavier nuclei into the surrounding universe. This cast-off material from the older stars would then be mixed with the light elements in the portion of a galaxy where new stars would be formed. It would appear that our sun and solar system were formed after the basic material was contaminated by these heavier elements. Probably all of the heavier elements in our solar system at one time were in the center of some older stars.‡

It should be pointed out that there are no direct data that confirm that the fusion processes taking place within the stars are what we have suggested. However, an experiment has been proposed that would directly test some of the above processes. Notice that in the first step of the proton-proton process, a neutrino is emitted. Of all the particles involved in that process, only the neutrino escapes the vicinity of the sun unaltered. The proposed experiment is to measure the number of

†See the article by Donald D. Clayton referenced at the end of this chapter.

‡Human beings, of course, are made up of some of these heavier elements and thus a good fraction of each of us may have existed in some older star at some point in time.

neutrinos coming from the sun and compare the result with the number predicted. If we were to study the process of fusion in detail, we would find that there are neutrinos being emitted other than the ones predicted in the first step. An experiment has been set up to detect these other neutrinos.† The experiment involves looking for the interaction of a neutrino with a $^{37}_{17}Cl$ nucleus. Since the neutrino interacts only very very weakly, it is necessary to have a large quantity of chlorine. Such an experiment has been performed by R. Davis, in which a hundred thousand gallons of carbon-tetrachloride (C_2Cl_4) were housed in a deep gold mine. A chlorine nucleus interacts with a neutrino to form Ar, which becomes Cl in an excited state. The chlorine decays to the ground state with the emission of a characteristic photon. By counting these photons, Davis ascertained the number of neutrinos entering his detecting tank. From this information, the total number of neutrinos leaving the sun can be determined.

The total number of neutrinos as determined by Davis is far too few to be understood on the basis of what we know about the processes going on within our sun. This leads to a dilemma. Either we don't understand the processes in our sun as well as we think or something unusual is happening to the neutrinos after they are emitted. One possible explanation for the result is that some of the neutrinos change into another type of neutrino.‡ Another explanation may well be that the basic proton-proton reaction is behaving as we understand it but the interactions that yield the $^{7}_{4}Be$ are not.

12.16 FUSION IN THE LABORATORY

If fusion can take place in the stars, is it possible to cause fusion here on earth? Before Bethe proposed the carbon cycle for the sun, each of the reactions of that cycle had been produced on the earth. Thus by 1939 there had been fusion in the laboratory in a limited sense. But like many other exothermic nuclear reactions, there was no possibility of making practical use of the energy liberated as long as it took extensive apparatus using much *more* energy to make the reaction proceed. The key to practical nuclear power is a chain reaction that runs itself. The links of the fission chain are neutrons, while the links of a fusion chain are protons.§ Fission proceeds best with thermal *neutrons*, where thermal means at or near room temperature. Fusion proceeds best with thermal *particles*, where thermal means temperatures

† The neutrino that is detected is one that comes from the beta decay of $^{8}_{5}B$, which is formed when $^{7}_{4}Be$ interacts with a proton. $^{7}_{4}Be$ is formed by the fusion of two helium nuclei.

‡ In Chapter 13 we shall see that there are three different neutrinos — e-neutrino, μ-neutrino, and τ-neutrino. There is some slight evidence that neutrinos oscillate between these three types. If so, neutrinos cannot be massless particles. If they have mass, this has great significance for astrophysics.

§ There are fusion reactions in which deuterons, alpha particles, etc., provide the "links" of the chain.

measured in millions of degrees. If the solar fusion of hydrogen into helium is a slow process in the sun, where Eddington estimated the temperature to be 15,000,000 K, the prospect of fusion on earth looks dim.

Our best hopes lie in some of the possible fusion reactions listed below, together with their Q-values in MeV:

$$
\begin{array}{ll}
\mathrm{D(d, p)T} & 4.02 \\
\mathrm{T(p, \gamma)^4He} & 19.6 \\
\mathrm{T(d, n)^4He} & 17.6 \\
\mathrm{Li^6(n, \alpha)T} & 4.96 \\
\mathrm{Li^6(d, \alpha)^4He} & 22.4 \\
\mathrm{Li^7(p, \alpha)^4He} & 17.3 \\
\mathrm{Li^7(d, n)^8Be} & 14.9
\end{array}
\qquad \textbf{(12.23)}
$$

A fusion reactor might utilize several of the reactions in this list. For example, the fourth reaction could be initiated with a neutron. If the tritium product struck deuterium, the third reaction would provide a neutron. This neutron might excite the fourth reaction again, etc. Hydrogen isotopes have been ignited to fusion in a fission bomb. This so-called hydrogen bomb (fusion bomb) is a hideous thing because its energy release is "open-ended." Once the fusion is started, the fusion itself maintains the temperature to keep the process going. The energy liberated is a function of how much fusible material is present, and there is no theoretical limit. Recall that in a fission bomb, the parts, before detonation, must each be smaller than the critical size lest the parts explode spontaneously. A fusion bomb cannot explode until it is "ignited" and any amount of fusible material is safe until ignited. Thus the amount of fusible material in a fusion bomb is not limited.

The constructive utilization of the energy released from fusion is now a major field of research endeavor. Fusion on the sun is "contained" by the tremendous gravitational field of the sun, and therefore a continuous reaction takes place. Fusion in an H-bomb is a destructive explosion, since it is not contained. At first glance it appears entirely hopeless, not only to achieve the temperatures necessary for fusion, but at the same time to contain the reaction at a temperature many orders of magnitude above that at which earth materials vaporize.

One answer to the achievement of suitable temperature is already implied by our discussion. We do not need temperature as such. What we need is high-velocity particles. Although both the sun and the H-bomb produce these velocities with thermal energy, all the fusion reactions have been produced on a small scale by using particle accelerators. We have seen that the kinetic energy of a particle moving with the most probable speed in a Maxwellian distribution for a group of particles at room temperature, 293 K, is 0.025 eV. *By stretching the concept of temperature from a statistical property associated with the random motion of all the atoms or molecules in a piece of gross matter to a description of individual particle energy* we get a new concept, called *kinetic temperature*. Thus any particle with 0.025 eV of energy has a kinetic temperature of 293 K. Multiplying each of these quantities by 4×10^4, we find any particle having 1 keV of energy has a kinetic temperature of 11.6 million K. A baseball thrown by a fast-ball pitcher may have a kinetic temperature of 10^{20} electron volts or about 10^{24} K. Calculations show that it may be possible to maintain fusion continuously at a kinetic temperature as low as 45 million degrees. Charged particles may be given this kinetic temperature by accelerating them through a mere 4000 V potential difference.

If an electric arc is maintained by a potential of this order of magnitude, any gas of low atomic number in the arc becomes completely ionized. The space between the electrodes is then filled with a mixture of two gases, one consisting of negative electrons and the other consisting of positive nuclei. This mixture of charged-particle gases is called a *plasma*.

Even though the number of particles per unit volume in a plasma may be much smaller than in the atmosphere, a kinetic temperature of 100 keV may produce a pressure as high as 1000 atm.

If we assume that a plasma can be produced in which fusion can take place, how might a body of gas with a pressure of 1000 atm and a kinetic temperature of about a billion degrees be contained? Obviously, contact with any material container will result in the absorption of energy from the plasma, thereby reducing its kinetic temperature. One important mechanism that causes the plasma to lose energy rapidly is radiative, not thermal, in character. When a high-energy particle from the plasma strikes the wall of the container, some of the material (usually stainless steel) is vaporized. The atoms of high atomic number that are released diffuse into the plasma stream and cause a marked increase in the rate at which the moving, colliding, charged particles lose energy by bremsstrahlung. Even minute amounts of impurities cause a large increase because the power radiated by bremsstrahlung varies as the square of the atomic number of the atom causing deceleration. Since the presence of elements of high atomic number lowers the kinetic temperature of the plasma, a container must be fantastically clean before deuterium or some other gas is introduced. The best hope of preventing contact between a stream of charged particles and the walls of the container lies in the restraining effect that can be produced by a magnetic field. If the plasma arc is along the axis of a solenoid, any charges moving parallel to this axis experience no magnetic force. If charges are knocked out of their parallel paths, however, their paths become helices, since their velocities now have components perpendicular to the field (see Problem 2.25). Thus a strong enough magnetic field turns escaping charged particles back into the body of the plasma and constitutes a "lens."

The high kinetic temperature necessary to establish fusion conditions is reached by shrinking the plasma with what is known as the *pinch effect*. We know that two parallel wires carrying currents in the same direction attract each other. In a plasma carrying a current, we may think of the current as taking place in a bundle of wires each attracting all the others. If a large bank of highly charged capacitors is discharged through the plasma, there may be a transient current of many million amperes. Such a current, properly directed through the plasma parallel to the axis of the existing magnetic field, produces an intense pinch effect. Charges at the outside edges of the plasma are thrown toward its axis, where they collide, with high kinetic temperature. Fusion conditions have been produced momentarily in this way for deuterium nuclei. The energy released has been small compared with the energy required to operate the apparatus.

If a pinch-initiated fusion reaction were to sustain itself continuously, or long enough for the energy liberated to exceed the energy input to the apparatus, then electromagnetic radiation and uncharged particles (principally neutrons) could carry away energy to heat a boiler. There is also hope that because of the electrical nature of the fusion plasma it may be possible to generate electricity by pulsing the plasma and thereby inducing currents in a secondary coil.

The simplest magnetic confinement system is the mirror type cylindrical system. The plasma cavity is a cylindrically shaped region containing the magnetic field for confinement. At each end of the cylindrical chamber, separate magnets are used to generate a magnetic field to reflect the plasma particles back toward the primary chamber.

Another technique to confine the hot plasma makes use of a specially shaped magnetic bottle. The magnetic bottle is a toroidal- or doughnut-shaped container called a tokamak (Fig. 12.15). The machine consists of a plasma with a circular cross section where a toroidal magnetic field is provided for confinement and a toroidal electric field is provided to maintain current through the plasma. A circular magnetic field in the cross section of the toroid is generated by the electric current. The combination of this field and the toroidal field produces helical magnetic field lines that confine the plasma. The three main components of the machine are a toroidal plasma chamber, a coil to provide the toroidal magnetic field, and a transformer to induce the electric field. In Fig. 12.15, the plasma is shown confined by the doughnut-shaped inner cylinder. The outer ring of concentric coils provide the toroidal-shaped magnetic field in the doughnut-shaped region.[†]

Many scientists believe that the tokamak is the best prospect for the first fusion reactor to yield significantly more energy output than energy input. The ultimate fusion reactor, however, may come from a mirror machine or from a completely different confinement technique — for example, inertial confinement.

The newest activity on the fusion scene is a project to attempt to produce fusion by irradiation of tiny deuterium-tritium fuel pellets with very high energy lasers. As we pointed out in Chapter 4, a laser beam can be made almost perfectly parallel and can be focused to a spot with a diameter of only a few 100 millionths of an inch. Solid-state ruby or neodymium-glass lasers that deliver energies in excess of 1000 joules in less than a nanosecond are used to obtain the highest peak power. After

[†] For a review of magnetic confinement systems, see the articles referenced at the end of this chapter by Masanori Marakami and Harold P. Eubank, and by Francis F. Chen.

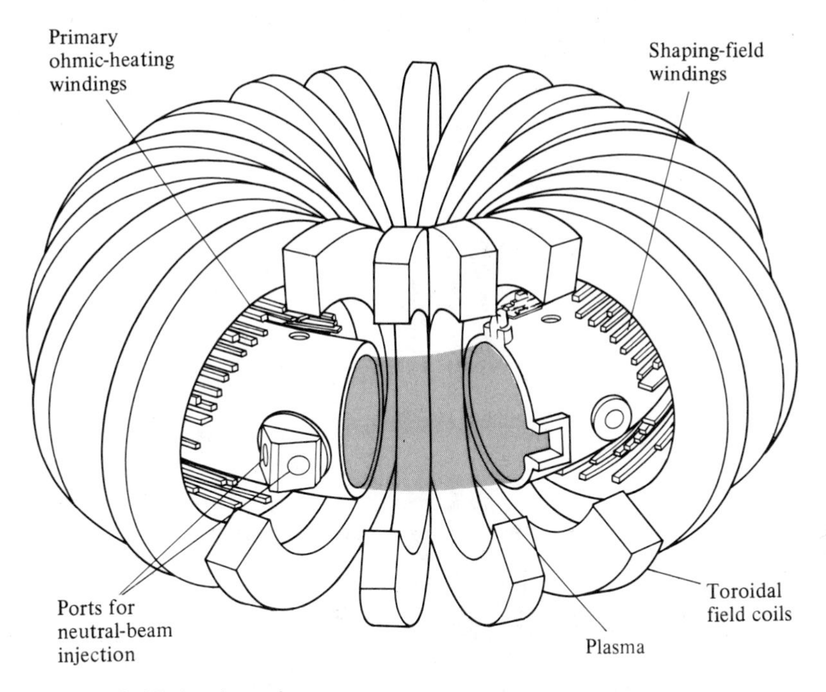

Figure 12.15 A tokamak.

From "Progress toward a Tokamak Fusion Reactor," by Harold P. Furth.

focusing the beam, the power is about 10^{17} watts per square centimeter, which results in local electric fields of 10^{10} volts per centimeter. For an appreciation of the magnitude of this field, consider a free electron. In this field, the electron would be accelerated to a velocity comparable to the velocity of light in a fraction of an optical wavelength. These very high power densities are believed to be approaching what is necessary to initiate the fusion reaction.

A proposed technique for extracting the energy from the fusion process is shown in Fig. 12.16. The fusion reactor would consist of a sphere 10 to 15 feet in diameter filled with lithium swirling fast enough to cause a vortex around the vertical axis. Deuterium-tritium pellets would be dropped from above, falling into the chamber through the vortex to the center of the sphere. A high-energy focused laser beam would cause the pellet to vaporize. Some of the deuterium-tritium ions would fuse, increasing the temperature of the rest of the pellet and causing about 5% of the remaining ions to fuse before the plasma expanded to terminate the reaction. Approximately 75% of the fusion energy would go into energetic neutrons that would yield their energy to the lithium as they slowed. The remaining 25% of the energy would go into kinetic energy of the alpha particles. As the alpha particles slowed, x-rays would be produced that would be absorbed in the first few millime-

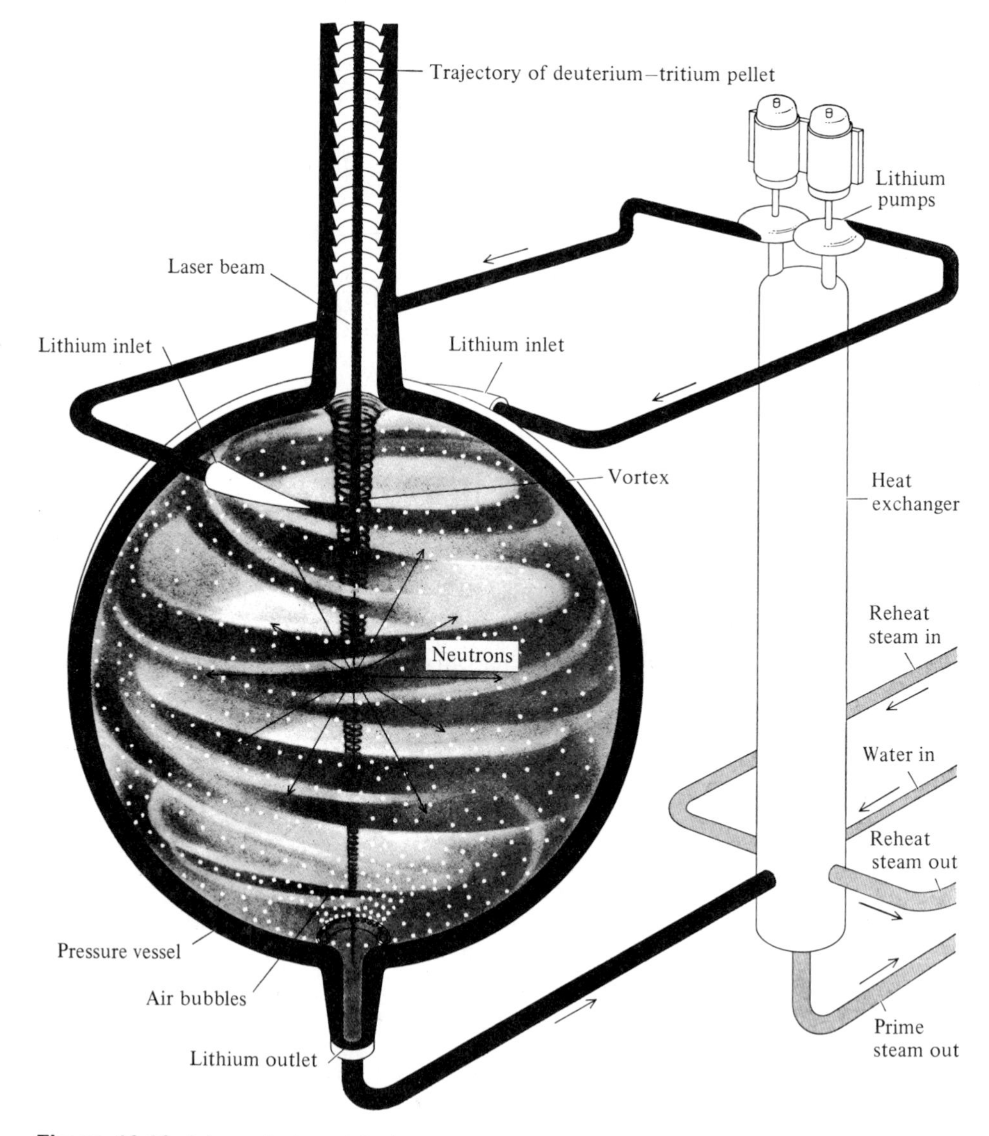

Figure 12.16 A laser-induced fusion reactor.

ters of the lithium. The temperature of the central region would increase by a few hundred degrees in about a millionth of a second, causing the lithium to expand sending a shock wave radially outward. Air bubbles injected from holes in the bottom of the chamber, hopefully, would diminish the damaging effect of the shock wave.

Hot lithium would be extracted from the bottom of the reactor sent through a heat exchanger and returned to the sphere at the top in a manner that would cause the swirling and the vortex. Water passing through the heat exchanger would be converted into steam, which would then drive a typical steam generator. Deuterium and tritium would be recaptured from the lithium and used for making more pellets.

Some of the questions yet to be answered about laser-induced fusion are (1) Can a sufficiently powerful laser be built to ignite the fusion process? (2) Can the blast wave be attenuated sufficiently by the gas bubbles? and (3) Will the fusion energy be sufficiently greater than the energy input to the laser?

It is easy to see why research on fusion continues in spite of its many discouraging aspects. Although fission presents a far greater energy source than our remaining fossil fuels, this source could be used up in a few centuries of modern civilization. Furthermore, fission produces radioactive wastes that are certain to be an increasingly serious problem, possibly requiring that these "poisons" be projected into space. On the other hand, the product materials of fusion are almost harmless. However, some problem can be generated with the fusion containment system. Its materials can become activated by the dense neutron flux. The known fuel supply for fusion is sufficient for our extravagant use for many, many years. If hydrogen nuclides are what we need, we are fortunate since every water molecule in the oceans has two of them.

FOR ADDITIONAL READING

Agnew, Harold M., "Gas-Cooled Nuclear Power Reactors," *Scientific American* **244** (June 1981).

Bekefi, G., and S. C. Brown, "Resource Letters PP-2 on Plasma Physics: Waves and Radiation Processes in Plasmas," *Am. J. Phys.* **34**, 1001 (1966).

Bethe, H. A., "The Necessity of Fission Power," *Scientific American* **234**, 21 (January 1976).

Brown, S. C., *Plasma Physics*, Resource Letter PP-1 and Selected Reprints. New York: American Institute of Physics, 1963.

The Bulletin of the Atomic Scientist. A monthly magazine of science and public affairs.

Bump, T. R., "A Third Generation of Breeder Reactors," *Scientific American* **216**, 25 (May 1967).

Chen, Francis F., "Alternate Concepts in Magnetic Fusion," *Physics Today* **32**, #5, 36 (May 1979).

Clauser, F. H., editor, *Plasma Dynamics*. Reading, Mass.: Addison-Wesley, 1960. An inclusive and authoritative book.

Clayton, Donald D., "The Origin of the Elements," *Physics Today* **22**, 28 (May 1969).
Cohen, Bernard L., "Environmental Hazards in Radioactive Waste Disposal," *Physics Today* **29**, 9 (January 1976).
Compton, Arthur H., *Atomic Quest, A Personal Narrative*. New York: Oxford University Press, 1956.
Coppi, Bruno and Jan Rem, "The Tokamak Approach in Fusion Research," *Scientific American* **227**, #1, 65, (July 1972).
Cottrell, Sir Alan, FRS., *How Safe is Nuclear Energy*? London-Exeter, N.H.: Heinemann, 1981.
Cowan, George A., "Natural Fission Reactor," *Scientific American* **235**, 36 (July 1976).
Fermi, E., et al., *Neutronic Reactor*, U.S. Patent No. 2,708,656; dated May 17, 1955; filed December 19, 1944 (57 pages). Reveals the state of the art of reactors at the time of filing. Well worth reading.
Fermi, Laura, *Atoms in the Family*. Chicago: University of Chicago Press, 1954. A delightful biography of Enrico Fermi.
Fetter, Steven A. and Kosta Tsipis, "Catastrophic Release of Radioactivity," *Scientific American* **244** (April 1981).
Furth, Harold P., "Progress Toward a Tokamak Fusion Reactor," *Scientific American* **241**, 51 (August 1979).
Garmon, Linda, "Radioactive Waste Disposal Options," *Science News* **120**, 398 (1981).
Garmon, Linda and Ivars Peterson, "International Aspects of Radwaste Storage," *Science News* **121**, 60 (1982).
Glackin, J. J., "The Dangerous Drift in Uranium Enrichment," *The Bulletin of the Atomic Scientist* **32**, No. 2, 22 (February 1976).
Hammond, Allen, et al., *Energy and the Future*. Washington, D.C.: American Association for the Advancement of Science, 1973. This has good concise discussions of fission and fusion reactors as well as other energy conversion systems.
Hughes, D. J., and R. B. Schwartz, *Neutron Cross Sections, BNL 325*. 2nd ed. Washington, D.C.: U.S. Government Printing Office, 1958. The most extensive data published.
Landa, Edward R., "The First Nuclear Energy," *Scientific American* **247**, 180 (November 1982).
Lipschutz, Ronnie, *Radioactive Waste: Politics, Technology, and Risk*. Cambridge, Mass.: Ballinger Publishing Company, 1980.
Murakami, Masanori and Harold P. Eubank, "Recent Progress in Tokamak Experiments," *Physics Today* **32**, #5, 25 (1979).
Nuckolls, J., J. Emmit, and L. Wood, "Laser-Induced Thermonuclear Fusion," *Physics Today* **26**, 46 (August 1973).
Peterson, Ivars, "Radwaste Storage," *Science News*, **121**, 9 (1982).
Post, R. F., "Prospects for Fusion Power," *Physics Today* **26**, 30 (April 1973).
Schroeer, Dietrich and John Dowling, "Resource Letter PNAR-1: Physics and the Nuclear Arms Race," *Am. J. Phys.* **50**, 786 (1982). An excellent list of books and articles on both the technical and political aspects of nuclear arms rivalry.
Smyth, Henry D., *Atomic Energy for Military Purposes*. Princeton, N.J.: Princeton University Press, 1945. The official report of the development of the nuclear bomb. A very good historical account.
Spitzer, Lyman, *Physics of Fully Ionized Gases*. 2nd ed. New York: Wiley, 1962.

Taylor, Denis, *Neutron Irradiation and Activation Analysis*. Princeton, N.J.: Van Nostrand, 1964.
Teller, Edward, *Energy From Heaven and Earth*. San Francisco: W. H. Freeman, 1979.
Vook, F. L., chairman, "Report to the American Physical Society by the Study Group on Physics Problems Relating to Energy Technologies. Radiation Effects on Materials," *Reviews of Modern Physics* **47**, Supplement No. 3, 51 (Winter 1975).
Weinberg, A. M., "Breeder Reactors," *Scientific American* **82** (January 1960).
United States Atomic Energy Commission, *The Effects of Nuclear Weapons*. Washington, D.C.: U.S. Government Printing Office, 1957. An extensive account.

REVIEW QUESTIONS

1. What are the conditions necessary to sustain a chain reaction?
2. What is meant by reactor criticality?
3. What is the function of a moderator in a nuclear reactor?
4. What is a breeder reactor?
5. What is the source of energy in our sun?

PROBLEMS

12.1 A crafty artisan offered a wealthy monarch an ornately carved chessboard, asking in return only one grain of rice for the first square of the board, 2 grains for the second square, 4 for the third, 8 for the fourth, 16 for the fifth, and so on. The monarch, unfamiliar with the nature of exponential growth, accepted. How many grains of rice were required for the first row of 8 squares? How many tons of rice were required for the first half of the board (32 squares)? (There are 50 grains of rice per gram and 10^6 grams/ton.) How many tons were required for the 64th square? (The world annual rice harvest is about 3.0×10^8 tons.)

12.2 Neutrons emerge from a reactor with an energy corresponding to the equipartition value for thermal equilibrium at 400 K. Find their velocity and de Broglie wavelength.

12.3 (a) If ^{90}Sr constitutes 5% of the fission-fragment yield of uranium (see Fig. 12.1), how many grams of this strontium are formed when a uranium bomb having the explosive equivalent of one megaton of TNT is exploded? (All the atoms in 50 kg of uranium must fission to produce the equivalent blast effect.) (b) If all this strontium is distributed uniformly throughout the atmosphere over the whole earth within a few months, what is the approximate radioactivity in microcuries above each square mile of the surface of the earth due to this one kind of fission fragment? (c) Comment on the likelihood of the uniform distribution assumed. (d) What happens to this strontium?

12.4 Show that the graph of the logarithm of the microscopic cross section of a "$1/v$-absorber" against the logarithm of the kinetic energy of the bombarding neutron is a straight line with a negative slope.

12.5 A neutron with nonrelativistic kinetic energy K has a head-on collision with a carbon-12 nucleus at rest. (a) What fraction of its energy does it lose? (Use conservation of energy and momentum.) (b) Assume the neutron strikes a hydrogen nucleus instead. What fraction of its energy does it lose?

12.6 The preceding problem assumed head-on collisions and thus gave the maximum possible energy lost per collision. Assume that the actual average energy lost per collision is half this value. How many collisions are required in a graphite moderator to reduce the energy of a neutron, which is initially 2 MeV, to thermal energy (0.025 eV)? Estimate the time required to reduce the energy from 2 MeV to 0.025 eV.

12.7 Assume a neutron causes fission in ${}^{235}_{92}U$ producing ${}^{97}_{40}Zr$, ${}^{134}Te$, and some neutrons. (a) Determine the atomic number of Te from the data. (b) How many neutrons are released?

12.8 A beam of 0.025 eV neutrons passes through an apparatus 20 cm long at the rate of 10^9/s. From the neutron half-life of 12 m, calculate the rate at which neutron decays are expected in the apparatus.

12.9 A nuclear reactor containing ${}^{235}U$ is operating at a power level of 2 W. Calculate the fission rate assuming that 200 MeV of useful energy is released in each fission.

12.10 Find the probability of fission, P_f, and the multiplication constant, k_∞, for slow neutron bombardment of ${}^{235}U$.

12.11 In a nuclear power reactor, 3.6% of the uranium is ${}^{235}U$ and the rest is ${}^{238}U$. What is the multiplication constant, k_p, for thermal neutron bombardment?

12.12 (a) What must be the ratio of ${}^{238}U$ to ${}^{235}U$ (N_8/N_5) in a reactor so that the multiplication constant, k_∞, equals unity for slow neutrons? (b) What is the enrichment of ${}^{235}U$ in that reactor? (c) The result in part (b) is less than 0.7%, the concentration of ${}^{235}U$ in natural uranium. Why doesn't natural uranium in ores or in solid pieces fission?

12.13 Small power stations in remote areas make use of the energy obtainable from the radioactive decay of Pu to Sn. This nucleus is an alpha emitter with an energy of 5.3 MeV with a half-life of 138 days. Calculate the power in watts per kilogram of Pu.

12.14 At present, the abundances of the isotopes in natural uranium are 0.72% of ${}^{235}U$, and 99.28% of ${}^{238}U$. What were the values 1.80×10^9 years ago? The half-lives are 7.13×10^8 y and 4.51×10^9 y, respectively. (See the George A. Cowan reference in this chapter.)

12.15 Assume that the two isotopes ${}^{235}U$ and ${}^{238}U$ were equally abundant at the time of the formation of the earth. Use the information in the preceding problem to estimate the number of years since the earth was formed. (Age $= 5.9 \times 10^9$ yrs.)

12.16 A beryllium target with a thickness of 6.5×10^{-5} cm is bombarded with a 20 μA beam of 3 MeV alpha particles. Neutrons are produced from the reaction ${}^9Be(\alpha, n)C^{12}$, which has a cross-section of 9.00×10^{-2} barns. What is the rate at which neutrons are produced?

12.17 A very thin sheet of gold foil (thickness $= 4 \times 10^{-3}$ cm) is irradiated for 2000 s in a neutron beam of flux 2×10^{10} neutrons/s. The gold is transmuted to ${}^{198}Au$ by an (n, γ) reaction. After the irradiation has ended, the measured activity of the ${}^{198}Au$ is 10^{10} disintegrations/h. Calculate the neutron capture cross section of gold for the neutrons used in this experiment.

12.18 A piece of ${}^{59}Co$ (relative density, 8.9) whose dimensions are 3 cm $\times$ 4 cm $\times$ 0.5 mm is to be irradiated by neutrons in a graphite pile until its total activity is 2 curies, due to the formation of ${}^{60}Co$ by the reaction ${}^{59}Co(n, \gamma){}^{60}Co$. (a) How many atoms of ${}^{60}Co$ does the activated piece of metal contain? (b) How many neutrons must interact with the ${}^{59}Co$, according to the reaction given, to produce the number of atoms in part (a)? (c) The microscopic activation cross section of ${}^{59}Co$ for the given reaction is 34 barns. If the irradiating beam is 2×10^{12} neutrons/cm^2/s normal to the 3×4 cm face, how long will it take to achieve the total activity of 2 curies? Assume that the radioactive disintegration of the ^{60}Co being formed during the time of bombardment is negligible. (d) Under what conditions is the assumption in part (c) reasonable?

12.19 (a) What fraction of the total number of atoms in the sheet of activated cobalt in Problem 12.18 changes into the daughter product in a day? (b) In 1975 the population of the United States was 215 million and the births were 3.2 million. What fraction of the total population became "daughter products" each day? (c) How much greater is this fraction than the fractional rate of radioactive disintegration in the cobalt sheet in part (a)? (d) Is radioactive "blowing up" a commonplace event from the viewpoint of the atoms in this piece of cobalt? (The cobalt sheet in this problem actually has a high activity for a radioactive material.)

12.20 A borated-steel sheet (relative density 7.81) that is used as a control "rod" in a reactor is 1/16 inch thick and contains 2% boron by weight. The microscopic absorption cross section of iron for neutrons is 2.5 barns and for boron, 755 barns. (a) Find the macroscopic absorption cross sections of each of these elements in the borated-steel sheet. (b) What fraction of a neutron beam is absorbed in passing through this sheet?

12.21 100 ml of a solution contains 2 g of boric acid, H_3BO_3, dissolved in 100 g of water. (a) What is the macroscopic neutron absorption coefficient of each element in the solution? (b) What is the total absorption cross section of the solution? (The microscopic absorption cross sections for these elements are given in Table 12.3.)

12.22 Using the same method as that used to obtain the average life of a radioactive atom, Eq. (10.8), show from Eq. (12.12) that the neutron mean free path in a material is $1/N\sigma$.

12.23 What is the absorption mean free path of (a) neutrons in iron ($\sigma_a = 2.5$ barns), (b) neutrons in borated-steel (Problem 12.20)? (c) What fraction of a beam of neutrons is transmitted by a sheet of material having a thickness equal to their mean free path?

12.24 (a) It is said that the effective neutron shielding property of concrete depends primarily on its hydrogen content but that its gamma-ray absorbing property depends mostly on its aluminum and silicon content. Do Eq. (6.15) and the data in Table 12.4 confirm this? (b) What would be the effect on the scattering property for neutrons and the absorption for gamma radiations if the concrete were made denser by using steel punchings in the mix instead of gravel?

12.25 The index of refraction of water at 20°C for red light is 1.33 and for blue light is 1.34. (a) What is the half-angle, θ, of the bow wave for red light and for blue light in the Cherenkov radiation produced by a beta particle emitted from a sheet of ^{60}Co with a speed of $0.8c$? (b) At what particle speed will (1) the red light and (2) the blue light in Cherenkov radiation in water disappear?

12.26 A beam of electrons traveling through a glass of index of refraction 1.50 is observed to emit Cherenkov radiation at an angle of 40° from the electron beam. Determine the kinetic energy of the electrons.

12.27 An Air Force jet traveling at 1400 miles per hour passes directly overhead and 45 seconds later the sonic boom is heard. Assume the velocity of sound to be 1088 feet per second and calculate how high the airplane is flying.

12.28 Each fission of a certain hypothetical nucleus produces two identical fragments, each with a kinetic energy of 50 MeV. What is the kinetic temperature of the "gas" of fission fragments?

12.29 The energy received from the sun by the earth and its surrounding atmosphere is 2.00 cal/(cm^2 · m) on a surface normal to the rays of the sun. (a) What is the total energy received in joules per minute by the earth and its atmosphere? (The diameter of earth-atmosphere system is 1.27×10^4 km.) (b) What is the total energy radiated in joules per minute by the sun to the universe? (Distance from sun to earth is 1.49×10^8 km.) (c) At what rate, in tons/m, must hydrogen be consumed in the fusion reaction, Eq. (12.22), to provide the sun with the energy it radiates? (d) At what rate, in tons/m, must the mass of the sun decrease to provide the radiated energy calculated in part (b)?

12.30 A deuterium reaction that occurs in experimental fusion reactors is D(d, p)T followed by a reaction of the tritium product with another deuterium atom. This second reaction is T(d, n)^{4}He. (a) Compute the combined Q-value of these successive reactions. (b) What is the energy released in the combined reactions per unit mass of the three initial deuterium particles? (c) What percent of the total rest mass-energy of these three initial particles is released in the combined reactions? (d) Compare the answers for parts (b) and (c) with the corresponding calculations for uranium fission in Problem 5.52. (e) Compare ^{235}U fission with deuterium fusion as a source of energy on the basis of the availability of the isotopes and of the energy released per unit mass.

12.31 It is believed that the microscopic absorption cross section of iron for neutrinos is of the order of 10^{-20} barn. In kilometers, what is the half-value layer of iron for neutrinos?

13 HIGH-ENERGY PHYSICS

13.1 INTRODUCTION

In this chapter, we return again to the early part of this century to describe the discovery and some of the properties of cosmic rays. We shall see how studies of these natural phenomena, together with studies of events produced by the modern accelerators, serve to clarify some matters already discussed. But this chapter will not tie up a neat package. In March 1966, Nobel Prize winner T. D. Lee stated, "The more we learn about symmetry operations — space inversion, time reversal, and particle-antiparticle conjugation — the less we seem to understand them. At present, although still very little is known about the true nature of these discrete symmetries, we have, unfortunately, already reached the unhappy state of having lost most of our previous understanding."† Our hope is that the reader will gain some comprehension of the problems that are most challenging to the physicists of today.

13.2 COSMIC RAYS

Cosmic rays were discovered during the search for a cause of ionization chamber leakage. Of course, no insulator is perfect, and some ionization chamber leakage was expected. When it was found that the leakage rate was greater than could be

† T. D. Lee, *Physics Today* **19**, #3, 23 (March 1966).

accounted for by imperfect insulation, it was assumed that the difference was due to ionization from traces of radioactivity in the earth. To test this point, ionization chambers were carried above the earth in balloons. During the first part of the ascent, the ionization did indeed decrease slightly. But in 1909, Gockel found that at greater heights the ionization began to increase, and at 2.5 miles the ionization was greater than at the surface of the earth. In 1910, Hess and Kohlhoerster carried a chamber even higher and, after extensive study, they proposed that there must be a very penetrating radiation coming from outside the earth — *cosmic rays*. It was estimated that these rays were at least ten times as penetrating as the most energetic gamma rays known.

Just as cosmic-ray research was begun by moving ionization chambers up into the air, so it was continued by moving them into the earth and over its surface. The presence of cosmic rays was traced deep inside mines, where the effect of any radioactive material was excluded with lead shields. They were found deep in lakes. Glacier-fed lakes were chosen because their water is "distilled" and free of radioactive materials. Recording ionization chambers were sent all over the world on ships.

As a result of these geographic explorations, it was concluded that cosmic rays are extremely penetrating, and that although they come to the earth from all directions, a few more reach the earth in the latitudes of the magnetic poles than at the equator.

The fact that the intensity is greater toward the polar regions is called the *latitude effect* and it suggests that at least some of the cosmic rays are charged. Charged particles approaching the earth at the equator must move perpendicular to the magnetic field of the earth. Depending on the sign of the charge, these particles experience a force either toward the east or toward the west, which may cause some rays to miss the earth that would otherwise have hit it. Charged particles approaching the high-latitude portions of the earth move nearly parallel to the magnetic field and experience a negligible deflecting force. If these particles are moving at an angle with respect to the magnetic axis of the earth, they will spiral along a helical path. Figure 13.1 summarizes the possible trajectories of cosmic-ray particles in the geomagnetic field.

Although this analysis indicated that at least some of the rays were charged, it did not indicate the sign of the charge. It also follows from the analysis that at the magnetic equator charged particles should strike the earth from a preferred direction, either east or west of the vertical, depending on the sign of the charge. Ionization chambers have no directional sensitivity; therefore clever arrangements of Geiger-Mueller tubes were used to investigate the east-west effect. A single G-M tube has no directional sensitivity but several of them, connected electronically so that no count is recorded unless they all count in coincidence, may be physically arranged like the rungs of a ladder. With such an apparatus, called a *cosmic-ray telescope*, the direction of a ray capable of firing all the tubes could be determined.

The east-west experiment was performed by observing the counting rate of a cosmic-ray telescope set at a variety of angles east and west of the vertical. A symmetrical distribution was found at 35° south latitude (Buenos Aires). At the

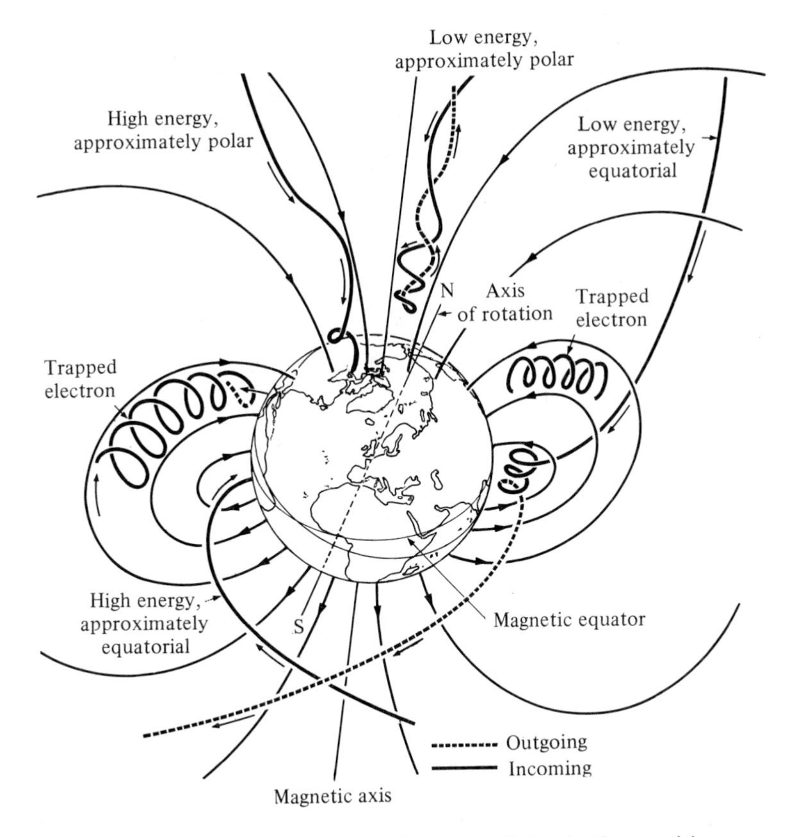

Figure 13.1 Trajectories of cosmic-ray particles in the earth's magnetic field.

From Marcelo Alonso and Edward J. Finn, *Physics*, © 1970. Addison-Wesley, Reading, Mass. Reprinted with permission.

magnetic equator, however, although the distribution had the same general shape, it skewed a little to the west. The slightly greater intensity from the west indicated a preponderance of positive charges, but it was impossible to conclude from this that there were no negative particles or uncharged particles present.

13.3 COSMIC-RAY SHOWERS

In 1928, Skobeltsyn photographed cosmic-ray tracks in a Wilson cloud chamber. He found that the tracks appeared in groups and that all the particles in a group seemed to originate from some point near and above the apparatus, such as the ceiling. It was later found that G-M tubes that were widely separated horizontally had more

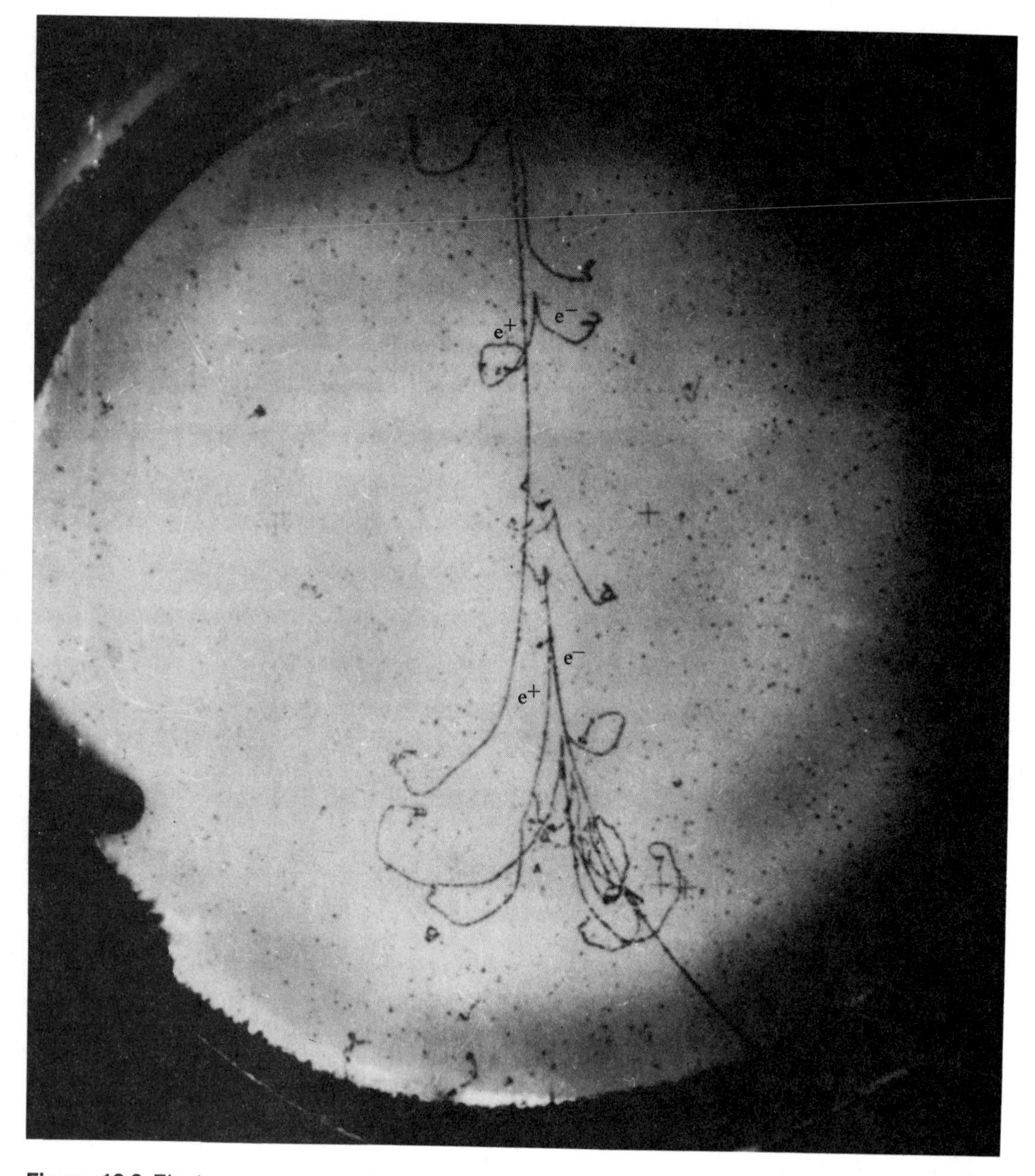

Figure 13.2 Electron-gamma ray shower formation. An electron with about 1 GeV of energy enters a methyl iodide propane bubble chamber, initiating a shower of gamma rays and electron-positron pairs. The entire development can be seen because of the high atomic number and high density of the medium.

Courtesy of the "Cambridge Bubble Chamber Group," Brown University, Harvard University, Massachusetts Institute of Technology, and Brandeis University.

coincidence counts than could be accounted for by chance. These observations led to the discovery of what are called *cosmic-ray showers*. The mechanism of these bursts of rays is well understood. Somewhere in the atmosphere a cosmic ray may produce a photon with several GeV of energy. This photon may produce an electron-positron pair. (Recall that positrons were discovered by Anderson while using a cloud chamber in cosmic-ray research.) These very high-energy electrons can next produce high-energy photons by the mechanisms of x-ray production and also by the eventual production of annihilation radiation. Consequently, one initiating particle (either photon or electron) may be multiplied into many shower particles, as shown in Fig. 13.2.

The discovery of showers points out how difficult it has been to establish the nature of the real primary cosmic radiation that comes to the earth from outer space. The nature of the rays changes as they come down through the atmosphere, and sea-level cosmic rays bear little resemblance to their initial state. Only with the recent use of satellite and rocket-borne detectors has it been shown that the primary cosmic particles are nuclei of hydrogen (protons), nuclei of helium (alpha particles), as well as heavier nuclei.

Primary cosmic rays come infrequently — just a few particles per square centimeter per second at the top of the atmosphere. The rates fall off for individual particles at roughly the inverse cube of their energy, with the extreme limit of the known spectrum at 10^{20} eV. After many years of study, it is still unclear where all cosmic ray particles come from. Those with energies below about 10^{17} eV seem to come from somewhere in our Milky Way. They are possibly accelerated in supernova explosions or by pulsars. However, astrophysicists have recently leaned toward a more indirect process. They believe the charged particles are accelerated by shock waves in the gas of the instellar media.

New experiments are underway to try to determine the origin of the highest energy cosmic rays. Scientists are making use of an optical cosmic-ray detector called the fly's eye because of its many segmented parts. The system uses 67 telescopes set in concentric rings. Each telescope, equipped with a set of photomultiplier tubes, scans different segments of the sky to observe scintillation of light left by the air shower of particles. The assembly is a giant optical scintillation counter that uses the atmosphere as its detector. It is anticipated that data accumulated over the next decade will yield evidence about the source of the high energy cosmic rays.

13.4 DISCOVERY OF THE MUON

In trying to explain the binding force between neutrons and protons in the nucleus, Fermi had attributed it to the "exchange" of electrons between the two types of particles. It developed that the magnitude of a force from this cause was too small to fit the experimental evidence. In 1935, Yukawa followed the same general idea but postulated that the exchange particle was not an electron but a new particle having a mass intermediate between those of an electron and a proton. Since this postulated particle, if it existed, would have the penetrating properties consistent with cosmic-ray data, cosmic-ray researchers set out to find it. In 1936, Anderson and Neddermeyer observed a cloud-chamber track of such a particle, now called a muon,† but it was shown some years later that it was not identical with the one assumed by Yukawa. In 1940, Leprince-Ringuet obtained the first good cloud-chamber picture that showed a collision between a muon and an electron. Since both tracks were curved by a magnetic field, the particle energies could be calculated. The mass of the muon was calculated from the mechanics of the collision; the modern accepted value is 207 electron rest masses.

The discovery of the *muon* accounted for the penetrating ability of cosmic rays, but it was soon evident that the muon could not be the primary radiation from outer space. It developed that the muon is itself radioactive, with a very short half-life. Measurement of this half-life is complicated by the fact that muons formed in the upper atmosphere by cosmic-ray bombardment are projected toward the earth with a speed close to that of light. In our discussion of relativity we saw that there is a time transformation equation that takes into account the relative velocity of two observers. An observer moving with a muon would conclude that its half-life was a few microseconds. To an observer on earth, the half-life appears considerably longer — long enough, in fact, to let a muon traverse the atmosphere but not long enough for it to have come from outer space. Thus muons cannot be primary radiation but must be created by something else at the top of the atmosphere. Present data indicate that most of the primary cosmic rays are protons, but there are also some positive nuclei or stripped atoms of elements up to $Z = 26$. These very high-energy primary particles produce a large variety of other particles by *spallation* (nuclear smashing) of the atoms in the outer atmosphere.

Although much more is known about the primary radiation than we have discussed here, there is still speculation going on. Earth satellites and instruments left on the moon by the astronauts are now sending us information from beyond the atmosphere.

13.5 DISCOVERY OF PI-MESONS

One of the first discoveries made with the nuclear emulsions that were discussed in Chapter 10 was that of the pi-meson. A block of emulsions was exposed for several weeks on a mountain. One mosaic of track pieces showed a meson slowing down and then emitting a new high-speed meson in a new direction. The second meson could be identified as the familiar muon. Since the first could create a muon and give it kinetic energy as well, this first meson had to be more massive. It is called the *pi-meson* (primary meson) and its mass is now known to be 270 electron masses.

† This particle first was incorrectly identified as a meson (mu-meson) but has since been included with the leptons.

Three types of pi-mesons, or *pions*, have been discovered. One has a positive charge, another is negative, and the third is neutral. All are radioactive. The charged ones have mean lives of the order of 10^{-8} s,† but the life of a neutral pion is only about 10^{-16} s.

In flight, the positive and negative pi-mesons decay into the positive and negative muons, respectively, and also into neutrinos. When negative pions are slowed down in matter, they are captured by atoms, make x-ray transitions down to the K level, and then are absorbed by the nucleus, causing violent explosions called "stars." Positive pi-mesons stopped by matter decay into muons, all of which have about the same energy. Experiments show that neither the positive nor the negative muon has an affinity for atomic nuclei. Because of this, it is difficult to assign to muons the role of providing the nuclear exchange forces of the Yukawa theory. According to present nuclear theory, all three types of pi-mesons are found in the nucleus, and it is believed that both protons and neutrons are continuously emitting or absorbing positive, negative, and neutral pi-mesons. The particular process depends on which nucleons are paired in experiencing exchange forces at the moment, that is, proton-neutron, proton-proton, or neutron-neutron. Thus the general view is that each nucleon has an associated meson field through which it interacts with other nucleons. This is analogous to the action through an electromagnetic field of one electrically charged body on another. The pions play the same role in the meson field as photons do in an electromagnetic field.

13.6 "ELEMENTARY" PARTICLES

Electrons, protons, and neutrons are familiar in our scheme of things. In earlier chapters we introduced positrons and neutrinos. The muon and pi-meson facilitate our understanding of the penetrating ability of cosmic rays, and pi-mesons fit into the theory of nuclear binding forces. But the total number of "elementary" particles now known is above 100. Those we have discussed in this book thus far are "useful" in the sense that they fit into a rational scheme for the structure of matter, and they "explain" events that are more or less common and natural. The plethora of new particles present an enigma. One can make an analogy between these particles and the transuranic elements. Elements with atomic numbers above 92 did not exist in nature in measurable amounts before they were "'made" in the laboratory.‡ They are all unstable with half-lives short compared to the age of the earth, so that if they ever existed, they have long since disappeared. Thus these elements are strange elements. Having made them, however, we seek to measure their chemical and physical properties, understand them, and find uses for them. The situation is similar for the elementary particles. They are created (or liberated) by collisions of the

† In describing the fundamental particles, mean lives rather than half-lives are usually stated. The definition of mean life is given in Section 10.10.

‡ The possible exception is the element with $Z = 94$ discussed in Section 10.11.

extremely energetic particles from the modern accelerators. They are all unstable with mean lives from about 10^{-8} to 10^{-16} s. They are not elementary in the sense that they are simple and basic like elementary algebra. They are elementary in the sense that they leave their tracks in bubble chambers and nuclear emulsions the way elementary electrons and protons do. Indeed, it is only from these tracks that we know of them at all. An analysis of these tracks — their density, their relative directions, their curvatures in magnetic fields, and their branchings — serves to specify their unique identities and their physical properties. See Fig. 13.3, which we will discuss in some detail shortly. As in the case of the transuranic elements, having made them we seek to understand them. Because we believe in the *oneness-of-knowledge*, until we understand these particles and can incorporate them into our

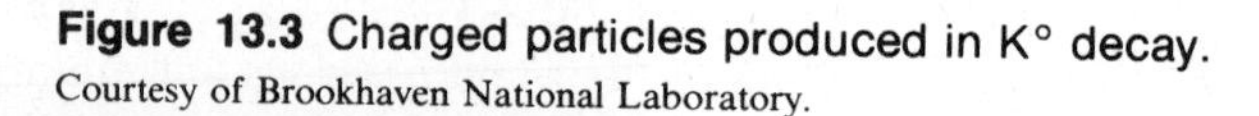

Figure 13.3 Charged particles produced in K° decay.
Courtesy of Brookhaven National Laboratory.

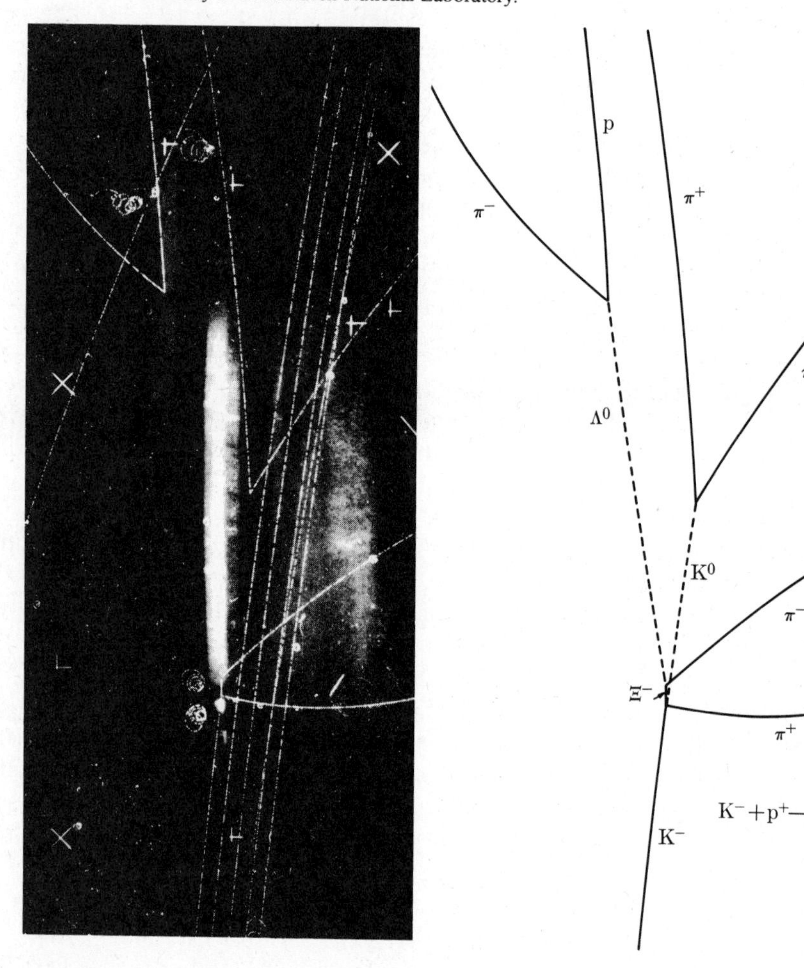

theoretical scheme, they constitute a threat to much of what we think we now understand about the physical world. Note again the statement of Professor Lee quoted at the beginning of this chapter.

Figure 13.4 represents the 30 particles scientists were aware of in 1957 in a kind of "periodic table." Masses increase upward from the bottom of the figure. Particles on the left are considered "ordinary" particles whereas those on the right are antiparticles. In Section 5.20, we related the discovery of the positron, which is the antiparticle of the electron. As the chart shows, nearly every particle has its antiparticle. There is no significance as to which is which except that the rarer particle is chosen to be the antiparticle. We have discussed the annihilation of positive and negative electrons, and whenever any particle and its antiparticle collide, they annihilate each other. This is the "particle-antiparticle conjugation" referred to by Professor Lee in the passage we quoted.

The 30 particles shown in Fig. 13.4 are divided into four groups: baryons, mesons, leptons, and the photon. The baryons are the most massive of these particles and include the proton, the antiproton, the neutron, the antineutron, and two lambda particles (Λ), six sigma particles (Σ), and four xi particles (Ξ). Excluding the photon, the least massive of these particles are called the leptons. This group is made up of the electron, the positron, two muons, and two neutrinos. There is a neutrino and an antineutrino associated with the electron and also a neutrino and antineutrino associated with the muon; however, this was not known in 1957. Between these two groups is a group of medium mass particles called mesons. The mesons include the three pi-mesons and the four K-mesons. Mesons have integral spin while baryons have half integral spin.

Although these new particles constitute one of the major puzzles of physics, we see that much information has been gathered and a few generalizations have been made that are appropriate to a book of this level. The masses, spins, and charges are known. The mean lifetimes of the unstable particles are known as are many of the decay schemes. Thus, for example, there is the decay

$$\Sigma^0 \rightarrow \Lambda^0 + \gamma.$$

Similarly, a wide variety of elementary particle reactions are known, which are produced by bombarding one particle with another. Thus we have

$$\pi^+ + p \rightarrow \Sigma^+ + K^+.$$

In this reaction mass-energy and charge are conserved as in any nuclear reaction. Translational momentum and spin are conserved. Another property of the particles, discovered by Murray Gell-Mann and called *strangeness*,† is also conserved. Figure

† The term strangeness comes from a group of particles called "strange" particles. Of the particles in Fig. 13.4 the Ξ, Σ, Λ, and K-meson belong to this group. The decay time of each of these particles is much longer than what was predicted by the theory at the time of the discovery. The time scale for the production of these particles is about 10^{-23} seconds and it was expected that they should decay on this same time schedule. However, the lifetime of these particles is from 10^{-8} to 10^{-10} seconds, which is 10^{15} to 10^{13} times longer than expected. Therefore, they have been called "strange" particles.

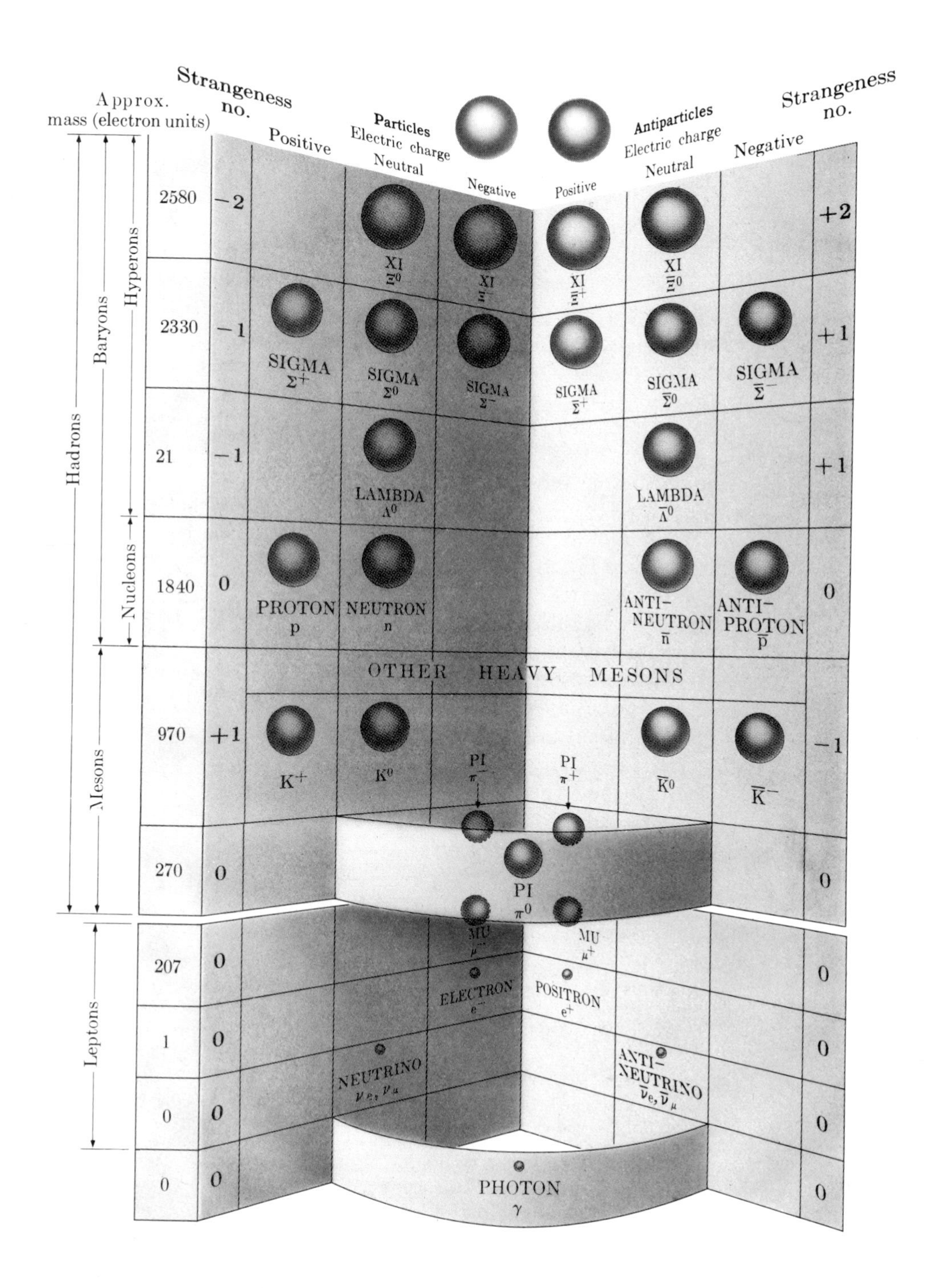
Approx. mass (electron units)
Strangeness no.
Particles
Electric charge
Positive
Neutral
Negative
Antiparticles
Electric charge
Positive
Neutral
Negative
Strangeness no.
Hadrons
Baryons
Hyperons
Nucleons
Mesons
Leptons
2580
−2
XI
Ξ0
XI
Ξ−
XI
Ξ̄+
XI
Ξ̄0
+2
2330
−1
SIGMA
Σ+
SIGMA
Σ0
SIGMA
Σ−
SIGMA
Σ̄+
SIGMA
Σ̄0
SIGMA
Σ̄−
+1
21
−1
LAMBDA
Λ0
LAMBDA
Λ̄0
+1
1840
0
PROTON
p
NEUTRON
n
ANTI-NEUTRON
n̄
ANTI-PROTON
p̄
0
OTHER HEAVY MESONS
970
+1
K+
K0
PI
π−
PI
π+
K̄0
K̄−
−1
270
0
PI
π0
0
207
0
MU
μ−
MU
μ+
0
ELECTRON
e−
POSITRON
e+
1
0
0
NEUTRINO
νe, νμ
ANTI-NEUTRINO
ν̄e, ν̄μ
0
0
0
0
0
PHOTON
γ
0

12.4 includes columns headed "Strangeness numbers." Gell-Mann found it possible to assign these numbers in such a way that for all known reactions (as distinguished from mere disintegrations) the total strangeness before and after the reaction are the same. Thus for the reaction given

$$0 + 0 = -1 + 1.$$

Having introduced the symbols of the particles, let us discuss the events depicted in Fig. 13.3 more fully. A high-energy heavy meson K^- entered the hydrogen-filled bubble chamber from below. Unlike its companions, which traversed the chamber without encountering a nuclear collision, this one struck a proton hydrogen nucleus. A reaction took place and the products were Ξ^-, K^0, and π^+, according to the reaction

$$K^- + p \rightarrow \Xi^- + K^0 + \pi^+. \tag{13.1}$$

Since the product particles have about 1000 more electron masses than the initial particles, we know the incident K^- must have had at least 500 MeV of energy. It is obvious from the length of the tracks that the product particles had great kinetic energy; therefore the incident K^- particle must have had far more than 500 MeV of energy.

The π^+ left the region photographed at the right (see Fig. 13.3). The charge-free K^0 left no track, but it evidently continued in the direction of the original K^- until it produced a π pair at the upper right. The Ξ^- went only a short distance before it disintegrated into a neutral Λ^0 and a π^-. The neutral Λ^0 proceeded upward and to the left leaving no track, but it disintegrated into a proton and a π^-. By sighting along the tracks, it is evident that the magnetic field curved the paths of positive particles one way and the paths of negative particles the other. Now it may seem that a great deal has been read into this photograph, but it must be realized that the photograph contains a vast amount of data. Using conservation of charge, conservation of translational momentum, conservation of angular momentum (spin), conservation of energy, and conservation of strangeness, no other interpretation of this diagram is reasonable. It is through such an analysis of nuclear emulsions and bubble chamber photographs that most of the new particles have been identified and their properties determined.

◄**Figure 13.4** The mirror symmetry of elementary particles. Only seven of the thirty particles shown are intrinsically stable: photons, protons, electrons, neutrinos, antineutrinos, antiprotons, and positrons. The last two, however, share the usual fate of antiparticles that wander into this world: They are annihilated when they meet their counterparts. Photons and neutral pi-mesons commute freely between both worlds since each is its own antiparticle — although the neutral pi-meson has little time for traveling because it is the second shortest-lived of all the known particles. A neutron is stable if bound in an atomic nucleus; otherwise it has a half-life of 12 minutes, decaying into a proton, an electron, and a neutrino. Similarly, all unstable particles decay, directly or indirectly, into two or more of the seven stable ones. Their mean lifetimes range from about 10^{-8} seconds to less than 10^{-16} seconds.

Max Gschwind for *Fortune Magazine*,

13.7

PARTICLE PROLIFERATION

More than 100 new particles have been discovered since the 30 of 1957, which has made necessary a new classification system. Particles are classified into groups according to the types of forces that they "feel" or interactions they undergo. There are four known forces or interactions in nature: gravitational, electromagnetic, strong, and weak. These forces and interactions will be discussed in the next section. Of these forces, the weak and strong are used to classify all particles except the photon. Those that feel the strong force are called hadrons and are divided into two groups: baryons and mesons. Those that do not feel the strong force are called leptons. The photon is in a completely different category since it has no mass and no charge and does not participate in either of these interactions.

The proliferation of particles has occurred almost exclusively in the strongly interacting particles. The number of hadrons has increased by more than 100, while the number of leptons has increased only by two. Strongly interacting particles known as of 1964 are shown in Fig. 13.5. These are only those with rest mass below 2000 MeV.

Classification of the particles is made easier by the assignment of quantum numbers that refer to individual properties of the particle. Quantum numbers assume only certain discrete values and a complete quantum number list uniquely identifies the particle and will define its characteristics. Some quantum numbers have already been discussed — for example, those of electric charge, spin, strangeness, etc.

A scheme that has been valuable not only for classifying particles but also for predicting the existence of previously undiscovered particles is a system called the "eightfold way." In 1961, this system was developed independently by Murray Gell-Mann and Y. Neemann. It has been referred to as the eightfold way because it involves eight quantum numbers associated with each particle. The choice of name may be partly due to a statement attributed to Buddha: "Now this, O monks, is noble truth that leads to the cessation of pain: this is the noble Eightfold way: namely, right views, right intention, right speech, right action, right living, right effort, right mindfulness, right concentration." One significant accomplishment for the eightfold way was the prediction of a new particle, the omega minus (Ω^-), with a mass of between 1676 and 1680 MeV. In January 1964, scientists at Brookhaven using the alternating gradient synchrotron that we discussed in Chapter 11 first detected this particle, thus confirming the prediction of its existence.

A more recent development of the eightfold way has been the prediction of quarks.† Gell-Mann and Zweig independently proposed that there were three particles that could be combined to form all of the hadrons. This quark model uses

†Gell-Mann took the word "quark" from James Joyce's novel, *Finnegan's Wake*. In that work, a barkeeper, H. C. Earwicker, says at intervals: "Three quarks for Muster Mark."

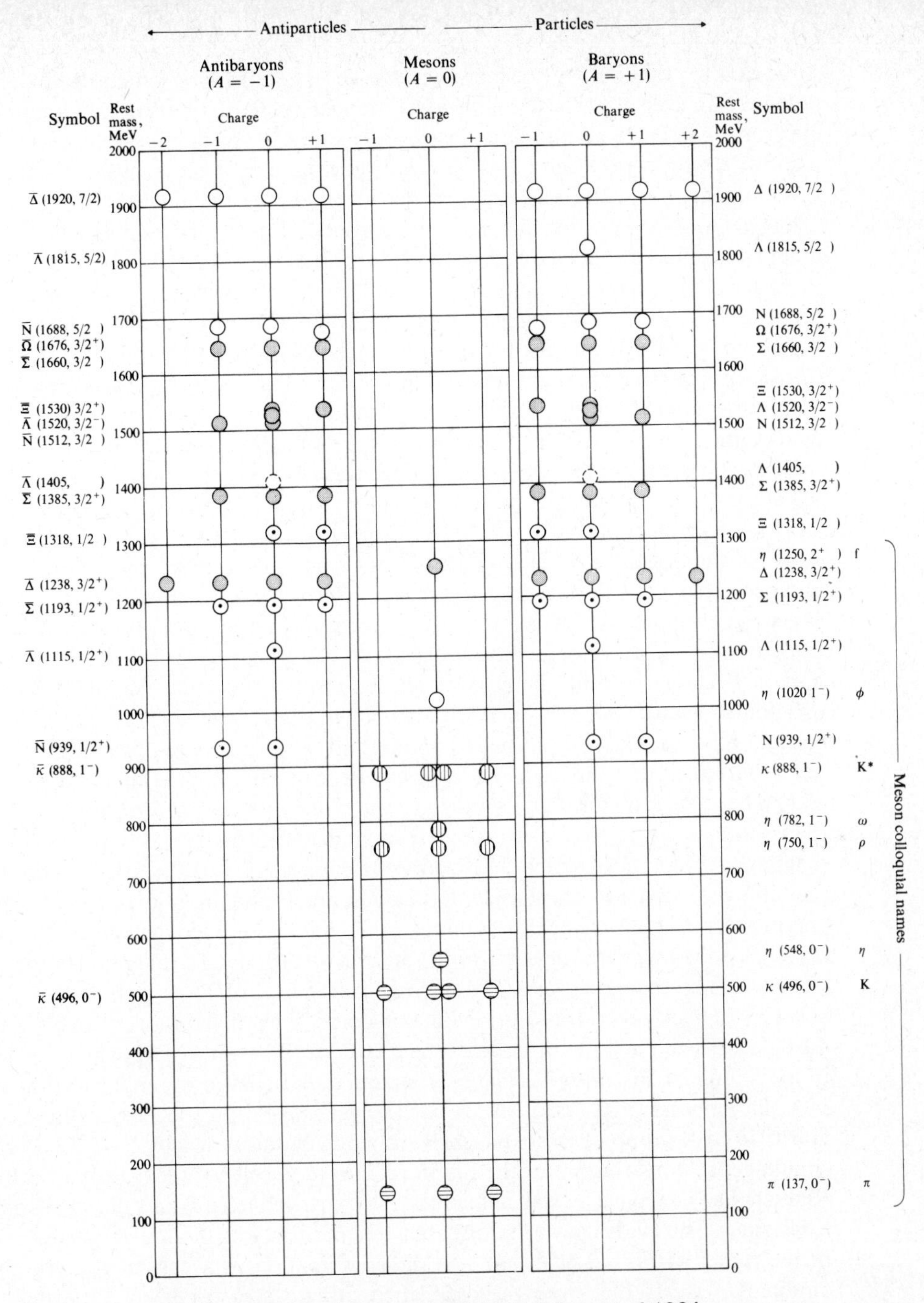

Figure 13.5 Strongly interacting particles as of 1964.

From "Strongly Interacting Particles," by Geoffrey F. Chew, Murray Gell-Mann, and Arthur H. Rosenfeld.

three quarks: one with the electronic charge $-\frac{1}{3}e$ and strangeness 0 (d quark), a second with electric charge $+\frac{2}{3}e$ and strangeness 0 (u quark), and a third with electronic charge $-\frac{1}{3}e$ and strangeness -1 (s quark). Each quark has an antiquark associated with it ($\bar{\mathrm{u}}$, $\bar{\mathrm{d}}$, and $\bar{\mathrm{s}}$). The magnitude of each of the quantum numbers for the antiquarks has the same magnitude as those for the quarks, but the sign is changed. All baryons are formed of three quarks. For example, the proton is made up of two u quarks and a d quark (uud). For these quarks, the electric charges are $+\frac{2}{3}$, $+\frac{2}{3}$, and $-\frac{1}{3}$ for a total value of $+1$. The baryon numbers are $+\frac{1}{3}$, $+\frac{1}{3}$, and $+\frac{1}{3}$, for a total of $+1$. The strangeness numbers are 0, 0, and 0 for a total strangeness of 0. All are in agreement with the quantum numbers for the proton. The mesons are formed of a quark and an antiquark. For example, the positively charged pi-meson is the combination of a u quark and a d antiquark (u$\bar{\mathrm{d}}$). Electric charges of these quarks are $+\frac{2}{3}$ and $+\frac{1}{3}$ for a total of $+1$. The baryon numbers are $+\frac{1}{3}$ and $-\frac{1}{3}$ for a total baryon number of 0. The strangeness numbers are 0 and 0 for a total of 0. All of these are in agreement with the quantum numbers for the pi-meson.

In recent years, a fourth quark has been added to the list; it has electric charge $+\frac{2}{3}e$ and strangeness 0. This quark differs from the u quark by a new quantum number given the arbitrary name "charm." The charmed quark was suggested to explain the suppression of certain decay processes that are not observed. With only three quarks, the processes would proceed at measurable rates and should have been observed. Until recently, there had been no physical evidence for the existence of this fourth quark. All of the hadrons known could be made by combining the original three quarks with no need for the fourth one. However, by using the fourth quark with the other three, the hadrons can be arranged in families as shown in Fig. 13.6. The planes of the figures indicate the values of charm. Positions within the planes are determined by isotropic spin and strangeness. Each point on these diagrams represents a particle formed by the combination of the quarks indicated in parentheses. All hadrons discussed thus far are found on the planes designated charm = 0 in the diagram. For example, in diagram (b) we see the kaons. In diagram (d) we see the nucleons [neutron (udd) and proton (uud)]. In diagram (e) the sigma particles, the cascade particles, and the omega minus particle are found.

The first evidence for charm did not come from the discovery of a charmed particle but from a study of the mesons ρ (u$\bar{\mathrm{u}}$), ω (d$\bar{\mathrm{d}}$), and ϕ (ss), which are shown in the middle of Fig. 13.6(b). The ρ, ω, and ϕ were believed to be particles made up of a quark bound to its antiquark. A particle with mass of 3.1 GeV was discovered in late 1974 by a group at Brookhaven National Laboratory led by S. C. C. Ting and simultaneously by a group at the Stanford Linear Accelerator Center led by Burton Richter. The new particle was named the "J" particle by Ting and the "Psi" particle by Richter. The evidence was strong that this particle was the fourth particle shown in the center of Fig. 13.6(b) and was the combination of a charmed quark and its antiquark (c$\bar{\mathrm{c}}$). If this were the case, then other states of the new particle were expected to exist and one such state was to be at about 3.7 GeV. This 3.7 GeV state should decay into an intermediate state with the emission of a photon and then the

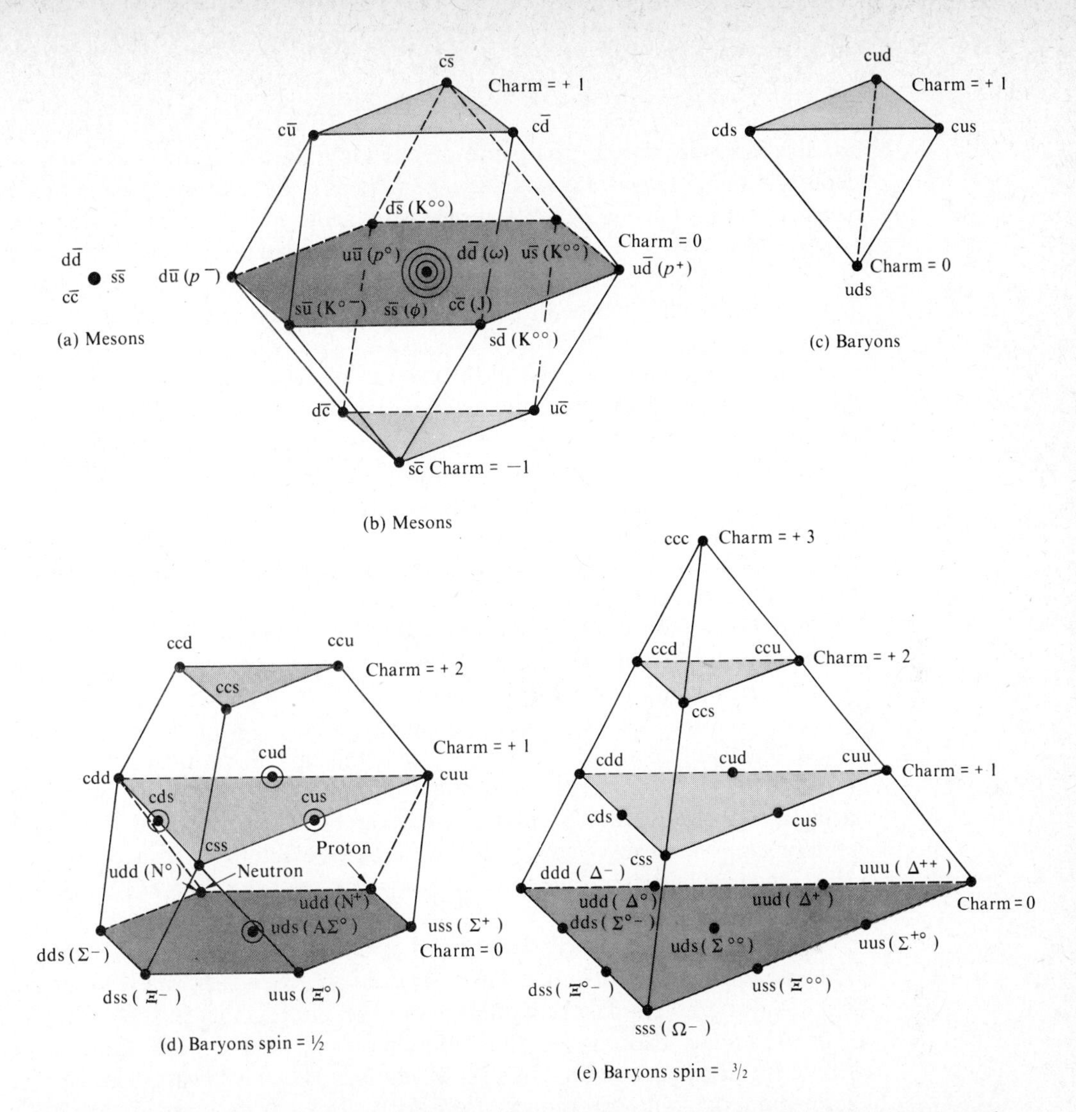

Figure 13.6 The hadrons can be arranged as polyhedrons. Each supermultiplet consists of particles with the same value of spin angular momentum. Within each supermultiplet the particles are assigned positions according to three quantum numbers: Positions on the shaded planes are determined by isotopic spin and strangeness; the planes themselves indicate values of charm. The mesons are represented by a point (a) and by an Archimedean solid called a cuboctahedron (b), which comprises 15 particles, including six charmed ones. The mesons shown are those with a spin of 1, but all mesons fit the same point and cuboctahedron representations. The baryons form a small regular tetrahedron (c) of four particles, a truncated tetrahedron (d) of 20 particles, and a larger regular tetrahedron (e) also made up of 20 particles. Both mesons and baryons are identified by their quark constitution, and for those particles that have been observed the established symbol is also given. Each figure contains one plane of uncharmed particles that are identical with earlier representations of the "eightfold way."

From "Quarks with Color and Flavor" by S. L. Glashow.

intermediate state should decay into the 3.1 GeV state with the emission of another photon. The 3.7 GeV state was discovered at SPEAR and the cascade process was recently observed at the DORIS storage ring facility in Hamburg, Germany, as well as at SPEAR. The J/ψ particle appears to be the particle formed by the combining of the charmed quark and its antiquark ($c\bar{c}$). However, the particle has charm equal to zero because the charmed quark has charm $+1$ and the charmed antiquark has charm -1.

A necessary test of the charm theory was, of course, the discovery of a particle with nonzero charm quantum number. As can be seen in Fig. 13.6, there are many mesons with charm of $+1$ and -1 and many baryons with charm of $+1$, $+2$, or $+3$. The first observation of a meson with charm was made in the summer of 1976. It was composed of a charmed quark and the antiparticle of a normal quark; thus the charm quantum number was $+1$. The first observation of a charmed baryon was made at the Fermi National Laboratory at Batavia, and is believed to be the (cud) baryon with charm $+1$, indicated in Fig. 13.6(c).

In 1977, a new particle was discovered at Fermi Lab that provided evidence for yet another quark. This particle, called the upsilon-meson, was thought to be made up of the new quark along with the associated antiquark. The fifth quark has been named the bottom quark (sometimes called beauty). As in the case of the J/psi meson, consisting of a charm-anticharm pair, this upsilon particle can be considered as a system of two quarks, bottom-antibottom, that orbit each other. Another particle has been discovered that is believed to be a member of this family and exhibiting naked beauty just as the D particles exhibited naked charm (see Fig. 13.6). This new particle formed by the decay of one of the upsilon particles is called a B-meson.

Things seem to be getting out of hand again. A few years ago everything seemed to be in a high state of disorder. There were a few leptons and many many hadrons. Then it was observed that three quarks could be put together in such a way to form all of the known hadrons. At that time, it seemed that leptons and three quarks could be the basic building blocks of nature. The need was then seen for the charmed quark. And now evidence for a fifth quark has occurred. So the question is how many quarks are there?

There were problems with the quark model, one of them being the omega-minus hyperon. It was believed to contain three identical s-quarks (see Fig. 13.6e). If this were true, it would violate the Pauli exclusion principle that prohibits two or more fermions from occupying identical quantum states.† The solution was the assignment of individual characteristics called "color" to quarks. The colors are red, blue, and yellow.

It appears to be developing that we have two separate families of particles, leptons and quarks. There are six known leptons and appear to be five known quarks, each with three different colors. For consistency a sixth quark has been

† The proton, neutron, and others with two identical quarks would violate this principle also.

Table 13.1

	Leptons		Quarks					
			Red		Blue		Yellow	
Third generation	ν_τ	0	t	+2/3	t	+2/3	t	+2/3
	τ^-	−1	b	−1/3	b	−1/3	b	−1/3
Second generation	ν_μ	0	c	+2/3	c	+2/3	c	+2/3
	μ^-	−1	s	−1/3	s	−1/3	s	−1/3
First generation	ν_e	0	u	+2/3	u	+2/3	u	+2/3
	e^-	−1	d	−1/3	d	−1/3	d	−1/3

predicted but not observed in any way. The families of particles are broken down by generation and there appear to be three generations. Table 13.1 shows how these basic particles are presently classified.

We see that the first generation contains two leptons, the electron and the electron neutrino, and two quarks, up and down. The second generation includes the muon and muon-neutrino and the strange and charmed quarks. And the third generation includes the tau and the tau-neutrino and the bottom and perhaps the top quarks. Note that in ordinary matter, made up of atoms, only the first generation particles are necessary. The other basic particles are only observed in extremely high-energy collisions. One puzzle in nature is why the second and third generation particles appear in nature at all.

There are basic differences in the families of particles shown in Table 13.1. The first is that the leptons have charges of absolute value zero or one, while the quarks have charges of absolute value one third or two thirds. Combinations of quarks to form particles observed in nature yield charges that are integral. The second significant difference is that the leptons appear as free particles in nature but quarks are only observed as constituents of observed mesons or hadrons.

The fact that no free quarks have been observed is predicted by the accepted theory of the strong interactions. However, many searches have been carried out to try to observe one of these particles. Searches have been conducted in large accelerator facilities throughout the world. In at least ten countries, many groups using cosmic-ray techniques have contributed to the search for quarks. Other searches have been carried out in material samples, using several different methods including mass spectroscopy, improved versions of the Millikan oil-drop experiment, and optical spectroscopy. All have been negative with the possible exception of the Millikan oil-drop technique. William M. Fairbank and co-workers using niobium spheres in a magnetic field at extremely low temperatures believe they have observed a particle with fractional charge. If these data are indeed due to free quarks, it will be an exciting new experimental discovery. However, it will have serious effects on the standing of the theory for these interactions.

Our recipe for forming particles by combining quarks is now modified a small amount. Quarks can be combined in two ways to form hadrons. Three quarks bound together yield a baryon. One quark bound with an antiquark yields a meson. However, the rule now is that the combination of quark colors must result in white in analogy to the mixing of the three primary colors to obtain white. A particle is made white (or colorless†) by combining a red quark, a blue quark, and a yellow quark (baryon) or a colored quark and its antiquark (meson). Combining three different colored quarks or a quark and its antiquark gives rise to a color neutral state, much as combining an electron and a proton creates a charge neutral state (hydrogen atom).

13.8 BASIC INTERACTIONS

As was pointed out in the last section, there are four basic interactions or forces in nature — gravitational, electromagnetic, strong, and weak. These four interactions vary in strength and range over wide margins, as can be seen in Table 13.2, which summarizes the characteristics.

The theory that describes the electromagnetic interaction is quantum electrodynamics, QED. This theory, which is one of the most accurate theories ever devised, was finalized in the 1950s after some 25 years of work. The electromagnetic interaction between two charged particles is understood on the basis of this theory as the exchange of a third particle. The third particle is identified as a photon, the quantum of electromagnetic radiation. As we have pointed out, the photon is a massless particle that has no electric charge. QED theory dictates that the range of interaction is inversely proportional to the mass of the particle exchanged. Thus the theory is consistent with the mass of the photon being zero and the range of the electromagnetic interaction being infinite.

The theory for the strong interaction is modeled after quantum electrodynamics. Whereas in the electromagnetic interaction the forces are between particles that have a charge, in the strong interaction the forces are between particles that have color. Particles that have no color are not subject to the force. This theory is referred to as quantum chromodynamics, QCD. If we arbitrarily take the strength of the electromagnetic interaction to be one, then the relative strength of the strong interaction is 100. The strong interaction is 100 times as strong as the electromagnetic.

As you might guess, this theory is a much more complicated theory because there are three colors involved, whereas in the electromagnetic theory there is only

† Basic colors combine to give white, which is referred to as "colorless' in some cases.

Table 13.2 Characteristics of the Four Fundamental Forces in Nature

	Gravitational	Electromagnetic	Strong	Weak
Range of action	∞	∞	10^{-13} cm	$< 10^{-15}$ cm
Relative strength	10^{-37}	1	100	10^{-11}
Particles acted upon	All particles and energy	All charged particles	Hadrons	Hadrons and leptons
Carrier (or mediator) of force	Graviton	Photon	Eight gluons	Three intermediate vector bosons

one charge. In this theory, the force is also transmitted by the exchange of a third particle. However, in this case more than one exchange particle is needed and, in fact, eight are necessary. These particles, called gluons, are massless and some carry color charge. The significance of this characteristic is that some quarks can emit a gluon carrying color and change to another type quark.

The weak interaction has a special charge associated with it and is 10^{-11} times as strong as the electromagnetic. Three particles are associated with the transmission of the weak interaction. They are the W^+, which also has a positive electric charge, the W^-, which has a negative electric charge, and the W^0, which is electrically neutral.

The remaining interaction is gravitational, the weakest of them all. This interaction is 10^{-37} times as strong as the electromagnetic and has a range of infinity. It is speculated that the carrier of this force is a particle called a graviton.

For many years physicists have been working on the unification of interactions in nature.† The first major success in this work occurred in 1967 when Sheldon Glashow and Stephen Weinberg formulated a theory that combined the weak and the electromagnetic interactions, and a year later, when Abdus Salam independently formulated the same theory. This theory combined these two interactions into an electro-weak force mediated by a family of four particles — the photon, two charged W mesons, and a new neutral meson Z^0. In 1978, a group of experimenters at SLACK conducted an elegant experiment that precisely tested and confirmed predictions of the combined theory. Since that time, many other experiments have supported the theory. For this work, Glashow, Weinberg, and Salam were awarded the 1979 Nobel Prize in physics.

† Einstein spent his last 30 years working to unify gravity and electromagnetism and published his negative results every time "to save another fool from wasting time on the same idea." Wolfgang Pauli, who was skeptical that these different forces could stem from a single source, once stated that "what God hath put asunder no man shall ever join."

Now physicists are trying to incorporate the strong interaction with the successful theory for the electro-weak interaction. Several theories have been proposed. One called the grand unified theory combines the electromagnetic, weak, and strong forces within a framework using the following particles to carry the force: the photon, the weak bosons, the eight gluons, and twelve other very heavy bosons with masses of approximately 10^{15} GeV. In this theory it would be possible for a quark to change into a lepton. One consequence of this change would be the decay of a proton with a predicted lifetime of the order of 10^{33} years. The success of this grand unified theory will have to wait the results of experiments presently being performed.

Finally, the successful unification of the four interactions, the ultimate goal of physicists, would be a tremendous triumph of the great intellectual adventure we call science. It would establish a single set of uniform rules for all phenomena everywhere in the universe on scales from inconceivably small to the inconceivably large. When it occurs it will stand with the other great steps taken in science — Copernicus's helioconcentric universe, Newton's law of gravity, Einstein's general theory of relativity and quantum mechanics.

13.9 CONSERVATION OF PARITY

We mentioned various conservation laws, for example, those of conservation of mass, conservation of momentum, conservation of strangeness, etc. Some of these laws are good for all interactions. However, some do not hold for one or more of the four types. For instance, the total strangeness value remains constant for the strong and electromagnetic interactions, but for the weak interaction this law is violated. Another limited conservation law is conservation of parity. Parity is conserved in the strong and electromagnetic interactions and was thought to be conserved in all interactions until a few years ago. It is now known that parity is not conserved in the weak interactions.

In the mathematical description of particles and their interactions, it is possible to describe particles so that if all of the coordinates (x, y, and z) are reversed and made negative, the resulting description will either be identical to or have exactly the negative value of the previous description. This property, called parity, is even or odd depending on whether the description remains positive or turns negative. The algebraic function $y = \cos x$ is symmetrical and has the same form when reversed right to left, that is $f(x) = +f(-x)$. Its parity is even $(+)$ (see Fig. 13.7). The function $y = \sin x$ is antisymmetrical. When it is reversed right to left, $f(x) = -f(-x)$ and its parity is odd $(-)$. The quantum-mechanical wave functions that represent particles or systems of particles can be described in the same way. They are either symmetric (even parity $+$) or antisymmetric (odd parity $-$). Thus the

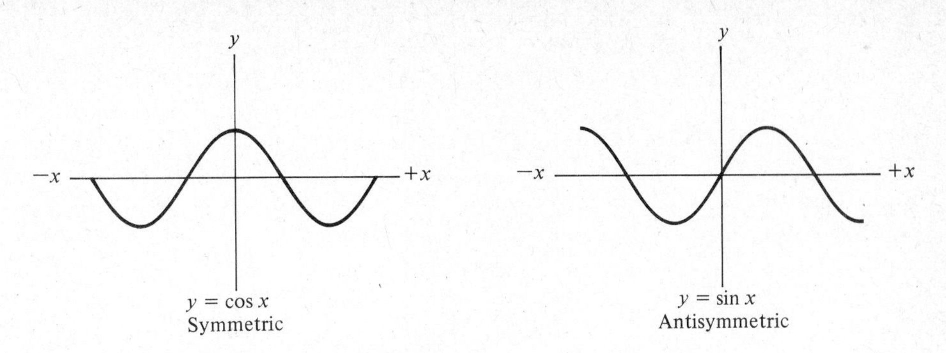

Figure 13.7 Symmetric and antisymmetric algebraic functions.

intrinsic parity of a particle described by these wave functions can be either even or odd.

Until 1956, it was believed that there was an absolute conservation of parity law. If the description of a system had even parity before an interaction, it would have even parity after. The physical implication of this was that there would be no way of telling a mirror image of a real process from a real world process. In other words, the mirror image would be as good a physical process as the real process. Figure 13.8(a) shows a spinning ball that ejects particles equally from the top and bottom. The mirror image is an equally good physical process and, in fact, is exactly equal to the real ball turned upside down. On the other hand, if spinning balls were to emit particles *only* along one axis (or more along that axis than the other) as shown in Fig. 13.8(b) the mirror image would be an unreal process.

The challenge to the universal parity conservation law arose with what was called the θ-τ puzzle. These mesons (now called kaons) were being studied in the early 1950s. There appeared to be two types (θ and τ) that were identical except for the way in which they decayed. The θ decayed into two pions while the τ decayed into three pions:

$$\theta^+ \rightarrow \pi^+ + \pi^0$$
$$\tau^+ \rightarrow \pi^+ + \pi^+ + \pi^-.$$

The pions were known to have odd parity so in the decay into two pions the combination had to be even and the parent θ meson had to have even parity. In the decay into three pions, analysis indicated that the τ had to have odd parity. This led to a perplexing dilemma: (1) the two mesons, though indistinguishable in properties, were really two different particles, the θ meson with even parity and the τ meson with odd parity, or (2) conservation of parity did not hold for these interactions.

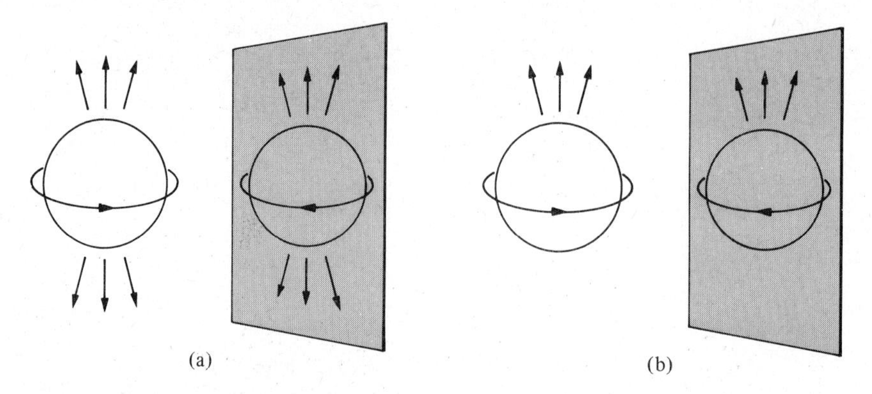

Figure 13.8 A hypothetical spinning ball ejecting particles. The shaded areas represent mirrors.

Most physicists found the second alternative unthinkable. It would have meant admitting that the left-right symmetry of nature was being violated and that nature was showing a bias for one type of "handedness." T. D. Lee and C. N. Yang in 1956 while searching for an answer to this puzzle reexamined the basis for the generally held conviction that parity is always conserved and found no evidence for this conclusion for weak interactions. They suggested that parity need not be conserved in weak interactions and that the θ and τ mesons were the same particle with two different decay modes.

Lee and Yang suggested experiments that would prove one way or the other whether parity was conserved in all weak interactions. One of these experiments was the study of the emission of β particles from spin-polarized radioactive nuclei. C. S. Wu and a group from the National Bureau of Standards used this method to first verify the nonconservation of parity. They cooled radioactive cobalt to very low temperatures using liquid helium and polarized the nuclei with a strong magnetic field. Thus, the cobalt nuclei had their spins aligned in the direction of the magnetic field. The question to be answered was whether the β particles were emitted in equal numbers along the direction of spin and opposite to that direction or whether they came off preferentially in one direction. If they were emitted uniformly along both the direction of spin and opposite to it, then we would have the analogous situation to Fig. 13.8(a). However, if more β particles were emitted along one direction than along the other, we would have the situation analogous to that shown in Fig. 13.8(b). What Wu and her associates found was that the β particles were emitted preferentially in one direction. More electrons were observed to come out in a direction opposite to the direction of the magnetic field than along the direction of the magnetic field. Thus, a spinning cobalt nucleus is more likely to emit a particle opposite to the direction of the magnetic moment. Wu's experiment provided for the first time in the history of science a method of labeling the ends of a magnetic axis in a way that is not at all conventional. The south end is the end of a cobalt-60 nucleus that is most likely to fling out an electron.

13.10 CONCLUSION

One way of summarizing what we have said in this book is to point out that we have considered events of higher and higher energy. We started with molecules, whose thermal energies are about 0.025 eV. We discussed the electronic structure of atoms, where energies have magnitudes of a few electron volts. The x-rays that interact with inner atomic electrons have energies of thousands of electron volts, and the energies of nuclear transformations fall in the MeV range. With our discussion of cosmic rays and events artificially produced with the accelerators, we introduced super-high-energy phenomena in the GeV range. The highest of these energies is 10^{11} times the lowest.

The story we have told has, in general, followed a familiar pattern: A large body of information had been gathered before some genius saw that the data led to new concepts and new generalizations. These interpretations of data have lifted the human intellect so that the past could be seen in perspective and the future could present an exciting challenge. Avogadro, Planck, Einstein, Rutherford, Bohr — all stood on the strong shoulders of experimentalists and breathed the heady air of speculation. De Broglie, on the other hand, exemplified the genius who speculated before the fact.

Just as Rutherford discovered the nucleus of the atom by probing with high-energy alpha particles, today the interior of the nucleus is being probed with particles of still higher energy. A great deal of information has been gathered, and much of it is reasonably well understood. But there are far more questions than answers and the time is ripe for some geniuses to put the house in order.

Except for mention of the mesons and the neutrino, we have discussed the nucleus as though it involved only protons and neutrons, whereas there now have been identified many nuclear particles. We spoke of short-range attractive nuclear forces, potential barriers, and binding energies, but we said almost nothing about the nature of the forces within the nucleus. There are partial theories about these nuclear forces and the many new particles certainly have a lot to do with the ways in which these forces act. But the impressive thing about nuclear forces (if indeed the term force is to be retained) is not what we know about them, but what we do not know about them.

Physicists have turned away from devising nuclear reactors, and are back at the task of unraveling the basic problems of nature. At this point in time with the many existing riddles of nature we may be likened to those who knew only Ptolemy's complex description of the solar system. What we need is a Copernicus to assimilate and interpret the data with a generalization that will not only solve the riddles but lift our sights to levels we cannot now foresee.

FOR ADDITIONAL READING

Chew, G. F., M. Gell-Mann, and A. H. Rosenfeld, "Strongly Interacting Particles," *Scientific American* **210**, 74 (February 1964).

Drell, S. D., "Electron Positron Annihilation and the New Particles," *Scientific American* **50** (June 1975).

Fowler, W. B., and N. P. Samios, "The Omega-Minus Experiment," *Scientific American* **36** (October 1964).

Freedman, Daniel Z., and Peter Van Nieuwenhuizen, "Super Gravity and the Unification of the Laws of Physics," *Scientific American* **238**, 126 (February 1978).

Gell-Mann, M., and E. P. Rosenbaum, "Elementary Particles," *Scientific American*, **197**, 72–88 (1957). A very good article.

Georgi, Howard, "A Unified Theory of Elementary Particles and Forces," *Scientific American* **244**, 48 (April 1981).

Glashow, S. L., "Quarks with Color and Flavor," *Scientific American* **38** (October 1975).

Heisenberg, W., "The Nature of Elementary Particles," *Physics Today* **29**, 32 (1976).

Hofstadter, R., and L. I. Schiff, *Nucleon Structure*. Stanford, Ca.: Stanford University Press, 1964.

Ishikawa, Kenzo, "Glueballs," *Scientific American* **247**, 142 (November 1982).

Larue, G. S., W. M. Fairbank, and A. F. Hebard, "Evidence for the Existence of Fractional Charge on Matter," *Phys. Rev. Lett.* **38**, 1011 (1977).

Lederman, Leon M., "The Upsilon Particle," *Scientific American* **239**, 72 (October 1978).

Preston, M. A., *Nuclear Structure*, Resource Letter NS-1 and Selected Reprints. New York: American Institute of Physics, 1965.

Rochester, G. D., and J. D. Wilson, *Cloud Chamber Photographs of the Cosmic Radiation*. London: Pergamon Press, 1952. An excellent collection of photographs.

T'Hofft, Gerard, "Gauge Theories of the Forces between Elementary Particles," *Scientific American* **242**, 104 (June 1980).

Weinberg, Steven, "The Decay of the Proton," *Scientific American* **244**, 64 (June 1981).

Winckler, J. R., and D. J. Hofmann, "Resource Letter CR-1 on Cosmic Rays," *Am. J. Phys.* **35** (January 1967).

REVIEW QUESTIONS

1. What are the primary cosmic ray particles?
2. What is meant by the term elementary particle?
3. What are the characteristics of the known quarks?

4. What are the characteristics of the known leptons?
5. What are the four basic interactions found in nature? What are the relative strengths and ranges of these interactions?

PROBLEMS

13.1 In Chapter 7, we discussed the uncertainty principle for energy and time. An unstable particle with a mean life τ will have an uncertainty of its total energy given from this principle. What is the uncertainty in the mass of this particle?

13.2 A π^0-meson (mass = 264 m_e) at rest decays into two gamma-ray photons. (a) What is the energy of each photon? (b) Why must each photon have the same energy? (c) What is the direction of emission of one photon with respect to the other?

13.3 Calculate the kinetic energy of the μ^--meson (mass = 207 m_e) emitted in the decay of a stationary π^--meson (mass = 270 m_e) according to the equation $\pi^- \rightarrow \mu^- + \bar{\nu}$.

13.4 A proton-antiproton pair is created from a photon. If the kinetic energy of each particle is 15 MeV, what were the energy and wavelength of the photon?

13.5 For photons striking a tungsten target, what minimum energy is needed to have a proton-antiproton pair produced? How might the photons be produced?

13.6 The cross section for neutrino interaction in lead is approximately 10^{-40} cm^2. What is the mean free path for neutrinos in lead? Compare your answer with the earth-sun distance.

13.7 A neutron star with mass of 2×10^{33} kg has a radius of 10 km. Given that the neutrino-neutron cross section is 10^{-43} cm^2, what is the mean free path of the neutrino in this star?

13.8 What is the sum of the strangeness numbers on each side of Eq. (13.1)?

13.9 Determine the change in strangeness number for the following decays: (a) $\Sigma^0 \rightarrow \Lambda + \gamma$, and (b) $\Lambda \rightarrow p^+ + \pi^-$.

13.10 (a) List the upper-case letters of the Latin alphabet that remain unchanged under the operation of parity. (b) Of the ten digits 0–9, which remain unchanged under the operation of parity?

13.11 Assume that you are in communication with scientists on a planet that is in orbit around a distant star. How would you communicate to them what you mean by left or right?

13.12 When a stationary target containing protons is bombarded by a proton beam, the reaction

$$p + p \rightarrow n + p + \pi^+$$

takes place. What is the threshold energy for the incident proton? (Mass of the π^+-meson = 273 m_e.)

13.13 Calculate the total kinetic energy of the particles created in the decay of a neutron at rest.

13.14 A π^+ meson at rest decays into a μ^+ meson and a neutrino: $\pi^+ \rightarrow \mu^+ + \nu$. The μ^+ has a mass of 207 m_e, where m_e is the electron rest mass and it has a kinetic energy of 4 MeV. The rest mass of the π^+ is 270 m_e. What is the kinetic energy of the neutrino?

13.15 A π^- meson at rest interacts with a proton and produces a neutron and a gamma ray: $\pi^- + p \rightarrow n + \gamma$. Calculate the energy of the gamma ray emitted.

APPENDIXES

APPENDIX 1: A CHRONOLOGY OF THE ATOMIC VIEW OF NATURE

c. 550 B.C. **Thales of Miletus** (Greece, c. 640–546 B.C.) recorded the attractive properties of rubbed amber and of lodestone.

c. 450 B.C. **Leucippus** (Greece) proposed an atomic concept of matter.

c. 400 B.C. **Democritus of Abdera** (Greece, c. 460–357 B.C.), pupil of Leucippus, was the most famous of the atomists in ancient times. He taught: "The only existing things are the atoms and empty space; all else is mere opinion."

c. 335 B.C. **Aristotle** (Greece, 384–322 B.C.) held that all matter was basically composed of the same continuous primordial stuff.

c. 300 B.C. **Epicurus of Samos** (Greece, c. 342–270 B.C.) founded a philosophical system based on the atomism of Democritus.

c. 300 B.C. **Zeno of Cition** (Greece, c. 336–264 B.C.) founded the Stoic school of philosophy, which held that matter, space, etc., were continuous.

c. 60 B.C. **Titus Lucretius Carus** (Rome, c. 96–55 B.C.) attempted to formulate a rational explanation of natural phenomena by extending the beliefs of Democritus and Epicurus. His poem, *De Rerum Natura*, is the most complete record of Greek atomism extant. The atomism of antiquity was primarily a system of metaphysics. The atomic view of matter in the modern sense was barely introduced in its most elementary form by the beginning of the nineteenth century.

c. 400 **Saint Augustine** (Aurelius Augustinus) (North Africa, 354–430) was the first to report that the forces exerted by rubbed amber and by lodestone are different properties.

c. 1600 **William Gilbert** (England, 1540–1603) made the first detailed study of magnetism and also showed that, in addition to amber, many other materials can be electrified.

1638 **Galileo Galilei** (Italy, 1564–1642) published *Discorsi e Dimostrazioni Matematiche intorno a due nuove Scienze attenti alla Mecanica e Movimenti locali* (*Discourses and Mathematical Demonstrations Concerning Two New Sciences Pertaining to Mechanics and Local Motions*, usually contracted to *Two New Sciences*). This account of Galileo's contributions to science establishes him as the founder of dynamics. He was the first to make extensive use of the experimental method to study natural phenomena. From this time on, induction from experiment replaced the teleology of the scholastics as a guiding principle in the organization of the natural sciences.

1650–1700 **Robert Boyle** (England, 1627–1691), **Robert Hooke** (England, 1635–1703), and **Isaac Newton** (England, 1642–1727) gave qualitative explanations of Boyle's law by assuming a kinetic theory of gases.

1675 **Jean Picard** (France, 1620–1682) observed the luminous glow in the Torricellian vacuum of a barometer produced by motion of the mercury when the instrument was carried from place to place.

1675 **Isaac Newton** developed a corpuscular theory of light.

1676 **Ole Christensen Roemer** (Denmark, 1644–1710) was the first to show that the velocity of light is finite. His conclusion was based on the variations of the time intervals between consecutive eclipses of one of the moons of Jupiter during the course of the revolution of the earth around the sun.

1678 **Christian Huygens** (Netherlands, 1629–1695) developed a wave theory of light in which light was regarded as composed of longitudinal "pulses" consisting of compressions and rarefactions, similar to sound, in an extremely thin, all-pervading medium that he called the *aether*. The concept that light is a periodic wave motion was introduced in about 1750 by **Leonard Euler** (Switzerland, Germany, Russia, 1707–1783).

Not only did Huygens correctly account for the refraction of light by transparent bodies by means of spherical emanations (wavelets), but also, by using both spherical and spheroidal wavelets, he became the first one to explain double refraction, a phenomenon that had been discovered in 1669 by **Erasmus Bartholinus** (Denmark, 1625–1692).

1687 **Isaac Newton** published *Philosophiae Naturalis Principia Mathematica* (*Mathematical Principles of Natural Philosophy*), which contains the fundamental laws of classical dynamics and the law of gravitation. The synthesis involved in obtaining these laws is one of the greatest achievements of the human mind.

1705 **Francis Hauksbee** (England, d. 1713) made a "powerful" electrostatic generator and discovered the conditions for producing luminous electric discharges in gases.

1728 **James Bradley** (England, 1693–1762) explained the aberration of light from stars by taking the vector sum of the orbital velocity of the earth v and the free-space velocity of light c, and showed that the angle of aberration was a function of the ratio of these velocities, v/c. On this basis, he also showed that the revolution of the earth around the sun correctly accounted for the observed cyclic change in the aberration of starlight. (The Anti-Copernicans, still numerous in the first half of the eighteenth century, were unable to refute this explanation of the change.) Bradley's work is the first of many instances that *seemed* to show that the value of the velocity of light depends on the motion of the observer.

1731 **Stephen Gray** (England, c. 1666–1736) discovered the conduction of electricity.

1734 **Charles François de Cisternay Dufay** (France, 1698–1739) showed that there are two kinds of electrification, resinous and vitreous, and then proposed a two-fluid theory of electric discharge. He also found that the air in the vicinity of a hot body is conducting.

1738 **Daniel Bernoulli** (Switzerland, 1700–1782) was the first to devise a quantitative kinetic theory of gases.

1745 **Ewald Jurgen Von Kleist** (Germany, d. 1748) and **Peiter Van Musschenbroek** (Netherlands, 1692–1761) independently made the first capacitors, called Leyden jars.

1752 **Benjamin Franklin** (USA, 1706–1790) experimentally verified the electrical nature of lightning and introduced the one-fluid theory of flow of electricity — from surplus or positive to deficiency or negative. His theory contained the first clear statement of the law of conservation of electric charge.

1753 **John Canton** (England, 1718–1772) discovered the facts of electrostatic induction.

1766 **Henry Cavendish** (England, 1731–1810) discovered hydrogen. Within the next 20 years he found the inverse square law of force action between electric charges and other important laws of electricity but, because of excessive shyness, he withheld announcement of his experiments. The great extent of his work was not known until James Clerk Maxwell published Cavendish's papers in 1879.

1785 **Charles Augustin Coulomb** (France, 1736–1806) determined the law of force action between electric charges.

1789 **Antoine Laurent Lavoisier** (France, 1734–1794) published a book containing a well-founded concept of chemical elements and the verification of the law of conservation of matter in chemical reactions.

1791 **Bryan Higgins** (Ireland, 1737–1820) and **William Higgins** (Ireland, c. 1769–1825) reported the first of a series of experiments leading to the laws of chemical combination.

1799 **Joseph Louis Proust** (France, Spain, 1754–1826) established the law of definite proportions for chemical compounds.

1800 **Alessandro Guiseppe Antonio Anastasio Volta** (Italy, 1745–1827) made the first voltaic pile (battery), based on his discovery of the fundamental conditions necessary to produce the "animal electricity" that had first been observed in 1780 by **Aloisio Galvani** (Italy, 1737–1798).

1801 **Thomas Young** (England, 1773–1829) showed that his interference experiments verified the wave theory of light.

1803 **John Dalton** (England, 1766–1844) published the first of a series of papers introducing atomic weights, establishing the law of multiple proportions, and founding the atomic theory of matter.

1808 **Joseph Louis Gay-Lussac** (France, 1778–1850) discovered the law of combining volumes of gases.

1810–1875 **Etienne Louis Malus** (France, 1775–1812), **Dominique François Jean Arago** (France, 1786–1853), **Augustin Jean Fresnel** (France, 1788–1827), **Jean Bernard Leon Foucault** (France, 1819–1868), **Hippolyte Louis Fizeau** (France, 1819–1896), and **Marie Alfred Cornu** (France, 1841–1902) established conclusively, through many experiments, especially in physical optics, that light is a transverse wave. Several of these men made precise measurements of the velocity of light in various media. In 1818, Arago found that the refraction of a prism for starlight was the same for light incident in the direction of the earth's orbital velocity v as for that coming in the opposite direction. This unexpected null result was explained that same year by Fresnel's ether-drag theory, which assumed partial ether entrainment in transparent media by an amount depending on the *first* power of v. This theory appeared fully verified by the measurements of the speed of light in moving water by Fizeau in 1851 and, in 1871, by the observations of the aberration of starlight with a water-filled telescope by **George Biddell Airy** (England, 1801–1892).

1811 **Lorenzo Romano Amadeo Avogadro** (Italy, 1776–1856) introduced Avogadro's hypothesis and differentiated atoms and molecules.

1813 **Jons Jacob Berzelius** (Sweden, 1779–1848) introduced the present symbols for the chemical elements.

1815 **William Prout** (England, 1785–1850) proposed that all elements are composed of an integral number of hydrogen atoms.

1815–1820 **Joseph Fraunhofer** (Germany, 1787–1826) noted the spectral lines of several elements, obtained the first grating spectra, and observed the Fraunhofer (absorption) lines in solar spectra.

1819 **Pierre Louis Dulong** (France, 1785–1838) and **Alexis Therèse Petit** (France, 1791–1821) found the law of constancy of molar specific heat capacities of elements.

1820 **Hans Christian Oersted** (Denmark, 1777–1851) discovered that an electric current produces a magnetic field. This initiated the study of electromagnetism.

1821 **Thomas Johann Seebeck** (Russia, Germany, 1770–1831) discovered thermoelectricity.

1823 **André Marie Ampère** (France, 1775–1836) published his mathematical theory of electromagnetism and the laws of magnetic field produced by currents. Some of these laws were also discovered independently by **Jean Baptiste Biot** (France, 1774–1862) and **Felix Savart** (France, 1791–1841).

1826 **Georg Simon Ohm** (Germany, 1787–1854) discovered Ohm's law.

1827 **Robert Brown** (England, 1773–1858) discovered Brownian movement.

1831 **Michael Faraday** (England, 1791–1867) and **Joseph Henry** (USA, 1797–1878) independently discovered electromagnetic induction.

1833 **Michael Faraday** discovered the laws of electrolysis and introduced the terms "anode" and "cathode."

1835 **Joseph Henry** discovered self-induction and, in 1842, oscillatory electric discharge.

1842 **Johann Christian Doppler** (Austria, 1803–1853) deduced a relation that showed that the observed frequency of waves depends on the relative motion of the source and the observer.

1842 **Julius Robert Mayer** (Germany, 1814–1878) calculated the mechanical equivalent of heat theoretically from the specific heats of gases and vaguely proposed a law of conservation of energy based on "*ex nihilo, nihil fit*." His work was not published for several years.

1843 **James Prescott Joule** (England, 1814–1889) published the first of a series of reliable experimental results that showed the constancy of the relation between mechanical energy and heat — a basic step toward the law of conservation of energy.

1847 **Hermann Ludwig Ferdinand Von Helmoltz** (Germany, 1821–1894) proposed the law of conservation of "force" (energy).

1848 **William Thomson** (Lord Kelvin, 1st Baron) (Ireland, Scotland, 1824–1907) introduced absolute temperature.

1850 **Rudolph Julius Emanuel Clausius** (Germany, 1822–1888) announced the second law of thermodynamics. Lord Kelvin independently found the same law in 1852.

1850–1900 **August Karl Kroenig** (Germany, 1822–1879), **Rudolph Julius Emanuel Clausius, James Clerk Maxwell** (Scotland, England, 1831–1879), **Ludwig Boltzmann** (Austria, 1844–1906), and **Josiah Willard Gibbs** (USA, 1839–1903) developed the kinetic theory of gases and founded statistical mechanics. Maxwell derived his speed distribution law in 1860, Clausius introduced the concept of entropy in 1865, and Boltzmann related entropy to thermodynamic probability in 1877.

1858 **Stanislao Cannizzaro** (Italy, 1826–1910) resolved the conflicting values of atomic weights by clarifying the terms "atomic," "molecular," and "equivalent" weights.

1859 **Gustav Robert Kirchhoff** (Germany, 1824–1887) showed that the ratio of the emittance to the absorptance for a given wavelength of radiation is the same for all surfaces at the same temperature, and introduced the concept of cavity (Hohlraum) or blackbody radiation.

1859 **Heinrich Geissler** (Germany, 1814–1879) and **Julius Pluecker** (Germany, 1801–1868) discovered the "rays" (now called cathode rays) from the negative electrode in gaseous discharge tubes.

1863 **James Alexander Reina Newlands** (England, 1837–1898) stated the law of octaves, a limited and elementary form of the periodic table of the elements.

1864 **James Clerk Maxwell** wrote *A Dynamical Theory of the Electromagnetic Field*, a paper synthesizing electricity, magnetism, and light. This was probably the greatest work since Newton's *Principia*.

1865 **Joseph Loschmidt** (Germany, 1821–1895) used the equations of the kinetic theory of gases to make the first determination of Avogadro's number and of molecular diameters.

1869 **Dimitri Ivanovich Mendeleev** (Russia, 1834–1907) and **Julius Lothar Meyer** (Germany, 1830–1895) independently introduced the periodic table of the elements, a concise summary of years of experimental and theoretical chemistry. The table is both mnemonic and heuristic.

1869 **Johann Wilhelm Hittorf** (Germany, 1824–1914) observed the deflection of rays from the cathode in a discharge tube, by means of a magnetic field.

1871 **Cromwell Fleetwood Varley** (England, 1828–1883) found that the rays from the cathode are negatively charged.

1876 **Eugen Goldstein** (Germany, 1850–1930) introduced the name "cathode rays" and began experiments leading eventually to the discovery of the positive counterpart, Kanalstrahlen (channel or canal rays). In 1886, he suggested that the aurora is due to cathode rays from the sun.

1877 **William Ramsay** (England, 1852–1916) and, independently, **Joseph Delsaulx** (France, 1828–1891) and **Ignace J. J. Carbonelle** (France, 1829–1889) advanced the first rather complete qualitative explanation of Brownian movement by attributing it to molecular impact. Some years later Ramsay discovered several of the noble gases, and made important contributions to the study of radioactivity. He was awarded the Nobel prize for chemistry in 1904.

1879 **Edwin Herbert Hall** (USA, 1855–1938) discovered the existence of a potential difference between the opposite edges of a metal strip carrying a longitudinal electric current, when the plane of the strip is set normal to a magnetic field. This is called the Hall effect.

1879 **William Crookes** (England, 1832–1919) began a long series of brilliant experiments on the discharge of electricity through gases.

1879 **Josef Stefan** (Austria, 1835–1893) announced Stefan's law, which gives the total energy radiated by a blackbody. This was the first successful attempt to connect absolute temperature and radiation.

1881 **Julius Elster** (Germany, 1854–1920) and **Hans Geitel** (1855–1923) started a long, systematic investigation of electrical effects produced by incandescent solids.

1883 **Thomas Alva Edison** (USA, 1847–1931) discovered the Edison effect, the emission of negative electricity from incandescent filaments in a vacuum.

1884 **Johann Jakob Balmer** (Switzerland, 1825–1898) found an empirical wavelength relation for a spectral series of hydrogen. This was the first series equation found for any spectrum.

1887 **Svante August Arrhenius** (Sweden, 1859–1927) conclusively established the ion dissociation theory of electrolytes, which grew from suggestions made by Clausius in 1857. Arrhenius was awarded the Nobel prize for chemistry in 1903.

1887 **Albert Abraham Michelson** (Germany, USA, 1852–1931) and **Edward Williams Morley** (USA, 1838–1923) performed the first precision experiment that showed that the earth has no ether drift. In a letter to *Nature* in 1879 Maxwell pointed out that evidence of ether drift had to be sought in second-order effects — those depending on v^2/c^2. These are involved in interference methods. The first trial by Michelson in 1881 gave inconclusive results. Michelson was awarded the Nobel prize for physics in 1907.

1887 **Heinrich Rudolph Hertz** (Germany, 1857–1894) discovered the photoelectric effect while verifying the existence of the electromagnetic waves predicted by Maxwell.

1888 **Wilhelm Hallwachs** (Germany, 1859–1922) showed that only negative charges are emitted in the photoelectric effect.

1890 **Johannes Robert Rydberg** (Sweden, 1854–1919) found an empirical wavelength relation for complex series of spectral lines.

1891 **Johnstone Stoney** (England, 1826–1911) introduced the name "electron" for an elementary unit of negative charge in electrolysis.

1892 **George Francis Fitzgerald** (Ireland, 1851–1901) and **Hendrik Antoon Lorentz** (Netherlands, 1853–1929) independently made the *ad hoc* assumption of contraction of length to account for the null result of the Michelson-Morley experiment. As Lorentz successively refined his electric theory of matter to conform with the results of new experiments, he obtained the space and time transformations later derived by Einstein. For later work Lorentz was awarded the Nobel prize for physics jointly with P. Zeeman in 1902.

1893 **Wilhelm Wien** (Germany, 1864–1928) derived his blackbody radiation displacement law. His blackbody radiation law was announced in 1896. He was awarded the Nobel prize for physics in 1911.

1893 **Philipp Eduard Anton Von Lenard** (Hungary, Germany, 1862–1947) investigated cathode rays by passing them through a Lenard window (thin-window) tube into air. For this and later work he was awarded the Nobel prize for physics in 1905.

1895 **Jean Baptiste Perrin** (France, 1870–1942) demonstrated conclusively that cathode rays are negatively charged. For this and later work he was awarded the Nobel prize for physics in 1926.

1895 **Wilhelm Conrad Roentgen** (Germany, 1845–1923) discovered x-rays. He was awarded the Nobel prize for physics in 1901.

1896 **Antoine Henri Becquerel** (France, 1852–1908) discovered the radioactivity of uranium. He was awarded the Nobel prize for physics jointly with the Curies in 1903.

1896 **Pieter Zeeman** (Netherlands, 1865–1943) observed the splitting of spectral lines radiated by excited atoms in an intense magnetic field. The early theory of this effect was derived by **H. A. Lorentz** (Netherlands). They were jointly awarded the Nobel prize for physics in 1902.

1896 **Oliver Lodge** (England, 1851–1940) reported that, contrary to expectations, there was no detectable ether drag on light passing between two large closely spaced disks of steel rotating at enormous speeds even when the disks were strongly magnetized or electrified.

1897 **Joseph John Thomson** (England, 1856–1940) determined q/m for cathode rays. He was awarded the Nobel prize for physics in 1906.

1897 **Ernest Rutherford** (Lord Rutherford of Nelson, 1st Baron) (New Zealand, Canada, England, 1871–1937) showed that the radiation from uranium was complex, consisting of "soft" (alpha) and "hard" (beta) rays. He was awarded the Nobel prize for chemistry in 1908.

1898 **Pierre Curie** (France, 1859–1906) and **Marie Sklodowska Curie** (Poland, France, 1867–1934) isolated radium and polonium. They were awarded the Nobel prize for physics jointly with H. Becquerel in 1903. Marie Curie was also awarded the Nobel prize for chemistry in 1911.

1899 **Henri Becquerel**, **Stefan Meyer** (Austria, 1872–1949) and **Egon Von Schweidler** (Austria, 1873–1948), and **Frederick Otto Giesel** (Germany, 1852–1927) independently observed the magnetic deflection of alpha and beta rays.

1899 **Julius Elster** (Germany) and **Hans Geitel** (Germany) determined the law of radioactive decay experimentally.

1899 **Philipp Lenard** showed that photoelectric emission is due to electrons.

1899 **J. J. Thomson** showed that the Edison effect is due to electrons.

1899 **Otto Lummer** (Russia, Germany, 1860–1925) and **Ernst Georg Pringsheim** (Germany, 1881–1917), and also **Ferdinand Kurlbaum** (Germany, 1857–1927) and **Heinrich Rubens** (Germany, 1865–1922) made precise measurements of the intensity-wavelength distribution of blackbody radiation.

1900 **John William Strutt** (Lord Rayleigh, 3rd Baron), (England, 1842–1919) announced a blackbody radiation law. The derivation of this law was re-examined in collaboration with **James Hopwood Jeans** (England, 1877–1946) and, after publication in 1905, became known as the Rayleigh-Jeans law. Rayleigh was awarded the Nobel prize for physics in 1904.

1900 **Max Karl Ernst Ludwig Planck** (Germany, 1858–1947) introduced the quantum theory of radiation — a revolutionary concept. He was awarded the Nobel prize for physics in 1918.

1900 **Henri Becquerel** showed that beta rays are identical with cathode-ray corpuscles.

1900 **Paul Villard** (France, 1860–1934) discovered gamma rays.

1902 **Philipp Lenard** discovered photoelectric threshold frequency and also that the kinetic energy of photoelectrons is independent of the intensity of the incident light.

1903 **Frederic Thomas Trouton** (Ireland, England, 1863–1922) and **H. R. Noble** (England, 1833–1896) were unable to observe any orienting torque on a suspended, charged capacitor as predicted on the basis of an ether drift (a second-order effect).

1903 **Ernest Rutherford** and **Frederic Soddy** (England, 1877–1956) showed that every radioactive process is a transmutation of elements. Soddy was awarded the Nobel prize for chemistry in 1921.

1903 **William Crookes** and, independently, **Julius Elster** and **Hans Geitel** (Germany) found that the luminescence produced when alpha particles strike zinc sulfide consists of discrete flashes of light or scintillations. This led to a method of counting individual alpha particles.

1904 **William Ramsay** and **Frederic Soddy** discovered the remarkable occurrence of helium in all radium compounds.

1904 **Dewitt Bristol Brace** (USA, 1859–1905) found no trace of the double refraction predicted for an isotropic transparent body when it is rotated from parallel to the ether drift to normal to it (a second-order effect). This type of experiment had been suggested by the elder Lord Rayleigh.

1904 **John Ambrose Fleming** (England, 1849–1945) applied the Edison effect to make the first thermionic valve ("radio" tube).

1904 **Maryan Von Smoluchowski** (Austria, 1872–1919) proposed a statistical theory of Brownian movement.

1905 **Albert Einstein** (Germany, Switzerland, USA, 1879–1955) completed the statistical theory of Brownian movement, introduced the quantum explanation of the photoelectric effect, and announced the *special* theory of relativity. He was awarded the Nobel prize for physics in 1921. The citation stated that the award was "for his contributions to mathematical physics, and especially for his discovery of the law of the photoelectric effect."

1905 **Egon Von Schweidler** (Austria) derived the law of radioactive decay from probability theory — not obtainable from causality.

1906 **Owen Willans Richardson** (England, 1879–1959) began a long series of important investigations on the emission of electricity from hot bodies (thermionic emission). He was awarded the Nobel prize for physics in 1928.

1906 **Lee De Forest** (USA, 1873–1961) made the first audion (triode) by introducing a grid into a Fleming valve.

1907–1912 **J. J. Thomson** devised methods of positive-ray analysis. This was the beginning of mass spectroscopy.

1908 **Walter Ritz** (Switzerland, 1878–1909) announced the combination principle for computing the frequencies of spectral lines.

1908 **Louis Carl Heinrich Friedrich Paschen** (Germany, 1865–1947) experimentally verified the existence of a spectral series of hydrogen in the near infrared predicted by the Rydberg-Ritz relation.

1908 **Charles Glover Barkla** (England, 1877–1944) discovered from absorption experiments that the secondary x-rays of various elements are composed of groups of characteristic x-rays that he called the K, L, and M radiations, and demonstrated the polarization of x-rays. He was awarded the Nobel prize for physics in 1917.

1908 **Jean Perrin** (France, 1870–1942) verified experimentally the several equations for Brownian movement, obtained good values of Avogadro's number, and showed that equipartition of energy held for small particles suspended in a stationary liquid. He received the Nobel prize for physics in 1926.

1908 **Hermann Minkowski** (Lithuania, Germany, 1864–1909) developed a geometrical interpretation of the special theory of relativity in which time and the three space coordinates all had the same validity in a four-dimensional continuum.

1908–1910 **Alfred Heinrich Bucherer** (Germany, 1863–1927), **E. Hupka** (Germany), and **Charles Eugene Guye** (France, 1866–1942) and **Simon Ratnowsky** (Russia, Switzerland, 1884–1945) independently made precision measurements of the mass of an electron as a function of its velocity. The results verified the Lorentz-Einstein mass variation relation.

1908 **Heike Kamerlingh-Onnes** (Netherlands, 1853–1926) first liquefied helium. He received the Nobel prize for physics in 1913.

1909 **Guglielmo Marconi** (Italy, 1874–1937) and **Carl Ferdinand Braun** (Germany, 1850–1918) were jointly awarded the Nobel prize for physics — the former for combining the basic knowledge about Hertzian waves to produce wireless telegraphy, and the latter for the study, production, and use of electrical oscillators. Braun also developed the Braun tube, called the "cathode-ray" tube in the USA.

1909 **Ernest Rutherford** and **Thomas Royds** (England, 1884–1955) showed that alpha particles are doubly ionized helium atoms.

1909–1910 **T. Wulf** (France) observed the rate of leak of charge from a highly insulated electroscope placed at the top of the Eiffel Tower, and **Albert Gockel** (Switzerland, 1860–1927) studied the same effect in balloon ascents up to 4500 meters. Both found the leakage rate greater than at the surface of the earth. Their results were unexpected because the effect at ground level had been ascribed to local radioactivity of the soil.

1909–1911 **Robert Andrews Millikan** (USA, 1868–1953) established the law of multiple proportions for electric charges and made the first precise determination of the electronic charge. He was awarded the Nobel prize for physics in 1923.

1910–1912 **Victor Franz Hess** (Austria, USA, 1883–1964) and **Werner Kolhoerster** (Austria, 1887–1945) discovered cosmic rays. Hess was awarded the Nobel prize for physics jointly with C. D. Anderson in 1936.

1911 **Peter Joseph Wilhelm Debye** (Netherlands, Switzerland, Germany, USA, 1884–1966) used the quantum theory to obtain a rather complete theory of specific heats, and later applied the quantum concept to many problems in physical chemistry. He was awarded the Nobel prize for chemistry in 1936.

1911–1913 **Ernest Rutherford**, **Hans Geiger** (Germany, 1882–1945), and **Ernest Marsden** (England, b. 1889) showed that a nuclear model of the atom was required to explain their experiments on alpha-particle scattering by thin metal foils.

1911 **Charles Thomson Rees Wilson** (Scotland, England, 1869–1959) made the first expansion cloud chamber. This was the most important device in nuclear physics. He was awarded the Nobel prize for physics jointly with A. H. Compton in 1927.

1911 **Heike Kamerlingh-Onnes** discovered superconductivity.

1912 **Max Felix Theodor Von Laue** (Germany, 1879–1960) with **Walter Friedrich** (Germany, b. 1883) and **Paul C. M. Knipping** (Germany, 1883–1935) established the wave nature of x-rays by crystal diffraction. Laue was awarded the Nobel prize for physics in 1914.

1912 **Hans Geiger** (Germany, 1882–1945) and **John Mitchell Nuttall** (England, 1890–1958) obtained an empirical law relating the energy of an emitted alpha particle to the disintegration constant of the parent nucleus.

1913 **Hans Geiger** published a detailed description of the point-discharge counter tube that was developed from a simpler form first made in 1908. This instrument was greatly improved in 1928.

1913 **Frederick Soddy** and **Kasimir Fajans** (Poland, Germany, 1887–1975) announced the laws of displacement in the periodic table for elements undergoing radioactive decay. Soddy introduced the term "isotopes."

1913 **George Charles De Hevesy** (Hungary, Germany, Sweden, 1885–1966) and **Fritz Adolf Paneth** (Austria, 1887–1959) used radium-D, an isotope of lead, to study the solubility and the chemistry of lead compounds. This was the first use of an isotope as a tracer element. Hevesy was awarded the Nobel prize for chemistry in 1943.

1913 **Niels Henrik David Bohr** (Denmark, 1885–1962) developed the first successful theory of atomic structure. He was awarded the Nobel prize for physics in 1922.

1913 **Johannes Stark** (Germany, 1874–1957) observed the splitting of spectral lines radiated by excited atoms in an intense electric field. He was awarded the Nobel prize for physics in 1919.

1913 **James Franck** (Germany, USA, 1882–1964) and **Gustav Hertz** (Germany, 1887–1975) supported the Bohr atomic theory with their measurements of ionization and resonance potentials. They were awarded the Nobel prize for physics in 1925.

1913 **William Henry Bragg** (England, 1862–1942) and son, **William Lawrence Bragg** (Australia, England, 1890–1971), studied x-ray "reflection" from crystals and devised an x-ray spectrometer. They were awarded the Nobel prize for physics in 1915.

1914 **Henry Gwyn Jeffrey Moseley** (England, 1884–1915) made x-ray spectrograms of the elements and established the identity of the ordinal number of an element in the periodic table with its nuclear charge (atomic number).

1914 **Karl Manne Georg Siegbahn** (Sweden, 1886–1979) began a long series of pioneer researches in the theory and application of precision x-ray spectroscopy. He was awarded the Nobel prize for physics in 1924.

1915 **Arnold Johannes Wilhelm Sommerfeld** (Germany, 1868–1951) improved the Bohr atomic model by introducing elliptical orbits and relativistic effects.

1915 **William Duane** (USA, 1872–1935) and **Franklin Livingston Hunt** (USA, b. 1883) showed that the short-wavelength limit of emitted x-radiation is determined by the quantum theory.

1915 **Albert Einstein** announced the *general* theory of relativity. It considers the observations of phenomena on accelerated reference frames.

1916 **R. A. Millikan** experimentally verified Einstein's photoelectric equation.

1916 **P. J. W. Debye** (Netherlands, Switzerland, USA), **Paul Scherrer** (Switzerland, 1890–1969), and, independently, **Albert Wallace Hull** (USA, 1880–1966) obtained the first x-ray powder diffraction patterns.

1916 **Theodore Lyman** (USA, 1874–1954) found the Lyman series lines predicted by Bohr's theory of the hydrogen atom. Lyman had observed at least one of these lines as early as 1906.

1919 **Ernest Rutherford** produced hydrogen and oxygen by alpha-particle bombardment of nitrogen, the first transmutation of an element in the laboratory.

1919 **Francis William Aston** (England, 1877–1945) made the first high-precision determinations of isotopic masses. He was awarded the Nobel prize for chemistry in 1922.

1919 The observations made during a total solar eclipse in this year by an expedition from the Royal Astronomical Society and the Royal Society of London confirmed the deviation of starlight in the gravitational field of the sun as predicted by the general theory of relativity. Other support for this theory came later from the agreement between the calculated and observed values of the precession of the perihelion of Mercury.

1921 **Otto Stern** (Germany, USA, 1888–1969) and **Walter Gerlach** (Germany, b. 1889) verified the space quantization of silver atoms in a magnetic field and measured their magnetic moment. Stern was awarded the Nobel prize for physics in 1943.

1923 **Arthur Holly Compton** (USA, 1892–1962) discovered the Compton effect, which showed that a photon has momentum. He was awarded the Nobel prize for physics jointly with C. T. R. Wilson in 1927.

1924 **Edward Victor Appleton** (England, 1892–1965) began a series of experiments that established the existence and properties of ionized layers in the high atmosphere. Such layers had been postulated in 1902 by **Arthur Edwin Kennelly** (India, USA, 1861–1939) and, independently, by **Oliver Heaviside** (England, 1850–1925) to account for long-distance wireless telegraphy. Appleton was awarded the Nobel prize for physics in 1947.

1924 **Satyendranath Bose** (India, 1894–1974) and **A. Einstein** independently developed the statistics "obeyed" by bosons, a collective name for photons, nuclei of even mass number, and certain other particles.

1924 **Louis Victor, Duc de Broglie** (France, b. 1892) introduced the concept of de Broglie waves, the beginning of the wave theory of matter. He was awarded the Nobel prize for physics in 1929.

1925 **Walter M. Elsasser** (Germany, USA, b. 1904) predicted from de Broglie's theory that electrons could be diffracted by crystals.

1925 **Charles Drummond Ellis** (England, b. 1895) and **W. A. Wooster** (England, b. 1903) established that in a number of elements the emission of either an alpha or a beta particle precedes the radiation of gamma rays, and thus the latter should be associated with the daughter product, not with the parent.

1925 **Pierre Victor Auger** (France, b. 1899) discovered a type of energy transition in which an atom goes from a higher to a lower state by ejecting one of its own electrons, without the emission of electromagnetic radiation.

1925 **George Eugene Uhlenbeck** (Java, Netherlands, USA, b. 1900) and **Samuel Abraham Goudsmit** (Netherlands, USA, 1902–1979) introduced spin and magnetic moment of the electron into atomic theory.

1925 **Wolfgang Pauli** (Austria, Switzerland, 1900–1958) announced the exclusion principle. He was awarded the Nobel prize for physics in 1945.

1925 **Patrick Maynard Stuart Blackett** (England, 1897–1974) obtained the first cloud-chamber tracks of the induced transmutation of nitrogen and of other elements, and later made many cosmic-ray studies. He was awarded the Nobel prize for physics in 1948.

1925 **Max Born** (Germany, 1882–1970), **Werner Karl Heisenberg** (Germany, 1901–1976), and **Pascual Jordan** (Germany, b. 1902) developed quantum mechanics. Later, Born originated the statistical interpretation of wave mechanics, and he was awarded the Nobel prize for physics jointly with W. Bothe in 1954.

1926 **Erwin Schroedinger** (Austria, Ireland, 1887–1961) proposed the wave-mechanical theory of the hydrogen atom. He was awarded the Nobel prize for physics jointly with P. A. M. Dirac in 1933.

1926 **Enrico Fermi** (Italy, USA, 1901–1954) and **Paul Adrien Maurice Dirac** (England, b. 1902) independently developed the statistics "obeyed" by fermions, a collective name for nuclei of odd mass number, some particles, and electrons, particularly the electron gas in a conductor. Each was awarded the Nobel prize for work listed later in this chronology.

1926 **Eugene Paul Wigner** (Hungary, USA, b. 1902) published the first of a long series of important papers on the application of group theory in quantum mechanics. He was awarded the Nobel prize for physics jointly with Maria Mayer and J. H. D. Jensen in 1963.

1927 **Werner Heisenberg** (Germany, 1901–1976) announced the "Unbestimmtheit Prinzip" (indeterminacy or uncertainty principle). He was awarded the Nobel prize for physics in 1932.

1927 **Clinton Joseph Davisson** (USA, 1881–1958) and **Lester Halbert Germer** (USA, 1896–1971) obtained electron diffraction from single crystals, and **George Paget Thomson** (England, 1892–1975) obtained powder diffraction patterns using electrons. Their work verified the existence of de Broglie waves. Davisson and Thomson were awarded the Nobel prize for physics in 1937.

1928 **Edward Uhler Condon** (USA, b. 1902) and **Ronald Wilfrid Gurney** (England, USA, 1898–1953) and, independently, **George Gamow** (Russia, USA, b. 1904) solved the nuclear problem of alpha-particle emission by means of wave mechanics and derived the Geiger-Nuttall law.

1928 **P. A. M. Dirac** (England, b. 1902) developed relativistic quantum mechanics and predicted the existence of the positron. He was awarded the Nobel prize for physics jointly with E. Schroedinger in 1933.

1928 **Chandrasekhara Venkata Raman** (India, 1888–1970) discovered the Raman effect. This is the presence, in light scattered from molecules, of frequencies differing from that of the incident light by amounts characteristic of the scattering substance and independent of the incident frequency. He was awarded the Nobel prize for physics in 1930.

1928 **Dmitri Vladimirovich Skobeltsyn** (Russia, b. 1892) obtained the first cloud-chamber photographs of cosmic rays. These showed that the rays either were, or produced, many charged, high-energy particles.

1928 **Hans Geiger** and **W. Mueller** developed the Geiger point counter (1913) into a greatly improved form, called the Geiger-Mueller counter.

1928 **Walther Wilhelm Georg Franz Bothe** (Germany, 1891–1957) and **W. Kolhoerster** (Austria) applied G-M tubes to make coincidence counters and other ingenious devices for cosmic-ray study. Bothe was awarded the Nobel prize for physics jointly with M. Born in 1954.

1929 **Otto Stern** obtained crystal diffraction of a beam of helium atoms.

1930 **W. Bothe** and **H. Becker** observed a puzzling penetrating "radiation" from beryllium bombarded with alpha particles.

1930 **Jacob Clay** (Netherlands, b. 1882) discovered that cosmic-ray intensity decreased in going toward the geomagnetic equator. This latitude effect was investigated exhaustively by A. H. Compton, R. A. Millikan, and others.

1930–1934 **Fredrik Carl Muelertz Stoermer** (Norway, 1874–1957) applied his theory of the motion of charged particles in the magnetic field of the earth, originally developed to account for the aurora borealis, to cosmic rays. This theory of the cause of geomagnetic effects was greatly expanded in 1933 by **Georges Lemaitre** (Belgium, 1894–1966) and **Manuel Sandoval Vallarta** (Mexico, 1899–1981). Further theoretical work was done by **William Francis Gray Swann** (England, USA, 1884–1962) and others.

1930–1933 **Ira Forry Zartman** (USA, b. 1899) and **C. C. Ko** (China, USA) experimentally verified the Maxwell distribution law of molecular speeds.

1931 **Thomas Hope Johnson** (USA, b. 1899) obtained crystal diffraction of a beam of hydrogen atoms.

1931 **Robert Jemison Van De Graaff** (USA, 1901–1967) constructed the first reliable, high-voltage, electrostatic generator for nuclear research.

1931 **Wolfgang Pauli** (Austria, Switzerland, 1900–1958) proposed a hypothesis of beta-decay processes postulating that a "new," small, neutral particle was emitted simultaneously with the electron. This particle was later named the neutrino by **Enrico Fermi**. Pauli received the Nobel prize for physics in 1945.

1931 **Harold Clayton Urey** (USA, 1893–1981), **Ferdinand Graft Brickwedde** (USA, b. 1903), and **George Moseley Murphy** (USA, b. 1903) discovered deuterium and made the first heavy water. Urey was awarded the Nobel prize for chemistry in 1934.

1932 **Roy James Kennedy** (USA, b. 1897) and **Edward Moulton Thorndike** (USA, b. 1905) sought to detect the ether drift with a very stable and refined form of interferometer having arms of unequal length. Both the length and the time transformations of the special theory of relativity had to be used to account for the null result.

1932 **Ernest Orlando Lawrence** (USA, 1901–1958) and **Milton Stanley Livingston** (USA, b. 1905) made the first cyclotron. Lawrence was awarded the Nobel prize for physics in 1939.

1932 **John Douglas Cockcroft** (England, 1897–1967) and **Ernest Thomas Sinton Walton** (Ireland, b. 1903) accomplished the transmutation of lithium by bombarding it with high-energy protons, and so obtained the first direct verification of Einstein's law of mass-energy equivalence. This was also the first time a high-voltage accelerator was used successfully to produce a nuclear reaction. They were awarded the Nobel prize for physics in 1951.

1932 **James Chadwick** (England, 1891–1974) discovered the neutron. This particle accounted for Bothe and Becker's penetrating "radiation." He was awarded the Nobel prize for physics in 1935.

1932 **Bruno Benedetto Rossi** (Italy, USA, b. 1905) found an initial increase with thickness in the cosmic-ray intensity "transmitted" by an absorber and explained this by cosmic-ray showers. A transition to decreasing intensities was observed beyond a certain thickness.

1932 **Carl David Anderson** (USA, b. 1905) discovered the positron during cosmic-ray research. He was awarded the Nobel prize for physics jointly with V. F. Hess in 1936.

1932 **Karl G. Jansky** (USA, 1905–1950) found, while investigating factors causing disturbances of radio transmission in the HF band, that the Milky Way is a strong source of continuous radio emission at a wavelength of 14.6 meters. This discovery gave birth to the new science of radio astronomy.

1933 **P. M. S. Blackett** and **G. P. S. Occhialini** obtained the first cloud-chamber photographs of electron-positron pair production.

1933 **Jean Valentin Thibaud** (France, 1901–1960) and **Frederic Joliot** (France, 1900–1958) observed the radiation produced by electron-positron annihilation. They also showed that the mass of the positron is equal to that of the electron.

1933 **T. H. Johnson** and **Jabez Curry Street** (USA, b. 1906) observed that the cosmic-ray intensity from the west exceeded that from the east. This east-west asymmetry shows that there is an excess of positively charged particles in the primary cosmic-ray beam.

1933 **W. Meissener** (Germany, b. 1882) discovered a new property of superconducting materials not explained by infinite conductivity. His results show that within a solid superconductor the magnetic field is everywhere zero.

1934 **Pavel Aleksejevic Cherenkov** (Russia, b. 1904) observed the weak, bluish glow in transparent substances when irradiated with high-energy beta particles. The theory of this Cherenkov radiation was given by **Igor Jevgenevic Tamm** (Russia, 1895–1971) and **Ilya Michajlovic Frank** (Russia, b. 1908) three years later. These three scientists were jointly awarded the Nobel prize for physics in 1958.

1934 **S. Mohorovicic** (Yugoslavia) predicted the existence of a transitory non-nuclear "element" (later called positronium) preceding electron-positron annihilation.

1934 **Irene Joliot–Curie** (France, 1897–1956) and **Frederic Joliot** (France) discovered artificial (induced) radioactivity. They were awarded the Nobel prize for chemistry in 1935.

1934 **Enrico Fermi** developed Pauli's theory of beta decay and named the "new" particle the neutrino (little neutron). It is postulated in Fermi's theory that the neutron is radioactive, disintegrating into a proton with the formation of an electron and a neutrino just before beta emission. He also began a series of experiments in collaboration with **Eduardo Amaldi** (Italy, b. 1908), **Oscar D'Agostino, Franco Rasetti** (Italy, USA, b. 1901), and **Emilio Gino Segrè** (Italy, USA, b. 1905) to produce transuranic elements by irradiating uranium with neutrons. They were granted a patent on the graphite moderator in 1955. Fermi was awarded the Nobel prize for physics in 1938.

1935–1939 **Isidor Isaac Rabi** (Austria, USA, b. 1898) made precise determinations of nuclear magnetic moments in beams of atoms by his radiofrequency resonance method. He was awarded the Nobel prize for physics in 1944.

1935 **Hideki Yukawa** (Japan, 1907–1981) announced his theory of nuclear binding forces involving the postulate of a particle having a mass intermediate between that of the electron and the proton. He was awarded the Nobel prize for physics in 1949.

1936 **C. D. Anderson** and **Seth Henry Neddermeyer** (USA, b. 1907) discovered, during cosmic-ray research, a particle of the type postulated by Yukawa. They called it the "mesotron" (later changed to "meson").

1936 **Marietta Blau** (Austria, 1894–1970) was the first to use nuclear track plates.

1937 **Niels Bohr** (Denmark) introduced the liquid-drop model of the nucleus.

1938 **Irene Joliot–Curie** and **Pavle Savitch** (Yugoslavia, b. 1908) found indications of the existence of lanthanum in uranium after it was irradiated with neutrons.

1938 **Otto Hahn** (Germany, 1879–1968) and **Fritz Strassmann** (Germany, 1902–1981) discovered that bombarding uranium with neutrons produces alkali earth elements. Hahn was awarded the Nobel prize for chemistry in 1944.

1939 **Lise Meitner** (Austria, Germany, Sweden, 1878–1968) and **Otto Robert Frisch** (Austria, Germany, England, 1904–1980) proposed nuclear splitting to explain Hahn's results on the disintegration of uranium by neutrons and predicted the release of an enormous amount of energy per fission.

1939 **Niels Bohr** and **John Archibald Wheeler** (USA, b. 1911) developed the theory of nuclear fission.

1939 **Hans Albrecht Bethe** (Germany, USA, b. 1906) and **Carl Friedrich Von Weizsaecker** (Germany, b. 1912) independently proposed two sets of nuclear reactions to account for stellar energies: the carbon cycle and the proton-proton chain. Bethe was awarded the Nobel prize for physics in 1967.

1939 **J. Robert Oppenheimer** (USA, 1904–1967) and **Hartland S. Snyder** suggested the possibility of black holes in the cosmos.

1940 **John Ray Dunning** (USA, 1907–1975), **Eugene Theodore Booth** (USA, b. 1912), and **Aristid V. Grosse** (Russia, USA, b. 1905) showed that it is ^{235}U, the less abundant isotope of uranium, that is fissioned by slow neutrons.

1940 **Louis Leprince–Ringuet** (France, b. 1901) obtained the first cloud-chamber photograph of a meson-electron collision, from which the mass of the meson could be deduced.

1940 **Donald William Kerst** (USA, b. 1911) made the first betatron, an induction-type accelerator.

1940 **Edwin Mattison McMillan** (USA, b. 1907) and **Philip Hague Abelson** (USA, b. 1913) produced the first transuranic element, neptunium; and **Glenn Theodore Seaborg** (USA, b. 1912), **Edwin Mattison McMillan**, **Joseph William Kennedy** (USA, b. 1917), and **Arthur Charles Wahl** (USA, b. 1917) prepared the second transuranic element, plutonium. McMillan and Seaborg were awarded the Nobel prize for chemistry in 1951.

1940 **Grote Reber** (USA, b. 1912) constructed the first parabolic antenna, 30 ft in diameter, to receive radio signals from space, and made a radio contour map of the Milky Way at a wavelength of one to two meters.

1942 **Enrico Fermi**, **Leo Szilard** (Hungary, Germany, USA, 1898–1964), and associates built the first successful self-sustaining fission reactor. It was first put into operation on December 2 and operated at a power level of one-half watt. It was located in Chicago, Illinois.

1945 **J. Robert Oppenheimer** served as director of the Los Alamos Scientific Laboratory and the "crash" program of basic nuclear research and development that saw as its culmination the detonation of the first nuclear bomb at Almagordo, New Mexico, on July 16.

1945 **E. M. McMillan** and **Vladimir Iosifovich Veksler** (Russia, 1907–1966) independently proposed the principle of the synchrotron, the type of accelerator that produces very high-energy particles, as in the Cosmotron, Bevatron, etc.

1946 **Felix Bloch** (Switzerland, USA, b. 1905) devised the magnetic induction method and, independently, **Edward Mills Purcell** (USA, b. 1912) originated the magnetic resonance absorption method for determining nuclear magnetic moments, using liquids or solids in bulk (not beams). This led to the nuclear resonance spectrometer. They were awarded the Nobel prize for physics in 1952.

1947 **Polykarp Kusch** (Germany, USA, b. 1911) made high-precision determinations of the magnetic moment of the electron and found a small but theoretically significant difference between the predicted value and the experimental results. Kusch was awarded the Nobel prize for physics jointly with W. R. Lamb in 1955.

1947 **Willis Eugene Lamb, Jr.,** (USA, b. 1913) and **Robert E. Retherford** observed, during the course of spectral measurements of the fine structure of hydrogen in the microwave region, a small displacement (the "Lamb shift") of an energy level from its theoretical position as predicted by Dirac's quantum theory of the electron. Lamb was awarded the Nobel prize for physics jointly with P. Kusch in 1955.

1947 **H. A. Bethe** and, independently, **Julian Seymour Schwinger** (USA, b. 1918) explained the discrepancies found by Kusch and by Lamb as resulting from an interaction of electrons with the radiation field. Schwinger was awarded the Nobel prize for physics jointly with R. P. Feynman and S. Tomonaga in 1965.

1947 **Hartmut Paul Kallmann** (Germany, USA, 1896–1978) and, independently, **John Wesley Coltman** (USA, b. 1915) and **Fitz-Hugh Ball Marshall** (USA, 1912–1966) developed scintillation counters.

1947 **Cecil Frank Powell** (England, 1903–1969), **G. P. S. Occhialini**, and **Cesare Mansueto Giulio Lattes** (Brazil, England, b. 1924) discovered the pi-meson. Powell was awarded the Nobel prize for physics in 1950.

1947 **George Dixon Rochester** (England, b. 1908) and **Clifford Charles Butler** (England, b. 1922) discovered V-particles and hyperons.

1948 **Eugene Gardner** (USA, 1913–1950) and **C. M. G. Lattes** (Brazil, England, USA) were the first to produce mesons in the laboratory.

1948–1950 **Willard Frank Libby** (USA, 1908–1980) and collaborators developed the techniques of radiocarbon dating. He was awarded the Nobel prize for chemistry in 1960.

1949 **Maria Goeppert Mayer** (Germany, USA, 1906–1972) and, independently, **Otto Haxel** (Germany, b. 1909), **Johannes Hans Daniel Jensen** (Germany, b. 1907), and **Hans Eduard Suess** (Austria, Germany, USA, b. 1909) developed the shell theory of the nucleus, which assumes a spherical distribution of nucleons. Mayer and Jensen were awarded the Nobel prize for physics jointly with E. P. Wigner in 1963.

1950 **Arthur Hawley Snell** (Canada, USA, b. 1911) and associates at the Oak Ridge National Laboratory and **John Michael Robson** (England, Canada, b. 1920) and his associates at the Chalk River Laboratory experimentally verified that the free neutron is radioactive.

1950 Scientists began intensified research on light-element fusion reactions.

1951 **Martin Deutsch** (Austria, USA, b. 1917) experimentally confirmed the prediction of the existence of positronium.

1952 Brookhaven National Laboratory was the first to achieve the acceleration of particles to the giga-electron-volt energy range: 2.3-GeV protons.

1952 **Aage Bohr** (Denmark, b. 1922) and **Ben Roy Mottelson** (Denmark, b. 1926) developed the unified (collective) shell model of the nucleus, which assumes a nonspherical nuclear core.

The possibility of a distorted core had been suggested in 1950 by **Leo James Rainwater** (USA, b. 1917). These three scientists were awarded the Nobel prize for physics in 1975.

1952 **Donald Arthur Glaser** (USA, b. 1926) made the first bubble chamber. He was awarded the Nobel prize for physics in 1960.

1952 The first large-scale, terrestrial thermonuclear reaction was produced when a "hydrogen fusion device" was tested at Einewetok Atoll on November 1.

1953 **Murray Gell-Mann** (USA, b. 1929) introduced the strangeness numbers for nucleons, mesons, and hyperons, and found that strangeness is conserved in strong interactions. He was awarded the Nobel prize for physics in 1969.

1953 **Robert Hofstadter** (USA, b. 1915) and collaborators started a series of experiments on the scattering of high-energy electrons by atoms. The results led to the determination of the charge distribution and structure of nuclei and nucleons. Hofstadter was awarded the Nobel prize for physics jointly with R. L. Mössbauer in 1961.

1954 **James Power Gordon** (USA, b. 1928), **H. J. Zeiger** (USA, b. 1925), and **Charles Hard Townes** (USA, b. 1915) made the first *maser* [molecular (formerly, microwave) amplification by stimulated emission of radiation]. In this device, many molecules that have been put into high energy states are induced to emit their energy as radiation by a weak incoming signal of the same frequency. Townes was awarded the Nobel prize for physics jointly with N. Basov and A. Prokhorv in 1964.

1955 **Owen Chamberlain** (USA, b. 1920), **Emilio Gino Segrè** (Italy, USA, b. 1905), **Clyde Edward Wiegand** (USA, b. 1915), and **Thomas John Ypsilantis** (USA, b. 1928) created proton-antiproton pairs. Chamberlain and Segrè were awarded the Nobel prize for physics in 1959.

1956 **Luis Walter Alvarez** (USA, b. 1911) and collaborators accomplished cold fusion of deuterium with the negative mu-meson as a catalyst. He was awarded the Nobel prize for physics in 1968.

1956 **John Bardeen** (USA, b. 1908), **Walter Houser Brattain** (China, USA, b. 1902), and **William Shockley** (England, USA, b. 1910) were awarded the Nobel prize for physics in recognition of their work in theory of the solid state, particularly semiconductors.

1956 **Frederic Reines** (USA, b. 1918) and **Clyde Lorrain Cowan, Jr.** (USA, b. 1919) and collaborators experimentally confirmed the existence of the neutrino.

1956 The world's first full-scale nuclear power plant was put into operation on October 17 at Calder Hall, England. The gas-cooled reactors develop 360 megawatts of thermal power to deliver 78 megawatts of electrical power.

1956 **Tsung Dao Lee** (China, USA, b. 1926) and **Chen Ning Yang** (China, USA, b. 1922) deduced theoretically that the law of conservation of parity (the invariance of spatial inversion) is invalid for weak interactions. They were awarded the Nobel prize for physics in 1957.

1956 **Chien-Shiung Wu** (China, USA, b. 1915) and collaborators performed the first experiment that demonstrated the violation of conservation of parity. They observed the beta emission from ^{60}Co at very low temperatures.

1957 **John Bardeen, Leon N. Cooper** (USA, b. 1930), and **John Robert Schrieffer** (USA, b. 1931) announced the first comprehensive theory of superconductivity. These three scientists were awarded the Nobel prize for physics in 1972. This was Bardeen's second Nobel prize.

1958 **C. H. Townes, John Perry Cedarholm** (USA, b. 1927), **George Francis Bland** (USA, b. 1927), and **Byron Luther Havens** (USA, b. 1914) employed maser beams in the most precise ether-drift experiment yet performed. The results showed that if the effect exists, it is less than one-thousandth of the earth's orbital speed or less than one ten-millionth of the speed of light. The precision in the comparison of the frequencies of the masers was about one part in a million million.

1958 **Rudolph L. Mössbauer** (Germany, b. 1929) predicted and found an extremely small frequency spread in the emission of low-energy gamma rays from nuclei bound in a crystal lattice. This effect results from giving the gamma-ray recoil momentum to the whole lattice instead of to an individual nucleus. The effect provides a very high-precision frequency standard suitable for testing several predictions of the special and general theories of relativity. He was awarded the Nobel prize for physics jointly with R. Hofstadter in 1961.

1959 **James Alfred Van Allen** (USA, b. 1914) showed from the data obtained from instruments carried by artificial satellites that the earth is encircled by two zones, called Van Allen radiation belts, of high-energy charged particles that are trapped by the earth's magnetic field.

1960 **Theodore Harold Maiman** (USA, b. 1927) made the first ruby laser.

1960 **Ali Javan** (Iran, USA, b. 1926) made the first helium-neon laser.

1960 **Vernon Willard Hughes** (USA, b. 1921), **D. W. McColm** (b. 1933), **Klaus Otto Ziock** (Germany, USA, b. 1925), and **R. Prepost** (b. 1935) made and studied muonium, a short-lived atom having a positive mu-meson nucleus and an orbiting electron.

1961 **M. Gell-Mann** (USA, b. 1929) and **Y. Ne'eman** (Israel, b. 1925) predicted the existence of a particle of mass about 1675 MeV, strangeness minus 3, spin three halves, positive parity and negative charge. This particle was later detected and named the omega minus. Gell-Mann was awarded the Nobel prize for physics in 1969.

1961 **M. Gell-Mann** introduced the idea of the quark.

1962 **B. D. Josephson** (England, b. 1940) discovered and theoretically analyzed a number of unexpected phenomena occurring at a "Josephson junction," an arrangement consisting of two superconductors separated by a very thin layer of insulating material. He was awarded the Nobel prize for physics along with **Ivar Giaever** (Norway, USA, b. 1929) and **Leo Esaki** (Japan, USA, b. 1925) in 1973 for their discoveries regarding tunneling phenomena.

1964 **G. K. O'Neill** (USA, b. 1927) and associates produced the first electron-electron colliding beam reactions at Princeton.

1964 A large group at the Brookhaven National Laboratory verified the existence of the omega minus particle.

1964 **Allan R. Sandage** (USA, b. 1926) first discovered quasars.

1964 **James Cronin** (USA, b. 1931) and **Val L. Fitch** (USA, b. 1923) showed that in a certain rare decay of the K meson time-reversal invariance was violated.

1965 **Jerome V. V. Kasper** and **George Claude Pimentel** (USA, b. 1922) made the first chemical laser, a device in which pumping energy is supplied by chemical reactions instead of by an external source of power.

1966 **Peter Sorokin** (USA, b. 1931), an investigator in quantum optics and lasers, made the first organic dye laser.

1967 **Jocelyn Bell** (England, b. 1943), a graduate student at Cambridge University, started a detailed radiation survey of the sky with a high-resolution, quick-response antenna array developed by **Antony Hewish** (b. 1924). She discovered cosmic, weak, pulselike signals having a repetition period of 1.337 seconds. Subsequently, several other such sources were found. In accordance with their dominant characteristic, these sources are now called pulsars and the signals are due to rapidly rotating neutron stars.

1968 **Irwin I. Shapiro** (USA, b. 1929) of MIT reported the first use of radar for a test of the general theory of relativity. By bouncing radar beams off Venus and Mercury as they passed the opposite side of the sun, he was able to show that the delay of the signal was just that predicted by Einstein's theory.

1969 **Neil A. Armstrong** (USA, b. 1930) and **Edwin Eugene Aldrin** (USA, b. 1930) landed on the surface of the moon.

1970 **Terrill A. Cool** (USA, b. 1936) produced the first continuous laser by purely chemical means. No external energy source was required. The laser operated only by the simple mixing of bottled gas.

1972 **Kenneth M. Evenson** (USA, b. 1932) and colleagues measured the frequency and wavelength of infrared radiation emitted from a stabilized helium neon laser and obtained the value of the speed of light $299{,}792.4562 \pm 0.0011$ km/sec. (The accuracy is one part in 300 million.)

1972 **E. C. Silverberg** (USA) of McDonald Observatory, Texas, and **D. C. Currie** (USA) of the University of Maryland, in a lunar-ranging experiment, measured the distance to the moon to an accuracy of 15 cm.

1972 **Robert Rathbun Wilson** (USA, b. 1914) and associates in March produced the first particle beam at the Fermi National Accelerator Laboratory. In August, the first experiments were carried out using the beam.

1972 **David Morris** (USA, b. 1931) and **Robert Coleman Richardson** (USA, b. 1937) with their graduate students Osheroff and Gully made the first measurements giving evidence that liquid helium isotope 3 is a superfluid.

1973 **J. R. Gavaler** (USA) fabricated a superconducting sample (Nb_3GE) with a transition temperature of 22.3 kelvin. This is the first superconductor with the transition temperature above the boiling point of hydrogen.

1974 **Burton Richter** (USA, b. 1931) and associates at the Stanford Linear Accelerator and **Samuel C. C. Ting** (USA, b. 1936) and associates at the Brookhaven National Laboratory independently discovered the subatomic particle called the "J" or "psi" particle. They were awarded the Nobel prize for physics in 1976.

1975 **Richard Anthony Sramek** (USA, b. 1943) and **E. B. Fomalong** (USA) have been able to show that radio waves emanating from a distant quasar are bent when passing near the sun by exactly the amount predicted by Einstein's theory of general relativity.

1975 **Martin Perl** (USA, b. 1927) and his colleagues at the Stanford Linear Accelerator discovered a new lepton and named it the tau particle.

1976 Jet Propulsion Laboratory scientists landed Viking on the surface of Mars.

1979 **D. Walsh, R. F. Carswell,** and **R. J. Weyman** discovered a galaxy or a group of galaxies located between a distant quasar and earth that is performing the function of a gravitational lens. Two images of the distant quasar are observed due to this effect.

APPENDIX 2: NOBEL PRIZE WINNERS

Physics		Chemistry
W. C. Roentgen (Germany)	1901	J. H. van't Hoff (Netherlands)
H. A. Lorentz (Netherlands) Pieter Zeeman (Netherlands)	1902	Emil Fischer (Germany)
Henri Becquerel (France) Pierre Curie (France) Marie S. Curie (France)	1903	S. A. Arrhenius (Sweden)
J. W. Strutt (Lord Rayleigh) (England)	1904	William Ramsay (England)
Philipp Lenard (Germany)	1905	Adolf von Baeyer (Germany)
J. J. Thomson (England)	1906	Henri Moissan (France)
A. A. Michelson (USA)	1907	Eduard Buchner (Germany)
Gabriel Lippman (France)	1908	Ernest Rutherford (England)
Guglielmo Marconi (Italy) K. F. Braun (Germany)	1909	Wilhelm Ostwald (Germany)
J. D. van der Waals (Netherlands)	1910	Otto Wallach (Germany)
Wilhelm Wien (Germany)	1911	Marie S. Curie (France)
N. G. Dalen (Sweden)	1912	Victor Grignard (France) Paul Sabatier (France)
Kamerlingh-Onnes (Netherlands)	1913	Alfred Werner (Switzerland)
Max von Laue (Germany)	1914	T. W. Richards (USA)
W. H. Bragg (England) W. L. Bragg (England)	1915	Richard Willstaetter (Germany)
No Award	1916	No Award
C. G. Barkla (England)	1917	No Award
Max Planck (Germany)	1918	Fritz Haber (Germany)
Johannes Stark (Germany)	1919	No Award
C. E. Guillaume (Switzerland)	1920	Walther Nernst (Germany)
Albert Einstein (Germany)	1921	Frederick Soddy (England)
Niels Bohr (Denmark)	1922	F. W. Aston (England)
R. A. Millikan (USA)	1923	Fritz Pregl (Austria)
K. M. G. Siegbahn (Sweden)	1924	No Award
James Franck (Germany) Gustav Hertz (Germany)	1925	Richard Zsigmondy (Germany)
Jean Perrin (France)	1926	Theodor Svedberg (Sweden)
A. H. Compton (USA) C. T. R. Wilson (England)	1927	Heinrich Wieland (Germany)
O. W. Richardson (England)	1928	Adolf Windaus (Germany)

Physics		Chemistry
Louis de Broglie (France)	1929	Arthur Harden (England) H. von Euler-Chelpin (Sweden)
C. V. Raman (India)	1930	Hans Fischer (Germany)
No Award	1931	Carl Bosch (Germany) Friedrich Bergius (Germany)
Werner Heisenberg (Germany)	1932	Irving Langmuir (USA)
Erwin Schroedinger (Austria) P. A. M. Dirac (England)	1933	No Award
No Award	1934	H. C. Urey (USA)
James Chadwick (England)	1935	Frederic Joliot (France) Irene Joliot-Curie (France)
V. F. Hess (Austria) C. D. Anderson (USA)	1936	Peter J. W. Debye (Germany)
C. J. Davisson (USA) G. P. Thomson (England)	1937	W. N. Haworth (England) Paul Karrer (Switzerland)
Enrico Fermi (Italy)	1938	Richard Kuhn (Germany)
E. O. Lawrence (USA)	1939	Leopold Ruzicka (Switzerland) Adolf Butenandt (Germany)
No Award	1940	No Award
No Award	1941	No Award
No Award	1942	No Award
Otto Stern (Germany)	1943	George de Hevesy (Hungary)
I. I. Rabi (USA)	1944	Otto Hahn (Germany)
Wolfgang Pauli (Austria)	1945	A. I. Virtanen (Finland)
P. W. Bridgman (USA)	1946	J. B. Sumner (USA) J. H. Northrop (USA) W. M. Stanley (USA)
E. V. Appleton (England)	1947	Robert Robinson (England)
P. M. S. Blackett (England)	1948	Arne Tiselius (Sweden)
Hideki Yukawa (Japan)	1949	W. F. Giauque (USA)
C. F. Powell (England)	1950	Otto Diels (Germany) Kurt Alder (Germany)
J. D. Cockcroft (England) E. T. S. Walton (Ireland)	1951	E. M. McMillan (USA) G. T. Seaborg (USA)
Felix Bloch (USA) E. M. Purcell (USA)	1952	A. J. P. Martin (England) E. L. M. Synge (England)
Fritz Zernike (Netherlands)	1953	Herman Staudinger (Germany)

Physics		Chemistry
Max Born (Germany) Walter Bothe (Germany)	1954	Linus Pauling (USA)
Polykarp Kusch (USA) W. E. Lamb (USA)	1955	Vincent du Vigneaud (USA)
John Bardeen (USA) W. H. Brattain (USA) William Shockley (USA)	1956	C. N. Hinshelwood (England) N. N. Semenov (USSR)
C. N. Yang (China, USA) T. D. Lee (China, USA)	1957	Alexander Todd (England)
P. A. Cherenkov (USSR) I. Y. Tamm (USSR) I. M. Frank (USSR)	1958	Frederic Sanger (England)
Owen Chamberlain (USA) E. G. Segrè (USA)	1959	Jaroslav Heyrovsky (Czechoslovakia)
D. A. Glaser (USA)	1960	W. F. Libby (USA)
Robert Hofstadter (USA) R. L. Mössbauer (Germany)	1961	Melvin Calvin (USA)
L. D. Landau (USSR)	1962	J. C. Kendrew (England) M. F. Perutz (England)
J. H. D. Jensen (Germany) Maria G. Mayer (USA) E. B. Wigner (USA)	1963	Karl Ziegler (Germany) Giulio Natta (Italy)
Nikolai Basov (USSR) Alexander Prokhorv (USSR) C. H. Townes (USA)	1964	Dorothy C. Hodgkin (England)
R. P. Feynman (USA) J. S. Schwinger (USA) Sin-itiro Tomonaga (Japan)	1965	R. B. Woodward (USA)
Alfred Kastler (France)	1966	R. S. Mullikan (USA)
Hans Albrecht Bethe (USA)	1967	M. Eigen (Germany) R. G. W. Norrish (Great Britain) George Porter (Great Britain)
Luis W. Alvarez (USA)	1968	L. Onsager (USA)
Murray Gell-Mann (USA)	1969	D. H. R. Barton (Great Britain) O. Hassel (Norway)
Louis Néel (France) Hannes Alfven (Sweden)	1970	L. F. Leloir (Argentina)
Dennis Gabor (England)	1971	G. Herzberg (Canada)
John Bardeen (USA) Leon Cooper (USA) John R. Schrieffer (USA)	1972	C. B. Anfinsen (USA) Stanford Moore (USA) W. H. Stein (USA)

Physics		Chemistry
Brian Josephson (England) Ivar Giaever (USA) Leo Esaki (USA)	1973	E. Fisher (Germany) G. Wilkinson (Great Britain)
Martin Ryle (England) Antony Hewish (England)	1974	P. J. Flory (USA)
L. James Rainwater (USA) Aage Bohr (Denmark) Ben R. Mottelson (Denmark)	1975	J. W. Cornforth (Great Britain) V. Prelog (Switzerland)
Burton Richter (USA) Samuel C. C. Ting (USA)	1976	William N. Lipscomb (USA)
P. W. Anderson (USA) N. F. Mott (England) J. H. Van Vleck (USA)	1977	I. Prigogine (Belgium)
Arno A. Pencias (USA) Robert W. Wilson (USA) Peter L. Kapitza (USSR)	1978	Peter Mitchell (Great Britain)
Sheldon Glashow (USA) Steven Weinberg (USA) Abdus Salam (Pakistan)	1979	Herbert C. Brown (USA) Georg Wittig (Germany)
Val Fitch (USA) James Cronin (USA)	1980	Paul Berg (USA) Walter Gilbert (USA) Frederick Sanger (Great Britain)
Nicolaas Bloembergen (USA) Arthur Schawlow (USA) Kai Siegbahn (Sweden)	1981	Kenichi Fukni (Japan) Ronald Hoffman (USA)
Kenneth G. Wilson (USA)	1982	Aaron Klug (Great Britain)
W. A. Fowler (USA) S. Chandrasekhar (USA)	1983	H. Taube (USA)

APPENDIX 3: PERIODIC TABLE OF THE ELEMENTS

↓ Period	Group →	I	II													III	IV	V	VI	VII	VIII
1	1s	1 **H** 1.00 $1s^1$																			2 **He** 4.00 $1s^2$
2	2s	3 **Li** 6.94 $2s^1$	4 **Be** 9.01 $2s^2$												2p	5 **B** 10.81 $2p^1$	6 **C** 12.01 $2p^2$	7 **N** 14.01 $2p^3$	8 **O** 16.00 $2p^4$	9 **F** 19.00 $2p^5$	10 **Ne** 20.18 $2p^6$
3	3s	11 **Na** 22.99 $3s^1$	12 **Mg** 24.31 $3s^2$												3p	13 **Al** 26.98 $3p^1$	14 **Si** 28.09 $3p^2$	15 **P** 30.98 $3p^3$	16 **S** 32.07 $3p^4$	17 **Cl** 35.46 $3p^5$	18 **Ar** 39.94 $3p^6$
4	4s	19 **K** 39.10 $4s^1$	20 **Ca** 40.08 $4s^2$	3d	21 **Sc** 44.96 $3d^1$	22 **Ti** 47.90 $3d^2$	23 **V** 50.94 $3d^3$	24 **Cr** 52.00 $4s^1 3d^5$	25 **Mn** 54.9 $4s^2 3d^5$	26 **Fe** 55.85 $4s^2 3d^6$	27 **Co** 58.93 $4s^2 3d^7$	28 **Ni** 58.71 $4s^2 3d^8$	29 **Cu** 63.54 $4s3d^{10}$	30 **Zn** 65.37 $4s^2 3d^{10}$	4p	31 **Ga** 69.72 $4p^1$	32 **Ge** 72.59 $4p^2$	33 **As** 74.92 $4p^3$	34 **Se** 78.96 $4p^4$	35 **Br** 79.91 $4p^5$	36 **Kr** 83.8 $4p^6$
5	5s	37 **Rb** 85.47 $5s^1$	38 **Sr** 87.66 $5s^2$	4d	39 **Y** 88.91 $5s^2 4d^1$	40 **Zr** 91.22 $5s^1 4d^2$	41 **Nb** 92.91 $5s^1 4d^4$	42 **Mo** 95.94 $5s^1 4d^5$	43 **Tc** (99) $5s^2 4d^5$	44 **Ru** 101.1 $5s^1 4d^7$	45 **Rh** 102.91 $5s^1 4d^8$	46 **Pd** 106.4 $5s^0 4d^{10}$	47 **Ag** 107.87 $5s^1 4d^{10}$	48 **Cd** 112.40 $5s^2 4d^{10}$	5p	49 **In** 114.82 $5p^1$	50 **Sn** 118.69 $5p^2$	51 **Sb** 121.75 $5p^3$	52 **Te** 127.60 $5p^4$	53 **I** 126.90 $5p^5$	54 **Xe** 131.30 $5p^6$
6	6s	55 **Cs** 132.91 $6s^1$	56 **Ba** 137.34 $6s^2$	5d	57–71 *	72 **Hf** 178.49 $6s^2 5d^2$	73 **Ta** 180.95 $6s^2 5d^3$	74 **W** 183.85 $6s^2 5d^4$	75 **Re** 186.2 $6s^2 5d^5$	76 **Os** 190.2 $6s^2 5d^6$	77 **Ir** 192.2 $6s^2 5d^7$	78 **Pt** 195.09 $6s^1 5d^9$	79 **Au** 197.0 $6s5d^{10}$	80 **Hg** 200.59 $6s^2 5d^{10}$	6p	81 **Tl** 204.37 $6p^1$	82 **Pb** 207.19 $6p^2$	83 **Bi** 208.98 $6p^3$	84 **Po** (210) $6p^4$	85 **At** (210) $6p^5$	86 **Rn** 222 $6p^6$
7	7s	87 **Fr** (223) $7s^1$	88 **Ra** 226.05 $7s^2$	6d	89–103 †																

4f	57 **La** 138.91 $6s^2 5d^1$	58 **Ce** 140.12 $5d^1 5f^2$	59 **Pr** 140.91 $5d^0 4f^3$	60 **Nd** 144.24 $5d^0 4f^4$	61 **Pm** (145)	62 **Sm** 150.35 $5d^0 4f^6$	63 **Eu** 152.0 $5d^0 4f^7$	64 **Gd** 157.25 $5d^1 4f^7$	65 **Tb** 158.92 $5d^1 4f^8$	66 **Dy** 162.50	67 **Ho** 164.92	68 **Er** 167.26	69 **Tm** 168.93 $5d^0 4f^{13}$	70 **Yb** 173.04 $5d^0 4f^{14}$	71 **Lu** 174.97 $5d^1 4f^{14}$
5f	89 **Ac** 227 $7s^2 6d^1$	90 **Th** 232.04 $6d^2 5f^0$	91 **Pa** 231	92 **U** 238.03 $5d^1 5f^3$	93 **Np** (237)	94 **Pu** (242)	95 **Am** (243) $6d^0 5f^7$	96 **Cm** (247)	97 **Bk** (249)	98 **Cf** (251)	99 **Es** (254)	100 **Fm** (253)	101 **Md** (256)	102 **No** (254)	103 **Lr** (257)

*Lanthanides (rare earths).
†Actinides.

APPENDIX 4: PROPERTIES OF ATOMS IN BULK

Element	Symbol	Atomic number, Z	Atomic mass	Density, g/cm^3 at 20°C	Melting point, °C	Boiling point, °C	Specific heat, cal / (g · C°) at 25°C
Actinium	Ac	89	(227)	...	1323	(3473)	(0.022)
Aluminum	Al	13	26.9815	2.699	660	2450	0.215
Americium	Am	95	(243)	11.7	1541	...	...
Antimony	Sb	51	121.75	6.62	630.5	1380	0.049
Argon	Ar	18	39.948	1.6626×10^{-3}	−189.4	−185.8	0.125
Arsenic	As	33	74.9216	5.72	817 (28 at.)	613	0.079
Astatine	At	85	(210)	...	(302)	...	...
Barium	Ba	56	137.34	3.5	729	1640	0.049
Berkelium	Bk	97	(247)	...	...	...	...
Beryllium	Be	4	9.0122	1.848	1287	2770	0.436
Bismuth	Bi	83	208.980	9.80	271.37	1560	0.0292
Boron	B	5	10.811	2.34	2030	...	0.265
Bromine	Br	35	79.909	3.12 (liquid)	−7.2	58	0.070
Cadmium	Cd	48	112.40	8.65	321.03	765	0.054
Calcium	Ca	20	40.08	1.55	838	1440	0.149
Californium	Cf	98	(251)	...	...	...	...
Carbon	C	6	12.01115	2.25	3727	4830	0.165
Cerium	Ce	58	140.12	6.768	804	3470	0.045
Cesium	Cs	55	132.905	1.9	28.40	690	0.058
Chlorine	Cl	17	35.453	3.214×10^{-3} (0°C)	−101	−34.7	0.116
Chromium	Cr	24	51.996	7.19	1857	2665	0.107
Cobalt	Co	27	58.9332	8.85	1495	2900	0.101
Copper	Cu	29	63.54	8.96	1083.40	2595	0.092
Curium	Cm	96	(247)	...	...	...	...
Dysprosium	Dy	66	162.50	8.55	1409	2330	0.041
Einsteinium	Es	99	(254)	...	...	...	...
Erbium	Er	68	167.26	9.15	1522	2630	0.040
Europium	Eu	63	151.96	5.245	817	1490	0.039
Fermium	Fm	100	(257)	...	...	...	...
Fluorine	F	9	18.9984	1.696×10^{-3} (0°C)	−219.6	−188.2	0.18
Francium	Fr	87	(223)	...	(27)	...	...
Gadolinium	Gd	64	157.25	7.86	1312	2730	0.056
Gallium	Ga	31	69.72	5.907	29.75	2237	0.090
Germanium	Ge	32	72.59	5.323	937.25	2830	0.077
Gold	Au	79	196.967	19.32	1064.43	2970	0.0312
Hafnium	Hf	72	178.49	13.09	2227	5400	0.0345
Helium	He	2	4.0026	0.1664×10^{-3}	−269.7	−268.9	1.25

(Continued)

Element	Symbol	Atomic number, Z	Atomic mass	Density, g / cm^3 at 20°C	Melting point, °C	Boiling point, °C	Specific heat, cal / (g · C°) at 25°C
Holmium	Ho	67	164.930	8.79	1470	2330	0.0394
Hydrogen	H	1	1.00797	0.08375×10^{-3}	−259.19	−252.7	3.45
Indium	In	49	114.82	7.31	156.634	2000	0.0556
Iodine	I	53	126.9044	4.94	113.7	183	0.052
Iridium	Ir	77	192.2	22.5	2447	(5300)	0.0311
Iron	Fe	26	55.847	7.87	1536.5	3000	0.1069
Krypton	Kr	36	83.80	3.488×10^{-3}	−157.37	−152	(0.059)
Lanthanum	La	57	138.91	6.189	920	3470	0.0466
Lawrencium	Lw	103	(257)	...	...	...	...
Lead	Pb	82	207.19	11.36	327.45	1725	0.0307
Lithium	Li	3	6.939	0.534	180.55	1300	0.856
Lutetium	Lu	71	174.97	9.849	1663	1930	0.037
Magnesium	Mg	12	24.312	1.74	650	1107	0.245
Manganese	Mn	25	54.9380	7.43	1244	2150	0.115
Mendelevium	Md	101	(256)	...	...	...	...
Mercury	Hg	80	200.59	13.55	−38.87	357	0.033
Molybdenum	Mo	42	95.94	10.22	2617	5560	0.060
Neodymium	Nd	60	144.24	7.00	1016	3180	0.045
Neon	Ne	10	20.183	0.8387×10^{-3}	−248.597	−246.0	(0.246)
Neptunium	Np	93	(237)	19.5	637	...	0.30
Nickel	Ni	28	58.71	8.902	1453	2730	0.106
Niobium	Nb	41	92.906	8.57	2468	4927	0.063
Nitrogen	N	7	14.0067	1.1649×10^{-3}	−210	−195.8	0.247
Nobelium	No	102	(255)	...	...	...	...
Osmium	Os	76	190.2	22.57	3027	5500	0.031
Oxygen	O	8	15.9994	1.3318×10^{-3}	−218.80	−183.0	0.218
Palladium	Pd	46	106.4	12.02	1552	3980	0.058
Phosphorus	P	15	30.9738	1.83	44.25	280	0.177
Platinum	Pt	78	195.09	21.45	1769	4530	0.0320
Plutonium	Pu	94	(244)	...	640	3235	0.031
Polonium	Po	84	(210)	9.24	254	...	...
Potassium	K	19	39.102	0.86	63.20	760	0.181
Praseodymium	Pr	59	140.907	6.769	931	3020	0.047
Promethium	Pm	61	(145)	...	(1027)	...	
Protactinium	Pa	91	(231)	...	(1230)	...	...
Radium	Ra	88	(226)	5.0	700	...	...
Radon	Rn	86	(222)	9.96×10^{-3} (0°C)	(−71)	−61.8	(0.022)

Element	Symbol	Atomic number, Z	Atomic mass	Density, g / cm^3 at 20°C	Melting point, °C	Boiling point, °C	Specific heat, cal / (g · C°) at 25°C
Rhenium	Re	75	186.2	21.04	3180	5900	0.032
Rhodium	Rh	45	102.905	12.44	1963	4500	0.058
Rubidium	Rb	37	85.47	1.53	39.49	688	0.087
Ruthenium	Ru	44	101.107	12.2	2250	4900	0.057
Samarium	Sm	62	150.35	7.49	1072	1630	0.047
Scandium	Sc	21	44.956	2.99	1539	2730	0.136
Selenium	Se	34	78.96	4.79	221	685	0.076
Silicon	Si	14	28.086	2.33	1412	2680	0.170
Silver	Ag	47	107.870	10.49	960.8	2210	0.056
Sodium	Na	11	22.9898	0.9712	97.85	892	0.293
Strontium	Sr	38	87.62	2.60	768	1380	0.176
Sulfur	S	16	32.064	2.07	119.0	444.6	0.169
Tantalum	Ta	73	180.948	16.6	3014	5425	0.033
Technetium	Tc	43	(99)	11.46	2200	...	(0.050)
Tellurium	Te	52	127.60	6.24	449.5	990	0.048
Terbium	Tb	65	158.924	8.25	1357	2530	0.043
Thallium	Tl	81	204.37	11.85	304	1457	0.031
Thorium	Th	90	(232)	11.66	1755	(3850)	0.028
Thulium	Tm	69	168.934	9.31	1545	1720	0.038
Tin	Sn	50	118.69	7.2984	231.868	2270	0.054
Titanium	Ti	22	47.90	4.507	1670	3260	0.125
Tungsten	W	74	183.85	19.3	3380	5930	0.032
Uranium	U	92	(238)	19.07	1132	3818	0.028
Vanadium	V	23	50.942	6.1	1902	3400	0.117
Xenon	Xe	54	131.30	5.495×10^{-3}	−111.79	−108	(0.038)
Ytterbium	Yb	70	173.04	6.959	824	1530	0.037
Yttrium	Y	39	88.905	4.472	1526	3030	0.071
Zinc	Zn	30	65.37	7.133	419.58	906	0.0930
Zirconium	Zr	40	91.22	6.489	1852	3580	0.066

The values in parentheses in the column of atomic masses are the mass numbers of the longest-lived isotopes of those elements that are radioactive. Melting points and boiling points in parentheses are uncertain. Specific heats in parentheses are calculated values.

All the physical properties are given for a pressure of one atmosphere except where otherwise specified.

The data for gases are valid only when these are in their usual molecular state, such as H_2, He, O_2, Ne, etc. The specific heats of the gases are the values at constant pressure.

Source: The data for this table were obtained from the "Key to Periodic Chart of the Atoms," 1963 edition, by William F. Meggers (courtesy of the Welch Scientific Company, Skokie, Illinois) and updated using "Selected Values of the Thermomagnetic Properties of the Elements," by R. Hultgren, P. D. Desai, D. T. Hawkins, M. Gleiser, K. K. Kelley, and D. D. Wagman, published by the American Society for Metals in 1973.

APPENDIX 5: PARTIAL LIST OF ISOTOPES

The values of the isotopic masses are based on carbon-12. Naturally occurring radioactive isotopes are indicated by (NR). The radioactive ones indicated by (R) are not found in nature. The mass numbers given for the radioactive elements are those of the longest-lived isotopes.

The data for this table were obtained from the Chart of the Nuclides, 8th edition, revised to March 1965 by David T. Goldman. (Courtesy of Knolls Atomic Power Laboratory, Schenectady, N.Y., operated by the General Electric Company for the United States Atomic Energy Commission) and revised with data from *Handbook of Chemistry and Physics*, 55th ed., CRC Press (1974–1975).

At. no. Z	Element	Symbol	Mass no., A	Isotopic mass, u	Relative abundance, %	No. of Isotopes: Stable	No. of Isotopes: Radio-active
0	Neutron	n				0	1
			1 (R)	1.008665			
1	Hydrogen	H				2	1
			1	1.007825	99.985		
		D					
			2	2.0140	0.015		
		T					
			3 (NR)	3.01605			
2	Helium	He				2	3
			3	3.01603	0.00013		
			4	4.00260	100		
3	Lithium	Li				2	3
			6	6.01512	7.42		
			7	7.01600	92.58		
4	Beryllium	Be				1	6
			9	9.01218	100		
5	Boron	B				2	4
			10	10.01294	19.78		
			11	11.00931	80.22		
6	Carbon	C				2	6
			12	12.00000	98.89		
			13	13.00335	1.11		
7	Nitrogen	N				2	5
			14	14.00307	99.63		
			15	15.00011	0.37		
8	Oxygen	O				3	5
			16	15.99491	99.759		
			17	16.99914	0.037		
			18	17.99916	0.204		
9	Fluorine	F				1	5
			19	18.99840	100		

PARTIAL LIST OF ISOTOPES

At. no. Z	Element	Symbol	Mass no., A	Isotopic mass, u	Relative abundance, %	No. of Isotopes: Stable	No. of Isotopes: Radio-active
10	Neon	Ne				3	5
			20	19.99244	90.92		
			21	20.99395	0.257		
			22	21.99138	8.82		
11	Sodium	Na				1	6
			23	22.98977	100		
12	Magnesium	Mg				3	5
			24	23.98504	78.70		
			25	24.98584	10.13		
			26	25.98259	11.17		
13	Aluminum	Al				1	7
			27	26.98153	100		
14	Silicon	Si				3	5
			28	27.97693	92.21		
15	Phosphorus	P				1	6
			31	30.97376	100		
16	Sulfur	S				4	6
			32	31.97207	95.0		
17	Chlorine	Cl				2	7
			35	34.96885	75.53		
			37	36.96590	24.47		
18	Argon	Ar				3	6
			40	39.96238	99.60		
19	Potassium	K				2	9
			39	38.96371	93.10		
			40 (NR)	39.974	0.0118		
			41	40.96184	6.88		
20	Calcium	Ca				6	8
			40	39.96259	96.947		
			44	43.95549	2.083		
21	Scandium	Sc				1	11
			45	44.95592	100		
22	Titanium	Ti				5	5
			48	47.94795	73.94		
23	Vanadium	V				1	9
			50 (NR)	49.9472	0.24		
			51	50.9440	99.76		
24	Chromium	Cr				4	7
			52	51.9405	83.76		
			53	52.9407	9.55		
25	Manganese	Mn				1	8
			55	54.9381	100		
26	Iron	Fe				4	6
			56	55.9349	91.66		
			57	56.9354	2.19		

(Continued)

At. no. Z	Element	Symbol	Mass no., A	Isotopic mass, u	Relative abundance, %	No. of Isotopes: Stable	No. of Isotopes: Radio-active
27	Cobalt	Co				1	10
			59	58.9332	100		
28	Nickel	Ni				5	7
			58	57.9353	68.274		
			60	58.9332	26.095		
29	Copper	Cu				2	9
			63	62.9298	69.09		
			65	64.9278	30.91		
30	Zinc	Zn				5	8
			64	63.9291	48.89		
			66	65.9260	27.81		
			68	67.9249	18.57		
31	Gallium	Ga				2	12
			69	68.9257	60.4		
			71	70.9249	39.6		
32	Germanium	Ge				4	10
			70	69.9243	20.52		
			72	71.9217	27.43		
			74	73.9219	36.54		
33	Arsenic	As				1	14
			75	74.9216	100		
34	Selenium	Se				6	11
			78	77.9173	23.52		
			80	79.9165	49.82		
35	Bromine	Br				2	16
			79	78.9183	50.54		
			81	80.9163	49.46		
36	Krypton	Kr				6	16
			82	81.9135	11.56		
			83	82.9141	11.55		
			84	83.9115	56.90		
			86	85.9106	17.37		
37	Rubidium	Rb				1	16
			85	84.9117	72.15		
			87		27.85		
38	Strontium	Sr				4	12
			88	87.9056	82.56		
39	Yttrium	Y				1	14
			89	88.9054	100		
40	Zirconium	Zr				5	9
			90	89.9043	51.46		
			92	91.9046	17.11		
			94	93.9061	17.40		

At. no. Z	Element	Symbol	Mass no., A	Isotopic mass, u	Relative abundance, %	No. of Isotopes Stable	No. of Isotopes Radio-active
41	Niobium (or Columbium, Cb)	Nb				1	13
			93	92.9060	100		
42	Molybdenum	Mo				7	10
			92	91.9063	15.84		
			95	94.90584	15.72		
			96	95.9046	16.53		
			98	97.9055	23.78		
43	Technetium	Tc				0	14
			99 (R)				
44	Ruthenium	Ru				7	9
			102	101.9037	31.61		
			104	103.9055	18.58		
45	Rhodium	Rh				1	14
			103	102.9048	100		
46	Palladium	Pd				6	12
			105	104.9046	22.23		
			106	105.9032	27.33		
			108	107.90389	26.71		
47	Silver	Ag				2	14
			107	106.90509	51.82		
			109	108.9047	48.18		
48	Cadmium	Cd				8	11
			110	109.9030	12.39		
			111	110.9042	12.75		
			112	111.9028	24.07		
			113	112.9046	12.26		
			114	113.9036	28.86		
49	Indium	In				1	18
			113	112.9043	4.28		
			115 (NR)	114.9041	95.72		
50	Tin	Sn				10	15
			116	115.9021	14.30		
			118	117.9018	24.03		
			119	118.9034	8.58		
			120	119.9022	32.85		
51	Antimony	Sb				2	22
			121	120.9038	57.25		
			123	122.9041	42.75		
52	Tellerium	Te				8	16
			123 (NR)	122.9042	0.87		
			126	125.9032	18.71		
			128	127.9047	31.79		
			130	129.9067	34.48		

(Continued)

At. no. Z	Element	Symbol	Mass no., A	Isotopic mass, u	Relative abundance, %	No. of Isotopes Stable	No. of Isotopes Radio-active
53	Iodine	I				1	22
			127	126.9004	100		
54	Xenon	Xe				9	16
			129	128.9048	26.44		
			131	130.9051	21.18		
			132	131.9042	26.89		
55	Cesium	Cs				1	20
			133	132.9051	100		
56	Barium	Ba				7	14
			137	136.9056	11.32		
			138	137.9050	71.66		
57	Lanthanum	La				1	20
			138 (NR)	137.9068	0.089		
			139	138.9061	99.911		
58	Cerium	Ce				3	16
			140	139.9053	88.48		
			142 (NR)	141.9090	11.07		
59	Praseodymium	Pr				1	14
			141	140.9074	100		
60	Neodymium	Nd				6	8
			142	141.9075	27.11		
			144 (NR)	143.9099	23.85		
			146	145.9127	17.62		
61	Promethium	Pm				0	14
			145 (R)				
62	Samarium	Sm				4	14
			147 (NR)	146.9146	14.97		
			148 (NR)	147.9146	11.24		
			149 (NR)	148.9169	13.83		
			152	151.9195	26.72		
			154	153.9220	22.71		
63	Europium	Eu				2	16
			151	150.9196	47.82		
			153	152.9209	52.18		
64	Gadolinium	Gd				6	12
			152 (NR)	151.9195	0.20		
			156	155.9221	20.47		
			158	157.9241	24.87		
			160	159.9271	21.90		
65	Terbium	Tb				1	17
			159	158.9250	100		
66	Dysprosium	Dy				6	13
			156 (NR)	155.9238	0.052		

At. no. Z	Element	Symbol	Mass no., A	Isotopic mass, u	Relative abundance, %	No. of Isotopes: Stable	No. of Isotopes: Radio-active
			162	161.9265	25.53		
			163	162.9284	24.97		
			164	163.9288	28.18		
67	Holmium	Ho				1	18
			165	164.9303	100		
68	Erbium	Er				6	12
			166	165.9304	33.41		
			167	166.9320	22.94		
			168	167.9324	27.07		
69	Thulium	Tm				1	17
			169	168.9344	100		
70	Ytterbium	Yb				7	10
			172	171.9366	21.82		
			174	173.9390	31.84		
71	Lutetium	Lu				1	15
			175	174.9409	97.41		
			176 (NR)		2.59		
72	Hafnium	Hf				5	13
			174 (NR)	173.9403	0.18		
			178	177.9439	27.14		
			180	179.9468	35.24		
73	Tantalum	Ta				2	13
			181	180.9480	99.988		
74	Tungsten (Wolfram)	W				5	10
			182	181.9483	26.41		
			184	183.9510	30.64		
			186	185.9543	28.14		
75	Rhenium	Re				1	14
			185	184.9530	37.50		
			187 (NR)	186.9560	62.50		
76	Osmium	Os				7	8
			190	189.9586	26.4		
			192	191.9612	39.952		
77	Iridium	Ir				2	15
			191	190.9609	37.4		
			193	192.9633	62.6		
78	Platinum	Pt				5	16
			190 (NR)	189.9600	0.0127		
			194	193.9628	32.9		
			195	194.9648	33.8		
			196	195.9650	25.3		
79	Gold	Au				1	18
			197	196.9666	100		

(Continued)

At. no. Z	Element	Symbol	Mass no., A	Isotopic mass, u	Relative abundance, %	No. of Isotopes Stable	No. of Isotopes Radio-active
80	Mercury	Hg				7	14
			199	198.9683	16.84		
			200	199.9683	23.13		
			202	201.9706	29.80		
81	Thallium	Tl				2	18
			203	202.9723	29.50		
			205	204.9745	70.50		
			207 (NR)				
82	Lead	Pb				3	18
			204 (NR)	203.9731	1.48		
			206	205.9745	23.6		
			207	206.9759	22.6		
			208	207.9766	52.3		
83	Bismuth	Bi				1	18
			209	208.9804	100		
			210 (NR)				
84	Polonium	Po				0	27
			210 (NR)	209.9829			
85	Astatine	At				0	20
			210 (NR)				
			211 (NR)	210.9875			
86	Radon	Rn				0	20
			222 (NR)	222.0175			
87	Francium	Fr				0	20
			223 (NR)	223.0198			
88	Radium	Ra				0	16
			226 (NR)	226.0254			
89	Actinium	Ac				0	11
			227 (NR)	227.0278			
90	Thorium	Th				0	13
			232 (NR)	232.0382			
91	Protactinium	Pa				0	13
			231 (NR)	231.0359			
92	Uranium	U				0	14
			234 (NR)	234.0409	0.0057		
			235 (NR)	235.0439	0.72		
			238 (NR)	238.0508	99.27		
93	Neptunium	Np				0	13
			237 (R)	237.0480			
94	Plutonium	Pu				0	15
			239 (R)	239.0522			
			242 (R)	242.0587			
			244 (R)				

At. no. Z	Element	Symbol	Mass no., A	Isotopic mass, u	Relative abundance, %	No. of Isotopes Stable	No. of Isotopes Radio-active
95	Americium	Am				0	11
			241 (R)	241.0567			
			243 (R)	243.0614			
96	Curium	Cm				0	13
			243 (R)	243.0614			
			247 (R)				
97	Berkelium	Bk				0	8
			247 (R)	247.0702			
98	Californium	Cf				0	12
			251 (R)				
99	Einsteinium	Es				0	11
			254 (R)	254.0881			
100	Fermium	Fm				0	11
			257 (R)				
101	Mendelevium	Md				0	3
			256 (R)				
102	Nobelium	No				0	7
			255 (R)				
103	Lawrencium	Lw				0	2
			257 (R)				
104	Rutherfordium	Ru				0	1
			260 (R)				
105	Hahnium	Ha				0	3
			260 (R)				

APPENDIX 6: PARTIAL LIST OF RADIOISOTOPES

Element	Nuclide	Half-life, T	Decay constant λ, s^{-1}	Principal particle energy, MeV
Antimony	$^{122}_{51}Sb$	2.80 d	2.87×10^{-6}	β^-, 1.40; γ, 0.56, 0.70
Argon	$^{37}_{18}Ar$	35.1 d	2.29×10^{-7}	K-capture
Arsenic	$^{76}_{33}As$	26.5 h	7.26×10^{-6}	β^-, 2.97; γ, 0.56, 1.21
Barium	$^{140}_{56}Ba$	12.8 d	6.26×10^{-7}	β^-, 1.02; γ, 0.03, 0.54
Bismuth	$^{210}_{83}Bi$ (NR)	5.01 d	1.60×10^{-6}	β^-, 1.16
Cadmium	$^{115}_{48}Cd$	53.5 h	3.60×10^{-6}	β^-, 1.11; γ, 0.52
Calcium	$^{45}_{20}Ca$	165 d	4.87×10^{-8}	β^-, 0.25
Carbon	$^{14}_{6}C$	5730 y	3.83×10^{-12}	β^-, 0.156
Cerium	$^{141}_{58}Ce$	33 d	2.43×10^{-7}	β^-, 0.44; γ, 0.15
Cesium	$^{134}_{55}Cs$	2.05 y	1.07×10^{-8}	β^-, 0.65; γ, 0.60
	$^{137}_{55}Cs$	30.23 y	7.27×10^{-10}	β^-, 0.51; γ, 0.66
Chlorine	$^{36}_{17}Cl$	3.1×10^5 y	7.09×10^{-14}	β^-, 0.71
Chromium	$^{51}_{24}Cr$	27.8 d	2.89×10^{-7}	γ, 0.32; K-capture
Cobalt	$^{58}_{27}Co$	71.3 d	1.125×10^{-7}	β^+, 0.48; γ, 0.81, 1.65
	$^{60}_{27}Co$	5.26 y	4.17×10^{-9}	β^-, 0.31; γ, 1.17, 1.33
Gold	$^{198}_{79}Au$	2.693 d	2.98×10^{-6}	β^-, 0.96; γ, 0.41, 0.67
Hafnium	$^{181}_{72}Hf$	42.4 d	1.89×10^{-7}	β^-, 0.41; γ, 0.48
Hydrogen	$^{3}_{1}H$	12.26 y	1.79×10^{-9}	β^-, 0.018
Iodine	$^{131}_{53}I$	8.070 d	9.94×10^{-7}	β^-, 0.61; γ, 0.36, 0.72
Iron	$^{59}_{26}Fe$	45.1 d	1.78×10^{-7}	β^-, 0.46; γ, 1.10, 1.29
Krypton	$^{85}_{36}Kr$	10.76 y	2.04×10^{-7}	β^-, 0.67; γ, 0.52
Lanthanum	$^{140}_{57}La$	40.22 h	4.79×10^{-6}	β^-, 1.34; γ, 1.60, 0.49
Mercury	$^{203}_{80}Hg$	46.57 d	1.72×10^{-7}	β^-, 0.21; γ, 0.28
Molybdenum	$^{99}_{42}Mo$	66.69 h	2.89×10^{-6}	β^-, 1.23; γ, 0.04, 0.14
Neptunium	$^{237}_{93}Np$	2.14×10^6 y	1.03×10^{-14}	α, 4.50; γ, 0.03, 0.09
	$^{239}_{93}Np$	2.35 d	3.42×10^{-6}	β^-, 0.72; γ, 0.05, 0.33
Nickel	$^{63}_{28}Ni$	92 y	2.39×10^{-10}	β^-, 0.07
Phosphorus	$^{32}_{15}P$	14.3 d	5.61×10^{-7}	β^-, 1.71
Plutonium	$^{239}_{94}Pu$	2.44×10^4 y	9.01×10^{-13}	α, 5.15; γ, 0.013, 0.038
Potassium	$^{40}_{19}K$ (NR)	1.28×10^9 y	1.72×10^{-17}	β^-, 1.32; γ, 1.46
	$^{42}_{19}K$	12.4 h	1.55×10^{-5}	β^-, 3.53; γ, 1.52
Selenium	$^{75}_{34}Se$	120.4 d	6.66×10^{-8}	γ, 0.27; K-capture
Silver	$^{111}_{47}Ag$	7.5 d	1.07×10^{-6}	β^-, 1.05; γ, 0.34
Sodium	$^{24}_{11}Na$	15.0 h	1.28×10^{-5}	β^-, 1.39; γ, 1.37, 2.75

Element	Nuclide	Half-life, T	Decay constant λ, s^{-1}	Principal particle energy, MeV
Strontium	$^{89}_{38}Sr$	52 d	1.54×10^{-7}	β^-, 1.46
	$^{90}_{38}Sr$	28.1 y	7.82×10^{-10}	β^-, 0.54
Sulfur	$^{35}_{16}S$	88 d	9.12×10^{-8}	β^-, 0.17
Tantalum	$^{182}_{73}Ta$	115 d	6.98×10^{-8}	β^-, 0.51; γ, 0.10, 1.12
Xenon	$^{135}_{54}Xe$	9.2 h	2.09×10^{-5}	β^-, 0.91; γ, 0.25, 0.61
Zinc	$^{65}_{30}Zn$	243.6 d	3.29×10^{-8}	β^-, 0.33; γ, 1.12

Source: From *Handbook of Chemistry and Physics*, 63rd ed., C. D. Hodgeman, editor. Cleveland, Ohio: Chemical Rubber Publishing Co., 1982. Copyright CRC Press, Inc., Boca Raton, Fla. Reprinted by permission.

APPENDIX 7: CONVERSION FACTORS

CONVERSION FACTORS

A conversion factor is a dimensionless ratio used to make a change in units. Thus the conversion factor

$$\frac{12 \text{ in.}}{1 \text{ ft}}$$

(read, "There are 12 inches in 1 foot") may be used to convert 10 feet to inches by direct multiplication and cancellation of the units, feet. To convert 15 inches to feet we evidently must invert the conversion factor to produce the desired result and multiply by

$$\frac{1 \text{ ft}}{12 \text{ in.}}$$

(read, "In 1 foot there are 12 inches"). When the conversion factor is used incorrectly we notice immediately that the desired cancellation of units is not obtained and this is the signal to invert the factor.

To convert speed in $\text{mi} \cdot \text{h}^{-1}$ to $\text{m} \cdot \text{s}^{-1}$ we proceed as follows:

$$1 \frac{\text{mi}}{\text{h}} \times \frac{1 \text{ h}}{3600 \text{ s}} \times \frac{5280 \text{ ft}}{1 \text{ mi}} \times \frac{12 \text{ in.}}{1 \text{ ft}} \times \frac{2.540 \text{ cm}}{1 \text{ in.}} \times \frac{1 \text{ m}}{100 \text{ cm}} = 0.4470 \text{ m} \cdot \text{s}^{-1}.$$

The conversion factor then is

$$\frac{0.4470 \text{ m} \cdot \text{s}^{-1}}{1 \text{ mi} \cdot \text{h}^{-1}}.$$

The conversion factors in the following table can be considered exact except where they are given to five significant figures.

	In one unit	There are	
Length s	cm	10^{-2}	meter, m
	in	2.54×10^{-2}	
	ft	0.3048	
	mi (U.S. statute)	1609.3	
	μ (micron)	10^{-6}	
	nm (nanometer)	10^{-9}	
	Å (angstrom)	10^{-10}	
	f (fermi)	10^{-15}	
Wave number $\frac{1}{\lambda}, \bar{f}, \bar{\nu}$	kayser	10^{-2}	m^{-1}
Time t	d (day, mean solar)	8.64×10^4	second, s
	y (year, calendar)	3.1536×10^7	
	shake	10^{-8}	
Frequency f, ν	$\text{cycle} \cdot \text{s}^{-1}$	1	hertz, Hz
Velocity u, v	$\text{ft} \cdot \text{s}^{-1}$	0.3048	$\text{m} \cdot \text{s}^{-1}$
	$\text{mi} \cdot \text{h}^{-1}$	0.44704	

	In one unit	There are	
Acceleration a	g (free fall, standard)	9.8067	$m \cdot s^{-2}$
	gal	10^{-2}	
Mass m	g (gram)	10^{-3}	kilogram, kg
	slug	14.594	
Force F	dyne	10^{-5}	newton, N
and weight w, F	poundal	0.13826	
	lb (avoirdupois)	4.4482	
Energy W, E	erg	10^{-7}	joule, J
	eV	1.6021×10^{-19}	
	kWh	3.6×10^{6}	
	cal (thermochemical)	4.184	
	kcal	4.184×10^{3}	
	$ft \cdot lb$	1.3558	
	Btu (thermochemical)	1054.4	
Power P	$erg \cdot s^{-1}$	10^{-7}	watt, W
	$cal \cdot s^{-1}$	4.184	
	$Btu \cdot h^{-1}$	0.29288	
	$ft \cdot lb \cdot s^{-1}$	1.3558	
	hp (electric)	746.00	
Pressure p	$dyne \cdot cm^{-2}$	10^{-1}	$N \cdot m^{-2}$ or pascal
	$lb \cdot in^{-2}$ (psi)	6.8948×10^{3}	
	bar	10^{5}	
	atm (normal)	1.0133×10^{5}	
	cm-mercury (0°C)	1.3332×10^{3}	
	torr (0°C)	1.3332×10^{2}	
Density ρ	$g \cdot cm^{-3}$	10^{3}	$kg \cdot m^{-3}$
	$lb \cdot ft^{-3}$	16.018	
Specific heat c	$cal \cdot g^{-1} \cdot (C°)^{-1}$	4.184×10^{3}	$J \cdot kg^{-1} \cdot (C°)^{-1}$
	$Btu \cdot lb^{-1} \cdot (F°)^{-1}$	4.184×10^{3}	
Charge q, Q	statcoulomb (esu)	$1/(3 \times 10^{9})$	coulomb, C
	abcoulomb (emu)	10	
Potential difference V	statvolt (esu)	300	volt, V
	abvolt (emu)	10^{-8}	
Capacitance C	statfarad (esu)	$1/(9 \times 10^{11})$	farad, F
	abfarad (emu)	10^{9}	
Permittivity ε_0	esu	$1/(36\pi \times 10^{9})$	$F \cdot m^{-1}$
Electric field intensity E	$V \cdot cm^{-1}$	10^{2}	$V \cdot m^{-1}$
	dyne · stat-coulomb^{-1} (esu)	3×10^{4}	
Electric flux density or electric displacement D	esu	$1/(12\pi \times 10^{5})$	$C \cdot m^{-2}$
	emu	$10^{5}/4\pi$	

(Continued)

	In one unit	There are	
Current i, I	statampere (esu)	$1/(3 \times 10^9)$	ampere, A
	abampere (emu)	10	
Resistance r, R	statohm (esu)	9×10^{11}	ohm, Ω
	abohm (emu)	10^{-9}	
Resistivity ρ	ohm · cm	10^{-2}	ohm · m, $\Omega \cdot$ m
	ohm · $(\text{mil-ft})^{-1}$	1.6624×10^{-9}	
Magnetic flux ϕ	maxwell (emu)	10^{-8}	weber, Wb
	esu	3×10^2	
Magnetic flux density or magnetic induction B	gauss (emu)	10^{-4}	Wb · m^{-2} or tesla, T
	gamma	10^{-9}	
	line · in^{-2}	1.5500×10^{-5}	
	esu	3×10^6	
Inductance L	abhenry (emu)	10^{-9}	henry, H
	stathenry (esu)	9×10^{11}	
Permeability μ_0	emu	$4\pi \times 10^{-7}$	H · m^{-1}
Magnetic field intensity H	ampere (turn) · cm^{-1}	10^2	A · m^{-1}
	oersted (emu)	$10^3/4\pi$	

MULTIPLES AND SUBMULTIPLES

The names of multiples and submultiples of the units are formed with the following prefixes:

Factor by which unit is multiplied	Prefix	Symbol	Factor by which unit is multiplied	Prefix	Symbol
10^{12}	tera	T	10^{-2}	centi	c
10^9	giga	G	10^{-3}	milli	m
10^6	mega	M	10^{-6}	micro	μ
10^3	kilo	k	10^{-9}	nano	n
10^2	hecto	h	10^{-12}	pico	p
10	deka	da	10^{-15}	femto	f
10^{-1}	deci	d	10^{-18}	atto	a

Source: The material in Appendix 7 is from *Handbook of Chemistry and Physics*, 63rd ed., C. D. Hodgeman, editor. Cleveland, Ohio: Chemical Rubber Publishing Co., 1982. Copyright CRC Press, Inc., Boca Raton, Fla. Reprinted by permission.

APPENDIX 8: PHYSICAL CONSTANTS

Constant	Symbol	Système International (SI)	Centimeter-gram-second (cgs)
Speed of light in vacuum	c	3.00×10^{8} m · s^{-1}	3.00×10^{10} cm · s^{-1}
Avogadro constant	N_A	6.02×10^{26} k mole^{-1}	6.02×10^{23} mole^{-1}
Faraday constant	F	9.65×10^{4} C · mole^{-1}	9.65×10^{3} cm$^{1/2}$ · g$^{1/2}$ · mole^{-1} (emu)
Planck constant	h	6.63×10^{-34} J · s	6.63×10^{-27} erg · s
Elementary charge	e	1.602×10^{-19} C	1.602×10^{-20} cm$^{1/2}$ · g$^{1/2}$ (emu)
			4.803×10^{-10} cm$^{3/2}$ · g$^{1/2}$ · s^{-1} (esu)
Unified atomic mass unit, 1/12 mass $C^{12} = 1/N_A$	u	1.6604×10^{-27} kg	1.6604×10^{-24} g
		931 MeV/c^2	931 MeV/c^2
Electron rest mass	m_e	9.11×10^{-31} kg	9.11×10^{-28} g
		5.49×10^{-4} u	5.49×10^{-4} u
		0.511 MeV/c^2	0.511 MeV/c^2
Proton rest mass	m_p	1.673×10^{-27} kg	1.673×10^{-24} g
		1.0073 u	1.0073 u
		938.3 MeV/c^2	938.3 MeV/c^2
H-atom rest mass	m_H	1.673×10^{-27} kg	1.673×10^{-24} g
		1.0078 u	1.0078 u
		938.8 MeV/c^2	938.8 MeV/c^2
Neutron rest mass	m_n	1.675×10^{-27} kg	1.675×10^{-24} g
		1.0087 u	1.0087 u
		939.6 MeV/c^2	939.6 MeV/c^2

(Continued)

Constant	Symbol	Système International (SI)	Centimeter-gram-second (cgs)
Charge to mass ratio for electron	e/m_e	1.760×10^{11} C · kg^{-1}	1.760×10^{7} cm$^{1/2}$ · g$^{1/2}$ (emu)
Quantum-charge ratio	h/e	4.14×10^{-15} J · s · C^{-1}	1.38×10^{-17} cm$^{3/2}$ · g$^{1/2}$ · s^{-1} (esu) 4.14×10^{-7} cm$^{3/2}$ · g$^{1/2}$ · s^{-1} (emu)
Compton wavelength of electron $(h/m_e c)$	λ_C	2.426×10^{-12} m	2.426×10^{-10} cm
Compton wavelength of proton $(h/m_p c)$	$\lambda_{C,P}$	1.321×10^{-15} m	1.321×10^{-13} cm
Rydberg constant	R_∞	1.097×10^{7} m^{-1}	1.097×10^{5} cm^{-1}
Bohr radius	a_0	5.29×10^{-11} m	5.29×10^{-9} cm
Wien displacement constant $(\lambda_{max} T)$	b	2.90×10^{-3} m · K	2.90×10^{-1} cm · °K
Stefan-Boltzmann constant	σ	5.67×10^{-8} W · m^{-2} · K^{-4}	5.67×10^{-5} erg · cm^{-2} · s^{-1} · K^{-4}
Boltzmann constant	k	1.38×10^{-23} J · K^{-1}	1.38×10^{-16} erg · K^{-1}
Normal volume, perfect gas	V_0	22.4 m^{3} · kmole^{-1}	2.24×10^{4} cm^{3} · mole^{-1}
Gas constant	R	8.31×10^{3} J · K^{-1}kmole^{-1}	8.31×10^{7} erg · K^{-1} · mole^{-1}
Gravitational constant	G	6.67×10^{-11} N · m^{2} · kg^{-2}	6.67×10^{-8} dyn · cm^{2} · g^{-2}
Acceleration by gravity (sea level, 45°)	g	9.8062 m · s^{-2}	980.62 cm · s^{-2}
Permittivity of free space	ε_0	8.85×10^{-12} C^{2} · J^{-1} · m^{-1}	

The values are within ±0.1 percent of the best ones presently known. For the more precise values consult the references given in the footnote in Section 1.3, Chapter 1.

Source: From *Handbook of Chemistry and Physics*, 63rd ed., C. D. Hodgeman, editor. Cleveland, Ohio: Chemical Rubber Publishing Co., 1982. Copyright CRC Press, Inc., Boca Raton, Fla. Reprinted by permission.

ANSWERS TO SELECTED ODD-NUMBERED PROBLEMS

Chapter 1

1.5 (a) 6.2 cal/mole · K $\simeq$ 6 cal/mole · K; (b) 1.2 cal/mole · K $\neq$ 6 cal/mole · K

1.7 $p = mvr/A$

1.11 0.03 mm

1.13 (a) 10 m/s, 10 m/s; 6.25 m/s, 6.72 m/s; 7.62 m/s, 8.04 m/s; 5 m/s, 7.07 m/s; (b) No

1.15 (b) $2N/V^2$; (c) $2V/3$, $V/\sqrt{2}$, V; (d) 56%, 50%

1.17 (a) 11.2×10^3 m/s

1.19 1.004

1.21 (a) 927°C; (b) −10.5°C

1.23 (a) and (b) 1.91×10^3 m/s, 1.76×10^3 m/s, 1.56×10^3 m/s

1.25 0.66 cm

1.27 (a) 8.47×10^6 m/s; (b) 5.8×10^9 K

1.29 (a) 7.479×10^3 J; (b) 7.504×10^3 J; (c) 25 J

1.31 (b) $2.69 \times 10^{25}/\text{m}^3$; (c) $4.59 \times 10^{25}/\text{m}^3$

1.35 (a) 2.24×10^{-7} m; (b) 2.99×10^{-7} m; (c) 2.24×10^{-9} m; (d) 7.84×10^9/s, 6.82×10^9/s, 7.84×10^{11}/s; (e) 1 : 2.45

1.39 (a) 4.00×10^{-21} J; (b) 2.19×10^3 m/s

1.41 0.015, 0.035

1.43 (a) 5.64×10^{-21} J; (b) 3.76×10^{-21} J; (c) 9.40×10^{-21} J

1.45 (a) 4.74×10^{-17} kg; (b) 6.73×10^{23}/mole; (c) 3.19×10^{10}

Chapter 2

2.1 (a) 3.58×10^{13}; (b) 1.94×10^{10}; (c) Yes

2.3 3.67×10^6 V/m downward

2.5 (a) 0.56 mm, $\phi_x = 4°2'$; (b) 1.11 cm

2.7 $5.34 \times 10^{-6} N$

2.9 (a) 7.95 ns; (c) 1.67 cm; (c) No

2.11 (b) 2.00×10^7 m/s; (c) 5.69 mm

2.13 4.8×10^7 C/kg

2.15 $x = 0$; $y = 3.65$ mm

2.19 2.24×10^7 m/s

2.21 (b) No

2.25 (a) 4.0×10^6 m/s; (b) 69.6 cm

2.27 (a) 1.61×10^{-6} m; (b) 1.41×10^{-14} kg; (c) 0.23

2.29 (a) 2.50×10^{-4} m/s; (b) 1.11×10^{-4} m/s, 3.48×10^{-4} m/s, 2.30×10^{-4} m/s, 5.30×10^{-4} m/s, 1.67×10^{-4} m/s; (c) 3.61×10^{-4} m/s, 5.98×10^{-4} m/s, 4.80×10^{-4} m/s, 7.80×10^{-4} m/s, 4.17×10^{-4} m/s; (d) 6, 10, 8, 13, 7; (e) 1.63×10^{-19} C

2.31 (a) 9.00×10^{21}; (b) 4.50×10^{21}; (c) 1.05×10^{-22} g; (d) 6.04×10^{23}; (e) 6.04×10^{23}; (f) 9.67×10^4 C; (g) 1.65×10^{-24} g

2.33 (a) Inner segment; (b) outer segment

2.35 $v_x = \infty$

2.37 (a) 5×10^4 m/s; (b) 0.52 cm; (c) No

2.39 14.6 cm

2.41 70%, 30%

2.43 (a) 3.2×10^{-2} T; (b) 309 cm, 311 cm

Chapter 3

3.1 (a) 4.02×10^{17}; (b) 5.90×10^{-15} J

3.3 5690 K

3.5 (a) 1500%; (b) 46%; (c) 4%; (d) 0.4%

3.7 9.67×10^4 Å, 2.90×10^4 Å, 2.9×10^3 Å

3.9 $R_{\text{star}} = 80\ R_{\text{sun}}$

3.11 875 K

3.15 0.218 cm^2

3.17 1.62×10^3/s

3.23 4.14 eV

3.25 $1.44 \times 10^{30}/s$

3.29 (a) $n_3/n_1 = 0.135$, $n_2/n_1 = 0.368$; (b) 0.036 eV

3.31 7000 atoms

3.33 (a) 1.05 g; (b) 1265 tons; (c) 110 lb

3.35 (a) 1.04 eV; (b) 11,900 Å; (c) 2.52×10^{14} Hz; (d) 4.15×10^{-7} eV

3.37 (a) 1.44 V; (b) 1.44 eV; (c) 7.10×10^5 m/s

3.39 (a) 1.2 V; (b) 4.3 V; (c) no photoemission, 2.4 V

3.41 2260 Å

Chapter 4

4.3 (a) 0.529 Å, 2.11 Å, 4.75 Å; (b) 1.06 Å

4.5 (a) 8.23×10^{-8} N; (b) 3.62×10^{-47} N; (c) 2.27×10^{39}

4.7 (a) $r^2 = n\hbar/\sqrt{km}$; (b) $E = n\hbar\sqrt{k/m}$

4.9 21.8×10^{-19} J, 13.6 eV; 5.44×10^{-19} J, 3.40 eV; 2.42×10^{-19} J, 1.51 eV; 0, 0

4.11 796 Å

4.13 $1.097 \times 10^7/m$

4.15 (a) 6.16×10^{14} Hz; (b) 4861 Å; (c) $2.06 \times 10^6/m$

4.17 (a) -74.5 keV; (b) 55.9 keV, 0.222 Å; (c) x-ray

4.19 1836

4.21 (a) 15.6 V; (b) 2340 Å; (c) 12.5 V; (d) $1.013 \times 10^7/m$; (e) 6 eV, 0.7 eV

4.23 (a) 12.75 eV; (b) 6

4.25 973 Å, 1026 Å, 1216 Å, 4860 Å, 6560 Å, 18,800 Å

4.27 (a) -18 eV, -4.5 eV, -2.0 eV, -1.12 eV; (b) 13.5 V; (c) 918 Å, 776 Å, 4970 Å; (d) no; (e) 690 Å

4.29 (a) 1216 Å; (b) 3650 Å; (c) 13.6 V

4.31 (c) 51 eV

4.33 (a) 5.28×10^{45}; (c) 3.03×10^{-39} m

4.35 9.6×10^{139} mi^3

4.37 12.5 T

4.43 $3.18 \times 10^{15}/s$

4.45 (a) 0.65 cm; (b) 0.36 mi

Chapter 5

5.5 231 s, 266.7 s, 1.15

5.7 (a) 6.28×10^5 mi/s; (b) 1.86×10^5 mi/s; (c) 0; (d) No

5.11 (a) 0.0447 c; (b) 0.14 c; (c) 0.435 c

5.13 64.7 s

5.15 $d = 1.35 \times 10^{11}$ m

5.17 (a) 15 s; (b) 15 s

5.21 (a) 1.3 c; (b) 0.93 c

5.23 (a) 1.0 c; (c) 21°48′

5.25 0.87 c

5.29 (a) 1.80×10^5 eV; (b) 12.3×10^{-31} kg; (c) 2.02×10^8 m/s; (d) 1.30×10^{11} C/kg; (e) 2.52×10^8 m/s

5.31 2.70×10^{11} kg

5.33 (a) 1200 MeV; (b) 8.82×10^{-27} kg; (c) 1.96×10^8 m/s

5.37 (a) 5; (b) $\sqrt{15}$

5.39 (a) -2.8×10^{-13}; (b) $+2.8 \times 10^{-13}$

5.41 9×10^{16} m

5.43 2.60×10^8 m/s

5.47 (a) 1.95×10^3; (b) 2.9999996×10^8 m/s

5.49 (a) 5.11 kV; (b) 9.38 MV

5.51 (a) 2.35×10^8 m/s; (b) No

5.53 9×10^8 gal

5.55 (a) 8.03 u; (b) 7.47×10^3 MeV; (c) 17.3 MeV, 1.86×10^{-2} u

5.57 (a) 1.73 MeV

5.59 (a) 25.8×10^{-31} kg; (b) 2.81×10^8 m/s; (c) 0.038 T normal to the trajectories

5.63 $\sqrt{6}$ MeV/c; $\sqrt{6}$ MeV

Chapter 6

6.1 (a) 3.53 Å; (b) 7.06 Å

6.3 (a) 1.24 Å; (b) 0.413 Å; (c) x-ray

6.5 (a) 355 cal/s; (b) 33.8°C/s

6.7 (a) 24, 29; (b) Cr, Cu

6.9 (a) 10.4 keV, 69.4 keV, 77.7 keV; (b) 79.6 to 80 keV

6.11 27, 28, 29, 30

6.19 (a) 70.7%, 0.098%; (b) 10 cm

6.21 (a) 0.664; (b) 0.091; (c) 0.464

6.23 (a) 0.023 ft^2/lb, 0.022 ft^2/lb, 0.021 ft^2/lb, 0.022 ft^2/lb; (b) 154 lb/ft^2, 156 lb/ft^2, 163 lb/ft^2, 156 lb/ft^2

6.27 (a) 2.28; (b) 3.41×10^{-2}; (c) 4.85×10^{-6}; (d) no significant change for (a) and (b). Impossible for (c), but for whole atom, 2.19×10^{-10}.

6.29 (a) 570 eV; (b) 0; (c) 44°2′, no recoil

6.31 (a) 5.44×10^{-27} kg · m/s; (b) 5.42×10^{-7}%

6.35 1.56×10^{-22} kg · m/s

6.39 2.66×10^{-13} A

6.41 45.8°, 40.4°, 40.0°

6.43 (a) 1.80 Å, 1.56 Å, 1.11 Å, 0.95 Å, 0.90 Å; (b) (111), (200), (220), (311), (222); (c) 3.12 Å, 3.12 Å, 3.13 Å, 3.14 Å (average 3.13 Å); (d) 6.26 Å

Chapter 7

7.3 (a) 1.23 Å; (b) 0.14 Å

7.5 (a) 1.47 Å; (b) 1.30 Å; (c) 20.5°, 18.0°

7.7 7.26×10^6 m/s, 3.96×10^3 m/s, 151 eV, 0.082 eV

7.9 (a) 2.00×10^{-23} Å; (b) 1.77×10^{-25} Å; (c) 4.00×10^{-31} cm, 3.54×10^{-33} cm

7.11 0°55′

7.13 (a) Less; (b) Phase; equal; (c) No

7.19 3.52×10^{-19} Å, 23.2 Å

7.21 $\Delta X_e = 1.16$ cm, $\Delta x_{\text{bullet}} = 5.28 \times 10^{-31}$ m

7.23 6.59×10^{-8} eV; no

7.25 9.5 cm

Chapter 8

8.1 (a) 2.5, 7.5; (b) 2.5, 6.83

8.3 9.45 eV; 37.8 eV, 84.9 eV

8.5 Zero

8.11 (a) 1.35×10^{-3} J; (b) 5.41×10^{31}; (c) 4.97×10^{-35}

8.13 9.4×10^4 eV

8.15 (a) $(\pi a^2)^{-1/4}$; (b) $0, 1/2a^2$; (c) $0, \hbar^2 a^2/2$

8.17 $h\nu/4$

8.19 (b) $E = \frac{3\hbar}{2}\sqrt{k/m}$

Chapter 9

9.5 (a) 2.31×10^4 K; (b) 3.4×10^{-3}; no

9.7 3.1

9.13 1.66×10^4 Å

Chapter 10

10.1 (a) 40 d; (b) 60 d; (c) 0.0346/d, 28.9 d

10.3 (a) 3.62×10^4/s; (b) 2.92×10^4/s; (c) 1.46×10^5/s

10.5 10 m; 0.069 m^{-1}; 14.43 m

10.7 6.04 *y*

10.9 (a) 51, 59; (b) 51, 58; (c) 52, 58

10.11 $^{122}_{48}E$

10.13 6.48×10^{-4} g

10.17 (a) 0.368; (b) 0.632

10.19 7α, 4β

10.23 (b) 2.94×10^{-4}/s; (c) 39.2 m

10.25 4.25×10^9 y

10.27 6.50×10^{-6} g, 0.655 mm^3

10.29 6.23×10^{-15} m

10.31 1.55×10^{-18}/s, 1.42×10^{10} y

10.33 48 min

10.35 (a) 1.05×10^{-4} g, 6.65×10^{-10} g; (b) 0.103 mCi, 3.81 Rd

10.37 (a) 6.64 T_2; (b) 9.96 T_2

10.39 3.62×10^5 ft^3

10.41 0.12 μCi

10.45 (a) 93%; (b) $^{222}_{86}Rn$; (c) 86%

10.47 With calcium

10.49 6840 mR/h

Chapter 11

11.3 (a) 1.92×10^7 m/s; (b) 2.26×10^5 ip; (c) 1.34×10^{-9} A

11.5 4

11.7 (c) 100%

11.9 (a) 0.039 eV; (b) 1.46×10^{-10} m

11.11 1.59×10^{-5} MeV; 7.14×10^3 m/s

11.13 (a) 17.3 MeV; (b) 8.65 MeV

11.17 5.12×10^{-5} miles; 0.08 meter

11.19 (a) 34 GeV/s; (b) 714 eV/cycle; (c) 2.38×10^{-6}

11.21 (a) 1.54×10^6 J; (b) 6.17×10^6 J

11.23 (a) ^7Li; (b) ^{17}F; (c) ^{30}P; (d) ^7Be; (e) β^-; (f) ^{28}Al; (g) ^{32}P; (h) ^8Be; (i) n; (j) β^+; (k) ^{120}Te; (l) γ

11.25 (a) 0.087 MeV; (b) 4.84 MeV

11.27 (a) +0.781 MeV; (b) −1.80 MeV

11.29 10.01354 u

11.31 β-decay

11.33 (a) $^{23}_{11}Na$; (b) $^{23}_{10}Ne$; (c) $^{20}_{9}F$

11.35 4.1 MeV

11.37 (b) 34.6 MeV; (c) 8.65 MeV for each

11.39 (b) 4.8 MeV; (c) 2.06 MeV

11.41 (a) 0.63 MeV, −3.99 MeV; (c) radioactive, stable

11.45 ± 125 Y

11.47 2.5×10^6 y

11.49 1.77 mg

11.51 Supply, release, release

11.53 (a) 4.15×10^{-8} eV; (b) 2.88×10^{-12}; (c) 2.88×10^{-12}; (d) 11.8×10^6 Hz, 7.45×10^{-4} Å

11.55 0.03 m/s, away from emitter

Chapter 12

12.1 255, 85.9 tons, 1.84×10^{11} tons

12.3 (a) 945 g; (b) 663 $\mu Ci/mi^2$

12.5 (a) 0.28; (b) ≈ 1

12.7 (a) 52; (b) 5

12.9 6.24×10^{10}/s

12.11 1.90

12.13 1.245×10^5 W

12.15 5.9×10^9 y

12.17 99.5 barns

12.19 (a) 1.17×10^{-7}; (b) 4.08×10^{-5}; (c) 348

12.21 (a) H, 0.022/cm; B, 0.147/cm; 0, 6.82×10^{-6}/cm; (b) 0.169/cm

12.23 (a) 4.75 cm; (b) 0.15 cm; (c) 0.368

12.25 (a) Red, 69°38′; blue, 68°53′; (b) Red, 0.75c; blue, 0.746c

12.27 41,518 ft

12.29 (a) 1.07×10^{19} J/m; (b) 2.36×10^{28} J/m; (c) 4.10×10^{10} tons/m; (d) 2.88×10^8 tons/m

12.31 1.18×10^{16} km

Chapter 13

13.1 $\hbar/\Delta\tau c^2$

13.3 3.77 MeV

13.5 1876.6 MeV

13.7 340 m

13.9 (a) 0; (b) +1

13.13 0.79 MeV

13.15 128 MeV

INDEX